# Microbes

# Microbes

## A Source of Energy for 21st Century

**S.K. Soni**
Reader, Department of Microbiology
Panjab University, Chandigarh - 160 014

**2007**

New India Publishing Agency
Pitam Pura, New Delhi- 110 088

ISBN 81-89422-14-6

*Typeset at:*

**Laxmi Art Creations**
Shiva Mkt., Pitam Pura, New Delhi
Mobile : 9811482328

*Printed at :*

**Jai Bharat Printing Press**
Rohtas Nagar, Shahdara, Delhi

*Published by:*
Sumit Pal Jain *for*

New India Publishing Agency

101, Vikas Surya Plaza, CU Block, L.S.C. Mkt.,
Pitam Pura, New Delhi- 110 088, (INDIA)
Phone: 011-27341717, Fax: 011-27341616
E-mail: newindiapublishingagency@gmail.com
Web: www.bookfactoryindia.com

*This book is dedicated to the memory of my mother*
*Mrs. Urmil Soni*
*who expired during its preparation*

# FOREWORD

*"Not only will atomic power be released, But someday we will harness the rise and fall of the tides and imprison the rays of the sun".*

***— Thomas Edison***

Crisis always brings change - change in government policies, individual attitudes and society's concerns for its future. The Oil crisis of 1973 coupled with the depleting energy resources compelled one and all to think in terms of inventing renewable resources of energy to cope with the future needs of society. The Governments around the world are placing considerable faith in renewable energy resources as important technologies are being made available for reducing energy related environmental problems and thereby looking for renewable energy resources. One of the most notable features of renewable forms of energy is the diversity of available technologies and resources. There is little doubt that the ultimate size of the renewable energy resource is large and could, in principle, make a very substantial contribution to world energy demands.

In consequence, there is an urgent need to develop alternate energy sources by involving microbes, in particular. Biological fuel generation appears to have become increasingly important, especially as it can provide both liquid and gaseous fuels. Currently, a number of liquid bio-fuels, in the form of alcohols, esters, ethers, and other chemicals made from biomass, are in focus with respect to their substitution for fossil fuels. Two major types of fuels - first, via fermentation, ethanol to be obtained from a large number of crops: corn, sugar beet, sugarcane, and cellulosic materials and the second from a second group of crops with a substantial fraction of triglycerides, such as rapeseed and soy, methyl esters produced through chemical trans-esterification - are receiving growing attention as bio-fuels for transportation purposes.

It is, perhaps, in this background that the author, while looking at the current needs of the society, depleting energy resources and the above observations made by Thomas Edison, has thought it fit to put up concerted efforts in the form of present publication to create needed awareness and knowledge amongst the public in general, the academics and student community in particular for whom, I feel, it will prove to be an asset. I wish this publication all success.

KNPathak

**(K.N. Pathak)**
Vice-Chancellor
Panjab University, Chandigarh.

July 14, 2006

# MESSAGE

The increasing global population, industrialization and dependence upon fossil fuels has become a matter of concern for the whole world during the 20th century and have lead to the discovery of a number of alternate fuels, such as ethanol, methanol, natural gas and electricity, each having its own merits & demerits. The focus of researchers world over is to develop the cost effective technologies as well as searching new sources of energy which are eco-friendly, cheap and easy to prepare. One such class of fuels which has been identified by scientists include the biofuels which can be produced from renewable energy resource - the biomass, by the action of microorganisms which are ubiquitous and easy to handle.

This book - "Microbes: A source of energy for 21st century" highlights potential of microorganisms in solving the global energy crises and provides comprehensive information on the various alternate fuels. The contents of the book are well defined to provide the readers and researchers the fundamental knowledge about different types of microorganisms and their diversity in energy generating path. The volume makes the readers familiar with different types of biomass energy sources available on this planet and various possibilities which can be exploited for converting them into alternate energy sources using microorganisms. In this volume, efforts have also been made to provide comprehensive information about different alternate fuels from microbes for the 21st century. The use of various biotechnological tools for developing novel microorganisms and guidelines for protection of intellectual property rights have also been discussed. This manuscript is expected to assist all those who grapple with these areas and I hope it will be well received.

**S.S. Marwaha**
Director (Biotechnology),
Punjab State Council for Science & Technology &
Chief Executive Officer, Punjab Biotechnology Incubator

# PREFACE

Energy is the lifeline of modern world and this need is mainly fulfilled by the fossil fuels like petrol, diesel, natural gas etc. Due to increasing global population and industrialization, our dependence upon fossil fuels increased heavily during the last century. But today, India has 17% of the world's population, and just 0.8% of the world's known oil and natural gas resources. Transportation is the fastest growing energy consumer in India, now consuming nearly 112 million tons of oil annually. The complete substitution of oil imports for the transportation sectors and the depleting petroleum reserves in the world is the biggest and toughest challenge for the developing nations. The focus of researchers world over is to develop the cost effective technologies as well as searching alternative sources of energy which are eco-friendly, and easy to prepare to become energy independent in the years to come. One such class of fuels which have gained world-wide significance due to many advantages over the present fuels includes the biofuels which can be produced from renewable energy resource – the biomass, by the action of microorganisms.

Microorganisms are ubiquitous and indispensable for the existence of mankind. They show diversity in size, shape, metabolism and the range of positive functions they perform for sustaining the life on this planet. Bacteria have been exploited by the mankind since times immemorial for the production of various foods and enzymes. They reveal several types of metabolic reactions which are absent in higher organisms. The scientific understanding of metabolic diversity in microorganisms has made it possible to exploit the selected strains of microbes for the commercial production of enormous useful commodities. Further, the biotechnological tools have made it possible to tailor useful strains of microorganisms for the new millennium which foresees upcoming applications for the sustainability of the society.

The present book "Microbes: a source of energy for 21st century" highlights the potential of microorganisms in solving the global energy crisis and making this world as energy independent. The volume has been organized in 13 chapters which have been prepared to provide the readers with both an in-depth study and a broad perspective of microorganisms for sustainability of mankind. The book provides the fundamental knowledge about different types of microorganisms, overviews their distribution in nature and highlights their various applications and commercial uses. Further, it makes the readers familiar with the diversity in energy generating pathways among different groups of microbes and various possibilities which can be exploited for converting global biomass resources in to alternate biofuels. A great effort has been made to provide the readers a comprehensive knowledge about different alternative fuels from microbes for the 21st century. The use of various biotechnological tools for developing novel microorganisms and guidelines for the protection of intellectual property rights have also been discussed in this volume.

I am indebted to all other co-authors including Dr. (Mrs) Raman Soni, Dr. D.K. Rahi, Ms. Nidhi Goyal and Ms. Bhavna Lumba who have contributed for this book. The assistance rendered by my research students, Ms. Sushma, Ms. Tammana Toor, Ms. Jasmine Sidhu, Ms. Garima Vasudev, my loving son, Dikshit and daughter, Lubna in the compilation of this manuscript is gratefully acknowledged. It is hoped that this volume will prove useful to the students and professionals who are pursuing their career in Microbiology, Biotechnology, Biochemistry, Environmental sciences and Energy studies related to the alternate biofuels to solve the global energy crisis.

At the end, it is suggested that biofuel research should be extended in collaboration with R&D Laboratories, academic institutions and automobile industry to make it a "full fledged fuel" for the fleet running in the country in a time bound manner. The developing countries like India also need to evolve a comprehensive renewable energy policy for energy independence during the 21st century.

July 2006

**SANJEEV KUMAR SONI**
Chandigarh

# Contents

# 1

# Introduction and Fundamentals of Microbiology

**S.K. SONI[1] AND R. SONI[2]**
[1]*Department of Microbiology, Panjab University, Chandigarh-160 014*
[2]*Department of Biotechnology, D.A.V. College, Chandigarh-160 011*

---

# 1. INTRODUCTION

Microbiology is the study of the organisms which are too small to be clearly seen with unaided human eye. Organisms with diameter of 1mm or less than 1 mm are microorganisms which can be divided into six main types: Archaea, Bacteria, Protista, Fungi, Viruses, and Microbial Mergers. They are the oldest form of life on earth. Microbe fossils date back more than 3.5 billion years to a time when the Earth was covered with oceans that regularly reached the boiling point, hundreds of millions of years before dinosaurs roamed the earth. There are two main branches of microbiology, one which is concerned with the form, structure reproduction, physiology, metabolism and classification of microorganisms, is called. "General Microbiology". It includes the study of their distribution in nature, relationship to each other and other organisms, their abilities to make physical and chemical changes in the environment, their reactions to physical and chemical agents. The second one which is concerned with the association of microorganisms with the health and welfare of human beings is called "Applied Microbiology". The various general and applied disciplines of microbiology are mentioned in Table 1.

**Table 1.** Major disciplines of Microbiology

| Branch/ Area of Microbiology | Scope |
|---|---|
| **General Microbiology** | Gives the fundamental knowledge about various microorganisms |
| Bacteriology | Deals with the study of bacteria |
| Mycology | Deals with the study of fungi |
| Phycology | Deals with the study of algae |
| Protozoology | Deals with the study of protozoa |
| Virology | Deals with the study of viruses |
| Parasitology | Deals with the study of parasites |
| Microbial Physiology | Deals with the physiology of various microorganisms |
| Microbial Metabolism | Deals with the general metabolism of various microorganisms |
| Microbial Ecology | Deals with the study of microorganisms in their natural habitat |
| Microbial Systematics | Deals with the identification and classification of various microorganisms |
| Microbial Genetics | Deals with the genetic setup of various microorganisms |
| **Applied Microbiology** | Concerned with the association of microorganisms with the welfare of the mankind |
| Industrial Microbiology | Deals with the exploitation of microorganisms for the production of various useful products including antibiotics, organic acids, alcoholic beverages, amino acids, vaccines, enzymes etc at industrial scale |
| Food Microbiology | Deals with the study of food spoilage with various microorganisms, food borne illness, preservation of foods, preparation of various fermented foods |
| Medical Microbiology | Deals with the study of causative agents of disease in animals and their preventive measures |

*Contd...*

*Table 1 Contd...*

| Branch/ area of Microbiology | Scope |
|---|---|
| Agricultural Microbiology | Deals with the role of microorganisms in increasing soil fertility, geochemical cycles of nature |
| Plant Pathology | Deals with the study of causative agents of disease in plants and their preventive measures |
| Aquatic Microbiology | Deals with microbiological degradation of sewage, water purification with microorganisms, role of water as a vector for the transmission of various pathogens and disinfection of water |
| Aeromicrobiology | Deals with the role of air microflora in spoilage and transmission of various pathogens |
| Exomicrobiology | Deals with the exploration of microbial life in outer space |
| Geochemical Microbiology | Deals with the role of microorganisms in coal, mineral and gas formation, recovery of minerals with the help of microorganisms from low grade ores |

## 2. HISTORY AND SCOPE OF MICROBIOLOGY

The era of Microbiology began with the discovery of world of microorganisms, when people learned to grind lenses from pieces of glass and combine them to produce magnifications great enough to enable microbes to be seen. Zaccharias Jensen, (1591-1608) was the first one to make a crude, 'Compound' microscope but it had no provision for focusing. Later, Galleo Gallei in 1610, made a compound microscope, which he called "occiale" having a focusing arrangement. The name microscope was first proposed by Fabri in 1625. First compound microscope with all types of focusing was invented in 1820.

The first person to report seeing microbes under the microscope was an Englishman, Robert Hooke. Working with a crude compound microscope, he saw the cellular structure of plants around 1665. He also saw fungi which he drew. However, because his lens were of poor quality he was apparently unable to "see" bacteria. With these developments, the stage was set for revealing the world of microorganisms. It was about 1676, when Antony Van Leeuwenhoek (1632-1732), a Dutch cloth merchant who had a habit of grinding lenses, reported his observations on protozoa with accurate descriptions and drawings. He invented a 'Simple' microscope with magnification power of 300-500X and used the term 'animalcules' for microorganisms. While Leeuwenhoek may not have been the first to see bacteria under the microscope, he was certainly the first to describe these forms accurately and was probably, the first to recognize the value of the microscope as a laboratory instrument.

Although not the first to discover the microscope or to use magnifying lens, Leeuwenhoek was the first to see and describe bacteria. He was a "cloth merchant" living in Delft Holland and used magnifying lens to view the quality of the weave of the merchandise he purchased. He travelled to England in 1668 to view English cloth and there he saw drawings of magnifications of cloth much greater than any of the current lens available in Holland would do. He returned to Holland and took up lens grinding. Being meticulous, he developed his lens grinding to an art and in the process tested them by seeing how much detail he could observe with a given lens. One can guess that he chanced to look at a sample of pond water or other source rich in microbes and was amazed to see distinct, uniquely shaped organisms going, apparently purposefully, about their lives in a tiny microcosm. He made numerous microscopes from silver and gold and viewed everything he could including the scum on his teeth and his semen. His best lens could magnify ~300-500 fold which allowed him to see microscopic algae, protozoa and larger bacteria. He clearly had excellent eyesight because he accurately drew pictures of microbes that were at the limit of the magnification of his lens. He used only 'single lens' and not the compound lens of the true microscopes we employ today; which makes his observations all the more amazing. He wrote all his observations to the Royal Society of London in 1676 and included numerous drawings. He astonished everyone by claiming that many of the tiny things he saw with his lens were 'alive' because he saw them swimming purposefully about. This caused no end of shock and wonderment and numerous people hurried to Delft to see if this Dutchman was "in his cups" or if he was really onto something new and wonderful. A few minutes with one of his numerous microscopes was all it took to convert his visitors to enthusiastic believers in the existence of these tiny *beasties* living all around them.

## 2.1. Controversy over the origin of microorganisms

Immediately after the discovery of microorganisms, people debated on the origin of microorganisms. There were two schools of thought from the very beginning, some believed that the animalcules were formed spontaneously from the non-living matter (abiogenesis) while others including Leeuwenhoek believed that they were formed from the seeds or germs of these animalcules which were always present in air (biogenesis).

Before the discovery of microorganisms a controversy was going on regarding the spontaneous origin of plants and animals. It was just before the discovery of microorganisms, it was made evident by an Italian Physician – Francesco Redi in 1665 that the spontaneous origin of plants and animals is not possible. John Needham in 1749 experienced the appearance of microorganisms not previously present in meat and concluded that they appeared spontaneously. An Italian Naturalist Lazzaro Spallanzani conducted a long series of experiments in the middle of eighteenth century by heating the meat infusions for a variable period of time and then hermetically sealing the flasks. He showed no spontaneous development of microbes in his experiments. Proposers of abiogenesis objected to the hermetic sealing of Spallanzani suggesting that oxygen may be necessary for the spontaneous origin of microorganisms. The challenge was taken up by Theodor Schwann in 1837 who modified the experiment of Spallanzani and supplied the previously heated air to the boiled meat infusion and demonstrated no spontaneous development of microorganisms. To give some more weightage, H. Schroeder and T. Von Dusch modified the experiment of Schwann and supplied the air after passing through a long tube packed with cotton wool. Louis Pasteur in around 1861 demonstrated for the first time that air contains microscopically observable organized bodies. He passed the air through a tube containing a plug of gun cotton and then added it in a mixture of alcohol and ether which on examination under microscope revealed the presence of small round or oval bodies indistinguishable from microorganisms. Pasteur conducted an experiment in favour of biogenesis which was widely accepted and made way for the abandonment of the doctrine of spontaneous origin. He used special bent-necked flasks. Last experiment in favour of biogenesis was conducted by John Tyndall towards the end of nineteenth century. He used the concept of optically empty air and used hay infusions in his experiment. He concluded that bacteria has two phases in its life cycle – one relatively thermo-labile (destroyed by boiling for 5 min) and one thermo-resistant. A German Botanist Ferdinand Cohn used the term endospores to account for the resting bodies to be responsible for the thermo resistant phase. He developed a method of sterilization 'tyndallization' – discontinuous boiling for 1 min on five successive occasions.

## 2.2. Germ theory of fermentation

Leeuwenhoek's lucid reports on the ubiquity of microbes enabled Louis Pasteur approximately 200 years later to discover the involvement of these creatures in fermentation reactions. He found that fermentation of fruits and grains, resulting in alcohol, was brought about by microbes. By 1860 Louis Pasteur began to realize that there is some relationship between the development of microbes in organic infusions and chemical changes brought about by them. He observed two types of changes in organic infusions, one occurring in carbohydrates leading to the formation of alcohol and was called 'Fermentation' and the other one occurring in proteins leading to the formation of foul smelling compounds and was called 'Putrefaction'.

Louis Pasteur eventually convinced the scientific world that all fermentation processes are the result of microbial activity. He spent 20 years from 1857-1876 on microbial fermentations. He also discovered lactic acid and butyric acid fermentations. He postulated a germ theory of fermentation and discovered anaerobic life during butyric acid fermentation. Louis Pasteur also studied the physiological significance of fermentation and defined it as life without oxygen. He also observed the effect of $O_2$ on ethanol fermentation with yeast which brings about fermentation in absence of oxygen and if some oxygen is made available in between, the yeast stops fermentation and shifts the energy generation mechanism to respiration. This effect was termed as 'Pasteur effect'. Further knowledge about the nature of fermentation resulted from an accidental observation made in 1897 by H. Buchner. He tried to preserve an extract of yeast prepared by grinding yeast cells with sand and by adding excess of sugar to it. He observed the evolution of $CO_2$ with the formation of alcohol. A soluble enzymatic preparation, able to carry out fermentation was thus discovered and was established that living cells are not essential for fermentation instead enzymes are responsible.

Louis Pasteur during studies on fermentation observed that fruit juices, *Beer* and *Wines* get spoiled by the growth of undesirable microorganisms. He suggested that undesirable type of microbes can be removed by heating not-enough to spoil the flavour of these beverages, but enough to destroy a very high percentage of microbial population, (62.8ºC or 145ºF for half an hour). He described this method as "pasteurization" which is today widely used method in fermentation and dairy industries.

## 2.3. Germ theory of disease

After the discovery of the role of microorganisms in chemical tranformations another important contribution in the field of microbiology was by Robert Koch (1843-1885) who established the role of microorganisms in causing diseases in higher organisms including humans. The discovery that bacteria can cause diseases in higher animals was possible through the study of 'anthrax' – a serious infection in domestic animals that is transmissible to man. During his work on the etiology of 'anthrax', "C.J. Davaine" showed the presence of rod shaped bacteria in the blood of the diseased animals which were absent in healthy animals. The Infection was shown to be transmissible to healthy animals by the inoculation with blood containing the rod shaped bacteria. The demonstration of the bacterial cause of anthrax was provided by a German doctor, "Robert Koch" in 1876. The disease in mice was transmitted through a series of 20 successive inoculations and the characteristic symptoms appeared every time. Ultimately, Koch established the relationship between specific microorganism and specific disease and advocated four postulates:

1. Microorganism must be present in every case of disease.
2. Microorganism must be isolated from the diseased host and grown in pure culture.
3. Specific disease must be reproduced when a pure culture of the organism is inoculated in healthy susceptible host.
4. Microorganism must be recoverable once again from experimentally infected host.

## 2.4. Development of laboratory techniques and pure cultures

Both Pasteur and Koch believed that there is a great variation in the morphological and physiological functions of the microorganisms occurring in nature. First pure culture of bacteria was isolated with Lactic acid bacteria by Joseph Lister using serial dilution method which was very tedious. Koch realized the need for obtaining pure culture of bacteria by some other method and thought of using solid media for which he used cut surface of potato which was found to have some disadvantages and was soon abandoned. Koch then thought of solidifying a well defined liquid medium with some clear substance and added gelatin as solidifying agent. He deviced a universal culture medium, the nutrient broth and its solid counter part, nutrient agar. He also contributed to the development of various laboratory techniques including streaking, pour plating and spreading for the isolation of pure culture of microorganisms on the solid medium. He added gelatin as solidifying agent in the liquid medium but these had some drawbacks. Koch met a young German doctor, Walther Hesse who became interested in Microbiology and thus joined Koch's laboratory in 1881. The solution to solidification of culture media came in 1883 by W. Hesse's house wife, Fannie E. Hesse. She suggested the use of "agar" as a substitute for gelatin in microbiological media. Agar goes into solution at 100°C and soldifies at 45°C upon jelling, it remains solid at elevated temperatures just below 100°C.

## 2.5. Discovery of immunization

Smallpox was one of the greatest scourges of mankind. For thousands of years it swept through human populations, killing up to 40% of its victims and leaving many of the survivors horribly scarred for life; their faces covered with deep red pits. The ancient Chinese recognized that those who recovered from a case of "the pox" were immune to smallpox. Some unknown person in China, perhaps noting that even people who only developed a few scabs were as immune as those whose bodies were covered with scars, took material from a dried scab and scratched it into an uninfected person in an attempt to immunize them. It worked and the process was repeated by others, with the technique eventually reaching India. From India it travelled by various routes to Europe in the 17th century where it came into common usage. The only problem was that the scab from a victim contains fully virulent virus capable of producing the clinical disease. Thus, while one person might have only a mild case and become immune, they shed the virulent virus and were capable of starting their own epidemic. This was the situation when Edward Jenner entered the picture. Through a series of serendipitous events, Jenner led to the discovery of immunization and to the eventual elimination of the scourge of smallpox from the earth. As a young man he had lived in the country and had been told by a milkmaid that "she never had to worry about catching smallpox because she had had "cowpox", a mild chronic disease of cows that milk-maids usually contracted as a rash on their hands. Later, after Jenner became a physician and took up a country practice, he remembered the milkmaid's story. He began asking questions and was told by local men that "if you want to marry a woman who will never be scarred by the 'POX', marry a milkmaid". By 1796 he became convinced that the story was true so he inoculated an 8-yr. old boy with cowpox and 8 weeks later inoculated the same boy with the pus from a smallpox lesion. The boy showed no effects and Jenner repeated the experiment. As word of his results spread, others began to test it and by 1803 it was an established medical procedure in England. Shortly thereafter, Ben Franklin encouraged American doctors to adapt this technique in view of the dangers inherent in the older technique.

## 2.6. Discovery of protection against infection

In 1880 Louis Pasteur isolated a bacterium responsible for chicken cholera and grew it in pure culture. He made a public demonstration to prove Koch's postulates by taking the culture which he isolated 8 weeks ago. On inoculating in a healthy chicken the animal failed to become sick and die. He then used a fresh culture and inoculated two sets of the healthy animals, one which was previously inoculated with 8 weeks old culture and the other one which was not previously inoculated. The animals of the second set developed symptoms and died while those of first set remained healthy. He used the term "attenuated culture" for the 8 week old culture which failed to make the animal sick and defined it as the culture which had lost the ability to cause a disease but retained the ability to stimulate the host to produce the substances, "antibodies" that protect the host against subsequent exposure to virulent organisms. Attenuated cultures were called as vaccines. Pasteur also developed a vaccine against rabies, a viral infection. He inoculated the rabbits with saliva from mad dogs and then removed the brain and the spinal cord from the infected rabbits. He dried it, pulverized and mixed with glycerin and injected into dogs for protection against rabies.

## 2.7. Discovery of chemotherapeutic agents

Paul Ehrlich worked in Koch's lab where he learned to study bacteria. While considering the phenomenon of differential staining of different bacteria and of different components of eukaryotic cells, he speculated that if a dye chemical could bind to one cell and not another or to one substance within a cell and not others, perhaps you could find chemicals that would selectively kill certain pathogens without harming the surrounding host cells; this would act like a 'Magic bullet' selectively killing the villain and sparing the innocent victim. He embarked on a search for a magic bullet to cure syphilis, which in the late 19th century was a scourge as terrible as AIDS is today. In the final stages of syphilis, a sexually transmitted disease, its victim suffered horribly and eventually died insane as the brain was destroyed by the infection. Over many years he tested 100's of chemicals and finally in 1910 he found one, he named 'Salvarsan' or 'compound 606', that killed the syphilis organisms without killing the host (usually). This discovery laid the ground for the discovery of antibiotics and other chemotherapeutic agents.

### 2.8. Microorganisms as geochemical agents

In contrast to plants and animals, microorganisms show a wide range of physiological diversity. The role that microorganisms play in the biologically important cycles of matter on earth, the cycles of C, N and S was established by S. Winogradsky and W.M. Beijerinck towards the end of nineteenth century and the beginning of twentieth century. Winogradsky also discovered autotrophic bacteria capable of growing in completely inorganic environments obtaining energy by oxidation of reduced inorganic compounds and using $CO_2$ as the source of their cellular carbon.

Both Winogradsky and Beijerinck contributed towards the discovery of N-fixing microorganisms including bacteria and blue green algae and developed a new 'enrichment culture technique' involving an application of the principle of natural selection for the isolation of various physiological types of microorganisms.

## 3. DISTRIBUTION OF MICROORGANISMS

Microorganisms flourish in an extremely wide range of habitats and occur nearly every where in nature. They are carried by air currents from the earth's surface to the upper atmosphere. Even those indigenous to ocean may be found many miles away on mountain tops. Microorganisms occur most abundantly where they find food, moisture and a temperature suitable for their growth and multiplication. Since the conditions that favour the survival and growth of many microorganisms are those under which people normally live, it is inevitable that we live among a multitude of microbes. They are in the air we breathe and the food we eat. They are on the surfaces of our bodies, in our alimentary tract, in our mouth, nose and other body orifices. Bacteria are found in nature at pH ranging 1 to 10, temperatures from 0-80°C, at atmospheric pressure and at 10,000 m in ocean depths. They occur in air as well as in oxygen-free environments. Fortunately, most of them are harmless, our body also resists the invasion of pathogens and fights against them by various defense mechanisms.

## 4. DIVERSITY OF MICROORGANISMS AND THEIR PLACE IN THE LIVING WORLD

Until the eighteenth century the classification of living organisms placed all the organisms into two kingdoms-plants and animals. Some microorganisms have plant like characters and fell into plant kingdom. Some have animal like characters and fell into animal kingdom. Some have characters common to both plants and animals and thus could not be adjusted. A third kingdom "Protista" was formed in 1866 by a German Zoologist – E.H. Hackel to include those unicellular microorganisms that are typically neither plants nor animals, bacteria, algae, fungi and protozoa. Most microorganisms are unicellular where all the life processes are performed by a single cell. All unicellular microorganisms belong to the kingdom of Protista. In the so-called higher forms of life, organisms are composed of many cells that are arranged in tissues and organs to perform specific functions. Regardless of the complexity of an organism, the cell is the basic structural and functional unit of life. All the living cells are fundamentally similar.

Figure1 shows a broad classification of the protists. Microorganisms are classified into two large groups: lower protists, the prokaryotes and higher protists, the eukaryotes. Prokaryotic cells are characterized by having no nuclear membrane, having a closed loop of double stranded deoxyribonucleic acid (DNA) where the genetic information is encoded and cytoplasm in which ribonucleic acids (RNA) concerned with transcription and translation of the genetic message and ribosomes, which control and direct the synthesis of protein molecules (enzymes), reside. Eukaryotes, a large class which includes yeasts, fungi, animal and plant cells, have more than one DNA molecule enclosed in a nuclear membrane constituting the nucleus. Eukaryotes are characterized by membrane structures: vacuoles, mitochondria, chloroplasts and endoplasmic reticulum. Typical features of prokaryotic and eukaryotic cells are shown in Table 2. Another system of classification of living organisms, based on three modes of nutrition – photosynthesis, absorption and ingestion, has five kingdoms and was proposed by R.H. Whittaker in 1969 and is indicated in Figure 2. The main features of each of these kingdoms are given below:

1. *Monera:* Prokaryotes – lack ingestive mode of nutrition – bacteria and cyanobacteria
2. *Protista:* Unicellular eukaryotes – all three nutritional types – microalgae and protozoa
3. *Plantae:* Multicellular and multinucleate eukaryotes – green plants and higher algae
4. *Animalia:* Multicellular animals – ingestive
5. *Fungi:* Multinucleate higher fungi – absorptive – yeasts and moulds

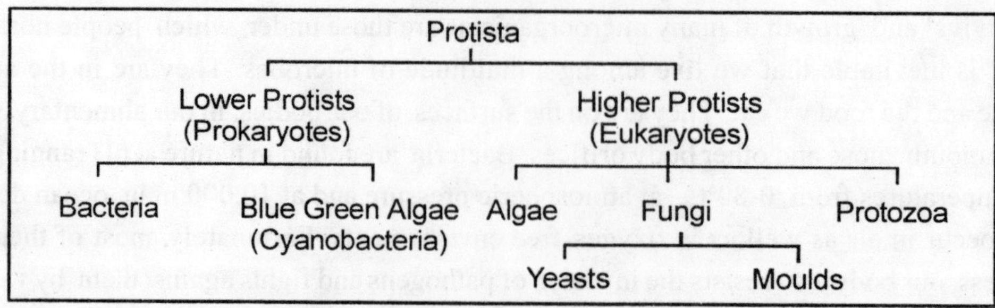


**Figure 1** Broad classification of Protists

**Table 2.** Some features distinguishing prokaryotic cells from eukaryotic cells

| Feature | Prokaryotic cells | Eukaryotic cells |
|---|---|---|
| Microorganisms where found as unit of structure | Bacteria | Fungi, algae and protozoa |
| Size range of microorganisms | 1-2 × 1-4 μm | More than 5 μm |
| Location of genetic system | Nucleoid, chromatin body or nuclear material | Nucleus, mitochondria, chloroplasts |
| Structure of nucleus | Not bounded by nuclear membrane | Bounded by nuclear membrane |
| Number of chromosomes | One circular chromosome not containing histones | More than one chromosome having histones |
| Nucleolus | Absent | Present |
| Sexuality | Zygote is partial diploid (merozygotic) | Zygote is diploid |
| Cytoplasmic streaming | Absent | Present |
| Pinocytosis | Absent | Present |
| Gas vacuoles | May be present | Absent |
| Mesosomes | Present | Absent |
| Ribosomes | 70S, distributed in cytoplasm | 80S, arrayed on membranes in cytoplasmic reticulum, 70S in mitochondria and chloroplasts |
| Mitochondria | Absent | Present |
| Chloroplasts | Absent | Present |
| Golgi apparatus | Absent | Present |
| Endoplasmic reticulum | Absent | Present |
| Membrane bound vacuoles | Absent | Present |
| Structure of cytoplasmic membrane | Contains part of respiratory and in some cases photo-synthetic machinery. Generally donot contain sterols | No part of respiratory or photosynthetic machinery. Contain sterols |
| Cell wall structure | Peptidoglycan as the main component | Peptidoglycan absent |
| Flagellar structure | Simple fibril | Multifibrilled with 9+2 microtubules |
| Metabolic mechanisms | Wide variation particularly in anaerobic energy-yielding reactions, some fix atmospheric nitrogen, some accumulate poly-ß-hydroxybutyrate as reserve material | Glycolysis is the pathway for anaerobic energy-yielding mechanism. Not able to fix nitrogen, not able to accumulate poly-ß-hydroxybutyrate as reserve material |
| DNA base ratio (G+C%) | 28-73 | About 40 |

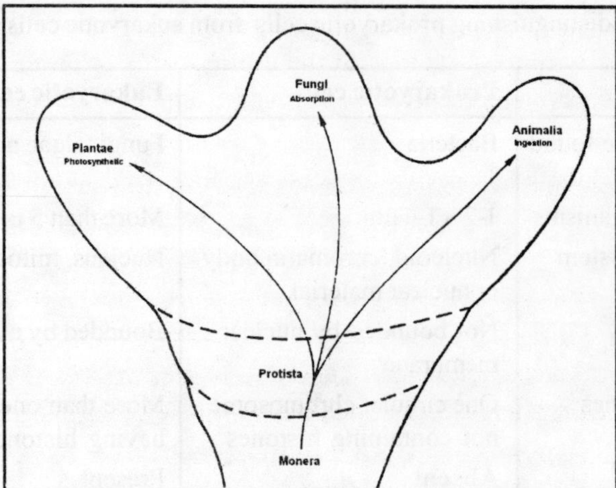


**Figure 2** A schematic representation of Whittaker's five-kingdom system

Another way of classification of microorganisms, giving prominence to more primitive and ancient forms of life, divides prokaryotes into archaebacteria and eubacteria. Archaebacteria are distinguished from other bacteria by profound differences and are found in rather extreme (special) habitats. These extreme conditions seem to resemble those supposed to have prevailed in the earliest times of the earth's development i.e. archaeic times. The archaebacteria include lithoautotrophic and heterotrophic aerobes and anaerobes. The members of this group have many properties in common that distinguish them from eubacteria. According to the present state of reserch the archaebacteria can be subdivided into three groups: methanogenic bacteria, halophilic bacteria and thermo-acidiphilic bacteria. The archaebacterial cell wall does not contain a peptidoglycan skeleton, only proteins and polysaccharides are present. This explains why archaebacteria are not sensitive to antibiotics that act on eubacterial cell wall. Eubacteria include all other well known bacteria such as *Bacillus subtilis* and *Escherichia coli.* The microorganisms with which we are most concerned are bacteria, yeasts and microfungi or moulds employed for various fermentations.

## 4.1. Types of microorganisms and their distribution

Microorganisms can be divided into six main types: Bacteria, Archaea, Fungi, Protista, Viruses and Microbial Mergers. The images of some representative microbial types are exhibited in Figure 3.

**4.1.1.** *Bacteria:* Bacteria are unicellular prokaryotic organisms or simple associations of similar cells which are relatively small ranging from 0.3-8 μm in size. Originally the bacteria were considered microscopic fungi (called Schizomycetes), except for the photosynthetic cyanobacteria, which were considered as a group of algae (called Cyanophyta or blue-green algae). It was only with the study of detailed cell structure it was realized that they form a fundamental group, separate from the other organisms. In 1956, Copeland gave them their own separate kingdom Mycota, later renamed Monera, Prokaryota, or Bacteria.

During the 1960s the concept was refined and bacteria (now including cyanobacteria) were recognized as one of two major divisions of the living world, together with the eukaryotes. Eukaryotes were generally believed to have evolved from bacteria, later from assemblies of bacteria.

The advent of molecular systematics challenged this view. In 1977, Woese divided the prokaryotes into two groups, based on 16S rRNA sequences, called the kingdom Eubacteria and Archaebacteria. He argued that each of these and the eukaryotes all evolved separately and in 1990 emphasized this by promoting them to domains, which were renamed the Bacteria, Archaea, and Eucarya. This redefinition has generally been accepted by molecular biologists but criticized by some others, who maintain that he over-emphasized a few genetic differences and that both archaebacteria and eukaryotes probably developed from within the eubacteria.

Bacteria come in a wide variety of shapes:

- Rod-shaped
- Round-shaped or spherical.
- Round-shaped in clusters.
- Round-shaped in twos.
- Spiral-shaped.
- Comma-shaped.

Bacteria come in a variety of different shapes. Most are rod-shaped, sphere-shaped, or helix-shaped; these are respectively referred to as bacilli, cocci, and spirilla. An additional group, vibrios, are comma-shaped. Shape is no longer considered a defining factor in the classification of bacteria, but many genera are named for their shape (e.g. *Bacillus*, *Streptococcus*, *Staphylococcus*) and it is an important part in their identification.

Cell multiplication in bacteria is by binary fission and under favourable conditions they double in size, mass and number every 30-50 min. They exist in three morphological forms, spirals (spirillae), spheres (cocci) and rods (bacilli) and are sometimes filamentous e.g. cyanobacteria and streptomycetes. Most are capable of motion (motile) and reproduce by division into two daughter cells (binary fission). All bacteria incorporate N-acetylglucosamine (NAG) molecules into their cell wall. The peptidoglycan murein gives strength to the cell wall and shape to the cells. A subdivision of bacteria may be distinguished by a characteristic 'Gram reaction'. The term Gram- positive refers to those bacteria rich in murein which are stained by crystal violet whereas cells poor in murein donot stain and are thus Gram-negative. Endospores are a dormant form of cell, able to resist heat, radiation and poisonous chemicals. Under adverse conditions some bacteria are capable of endospore formation. In a favourable environment endospores germinate to restore the biologically active cell state, the vegetative form.

Another important tool is Gram staining, named after Hans Christian Gram who developed the technique. This separates bacteria into two groups, based on the composition of their cell wall. The first formal grouping of bacteria into phyla was based largely on this test:

- Gracilicutes - bacteria with a second cell membrane containing lipids, giving them Gram-negative stains
- Firmicutes - bacteria with a single membrane and thick peptidoglycan wall, giving them Gram-positive stains
- Mollicutes - bacteria with no second membrane or wall, giving them Gram-negative stains

The archeabacteria were originally included as the Mendosicutes. As given, these phyla are no longer believed to represent monophyletic groups. The Gracilicutes have been divided into many different phyla. Most gram-positive bacteria are placed in the phyla Firmicutes and Actinobacteria, which are closely related. However, the Firmicutes have been redefined to include the mycoplasmas (Mollicutes) and certain gram - negative bacteria.

**Figure 3** Images of some representative microbial types Clockwise from upper left: *Amoeba proteus*, a member of the protista; a close-up view of a fog-desert *Niebla,* lichen [mergers]; basidiospores of the shiitake mushroom, *Lentinula edodes* [fungi]; *Trypanosoma brucei rhodesiense* [bacteria]; rotavirus [viruses]. *(For a coloured version of this figure, see plate section, page 555)*

Bacteria reproduce only asexually, not sexually. Specifically they reproduce by binary fission, or simple cell division. During this process, one cell divides into two daughter cells with the development of a transverse cell wall.

However, genetic variations can occur within individual cells through recombinant events such as mutation (random genetic change within a cell's own genetic code). Similar to more complex organisms, bacteria also have mechanisms for exchanging genetic material. The end result is that a bacterium contains a combination of traits from two different parental cells. Three different modes of exchange have thus far been identified in bacteria:

1. Transformation (the transfer of naked DNA from one bacterial cell to another in solution, this can include dead bacteria),
2. Transduction (the transfer of viral, bacterial, or both bacterial and viral DNA from one cell to another via bacteriophage) and;
3. Bacterial conjugation (the transfer of DNA from one bacterial cell to another via a special protein structure called a conjugation pilus).

Bacteria, having acquired DNA from any of these events, can then undergo fission and pass the recombined genome to new progeny cells. Many bacteria harbor plasmids that contain extrachromosomal DNA. Under favourable conditions, bacteria may form aggregates visible to the naked eye, such as bacterial mats.

Bacteria show a wide variety of different metabolisms and can accordingly be classified into primary nutritional groups. Heterotrophs depend on an organic source of carbon, while autotrophs are able to synthesize organic compounds from carbon dioxide and water. Autotrophs that obtain energy by oxidizing chemical compounds are called chemotrophs, and those that obtain their energy from light, via photosynthesis, are called phototrophs. There are many variations on this terminology such as chemoautotrophs and photosynthetic autotrophs and so on. In addition, bacteria are distinguished based on the source of reducing equivalents they are using. Those using inorganic compounds (e. g. water, hydrogen sulfide or ammonia) for this purpose are called lithotrophs and others needing organic compounds (e. g. sugars or organic acids) are called organotrophs. The metabolic modes of energy metabolism (phototrophy or chemotrophy), reducing equivalent sources (lithotrophy or organotrophy) and carbon sources (autotrophy or heterotrophy) can be combined differently in any single microorganism, and even shifting between different modes frequently occurs in many species.

The photolithoautotrophs include the cyanobacteria, which are some of the oldest organisms known from the fossil record and probably played an important role in creating the earth's oxygen atmosphere. They apparently pioneered the use of water as (lithotrophic) electron source and were the first to use the photosynthetic water splitting apparatus. Other photosynthetic bacteria use different electron sources and therefore do not produce oxygen. These anoxygenic phototrophs fall into four phylogenetic groups: the green sulfur bacteria, green non-sulfur bacteria, purple bacteria, and heliobacteria.

Other nutritional requirements include nitrogen, sulfur, phosphorus, vitamins and metallic elements such as sodium, potassium, calcium, magnesium, manganese, iron, zinc, cobalt, copper and nickel for normal growth. For some species, additional trace elements such as selenium, tungsten, vanadium or boron are needed.

Based on their response to oxygen, most bacteria can be placed into one of three groups: Some bacteria can grow only in the presence of oxygen and are called aerobes; others can grow only in the absence of oxygen and are called anaerobes; and some can grow in the presence or absence of oxygen and are called facultative anaerobes. Bacteria that do not utilize oxygen for respiration but still grow in its presence are called aerotolerant. Bacteria also thrive in environments that are considered extreme for mankind. These organisms are called extremophiles. Some bacteria inhabit hot springs and are called thermophiles; others inhabit highly saline lakes and are called halophiles; yet others inhabit acidic or alkaline environments and are called acidophiles and alkalophiles, respectively; and still others inhabit alpine glaciers and are called psychrophiles.

Often dismissed as "germs" that cause illness, bacteria help us do an amazing array of useful things, like make vitamins, break down some types of garbage, and maintain our atmosphere.

Bacteria can be found virtually everywhere. They are in the air, the soil, and water, and in and on plants and animals, including us. A single teaspoon of topsoil contains about a billion bacterial cells. The human mouth is home to more than 500 species of bacteria. Some bacteria thrive in the most forbidding, uninviting places on Earth, from nearly-boiling hot springs to super-chilled Antarctic lakes buried under sheets of ice. Microbes that dwell in these extreme habitats are aptly called extremophiles.

Bacteria are both harmful and useful to the environment, and animals, including humans. The role of bacteria in disease and infection is important. Some bacteria act as pathogens and cause tetanus, typhoid fever, pneumonia, syphilis, cholera, foodborne illness and tuberculosis. Sepsis, a systemic infectious syndrome characterized by shock and massive vasodilation, or localized infection, can be caused by bacteria such as *Streptococcus, Staphylococcus*, or many gram-negative bacteria. Some bacterial infections can spread throughout the host's body and become systemic. In plants, bacteria cause leaf spot, fireblight, and wilts. The mode of infection includes contact, air, food, water and insect-borne microorganisms. The hosts infected with the pathogens may be treated with antibiotics, which can be classified as bacteriocidal and bacteriostatic, which at concentrations that can be reached in bodily fluids either kill bacteria or hamper their growth, respectively. Antiseptic measures may be taken to prevent infection by bacteria, for example, prior to cutting the skin during surgery or swabbing skin with alcohol when piercing the skin with the needle of a syringe. Sterilization of surgical and dental instruments is done to make them sterile or pathogen-free to prevent contamination and infection by bacteria. Sanitizers and disinfectants are used to kill bacteria or other pathogens to prevent contamination and risk of infection.

In soil, microorganisms help in the transformation of nitrogen to ammonia with enzymes secreted by these microbes, which reside in the rhizosphere (a zone that includes the root surface and the soil that adheres to the root after gentle shaking). Some bacteria are able to use molecular nitrogen as their source of nitrogen, converting it to nitrogenous compounds, a process known as nitrogen fixation. Many other bacteria are found as symbionts in humans and other organisms. For example, their presence in the large intestine can help prevent the growth of potentially harmful microbes.

The ability of bacteria to degrade a variety of organic compounds is remarkable. Highly specialized groups of microorganisms play important roles in the mineralization of specific classes of organic compounds. For example, the decomposition of cellulose, which is one of the most abundant constituents of plant tissues, is mainly brought about by aerobic bacteria that belong to the genus *Cytophaga*. This ability has also been utilized by humans in industry, waste processing, and bioremediation. Bacteria capable of digesting the hydrocarbons in petroleum are often used to clean up oil spills. Some beaches in Prince William Sound were fertilized in an attempt to facilitate the growth of such bacteria after the infamous 1989 Exxon Valdez oil spill. These efforts were effective on beaches that were not too thickly covered in oil.

Bacteria, often in combination with yeasts and molds, are used in the preparation of fermented foods such as cheese, pickles, soy sauce, sauerkraut, vinegar, wine, and yoghurt. Using biotechnology techniques, bacteria can be bioengineered for the production of therapeutic drugs, such as insulin, or for the bioremediation of toxic wastes.

Bacteria and their microbial cousins, the archaea were the earliest forms of life on earth and may have played a role in shaping our planet into one that could support the larger life forms we know today by developing photosynthesis. Cyanobacteria fossils date back more than 3 billion years. These photosynthetic bacteria paved the way for today's algae and plants. Cyanobacteria grow in the water, where they produce much of the oxygen that we breathe. Once considered a form of algae, they are also known as blue-green algae. Like dinosaurs, bacteria left behind fossils (Figure 4). The main difference is that it takes a microscope to see them.

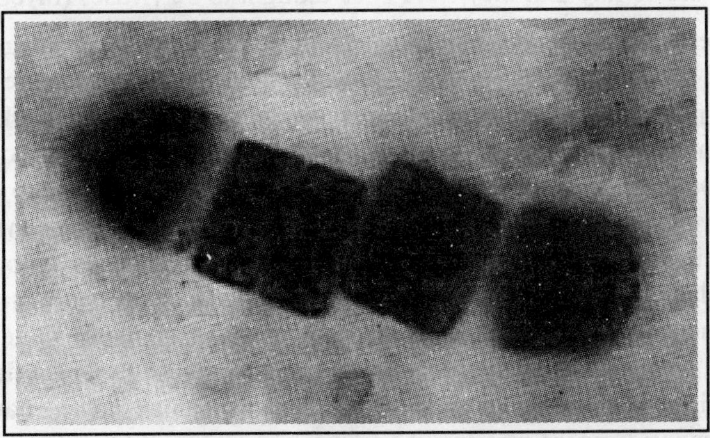

**Figure 4** Cyanobacteria fossil *(For a coloured version of this figure, see plate section, page 555)*

A bacterium consists of a single cell whose genetic information is contained in a single DNA molecule suspended in a jelly-like substance called cytoplasm (Fig 5). In most cases, this and other cell parts are surrounded by a flexible membrane that is itself surrounded by a tough, rigid cell wall. A few species, such as the mycoplasmas, don't have cell walls. Even though bacteria have only one cell each, they come in a wide range of shapes, sizes, and colors.

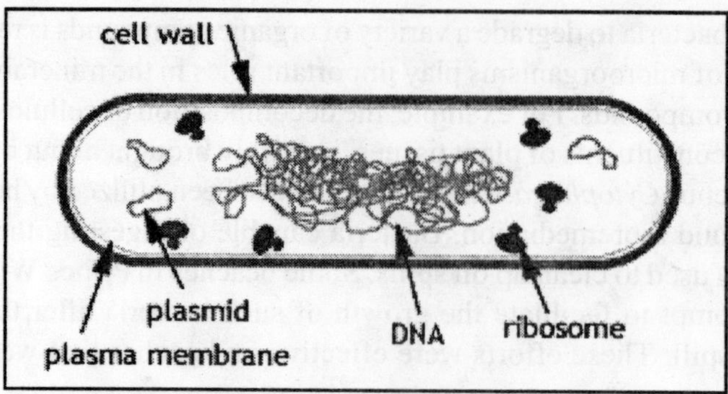


**Figure 5** Generalised diagram of a bacterial cell

Some bacteria look like little balls (*Micrococcus*) while others appear like tangled strings or corkscrews (*Leptospira*) under a microscope. Others look like medicine capsules (*Salmonella*) or segmented ribbons (cyanobacteria) or sticks. Still others look like fat commas (vibrios). Some bacteria are stalked (*Caulobacter*) while others have buds (*Rhodomicrobium*). Some have sheaths (*Sphaerotilus*) while others don't. Bacteria like mycoplasmas that lack a hard cell wall don't have any particular shape at all.

Just like in animals, where size ranges from the giant blue whale to the tiny gnat, bacteria vary from 1 millimeter in diameter at the largest end of the scale to 20 nanometers in length at the smallest. The largest bacteria found so far can actually be seen without the use of a microscope (*Thiomargarita namibiensis* and *Epulopiscium fischelsoni*). The smallest known bacteria are so tiny that they were once thought to be viruses (mycoplasmas).

Some bacteria have hair- or whip-like proteinaceous helical appendages, 20-30 nm in diameter and 15 μm long, emerging from cell wall called flagella which are used to 'swim' around. Others produce thick coats of slime and 'glide' about. Some stick out thin, rigid spikes called fimbriae to help 'hold' them to surfaces. Motile bacteria can move about, either using flagella, bacterial gliding, or changes of buoyancy. A unique group of bacteria, the spirochaetes, have structures similar to flagella, called axial filaments, between two membranes in the periplasmic space. They have a distinctive helical body that twists about as it moves. Bacterial flagella are arranged in many different ways. Bacteria can have a single polar flagellum at one end of a cell (monotrichous), clusters of many flagella at one end (lophotrichous), single or clusters at both the cell poles (amphitrichous) or flagella scattered all over the cell, as with peritrichous. Many bacteria (such as *E.coli*) have two distinct modes of movement: forward movement (swimming) and tumbling. The tumbling allows them to reorient and introduces an important element of randomness in their forward movement.

Motile bacteria are attracted or repelled by certain *stimuli*, behaviours called *taxes* - for instance, chemotaxis, phototaxis, mechanotaxis and magnetotaxis. In one peculiar group, the myxobacteria, individual bacteria attract to form swarms and may differentiate to form fruiting bodies. The myxobacteria move only when on solid surfaces, unlike *E. coli* which is motile in liquid or solid media.

Some contain little particles of minerals that orient with the planet's magnetic fields to help the bacteria figure out whether they're swimming up or down. Some bacteria do have natural colors. Certain species contain pigments, such as various chlorophylls, that make them naturally green, yellow, orange, or brown. Colonies of millions of bacteria may appear pink, yellowish, or white.

Cell wall in gram-ve is thinner (10-15nm) than in gram+ve bacteria (20-25nm) and accounts for 10-40% of the dry weight of the cell. Peptidoglycan , also termed as murein, a polymer of N-acetylglucosamine, N-acetylmuramic acid, L-alanine, D-alanine, D-glutamate, and a diamino acid (L,L-diaminopimelic acid, L-lysine, L-ornithine or diaminobutyric acid) is the main rigid component in the bacterial cell wall and accounts for 50% and 10% of cell wall components in gram+ve and in gram-ve cells respectively. Other major substances in cell wall of gram+ve bacteria include polysaccarides, covalently linked to murein (*Streptococcus pyogenes*), teichoic acids, acidic polymers of ribitol phosphate or glycerol phosphate covalently linked to peptidoglycan (*Staphylococcus aureus*, *Streptococcus faecalis*) – bind magnesium ions – protect the bacteria from thermal injury.

The cell wall in gram-ve bacteria is in the form of two layers, the outer one similar to cell membrane and termed as outer membrane which is attached to peptidoglycan, present in inner layer, by Braun's lipoprotein. The outer membrane is bilayered having phospholipids, proteins and lipopolysaccharides (LPS) which has endotoxic properties. LPS has three covalently linked parts, Lipid-A :firmly embedded in membrane, Core polysaccharide : located at membrane surface, O-antigens: which extend like whiskers from the membrane surface into the surrounding medium. Most of serological properties of gram-ve bacteria are due to O-antigens. The cell walls of gram-ve bacteria also contain special proteins having channels in them for transporting smaller molecules of nucleosides, oligosaccharides, monosaccharides, peptides and amino acids. Many porins also serve as receptors for bacteriophages. The two membranes in gram-ve bacteria have some points of direct contact known as adhesions.

Some bacterial cells are surrounded by extracellular polymeric substance (EPS), known as capsule which forms an envelope around the cell wall. The gelatinous polymer is composed of polysaccharide (*Klebsiella pneumoniae*) or polypeptide (*Bacillus anthracis*) or both. Capsule is antiphagocytic, prevents attachment of bacteriophages, protects cell against desiccation and helps the cell in attachment on to various supports.

Some gram-ve bacterial cells have proteinaceous hair like structures in the surface of cell wall which are smaller in length (1-2 µm long) and breadth (250A° in dia) than flagella but are more in number. These are called fimbriae or pili which have antigenic properties, help in the attachment of cells and exchange of genetic material between the two bacterial cells during conjugation.

Inner to the bacterial cell wall is plasma membrane which is 7.5 nm thick and is composed of 60-70% proteins, 20-30% phospholipids, 1-5% oligosaccharides and 20% water. Phospholipids form a bi-layer in which most of the proteins are held (integral proteins). Other proteins are only loosely attached (peripheral proteins). The plasma membrane helps in the transport of inorganic and organic molecules, contains enzymes for peptidoglycan synthesis, teichoic acid synthesis, polysaccharide synthesis, lipopolysacharide and phospholipids synthesis and has attachment sites for bacterial chromosome and plasmid DNA. Inner membrane is invaginated to form infoldings, known as mesosomes which play role in respiration and also in reproduction. The generalized diagrams of typical gram+ve and gram-ve envelopes are depicted in Figure 6.

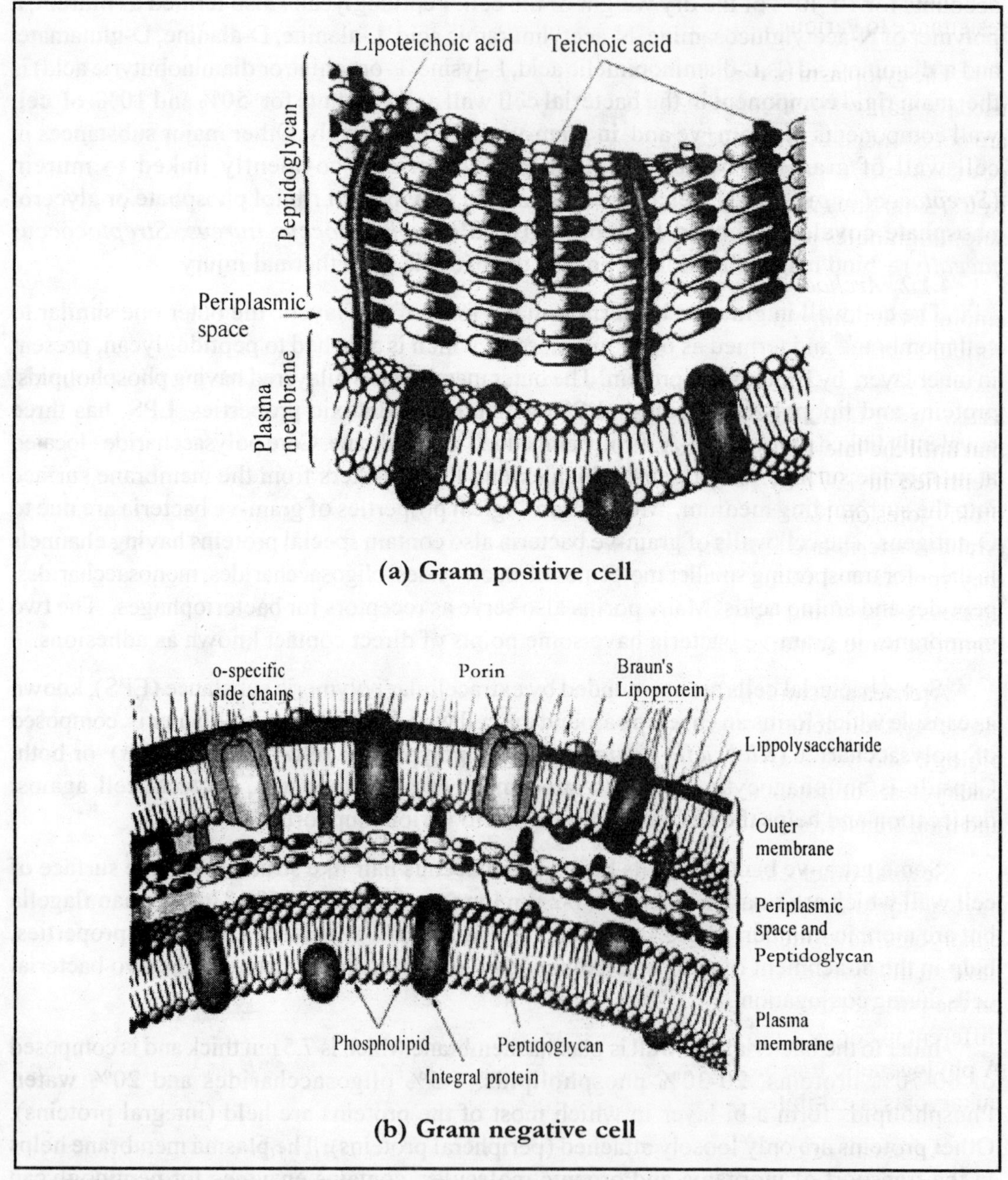


**Figure 6 (a,b)** Generalised diagrams of typical gram +ve and gram -ve bacterial envelopes

The internal matrix inside plasma membrane, known as cytoplasm, is composed of 80% water with proteins, carbohydrates, lipids, inorganic ions, DNA and ribosomes. Each bacterial cell has approximately 10,000 ribosomes which account for 30% of total dry weight of cell. High number of ribosomes gives cytoplasm a granular appearance. The cytoplasm has a single chromosome which is in the form of nucleoid, without a nuclear membrane and histones. Some of the bacterial cells have extrachromosomal covalently closed circular segments of DNA, knowm as plasmids, which have markers for sex factor, resistance to heavy metals, resisitance to various antibiotics and formation of proteinaceous toxins, colicins etc.

The cytoplasm of the cell also contains various granules of organic or inorganic materials, used for storage of carbon compounds, inorganic substances and energy. Some of the granules are not bounded by a membrane and lie freely in cytoplasm including polyphosphate granules or volutin granules, cyanophycean granules and some glycogen granules. Others are enclosed by a 2-4 nm thick membrane including poly-β-hydroxybutyrate granules, some glycogen and sulphur granules and some gas vacuoles

**4.1.2.** *Archaea:* The Archaea are a major group of prokaryotes. Their classification among living things was difficult because they are similar to *bacteria* in some respects but similar to eukaryotes in others. Archaea look and act a lot like bacteria and are the primitive bacteria which have a defective cell wall which is without peptidoglycan. These are considered as the living fossils that are providing clues to the earliest forms of life on Earth. So much so that until the late 1970s, scientists assumed they were a kind of "weird" bacteria. They were identified in 1977 by Carl Woese and George Fox based on their separation from other prokaryotes on 16S rRNA phylogenetic trees. These two groups were originally named the Archaebacteria and Eubacteria, treated as kingdoms or subkingdoms. Woese argued that they represented fundamentally different branches of living things. He later renamed the groups Archaea and Bacteria to emphasize this, and argued that together with Eukarya they comprise three domains of living things.

Archaea are similar to other prokaryotes in most aspects of cell structure and metabolism. However, their genetic transcription and translation - the two central processes in molecular biology - do not show the typical bacterial features, but are extremely similar to those of eukaryotes. For instance, archaean translation uses eukaryotic initiation and elongation factors, and their transcription involves TATA-binding proteins and TFIIB as in eukaryotes.

Several other characteristics also set the Archaea apart. Unlike most bacteria, they have a single cell membrane that lacks a peptidoglycan wall. Further, both bacteria and eukaryotes have membranes composed mainly of glycerol-ester lipids, whereas archaea have membranes composed of glycerol-ether lipids. These differences may be an adaptation on the part of Archaea to hyperthermophily. Archaeans also have flagella that are notably different in composition and development from the superficially similar flagella of bacteria. A phylogenetic tree based on *rRNA* data, showing the separation of bacteria, archaea, and eukaryotes is exhibited in Fig 7.

Archaea are divided into two main groups based on rRNA trees, the Euryarchaeota and Crenarchaeota. Two other groups have been tentatively created for certain environmental samples and the peculiar species *Nanoarchaeum equitans*, discovered in 2002 by *Karl Stetter*, but their affinities are uncertain. Woese argued that the bacteria, archaea, and eukaryotes each represent a primary line of descent that diverged early on from an ancestral progenote with poorly developed genetic machinery. This hypothesis is reflected in the name Archaea, from the Greek *archae* or ancient. Later he treated these groups formally as domains, each comprising several kingdoms. This division has become very popular, although the idea of the progenote itself is not generally supported. Some biologists, however, have argued that the archaebacteria and eukaryotes arose from specialized eubacteria.

The relationship between Archaea and Eukarya remains an important problem. Aside from the similarities noted above, many genetic trees group the two together. Some place eukaryotes closer to Euryarchaeota than Crenarchaeota are, although the membrane chemistry suggests otherwise. However, the discovery of archaean-like genes in certain bacteria, such as *Thermotoga*, makes their relationship difficult to determine. Some have suggested that eukaryotes arose through fusion of an archaean and eubacterium, which became the nucleus and cytoplasm, which accounts for various genetic similarities but runs into difficulties explaining cell structure.

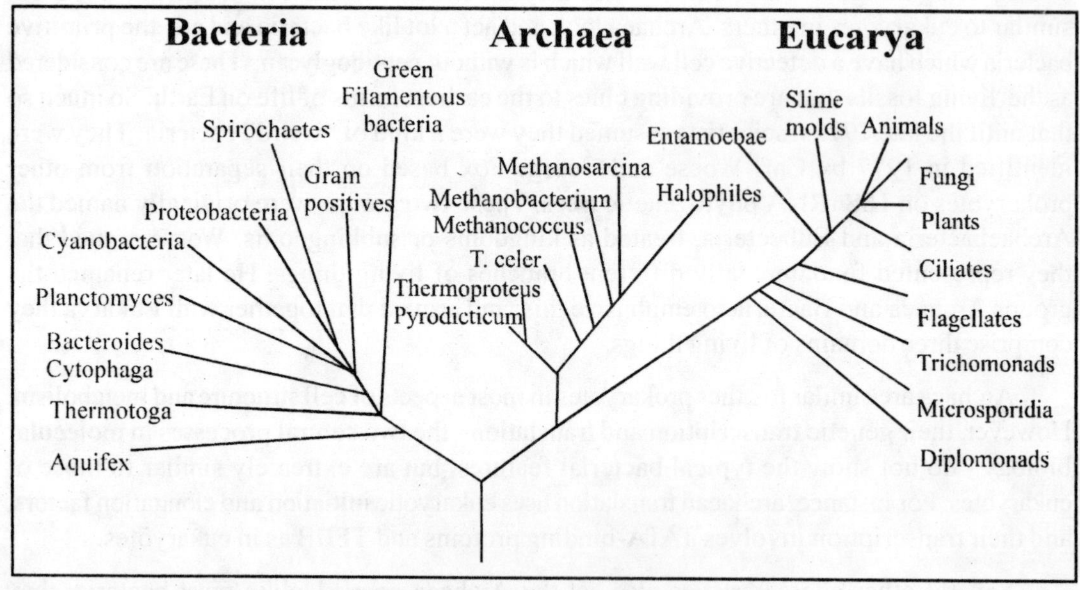


**Figure 7** Phylogenetic tree of life

Archaea comes from the Greek word meaning "ancient." An appropriate name, because many archaea thrive in conditions mimicking those found more than 3.5 billion years ago. Back then, the earth was still covered by oceans that regularly reached the boiling point — an extreme condition not unlike the hydrothermal vents and sulfuric waters where archaea are found today. Archaea can be found in many "extreme environments". In addition to superheated waters, some archaea have been found in acid-laden streams around old mines, in frigid Antarctic ice and in the super-salty waters of the Dead Sea. The extremophilic archaea, which are also sometimes termed as extremists, have been classified into various categories. These include i) Thermophiles which grow at unusually hot temperatures. A few species have been found to survive even above 110°C, ii) Psychrophiles which grow at extremely cold temperatures (even down to -10°C, iii) Halophiles which thrive in unusually salty habitats. Some can thrive in water with 9% salt, iv) Acidophiles which prefer acidic conditions and v) Alkalophiles which prefer very alkaline environments. However, other archaeans are mesophiles, and have been found in environments like marshland, sewage, and soil. Many methanogenic archaea are found in the digestive tracts of animals such as ruminants, termites, and humans. Archaea are usually harmless to other organisms and none are known to cause disease.

Individual archaeans range from 0.1 to over 15 μm in diameter, and some form aggregates or filaments up to 200 μm in length. They occur in various shapes, such as spherical, rod-shaped, spiral, lobed, or rectangular. They also exhibit a variety of different types of metabolism. The halobacteria can use light to produce ATP, although no Archaea conduct photosynthesis with an electron transport chain, as occurs in other groups.

Archaea that populate extreme environments (along with some of their bacterial cousins) have developed some clever tricks and tools to do so. For example, they produce special enzymes that help keep all the parts of their cells intact even in conditions that would have our human skin falling apart. Some archaea look like little rods or tiny balls, and some even get around like bacteria, using long hair- or whip-like appendages called flagella that stick out of their cell walls and act like a microscopic outboard motor to get them where they are going. Like bacteria, archaea lack a true nucleus. Both bacteria and archaea usually have one DNA molecule suspended in the cell's cytoplasm contained within a cell membrane. Most, but not all, have a tough, rigid outer cell wall. Although many archaea have tough outer cell walls, these walls contain different kinds of amino acids and sugars than those found in bacteria. Archaeal cell membranes also are chemically distinct from bacterial membranes with differing lipid structures and chemical links. This means that drugs that slow or kill bacteria by interfering with their ability to produce certain key proteins have no effect on archaea.

**4.1.3.** *Fungi:* Fungi are a group of organisms which belong to one of the four kingdoms of eukaryotes, the others being animals, plants and protists. Long counted among the plantae they are nowadays considered to be more closely related to the animals and grouped together with these in the taxon of opisthokonts. Notable differences from animals include the mostly saprobiontic nutrition of fungi and in many cases the presence of a cell wall. This cell wall comprises chitin as a structural component, which together with their chemo-organo-heterotrophy distinguishes fungi from the photoautotrophic plants. Chitin cell walls and haploid nuclei differentiate them from structurally similar protists such as water moulds.

Some also consider fungi as the eukaryotic lower plants without chlorophyll. These may be unicellular or multicellular but not differentiated into roots, stems and leaves. True fungi are composed of filaments and masses of cells - mycelium. Reproduction occurs by budding or fission or by means of spores borne on fruiting structures. They can do everything from helping to bake bread, to make wine and to decompose waste. These are spore forming microscopic organisms playing important role in industrial fermentations for beer, wine, and antibiotics production. They range in size from the single-celled organism, the yeast, to the largest known living organism on earth, a 3.5-mile-wide mushroom. Dubbed "the humongous fungus," this honey mushroom (*Armillaria ostoyae*) covers some 2,200 acres in Oregon's Malheur National Forest, USA (Figure 8).

Fungi occur in all environments on the planet and include important decomposers and parasites. Parasitic fungi infect animals, including humans, other mammals, birds, and insects, with consequences varying from mild itching to death. Other parasitic fungi infect plants, causing diseases such as butt rot and making trees more vulnerable to toppling. The vast majority of vascular plants are associated with mutualistic fungi, called mycorrhizae, which assist their roots in absorption of nutrients and water.

Fungi can be found in rising bread, moldy bread, and old food in the refrigerator, and on forest floors. Most decompose non-living things, but some damage crops and plants. A few cause problems in people, such as *Candida*, which causes yeast infections. Many decades ago, scientists thought that fungi were primitive kinds of plants. New studies looking at the DNA of fungi have confirmed that these organisms are not plants. Unlike plants, fungi do not make their own food energy via photosynthesis, but dine on organic matter like rotting leaves, wood, and other debris, or upon the tissues of living plants and animals. Fungi, along with bacteria, are the planet's major composters and recyclers.

**Figure 8. (a)** The only above-ground signs of the humongous fungus are patches of dead trees and **(b)** the mushrooms that form at the base of infected trees *(For a coloured version of this figure, see plate section, page 555)*

Although fungi may seem like a nuisance when they dine in your fruit bowl or refrigerator, their ability to degrade some of the toughest organic materials, including tree wood and insect exoskeletons, means that our planet is not cluttered with a mass of debris. Fungi dine at home, eating whatever they're growing on. Fungi secrete digestive enzymes in order to break down complex food sources, such as animal corpses and tree stumps, into smaller components they can absorb. Most fungi can best be described as grazers, but a few are active hunters. Hunter fungi prey on tiny protozoa and worm-like creatures called nematodes. Some produce a sticky substance on their hyphae, which then act like flypaper, trapping passing prey. A species called *Arthrobotrys dactyloides* sets snares made out of loops formed by its hyphae. When a nematode comes into contact with the loops, the movement triggers the fungal cells to swell with fluid, constricting the loop like a noose around the hapless nematode. Other hyphae then grow toward the trapped prey, eventually punching through its body where they begin absorbing its fluids.

Fungi usually reproduce without sex. Single-celled yeasts reproduce asexually by budding. A single yeast cell can produce up to 24 offsprings. Fungi that make hyphae can reproduce asexually as well. Bits of the hyphae can break off and continue to grow as separate entities, or can form stalks containing seed-like spores. Although less common, fungi can produce spores sexually. Two mating cells from hyphae of different strains of fungi can mate by fusing together and forming a spore stalk.

Yeasts which form one of the most important subclasses of fungi are rather more complex and usually larger than bacteria. They are distinguished from most of the fungi by their usual existence as single ovoid, elongate, elliptical or spherical cell shapes and by the production of buds during the process of division of cells which are about 8μm long and 5 μm in diameter, doubling every 1-3 hours in favourable media. Yeasts can grow over wide range of pH, alcohol and sugar concentrations. The true yeasts or ascosporogenous yeasts reproduce by sexual reproduction involving copulation to varying degrees. True yeasts can also produce asexual spores and chlamydospores. The latter are durable and tend to be produced when a culture encounters unfavourable conditions of growth. Most yeasts used industrially are hemiascomycetes and most are in the genus *Saccharomyces*. Yeasts are significant in wine and beer making because they conduct the alcoholic fermentation, they can spoil wines during conservation in the cellar and after packaging and they affect wine quality through autolysis.

Yeasts are often placed into groupings based upon some particular function or activity, for example, film yeasts are those that grow at the surface of certain acid products such as sauerkraut and pickles and include the species and strains of genera *Candida* and *Hansenula*. These organisms are capable of oxidising acids and alcohols as sources of energy. Top yeasts are those which carry out the conversion of sugars to alcohol at the top of a vessel while bottom yeasts are those that can do the same but from the bottom of the vessel. Apiculate or lemon shaped yeasts are undesirable in wine fermentation where they produce off-flavours.

Moulds are the heterotrophic multicellular higher fungi with branching vegetative structures, the mycelia. Mycelia are formed from thin filaments of cells, the hyphae. When grown on solid surfaces, mycelia ramify as a delicate branching, dense web absorbing nutrients and produce spores for reproduction, survival and dissemination purposes. The hyphae are multinucleate with the genetic material contained in nuclear membranes and the cell walls enclose mitochondria and endoplasmic reticulum. Their usual habitat is the soil. Generally speaking they are strict aerobes and the earlier industrial processes based on moulds, the fermentation production of antibiotics, were conducted by growing moulds now used for industrial production of enzymes including cellulase, amylase, protease and lipase enzymes.

The fungal cell wall is made of hemicellulose or chitin with some cellulose, proteins and lipids also embedded. The growth is distal where young hypha may get divided in to cells by cross walls which are formed by centripetal invagination. The cross walls grow inwards to form an incomplete septum that has a central pore which allows the protoplasm streaming. Even nuclei may migrate from cell to cell in hypha. There are 3 forms of hyphae (Fig 9): i) Non septate or coenocytic, ii) Septate with uninucleate cells, and iii) Septate with multinucleate cells. Mycelia can be either vegetative or reproductive. Some hyphae of the vegetative mycelium penetrate into the medium in order to obtain nutrients – soluble by absorption and insoluble by digestion with extracellular enzymes. Reproductive mycelium are responsible for spore production and usually extend from the medium into the air. The mycelium of a mould may be loosely woven network or it may be organized, compact structure as in mushrooms (Figure 10).

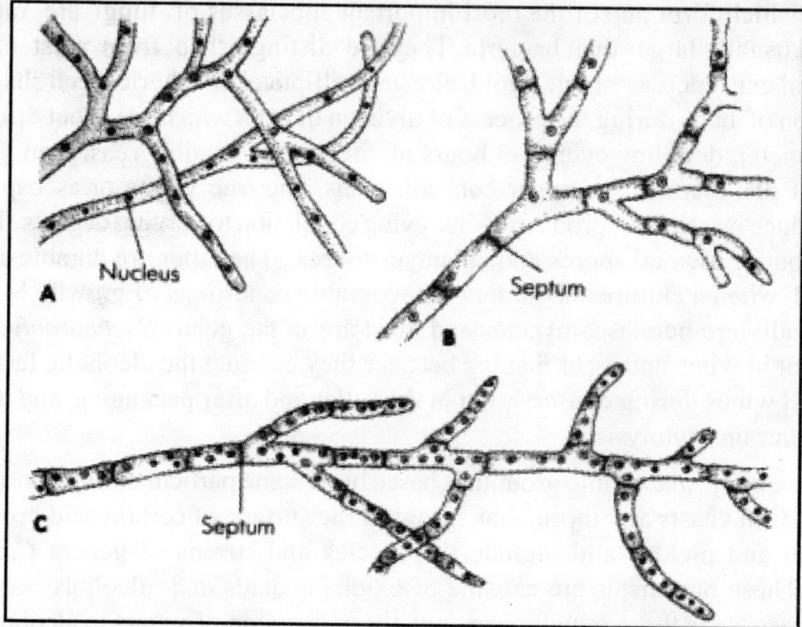


**Figure 9** Three types of hyphae in fungi. **(a)** Nonseptate, **(b)** Septate with uninucleate cells and **(c)** Septate with multinucleate cells

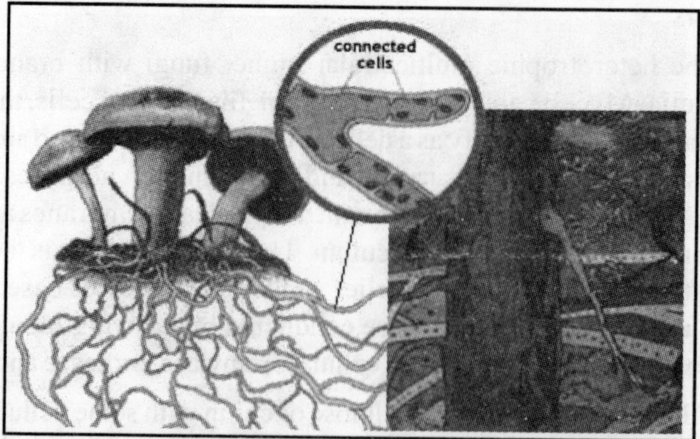


**Figure 10** Multicellular nature of mushrooms

The arrangement of the various fungal taxa has undergone substantial changes in the last several decades, especially as techniques for comparing biochemical characteristics (such as ribosomal RNA and DNA) have become increasingly more sophisticated. The phylogeny presented here is for the Eumycota (true fungi) and recognizes four divisions: the Chytridiomycota (chytrids), Zygomycota (conjugating fungi including bread moulds), Ascomycota (sac fungi, including common moulds, cup fungi) and Basidiomycota (club fungi). Most mushrooms are classified in the Basidiomycota.

Fungi reproduce both by asexual and sexual means. Asexual reproduction does not involve the fusion of nuclei or sex cells. It may be accomplished by i) fission of somatic cells yielding two similar daughter cells ii) budding of somatic cells or spores iii) fragmentation or disjoining of hyphal cells iv) spore formation – many kinds of asexual spores (Figure 11): a) Sporangiospores, single celled spores formed within sacs, sporangia which are formed at the end of special hyphae, sporangiophores. Such spores may be nonmotile (Aplanospores) or motile, having flagella (Zoospores), b) Conidiospores or conidia which may be single celled (Microconidia) or large, multicelled (Macroconidia) and are formed at the tip or side of a hypha, conidiophore, c) Oidia or arthrospores, single celled spores formed by disjoining of hyphal cells, d) Chlamydospores, thick walled single-celled spores are highly resistant to adverse conditions and are formed from cells of vegetative hyphae, e) Blastospores, spores formed by budding.

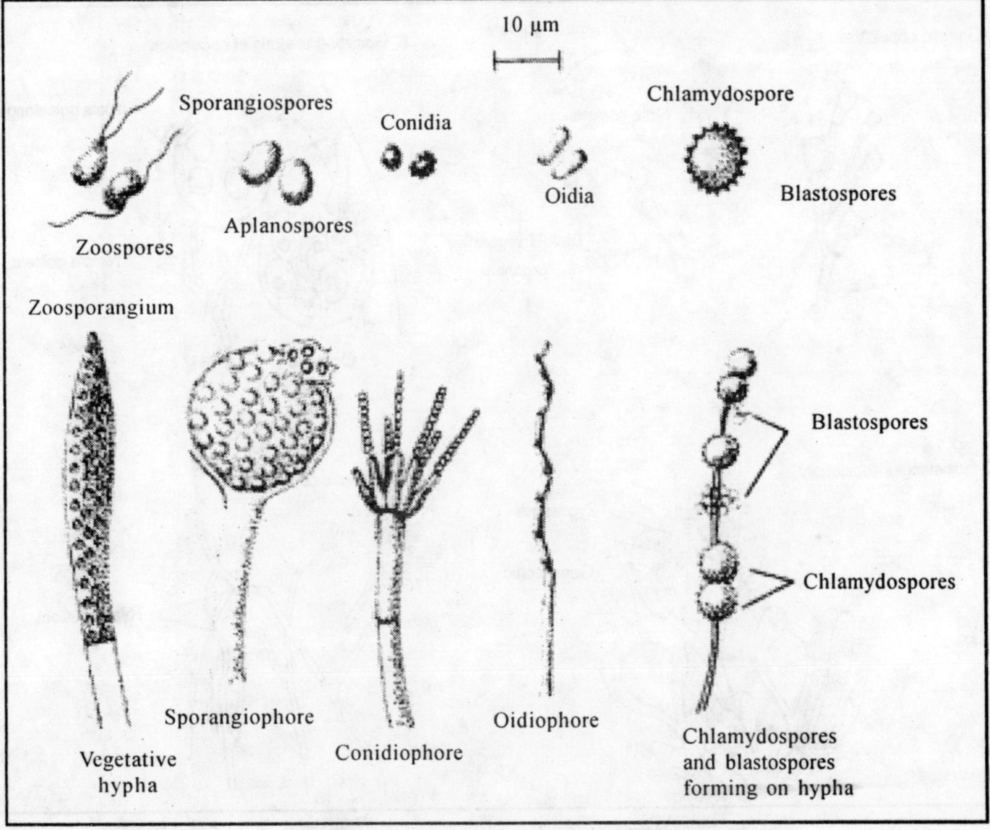


**Figure 11** Asexual spore types in fungi

Sexual reproduction in Fungi takes place by fusion of the compatible nuclei from two parent cells followed by meiosis and again reduces the number of chromosomes to haploid number. The sex organelles in fungi, if present, are called gametangia which form differentiated sex cells known as gametes. The male and female gametangia may be morphologically different and known as antheridium and oogonium respectively. There are various methods of sexual reproduction where the compatible nuclei are brought together (Figure 12): i) Gametic copulation: fusion of naked gametes – one or both may be motile, ii) Gamete-gametangial copulation: two gametangia come in contact but donot fuse – male nucleus migrates through a pore or fertilization tube into female gametangium, iii) Gametangial copulation: two gametangia or their protoplasts fuse and give rise to a zygote, iv) Somatic copulation: fusion of somatic cells or vegetative cells, and v) Spermatization: union of a special male structure – spermatium with a female receptive structure.

Sexual spores occur less frequently and in small numbers than asexual spores. These are of different types (Figure 13) and include: i) Ascospores: single celled – produced in a sac – ascus with usually 8 ascospores, ii) Basidiospores: single celled – borne on a club shaped structure – basidium, iii) Zygospores: large – thick walled spores formed when the tips of two gametangia fuse together, iv) Oospores: formed within a special female structure – oogonium where fertilization of eggs – oospheres by male gametes gives rise to oospores.

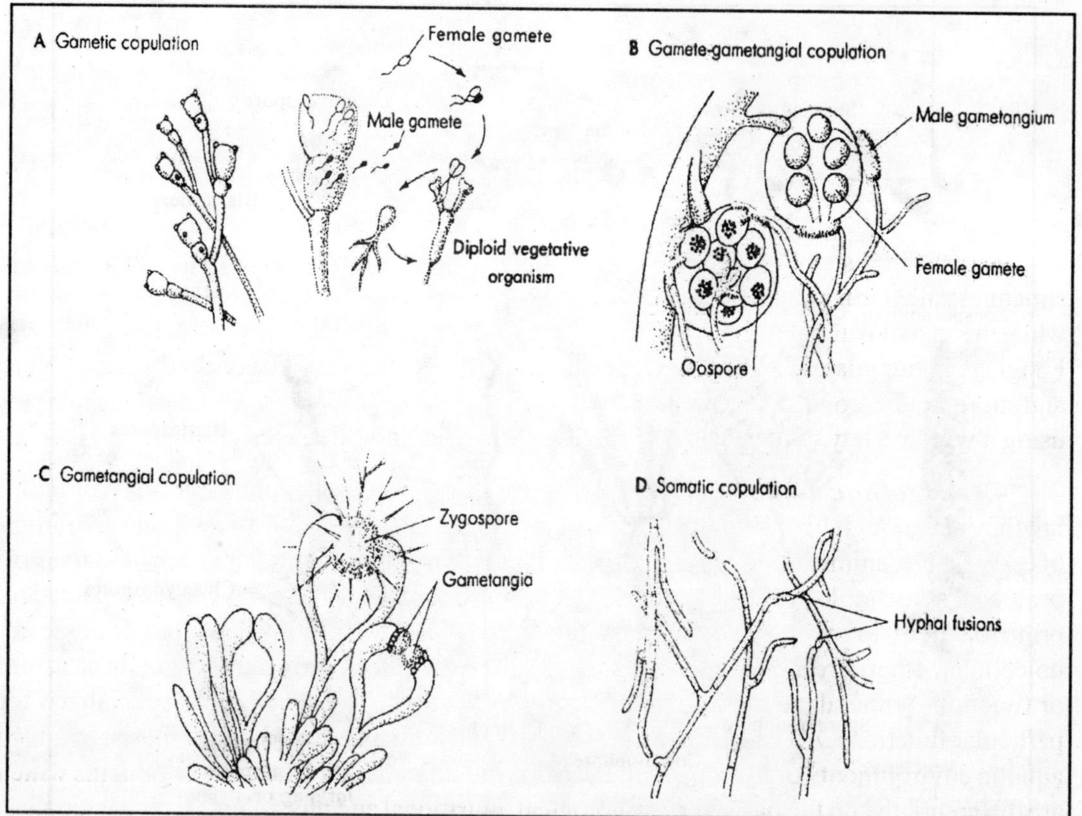


**Figure 12** Various sexual mechanisms in fungi

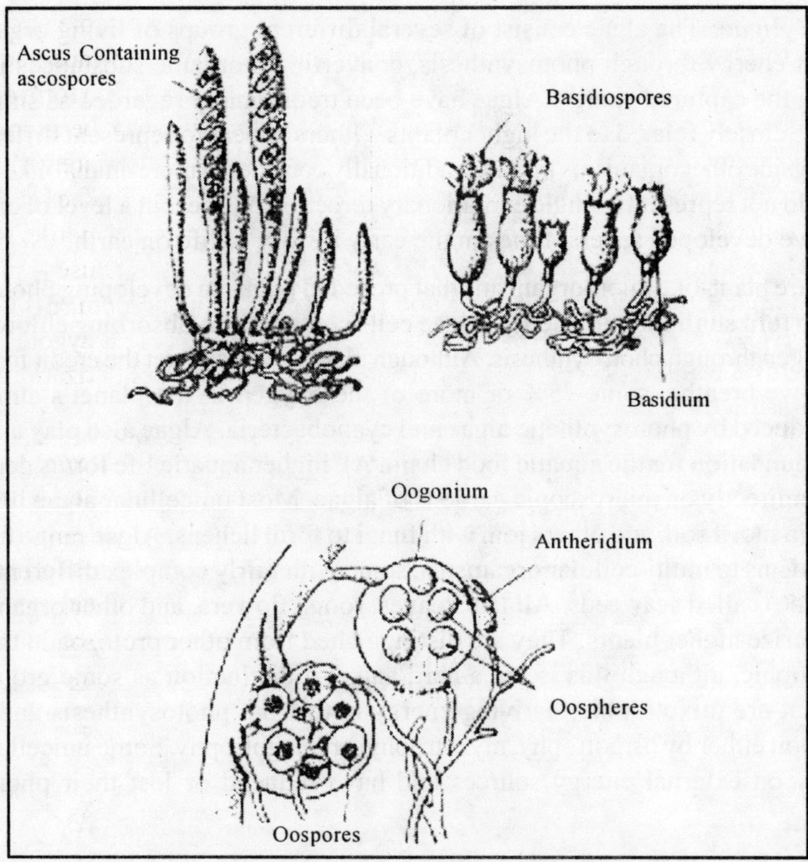


**Figure 13** Some sexual spore types in fungi

Asexual and sexual spores in fungi may be surrounded by highly organized protective structures called fruiting bodies. Asexual fruiting bodies may be called acervulus and pycnidium while the sexual fruiting bodies may be called as perithecium, apothecium and cleistothecium. Fungi are better adapted to certain extreme conditions including high concentration of sugar and more acidic conditions. Some yeasts are facultative anaerobes. Fungi are capable of using a wide variety of materials for nutrition however, they are heterotrophic.

**4.1.4.** *Protista:* Protists are eukaryotic creatures which are not plants, animals or fungi, but they act enough like them that scientists believe protists paved the way for the evolution of early plants, animals, and fungi. Protists fall into four general subgroups: unicellular algae, protozoa, slime molds, and water molds. Algae produce much of the oxygen we breathe and protozoa help to maintain the balance of microbial life. Most primitive types of algae are unicellular, others are aggregations of similar cells with little or no differentiation in structure or function. Some algae have a complex structure with different cell types specialized for particular functions. All cells contain chlorophyll and are capable of photosynthesis. Found in aquatic environments or in damp soil. Protozoa are unicellular eukaryotic organisms which are differentiated on the basis of morphological, nutritional and physiological characteristics. Some cause diseases in animals and humans.

**4.1.4.1.** *Algae:* The algae consist of several different groups of living organisms that capture light energy through photosynthesis, converting inorganic substances into simple sugars using the captured energy. Algae have been traditionally regarded as simple plants, and some are closely related to the higher plants. Others appear to represent different protist groups, alongside other organisms that are traditionally considered more animal-like (protozoa). Thus algae do not represent a single evolutionary direction or line, but a level of organization that may have developed several times in the early history of life on earth.

Algae are plant-like microorganisms that preceded plants in developing photosynthesis, the ability to turn sunlight into energy. Algae cells contain light-absorbing chloroplasts and produce oxygen through photosynthesis. Although plants generally get the credit for producing the oxygen we breathe, some 75% or more of the oxygen in the planet's atmosphere is actually produced by photosynthetic algae and cyanobacteria. Algae also play an important role as the foundation for the aquatic food chain. All higher aquatic life forms depend either directly or indirectly on microscopic gardens of algae. Most unicellular algae live in water, some dwell in moist soil, and others join with fungi to form lichens. Algae range from single-celled organisms to multi-cellular organisms, some with fairly complex differentiated form and (if marine) called seaweeds. All lack leaves, roots, flowers, and other organ structures that characterize higher plants. They are distinguished from other protozoa in that they are photoautotrophic, although this is not a hard and fast distinction as some groups contain members that are mixotrophic, deriving energy both from photosynthesis and uptake of organic carbon either by osmotrophy, myzotrophy, or phagotrophy. Some unicellular species rely entirely on external energy sources and have reduced or lost their photosynthetic apparatus.

All algae have photosynthetic machinery ultimately derived from the cyanobacteria, and so produce oxygen as a by-product of photosynthesis, unlike the non-cyanobacterial photosynthetic bacteria.

Algae are usually found in damp places or bodies of water and thus are common in terrestrial as well as aquatic environments. However, terrestrial algae are usually rather inconspicuous and far more common in moist, tropical regions than dry ones, because algae lack vascular tissues and other adaptions to live on land. Algae can endure dryness and other conditions in symbiosis with a fungus as lichen. The various sorts of algae play significant roles in aquatic ecology. Microscopic forms that live suspended in the water column are called phytoplankton which provide the food base for most marine food chains. In very high densities (so-called algal blooms) they may discolor the water and out compete or poison other life forms. The seaweeds grow mostly in shallow marine waters. Some are used as human food or are harvested for useful substances such as agar or fertilizer. The study of algae is called phycology or algology.

**4.1.4.1.1.** *Green algae:* The most clearly plant-like algae, this species gets its namesake hue from high levels of chlorophyll. Their cell walls are made up of cellulose, the same material that makes up the cell walls in larger, multicellular plants. Like plants, they store the food they make through photosynthesis as starches. Growing in large masses, these algae can form visible layers of slick, green scum on the surfaces and sides of ponds, puddles or damp soil. Fossil records suggest that the first green algae originated 500 to 600 million years ago. Early algae probably gave rise to multicellular plants.

**4.1.4.1.2.** *Diatoms:* These algae hardly look like plants, but more like flying saucers, tiny canoes or cigars, lobed leaves, the undersides of mushroom caps, striated ribbons, or filigreed Christmas ornaments. Whatever their shape, all diatoms have shell-like, brittle cell walls made out of silica (glass) and pectin. The walls are two interlocking halves or shells that fit together like a pillbox. Because they depend on sunlight for photosynthesis, diatoms generally live in the upper 200 meters of oceans and bodies of fresh water.

Some species of diatoms simply float in the water currents near the surface. Others attach themselves to larger floating objects or to the sea floor, some are capable of living in frigid Antarctic waters (Figure 14) . When diatoms die, they slowly sink to the sea floor. The buildup of trillions of these shells forms a crumbly white sediment known as diatomaceous earth or diatomite, which is used in manufacturing pool filters and abrasives, including toothpaste.

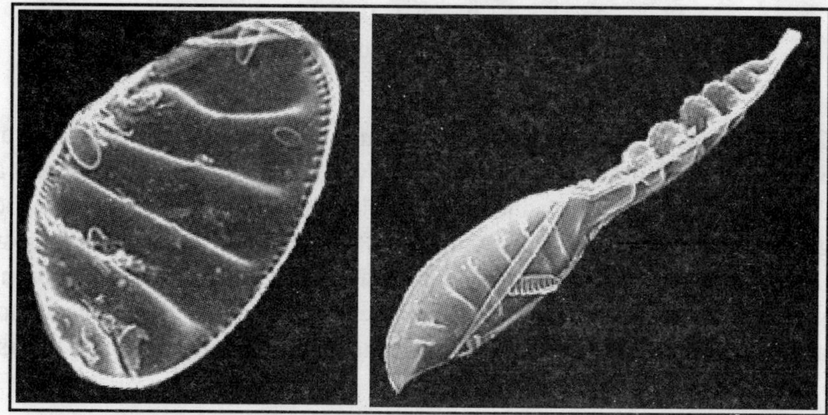

**Figure 14** Diatoms, *Cymatopleura* sp. (left) and *Entomoneis* sp. (right) capable of living in frigid Antarctic waters

**4.1.4.1.3.** *Dinoflagellates:* Dinoflagellates have long whip-like structures called flagella that let them turn, maneuver and spin about through the water (Figure 15). About 90% of these algae dwell in the ocean. Some species glow in the dark in a process called bioluminescence. These species contain a compound called luciferin (the same compound found in fireflies). The glow increases markedly if the algae cells are agitated, as when a ship churns through the water.

About half the species of dinoflagellates are photosynthetic; the other half are predators that attack bacteria, algae, and even fish. Dinoflagellate neurotoxins can concentrate in the bodies of shellfish and fish that eat the algal cells, in turn causing people who eat these seafoods to come down with illnesses such as paralytic shellfish poisoning and ciguatera (a combination of gastrointestinal, neurological, and cardiovascular disorders.) so-called “red tides” occur when enormous blooms of trillions of dinoflagellates are triggered by an upwelling of nutrients from the water's depths during warmer seasons. The population of dinoflagellates can jump to more than 20 million cells per liter of sea water along some coasts during these blooms, giving the water a reddish hue.

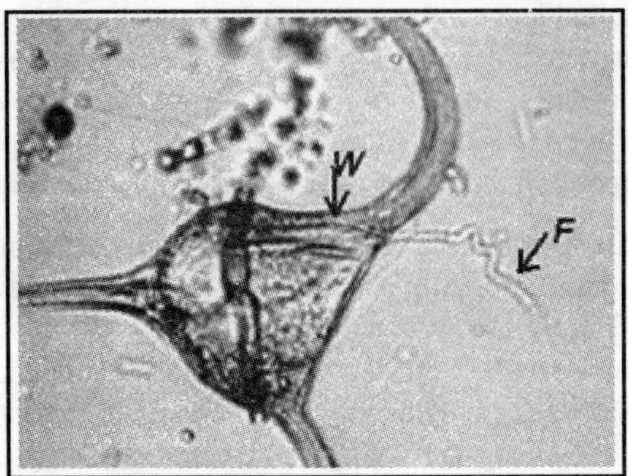


**Figure 15** The dinoflagellate *Ceratium* completely retracts its flagellum (F) into a sort of well or depression (W)

**4.1.4.2.** *Protozoa:* The name protozoa means "first animals." Protozoa are single-celled eukaryotes (organisms with nuclei) that show some characteristics usually associated with animals, most notably mobility and heterotrophy. They are often grouped in the kingdom Protista together with the plant-like algae. In some newer schemes, however, most algae are classified in the kingdoms Plantae and Chromista, and in such cases the remaining forms may be classified as a kingdom Protozoa. The name is misleading, since they are not animals. Protozoa have traditionally been divided on the basis of locomotion, although this is no longer believed to represent genuine relationships:

- Ciliates
- Flagellates
- Sarcodina: Amoeboids
- Apicomplexans : Sporozoans

Most protozoans are too small to be seen with the naked eye - most are around 0.01-0.05 mm, although forms up to 0.5 mm are still fairly common - but can easily be found under a microscope. Protozoa are ubiquitous throughout aqueous environments and the soil, and play an important role in their ecology. A few are also important parasites.

As the principal hunters and grazers of the microbial world, protozoa play a key role in maintaining the balance of bacterial, algal, and other microbial life. They also are themselves an important food source for larger creatures and the basis of many food chains. Protozoa have been found in almost every kind of soil environment from peat bogs to arid desert sands. They teem in the deep sea as well as near the surface of waters, and can be found even in frigid Arctic and Antarctic waters. Some species of protozoa are part of the normal microbial flora of animals, and live in the guts of insects and mammals, helping to break down complex food particles into simpler molecules. A very small number of species cause disease in people, including *Plasmodium vivax*, which causes malaria. The four main subgroups of protozoa are the ciliates, the flagellates, the sarcodina, and the apicomplexans.

**4.1.4.2.1.** *Ciliates:* Ciliates are covered in part or entirely with what look like little bristles called cilia (the Latin word for eyelash). The cilia are used for locomotion, and to snag bacteria, algae and other food and direct it into the ciliate's mouth-like opening. Ciliates include both grazers that dine on algae and bacterial cells and predators that attack and gulp down other protozoa. Grazers include *Paramecium* and *Vorticella*. An example of a predatory ciliate is *Didinium which* is a voracious hunter of live food such as *Paramecium* (Figure 16). Ciliates are among the most complex of all single-celled creatures, with a diverse array of structures and organelles that perform a range of activities, from finding and catching food, digesting it, excreting it, moving about, respiring, sensing environmental conditions, and balancing the fluids inside their cells. A few ciliates can grow up to 2 mm in length, big enough to be seen without a microscope.

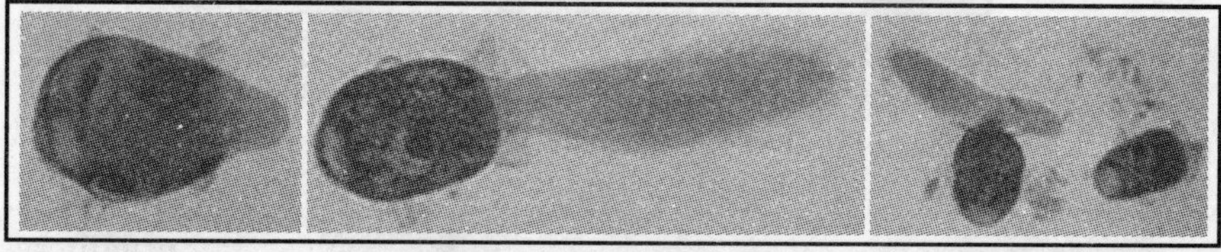

**Figure 16** *Didinium*, a ciliate that lives in fresh water, is a voracious hunter of live food such as Paramecium. A *Didinium* makes contact with a *Paramecium* and begins to ingest it *(For a coloured version of this figure, see plate section, page 556)*

**4.1.4.2.2.** *Flagellates:* Similarly complex single-celled organisms, flagellates have whip-like appendages called flagella sticking out of their cells. The flagella are used for locomotion and to direct food particles or cells into the organism's mouth-like opening. Flagellates dine on bacteria, algae, and other protozoa. Several well-known flagellates cause parasitic diseases, such as trypanosomes that cause sleeping sickness, and *Giardia lamblia*, a parasite found in mountain streams and rivers that causes severe gastrointestinal distress.

**4.1.4.2.3.** *Sarcodina:* This subgroup of protozoa includes the familiar shape-shifting amoebas, as well as heliozoa, radiozoa, and foraminifera (or forams for short). Sarcodina are best known for their pseudopods ("false feet") used for locomotion and feeding. In amoebas, these pseudopods are generally lobe-like bulges that extend from the cell membrane. In heliozoa, radiozoa, and forams, the pseudopods more resemble needles or spikes sticking out from the cells. Sarcodinas use their pseudopods to engulf or latch onto prey, which may include bacteria, algae or other protozoa. Many amoebas are active predators, oozing about on their false feet until they come into contact with a suitable meal. At that point, the pseudopod flows around and engulfs the hapless prey, until it is completely "swallowed". Heliozoa, radiozoa, and forams tend to be passive grazers and predators, relying on suitable food swimming or drifting past to come into contact with their pseudopods. Several species in the sarcodina group, including some species of amoebas, cover themselves with protective shell-like coverings called tests. These tests are stippled with many small and large openings through which water can flow in and out and through which the pseudopods protrude. The tests of radiozoa are made up of silica (the same substance in diatom cell walls) and can form very intricate, lacy designs that may be studded with long spines that increase buoyancy and ward off predators.

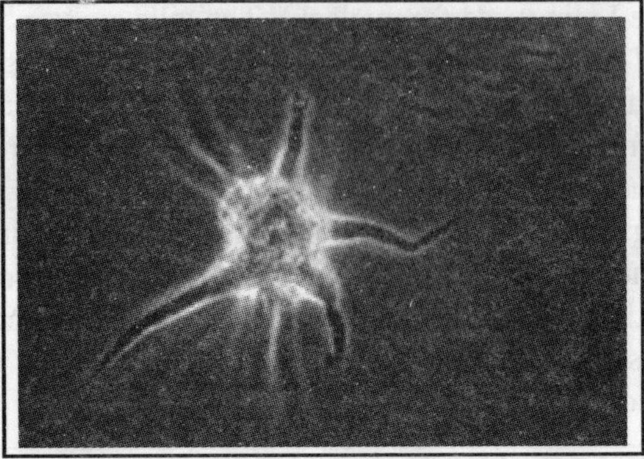

**Figure 17** *Amoeba radiosa* with its pseudopods which are used to engulf or latch onto prey which in the form of a Spirochaete is just visible at top right

The tests of forams are made up of sand grains or organic compounds or chalk extracted from sea water (Figure 18) . These can become quite large, the biggest reaching a little over 2 inches in diameter, making them some of the largest single-celled organisms known. When forams die, their tests sink and accumulate in large batches; the Great Pyramids of Egypt are built from sandstone composed largely of fossilized giant *Nummulites*, an ancient kind of foram. The famous White Cliffs of Dover are limestone cliffs formed from the skeletal remains of forams

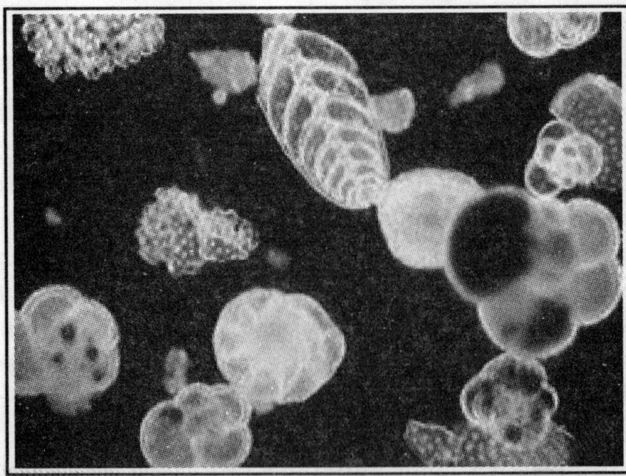

**Figure 18** Foraminifera bodies made up of chalk extracted from sea water
*(For a coloured version of this figure, see plate section, page 556)*

**4.1.4.2.4.** *Apicomplexans:* These protozoa are obligatory intracellular parasites: they must spend atleast a part if not all of their life cycle in a host animal. Apicomplexans are characterized by the presence of special organelles (tiny organ-like structures) located at the tips (apices) of the cells. These organelles contain enzymes that punch through, slice open and otherwise penetrate host tissues.

The best known apicomplexan is *Plasmodium*, the agent that causes malaria. *Plasmodium* spends a part of its life cycle in mosquitoes and the other part in human hosts where it ultimately infects and ruptures blood cells in large numbers. Another familiar apicomplexan is *Cryptosporidium parvum*. This water-borne parasite forms extremely durable cyst-like structures that enable it to survive UV radiation and sometimes chlorine in swimming pools and treated water. An outbreak of *Cryptosporidium* in Milwaukee's drinking water supply in 1993 killed 50 people and sickened more than 4,00,000.

**4.1.4.3.** *Slime molds:* In 1973, a Dallas resident went out to the backyard only to stumble upon a reddish, jelly-like mass pulsating in the grass. News reports on the discovery claimed that a "new life form" had been found, and many people couldn't help recalling the cult classic scientific thriller *The Blob*.

Scientists called to the scene, however, put any fears of menacing goo or alien creatures to rest by identifying the mass as an unusually large (46 cm or more than 14 inches in diameter) plasmodial slime mold. Slime molds were once considered fungi, but unlike fungi, they can move, and their cell membranes are made of different stuff. Slime molds are made up of individual cells that form an aggregate mass. In their visible, aggregate states, they look like blobs, gooey or foamy masses, spilled jelly, or even dog vomit. They may be bright orange, red, yellow, brown, black, blue, or white. These large masses act like giant amoebas, creeping slowly along and engulfing food particles along the way. If a slime mold aggregate is diced up, the pieces will pull themselves back together (Figure 19). The blobs can navigate and avoid obstacles and if a food source is placed nearby, they seem to sense it and head unerringly for it.

**Figure 19** Slime molds spend most of their lives independently, but during food shortages, they swarm and aggregate into an enormous single cell *(For a coloured version of this figure, see plate section, page 556)*

There are two kinds of slime molds. Plasmodial slime molds (the most common kind) share one big cell wall that surrounds thousands or millions of nuclei. Proteins called microfilaments act like tiny muscles that enable the mass to crawl at rates of about 1/25th of an inch per hour. As long as there is enough food and moisture, the mass thrives. But when food and water are scarce, the mass separates into smaller blobs. The *Plasmodium* forms stalks topped by sphere-like fruiting bodies that contain spores that are carried by the rain or wind to new locations. Cellular slime molds also produce spores, but these germinate into amoeba-like cells. The cells happily go their individual ways, as long as food and water are available. When nutrients and moisture are scarce, individual cells send out a chemical beacon to attract other cells of the same species. The cells join up to form a mass that looks and acts like a slug to take them to a more favorable location. Cells in cellular slime molds retain their individual cell walls when they form a mass, so the visible slug is actually a collection of hundreds of thousands of individual cells joined together. Slime molds eat decaying vegetation, bacteria, fungi, and even other slime molds. They are most commonly found in forests.

**4.1.4.4.** *Water molds:* Water molds are always found in wet environments, especially in fresh water sources and near the upper layers of moist soil. Officially named Oomycota, they are also known as downy mildews and white rusts. Water molds were long considered fungi because they produce fungi-like filamentous hyphae and feed on decaying tissue like rotting logs and mulch. The Oomycota species *Phytophthora infestans* caused the Great Potato Famine that killed nearly a million people in Ireland in 1846–1847. The water mold virtually wiped out the country's potato crops, which were an essential staple in the Irish diet (sometimes the only food on the table.) In addition to widespread starvation and malnutrition, the potato blight led more than 1.5 million Irish to flee the country. Because nearly all of the country's potato crops were clones of a few original imports from South America, they had no natural ability to resist the pathogen. Another water mold nearly wiped out the entire French wine industry. *Plasmopara viticola* (also known as downy mildew of grapes) was brought to Europe in the late 1870s on vines from America meant to be bred with French vines in hopes of yielding hybrids with a greater ability to ward off attacks by aphids. The infestation led to the use of the first fungicide, a mixture of copper sulfate and lime, which became known as the Bordeaux mixture for its role in saving the French vines. Other water mold species can cause disease in fish.

**4.1.5.** *Viruses:* These are very small non-cellular parasites which are included in the microbial world for two reasons: i) they are very small and the techniques used to study these are microbiological in nature, ii) These are the causative agents of many diseases and the diagnostic procedures for their identification are employed in clinical microbiological laboratory or the plant pathology laboratory. Unable to do much of anything on their own, viruses go into host cells to reproduce, often wreaking havoc and causing disease (Figure 20). Their ability to move genetic information from one cell to another makes them useful for cloning DNA and could provide a way to deliver gene therapy.

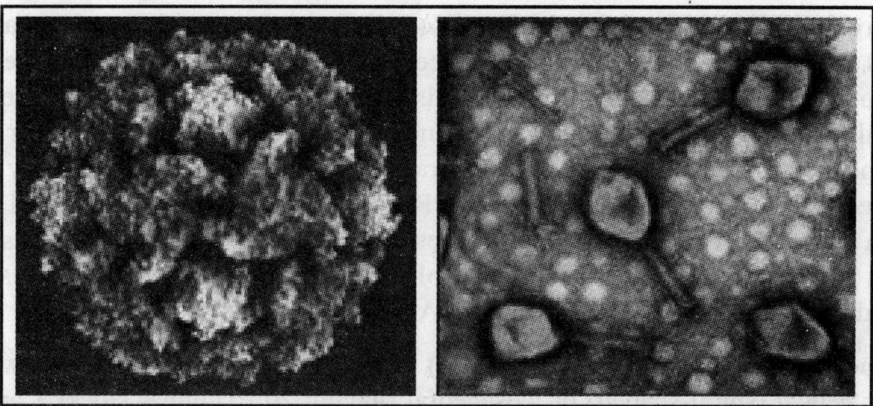

**Figure 20** The *polio virus* (left) once crippled millions. The *T4 bacteriophage* (right), is a virus that invades bacterial cells *(For a coloured version of this figure, see plate section, page 556)*

Viruses have played key roles in shaping the history of life on our planet by shuffling and redistributing genes in and among organisms and by causing diseases in animals and plants. These have been the culprits in many human diseases, including smallpox, flu, AIDS, certain types of cancer, and the ever-present common cold. Viruses can infect virtually all types of cells: bacteria, fungi, protozoa, plants, animals, and human. True parasites, viruses are basically little more than molecular syringes moving genetic information from one cell to another. Some viruses enter a host and leave virtually unnoticed. Others cause disease and destroy the host.

Viruses, the infectious agents are found in virtually all life forms, including humans, animals, plants, fungi, and bacteria. Viruses consist of genetic material in the form of either deoxyribonucleic acid (DNA) or ribonucleic acid (RNA), surrounded by a protective coating of protein, called a capsid, with or without an outer lipid envelope. Viruses are the simplest and tiniest of microbes; they can be as much as 10,000 times smaller than bacteria and hence are too small to be seen by light microscopy. Viruses vary in size from the largest poxviruses of about 450 nm in length to the smallest polioviruses of about 30 nm. Viruses are not considered free-living, since they cannot reproduce outside of a living cell; they have evolved to transmit their genetic information from one cell to another for the purpose of replication.

Viruses often damage or kill the cells that they infect, causing disease in infected organisms. A few viruses stimulate cells to grow uncontrollably and produce cancers. Although many infectious diseases, such as the common cold, are caused by viruses, there are no cures for these illnesses. The difficulty in developing antiviral therapies stems from the large number of variant viruses that can cause the same disease, as well as the inability of drugs to disable a virus without disabling healthy cells. However, the development of antiviral agents is a major focus of current research, and the study of viruses has led to many discoveries important to human health.

Individual viruses, or virus particles, also called virions, contain genetic material, or genomes, in one of the several forms. Unlike cellular organisms, in which the genes always are made up of DNA, viral genes may consist of either DNA or RNA. Like cell DNA, almost all viral DNA is double-stranded, and it can have either a circular or a linear arrangement. Almost all viral RNA is single-stranded; it is usually linear, and it may be either segmented (with different genes on different RNA molecules) or non segmented (with all genes on a single piece of RNA). Retroviruses are among the infectious particles that use RNA as their hereditary material. Probably the most famous retrovirus is human immunodeficiency virus (HIV), the cause of AIDS.

Viruses come in many shapes (Figure 21) with a protective shell, or capsid, which can be either helical (spiral-shaped) or icosahedral (having 20 triangular sides). Capsids are composed of repeating units of one or a few different proteins. These units are called protomers or capsomers. The proteins that make up the virus particle are called structural proteins. Viruses also carry genes for making proteins that are never incorporated into the virus particle and are found only in infected cells. These viral proteins are called nonstructural proteins; they include factors required for the replication of the viral genome and the production of the virus particle.

Capsids and the genetic material (DNA or RNA) are together referred to as nucleocapsids. Some virus particles consist only of nucleocapsids, while others contain additional structures. Some icosahedral and helical animal viruses are enclosed in a lipid envelope acquired when the virus buds through host-cell membranes. Inserted into this envelope are glycoproteins that the viral genome directs the cell to make; these molecules bind virus particles to susceptible host cells. The most elaborate viruses are the bacteriophages, which use bacteria as their hosts. Some bacteriophages resemble an insect with an icosahedral head attached to a tubular sheath. From the base of the sheath extend several long tail fibers that help the virus attach to the bacterium and inject its DNA to be replicated and to direct capsid production and virus particle assembly inside the cell.

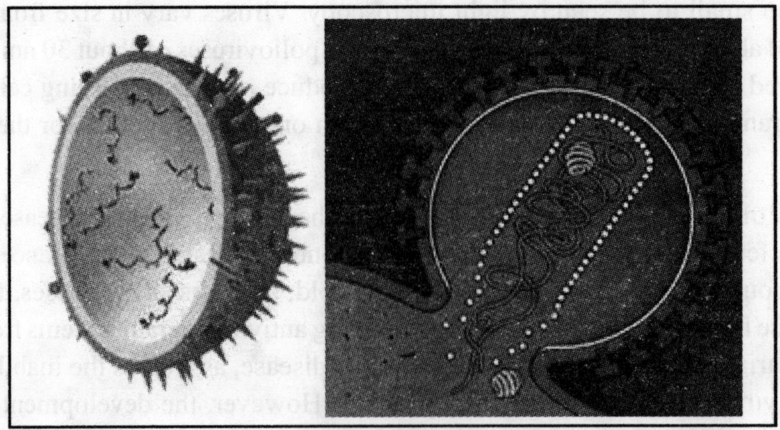

**Figure 21** An illustration of influenza virus (left) and human Immuno-deficiency Virus (HIV) (right) *(For a coloured version of this figure, see plate section, page 556)*

Viruses are classified according to their type of genetic material, their strategy of replication, and their structure. The International Committee on Nomenclature of Viruses (ICNV), established in 1966, devised a scheme to group viruses into families, subfamilies, genera, and species. The ICNV report published in 1995 assigned more than 4000 viruses into 71 virus families. Hundreds of other viruses remain unclassified because of the lack of sufficient information.

When viruses come into contact with host cells, they trigger the cells to engulf them, or fuse themselves with the cell membrane so that they can release their DNA into the cell. Once inside a host cell, viruses take over its machinery to reproduce. Viruses override the host cell's normal functioning with their own set of instructions that shut down production of host proteins and direct the cell to produce viral proteins to make new virus particles. Some viruses insert their genetic material into the host cell's DNA, where they begin directing the copying of their genes or simply lie dormant for years or a lifetime. Either way, the host cell does all the actual work: the viruses simply provide the instructions. Viruses may be able to infect and reproduce in more than one kind of animal, but the same virus can cause different reactions in different hosts. For example, flu viruses infect birds, pigs, and humans. While some types of flu viruses don't harm birds, they can overwhelm and kill humans. Plant viruses do not infect animals or vice versa. Viruses that infect bacteria do nothing to animal or plant cells. The first contact between a virus particle and its host cell occurs when an outer viral structure docks with a specific molecule on the cell surface. For example, a glycoprotein called gp120 on the surface of the human immunodeficiency virus (HIV, the cause of acquired immunodeficiency syndrome, or AIDS) virion specifically binds to the CD4 molecule found on certain human T lymphocytes (a type of white blood cell). Most cells that do not have surface CD4 molecules generally cannot be infected by HIV.

After binding to an appropriate cell, a virus must cross the cell membrane. Some viruses accomplish this goal by fusing their lipid envelope to the cell membrane, thus releasing the nucleocapsid into the cytoplasm of the cell. Other viruses must first be endocytosed (enveloped by a small section of the cell's plasma membrane that pokes into the cell and pinches off to form a bubblelike vesicle called an endosome) before they can cross the cell membrane. Conditions in the endosome allow many viruses to change the shape of one or more of their proteins. These changes permit the virus either to fuse with the endosomal membrane or to lyse the endosome (cause it to break apart), allowing the nucleocapsid to enter the cell cytoplasm.

Once inside the cell, the virus replicates itself through a series of events. Viral genes direct the production of proteins by the host cellular machinery. The first viral proteins synthesized by some viruses are the enzymes required to copy the viral genome. Using a combination of viral and cellular components, the viral genome can be replicated thousands of times. Late in the replication cycle for many viruses, proteins that make up the capsid are synthesized. These proteins package the viral genetic material to make newly formed nucleocapsids.

To complete the virus replication cycle, viruses must exit the cell. Some viruses bud out of the cell's plasma membrane by a process resembling reverse endocytosis. Other viruses cause the cell to lyse, thereby releasing newly formed virus particles ready to infect other cells. Still other viruses pass directly from one cell into an adjacent cell without being exposed to the extracellular environment. The virus replication cycle can be as short as a couple of hours for certain small viruses or as long as several days for some large viruses.

Some viruses kill cells by inflicting severe damage resulting in cell lysis; other viruses cause the cell to kill itself in response to virus infection. This programmed cell suicide is thought to be a host defense mechanism to eliminate infected cells before the virus can complete its replication cycle and spread to other cells. Alternatively, cells may survive virus infection, and the virus can persist for the life of its host. Virtually all people harbor harmless viruses.

Retroviruses, such as HIV, have RNA that is transcribed into DNA by the viral enzyme reverse transcriptase upon entry into the cell. (The ability of retroviruses to copy RNA into DNA earned them their name because this process is the reverse of the usual transfer of genetic information, from DNA to RNA.) The DNA form of the retrovirus genome is then integrated into the cellular DNA and is referred to as the provirus. The viral genome is replicated every time the host cell replicates its DNA and is thus passed on to daughter cells.

Hepatitis B virus can also transcribe RNA to DNA, but this virus packages the DNA version of its genome into virus particles. Unlike retroviruses, hepatitis B virus does not integrate into the host cell DNA.

Viruses can act as miniature couriers. When they infect, they may inadvertently take up a bit of their host's DNA and have it copied into their progeny. When the offspring viruses move on to infect new cells, they may insert this bit of accidentally pilfered DNA into the new host's genome. This process is called transduction. This can sometimes create a happy outcome. For example, the soil-dwelling bacterium *Bacillus subtilis* has viral genes that help protect it from heavy metals and other harmful substances in the soil. Other times, viruses can wreak havoc when they bring in new genes. For example, *Vibrio cholerae*, the bacterium that causes cholera, is harmless in itself. The disease-causing toxin that causes illness is actually made by a virus that at some point smuggled itself into its host's genome. Viruses can also influence host genes by where they insert themselves into their host's DNA. Recent decoding of the human genome shows that viral DNA sequences have been reproducing jointly with our genes for ages. Some of these DNA sequences stay out, but others seem to move about our genome, jumping from place to place on a chromosome or from chromosome to chromosome. These "mobile elements" take up nearly half of the human genome. Hemophilia and muscular dystrophy are two human diseases that researchers now believe resulted from mobile elements that, while skipping about the genome, ungraciously barged right into the middle of key human genes. Scientists in the 1970s began discovering even simpler and smaller virus-like organisms that can cause disease. Viroids contain only RNA, but lack an envelope and capsid. They cause problems in potatoes, tomatoes, and some fruit trees, and recently a viroid has also been linked to hepatitis D.

Unlike other infectious agents such as bacteria, viruses, and viroids, prions do not have the nucleic acids DNA or RNA. Prions are proteins that have the ability to transmit diseases, a finding that defied scientific expectations. There is still much debate about how they work, but scientists think these rogue proteins direct the host to create abnormal proteins that can cause serious neurological disease in animals and humans. Prions are blamed for scrapie in sheep, and bovine spongiform encephalopathy ("Mad Cow Disease") in cattle, and its human variant, Creutzfeldt-Jakob disease. Some scientists suspect that prions may be responsible for Alzheimer's disease.

**4.1.6.** *Microbial Mergers:* Mergers and collaborations on a minute scale paved the way for higher life forms. Today, symbionts, the scientific terms for these mergers, help fertilize plants (Figure 22), construct coral reefs, and help us digest food.

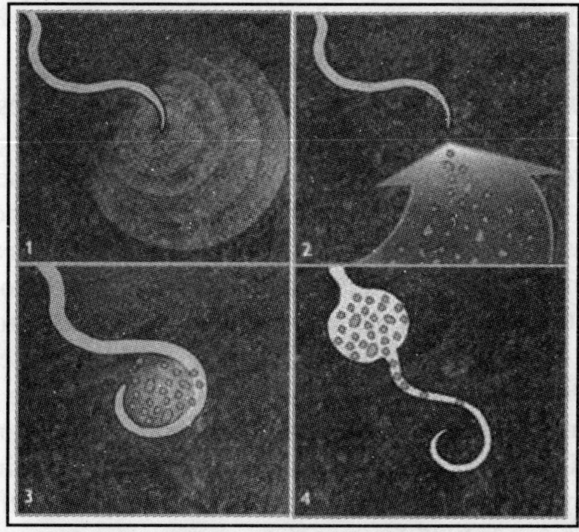


**Figure 22** The pea root sends a chemical message that attracts *Rhizobia*, then surrounds the bacteria, which set up housekeeping inside the root's cells *(For a coloured version of this figure, see plate section, page 557)*

Over millions of years of evolution, we humans have worked out a mutually beneficial partnership with the microbes that came to inhabit our guts. In return for their aid in digestion, we give them a stable, protected home and plenty of nutrients via the food we eat. We need them as much as they need us. Microbes break down food molecules that our body's enzymes and acids can't dissolve, helping us squeeze all the nutrients out of that our food. Some make valuable vitamins that our body needs. Many microbial species have proved to be consummate evolutionary wheelers and dealers, arranging collaborations, mergers, and acquisitions that usually serve both partners well.

Mitochondria and chloroplasts: the partnerships that led to higher life. Mitochondria are the energy factories found in each cell of fungi, protozoa, insects, and animals. Once nutrients are absorbed or digested, they move in the form of minuscule molecules into the mitochondria, which convert the molecules into chemical energy to power the cell. Chloroplasts undertake a similar function in the cells of plants, algae, and some protozoa. They capture sunlight and, through a series of chemical reactions called photosynthesis, use the light to make energy. These organelles are absolutely essential to the existence of all higher life forms on Earth. If all of the mitochondria in our bodies were to suddenly shut down, we would die. The same is true for plants were they to lose their chloroplasts. As microscopes improved over the years, scientists began noticing striking similarities in the appearances of mitochondria, chloroplasts, and bacterial cells. They discovered that these two organelles contain their own DNA, or gene set, organized very much like the DNA in bacterial cells. Mitochondria and chloroplasts also reproduce independently from the cells in which they reside, in a manner very like bacterial fission. Many microbiologists think it is likely that mitochondria and chloroplasts were once free-living prokaryotes (cells that lack a nucleus and organelles) that somehow took up residence in larger cells. An invading bacterium may have infected a cell, then became a permanent resident as it adapted to become less virulent, or the target cell became less susceptible. Or both chloroplasts likely first entered a host cell as food before establishing a successful merger with the cell so they were not digested.

**4.1.6.1.** *Fungi and bacteria: the fundamental fertilizers:* Ages ago, as land plants were evolving, they ran into a few impediments. Soil can sometimes prove a nutrient-poor and inhospitable environment. In order to grow and thrive, plants need nitrogen to make proteins, but they lack the chemistry set to convert free nitrogen in the air into a form their cells can use. To overcome these obstacles, early plants struck deals with co-evolving bacteria and fungi. Some early bacteria developed the chemical tools to harvest free nitrogen from the air and convert it into forms such as ammonia and nitrate through a process called nitrogen fixation. The catch is that they need sufficient amounts of energy in the form of carbohydrates to power these conversions, and the supply of carbohydrates in the soil can be limited. On the other hand, plants produce copious amounts of carbohydrates as a product of photosynthesis. And ammonia and nitrate are perfect protein-building nitrogen forms for plants (Figure 23). A group of enterprising plants called the legumes (these include all beans and peas as well as clover and alfalfa) entered into a merger agreement with nitrogen-fixing bacteria called *Rhizobium*. The bacteria moves into the plants' roots, forming bumps on the roots called nodules that supply the fixed nitrogen plants need. In exchange, the plants supply the bacteria with the carbohydrates they require. Because Rhizobia can still dwell independently in the soil, plants are more dependent upon them than the microbes are on plants.

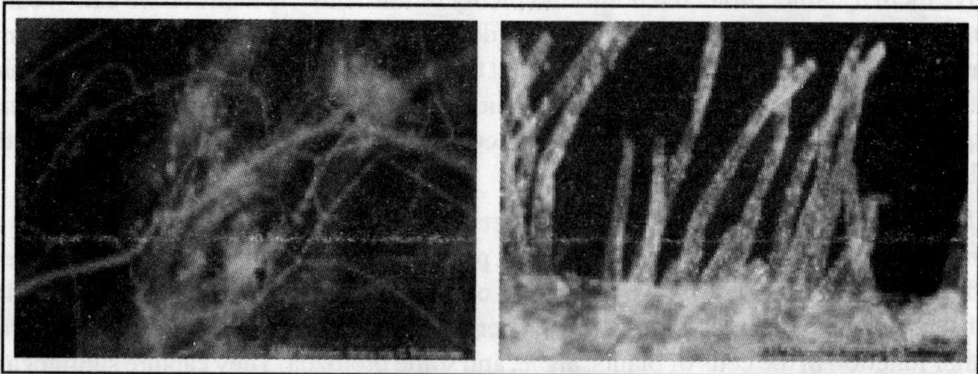

**Figure 23** Rhizobia living as a merger agreement with legumes (including beans and peas as well as clover and alfalfa) *(For a coloured version of this figure, see plate section, page 557)*

Animals need nitrogen for protein building, too. We humans get our nitrogen by eating plants (or by eating animals that eat plants). Another partnership teams plants with soil-dwelling fungi called mycorrhizae. Virtually all plants from flowers to towering trees like Sequoias have partner mycorrhizae. Some species of mycorrhizae cover the surface of plants' root hairs; others settle down inside the plant roots. The fungi act as extensions of the plants' roots, vastly increasing the surface space of their nutrient-absorbing network. Mycorrhizae increase the plants' uptake of water and essential nutrients, particularly phosphorous, which doesn't spread readily in soil. In exchange, the plants provide the fungi energy in the form of carbohydrates. This partnership enables both plants and fungi to survive in nutrient-poor places where they otherwise might die.

**4.1.6.2.** *Mouthless, gutless worms thrive thanks to bacterial feeders:* In the deepest sea, where not a single photon of sunlight ever penetrates, life persists in eternal darkness, crowded around chemical- and lava-spewing fissures in the ocean's floor. Life around these hydrothermal vents includes shrimp, crabs, and tall, slender tubeworms. These worms have no mouths, no stomachs, and no intestines, yet they clearly thrive, some species growing up to eight feet in length. They need no eating apparatus because their bodies house billions of bacteria that feed them (Figure 24). The tubeworm's tissues are lined with bacteria that convert hydrogen sulfide from the hydrothermal vents into molecules that serve as usable nutrients for the worms. The worm's gills look like a red plume sticking out from its protective tube. Its blood contains a special type of hemoglobin that transports sulfides, oxygen, and carbon dioxide into its tissues, where the bacteria turn them into food for themselves and their host. In return, the worms provide the bacteria with a protected and stable home.

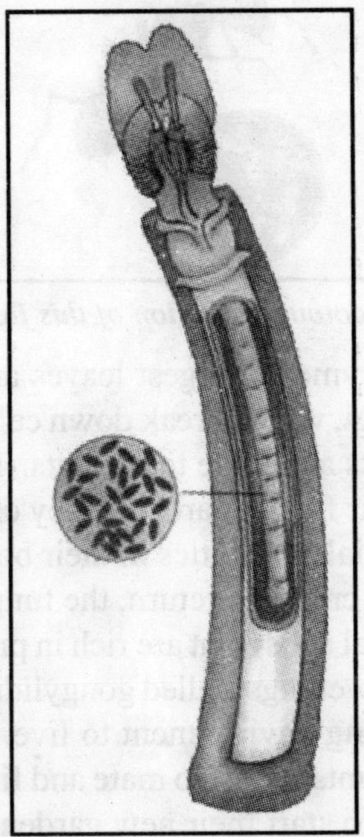

**Figure 24** Bacteria and their tubeworm host *(For a coloured version of this figure, see plate section, page 557)*

**4.1.6.3.** *Algae: the invisible partner in global reef construction:* Major development projects are taking place in oceans across the globe all the time, enterprises that will provide shelter and food for a vast number of fish, mussels, urchins, and other marine life. While credit is regularly and duly given to the visible construction crew - coral polyps - recognition is also due to the polyps' invisible, but very active algal partners, the zooxanthelle. These algae (a type called dinoflagellates) live inside the body tissues of coral polyps. Coral polyps take care of some of their nutritional needs on their own by catching tiny protists and organic matter that drift past their tentacles. Their partnership with the chlorophyll-containing algae enable them to get food from sunlight as if they were plants.

The zooxanthelle do the actual work of converting the sunlight into energy via photosynthesis. The by-products they generate (organic carbons such as glycerol and sugars) are excellent nutrients for their polyp hosts. The zooxanthellae supply much of the polyps' energy needs. In turn, the polyps provide the algae a protected, stable environment and nutrients they need for growth, such as nitrates and phosphates. Zooxanthellae also increase the rate of coral calcification, the growth of the hard shells that form the actual reef structures.

**4.1.6.4.** *Ant farmers and their gardens of fungi:* It's not clear precisely how and why some ancestral species of ant first took up fungus farming, but scientists have determined by genetic testing that it happened about 50 million years ago. Today, several species of ant (including the leaf-cutters of Central and South America) carefully tend and nurture the gardens of fungi in vast underground nests (Figure 25).

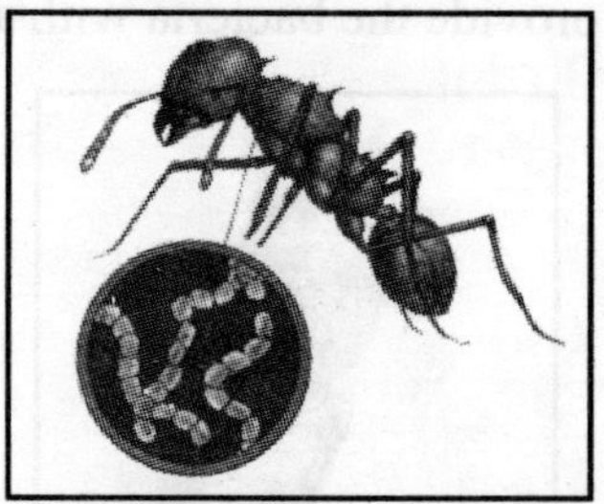

**Figure 25** Leafcutter ant *(For a coloured version of this figure, see plate section, page 558)*

Ants lack the necessary enzymes to digest leaves and stems themselves, so they cut them up and feed them to the fungi, which break down cellulose (the tough fibrous material in plant tissues), making nutrients available to the ants. The ants excavate nests and build nice, cozy, safe chambers for their fungal gardens; they clean off debris and parasites from the fungi; and even produce special antibiotics in their bodies to ward off or kill infectious organisms that might attack their crops. In return, the fungi produce swellings at the tips of their hyphae (long strings of fungal cells) that are rich in proteins, sugars and other nutrients. The ants dine on these nutritious swellings, called gongylidia. The fungi are ensured a copious food supply and a stable, nurturing environment to live. The fungi rely upon the ants for reproduction. Before new queen ants fly off to mate and find their own colonies, they tuck a bit of fungus in their mandibles to start their new gardens. The fungi growing in virtually every leaf-cutter garden are actually clones of the same fungus farmed by ants 25 million years ago.

**4.1.6.5.** *Lichens: when fungi and algae (or cyanobacteria) merged:* Fungi feed themselves quite ably, absorbing nutrients from organic materials. Algae and cyanobacteria fulfill their nutritional needs by turning sunlight into energy through photosynthesis. Yet many thousands of years ago, some fungi merged with some algae (or cyanobacteria in some cases) to create a new kind of partnership called a lichen. Some 20,000 different kinds of lichens live in such diverse habitats as the surfaces of rocks in arctic tundra and desert sands as well as the bark of trees in bayous and the sides of buildings.

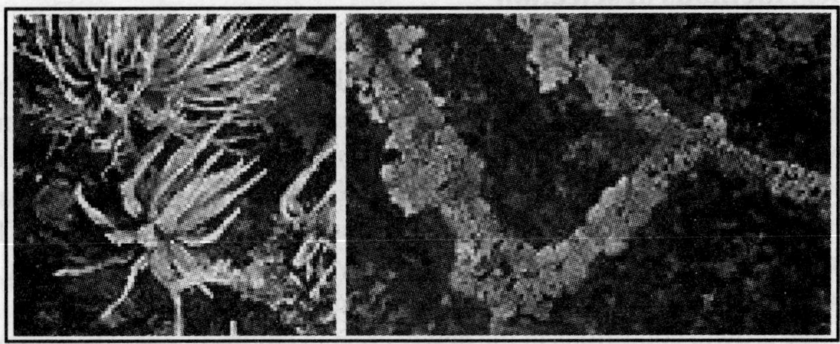

**Figure 26** A close-up view of a fog-desert *Niebla* lichen (left), one of the common kinds of lichens found on rock and a lichen following cracks in the rock (right) *(For a coloured version of this figure, see plate section, page 558)*

Lichen formation enables each of the partners to expand into habitats and environments they might not survive alone, such as wind-whipped tundra or sun-scorched desert. Fungi provide the shelter and stability while the algae or cyanobacteria provide the food. The fungal filaments surround the algal or bacterial cells and make up the majority of the lichen's bulk and shape. Meanwhile, the photosynthetic algae or bacteria churn out nutrients in the form of carbons, with up to 60% of what they produce being used by the fungal cells. Lichens have an amazing capacity to withstand drought. Like very efficient sponges, lichens absorb moisture from fog, dew and even humid air. They can take in as much as 35 times their weight in water. They also retain water well, drying out very slowly. This ability enables them to survive in places like bare rock surfaces, deserts and tundra. Some lichen partners are dependent upon one another for survival, but in many cases, the fungus and the algal or cyanobacterial species can each be found living independently.

## 5. SCREENING OF USEFUL MICROORGANISMS

One of the most fundamental tasks of a microbiologist is to isolate an organism with particular properties in a pure state, free of contamination, i.e. the selection of one type of organism from numerous others with which it is normally associated in nature. The ultimate sources for fresh isolates of microorganisms when contemplating a new process for industrial use are water, soil, plant material, sewage, spoiled foods and other natural habitats where they exist in complex mixtures. The detection and isolation of micro organisms form the natural sources is called screening which requires a selective method for isolation of a pure culture from natural environments. The desired organism may be less rapidly growing on the ordinary culture medium than other organisms. Selective methods favor the growth of desired species while discouraging the other organisms. There are three main types of selective methods.

- Chemical methods
- Physical methods
- Biological methods

## 5.1. Chemical methods of selection

These involve the use of some chemical compounds in the culture media which may favour the growth of the desirable organisms or may inhibit the undesirable ones. One approach used in chemical methods of selection involves the incorporation of a single carbon or nitrogen source that can be used only by the species to be isolated. This particular kind of selection is also known as "enrichment culture technique" (Figure 27) which is a valuable tool in many screening programmes meant for isolating industrially important strains. This involves the alterations in medium so as to favour the growth of the desired micro organisms. On the other hand, unwanted micro organisms are eliminated or develop poorly since they do not find suitable growth conditions in the newly created environment. The second approach involves the use of "dilute media" for the isolation of certain aquatic bacteria which are capable of growing with very low levels of carbon or nitrogen sources. This involves the use of a medium with low levels of the nutrients at which only the desired species are able to grow well while others are not. The third approach involves the supplementation of the culture media with low levels of certain chemicals including dyes, antibiotics, bile salts, salts of heavy metals which are inhibitory or toxic to the undesirable organisms so that only the desirable species, which are not inhibited by these chemicals are able to grow. Many gram -ve bacteria can grow in presence of low concentrations of various dyes that inhibit gram +ve bacteria. Many intestinal bacteria can grow in presence of bile salts whereas non-intestinal bacteria are inhibited. A medium containing crystal violet and sodium deoxycholate will allow gram negative intestinal bacteria to grow but will inhibit most other kinds of bacteria.

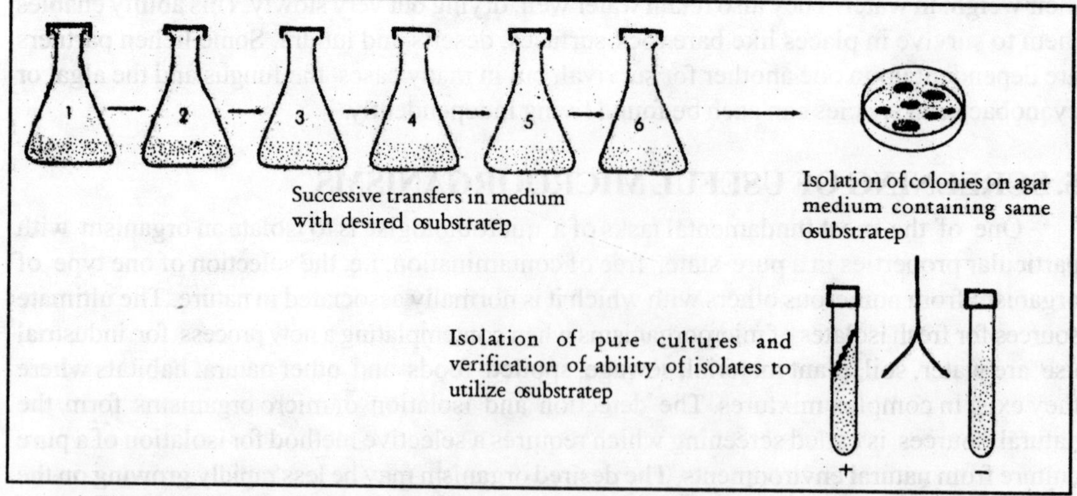


**Figure 27** Enrichment method for bacteria capable of utilizing a particular substrate

### 5.2. Physical methods of selection

One approach in this involves the use of heat treatment at which the endospore producing bacteria can be selected by heating a mixed culture to 80°C for 10 min. The second approach involves the use of incubation temperature according to the temperature range of the organisms which are to be screened: low temperature is used for psychrophiles while high temperature for thermophiles. The third approach involves the use of a culture media with specific pH according to the pH range of the organisms to be screened: low pH for acidophiles, high pH for alkalophiles. The fourth approach makes use of small cell diameter or the motility of the culture to be screened. A membrane filter having a pore size of 0.15 μm is placed on the surface of an agar plate and the mixed microbial population is applied over the filter surface. The cells with a diameter less than 0.15 μm penetrates the pores of the filter to reach the underlying agar and grow on the agar while the organisms which are too large to penetrate the membrane filter are unable to grow on the agar. Those which are smaller in size and also have the ability to swim will penetrate through the pores and migrate away from the point thus forming hazy zone within the agar.

### 5.3. Biological methods of selection

These involve the exploitation of negative interactions of the organisms towards the other organisms. One approach involves the selection of disease-producing species by taking advantage of pathogenic properties. If a mixed sample is injected in to a mouse the pathogens will multiply extensively while non-pathogens will be inhibited or killed by defense mechanism of the animal. The crowded plate technique is the other approach involving the detection and isolation of antibiotic producers. It consists of preparing a series of dilutions of the soil or other source material followed by spreading of dilutions on the nutrient agar plates. The agar plates having 300 to 400 or more colonies per plate are considered. The ability of a colony to exhibit antibiotic activity is indicated by the presence of a zone of growth inhibition surrounding the colony if one is interested to isolate.

## 6. ISOLATION OF PURE CULTURES

Once the desired microorganisms are enriched and screened by the selective methods, the same are purified, súb cultured and the strains of the same are preserved for future use. A variety of techniques have been developed for the isolation of pure cultures. One of the methods, "Streak-plate technique", involve the streaking of the culture on agar plate across the surface and colony developing from one cell will be sufficently apart from another. A single streak plate does not automatically assure purity – slime or chain producing bacteria may adsorb some contaminants. It is advisable to streak a culture several times in succession, preferably on a non-selective medium. A modification of streaking method, "Roll-tube technique" has been developed for the isolation of strict anaerobes. This involves the use of stoppered anaerobic culture tubes whose inner walls have been coated with a prereduced agar medium (Fig 28a). The tube containing an atmosphere of $O_2$-free $CO_2$ is rotated by a motor and inoculated with a transfer loop held against the agar surface starting at the bottom and drawing the loop gradually upwards (Fig 28b). After inoculation the tube is re-stoppered and incubated (Fig. 28c).

In the other two methods, "Pour plate" and "Spread plate techniques", the mixed culture is first diluted (only few cells/ml) before used to inoculate media. In pour plate, mixed culture is diluted directly in tubes of cooled, liquid agar medium and then dispensed into petridishes, allowed to solidify and then incubated. In spread plate, the culture is not diluted in medium but in sterile saline. Small amount of sample (diluted culture) is placed on to the surface of agar plate and spread evenly with the help of bent, sterile glass rod. Here, only surface colonies develop. Another method for obtaining pure culture involves the use of a device called the micromanipulator in conjuction with a microscope to pick a single bacterial cell from a mixed culture. Schematic illustration of a single bacterium from a mixture of cells is shown in Fig 29. The microscopic field (1) shows the point of the microprobe touching a bacterial cell and causing the organism to float in a small amount of water. The microscopic stage is then moved to the left (2) while the microscope remains fixed. This causes the bacterium to follow the probe and becomes separated from the other cells.

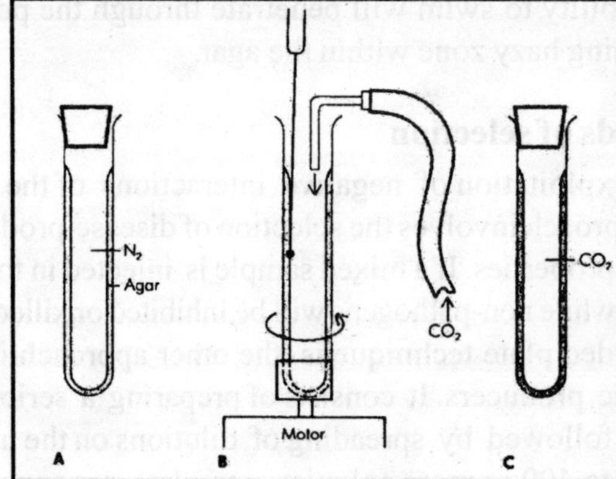

**Figure 28** Roll-tube method for isolating stringent anaerobes

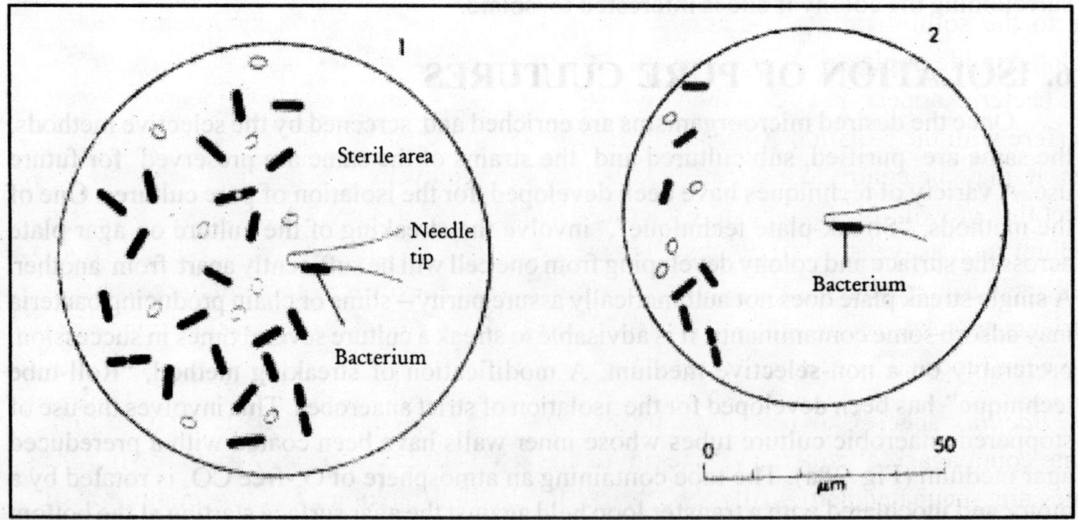

**Figure 29** Isolation of a single bacterial cell with the help of a micromanipulator equipment and microscope

### 6.1. Isolation and enumeration of bacteria

Lactic acid bacteria have been isolated by either of three procedures: (i) enrichment of the sample in broth culture followed by plating of this culture onto agar medium; (ii) direct inoculation of the sample onto an agar medium, and (iii) membrane filtration of the sample followed by incubation of the membrane onto an agar medium. The most useful and preferred procedure is direct inoculation (spread plating) onto an agar medium because the colonies which develop can be enumerated, differentiated by their appearance and isolated for identification.

Numerous media have been used for the isolation and enumeration of lactic acid bacteria from various food samples and these include de Man Rogosa Sharpe (MRS) agar, tomato juice (TJ) agar, Irrmann agar, tomato juice wine agar, yeast extract agar and various modifications of MRS agar to incorporate either grape juice, tomato juice, cysteine, malic acid and various sugars. Nakagawa agar and sucrose agar, developed for the isolation of brewery lactic acid bacteria could be useful for wine's lactic acid bacteria. The addition of cycloheximide (100 mg/l) or pimaricin (50 mg/l) to the plating medium is necessary to inhibit the growth of yeasts and moulds. Incubation of plates under an atmosphere enriched in carbon dioxide or nitrogen can encourage faster growth of the colonies of lactic acid bacteria.

Acetic acid bacteria are isolated by direct (spread) inoculation of samples over the surface of an appropriate agar medium. A range of media have been used to isolate acetic acid bacteria. Glucose-yeast extract-calcium carbonate-agar has been used most effectively to survey the presence of Acetic acid bacteria. The calcium carbonate gives the medium milky white appearance and serves to neutralise the acid produced by colonies of acetic acid bacteria. These colonies are recognised as they are surrounded by clear zones due to the solubilization of calcium carbonate. It is necessary to incorporate either cycloheximide or pimaricin into the medium to inhibit the growth of yeasts. Wine lactic acid bacteria appear not to grow very well on this medium and consequently donot interfere with the isolation of acetic acid bacteria. If their growth becomes a problem then it may be controlled by the addition of streptomycin (25 mg/l) which does not affect the growth of acetic acid bacteria. Alternatively, it may be necessary to use a medium such as yeast extract-ethanol agar which will selectively support the growth of acetic acid bacteria but not the lactic acid bacteria. The use of an agar medium comprising nutrients, glucose, acetic acid and ethanol for selectively isolating acetic acid bacteria has also been reported by certain workers.

*Bacillus* species can be isolated by plating the samples onto a nutrient agar medium (plate count agar) to which cycloheximide has been added to prevent yeast growth. The plates are incubated aerobically at 25-30°C for 24-48 h. Colonies that are comprised of Gram positive rods, are catalase positive and produce endospores are considered as the species of *Bacillus*.

## 6.2. Isolation and enumeration of yeasts

Both pour and spread plating techniques are used as the standard procedures for isolating and enumerating yeasts from various sources. However, there is increasing evidence that spread plating gives more accurate data. Other useful methods for enumerating yeasts are membrane filtration and microscopic examination. A variety of plating media have been used for the isolation of yeasts from various *sources* sources and these include grape juice agar, malt extract agar, yeast-malt (YM) agar and other nutrient agars. These are non selective media that allow the growth of all yeast species associated with the fermentation. A limitation in their use, however, occurs with samples containing mixtures of yeast species of significantly different populations. In these circumstances colonies of the most prevalent species dominate on the plate preventing observation of those species that are present in the sample at lesser populations. Lysine agar is used most effectively to selectively isolate and enumerate the populations of non-*Saccharomyces* species including *Kloeckera apuculata*, *Candida stellata* during wine fermentations. Predominating *Saccharomyces cerevisiae* does not interfere with the growth of these yeasts on this medium because it is unable to use lysine as a nitrogen source and consequently does not grow. The successful use of lysine agar to specifically monitor the growth of non-*Saccharomyces* yeasts during wine fermentation has now been demonstrated by many researchers. Nutrient medium containing ethanol (12%) and sodium metabisulphite (0.015%) is also for the selective enumeration of *Saccharomyces* species during wine fermentation, because of their lesser tolerance to ethanol and sulphur dioxide. Species of *Kloeckera* and *Hanseniaspora* in particular and other non-*Saccharomyces* yeasts have difficulty in growing in this medium therefore making it useful to monitor the growth of *S. cerevisae* during the early stages of fermentation. Use of a medium containing elevated concentrations of sorbate or benzoate for the selective isolation of *Zygosaccharomyces bailii*, a species mostly associated with wine spoilage has also been reported. Biggy agar can be used to monitor the presence of hydrogen sulphide ($H_2S$) producing yeasts. This medium contains bismuth sulphite which reacts with $H_2S$-producing colonies causing them to appear brown-black.

Bacterial growth on yeast isolation plates may be a problem when examining samples of grapes or musts. Such growth can be controlled without detriment to yeast development, by incorporating oxytetracycline (100 mg/l) or other bacterial antibiotic into the medium composition. Fungal contamination of yeast isolation plates can also be a problem with some grape samples. Some authors have added sodium propionate (0.1-0.2%) or biphenyl (0.02%) to the medium to control this interference.

The detection of killer toxin producing yeast is done by the seeded agar plate technique using malt extract agar buffered at pH 4.5 with 0.5M phosphate citrate buffer. Approximately $10^5$ cells/ml of the senstitive strain are suspended in 15 ml of pre-sterilized, molten (45°C) medium and then the mixture is poured into the petridish and allowed to solidify. Strains of the killer yeast are streak inoculated onto the surface of the seeded agar which is then incubated at 20°C for 48 h. Killer activity is recognised by inhibition of growth (clear zones) of the seeded strain in the region surrounding growth of the killer strain.

Although enumeration of yeasts by agar plating is suitable for ecological studies, it is too slow to meet the requirements of modern quality assurance programs. Generally, it is necessary to incubate agar plates for upto four days before reliable data on counts is obtained. More rapid, automated methods are being developed for the enumeration of yeast populations and these include the application of impedance, ATP-bioluminescence and fluorescence microscopy techniques. In some cases these techniques have been applied to wine yeasts but further research and development of these applications is required.

### 6.3. Isolation and enumeration of moulds

Procedures for the isolation and enumeration of fungi from foods and beverages have undergone extensive revision in recent years and are discussed in detail by many workers.. Direct plating of sample suspensions onto dichloran-rose bengal-chloramphenicol (DRBC) agar is now widely accepted as the preferred method for isolation of fungi from foods and beverages. The plates are incubated aerobically at 20-25°C for 72-96 h. The advantage of this medium over others that have been used in the past (acidified potato dextrose agar) lies in the combined use of dichloran and rose bengal to limit the spreading-lateral growth of fungal colonies without inhibiting their initial development.

## 7. CLASSIFICATION AND IDENTIFICATION OF MICROORGANISMS

Upon the isolation of a particular microorganism, its identity as well as its cultural characteristics are necessary to ensure that the culture bears the specific epithet (genus + species), and has the properties cited for it in a scientific journal or patent, or in a taxonomic text such as Bergey's manual of systematic bacteriology, The Yeasts: characteristics and identification. With bacteria and yeasts, apart from microscopic examination, complete identification of individual types needs a series of further tests. The sum of the various test amounts to a character profile, and enables one to compare the isolate with previously reported micro-organisms. The description, characteristic morphology, the results of biochemical tests for bacteria are collected together and compared with the data in standard taxonomic texts. Since microfungi are relatively complex in structure, the individual types are identified on the basis of appearance, their growth on specified media and the exhibition of certain macroscopic and microscopic properties and physiological tests.

Alternatively, the authenticity and the identity of the isolate can be verified by one of the large culture collections held by national organizations such as The American Type Culture Collection (ATCC), Rockville, Maryland, Central Bureau voor Schimmelcultures (CBS), Delft, The Netherlands, Northern Regional Research Laboratory (NRRL), US Department of Agriculture, Peoria, Illinois, National Collection of Yeast Cultures (NCYC), Food Research Institute, Colney Lane, Norwich, UK. National Collection of Industrial Microorganisms (NCIM), NCL, Pune, India, Microbial Type Culture Collection (MTCC), IMT, Chandigarh, India.

Yeasts are classified into genera and species on the basis of their morphological, physiological, biochemical and more recently, genetic properties. They are identified by conducting tests and following the classification schemes. Generally it is necessary to conduct about 50-100 tests to reliably identify most yeasts to species level. Consequently the work load is demanding and one to two weeks are often needed to obtain a final result. Moreover accurate interpretation of the data requires considerable expertise and is further complicated by many changes to the generic and specific names of yeasts that have occurred over the years. These changes have arisen as a result of the newer molecular approaches (nucleic acid hybridization) to taxonomy and from the desire of taxonomists to describe the most realistic phylogenetic classifications.

The laborious, time-consuming and costly nature of identifying yeasts by conventional procedures has led to the development of shortened, more convenient identification schemes that are based on yeast response to a few carefully selected tests and on the use of commercially available miniaturized diagnostic kits Some commonly used identification systems for food spoilage yeasts are based on their responses to some 15-25 key tests. Although the reliability of these systems needs further verification they could be of use for rapidly identifying wine yeasts. A simplified system for identifying wine spoilage yeasts, based on their growth in lysine agar, pantothenate-free agar and two other media containing either cycloheximide or 5% sodium chloride is also used.

Several ready-to-use diagnostic kits have been developed for the rapid identification of clinically significant yeast species. Generally they exploit yeast growth response to a very limited range of carbon and nitrogen compounds and give results within 48 h. Computer assisted interpretation of data is available for some kits. Of these systems, the API 20C and API 20C Auxanogram kits have been used with considerable advantage and success for the identification of wine yeasts alongwith some additional tests in order to obtain reliable identification of wine yeasts. The development and evaluation of a novel system that enables all of the fermentation and assimilation tests in the wells of a microtitre tray, alongwith vitamin growth tests and sporulation tests has been very reliable in the identification of wine yeasts within 48 h.

## 7.1. Characterization of microorganisms

The organisms are first characterized under various categories before identification and classification. It is not easy to study the characters of a single microbial cell due to its smaller size so for the purpose of characterization we generally obtain the population of microorganisms of same type, the pure culture. Some of the important characteristics which are taken into consideration for identification and classification are as follows:

**7.1.1.** *Morphological characteristics:* Every organism has a characteristic cell shape, size, structure, cell arrangement, presence of special structures, staining reactions, motility, flagellar arrangement. These are studied with the help of light and electron microscopes

**7.1.2.** *Chemical composition:* Microbial cells consists of many organic compounds and show wide variation in qualitative and quantitative composition. Cells are broken apart and subjected to chemical analysis. Occurrence of lipopolysaccharides in cell walls of gram-ve bacteria, teichoic acid in gram+ve bacteria are some of the differences in the chemical composition of gram+ve and gram-ve bacteria. Fungal cell walls are also different from bacterial walls and viruses differ from each other on the basis of type of nucleic acid.

**7.1.3.** *Cultural characteristics:* Each microorganism has specific nutritional requirements and the physical conditions required for growth. Many can be grown in or on a culture medium, some grow in a medium containing only inorganic compounds while others require a medium containing organic compounds like sugars, amino acids, proteins, purines or pyrimidines, vitamins etc. Some require even complex natural substances like peptone, yeast extract, blood cells or blood serum. There are a few like rickettsias and viruses which cannot be grown on artificial culture media and can be cultivated only in the living cells. In addition to specific nutrients each organism requires specific physical conditions of temperature, gaseous atmosphere, light etc.

Each organism grows in a characteristic manner in liquid medium where the growth may occur as a sediment or as a film called pellicle. On solid medium the organisms form different types of colonies.

**7.1.4.** *Metabolic characteristics:* The microorganisms differ in the ways in which they obtain and use their energy for various metabolic reactions. Some obtain energy by absorbing light others by oxidizing organic or inorganic compounds, some by redistributing the atoms within certain molecules. The various metabolic reactions are catalyzed by enzymes which differ in different organisms.

**7.1.5.** *Antigenic characteristics:* Microorganisms have special large chemical components, acting as antigens of the cells which are distinctive for certain kinds of microorganisms. Antigens, along with cells when enter the animal body, the animal responds by forming specific blood serum proteins, the antibodies which bind with the antigen. Since different kinds of microorganisms have different types of antigens, antibodies are widely used as tools for the rapid identification of particular kinds of microorganisms.

**7.1.6.** *Genetic characteristics:* Characteristics of hereditary material of the cells (DNA) and occurrence and function of other kinds of DNA like plasmids show wide variation in different types of microorganisms. The double stranded chromosomal DNA of each organism has certain features that are constant and characteristic for that organism and generally used for identification. One of the genetic characteristic used for identification and classification is the DNA base composition which is made up of guanine-cytosine and adenine-thymine. The % of G+C (moles % G+C) is generally constant for different organisms and generally ranges from 23-75. The other characteristic includes the sequence of nucleotide bases in DNA which is unique for each kind of organism. In addition to chromosomal DNA, plasmid DNA may also be present in some organisms which give some organisms the ability to make toxins, to become resistant to antibiotics or to use unusual chemical compounds as nutrients.

**7.1.7.** *Pathogenecity:* The ability to cause a disease in other organisms is also variable in different groups of the microorganisms. Some may cause disease in plants, some in animals, some in human beings and some even in other microorganisms.

**7.1.8.** *Ecological characteristics:* Habitat and distribution of the microorganisms in nature and the interactions in natural environments is also important in the characterization. Some exist in soil, some in intestinal tract, some on plants, some on animal skin etc. They also exhibit different types of neutral, positive and negative interactions including with other micro and macroorganisms.

Once the characteristics of microorganisms have been determined, the process of classification begins

## 7.2. Classification

Taxonomic groups in Microbiology are constructed from 'Strains', the descendents of a pure culture and are usually a succession of culture from a single colony. Each strain has a specific history and designation. For example strain MTCC XXXX is a strain of ABC isolated originally from soil of Panjab University in Chandigarh in 1975 by XYZ and the cultures of this strain are maintained at MTCC, Chandigarh. Cultures of the same species that are isolated from other sources are considered different strains.

The basic taxonomic group is 'Species' which is a collection of strains having similar characteristics. Microbial species consist of a special strain, 'Type Strain' which is the reference strain together with all other strains that are considered similar to type strain are included in the 'Species'. Just as a microbial species is a collection of similar strains, a microbial 'Genus' is a collection of similar species. One of the species is designated as a 'Type Species', reference species, which serves as the permanent example of the genus. A group of similar genera - 'Family', a group of similar families - 'Order', a group of similar orders - 'Class' , a group of similar classes – 'Division' and a group of similar divisions – 'Kingdom' are the other taxonomic groups of higher rank.

**7.2.1.** *General methods of classifying microorganisms:* Once the characteristics are established, the process of the identification and classification starts. There are three methods generally used for the classification of microorganisms.

**7.2.1.1.** *The intuitive method:* Generally used by the persons who are thoroughly familiar with the properties of the organism, he has been studying for several years. One can classify on the basis of his experience.

**7.2.1.2.** *Numeral taxonomy:* One determines several characteristics (usually 100-200) for each strain and then by using a computer he calculates the % similarity (%S) of each strain.

$$\% S = NS/NS+ND$$

NS= number of similar characters

ND=number of different characters

**7.2.1.3.** *Genetic relatedness:* Most reliable method of classification. Initially mole % G+C was used but now two modern techniques have appeared.

1. DNA homology experiments: Double stranded DNA molecules from two organisms are heated to convert them to single strands. The single strands from two organisms are mixed and allowed to cool. If the two organisms are closely related, 'heteroduplexes' will be formed after pairing of the stands from different organisms
2. Ribosomal RNA homology experiments: Ribosomes – granular structures composed of proteins and RNA. The rRNA is coded by a small fraction of DNA molecule the 'rRNA cistrons'. The nucleotide sequence of these rRNA genes is highly conserved. This degree of similarity is used to determine the relatedness between organisms.

## 7.3. Nomenclature

After classification the microorganisms are given a common name and a binomial scientific name according to International code of nomenclature which is always written in Latin. The scientific name has two words, generic name and the specific name. Generic name always starts with capital letter while specific name always starts with a small letter but both are either written in Italics or are underlined.

## 7.4. Bergey's system of bacterial classification

This is a comprehensive manual of classification for bacteria. A comparable manual of classification does not exist for fungi, algae or protozoa. The first edition of Bergey's manual of determinative bacteriology was initiated by action of the Society of American Bacteriologists (now called the American Society for Microbiology) by appointment of an Editorial Board consisting of David H. Bergey, Chairman, Francis C. Harrison, Robert S. Breed, Bernard W. Hammer, and Frank M. Huntoon. This Board, under auspices of the Society of American Bacteriologists who, then as now, published the Journal of Bacteriology as a service to science, brought the first edition of the manual into print in 1923. The Board, with some changes in membership and Dr. David Bergey as Chairman, published a second edition of the manual in 1925 and a third edition in 1930. Since the creation of the Trust, the Trustees have published, successively, the fourth, fifth, sixth, seventh, and eighth editions of the manual (dated 1934, 1939, 1948, 1957, and 1974, respectively). In 1977 the Trust published an abbreviated version of the eighth edition, called The shorter Bergey's manual® of determinative bacteriology; this contained the outline classification of the bacteria, the descriptions of all genera and higher taxa, all of the keys and tables for the diagnosis of species, all of the illustrations, and two of the introductory chapters; however, it did not contain the detailed species descriptions, most of the taxonomic comments, the etymology of names, and references to authors. The manual has 4 Volumes containing internationally recognized names and description of bacterial species. The details of information in the above volumes are summarized below:

- Volume I : (Sections 1-11) 1984: Gram-ve bacteria
- Volume II: (Sections 12-17) 1986: Gram +ve bacteria, Phototrophic and other specialized bacteria including gliding bacteria
- Volume III: (Sections 18-25) 1989: Archaebacteria
- Volume IV: (Section 26-33) 1991: Actinomycetes and other filamentous bacteria

Bergeys' manual of systematic bacteriology places all bacteria in Kingdom Prokaryotae with 4 divisions.

| | | |
|---|---|---|
| Division I | : | Gracilicutes : Prokaryotes with a complex cell-wall structure characteristic of gram -ve bacteria |
| Division II | : | Firmicutes : Prokaryotes with cell wall structure characteristic of gram +ve bacteria |
| Division III | : | Tenericutes : Prokaryotes that lack a cell wall |
| Division IV | : | Mendosicutes : Prokaryotes having faulty cell wall, suggesting the lack of peptidoglycan |

## 7.5. General features and classification of some groups of microorganisms

**7.5.1.** *Rickettsias:* This group includes two subgroups, *Rickettsias* and *Chlamydias Ricketssias* mostly live in an obligate intracellular association – parasitic or mutualistic, with eukaryotic hosts (vertebrates or arthropods), a few can be grown on moderately complex bacteriological media containing blood. Glutamate is oxidized with ATP generation. Cell wall contains muramic acid. Mainly rod shaped, coccoid or even pleomorphic that stain gram –ve and lack flagella. Important genera include *Bartronella, Grahamella, Rochalimoea*. Chlamydias are non-motile, obligate parasites, multiply with in the membrane bound vacuoles in the cytoplasm of human cells, mammals, birds. Cell wall lacks muramic acid. Glucose is not oxidized with ATP generation, hence the parasite depends upon the host cell's ATP.

**7.5.2.** *Mycoplasma (Mollicutes):* The cell wall-less bacteria. This group consists of very small prokaryotes totally devoid of cell wall – consists of only plasma membrane and incapable of synthesizing peptidoglycan. Resistant to penicillins and sensitive to lysis by osmotic shock. Pleomorphic – varying their shape from spherical or pear shaped structures to branched or helical filaments. Generally non-motile but show gliding motility in liquid. Mostly facultative anaerobes but few strict anaerobes. Most species form characteristic fried egg like colonies. *Mycoplasma, Spiroplasma* important genera.

**7.5.3.** *Archaebacteria:* Unique feature is in the cell membrane structure. Membrane lipids are glycerol isopropyl ethers. Cell envelopes show resistance againt cell wall antibiotics and lytic agents as all lack murein. Some gram –ve and some gram +ve. Gram –ve composed of single layed or more complex crystalline protein or glycoprotein subunits while gram +ve have pseudomurein, hetero-polysaccharide. Single S-layers (three dimensional structure) are found in most organisms which help in withstanding extreme environmental conditions such as high salt, low pH and high temperature.

**7.5.4.** *Actinomycetes:* Filamentous bacteria due to mycelium like structures – less broad than fungal mycelium. A chain of sexual spores – conidia are produced on their hyphae and few bear the sporangium. Colonies are powdery mass over the surface of the culture media.

## 8. MAINTENANCE AND PRESERVATION OF MICROORGANISMS

Several methods are available for the preservations of the microorganisms but there are two criteria for selecting a method of preservation for a given culture. They are: i) period of preservation desired and ii) nature of the culture to be preserved. Cultures are preserved for several reasons including demonstration of specific properties, research, assay, comparison studies, type strains, production of fermentation products and starter cultures. It is most desirable to preserve the cultures in such a way that they can be handled and transported with ease, and remain physiologically active as near to initial performance as possible without loss of genetic unity or of biochemical characteristics. In general each preservation method can be assigned to one of the following groups:

### 8.1. Metabolically active methods

The cultures remain metabolically active by these methods and need to be maintained at low temperatures, in the refrigerators during the period of the preservation. There include periodic transfer on agar or in liquid medium or keeping agar cultures under mineral oil.

**8.1.1.** *Periodic transfer to fresh media:* The temperature and the type of medium chosen should support a slow rate of growth.

**8.1.2.** *Overlaying cultures with mineral oil:* Oil must cover the slant completely - It should be about ½ inch above the tip of the slanted surface.

Maintenance of strains by metabolically active methods are only for short-term preservation and should be used only in case a strain cannot be preserved by one of the metabolically inactive methods, or in addition to one of these methods

### 8.2. Metabolically inactive preservation techniques

The cultures are made metabolically inactive by these methods either by cryopreservation or by drying.

**8.2.1.** *Freezing and low temperature storage in or above liquid nitrogen:* Dense suspension containing some cryoprotective agent – glycerol, dimethyl sulfoxide (DMSO) is frozen to -150°C and then stored in liquid nitrogen refrigerator either by immersing in liquid nitrogen (-196°C) or in gas phase above liquid nitrogen (-150°C).

**8.2.2.** *Freezing and low temperature storage below -70°C:* Dense suspension is generally made in 10% glycerol and frozen in the deep freezers.

**8.2.3.** *Freeze drying (Lyophilization):* The advantages of freeze-drying are obvious. It is a convenient method for the preservation and long-term storage of a wide variety of microorganisms. However, special precautions are needed for the preservation of microorganisms sensitive to desiccation, light, oxygen, osmotic pressure, surface tension and other factors. Some effective protective agents for example skim milk and meso-inositol or honey or glutamate or raffinose, are used to suspend cells to be freeze-dried inorder to protect these against known freezing and drying injuries. Several anaerobic bacteria which are sensitive to aerobic freeze-drying, can successfuly be preserved using activated charcoal (5% w/v) in the suspending media along with the above protective agents. Freeze-drying involves the removal of water from frozen cell suspension by sublimation under reduced pressure. The outline of the freeze-drying procedure and the major steps involved are shown in Figure 30. Most bacteria die when the cultures become dry although spore formers remain viable for many years. Suspension is frozen at -60 to -70ºC and then connected to vacuum. Sublimation occurs causing dehydration of bacteria with minimum damage to the cell. Small cotton-plugged vials containing frozen suspensions of the microorganisms are placed in a flask, attached to a condenser which is connected to a vacuum pump (Figure 30a). The microbial cells become dessicated as the ice in the frozen suspension sublimes directly to water vapours which are trapped in the condenser thus preventing it from entering the vacuum pump (Figure 30b). After dessication the vials are removed and hermetically sealed under a vacuum.

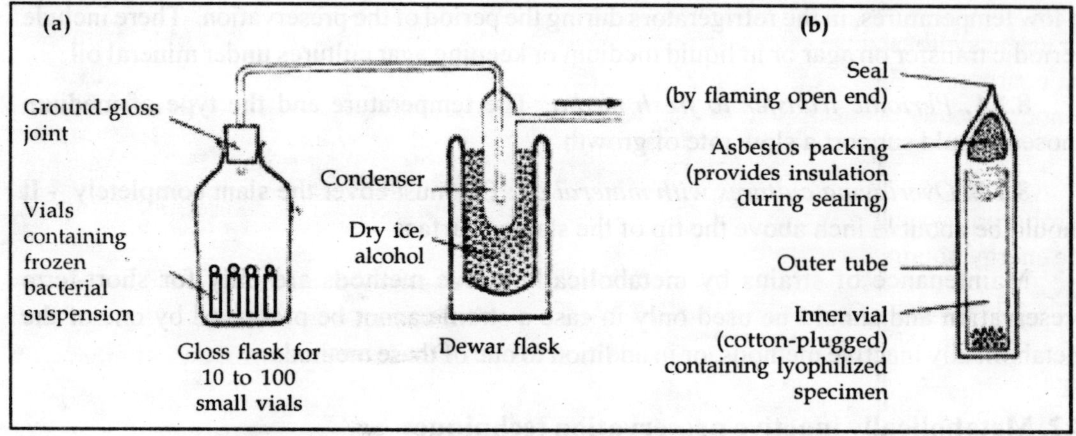


**Figure 30** Lyophilization process for preservation of microbial cultures

**8.2.4.** *Liquid drying:* Liquid-drying (L-Drying) involves vacuum-drying of samples from the liquid state without freezing. It is one of the sophisticated techniques used for the long-term preservation of microorganisms. Several microorganisms, which are sensitive to freezing or freeze-drying, can successfully be preserved by liquid drying. Liquid-drying has several advantages over freeze-drying and has been effectively used for preserving large collections of fragile microorganisms in various culture collections. However, specialized equipment is required for L-drying. Recently a simple and effective liquid-drying method has been described for the successful preservation of sensitive microorganisms (including various anerobes) which fail to survive freezing or freeze-drying.

Since micro-organisms can survive for years in a completely dry state, freeze drying (lyophilization), as well as soil tube storage are most effective and require the least maintenance for long-term storage. Cultures lyophilized in tubes sealed and stored in liquid nitrogen, at 4ºC in a cold room, or in a refrigerator have been known to survive storage for 30 or more years. Together with this, efficient record keeping and a precise inventory system asists in maintaining reliable records and eliminating costly mix-up. With proper care for nutrition, temperature, and inocula development, the lyophilized cultures are easily rehydrated and restored to vegetative growth. The verification of the culture after retrieval is usually checked microscopically for the presence of contamination, morphological variation and tested in shake flasks for authentication of the expected physiology.

## 9. NUTRITIONAL REQUIREMENTS AND CULTIVATION OF MICROORGANISMS

The substrates they utilize and transform for growth are highly diverse. All microrganisms require a source of energy. Some rely on chemical compounds for their energy and are distinguished as chemotrophs, others utilise light as soruce of energy and are called phototrophs. All require a source of electrons for their metabolism. Some use reduced inorganic compounds as electron donors, the lithotrophs (some may be chemolithotrophs, others photolithotrophs), others use organic compounds as electron donors, the organotrophs (some are chemo-organotrophs, others photoorganotrophs). All require carbon in some form for use in synthesizing cell components. Autotrophs require only carbon dioxide to satisfy their carbon needs whereas heterotrophs obtain their carbon supply from organic carbon. Among the autotrophs the phototrophic bacteria rely on light for energy and use either reduced inorganic compounds such as sulphur compounds or organic carbon as the source of reducing power under anaerobic conditions. Photosynthesis by the cyanobacteria is aerobic and water provides the reducing power. The chemotrophs fix carbon dioxide using the energy obtained by the oxidation of inorganic molecules.

All organisms require nitrogen. Some bacteria utilize atmospheric $N_2$ while others thrive on inorganic 'N' compounds or aminoacids. Among the heterotrophs some need simple source of organic carbon for growth whereas others have more complex growth requirements demanding a full complement of amino acids and vitamins. With most industrial heterotrophs carbon is supplied as carbohydrate and nitrogen is supplied as ammonium ion, urea or protein-rich byproducts. All living organisms require metal ions ($K^+$, $Ca^{+2}$, $Mg^{+2}$, $Fe^{+2}$) for their normal growth. Other metal ions are required in trace amounts and are generally present in medium as contaminants. All require oxygen, sulphur and phosphorus for cell components. Oxygen is provided in various forms such as water, component atom of various nutrients or molecular oxygen. Sulphur needed for synthesis of certain amino acids (cysteine, cystine and methionine) is used in the form of inorganic S-compounds. Some can even use elemental sulphur. All the nutrients must be present in aqueous solution form before they enter the cells. Also, water is a chemical reactant, which takes part in many hydrolytic reactions carried out by a cell.

Numerous media have been developed for bacterial cultivation. But as the nutritional requirements vary widely, there are great diffrences in the chemical compositions of the media used in laboratory cultivation.Generally, chemically defined media are employed for the cultivation of autotrophic organisms. For heterotrophs, simple liquid media like nutrient broth or solid media i.e., Nutrient agar is generally used. Yeast extract improves the nutrient quality of the media when added. For the cultivation of fastidious heterotrophs, blood, serum, extracts of plant and animal tissues are supplemented in nutrient broth or nutrient agar.

## 10. GROWTH AND REPRODUCTION

Growth can be considered as the increase in living substance, usually the number of cells or total mass of cells. The growth rate measures change in either cell number or cell mass per unit time. In unicellular organisms growth involves increase in cell numbers. Bacterial cells commonly multiply by transverse binary fission. At first, cell size doubles and the cell then divides into two daughter cells which have approximately the same size as the original mother cell. Some bacteria such as *Rhodopseudomonas acidophila* and *Hyphomicrobium* spp. reproduce by budding. Species of the genus *Streptomyces* and related bacteria produce many spores, conidiospores per organism by developing cross walls (septation) at the hyphal tips, each spore giving rise to a new organism. The time required for doubling of cell numbers is known as generation time while the time required for doubling of cell mass is referred to as the doubling time. Under conditions where cell mass and numbers double in the same time interval the generation time (g) and doubling time (td) are of course equal. The reciprocal value 1/g = number of doublings / h is called the growth rate, v. Yeasts reproduce asexually (budding and fission) or sexually producing resistant spores as part of their life cycle. In budding a small daugther cell begins to grow on the side of mother cell followed by physical separation of the mature offspring. The mother cell retains a bud scar. Under some conditions buds donot separate from the mother cell and a branched chain of cells called a pseudomycelium is formed. Sexual reproduction in yeasts occurs by the union of two cells of opposite mating types, each having a single set of chromosomes (haploid). The fusion results in the formation of a diploid cell (two sets of chromosomes per cell). The nucleus in the diploid cell undergoes one or several divisions and forms sexual spores called ascospores which are formed inside a sac or ascus, each eventually becomes an individual new haploid cell.

### 10.1 Methods of study of microbial growth

The term "growth" as commonly applied in microbiology refers to the magnitude of total population. Growth in this sense can be determined by numerous techniques based on one or more of the following types of measurements: 1) cell count directly by microscopy or by using electronic particle counter or indirectly by a colony count. 2) cell mass, directly by weighing or by a measurement of cell nitrogen or indirectly by turbidity. 3) cell activity, indirectly by relating the degree of biochemical activity to the size of the population.

Microscopic techniques may be used to count the total number of microbes present in the given specimen, however, there are several limitations to the accuracy of the method. One is that direct microscopic examination does not inform us whether the cells being counted are living or dead.

Since individual viable cells have the potential for growing on solid media to produce viable and visible clones, the procedure of plating can be adopted to count the number of viable microbes present in a specimen. A direct method often involves preparation of a serial dilution series using physiological saline or buffers, followed by plating on to nutrient medium agar. A common technique is the use of microporous membrane filters for counting microorganisms present in small numbers in large volume of sample (air or water). Passing the sample through the filter captures the microorganisms on the surface of the filters. Upon superimposing the filters on the surface of the solid nutrient media, incubation results in the development of viable cells to countable colonies. A wide range of selective indicator media are available, enabling a particular organism to develop in preference to others or particular type of colonies to be identified directly in the presence of others.

## 10.2. Factors affecting growth

Microorganisms adapt and grow under widely differing conditions. Very few places on earth are devoid of microbial life. Because of the ubiquity of microorganisms, when large number of individual cells are present in the environment and if at a particular place conditions suitable for microbial proliferation exist then that site will inevitably be colonized by the appropriate microorganisms. Thus establishment within the laboratory of a sterile environment free from living microbes and protected from contamination is of paramount importance.In addition to knowing the proper nutrients for the cultivation of bacteria, it is also becoming necessary to know the physical environment in which the organisms will grow best. Just as microorganisms vary greatly in their nutritional requirements so do they exhibit diverse responses to physical conditions such as temperature, gaseous condition and pH.

**10.2.1.** *Temperature:* Since all processes of growth are dependent on chemical reactions and since the rates of reactions are influenced by temperature, the pattern of microbial growth can be profoundly influenced by this condition. On the basis of temperature relationships, microorganisms are divided into three main groups: psychrophiles, which are able to grow at 0°C or lower but have optimum temperature of around 15°C or lower and a maximum temperature of about 20°C, mesophiles, growing best within a temperature range of approximately 25-40°C, thermophiles, growing best at temperature above 45°C. The growth range of many thermophiles extends into the mesophilic range but they always grow above 45°C. Factors which are implicated in the ability to grow at high temperature are increased thermal stability of ribosomes, membranes and various enzymes.

**10.2.2.** *Gaseous requirements:* The principal gases that affect the microbial growth are oxygen and carbon dioxide. Microorganisms display such a wide variety of responses to free oxygen that it is convenient to divide them into four groups. Aerobic, requiring oxygen for growth and can grow when incubated in an air atmosphere (21% oxygen). Anaerobic, which don't use oxygen to obtain energy. Moreover oxygen is toxic for them and cannot grow when incubated in an air atmosphere. Some can tolerate low levels of oxygen (non stringent or tolerant anaerobes) but others (stringent or strict anaerobes) cannot tolerate even low levels and may even die upon brief exposure to air. Facultative anaerobes, requiring oxygen for growth although they may use it for energy production if it is possible. They are not inhibited by oxygen and usually grow well under an air atmosphere as they do it in absence of oxygen. Microaerophilic, requiring low levels of oxygen for growth but cannot tolerate the level of oxygen present in atmosphere.

**10.2.3.** *Acidity or alkalinity (pH):* For most microorganisms, the optimum pH for growth lies between 6.5-7.5 and the limits generally lie somewhere between 5.0-9.0.

When organisms are cultivated in a medium originally adjusted to a given pH e.g. 7.0, it is very likely that this pH will change as a result of chemical activities of the organisms. If a carbohydrate is present it may be fermented or oxidised to organic acids thus decreasing the pH of the medium. If the salt of an organic acid is supplied as a carbon source (sodium malate) its oxidation will cause an increase in pH. Such shifts in pH may be so great that further growth of the organism is eventually inhibited.

Besides the requirement for temperature, gaseous environment and pH, some organisms have additional requirements for their growth. e.g. phototrophic bacteria must be exposed to a source of illumination. Since light is a source of energy, bacterial growth may also be influenced by hydrostatic pressure. Bacteria have been isolated from the deepest ocean trenches where the pressure is measured in tons per square inch and many of these organisms will not grow in laboratory unless the medium is subjected to similar pressure.

## 11. CONTROL OF MICROORGANISMS

The aim of control of microorganisms is aimed at reduction in numbers and / or activity of the total microflora. The microorganisms can be controlled by physical, chemical and chemotherapeutic agents which affect the microorganisms in any of the following ways: i) Damage to the cell wall or inhibition of cell wall synthesis, ii) Alteration of the permeability of the cytoplasmic membrane, iii) Alteration of the physical or chemical state of proteins and nucleic acids, iv) Inhibition of enzyme action, v) Inhibition of protein or nucleic acid synthesis. The antimicrobial action of various agents is influenced by some factors including number of microorganisms present, kinds of microorganisms and the physiological state of cells, environmental factors (temp, pH), dosage of agent used, presence of other inhibitory stuff (organic material). The relative resistance of some of the microorganisms is indicated in Table 3.

**Table 3** Relative resistance of Microorganisms

| | |
|---|---|
| Highest resistance | Spore forming organisms |
| Moderate resistance | Protozoan cysts |
| | Hepatitis B |
| | Poliovirus |
| | *Mycobacterium tuberculosis* |
| | *Staphylococcus aureus* |
| | *Pseudomonas* |
| Least resistance | Most bacteria |
| | Yeasts |

### 11.1 Control by physical agents

Microorganisms can be removed, inhibited or killed by various physical agents like high and low temperature, drying, osmotic pressure, radiations and filtration.

**11.1.1.** *High temperature:* This is one of the safest and most reliable methods used to destroy the microorganisms because of protein precipitation or denaturation. As the temperature increases beyond the maximum temperature for growth of a microorganism, lethal effects occur. The rate of death is a function of both the temperature and the time of exposure. Microbial death is an exponential function (death is linear when plotted on a log scale). Steam under pressure is most effective as it destroys all vegetative cells and spores. Some of the procedures involving high temperature include: a) *Canning*: generally used for the preservation of foods involving heating of the unsterilized foods sealed in unsterilized tin containers at 100°C for high acid foods, 121°C for low acid foods. b) *Pasteurization*: This involves heating of liquid foods at controlled temperatures of 145°F (62.8°C) for 30 min or 161°F (71.7°C) for 15 sec. This destroys all yeasts, molds, gram-ve bacteria and most gram+ve bacteria. Most heat resistant pathogen *Coxiella burnetti* is also killed c) Sterilization: This involves the use of 121°C using wet heat (steam under pressure) or 160-180°C using dry heat. To understand thermal destruction of microorganisms, it is necessary to understand certain basic concepts: i) Thermal death time (TDT): Time necessary to destroy a given population of microorganisms at a specified time, ii) Thermal death point (TDP): Temperature necessary to destroy a given population of microorganisms in a fixed time, usually 10 min, iii) Decimal reduction time, D value, Time necessary to destroy 90% of the organisms at a particular temperature, iv) Z value: Temperature in °F required to vary D value by 90%. D Value indicates resistance of microorganisms to a specified temperature. Z Value indicates relative resistance to different temperatures and is used to construct equivalent thermal processes.

**11.1.2.** *Low temperature:* This is used to lower or stop the activities of microorganisms. There are two ranges of lower temperature used for controlling microorganisms: i) refrigeration temperature (2-10° C), ii) freezing temperature ( below 0°C). The freezing temperature has some killing effect as the water in the cytoplasm expands on freezing leading to the bursting of the cells and also due to the mechanical injury caused by bigger ice crystals. The killing effect of the freezing temperature decreases as we go away from the freezing point of water due to less expansion and smaller size of ice crystals formed at that temperature. That is why the freezing temperature below -70°C is used for the preservation of microorganisms, however, the activities of the microorganisms at this temperature are stopped.

**11.1.3.** *Dehydration:* Drying reduces the water activity ($a_w$) and thus prevents the growth of microorganisms. Time of survival of microorganisms after desiccation depends upon factors – kind of microorganisms, material on which the organisms are dried, completeness of drying process, physical conditions to which the dried organisms are exposed – light, temperature, humidity.

**11.1.4.** *Osmotic pressure:* NaCl and sugars, when used in higher concentration, exert a high osmotic pressure which has the drying effect on the cells due to plasmolysis, ultimately causing the death. Halodurics, halophiles, osmophiles, however, tolerate the high osmotic pressure.

**11.1.5.** *Electromagnetic radiations:* There are two types of radiations used for the control of microorganisms: i) Ionising radiations like X rays and γ radiations which act by knocking out electrons from molecules and ionize them forming hydrogen radicals like hydroxyl radicals and peroxides which are toxic for cells, ii) Non-ionising radiations like UV radiations which are less energetic, absorbed specifically by different compounds causing excitation of their electrons and raising them to higher energy levels. UV radiations are absorbed by nucleic acids, more specifically by pyrimidine bases forming thymine dimers thus inhibiting DNA replication and resulting lethal mutations.

**11.1.6.** *Filters:* A variety of filters are used to remove the microorganisms from liquids and air. The mean diameter of the pores in these filters is generally less than 1 micrometer and are made of different materials like asbestos pas, diatomaceous earth, porcelain and sintered glass disks. These filters act as the mechanical sieves for the microorganisms and the efficiency of microbial filtration depends upon the electric charge of the filter and the organisms, the nature of the liquid being filtered. The high efficiency particulate air (HEPA) filters are generally used to provide germ-free air to an enclosure and the same together with a system of laminar airflow is used extensively to produce dust and germ-free air.

## 11.2. Control by chemical agents

A large number of chemical compounds have the ability to inhibit the growth and metabolism of microorganisms or to kill them. Some of chemicals are used for killing all organisms (sterilization), some are used for reducing pathogens till harmless (disinfection); examples include a) Antiseptic, used externally on tissue to destroy or inhibit organisms, b) Disinfectant - used on surfaces (don't kill spores), c) Sanitizer - used on food preparation equipment / utensils to reduce number, some are used for killing bacteria (bactericidal), some preventing bacterial growth (bacteriostatic). Some of the chemical compounds are also used for the preservation of foods and thus called as preservatives. There are two main categories of preservatives: i) class I preservatives which are a part of the food composition and are added in excess which include sugar and salt. These control the microorganisms by exerting a high osmotic pressure and thus lowering the water activity of the foods, ii) class II preservatives which are not a part of food composition and are added in very small quantity. These include organic acids, sulphites, metabisulphites, nitrites, nitrates etc. Though a wide range of chemical agents are available for controlling microorganisms but those generally used should fulfill the following conditions: i) fast acting , ii) work on bacteria without being poisonous to people (even solvent), iii) penetrate material without staining, iv) be stable when stored, v) inexpensive, vi) smell good. Some of the major groups of antimicrobial chemical agents including preservatives are as follows:

**11.2.1.** *Phenols and phenolic compounds:* Phenol was the first disinfectant used in 1880 by Joseph Lister to reduce the infection of surgical incisions. Used as a standard against which other disinfectants are compared to determine their antimicrobial activity. These control the microorganisms by precipitation of proteins, inactivation of enzymes, solubilization of membranes leading to leakage of amino acids from cells.

**11.2.2.** *Alcohols:* Ethyl alcohol in concentrations between 50 and 90% is effective against vegetative or non-spore forming cells. Methyl alcohol is less bactericidal, higher alcohols like propyl, butyl, amyl are more germicidal than ethyl alcohol but since the alcohols higher than propyl are not miscible with water, they are not commonly used in disinfectants. These control the microorganisms by acting as protein denaturants and also by damaging the lipid complex in cell membrane causing the solubilization of membranes.

**11.2.3.** *Halogens:* Iodine is one of the oldest and most effective germicidal agents. Since iodine is only slightly soluble in water, it is used as a germicidal agent as a tincture of iodine which is available in various preparations, generally having 2% iodine in 2% sodium or potassium iodide diluted in alcohol. Iodine is also used in the form of mixtures of iodine with surface-active agents like polyvivylpyrrolidone (PVP) which are known as iodophors. These oxidise and inactivate essential metabolic compounds including proteins with sulfhydryl groups. Other halogens which are widely used for control of microorganisms include chlorine and chlorine compounds like chlorine gas, hypochlorites, chloramines. Their major applications are in water treatment, in food industry as well as for domestic use. When chlorine is added in water it forms hypochlorous acid which is further decomposed to form nascent oxygen. Nascent oxygen being a strong oxidizing agent denatures major cellular constituents. Chlorine can also combine directly with proteins of cell membranes and enzymes of the microorganisms. Some of the commercial applications of halogens include the use of chlorine as water disinfectant and as an active ingredient in bleach, iodine in topical antisepsis and bromine in hot tubs as it doesn't smell like chlorine does.

**11.2.4.** *Oxidising agents:* $H_2O_2$ is used to kill anaerobes in wounds. These oxidise and inactivate essential metabolic compounds including proteins with sulfhydryl groups.

**11.2.5.** *Heavy metals and their compounds:* Most of the heavy metals including mercury, silver, gold and copper exert a detrimental effect on microorganisms by combining with cellular proteins especially containing sulfhydryl groups and inactivating them. Some of commonly used specific metal salts include silver nitrate against *Gonococcus*, copper sulfate in pipes to prevent algae growth, selenium in dandruff shampoo to kill fungi

**11.2.6.** *Dyes:* Two classes of dyes including triphenylmethane dyes like malachite green, brilliant green, crystal violet and acridine dyes like acriflavine, tryptoflavine have antimicrobial properties. Gram +ve bacteria are more susceptible than gram –ve. These control the bacteria by interfering with cellular oxidation processes.

**11.2.7.** *Soaps and detergents:* Detergents, the wetting agents and surface tension depressants are of two types: i) Anionic detergents with detergent property resident in the anion including soap, sodium lauryl sulphate, ii) Cationic detergents with detergent property resident in cation including cetylpyridinium chloride. These reduce the surface tension and increase the wetting power of water in which the microorganisms are dissolved and removed mechanically. Cationic detergents are more germicidal than anionic compounds. Soaps also solubilize membranes and are mildly alkaline.

**11.2.8.** *Quaternary ammonium compounds:* Most of germicidal cationic-detergent compounds are quaternary ammonium salts in which $R_1$, $R_2$, $R_3$ and $R_4$ groups are carbon groups linked to the nitrogen atom. These clean and disinfect the environment. The bactericidal power is high against gram+ve bacteria which are controlled by denaturation of proteins, interference with glycolysis and membrane damage.

**11.2.9.** *Aldehydes:* Most effective compounds of this category include formaldehyde and gluteraldehyde. These quickly inactivate enzymes and are good on most resistant organisms, however, these are toxic to tissues and thus are not good antiseptics. Formaldehyde is highly reactive chemical which combines readily with vital nitrogen compounds like proteins, nucleic acids.

**11.2.10.** *Gaseous agents:* Certain types of medical devices which need to be sterilized are made up of materials that are damaged by heat. These are sterilized by means of a gaseous agent, ethylene oxide which is a powerful sterilizing agent and the only accepted gaseous chemical sterilant which is liquid at $<10.8°C$ and is highly flammable. It destroys the microorganisms by alkylation reactions with organic compounds including enzymes and proteins. It is used in special autoclave like machine, generally used on spices, drugs, surgical supplies, plasticware and is very toxic.

**11.2.11.** *Benzoic acid and parabens:* $C_6H_5COOH$ and its sodium salt – $C_7H_5NaO_2$ along with esters of p-Hydroxybenzoic acid (Parabens) are used for the preservation of foods against fungal yeast and mold spoilage. Antimicrobial activity of benzoate is affected by pH with greatest activity at low pH and is ineffective at neutral pH. Antimicrobial activity resides in un-dissociated molecule whose dissociation is pH dependent. At pH 4.0, 60% compound is un-dissociated while at pH 6.0 only 1.5% compound is un-dissociated. Common permissible parabens are heptyl-, methyl-, propyl-, butyl-, ethyl-. These are less sensitive to pH than benzoate and are effective upto pH 8.0. Both benzoate, and parabens block oxidation of glucose to pyruvic acid and also inhibit uptake of substrate molecules.

**11.2.12.** *Sorbic acid:* $CH_3CH=CHCH=CHCOOH$, usually employed as Ca, Na or K salt are also active in acid foods than neutral foods and generally in-effective at $pH >6.5$. These are generally effective against molds and yeasts but also to certain bacteria and inhibit dehydrogenase enzyme system and the cellular uptake of substrate molecules including amino acids, phosphate, organic acids.

**11.2.13.** *Propionic acid:* $CH_3CH_2COOH$, usually employed as Ca and Na salts is mainly a mold inhibitor, used in breads, cakes, cheese. It also inhibits cellular uptake of substrate molecules

**11.2.14.** *Sulphur dioxide and Sulphites:* $SO_2$, Na or K salts of $SO_3$ (sulphite), $HSO_3$ (Bisulphite), $S_2O_5$ (metabisulphite), are generally effective against bacteria. The antibacterial effect is due to reducing power which reduces the $O_2$ tension to a point at which aerobes are not able to grow. These are also known as enzyme poison as these act on enzyme systems, generally with disulphide bonds.

**11.2.15.** *Nitrates and Nitrites:* Many bacteria are capable of utilizing nitrate as electron acceptor which is reduced to nitrite – highly reactive and capable of serving both reducing and oxidizing agent. Under acidic conditions it ionizes to form nitrous acid (HONO) – further decomposes to give nitric oxide (NO) – capable of reacting with catalase, peroxidases, cytochromes thus inhibiting aerobic bacteria. Nitrite also inhibits ferredoxin – thus prevents ATP synthesis from pyruvate.

## 11.3. Control by chemotherapeutic agents

Chemotherapeutic agents are the chemical substances used for the treatment of infectious diseases. These are prepared in the laboratory or obtained from microorganisms and some plants and animals. A commonly used example of chemotherapeutic agents is the antibiotics, most of which are prepared commercially by microbial biosynthesis while some are prepared synthetically. The major points of attack of antibiotics on the microorganisms include: inhibition of cell wall synthesis, damage to cytoplasmic membrane, inhibition of nucleic acid and protein synthesis, and inhibition of specific enzyme systems

**11.3.1.** *Inhibition of cell wall synthesis:* Penicillins, ampicillin, cephalosporins interfere with final stages of peptidoglycan synthesis by inhibiting transpeptidase reaction which involves cross-linking of the two linear polymers. Some including cycloserine and vancomycin inhibit the enzymes involved in the synthesis of pentapeptide side chains.

**11.3.2.** *Damage to cytoplasmic membrane:* Polymyxins, gramicidins, tyrocidines act on membrane adversely affecting the normal permeability. Polymyxins are particularly effective against gram-ve organisms while the gramicidins and tyrocidins are more effective against gram+ve organisms

**11.3.3.** *Inhibition of nucleic acid and protein synthesis:* Streptomycin, tetracycline interfere with binding of 30S ribosomes while chloramphenicol and erythromycin bind with 505 ribosome.

**11.3.4.** *Inhibition of specific enzyme systems:* Sulfonamides have the structure related to the compound p-aminobenzoic acid (PABA) which is required as a precursor by many bacteria for the synthesis of the essential coenzyme tetrahydrofolic acid (THFA). The selective action of sulfonamides is explained by the fact that the sulfonamide may enter the reaction in place of the PABA and block the synthesis of THFA, an essential cellular constituent whose functions include amino acid synthesis and thymidine synthesis

**11.3.5.** *Antifungal antibiotics:* These include nystatin, amphotericin and griseofulvin whose antimicrobial action is attributed to their ability to increase cell permealibility in fungal and animal cells having sterols in their cytoplasmic membrane.

**11.3.6.** *Synthetic chemotherapeutic agents:* These include Nitrofurans, generally effective against both gram+ve and gram-ve bacteria and Nalidixic acid which inhibits the DNA synthesis in gram-ve bacteria.

**11.3.7.** *Antiviral chemotherapeutic agents:* Antibiotics, discussed above are not effective against viruses, which are intracellular. Among the most promising chemotherapeutic agents for treating viral diseases is interferon which are glycoprotein substances whose antiviral action is attributed to interference of protein synthesis.

## 12. FURTHER READING

Alcamo, I. Edward (1997). Fundamentals of Microbiology. 5th ed. Menlo Park, Benjamin Cumming, California.

Atlas, Ronald M. (1995). Principles of Microbiology. Mosby, St. Louis, Missouri.

Dubey, R.C. and Maheshwari, D.K. (2004). A text book of Microbiology. S. Chand and Company Ltd., New Delhi.

**Holt, John.G. (1994). Bergey's manual of determinative bacteriology. 9th ed. Baltimore, Maryland: Williams and Wilkins.**

**Pelczar (Jr), M.J., Chan, E.C.S., and Kreig, N.R. (1986). Microbiology. 5th ed. McGraw Hill Book Company, Singapore.**

**Prescott, L.M., Harley, J.P. and Klein, D.A. (2005). Microbiology. 6th ed. McGraw Hill Companies, Inc. New York.**

**Stanier, R.Y., J. L. Ingraham, M. L. Wheelis, and P. R. Painter (1986). General Microbiology. 5th ed. Prentice Hall, Upper Saddle River, New Jersey.**

□□□

# 2

# Applications and Commercial Uses of Microorganisms

**D.K. RAHI AND S.K. SONI**
*Department of Microbiology, Panjab University, Chandigarh-160 014*

## 1. INTRODUCTION

Microorganisms are the most powerful creatures in existence and determine the life and death on this planet. They can kill mercilessly, but at the same time they can be harnessed to sustain life. Nature has provided us with a perfect balance in the oxygen, carbon, nitrogen, sulfur and phosphorus cycles with the help of microorganisms to sustain plant, animal and human life. We should therefore always keep in mind that it is the microbial world which determines the growth and existence of plant, animal and human on this planet and that, the microorganisms are much more flexible and adaptable to environmental changes than plants, animals and humans. Microorganisms are essential to our very existence. They are ubiquitous, found in common environments such as soil, water and air as well as exotic locales as diverse as deep sea hydrothermal vents and soda lime lakes. In these natural environments, microorganisms have very specific jobs. They are responsible for recycling nutrients in our soil and purifying our water. We also use microorganisms in constructed environments to serve our own functions. Microorganisms possess large diversities and can be employed for various commercial purposes. Some of the beneficial functions of microorganisms in various fields are as follows:

*Nutrient recycling:* When plants and animals take up nutrients, they are not available to other living organisms. When the plants and animals die, the nutrients remain in the carcass. If the dead material, or detritus, is not broken down by microbes, these nutrients will never become available to help sustain the life of other organisms. There are a finite amount of nutrients available in the environment and we cannot simply make more. The nutrients that are trapped in detritus must be released in order for life to continue. It is the microbial population of the environment that is responsible for this nutrient recycling.

*Food:* Microbes have been used for centuries to provide us with food. Bread is the result of a microbial fermentation of sugars to produce carbon dioxide that makes the bread to rise. We owe our beer and wine to similar little yeast that convert sugars into alcohol for our consumption. Yogurt and cheese are produced by bacterial fermentation of lactose, the sugar ín milk. Microbes such as phytoplankton also serve as the nutrient source that indirectly feeds all marine animals. Microbial symbioses with plants allow them to grow strong and increase productivity, sometimes they are even essential for plant survival.

*Health:* We depend on having a microbial population, called our microbiota, for our health. There are normally 10 times more microbial cells on/in a human body than human cells. The good bacteria that live on and in us, protect us from the bad invaders that might come our way. Without the good bacteria, the bad ones can slip in and easily cause us problems. Our own digestion has also evolved to use bacteria for assistance, allowing us to gain nutrition from plant polysaccharides that our own enzymes will not degrade. This is one of the reasons why taking antibiotics can lead to episodes of diarrhea.

*Biodegradation:* Microorganisms are responsible for getting rid of the waste generated by industry and households. They detoxify acid mine drainage and other toxins that we dump into the soil and water. The nutrients gained from the breakdown of these products then go to feed plants or algae, which in turn feed all animals.

*Wastewater treatment:* When we flush things down the drain or the toilet, they go to a septic system or waste water treatment plant at the end of the line. After primary mechanical treatment and aeration, microbes remove organic materials from the filthy waters that flow into these systems and, eventually, water can safely be returned to the rivers and streams. The methane produced during treatment can even be used to generate heat or electricity for the operation.

*Biosynthesis:* We have discovered that microorganisms are useful for making things as well. Products such as xanthan gum (a food thickener) are harvested from some bacterial producers. Most of our vitamin $B_{12}$, riboflavin, and vitamin C are produced by bacterial fermentation. Approximately 70% of antibiotics currently in use are also the product of microbial fermentation.

Because of a wide range of useful functions of microorganisms for the welfare of the mankind, they are exploited commercially for large scale applications, some of which are discussed below:

## 2. MICROORGANISMS IN FOOD INDUSTRY

Microorganisms are associated with all the foods that we eat and are responsible for the formation of certain foods through the process of fermentation and may also be directly used as a source of food in the form of single cell proteins and food supplements in the form of pigments, amino acids, vitamins, organic acids, enzymes and lipidic substances and also inhibit other microbes which may produce the toxins. The exploitation of microorganisms in food industry is as old as the history of civilization. The traditional application has been in the production of various indigenous fermented foods by involving the natural microflora coming from the environments. The current applications involve the production of various foods and food additives by the process of fermentation with genetically modified microbes.

Fermentation is one of the oldest methods of food preparation and originated centuries ago. The exact reports about the origin of fermented foods are lacking but it is believed that the discovery of food fermentations was purely by chance. Fermentation is a biological process in which a chemical substrate is transformed into different end products by the action of microbial cells. Fermentation processes are exploited by the mankind for the commercial production of various products and this is the basis of 'Fermentation Technology'. Alcohol fermentation was the first fermentation known to mankind. Distilleries began to appear in Europe in the middle of the seventeenth century. At first alcohol was used only for human consumption. Lately the demand for alcohol as a universal solvent and chemical raw material developed. Ethanol is the largest liquid industrial fermentation product world-wide which is produced to the extent of over 40 billion l / annum. The importance of ethyl alcohol as a future motor fuel alone or mixed to gasoline and its wide applications in chemical, pharmaceutical, potable and personal care product industry is well known.

Wine is the oldest product of fermentation and its origin is lost in antiquity. In 1863, Louis Pasteur established the microbial activity as the basis of wine fermentation and demonstrated that fermentation of grape musts and production of good quality wines is a complex biochemical process involving the sequential appearance of different microbial species whose development is affected by various intrinsic and extrinsic factors. The process includes the interaction of various yeasts, lactic acid and acetic acid bacteria with the yeast *Saccharomyces cerevisiae* as the primary organism in wine fermentations. There are several varieties of wines available around the world and are, generally, classified on the basis of the type of fruit used, source of yeast inocula, colour, alcohol content, presence or absence of carbon dioxide, processing of the fruit prior to fermentation etc. For the production of wines, various raw material including fruits, sugar, acid, nitrogen source, clarifying enzyme, filter aid etc. are required. The fruit constitutes the most important raw material. Most of the wines are generally prepared from over-ripened grapes but commercial wines are also made from several fruits other than grapes. In some cases the higher level of acidity in the fruit requires dilution, consequently lowering the sugar level which is compensated by the exogenous addition of table sugar to the must. Unlike grape which is a good culture medium for the growth of wine yeast needing no sugar supplementation, the juice/pulps from other fruits need sugar supplementation for successful alcoholic fermentation.

Indigenous fermented legume and cereal products form an important part of human diet in South-East Asia including India, the middle East and Africa. They are also becoming popular in the developed world due to their high nutritional value and organoleptic characteristics. People of Asia are the pioneers in the development of means of fermenting vegetable proteins to produce meat like flavours. Indonesians developed the fermentation methods for introducing meat like texture into vegetable products. Koreans are credited for the production of acid fermented vegetables. People of Egypt developed wheat breads leavened with yeasts while Indians discovered methods for souring and leavening cereal-legume batters.

Beer, wine and bread making by yeasts, the processing of milk to dairy products by lactic acid bacteria and the fermentation of vinegar by acetic acid bacteria show that these microbes are among the oldest domesticated microorganisms.

## 2.1. Microorganisms as the agents of fermentation in traditional fermented foods

The reports about the origin of fermented foods are lacking, but it is believed that the discovery of fermentations was purely by chance. The transition from hunting and gathering food to organized food cultivation and production is believed to have taken place some 10,000 to 15,000 years ago, in the Middle East. It is believed, that it was purely by chance that the fermentation became a method of preservation. Some of the common indigenous fermented foods consumed in different parts of the world are listed in Table 1 which indicates the geography, substrates employed, the nature and uses of various products. Indigenous fermented products form an important part of human diet in South-East Asia including India, the middle East and Africa. They are also becoming popular in the developed world due to their high nutritional value and organoleptic characteristics. People of Asia are pioneers in the development of means of fermenting vegetable proteins to produce meat like flavours. Indonesians developed the fermentation methods for introducing meat like texture into vegetable products. Koreans are credited for the production of acid fermented vegetables. People of Egypt developed wheat breads leavened with yeasts while Indians discovered methods for souring and leavening cereal-legume batters. Generally, the traditional methods of preparing indigenous fermented foods are simple and inexpensive. However, the ancient methods of making such foods are changing rapidly through modern microbial technology. Some of the fermented products including dairy products, coffee, cocoa and bread have been studied extensively for many decades and their production methods have been modernised keeping pace with the advancement in modern science. During the last few years much interest has been generated in the fermented foods of Asian and African countries including India where such foods are still being manufactured according to the traditional, technologically less advanced methods using simple equipment and are produced at cottage industry scale, by means of natural microflora, from the staples and the surroundings.

**Table 1** Some important indigenous fermented food of the world

| S. no | Fermented food | Country/place | Ingredient(s) | Nature of product | Product use |
|---|---|---|---|---|---|
| | | | **Legume products** | | |
| 1. | *Bhallae* | India | Black gram | Deep fried patties | Snack after soaking in curd, water |
| 2. | *Chee-fan* | China | Soybean, whey curd | Solid, cheese like | Eaten fresh |
| 4. | *Khaman* | Western India | Bengal gram | Spongy cake | Breakfast food |
| 5. | *Kenima* | Nepal, Sikkim, Darjeeling district of India | Soybeans | Solid | Snack |
| 6. | *Ketjap* | Indonesia | Black soybeans | Sirup | Seasoning agent |
| 7. | *Meitauza* | China, Taiwan | Soybean cake | Solid | Fried in oil or cooked with vegetables |
| 8. | *Meju* | Kotrea | Soybeans | Paste | Seasoning agent |
| 9. | *Miso* | Japan, China | Soybeans/ Soybeans and rice | Paste | Soup base, seasoning agent |
| 10. | *Natto* | Northern Japan | Soybeans | Solid | Roasted or fried in oil used as a meat substitute |
| 11. | *Oncom* | Indonesia | Peanut press cake | Solid | Roasted or fried in oil, used as a meat substitute |
| 12. | *Papadam* | India | Black gram and spices | Circular tortilla like wafers | Condiment |
| 13. | *Soybean milk* | China, Japan | Soybeans | Liquid | Drink |
| 14. | *Sufu* | China, Taiwan | Soybean, whey curd | Solid | Soybean cheese, condiment |
| 15. | *Soy sauce* | Japan, China, Philippines | Soybeans/ Soybeans and wheat | Liquid | Seasoning agent for meat, fish, cereals and vegetables |
| 16. | *Tempe* | Indonesia and vicinity | Soybeans | Solid | Fried in oil, roasted or used as meat substitute in soup |
| 17. | *Vadai* | India | Black gram | Deep fried patties | Snack |
| 18. | *Warries* | Northern India | Black gram and spices | Ball like hollow, brittle | Spicy condiment, eaten with vegetables, legumes, rice |

*Contd...*

| | | | | | |
|---|---|---|---|---|---|
| **Legume-cereal products** | | | | | |
| 19. | *Dhokla* | Western India | Rice, Bengal gram | Spongy cake | Condiment |
| 20. | *Dosa* | Southern India | Rice, black gram | Spony, pan cake | Staple food |
| 21. | *Hama natto* | Japan | Wheat flour, whole soybeans | Raisin like, | Flavouring agent for meat soft and fish, eaten as snack |
| 22. | *Idli* | Southern India | Rice, Black gram | Spongy | Breakfast food |
| 23. | *Kecap* | Indonesia and vicinity | Wheat, soybeans | Liquid | Condiment, seasoning agent |
| 24. | *Tao-si* | Philippines | Wheat flour, soybeans | Semi-solid | Seasoning agent |
| 25. | *Taotjo* | East Indies | Roasted wheat meal or glutinous rice, soybeans | Semi-solid | Condiment |
| **Cereal products** | | | | | |
| 26. | *Ang-Kak (Chinese red rice)* | China, Taiwan, Philippines, Thailand | Rice | Solid | Colouring other foods |
| 27. | *Ambali* | India | Millet Flour | Semi solid | All time food |
| 28. | *Bhatura* | India (North) | White wheat flour | Deep fried | Breakfast bread |
| 29. | *Fermented rice* | India | Rice | Semi solid | Breakfast |
| 30. | *Hopper (Appa)* | Sri Lanka | Rice or white wheat flour and coconut water | Semi solid | Breakfast |
| 31. | *Jalebie* | India and Pakistan | White wheat flour | Deep fried pretzel | Confection food |
| 32. | *Kenkey* | Ghana | Maize | Semi solid | Breakfast, lunch and supplement to fish stews |
| 33. | *Kisra* | Sudan | Sorgham flour | Spongy Bread | Staple food |
| 34. | *Kulcha* | India (North) and Pakistan | White wheat flour | Flat bread | Staple food |
| 35. | *Mahewu* | South Africa | Maize meal | Liquid | Beverage food |
| 36. | *Nan* | India (North), Pakistan, Iran and Afganistan | White wheat flour | Flat bread | Staple food |
| 37. | *Ogi* | Nigeria | Maize, Millet or Sorghum | Semi solid | Breakfast and Infant food |
| 38. | *Pozol* | Mexico | White maize | Liquid | Beverage or Porridge |
| 39. | *Puto* | Philippines | Rice | Semi solid | Breakfast and snack food |
| 40. | *Shamsy bread* | Egypt | Wheat flour | Spongy bread | Staple food |
| 41. | *Uji* | Kenya, Uganda and Tanzania | Maize, Sorghum or Millet flour | Semi solid | Breakfast and lunch |

*Contd...*

| Fermented Milk Products | | | | | |
|---|---|---|---|---|---|
| 42. | *Cultured cream* | United States, Russia Central Europe | Cow's milk cream | Sour gel/cream containing butter-like aromatic flavour | Toppling on vegetables, salads, fish, meats, |
| 43. | *Cultured butter milk* | United States | Cow's skim milk, cream | Viscous fluid milk | fruits, filling in cakes and in soups |
| 44. | *Yoghurt* | United States, Europe, Asia | Cow/goat/ sheep's milk, condensed skim milk, cream, non-fat dry milk, sucrose, fruits | Flavoured curd | Beverage food |
| 45. | *Acidophilus milk* | United States, Russia | Cow's milk | Acid fermented fluid milk | All time food |
| 46. | *Kefir* | Russia | Cow/goat/sheep 's milk | Acid and alcoholic milk | Beverage food with therapeutic value |
| 47. | *Koumiss* | Russia | Mare's milk | Acid and aicoholic milk | Beverage food |
| 48. | *Cottage cheese* | United States | Cow's milk | Natural unripened soft cheese | Beverage food, generally used in pulmonary tuberculosis, All time food |

Soyabeans and black grams are the principal legumes used in the preparation of a variety of fermented foods in different parts of the world. These are fermented separately or in combination with cereals. These are fermented as such, as a paste or as a liquid to produce *tempe*, *natto*, *miso*, *soy sauce* and *sufu*. Peanuts and locust bean are also fermented in some parts of the world to prepare *oncom* and *dawadawa*. *Tempe kedele*, a fermented product of Indonesia, is prepared by incubating the soaked soybeans at a warm place, which are knitted into compact cake by a fibrous mould mycelium for 1-3 days. The cake is then sliced thin and deep fried or used as a meat substitute in soups. Fermented whole soybeans are known as *natto* in Japan, *tu-si* in China, *tao-si* in Philippines and *tnua-nao* in Thailand. These are consumed with boiled rice or used as a seasoning agent with cooked meat, sea foods and vegetables. Fermented soybean pastes are known as *miso* in Japan, *chiang* in China, *tauco* in Indonesia, *doenjang* in Korea and believed to have been originated in China in 600 AD. In addition to soybeans and salts, most of these products also contain cereals. *Miso* has been manufactured in Japan for atleast 1,000 years and used as a base for soups, sauce served with meat, poultry, sea foods and vegetable dishes. Fermented soybean liquids, light brown to black in colour with meat like salty flavours, commonly known as *soy sauces*, are prepared by hydrolyzing soybeans (in China and Korea) and mixture of soybeans and wheat (in Japan) using enzymes produced by *Aspergillus oryzae*, in a strong salt solution. These are known as *shoyu* in Japan, *chiang-yu* in China, *kecap* in Indonesia and *kanjang* in Korea. Chinese have been using *soy sauce* for over 3,000 years. Production of *soy sauce* in Japan was probably a result of introduction of Buddhism from China and consequent changes in vegetarian diet in 552 AD. *Sufu*, a mold fermented soybean curd, which is highly creamy bean paste made by overgrowing soybean curd with mold, is popular among Chinese as *fu-zu*. It is known as *chao* in Vietnam, whereas in the West, *sufu* has been referred to as *Chinese cheese*. *Oncom* is a fermented peanut presscake product consumed largely in Indonesia as a roasted snack, covered with water and seasoned with salt or sugar. *Dawadawa*, a fermented locust bean product, consumed largely in West Africa, is also known as *daddowa* in Nigeria. The seeds after removing the powdery pulp are boiled in water and stored overnight. The swollen cotyledons are then boiled for 30 min, put into a hole in ground and left to ferment for 2-3 days. *Warries*, somewhat like japanese *miso* are spicy, hollow, brittle, friable balls 5-8 cm in diameter, very popular in Northern India, especially in Amritsar city and are used as condiment in cooking with vegetables, legumes or rice. The black gram paste after being spiced with asafoetida, caraway, cardamom, cloves, fenugreek, ginger and red pepper is moulded into small balls and fermented in open air for 2-8 days. *Papadam*, another condiment is similarly prepared from black gram paste but without fenugreek and ginger. It is allowed to ferment for 4-6 hours and rolled into tortilla like wafers which are served roasted or deep fried. *Bhallae* are prepared from unspiced black gram paste fermented for 6-24 hours. The paste is pan or deep fried and consumed as a snack as such or after briefly soaking in water or curd. Foods of acidified and leavened type like Indian *Idli*, constitute a very interesting group of cereal-pulse based foods of considerable potential importance in the developing and under-developed world. *Idli* is closely related to sour dough bread of western world but does not depend upon wheat or rye as a source of protein to retain $CO_2$ during leavening which is produced by bacterial rather than yeast activity. *Idlis* are spongy pancakes, steamed in a special container and consumed as a staple food primarily in South India. They have been in use since 1100 AD. Proportion of rice and black gram ranges from 1:4 to 4:1. *Dosa*, a thin, fairly crisp, fried pancake-like food of South India is prepared by spreading the fermented batter on hot greasy griddle. The batter is prepared in a similar way as that of *Idli* but the proportion of rice and black gram is 1:1 instead of 4:1.

In addition to the legume and cereal products, fermented dairy products also constitute a vital part of human diet in many parts of the world. The preservation and concomitant transformation of flavour and texture of milk by fermentation has been known centuries before the role of microorganisms in the process was appreciated. Historically, fermentation processes were based upon spontaneous souring of milk caused by inherent microflora. Modern processes affect milk fermentation under predictable, controllable and extracting conditions to yield cultured dairy products of high nutritional and sanitary standards. The origin of the fermented dairy products dates back to the dawn of civilization. The ancient Sanskrit scriptures of India, the Vedas, document the food value of dahi, a fermented milk product similar to modern *yoghurt*. Further evidence for the existence of soured milk as a food in early times is corroborated by the Bible. Milk is fermented as such or in combination of some additives to produce *cultured butter milk, cultured cream, yoghurt, acidophilus milk, kefir, koumiss* and *cheeses*. Cultured butter milk is obtained from pasteurised skim milk or part skim milk cultured with lactic acid and aroma producing organisms. The term buttermilk is also used for the phospholipid-rich fluid fraction obtained as a by-product during the churning of cream in butter manufacture. However, *cultured butter milk* is a viscous, cultured, fluid milk, containing a characteristic pleasing aroma and flavour. *Cultured cream* or *sour cream* is manufactured by ripening pasterurized cream of 18% fat content with lactic and aroma-producing bacteria. This product resembles *cultured butter milk* in terms of culturing procedure. However, in consistency it is an acid gel containing butter like aromatic flavour. *Cultured cream* is used as a topping on vegetables, salads, fish, meats and fruits, as a filling in cakes and in soups and cookery in place of butter-milk or sweet cream. *Yoghurt*, a fermented curd product, is prepared from a mix, standardized from whole, partially defatted milk, condensed skim milk, cream and non-fat dry milk. Alternatively, milk may be partly concentrated by removal of 15-20% water in a vacuum pan. *Kefir* and *Koumiss* belong to the class of acid and alcoholic fermented milks, very popular in Russia.

Pickles from vegetables and fruits like mango, gooseberry and lemons are prepared and used throughout India, More than their nutritive value, they act as food adjuncts and appetizers as they add palatability to Indian foods. Indian pickles is a very big industry now and the products are popular throughout the world.

A large number of bacteria, yeasts and yeast like fungi have been isolated from various fermented foods (Table 2). Filamentous fungi belonging to *Aspergillus, Mucor, Neurospora* and *Rhizopus*, bacteria belonging to *Leuconostoc mesenteroides,* lactobacilli, *Streptococcus faecalis,* pediococci and yeasts belonging to *Saccharomyces cerevisiae, Candida* sp., *Debaryomyces hansenii* and *Hansenula anomala* are most commonly associated with the traditional fermentations. These organisms are generally contributed by the ingredients, workers and the surroundings. Since the source of inoculum is not apparent, the study of biological agents actually associated with fermentations has attracted considerable attention over the years. Dehulled black gram grains, used for the production of various indigenous Indian products, have been reported to harbour *Leuconostoc mesenteroides* and other lactic acid bacteria in large numbers which play a major role in black gram fermentations. Ordinarily the microbes developing during the initial soaking of ingredients and later are sufficient to bring about the fermentations. Presoaking of the ingredients release the free sugars and non-protein nitrogen which support the growth of lactic acid bacteria. Spices, as added in some ingredients, also enhance the food fermentations by stimulating the development of certain lactic acid bacteria and accelerating lactic acid production. *Lactobacillus plantarum* and *Pediococcus cerevisiae* are able to grow in the presence of upto 10-12 g/l of the majority of spices including pepper cardamom, cinnamon, ginger, garlic, turmeric and celery seed with enhanced lactic acid production from carbohydrates. All the spices contain high manganese content which has been identified as a factor responsible for stimulatory effect in the enhancement of acid production . Seasonal variations affect qualitative and quantitative prevalence of bacteria and yeasts in different traditional fermented foods with summers favouring the former and winters favouring the latter.

*Tempe* has been known and produced by Indonesians for centuries and its production involves two distinct fermentations. The first one occurring during soaking is bacterial, resulting in the acidification of beans and the second fermentation results in the overgrowth of bean cotyledons by mold mycelium. The number of species of bacteria present in the commercial *tempe* are remarkably few, *Klebsiella pneumoniae* being predominant followed by *Bacillus* and *Micrococcus*, fungi belonging to *Aspergillus, Mucor, Penicillium* and *Rhizopus* and yeasts belonging to *Trichosporon.*

There are three main types of *natto* prepared in Japan. Itohiki *natto*, referred to simply as *natto*, is prepared by steaming the beans for about 15 min and inoculating with *Bacillus natto*, a variant strain of *Bacillus subtilis*. Yukiwari *natto*, a second type of fermented whole soybean product is made by mixing Itohiki *natto* with salt and rice koji, the source of enzymes to hydrolyze the soybean components in fermentation, produced by *A. oryzae*. In case of Hamma *natto*, the third major type of fermented whole soybean product in Japan, the soybeans after soaking in water for about 4 hours are steamed without pressure and are then inoculated after being cooled with koji. Bacteria such as *Micrococcus, Streptococcus* and *Pediococcus* are also reported to be widely distributed on the surface and in the inner part of hamma *natto* .

The essential microorganisms involved in fermentation of *miso* include a mold, *Aspergillus oryzae, Rhizopus oligosporus* and a yeast *Saccharomyces rouxii,* but under traditional conditions lactic acid bacteria such as *Pediococcus halophilus* or *Streptococcus faecalis* are also found to be present and contributing flavour through the production of acids.

The essential microorganisms taking part in Indonesian *tauco* fermentation inlude *Aspergillus oryzae, R. oligosporus, R. oryzae, Lactobacillus delbrueckii, Hansenula* sp. and *Zygosaccharomyces soyae* while *Mucor* sp., *Penicillium* sp., *Rhizopus* sp., *Aspergillus ozyzae, Bacillus subtilis, B. pumilis* and *Rhodotorula flava* or *Torulopsis dattila* and many other organisms are generally associated with Korean *Doenjang.*

*Soy sauce* fermentation involves two distinct basic processes. The first one includes the fermentation with microorganisms and the other chemical process involves the use of acids to promote the hydrolysis of the ingredients. The essential microorganisms in *soy sauce* fermentation are mould, *Aspergillus oryzae* or *A. soyae*, homofermentative lactic acid bacterium *Pediococcus cerevisiae* or *Lactobacillus delbrueckii* and yeast *Saccharomyces rouxii*. Some researchers have indicated that various groups of bacteria and yeasts predominate in sequence during the fermentation producing lactic acid from simple sugars and causing a fall in pH. Later, *Saccharomyces rouxii, Torulopsis* sp., *Zygosaccharomyces* sp. and other yeasts predominate. Moulds belonging to *Aspergillus niger, A. soyae, Monilia* sp., *Penicillium* sp. and *Rhizopus* sp. may appear on the surface of the mash but have no relation to proper fermentation. Koji, the source of enzymes used in *soy sauce*, is prepared by culturing a number of mixed strains of *Aspergillus oryzae* or *A. soyae* either on steam polished rice in Japan or a mixture of wheat bran and soybean flour in China. It is then mixed with saline water to form a mash. The mycelium of the koji mold is milled during the very early stages of mash preparation.

Various moulds belonging to mucoraceae especially *Mucor corticolus, M. hiemalis, M. prajni, M. racemosus, M. silvaticus, M. subtilissimus, Actinumucor elegans* and *Rhizopus chinensis* are generally isolated from *sufu*. *Oncom* is fermented with a mixed culture of microorganisms with *Rhizopus* or *Neurospora* spp. predominating. Depending upon the predominant mould, there are two types of *oncoms* viz. *Oncom hitam* (black *oncom*) having *Rhizopus* spp. as principal moulds and *Oncom merals* (red *oncom*) in which the principal moulds are *Neurospora sitophila, N. crassa* and *N. intermedia.*

The microorganisms responsible for *dawadawa* fermentation have not been determined but undoubtedly include spore forming bacilli, lactic acid bacteria and probably yeast. Microbial populations, whether mixed or pure, vary in their number and types with progress in fermentations.

The microbial population dynamics, their prevalence and succession in indigenous legume/cereal-legume Indian food fermentations suggest the involvement of bacteria alone or in combination with lesser proportion of yeasts during fermentations. Seasonal variations generally affect the development and prevalence of microorganisms during these fermentations; summers were found to favour a higher bacterial load ($10^{10}$ -$10^{12}$ /g) than winters ($10^{9}$-$10^{10}$ /g). *Bacillus subtilis* occurrs mainly during summers whereas *Leuconostoc mesenteroides, Streptococcus faecalis* and *Lactobacillus fermentum* are the principal bacteria associated during both the seasons. Yeasts *including Candida vartiovaarai, Kluyveromyces marxianus,* and *C. krusei* are largely encountered during winters while *Saccharomyces cerevisiae, Pichia membranaefaciens, Trichosporon beigelii* and *Hansenula anomala* are commonly involved during both the seasons. Moreover, since the source of inoculum is not apparent in such products, it is likely that legumes, cereals and spices mixture used in the preparation of dough/batters and the aeriel environment are contributing the initial inoculum for the fermentations. Several reports have indicated that black gram, the basic ingredient of all the Indian products, harbours *Leuconostoc mesenteroides* and many other lactic acid bacteria in large numbers which multiply rapidly during the initial soaking of the ingredients due to the release of free sugars and non-protein nitrogen. The less frequency of yeast occurrence in the traditional Indian fermentations is probably due to their less prevalence on the dry grains or ingredients. Since the grains are presoaked in water and ground to pastes, yeasts are likely to develop only after the softening of the ingredients. The biochemical and physiological characterization of predominant microorganisms associated with indigenous Indian foods indicates that both *Leuconostoc* and lactobacilli, the common bacteria associated with staples, presumably produce acid and gas from various carbohydrates thus making the environment unfit for other contaminants which gradually disappear with the progress in fermentation. *Streptococcus faecalis* perhaps comes from air in later stages and produces more acid causing further acidification, thus enhancing the shelf-life of the products. *Bacillus amyloliquefaciens* and *B. polymyxa*, which are generally associated with *Idli* and *Dosa* batter fermentation are likely to be contributed by rice and help in the degradation of starch into maltose and glucose by producing extracellular amylolytic enzymes. Some strains of *Saccharomyces cerevisiae* also produce acid and gas from starch itself and may thus play a significant role in adding to the acidification and leavening during fermentation.

All the milk products are generally produced by intense activity of lactic cultures. In fact milk is a natural habitat of number of lactic acid bacteria which may cause spontaneous souring. For desired fermentation processes, specially prepared cultures of harmless lactic acid bacteria are used which are called "starter culture" or "starters" and yield desirable flavor and texture to a particular fermented milk product. The lactic acid bacteria generally used in milk fermentation are from two groups viz. gram +ve cocci like *Streptococcus* spp. *(S. thermophilus, S.lactis, S.lactis subsp. diacetylactis, S.cremoris)* and gram +ve asporogenous rod shaped bacteria viz. *Lactobacillus spp. ( L. bulgaricus, L. acidophilous).*

The starter culture used for fermentation is generally a mixed culture i.e. consists of more than one strain. For commercial purposes, the starter cultures are generally preserved as frozen concentrates i.e. after growing the culture in optimum conditions in a fermenter, the culture is concentrated by centrifugation and its suspension containing about $10^{11}$ cells/ ml in the medium, is frozen in liquid nitrogen at -196°C. This eliminates the need for routine culture maintenance by repeated subculturing thereby reducing the chances of contamination.

The important microorganisms playing an important role in the preservation of vegetables and fruits for the production of pickles include salt-resistant lactic acid bacteria, initially heterofermentative *Leuconostoc* spp. and *Lactobacillus brevis*, are generally replaced by homofermentative *L. plantarum* and *Pediococcus* spp. As acids are produced and the pH drops, yeasts become more important.

**Table 2** Microorganisms associated with fermented foods

| S.No. | Organism | Fermented food |
|---|---|---|
| 1. | *Actinomucor elegans* | *Sufu* |
| 2. | *Aspergillus niger* | *Soy sauce, Tempe* |
| 3. | *Aspergillus oryzae* | *Doenjiang, Miso, Natto, Soyasauce, Tauco* |
| 4. | *Aspergillus soyae* | *Miso, Soyasauce* |
| 5. | *Bacillus* sp. | *Rabdi* |
| 6. | *Bacillus amyloliquefaciens* | *Dosa, Idli* |
| 7. | *Bacillus pumilis* | *Doenjang* |
| 8. | *Bacillus subtilis* | *Bhallae, Punjabi warri, Dosa, Doenjang, Kawai, Natto* |
| 9. | *Candida boidini* | *Dosa* |
| 10. | *C. cacaoi* | *Idli* |
| 11. | *C. curvata* | *Bhallae, Warri* |
| 12. | *C. famata* | *Bhallae, Warri* |
| 13. | *C. fragicola* | *Idli* |
| 14. | *C. glabrata* | *Dosa* |
| 15. | *C. glabrata* | *Idli* |
| 16. | *C. kefyr* | *Idli* |
| 15. | *C. krusei* | *Papadam, Punjabi Warri* |
| 17. | *C. membraneafaciens* | *Bhallae* |
| 18. | *C. parapsilosis* | *Punjabi Warri* |
| 19. | *C. pseudotropicalis* | *Idli* |
| 20. | *C. sake* | *Dosa, Idli* |
| 21. | *C. tropicalis* | *Idli* |
| 22. | *C. vartiovaarai* | *Bhallae, Warri* |

*Contd...*

| S.No. | Organism | Fermented food |
|---|---|---|
| 23. | *Cryptococcus humicolus* | *Bhallae, Warri* |
| 24. | *Debaryomyces hansenii* | *Bhallae, Papadam, Punjabi warri, Dosa, Idli* |
| 25. | *Deb. tamarii* | *Idli, Punjabi warri* |
| 26. | *Geotrichum candidum* | *Bhallae, Punjabi warri* |
| 27. | *Hansenula anomala* | *Bhallae, Idli, Papadam, Punjabi warri* |
| 28. | *H. polymorpha* | *Bhallae, Dosa* |
| 29. | *H. silvicola* | *Dhokla* |
| 30. | *Issatchenkia terricola* | *Dosa, Idli* |
| 31. | *Kluyveromyces marxianus* | *Bhallae, Warri* |
| 32. | *Klebsiella pneumoniae* | *Tempe* |
| 33. | *Lactobacillus* sp. | *Nan* |
| 34. | *L. acidophilis* | *Acidophilus milk* |
| 35. | *L. brevis* | *Sauerkraut, Pickles* |
| 36. | *L. bulgaricus* | *Yoghurt, Koumiss* |
| 37. | *L. buchneri* | *Jalebi* |
| 38. | *L. caucasicus* | *Kefir* |
| 39. | *L. coryniformis* | *Idli* |
| 40. | *L. delbrueckii* | *Dhokla, Idli, Soyasauce, Tauco* |
| 41. | *L. fermentum* | *Ambali, Dosa, Idli, Papadam, Bhallae, Punjabi warri, Dhokla, Jalebi* |
| 42. | *L. lactis* | *Idli, Dhokla, Jalebi* |
| 43. | *L. plantarum* | *Kawai, Sauerkraut, Pickles* |
| 44. | *Leuconostic cremoris* | *Butter milk, Sour cream* |
| 45. | *L. mesenteroides* | *Ambali, Bhallae, Dhokla, Dosa, Idli, Papadam, Punjabi warri, Sauerkraut* |
| 46. | *Micrococcus* sp. | *Rabdi* |
| 47. | *Mucor corticolus* | *Sufu* |
| 48. | *M. javenicus* | *Tempe* |
| 49. | *M. hiemalis* | *Sufu* |
| 50. | *M. praini* | *Sufu* |
| 51. | *M. racemosus* | *Sufu* |
| 52. | *M. rouxii* | *Tempe* |
| 53. | *M. silvaticus* | *Sufu* |
| 54. | *M. subtilissimus* | *Sufu* |
| 55. | *Neurospora crassa* | *Oncom* |
| 56. | *N. intermedia* | *Oncom* |
| 57. | *N. sitophila* | *Oncom* |
| 58. | *Pediococcus acidilactis* | *Idli, Rabdi, Soy sauce* |
| 59. | *P. cerevisae* | *Idli, Sauerkraut, Soy sauce* |

*Contd...*

| S.No. | Organism | Fermented food |
|---|---|---|
| 60. | *P. halophilus* | *Miso, Soy sauce* |
| 61. | *P. streptococcus* | *Idli* |
| 62. | *Pichia membranaefaciens* | *Bhallae, Papadam, Punjabi warri* |
| 63. | *Rhizopus achlamydosporus* | *Tempe* |
| 64. | *R. arrhizus* | *Tempe* |
| 65. | *R. chinensis* | *Sufu* |
| 66. | *R. formosaensis* | *Tempe* |
| 67. | *R. oligosporus* | *Miso, Oncom, Tauco, Tempe* |
| 68. | *R. oryzae* | *Tauco, Tempe* |
| 69. | *R. stolonifer* | *Tempe* |
| 70. | *R. graminis* | *Dosa, Idli* |
| 71. | *R. lactosa* | *Punjabi Warri* |
| 72. | *R. marina* | *Bhallae* |
| 73. | *Rhodotorula flava* | *Doenjang* |
| 74. | *Saccharomyces cerevisiae* | *Bhallae, Dosa, Idli,Jaleb, Papadam, Nan, Punjabi warri* |
| 75. | *S. kefir* | *Kefir* |
| 76. | *S. rouxii* | *Miso, Punjabi warri, Soy sauce* |
| 77. | *S. soyae* | *Miso* |
| 78. | *Staphylococcus sciuri* | *Kawai* |
| 79. | *Strepcoccus faecalis* | *Ambali, Bhallae, Dosa, Idli, Jelebi, Miso, Papadam, Punjabi warri* |
| 80. | *S. cremoris* | *Butter milk, sour cream, Cottage cheese* |
| 81. | *S. lactis* | *Yoghurt* |
| 82. | *S. thermophilus* | *Kefir, Koumiss, Cottage cheese, Jalebi* |
| 83. | *Torulopsis candida* | *Idli* |
| 84. | *T. dattila* | *Doenjang* |
| 85. | *T. holmii* | *Idli* |
| 86. | *Trichosporon beigelii* | *Bhallae, Dosa, Papadam, Punjabi warri* |
| 87. | *Tr. pullulans* | *Bhallae, Dosa, Idli, Punjabi warri* |
| 88. | *Wingea robertsii* | *Bhallae, Idli, Punjabi warri* |
| 89. | *Zygosaccharomyces soyae* | *Tauco* |

## 2.2. Microorganisms as the source of food

Though the every bite of food contains microbes, we are eating them all the time and, there is a constant stream of microbes into animal intestines. Some of the microorganisms or their fruiting bodies are used directly as a source of single cell proteins, mushrooms and truffles or are a source of proteins, generally termed as mycoproteins. The term 'single cell proteins' (SCP) refers to dead and dried cells of microorganisms such as bacteria, algae, molds, and yeasts. SCP from bacteria, yeasts, and fungi are produced by growing them on cheaper feedstocks such as molasses, methane, methanol, ethanol, cheese whey, cassava starch and a range of agricultural, forestry wastes and other industrial wastes. Generally four groups of microorganisms including bacteria, filamentous fungi, yeasts and algae are used as source of single cell proteins (Table 3). Of these yeasts are best due to their high

**Table 3** Microorganisms and substrates used for production of single cell proteins

| Microorganisms | Substrates for production |
|---|---|
| | **Bacteria** |
| *Aeromonas hydrophylla* | Dairy wastes |
| *Achromobacter delvacvate* | Hydrocarbons |
| *Acinetobacter calcoacetious* | Hydrocarbons |
| *Bacillus megaterium* | Meat industry wastes |
| *Bacillus subtilis* | Forest, agricultural, horticultural wastes |
| *Cellulomonas* spp. | Forest, agricultural, horticultural wastes |
| *Flavobacterium* spp. | Forest, agricultural, horticultural wastes |
| *Lactobacillus* spp. | Starch wastes |
| *Methylomonas methylotrophus* | Hydrocarbons |
| *Methylomonas clara* | Hydrocarbons |
| *Pseudomonas fluorescens* | Manures |
| *Rhodupseudomonas capsulata* | Starch wastes |
| *Thermomonospora fusa* | Lignocellulosic wastes |
| | **Filamentous fungi** |
| *Aspergillus niger* | Lignocellulosic wastes |
| *Aspergillus fumigatus* | Starch wastes |
| *Aspergillus oryzae* | Lignocellulosic wastes |
| *Cephalosporium eichorniae* | Starch wastes |
| *Chaetomium cellulolyticum* | Lignocellulosic wastes |
| *Penicillium cyclopium* | Dairy wastes |
| *Rhizopus chinencis* | Starch wastes |
| *Scytaldium acidiphilum* | Lignocellulosic wastes |
| *Trichoderma viridae* | Lignocellulosic wastes |
| *Trichoderma album* | Lignocellulosic wastes |
| | **Yeasts** |
| *Amoco torula* | Hydrocarbons |
| *Candida tropicalis* | Starch wastes |
| *Candida utilis* | Lignocellulosic wastes |
| *Candida novellas* | Hydrocarbons |
| *Candida intermedia* | Dairy wastes |
| *Saccharomyces cerevisae* | Lignocellulosic wastes |
| | **Algae** |
| *Chlorella pyrenidosa* | Utilizes $CO_2$ through photosynthesis |
| *Chlorella sorkiniana* | Utilizes $CO_2$ through photosynthesis |
| *Chondrus chrispum* | Utilizes $CO_2$ through photosynthesis |
| *Scenedesmus species* | Utilizes $CO_2$ through photosynthesis |
| *Spirulina species* | Utilizes $CO_2$ through photosynthesis |
| *Porphyrium species* | Utilizes $CO_2$ through photosynthesis |

nutritive values, ability to grow on low pH, vast substrate acceptability, less chances of contamination and are easy to harvest. Next to yeasts are the filamentous fungi. Bacteria may not be safe as food or feed as many of them produce toxins. While algae need large surface area for cultivation, economically and technically are difficult to grow.

A very important group of microorganisms used directly as a source of food are the mushrooms, most of which belong to Basidiomycetes, and a few belong to Ascomycetes. These fungi produce a fleshy fruiting structure containing the sexually produced spores, basidiospore or ascospores. The major, but relatively rare Ascomycete fruiting structure eaten are morels (*Morchella esculenta*). Most market mushrooms belong to group basidiomycetes. Mushrooms are consumed not only for their flavours and taste but also for their nutritive values. The nutritive value of edible mushrooms nearly equals to that of milk. The protein content of mushrooms are above most vegetables and somewhat less than most meats and milk. Protein content can vary from 10 to 50 % on dry weight basis. They contain high amount of amino acids (lysine, valine, and leucine), carbohydrates,vitamins, unsaturated fatty acids with a mineral content exceeding that in fish and meat and twice that in most vegetables. Fruiting bodies in general, on dry weight basis, contain 32% proteins, about 55% carbohydrate, 2% fat and rest is ash constituting the minerals.

**Table 4** Prominent cultivated edible mushrooms

| Mushroom species | Common name | Substrates for cultivation |
|---|---|---|
| *Agaricus bisporus, A. bitorquis* | Button mushroom | Wheat straw |
| *Pleurotus ostreatus, P. florida, P. sajor-caju* | Oyster mushroom | Wheat straw, bean straw, coir, corn cobs, paper wastes, cotton wastes (textile industry), crushed bagasse, etc. |
| *Volvariella volvacea* | Paddy straw mushroom | Paddy straw, oil palm pericarp waste, banana leaves, water hyacinth |
| *Lentinula edodes* | Shiitake or Oak mushroom | Sawdust, coffee pulp, sawdust-rice bran |
| *Pholiota nameko* | Nameko or viscid mushroom | Logs, sawdust-rice bran |
| *Auricularia spp.* | Jew's ear mushroom | Sawdust |
| *Hericium erinaceus* | Lion's head or Pom-Pom mushroom | Sawdust distillers grain waste |
| *Grifola frondosa* | Maitake or sitting hen mushroom | Sawdust |
| *Strophoria rugosonannulata* | - | Paper waste |
| *Morchella* sp. | Morels | *see footnote |
| *Flammulina velutipes* | Winter mushroom | Sawdust rice-bran |

*Commercial production of morels is currently not a practical option. Morels are being grown using a patented process, at only one production facility in North america (in Atlanta). The patent and facility are owned by Terry Farms and represent the only successful commercial process for cultivation of this ascomycetous mushroom.

Besides, mushrooms are a good source of vitamins like B-complex vitamins including riboflavin ($B_2$) niacin, pantothenic acid, thiamin ($B_1$) biotin, folate and vitamin $B_{12}$. Mushrooms are a particularly rich source of riboflavin. Mushrooms are unique in that they contain Vitamin $B_{12}$, something that vegetables can't produce at all. Since $B_{12}$ is mainly of animal origin, deficiency is commonly associated with vegetarian diets. Mushrooms were found to contain 0.32-0.65 mg per gram of $B_{12}$, Pro-vitamin D is present in some mushrooms, particularly shiitake, and can be converted to vitamin D by ultraviolet irradiation (sunlight).Vitamin A is uncommon although several mushrooms contain detectable amounts of pro-vitamin A measured as the β-carotene equivalent. Most cultivated mushrooms are believed to contain low amounts of the fat-soluble vitamins, K and E, and only a small contribution to the daily requirement of vitamin C. More than 2000 thousand species of mushrooms are reported to be edible throughout the world and about 150 species belonging to 30 genera are prominent among them. Over 60 species are cultivated commercially and 10 have reached an industrial scale in many countries. Table 4 enlists the prominent edible cultivated mushroom species through out the world.

There are several edible mycorrhizal mushrooms including the species of *Cantharellus, Russula, Lactarius* which cannot be cultivated on organic substrate due to their obligaory mycorrhizal nature and are known as mycorrhizal mushrooms

Truffles are a group of edible mycorrhizal (subterranean) fungi, the ascoma (fruiting body) of which is highly prized as food. These are generally served uncooked and shaved over steaming pasta or salads. The production of truffle is done by digging tree seedling in a mycorrhizal slurry before planting. After several years, under favourable growing conditions for both, the tree and the fungus, truffles form underground fruiting bodies that roughly resemble potatoes. These range from the size of a pea to that of a fist and give off a distinctive odor. Since these mushrooms do not completely emerge from the ground, they have traditionally been sniffed out by pigs or trained dogs. The requirements for growing the black truffle, *Tuber melanosporum* include choosing an appropriate host plant (usually oak or hazel nut), inoculating its roots with spawn and planting it.

## 2.3. Microorganisms as a source of food supplements

A food supplement, mycoprotein, is the processed cellular mass that is obtained from the filamentous fungus *Fusarium venenatum.* The filamentous nature of the cells gives the cellular mass a meat-like texture that makes it suitable for a variety of applications in food, including use as a muscle replacer in meat-alternative products. Mycoprotein also can be used as a fat replacer in certain dairy products and as a cereal replacer in products such as breakfast cereals or puffed snacks.The protein content of mycoprotein typically ranges from 42-50% percent on a dry weight basis. Mycoprotein is a high quality protein because mycoprotein contains all of the essential amino acids. The fat content of mycoprotein typically ranges from 12-14% on a dry weight basis and this fat content is more like vegetable fat than animal fat because it has a low proportion of saturated fatty acids and a high proportion of mono- and polyunsaturated fatty acids.The fungus is grown aerobically under steady-state conditions maintained by a continuous feed of nutrients and simultaneous removal of the culture. Quorn is the trademark of a fungus-based food product, sold (largely in Europe) as a meat substitute or imitation meat. It is marketed at the health-conscious, and to vegetarians. Quorn is the first commercial mycoprotein to be used in general food supplies.

A wide range of metabolites, produced by microorganisms, in the form of organic acids, amino acids, vitamins, lipids, proteins, enzymes etc. are also used as food supplements for the purpose of preservation, flavour enhancement, stabilization and an improvement in digestibility and nutritive value (Table 5).

Bacteria and fungi are producers of of a variety of organic acids including lactic acid, citric acid, acetic acid, gluconic acid and malic acid which are important in food industry.

**Table 5** Food additives of microbial origin

| Additive function | Product | Producing Microorganisms |
|---|---|---|
| Acidulants | Citric acid | *Aspergillus niger, Candida guilliermondii, Candida lipolytica* |
| | Acetic acid | *Acetobacter aceti, clostridia* |
| | Lactic acid | *Lactobacillus delbrueckii, L. bulgaricus* |
| | Malic acid | *Brevibacterium flavum* |
| Amino acids | Glutamic acid | *Corynebacterium glutamicum, Brevibacterium flavum* |
| | Lysine | *Corynebacterium glutamicum, Brevibacterium flavum* |
| | Tryptophan | *Klebsiella aerogenes* |
| Antimicrobial agents | Nisin | *Lactobaciluss lactis* |
| | Natamycin | *Streptomyces natalensis, S. chattanoogensis* |
| Antioxidans | Ascorbic acid | *Acetobacter suboxidans* |
| Colours | β - Carotene | *Blakeslea trispora, Dunaliella salina* |
| | Lycopene | *Blakeslea trispora, Streptomyces chattanoogensis* |
| | Zeaxanthin | *Flavobacterium species* |
| | Lutein | *Spongiococcum excentricum, Chorella pyrenoidosa* |
| | Cantaxanthin | *Cantharellus cinnabarinus, Rhodococcus maris* |
| | astaxanthin | *Mycobacterium lacticola, Xanthophyllomyces dendrorbous* |
| Emulsifiers | Various polysaccharides | *Many organisms* |
| | Glycolipids | *Rhodococcus erythropolis* |
| Flavours and flavor enhancer | Vanillin | *Saccharomyces species* |
| | Nucleosides | *Bacillus subtilis, Brevibacterium ammoniagenes* |

*Contd...*

| Additive function | Product | Producing Microorganisms |
|---|---|---|
| | Monosodium glutamate | *Corynebacterium glutamicum, Brevibacterium flavum* |
| Humectants | Glycols | *Bacillus species* |
| Lipids and fatty acids | Glycolipids | *Bacillus subtilis* |
| | γ – linoleic acid | *Mucor species* |
| Non nutritive sweetners | Aspartame | *From microbial production of aspartic acid and phenylalanine* |
| Stablizers and thickners | Glycan | *Saccharomyces species* |
| | Pulullan | *Auerobasidium* |
| | Xanthan | *Xanthomonas campestris* |
| Vitamins | Ascorbic acid (C) | *Acetobacter suboxidans* |
| | Cyanocobalamin($B_{12}$) | *Propionibacteria species, Pseudomonas denitrificans, Streptomyces griseus* |
| | Riboflavin ($B_2$) | *Ashbya gossypii, Ermothecium ashbyii, recombinant Bacillus subtilis* |
| Enzymes | α-Amylases | *Aspergillus flavus, A. oryzae* |
| | Amyloglucosidase | *Aspergillus oryzae, Rhizopus niveus* |
| | Catalase | *Micrococcus lysodeikticus* |
| | Cellulase | *Aspergillus niger, Trichoderma viridae* |
| | α-Galactosidase | *Mortierella vinacea* |
| | Glucose isomerase | *Actinoplanes missouriensis, Streptomy ces olivaceous, Bacillus coagulans* |
| | Glucose oxidase | *Aspergillus niger, Penicillium chrysogenum* |
| | Invertase | *Saccharomyces cerevisae* |
| | Lactase | *Candida pseudotropicalis, Kluyveromyces lactis* |
| | Lipase | *Aspergillus oryzae* |
| | Nuclease | *Penicillium citrinum* |
| | Pectinase | *Aspergillus niger* |
| | Urease | *Lactobacillus fermentum* |
| | Protease | *Bacillus licheniformis, Bacillus subtilis, Aspergillus niger* |
| | Rennet | *Bacillus cereus, Endothia parasitica, Rhizomucor miehei, Recombinant E. coli* |

The traditional organisms for citric acid fermentation are selected strains of *Aspergillus niger,* though citric acid has also been produced with several yeasts, mainly species of *Candida,* especially, *Candida lipolytica*. Citric acid provides a pleasant tartness and complements fruit, berry, and other flavors. Blends of citric acid combined with malic acid, lactic acid or fumaric acid are also used in beverage formulations. Other functions of citric acid in these applications rely on its ability to prevent metal-catalyzed off-flavors and the deterioration of color. This important food additive also serves as an antimicrobial preservative, retarding the growth of spoilage organisms. Acetic acid fermentation is applied in the aerobic, two-step process for the production of vinegar. However, in the last decade a significant amount of research has been devoted to the anaerobic production from biomass and its potential to compete with chemical synthesis. Vinegar is a liquid condiment used as a food flavoring and as a preservative. It is produced by the action of aerobic bacteria of the genus *Acetobacter* on dilute solutions of ethanol such as cider, wine, or diluted distilled alcohol solutions.The ethanol solutions utilized are also produced microbiologically by the ethanol fermentation of sugars by *Saccharomyces cerevisiae*. When starch materials are used (as in malt or rice vinegar), the enzymatic saccharification of starch is required prior to the alcohol fermentation. Lactic acid has been produced from various lactic acid bacteria in indigenous fermented foods including dairy products and is mainly responsible for the precipitation of milk proteins for yoghurt and cheese production. *Aspergillus niger* is used to produce gluconic acid at large scale. A portion of the gluconic acid is converted to δ-gluconolactone, a slow acting acidulant which is used with sodium bicarbonate for controlled leavening of baked goods. The lactone is also used in the dairy industry for cheese curd formation and for the improvement of the heat stability of milk. In addition, it serves as an additive in cured meat produts, luncheon meats, tofu, and packet desert mixes. L-malic acid obtained with fungal fermentations plays a role similar to that of citric acid as a food ingredient. As an acidulant, it has a smooth, long-lasting flavor profile, malic acid exhibits a synergy with aspartame permitting a 10% reduction in the usage of this sweetner in beverages.

Amino acids include an important class of compounds required by living cells as growth factors. Production of amino acids by fermentation processes was initiated in 1956 when L-glutamic acid was discovered as a product in the spent medium of *Corynebacterium glutamicum*. About 66% of amino acids are used in food industry and 31% used as feed additives. In food industry amino acids in the form of sodium glutamate, sodium aspartate, D,L-alanine, glycine are used to enhance flavours. L-cysteine improves quality of bread during baking process and acts as antioxidant in fruit juices. L-Tryptophan + L-Histidine act as antioxidants in powdered milk. L-phenylalanine and L-aspartic acid are used as the starting molecules for aspartame (L-aspartyl-L-phenylalanine methyl ester) which is used as a low calorie sweetner. Lysine is added in bread in Japan and in some countries soya products are enhanced by addition of methionine, L-Lysine. D,L-methionine used to upgrade animal feeds. The principal microorganisms capable of producing glutamic acid include *Corynebacterium, Brevibacterium, Microbacterium, Arthrobacter,* however, *Corynebacterium glutamicum* is mainly used commercially. L-methionine is synthesized with immobilized *Aspergillus oryzae* aminoacylase. L-Lysine is produced with L-amino-caprolactum hydrolase from *Cryptococcus laurentii* and L-phenylalanine is produced with immobilized *Rhodotorula rubra* cells bearing phenylalanine ammonia lyase. Phenylalanine is in demand for the synthesis of aspartame, a low-calorie, artificial sweetner used in soft drinks and desserts. Tryptophan, lysine and methionine all essential amino acids in the human diet, are used to enrich foods. Certain fungi are able to produce substantial amounts of these amino acids but are surpassed by improved industrial strains of bacteria. Strains of *Pichia anomala* are prolific producers of tryptophan. Strains of the yeasts *Candida utilis* and *Saccharomyces cerevisae* and the moulds *Ustilago maydis* and *Gliocladium* spp. produce significant quantities of lysine. Mutants of the yeasts *Candida tropicalis, Y. lipolytica, Pichia pastoris* and *C. utilis* have an increased methionine content and may have potential as food supplements.

Vitamins are essential mediators of metabolism and play an important role in the human diet. They are often supplements in foods and animal feeds. Many microorganisms are the potential producers of various vitamins. Many fungi and bacteria can produce riboflavin. The two ascomycetes *Ermothecium ashbyii* and *Ashbya gossypii* are the potential producers of riboflavin. Industrial fermentation with *A. gossypii* is now preferred and generates about 30% of world output. Sterols are widely distributed among the fungi. Ergosterol, a commercially important lipid, extracted from dehydrated yeast can be converted into vitamin D by irradiation with ultraviolet light. Such yeast strains may serve as alternative sources of vitamin D in future and can be used in various dairy products, often milk and margarine and in breakfast cereals and baby foods. There are certain fungal species for e.g., *Blakeslea trispora*, a heterothallic fungus of the Mucorales group and *Phycomyces blakesleeanus* are extremely rich in mycelial α-carotene, as a potential source of carotenoides which act as precursor for vitamin A production.

Microbial polysaccharides in the form of pullulan, scleroglucan and agar have attracted considerable interest as industrial fungal polymers with unique properties with food application. Pullulan is a neutral, linear homopolysaccharide composed of maltotriose units is produced by the dimorphic black yeast *Aureobasidium* (*Pullularia*). Its ability to form strong transparent films with low oxygen permeability enhances its usefulness as a food coating and packaging agent. Scleroglucan, a highly branched, neutral homopolysaccharide with a backbone of about 90 D-glucose units is produced by *Sclerotium glucanicum* or *S. rolfsii*. Scleroglucan exhibits pseudo-plasticity over a broad range of temperature, pH, and salt conditions, making it suitable for use as a food additives. Agar, a linear Galactan sulphate produced by red sea weed (*Gelidium algas*), having an ability to absorb water forming gels, is used as a stabilizer in foods.

About 70% of the world's supply of fats and oils is derived from plants and currently no food lipids are produced by microbiological means. It is notable that processes for fat production with yeasts *Rhodotorula gracilis, Lipomyces lipofer, Lipomyces starkeyi, Endomyces vernalis, Cryptococcus terricolus* and *Candida* species, which accumulate significant levels of fat, have been developed. The moulds *Mortierella vinacea, Mucor circinelloides, Aspergillus ochraceous, A. terreus*, and *Penicillium lilacinum* are also good producer of fat, as are certain fungi of genera *Hansenula, Chaetomium, Cladosporium, Malbranchea, Rhizopus* and *Pythium.* Fatty acids for the food, medical, and cosmetic industries could be standard products of fungal fermentation in coming years.

A wide range of microbial enzymes are also widely used in food processing and the manufacture of beverages (Table 6) which help to control texture, appearance and nutritive value of food as well as to generate desirable flavours and aromas.

Table 6 Applications of microbial enzymes in food industry

| Enzymes | Source | | Functions |
|---|---|---|---|
| | Bacteria | Fungi | |
| Proteases | *Bacillus subtilis, Lactobacillus acidophilus, L. plantanum, L. lactis, L. delbrueckii, L. helveticus* | *Aspergillus oryzae* | Baking, brewing, cheese production, meat tenderization, flavorants |
| Lipases | *Bacillus spp.* | *Aspergillus oryzae, a. niger, Rhizopus sp., Rhizomucor miehei, Rhodotorula sp.* | Flavorants, in dairy industry |
| α- amylases | *Bacillus spp.* | *Aspergillus spp. (Aspergillus oryzae)* | Baking, brewing, starch processing |
| α- amylases | *Bacillus spp.* | *Rhizopus sp.* | Brewing, maltose syrup production |
| Amyloglucosidases (glucoamylase) | - | *Aspergillus niger, Rhizopus niveus* | Brewing, baking, glucose syrup production |
| α- galactosidases (lactases) | *Lactobacillus spp.Bacillus spp., E. coli, Streptococcus lactis* | *Kluyveromyces sp., Candida spp.* | Processing of dairy products, whey-fructose syrup production |
| Glucose isomerases | *Actinoplane* sp., *Arthrobacter* spp., *Streptomyces* spp. | - | High-fructose syrup production |
| Glucose oxidases | - | *Aspergillus niger, Penicillium notatum* | As antioxidant, used alongwith catalase to remove oxygen from food products |
| Pullulanases | *Klebsiella pneumoniae, Bacillus acidopulluluticus* | - | Brewing, debranching of starch in sugar syrup manufacture |
| Pectinases | | *Aspergillus niger, Penicillium* spp. | Extraction of plant and fruit juices, coffee bean fermentation |
| Hemicellulases & α-glucanases/cellulases | *Bacillus* spp. | *Trichoderma reesei, Penicillium funiculosum, P. emersonii, Aureobasidium pullulans, Cryptococcus* sp., *Trichosporo*n sp. | Juice production, alcoholic beverages, baking, brewing, nutraceutics |
| Urease | *Lactobacillus fermentum* | - | Wine production |
| Invertase | - | *Saccharomyces cerevisae* | Variously in food processings |

Probiotics are an important part of the complex world of foods that are good for health. Probiotics are foods that contain live bacteria (Table 7). It is the bacteria and metabolites which they produce that give these probiotics their health promoting properties. The best known example of a probiotic is yoghurt. The consumption of yoghurt show increased milk digestibility, quicker recovery from certain types of diarrhea, enhanced immune function, reduction in certain cancers, and possible lowering of blood cholesterol levels. The probiotics exert the beneficial effect by producing certain metaboloites; known as prebiotics. Products that combine prebiotics with probiotics for enhanced health benefits are called as synbiotics and is the major new trend in the dairy sector. Synbiotic yoghurt has promising enhanced health benefits, such as improved immune function and increased calcium absorption in addition to overall health benefits. These products are at the forefront in providing consumers with a combination of health benefits in the tasty and convenient forms. Other emerging domestic products incorporating prebiotic fibers include kefir, yoghurt and other dairy drinks, sprouted products, functional waters, nutrition bars, weight loss products, soyamilk, green foods, probiotic supplements, mineral supplements, medical foods and pet foods.These products incorporate prebiotic fibers for various health benefits, including enhanced mineral and isoflavone absorption, fiber contribution, gut integrity, immune function and cholesterol control.

**Table 7** Important bacterial strains used as probiotics

| **Bacterial strains** | **Beneficial effects** |
|---|---|
| *Lactobacillus acidophilus*<br>*L. bulgaricus* | Balance intestinal flora, immune enhancement |
| *Mixture of L. lactophilus,*<br>*L. plantarum and Bacillus bifidum* | Prevention and treatment of antibiotic and rotavirus associated diarrhoea |
| *Lactobacillus casei* | Prevention of harmful intestinal microbiota |
| *Bacillus subtilis* | Used in bacteriotherapy, immunostimulatory agent |
| *Bifidobacterium bifidium* | Prevention of viral diarrhoea |
| *Propionibacterium fredenreichii* | Growth stimulation of other useful bacteria |

## 3. MICROORGANISMS IN PHARMACEUTICAL INDUSTRY

Microorganisms also find a wide application in the pharmaceutical industry and are the source of life saving drugs in the form of antibiotics, alkaloids, vaccines, bacteriophages, compounds of therapeutic value including steroids, recombinant proteins and peptides, microbial probiotics, prebiotics, enzymes and nutriceuticals.

### 3.1. Antimicrobial agents from microorganisms

Antibiotics are probably the most important group of compounds synthesized by microorganisms. Most antibiotics are secondary metabolites produced by actinomycetes, filamentous fungi and bacteria (Table 8).

**Table 8** Antibiotics of microbial origin

| Microbial genera | Microbial species | Antibiotics |
|---|---|---|
| | **Bacteria** | |
| *Pseudomonas* | *P. aeruginosa* | Pyocyanin |
| | *P. viscosa* | Viscosin |
| *Streptococcus* | *S. lactis* | Nisin |
| *Diplococcus* | *Diplococcus X-5* | Diplomycin |
| *Chromobacterium* | *C. prodigiosum* | Prodigiosin |
| *Escherichia* | *E. coli* | Coliformin |
| *Proteus* | *Pr. vulgaris* | Protaptin |
| *Bacillus* | *B. brevis* | Gramicidin |
| | *B. subtilis* | Subtilin |
| | *B. polymyxa* | Polymixin |
| | *B. licheniformis* | Bacitracin |
| **Actinomycetes** | | |
| *Streptomyces* | *S.griseus* | Streptomycin |
| | *S. aureofaciens* | Tetracycline |
| | *S. rimosus* | Tetracycline |
| | *S. spheroids* | Novobiocin |
| | *S. antibioticus* | Actinomycin |
| | *S. flavus* | Actinomycin |
| | *S. flaveolus* | Actinomycin |
| | *S. parvus* | Actinomycin |
| | *S.parvulus* | Actinomycin |
| | *S. chrysomallus* | Actinomycin |
| | *S.faradiae* | Actinomycin |
| | *S. melanochromogenes* | Actinomycin |
| | *S. erythreus* | Erythromycin |
| | *S. hygroscopicus* | Hygromycin |
| | *S. rimosus* | Oxytetracycline |
| | *S. griseofulvus* | Oxytetracycline |
| | *S. armilatus* | Oxytetracycline |
| | *S. aureofaciens* | Oxytetracycline |
| | *S. venezuelae* | Chloramphenicol |
| | *S. faradiae* | Neomycin |
| | *S. albogriseolus* | Neomycin |
| | *S. kanamyceticin* | Neomycin |
| *Nocardia* | *N. sulphurea* | Tetracycline |
| | **Fungi** | |
| **Imperfect fungi** | | |
| *Penicillium* | *P. chrysogenum* | Penicillin |
| | *P. griseofulvum* | Griseofulvin |
| *Trichothecium* | *T. roseum* | Trichothecin |
| **Basidiomycetes** | | |
| *Lenzites* (thermophilic fungi) | *L. thermophia* | Lenzitin |
| **Ascomycetes** | | |
| *Chaetomium* | *C. cochloides* | Chetomin |
| **Lichen** | *Lichen* | Usnic (Usninic) acid |
| | **Algae** | |
| *Chlorella* | *C. vulgaris* | Chlorellin |

Apart from bacteria, actinomycetes and fungi, algae also produces certain compounds which possess pharmaceutical properties such as anaesthetic, antipyretic, cough remedy, wound healing, thirst quenching and treatments for gout, gallstones, goitre, hypertension, bowel movement and skin diseases (Table 9).

**Table 9** List of the algal species with respective medicinal properties

| Organisms | Compound |
|---|---|
| *Lyngbya majuscula* | Terpenes, Carbohydrate |
| *Trichodesmium erythraeum* | Terpenes, Carbohydrate |
| *Fragilaria sp.* | Peptides |
| *Chaetoceros lauderi* | Polysaccharides |
| *Asterionella japonica* | Nucleotides fatty acids |
| *Gonyaulax tamarensis* | Terpenes, Carbohydrate |
| *Isochrysis sp.* | Terpenes, Carbohydrate |
| *Rhaeocystis pouchetii* | Acrylic acid |

Alkaloids are a diverse group of small nitrogen containing organic compounds produced by certain plants and microorganisms. Many are toxic, but some have various therapeutic properties. Species of filamentous fungus *Claviceps*, produce a range of alkaloids. These compounds are produced within the sclerotia (fruiting bodies) of *Claviceps purpurea* that develop naturally when this organism infects developing cereal grains. Infected grain become black and are referred to as ergots. These structures contain indole alkaloids, derived from a tertracyclic ergoline ring system, which are classified into two groups. The members of first group are based on clavin and contain no peptide component. Clavin-based alkaloids are also produced by other groups of fungi, including species of *Aspergillus* and *Penicillium*. Some possess antibiotic and anti tumor activity. The second group, based on lysergic acid (LSA) and containing a tripeptide or an amino alcohol, are found only in *Claviceps* species. Example include ergometrine, ergocristine, ergosine, and ergotamine. These LSA-based alkaloids have medical roles as analgesics in migraine therapy, as hallucinogens and for treating circulation problems. Others have particular uses in obstetrics, for inducing the smooth muscle of the uterus to contract during labour and after childbirth. Previously, the alkaloids were extracted from ergots that developed within the infected cereal crops, usually rye, or by chemical synthesis. Most are now produced by fermentation of *Claviceps fusiformis, C. paspali* or *C. purpurea* in surface, submerged or immobilized cell culture.

Many therapeutic steroids that are used for the treatment of allergies, inflammation, skin diseases and as oral coantraceptives are prepared by microbial transformations (Table 10). Initially, they were prepared by extraction from animal tissues or via complex chemical synthesis, both of which were extremely costly. Many steroids are now manufactured using combination of chemical and microbial transformation steps. These processes employ relatively cheap sterols as the starting materials, often diosgenin extracted from the Mexican yam (*Dioscorea composite*), or stigmasterol, a by product of soybean oil manufacture.The microorganisms involved are mostly filamentous fungi (*Rhizopus, Curvularia, Fusarium* and *Aspergillus* species), and mycobacteria, in the form of suspensions or immobilized growing cells, resting cells, spores and cell free extracts. They perform key reactions to modify the basic steroid structure, a cyclopentanoperhydrophenanthrene, including hydroxylations at positions 2 and 17, various side chain cleavages, hydrogenations and dehydrogenation and ring expansions, from a five –membered to a six-membered ring.

**Table 10** Some steroids whose production involves microbial transformation

| Biotransformation | Substrate | Product | Organisms |
|---|---|---|---|
| 1-Dehydrogenation | Hydrocortisone | Prednisolone | *Arthrobacter simplex* |
| 11-α-Hydroxylation | Progesterone | 11-α-Hydroxy-progesterone | *Rhizopus nigricans* |
| 11-β-Hydroxylation | Reichstein compounds | Hydrocortisone | *Curvularia lunata* |
| Side chain cleavage | β-sitosterol | Androstadienedione | *Mycobacterium* species |

Vaccines, another pharmaceutical preparations derived from suspensions of killed or attenuated microorganisms (bacteria, viruses, fungi, protozoa, or rickettsiae) or antigenic proteins derived from them are administered in the animals for the prevention, amelioration, or treatment of infectious and other diseases (Table 11).

**Table 11** Microbial vaccine for medical and veterinary use

| | |
|---|---|
| **Live attenuated bacterial vaccines** | |
| *Bacillus anthracis* (spore vaccine) | Anthrax |
| *Brucella abortus* S19 | Brucellosis |
| *Salmonella typhi* | Typhoid |
| *Shigella sonnei* | Shigelosis |
| **Inactivated bacterial vaccines** | |
| **Capsular antigens** | |
| *Escherichia coli* K88, K99 and 987P | Diarrhoea and food poisoning |
| *Neisseria meningitides* A & C | Meningitis |
| *Pasteurella multocida* | Pasteurellosis |
| **Cell wall antigens** | |
| *Bordetella pertussis* | Whooping cough (pertussis) |
| *Salmonella paratyphi* | Paratyphoid |
| *Salmonella typhi* | Typhoid fever |
| *Vibrio cholera* | Cholera |
| **Toxoid antigens** | |
| *Clostridium perfringens* | Food poisoning |
| *Clostridium tetani* | Tetanus |
| *Corynebacterium diphtheria* | Diptheria |
| **Recombinant vaccines** | |
| **Bacterial** | |
| *Chlamydia* | Chlamydia (pelvic inflammatory disease) |
| *Clostridium tetani* | Tetanus |

*Contd...*

| | |
|---|---|
| *Corynebacterium diphtheria* | Diptheria |
| **Protozoal** | |
| *Plasmodium vivax* | Malaria |
| **Viral** | Hepatitis B |
| | Herpes simplex |
| | Influenza |
| | Measels (rubella) |
| | Poliomyelitis |
| | Rabies |
| **DNA vaccine** | Malaria |

Recombinant DNA technology which employes the potential use of recombinant *E. coli*, has allowed the production of many recombinant therapeutic proteins and peptides with potential medical application (Table 12). Microbial vaccines are produced using either attenuated microbial cells, in which the organisms are adapted or engineered to delete pathogenicity while retaining immunogenicity, or live microbial vectors in which the microbial cells carrying added genes for an immunogenic protein are used.

**Table 12** Important recombinant proteins and peptides for therapeutic use

| |
|---|
| **Antibodies** |
| Tissue necrosis factor-α antibody (rheumatoid arthritis treatment) |
| **Cancer and viral diseases** |
| Interferons |
| Interleukins |
| Tissue necrosis factors |
| Transforming growth factors |
| **Cardiovascular diseases** |
| Erythropoietin (boosts red blood cell proliferation) |
| Hirudin (thrombin inhibitor from leech) |
| Urokinase (thrombosis treatment) |
| Tissue plasminogen activator (clot dissolution) |
| **Hormones** |
| Human growth hormone |
| Insulin |
| **Neurological diseases** |
| Endorphins |
| Nerve growth factors |
| Neuropeptides |
| **Vaccines** |
| Foot and mouth disease |
| Hepatitis B |
| **Wound healing and blood cloting factors** |
| Epidermal growth factors |
| Fibroblast growth factor |
| Clotting factor VIII (haemophilia treatment |

Due to increasing problems caused by antibiotic resistant bacterial pathogens, alternatives to antibacterial chemotherapy using antibiotics are now being sought. The benefits of bacteriophage (bacterial virus) therapy are that bacteria do not develop resistance to viruses, as they too evolve to continue invading their host bacteria. Also, they are specific and do not kill beneficial bacteria of the human or animal natural microflora. This therapy appears to be particularly useful for the treatment of cholera, dysentery, meningitis, septicaemia, tuberculosis and infections caused by *S. aureus* and species of *Campylobacter* and *Pseudomonas.*

## 3.2. Microorganisms as probiotics

A probiotic is an organism which contributes to the health and balance of the intestinal tract. A probiotic is also referred to as the 'friendly', 'beneficial', or 'good' bacteria which when ingested acts to maintain a healthy intestinal tract and help fight illness and disease. Probiotic is a live microbial feed supplement, which beneficially affects the host by improving its intestinal microbial balance. Probiotics enhance the immune system by favorably altering the gut micro-ecology and preventing unfriendly organisms from gaining a foothold in the body. They prevent the overgrowth of yeast and fungus and produce substances that can lower cholesterol. Probiotics are widely recommended for the treatment of candidal infection. Probiotics are also essential in the treatment and prevention of thrush, vaginal yeast infections, and athlete's foot. The important bacteria and yeasts used as probiotics include:

*Bifidobacterium:* They are saccharolytic organisms that produce acetic and lactic acids without generation of $CO_2$, except during degradation of gluconate. They are also classified as lactic acid bacteria (LAB). Bifidobacteria used as probiotics include *Bifidobacterium adolescentis, Bifidobacterium bifidum, Bifidobacterium animalis, Bifidobacterium thermophilum, Bifidobacterium breve, Bifidobacterium longum, Bifidobacterium infantis* and *Bifidobacterium lactis.*

*Lactobacillus: Lactobacilli* are normal inhabitants of the human intestine and vagina. *Lactobacilli* are gram-positive, facultative anaerobes, non-spore forming and non-flagellated rod or coccobacilli. *Lactobacilli* are also classified as lactic acid bacteria (LAB). *Lactobacilli* used as probiotics include *Lactobacillus acidophilus, Lactobacillus brevis, Lactobacillus bulgaricus, Lactobacillus casei, Lactobacillus cellobiosus, Lactobacillus crispatus, Lactobacillus curvatus, Lactobacillus fermentum, Lactobacillus gasseri, Lactobacillus johnsonii, Lactobacillus plantarum* and *Lactobacillus salivarus, Lactobacillus reuteri, Lactobacillus johnsonii, Lactobacillus delbrueckii* subspecies *delbrueckii, Lactobacillus delbrueckii* subspecies *bulgaricus* and *Lactobacillus paracasei* sub sp. *paracasei.*

*Lactococcus: Lactococci* are gram-positive facultative anaerobes. They are also classified as lactic acid bacteria (LAB). Lactococci that are used or are being developed as probiotics include *Lactococcus lactis, Lactococcus lactis* subspecies *cremoris (Streptococcus cremoris), Lactococcus lactis* subspecies *lactis.*

*Saccharomyces:* The principal probiotic yeast is *Saccharomyces boulardii. Saccharomyces boulardii* is also known as *Saccharomyces cerevisiae* Hansen CBS 5296 and is used to treat diarrhea associated with antibiotic use.

*Streptococcus:* One of the probiotic member of this genus is *Streptococcus thermophilus* which is a gram-positive facultative anaerobe and is nonmotile, non-spore forming and homofermentative. It is also classified as a lactic acid bacteria (LAB). It is found in milk and milk products. *Streptococcus salivarus* subspecies *thermophilus* is another probiotic strain.

*Enterococcus:* Enterococci are gram-positive, facultative anaerobic cocci of the Streptococcaceae family and are part of the intestinal microflora of humans and animals. *Enterococcus faecium* SF68 is a probiotic strain that has been used in the management of diarrheal illnesses.

The probiotic bacteria, exert various beneficial effects on health, both through their direct interactions with the gut, and in their metabolic products, known as prebiotics which enhance health in a number of ways through the support of probiotic bacteria in the gut. One important example of prebiotic metabolite is short chain fatty acids (SCFAs) which are the products of prebiotic fermentation by these gut microflora and are crucial for human health and well being. The probiotic bacteria and their SCFAs restrict the growth and activity of less beneficial species, including putrefactive and disease-causing bacteria and yeast. The effect of probiotics crowding out bad or less beneficial species is known as competitive exclusion which is essentially the good bacteria restricting the bad ones. The SCFAs produced by probiotic strains makes the gut environment less friendly to harmful strains. In addition, there is limited food and space in the colon, so the provision of a food source that is preferred by the good bacteria helps ensure they will have the competitive advantage. The SCFAs, in particular, are crucial to gut integrity and function, immune system modulation, calcium absorption and cholesterol maintenance. In fact, most of the energy required by the colon is provided directly by SCFAs. If the colon does not have enough SCFAs to provide energy needs, decreased integrity and function may follow. Various 'starved bowel' disorders, such as those known collectively as irritable bowel syndrome (IBS), are generally thought to be linked to inadequate SCFA production and a poor balance of gut microflora. The important point about SCFAs is that they are totally provided by probiotic fermentation of prebiotic fibers in the gut.

## 3.3. Microbial enzymes in pharmaceuticals

Enzymes as drugs have two important features that distinguish them from all types of drugs. First, enzymes often bind and act on their targets with great affinity and specificity. Second, enzymes are catalytic and convert multiple target molecules to desired products. These two features make enzymes specific and potent drugs that small molecules cannot. These characteristics have resulted in the development of many enzymes as drugs for a wide range of disorders (Table 13).

## 3.4. Mushroom nutriceuticals

Mushrooms were treated as a special kind of organisms which were known to possess all the essential curative ingredients in adequate quantities. The prominent health giving beneficial ingredients of mushrooms are mostly polysaccharides which are different in their composition, mode of action, degree of ramification and polymerization etc. yet they all are immunostimulants with no side effects and all of them functioning in strengthening health and increase immunity (Table 14). Thus, mushroom nutriceutical is a refined/partially defined food extractives which is consumed in the form of capsules or tablets as a dietary supplement (not a food), and which has potential therapeutic applications. A regular intake of mushroom nutriceutical enhance the immune response of the human body thereby increasing resistance against disease and in some cases causes regression of a disease state.

**Table 13** Microbial enzymes used for various therapeutic purposes

| Enzymes | Source | Applications |
|---|---|---|
| α –Amylase | *Aspergillus* and *Bacillus species* | Digestive aids |
| Alcohol dehydrogenase or alcohol oxidase | *Saccharomyces cervisae, Candida boidnii* and *Pichia pastoris* | Ethanol testing |
| Asparaginase | *Escherichia coli, Serratia marcesens* and *Erwinia carotovora* | Maintains low asparagines levels in treating cancers whose cells cannot synthesize the aminoacid |
| Catalase | *Aspergillus niger, Corynebacterium glutamicum and Micrococcus lysodeikticus* | Contact lens cleaning systems |
| Cholesterol esterase levels | *Pseudomonas fluorescens* | Monitoring serum cholesterol |
| Creatininase | *Pseudomonas putida* and *recombinant E. coli* | Determination of serum creatinine levels |
| Glucose oxidase | *Aspergillus niger* | Analysis of blood glucose levels |
| β- Lactamase | *Bacillus cereus and E. coli* | Treatment of penicillin allergy |
| Proteases (various) | *Several bacterial and fungal sources* | Used as digestive aids, for wound debridement and contact lens cleaning |
| Rhodanase (thiosulphate sulphur transferase) | *Trichoderma species* | Treatment of cyanide poisoning |
| Streptokinase (a cysteine protease) | *Various haemolytic streptococci* | To break down blood clots |
| Urease | *Lactobacillus fermentum* | Removal of bacteria from blood in renal failure |
| Uricase (ureate oxidase) | *Arthrobacter globiformis* and *Candida utilis* | Used in the treatment and diagnosis of gout |

**Table 14** Nutriceutically important mushroom with curative properties

| Mushroom species | Pharmacodynamics | Compounds |
|---|---|---|
| *Ganoderma lucidum, Lentinus edodes* | Immunomodulatory, Anticancer, Hepatoprotective, Hypoglycemic, Antihistamine release, Anti-angiogeinic, Antiviral | Polysaccharides/Proteins |
| *Ganoderma lucidum* | Cytotoxic tumor, Anti HIV, Hypolipidemic, Hypotensive, Antiplatelet aggregation, Cardioactive, Immunomodulatory | Triterpenes |
| *Ganoderma lucidum* | Anti-hypersensitivity, Anti-autoimmune diabetes, Anti-hepatitis B, Immunomodulatory | Amino acids |
| *Ganoderma lucidum* | Antiplatelet aggregation | Adenosine & derivatives |
| *Ganoderma lucidum* | Antitumor | Organic germanium |
| *Many species* | Antibacterial | Hirustic acid |
| *Odemansiella radiate* | Antibiotic effect | E-b methoxy acrylate |
| *Volvariella* | Cardiac tonic | Volvatoxin/Flammutoxin |
| *Collybia velutipes* | Decrease cholesterol | Eritadenine |
| *Ganoderma lucidum, Grifola frondosa* Ganodran/Glucan | Decrease blood sugar | Peptide/Glycogen/ |
| *Marasmius androsaceus* | Analgesic/Sedative effects | Marasmic acid |

# 4. MICROORGANISMS IN AGRICULTURAL INDUSTRY

Agriculture is dependent on microbes to maintain a biological balance of the soil; thus, they are essential for the growth of crops. More than 1,000,000,000 microbes can be found in only 1 gram of soil. Of this number, there may be more than 10,000 different species. Bacteria is the most numerous of the microbes. It breaks down organic matter and minerals into usable material for plants. Other types of microbes kill insects that are destructive to plants. Microorganisms are generally exploited in agriculture for controlling harmful insects and herbs by acting as bioinsecticides and bioherbicides as well as for improving the fertility of soil by way of acting as biofertilizers. These also play an important role in converting the complex polymers of agricultural residues into biological manures by the process of composting.

## 4.1. Microorganisms as biological control agents

One of the important category of biological agents used in control of insects and herbs in agriculture include microbial pathogens, like bacteria (*Bacillus thuringiensis* [Bt] for control of many catterpiller pests, beetles, mosquitoes), viruses, (nucleopolyhedrosis viruses for control of gypsy moth, European corn borer), fungi, (*Metarhizium* spp, for control of cockroach motels; *Beauvaria bassiana* for controlling Clorado potato beetle, corn root worms), protozoa, (*Nosema locustae* for controlling grasshoppers).

**4.1.1.** *Microbial insecticides:* Microbial insecticides are comprised of microscopic living organisms (viruses, bacteria, fungi, protozoa, or the toxins produced by these organisms (Table 15).They are formulated to be applied as conventional insecticidal sprays, dusts, liquid drenches, liquid concentrates, wettable powders, or granules. Each product's specific properties determine the ways in which it can be used most effectively. Microbial insecticides offer effective alternatives for the control of many insect pests. Their greatest strength is their safety, as they are essentially nontoxic and nonpathogenic to animals and humans. Although not every pest problem can be controlled by the use of a microbial insecticide, these products can be used successfully in place of more toxic insecticides to control many lawn and garden pests and several important field crop and forest insects.

**Table 15** Microbial insecticides: a summary of products and their uses

| Pathogen | Host range | Uses and comments |
|---|---|---|
| | | **Bacteria** |
| *Bacillus thuringiensis* var. *kurstaki* (*Bt*) | caterpillars (larvae of moths and butterflies) | Effective for foliage-feeding caterpillars (and Indian meal moth in stored grain). Deactivated rapidly in sunlight; apply in the evening or on overcast days and direct some spray to lower surfaces or leaves. Does not cycle extensively in the environment. Available as liquid concentrates, wettable powders, and ready to use dusts and granules. Active only if ingested. |
| *Bacillus thuringiensis* var. israelensis (*Bt*) | larvae of *Aedes* and *Psorophora* mosquitoes, black flies, and fungus gnats | Effective against larvae only. Active only if ingested. *Culex* and *Anopheles* mosquitoes are not controlled at normal application rates. Activity is reduced in highly turbid or polluted water. Does not cycle extensively in the environment. Applications generally made over wide areas by mosquito and blackfly abatement districts. |
| *Bacillus thuringiensis* var. *tenebrinos* | larvae of Colorado potato beetle, elm leaf beetle adults | Effective against Colorado potato beetle larvae and the elm leaf beetle. Like other *Bt*s, it must be ingested. It is subject to breakdown in ultraviolet light and does not cycle extensively in the environment. |
| *Bacillus thuringiensis* var. *aizawai* | wax moth caterpillars | Used only for control of was moth infestations in honeybee hives. |
| *Bacillus popilliae* and *Bacillus* | larvae (grubs) of Japanese beetle | The main Illinois lawn grub (the annual white grub, *Cyclocephala* sp.) is not susceptible to milky spore disease. The disease is very effective against |

*Contd...*

| | | |
|---|---|---|
| *lentimorbus* | | Japanese beetle grubs (not a major pest in Illinois) and cycles effectively for years in the soil. |
| *Bacillus sphaericus* | larvae of *Culex, Psorophora,* and *Culiseta* mosquitos, larvae of some *Aedes* spp. | Active only if ingested, for use against *Culex, Psorophora,* and *Culiseta* species; also effective against *Aedes vexans.* Remains effective in stagnant or turbid water. Commercial formulations will not cycle to infect subsequent generations. |
| | | **Fungi** |
| *Beauveria bassiana* | aphids, fungus gnats, mealy bugs, mites, thrips, whiteflies | Effective against several pests. High moisture requirements, lack of storage longevity, and competition with other soil microorganisms are problems that remain to be solved. |
| *Lagenidium giganteum* | larvae of most pest mosquito species | Effective against larvae of most pest mosquito species; remains infective in the environment through dry periods. A main drawback is its inability to survive high summertime temperatures. |
| | | **Protozoa** |
| *Nosema locustae* | European cornborer caterpillars, grasshoppers and mormon crickets | Useful for rangeland grasshopper control. Active only if ingested. Not recommended for use on a small scale, such as backyard gardens, because the disease is slow acting and grasshoppers are very mobile. Also effective against caterpillars. |
| | | **Viruses** |
| Gypsy moth nuclear plyhedrosis (NPV) | gypsy moth caterpillars | All of the viral insecticides used for control of forest pests are produced and used exclusively by the U.S. Forest Service. |
| Tussock moth NPV | tussock moth caterpillars | |
| Pine sawfly NPV | pine sawfly larvae | |
| Codling moth granulosis virus (GV) | codling moth caterpillars | Commercially produced and marketed briefly, but no longer registered or available. Future re-registration is possible. Active only if ingested. Subject to rapid breakdown in ultraviolet light. |

**4.1.2.** *Microbial herbicides:* Weeds or unwanted vegetation, in cut over or reforestation areas are competitive with the forest crop species and must be managed to raise the productivity and quality of the produce. Of the several options for control, including mechanical, incendiary and chemical methods, the herbicide option is the most efficient and cost effective. Several microorganisms of group fungi, bacteria, viruses and actinomycetes in this regard found quite potent. Fungi are considered to be one of the most important group of plant pathogen responsible for more than 75% of diseases in crop as well as in forest plants. Their pathogenic potential have been extensively exploited for the management of many weeds (Table 16).

**Table 16** Fungal pathogens used as mycoherbicides

| **Fungal pathogens** | **Target weeds** |
|---|---|
| *Phytophthora palmivora* | *Morrena odorata* |
| *Colletotrichum gloeosporides f sp. aeschynomene* | *Aeschynomene viginica* |
| *Colletotrichum gloeosporides f sp. malvae* | *Malva pusilla* |
| *Colletotrichum gloeosporides f sp. cuscutae* | *Cuscuta chinensis/ C. australis* |
| *Colletotrichumgloeosporides f sp. partheniae* | *Parthenium hysterophorus* |
| *C. orbicularae* | *Xanthium spinosum* |
| *C. dematium* | *Parthenium hysterophorus* |
| *C. coccodes* | *Abutilon theophrasti* |
| *Alternaria cassiae* | *Cassia esculentus* |
| *Puccinia canaliculata* | *Cyperus* |
| *Nectaria ditissima* | *Alnus rubra* |
| *Acremonium diospyri* | *Diospyros Virginiana* |
| *Sclerotium rolfsii* | *Parthenium hysterophorus* |
| *Cercospora rodmanii* | *Eichhornia crassipes* |
| *Bipolaris sorghicola* | *Sorghum halepense* |
| *Fusarium solani* | *Parthenium hysterophorus* |
| *F. oxysporum* | *Parthenium hysterophorus* |
| *Chondrostereum purpureum* | *Prunus serotina / Cassia obtusifolia* |
| *Phomopsis amaranthicola* | *Amaranthas sp.* |
| *Alternaria zinniae* | *Xanthium spinosum* |
| *Myrothecium verrucaria* | Broad spectrum |
| *Septoria passiflorae* | *Passiflora tripartita* |

Many potent plant pathogenic bacteria are also used as weedicides (Table 17). Bacteria are capable of causing devastating diseases viz., dwarfing wilts, lesions, blight, tumor, galls, scab, distortion and discoloration of various plant parts. In contrast fungal pathogens, bacteria are incapable of mechanically penetrating the cutinized plant surface tissues, periderm etc. They get into the plant through non-cutinised areas (root hairs, stigmas), natural openings (stomata, hydathodes) and incidental wounds or those incited by nematodes, insects and other microbes. On entry into the suitable host they establishes themselves, intracellularly, intercellularly or intracascularly or in more than one combination, and damage the plant.

Many viral pathogens are also known to incite severe infections and causes catastrophic losses in various categories of plants. Plant viruses do not contain any enzymes toxins or any substances necessary for pathogenecity. They are also unable to disseminate passively by way of wind, water or even sap of the infected hosts. They require a vector for transmission

**and entry into the host. Thus, requirement of vector and technical difficulties in in vitro mass production are the major constraints which discourage the use of these agents as microbial herbicides.**

**Table 17** Phytopathogenic bacterial pathogens used as herbicides

| **Phytopathogenic bacteria** | **Target weeds** |
|---|---|
| *Pseudomonas* sp. | *Acroptilum repens* |
| *Pseudomonas* sp. | *Avena fatua* |
| *Pseudomonas* sp. | *Calamagrostis canadensis var canadensis* |
| *Pseudomonas* sp. | *Poa annua* |
| *Pseudomonas* sp. | *Xanthium strumarium* |
| *Pseudomonas* sp. | *Bromus japonicus* |
| *Pseudomonas* sp. | *Abutilon theophrsti* |
| *Pseudomonas* sp. | *Bromus tectorum* |
| *Pseudomonas* sp. | *Parthenium hysterophorus* |
| *Pseudomonas* sp. | *Bromus tectorum* |
| *Agrobacterium* sp. | *Euphorbia esula* |
| *Xanthomonas* sp. | *Acroptilum repens* |
| *Xanthomonas* sp. | *Bromus secalinus* |
| *Enterobacter* sp. | *Bromus tectorum* |
| *Enterobacter* sp. | *Abutilon retroflexus* |
| *Erwinia herbicola* | *Ipomea* spp. / *Setaria viridis* |
| *Flavobacterium* sp. | *Euphorbia esula* |

**Actinomycetes constitutes a very important component of soil microbiota. They are not directly involed in causing diseases to plants and only one species, *Streptomyces scabies* is plant pathogenic causing potato scab disease in potato plants. Their toxins are used for management of weeds (Table 18).**

**Table 18** Toxins of Actinomycetes used as herbicides

| **Toxins** | **Producing Actinomycetes** |
|---|---|
| Gulphosinate / Anisomycin | *Streptomyces* spp. |
| Bialophos | *S. vridochromogens* |
| Hydantocidin Nigricin | *S. hygroscopicus* |
| Geldanomycin | *S. hygroscopicus* var. *gelanus* |
| Herbicidin | *S. saganonensis* |
| Toyacamycin/Sangivamycin/Tubercidin/Herbiphonin/ Gabaculinae | *S. toyacaenis* |

## 4.2. Microorganisms for increasing crop productivity

For a sustainable agriculture system, it is imperative to utilize renewable inputs which can maximize the ecological benefits and minimize the environmental hazards. One possible way of achieving this is to decrease dependence on use of chemical nitrogen fertilizers by harvesting the atmospheric nitrogen through biological processes.The value of organic fertilizer, not only provides plant nutrients but to help control soil borne diseases and maintain adequate soil structure, is now being widely recognized. It is living fertilizer compound of microbial inoculants or groups of micro-organisms which are able to fix atmospheric nitrogen or solubilize phosphorus, decompose organic material or oxidize sulphur in the soil. On application, it enhances the growth of plants increase in yield and also improve soil fertility and reduces pollution. There are variety of nitrogen fixing micro-organisms present in the nature. These are broadly divided into three categories: i) Symbiotic micro-organism including *Rhizobium,* ii) Asymbiotic or free living including *Azotobacter*, blue green algae, iii) Associative symbiosis including *Azospirillum.* These micro-organisms supply in addition to nitrogen, considerable amount of organic matter enriching structure of soil. Inoculants of these micro-organisms have proved their technical feasibility, economic viability and social acceptability. They are therefore called as 'biofertilizer'. Hence the term 'biofertilizer' or microbial inoculants may be defined as preparations containing living or latent cells of efficient strains of nitrogen fixing, phosphorous solubilizing or cellulolytic micro-organism. Biofertilizers have definite advantage over chemical fertilizers. Chemical fertilizers supply over nitrogen whereas biofertilisers provide in addition to nitrogen certain growth promoting substances like hormones, vitamins , amino acids, etc., crops have to be provided with chemical fertilizers repeatedly to replenish the loss of nitrogen utilised for crop growth. On the other hand biofertilizers supply the nitrogen continuously throughout the entire period of crop growth in the field under favourable conditions. Continuous use of chemical fertilisers adversely affect the soil structure whereas biofertilizers when applied to soil improve the soil structure. The deleterious effects of chemical fertilizers are that they are toxic at higher doses. Biofertilizers, however, have no toxic effects. It may be borne in mind that biofertilizers are no substitute for chemical fertilizers.Followings are some of the important types of biofertilizers which can be considered for agrobased industries. Some of the major microorganisms used as biofertilizers are discussed below:

*Rhizobium biofertilizers: Rhizobium* bacteria in association of legumunious plants fix atmospheric nitrogen in nodules formed on the roots of plants. These nodules are considered as miniature nitrogen production factories in the field.

*Azotobactor biofertilizers: Azotobacter* are free living microorganisms and they grow in the rhizosphere and fix atmospheric nitrogen non-symbiotically and make it available to particularly cereals. In addition, these bacteria produce growth promoting substances thereby enhancing the plant growth and finally yield.

*Azospirillum biofertilizers:* They are called as associative endosymbiont on roots of grasses and similar types of plants. They are also known to fix atmospheric nitrogen and benefit host plants by supplying growth hormones and vitamins.

*Blue green algal biofertilizers:* Blue green algae are considered as an important group of micro-organism capable of fixing atmospheric nitrogen non-symbiotically mostly in rice fields in heterocysts cells specially known as sites of nitrogen fixation. There is variation between species of blue green algae in fixation of nitrogen and other qualities of biofertilizers. Efficient strains of blue green algae (*Aulosira, Tolupothrix, Scytonema, Nostoc* and *Anabaena*) are used for multiplication on a large scale.

*Azolla biofertilizers: Azolla* is a water fern inside which grows the nitrogen fixing blue green algae *Anabaena*. It contans 2-3% nitrogen when wet and also produces organic matter in the soil. This type of biofertiliser is used all over the world. This can be grown in a cooler region. But there is a need to develop a strain tolerant to high temperature, tolerant to salinity and strains resistant to pests and diseases.

*Phosphorus-solubilizer:* Phosphorus is one of the important elements required for plant growth. This element is also essential for nodulation by *Rhizobium*. Phospho-microorganisms, mostly bacteria and fungi are used for commercial purposes. In addition, mycorrhizae (Plant roots-fungus association) have high potential of phosphorus accummulation in plants. There are two types of mycorrhiza viz :endo- and ecto-trophics. VA mycorrhiza are most popular and used for commercial purpose.

## 4.3. Microorganisms for composting

Composting is a microbially driven process that first breaks down organic matter, then builds that material into humus. Composting is a viable means of transforming various organic wastes into products that can be used safely and beneficially as biofertilizers and soil conditioners. During the composting process, organic wastes are decomposed, plant nutrients are mineralized into plant-available forms, pathogens are destroyed, and malodors are abated.Composting is a practice that farmers have used for centuries to convert organic wastes into useful biofertilizers and soil amendments. More specifically, composting is a microbiological process that depends on the growth and activity of mixed populations of bacteria, actinomycetes, and fungi (Table 19) that are indigenous to the wastes being composted. Composting can be conducted by either aerobic or anaerobic methods. However, the aerobic mode is generally preferred, since it proceeds more rapidly and provides greater pathogen reduction because higher temperatures are attained.

**Table 19** Major groups and succession of microorganisms occurring during composting

| Mesophiles | Thermotolerant (Succession of microorganisms→) | Thermophiles |
|---|---|---|
| | **Bacteria** | |
| *Flavobacterium* spp.<br>*Pseudomonas* spp.<br>*Serratia marcescens* | *Pseudomonas* spp.<br>*Bacillus licheniformis* | *Bacillus coagulans*<br>*B. stearothermophilis*<br>*B. subtilis* |
| | **Actinomycetes** | |
| *Streptomyces* spp.<br>*Nocardia* spp.<br>*Faeni* spp. | | *Thermoactinomycetes* spp.<br>*Thermomonospora* spp.<br>*Saccharomonospora viridis* |

*Contd...*

| | | *Streptomyces* spp. |
|---|---|---|
| | **Fungi** | |
| *Mucor* spp. | *Aspergillus fumigatus* | *Torula thermophila* (syn. *Scytalidium thermophilum)* |
| *Aspergillus* spp. | | *Chaetomium thermophile* |
| *Penicillium* spp. | | *Humicola insolens* |
| | | *Rhizomucor pusillis* |
| | | *Talaromyces thermophilus* |
| | | *Thermomyces lanuginosa* |

# 5. MICROORGANISMS IN ENERGY PRODUCTION

Microorganisms play an important role in fuel production, for energy generation, from biomass, a renewable resource. The primary source of the energy stored in biomass is solar energy which is trapped by green plants during photosynthesis and stored as potential energy in organic molecules. These biofuels are obtained from the terrestrial biomass (wood, forest litter), aquatic biomass (algae and water weeds like water hyacinth, *Pistia, Salvinia, Azolla*), various types of organic wastes (agricultural and industrial wastes, municipal and domestic wastes like sewage, garbage, animal dung, wastes from slaughter houses). Microorganisms convert biomass into liquid bioenergy fuels, gaseous fuels or into electricity by their metabolism. Some of the important fuels produced by microbes include biogas, ethanol, hydrogen and bioelectricity.

*Biogas:* It is known by various names such as biofuel, sewerage gas, Klar gas, sludge gas, digester gas and gobar gas. Biogas is often used for cooking and lighting purposes in the rural sector. It burns with a blue flame without smoke and is without odor. Biogas is formed under anaerobic conditions when organic materials are converted into gases (fuel), whose main constituents are methane (63%), carbon dioxide (30%) and organic fertilizer (sludge) through microbial reactions. Biogas can be produced from the biomass of plant origin (either terrestrial biomass or aquatic biomass) or animal origin (cattle dung, domestic sewage, organic manure, poultry, slaughter house and fishery wastes). Conversion of the biomass into biogas (methane) is a process called anaerobic digestion. It is completed in three stages by different groups of bacteria (Table 20).

**Table 20** Processess and microorganisms involved in production of biogas

| Stage | Process involved | Group of bacteria involved |
|---|---|---|
| 1 | Solubilization | Hydrolytic fermentative bacteria (mostly anaerobic). |
| 2 | Acetogenesis | Hydrogen producing acetogenic bacteria (facultative anaerobic) |
| 3 | Methanogenesis | Methanogens (methane producing anaerobic bacteria) |

*Bioethanol:* Ethanol is derived by fermentation from plant materials, particularly carbohydrates. Chemically it is the same molecule as found in potable alcohol. Bioethanol is by far the biggest fermentation product to date and is presently produced with the help of yeasts that convert easily fermentable carbohydrates obtained from sugar beet, sugar cane or cereals into ethanol. Currently, significant research efforts are directed towards producing alcohol from more difficult substrates such as agricultural residues or other organic waste. Agricultural co-products or waste such as straw, bran, corn cobs, corn stover, etc., products that are either being poorly valorised or left to decay on the land, are attracting increasing attention as an abundantly available and cheap renewable feedstock. Besides ethanol, a number of co-products are being produced during production such as protein fractions that are used as animal feed or as a fertilizer. The concomitant production of $CO_2$ is very significant, as there is as much production of $CO_2$ as ethanol during fermentation. This kind of $CO_2$ is very pure and is often recovered for use in the chemical industry and for the production of soft drinks. For use in motor fuels, the alcohol must be dewatered, mostly with the help of regenerable absorption agents. The dewatered alcohol can be used under different forms in motor fuels, usually in mixtures with normal gasoline.

*Hydrogen:* Hydrogen gas is an interesting alternative fuel because it completely eliminates carbon as an energy carrier.Therefore it obviously cannot contribute to the green house effect. Although hydrogen is currently being produced by either thermally decomposing natural gas or by electrolysis of water there is also a potential for biological hydrogen production. Although the capacity to produce hydrogen is generally acknowledged in anaerobic, fermentative bacteria like *Clostridium* , there are also examples of green algae who produce oxygen under anaerobic and dark conditions like *Chlamydomonas reinhardtii*. The generation of hydrogen using microorganisms, or cell free systems based on microbial components, is still very much in its infancy. However, there are three possible routes of production including biophotolysis of water using photosynthetic systems of cyanobacteria involving nitrogenase enzyme, or the algal phososynthetic systems involving hydrogenase enzyme, photoreduction involving the light dependent decomposition of organic compounds by the members of the Chlorobiaceae, Chromatiaceae and Rhodospirillaceae and fermentation of organic compounds by many bacteria.

*Bio-electricity:* Bio-electricity can be produced from biomass in a variety of ways. Typically, heat generated by burning solid biomass, biogas or biofuels can be converted into electricity by conventional means. Biogas can be used more directly in a gas motor that is linked to a generator for electricity production. More recently, methods for directly converting biomass into electricity with the help of microorganism or its enzyme based biofuel cells are emerging. Microbial fuel cells (MFC) are capable of converting energy, available in a bio-convertible substrate, directly into electricity with the help of bacteria including *Rhodoferax ferrireducens, Desulfuromonas acetoxidans, Proteus vulgaris*, *Geobacter* and *Shewanella* species. These bacteria switch from the natural electron acceptor, such as oxygen or nitrate, to an insoluble acceptor, such as the MFC anode (Figure 1). This transfer can either occur by membrane-associated components, or by soluble electron shuttles. The electrons then flow through a resistor to a cathode, at which the electron acceptor is reduced. In contrast to anaerobic digestion, a MFC creates electrical current and an off-gas containing mainly carbon dioxide instead of an energy-rich gas such as methane or hydrogen.

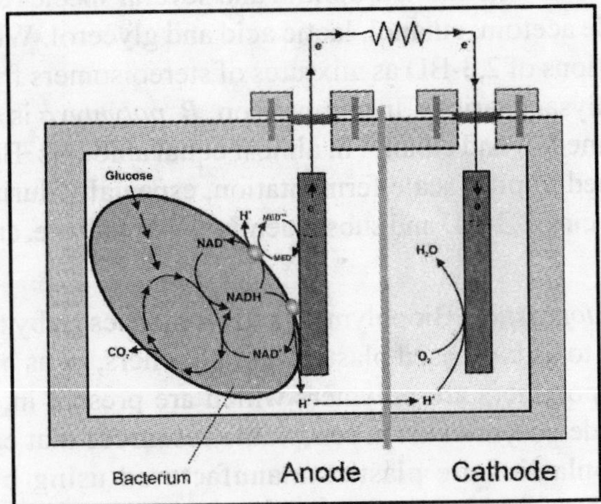


**Figure 1** A typical microbial fuel cell

## 6. MICROORGANISMS IN CHEMICAL INDUSTRY

Microorganisms are a source of many chemicals, commonly known as biochemicals, produced from organic matter. There are different types of biochemicals including solvents, fuel additives, lubricants, surfactants, adhesives, inks, enzymes, amino acids, vitamins, biosurfactants, biopolymers etc. Biochemicals can be made from the processing of plant sugars, starch hydrolysates, alkanes, acids etc. In the first step the substrates are fermented in a controlled environment, much like the fermentation of beer and wine in which the microorganisms convert sugars to the specific chemicals. In the next step, a purification process is undertaken where the fermented liquid is purified for the target biochemical. Some of the important biochemicals produced from microorganisms are discussed below:

*Acetone-Butanol:* Louis Pasteur observed butanol fermentation by *Clostridium acetobutylicum* in the 19$^{th}$ century in which acetone and ethanol are produced in addition to butanol. During world war I the product of interest was acetone, used in the production of explosive, trinitrotoluene (TNT) but after the war butanol became more important for its use as a solvent for rapid drying of nitrocellulose paints in automobile industry. Butryric acid, butanol, acetone, isopropanol are obtained through clostridial fermentation of starch, molasses, sucrose, wood hydrolysates and pentoses. The relative proportions of each of these products in the fermentation depends on the bacterial strain used and the fermentation conditions:

- Acetone-butanol fermentation with *Clostridium acetobutylicum*
- Butanol-isopropanol fermentation with *Cl. butylicum*
- Butyric acid-acetic acid fermentation with *Cl. butyricum*

$H_2$ and $CO_2$ are also produced as byproducts. In addition, ethanol is also produced through reduction of acetaldehyde.

*2-3, Butanediol:* Fermentation of xylose as well as glucose by *Klebsiella oxytoca,* (formerly known as *Klebsiella pneumoniae* and *Aerobacter aerogenes)* yields 2,3-butanediol (2, 3 BD) as the major product. Other microorganisms capable of producing 2,3-butanediol include *Bacillus subtilis*, *Aeromonas hydrophilia* and several species of *Serratia*. Secondary products formed include acetoin, ethanol, lactic acid and glycerol. While *K. oxytoca* is able to yield high concentrations of 2,3-BD as mixtures of stereoisomers from monosaccharides, it is unable to utilize polysaccharides. In comparison, *B. polymyxa* is able to ferment starch directly giving 2,3-butanediol and ethanol in almost equal amounts. Both *K. oxytoca* and *B. polymyxa* have been used in pilot scale fermentation, especially during World War II, as a possible means of producing 2,3-BD and subsequently 1,3-butadiene, an organic intermediate for rubber production.

*Biopolymers and Bioplastics:* Biopolymers and bioplastics go by many different names. They are often referred to as bio-based plastics and polymers, or as biodegradable plastics or polymers. Thus, biopolymers are polymers which are present in, or created by, living organisms. These include polymers from renewable resources that can be polymerized to create bioplastics. Bioplastics are plastics manufactured using biopolymers, and are biodegradable. Some polymerizable molecules which come from renewable natural resources, and can be converted into biodegradable plastics.

There are two ways fermentation can be used to create biopolymers for bioplastics. These include bacterial fermentation in which the bacteria use the sugar to fuel their cellular processes and produce the polymers as by-product of these cellular processes which are then separated from the bacterial cells. The other way converts sugars in to lactic acid by fermentation which is then converted to polylactic acid using traditional polymerization processes.

Another important category of biopolymers is polysaccharides which are produced by a wide range of microorganisms including bacteria, yeasts and fungi and algae. These are either homopolymers or heteropolymers and include cellulose, alginates, chitin, curdlan, dextran, glycans and phosphomannans, pullulan, scleroglucan, xanthan and gellan. Their commercial potential is now being realized and they have proved to be useful industrial products. They are mainly used in food industry as thickeners and stabilizers. Besides, they also possess several non food applications in pharmaceutical and chemical industries.

*Biosurfactants:* Surfactants are surface-active compounds capable of reducing surface and interfacial tension at the interfaces between liquids, solids and gases, thereby allowing them to mix or disperse readily as emulsions in water or other liquids. The enormous market demand for surfactants is currently met by numerous synthetic, mainly petroleum-based, chemical surfactants. These compounds are usually toxic to the environment and non-biodegradable.They may bio-accumulate and their production, processes and by-products can be environmentally hazardous. Tightening environmental regulations and increasing awareness for the need to protect the ecosystem have effectively resulted in an increasing interest in biosurfactants as possible alternatives to chemical surfactants. Biosurfactants are amphiphilic compounds of microbial origin with considerable potential in commercial applications within various industries. They have advantages over their chemical counterparts in biodegradability and effectiveness at extreme temperature or pH and in having lower toxicity. Biosurfactants are beginning to acquire a status as potential performance-effective molecules in various fields. At present biosurfactants are mainly used in studies on enhanced oil recovery and hydrocarbon bioremediation. Biosurfactants also have potential applications in agriculture, cosmetics, pharmaceuticals, detergents, personal care products, food processing, textile manufacturing, laundry supplies, metal treatment and processing, pulp and paper processing and paint industries (Table 21).

**Table 21** Biosurfactant producing microorganisms

| Microorganism | Biosurfactant |
|---|---|
| | **Bacteria** |
| *Acinetobacter* spp | Emulsan (lipoheteropolysaccharide); whole cell lipopeptide; fatty acids; mono- and di-glycerides |
| *Acinetobacter radioresistens* | Alasam |
| *Alcanivorax borkumensis* | Glycolipid |
| *Arthrobacter* spp. | Glycolipids; lipopeptides; heteropolysaccharides |
| *Bacillus* spp. | Surfactin; rhamnose lipids; hydrocarboprotein complex; polymyxin; gramicidin antibiotics |
| *Corynebacterium* spp. | Acyl glucoses ; polysaccharide-protein complex ; phospholipids ; corynemycolic acids ; fatty acids |
| *Clostridium* spp. | Neutral lipids |
| *Nocardia* spp. | Glycolipidic; whole cell |
| *Rhodococcus* spp. | Deemulsifiers; neutral lipids and fatty acids; trehalose dimycolates and dicorynomycolates; polysaccharide |
| *Pseudomonas* spp. | Viscosin; Ornithin; glycolipids |
| *Serratia* spp. | Serrawettin; Rubiwettin |
| *Thiobacillus* spp. | Phospholipids |
| | **Fungi** |
| *Candida bombicola* and *Candida atarctica* | Liposan; glycolipids – sophorolipids, mannosylerythrisol lipids; peptidolipid; polysaccharide-fatty acid complex |
| *Torulopsis* spp. | Glycolipids; proteins |
| *Ustilago* spp. | Cellobiolipids |

*Microbial pigments :* Pigments from natural sources has been obtained since long time ago, and their interest has increased due to the toxicity problems caused by those of synthetic origin. In this way the pigments from microbial sources are a good alternative. Some of more important natural pigments, are the carotenoids, flavonoids (anthocyanins) and some tetrapirroles (chlorophylls, phycobilliproteins). Another group less important are the betalains and quinones. The carotenoids are molecules formed by isoprenoids units and the most important used as colorant are the alpha and beta carotene which are precursors of vitamin A, and some xanthophylls as astaxanthin. The pigment more used in the industry is the beta-carotene which is obtained from some microalgae and cyanobacteria. The astaxanthin another important carotenoid is a red pigment of great commercial value, and it is used in the pharmaceutical, feed and aquaculture industries. This pigments is mainly obtained from *Phaffia rhodozyma* and *Haematococcus pluvialis* and other organisms. The phycobilliproteins obtained from cyanobacteria and some group of algae, have recently been increased on the food industries.

# 7. MICROORGANISMS IN POLLUTION CONTROL

Microorganisms play an important role in pollution control by treatment of polluted soils, waters by bioremediation, waste water treatment, biofiltration, biosorption and bioaccumulation.

## 7.1. Bioremediation

Bioremediation, also known as biological treatment or biotreatment, uses microorganisms including bacteria and fungi to biologically degrade hydrocarbon-contaminated waste, pesticide contaminated soils or waters into nontoxic residues. These microorganisms are naturally occurring and ubiquitous bacteria, fungi, and other one-celled creatures that eat some types of contaminants, especially petroleum hydrocarbons (gas and oil). Microbes use it as a food source. Thus, Bioremediation is the completely safe and natural process of cleaning up organic contaminants through the use of microbes.

Active principle involved in bioremediation is the microbial metabolism which refers to all the chemical reactions that happen in a cell or organism. All living processes are based on a complex series of chemical reactions. Metabolic processes fall into two types – those that build complex molecular structures from simpler molecules, called anabolism and those that breaks down complex molecules into simpler molecules, called catabolism. Chemicals present in contaminated sites can be remediated through either, or both, of these processes.

Microorganisms involved in bioremediation can be isolated from almost any environmental conditions and can be divided into the following groups:

*Aerobic:* These live in the presence of oxygen. Examples of aerobic bacteria recognized for their degradative abilities are *Pseudomonas, Alcaligenes, Sphingomonas, Rhodococcus,* and *Mycobacterium.* These microbes have often been reported to degrade pesticides and hydrocarbons, both alkanes and polyaromatic compounds. Many of these bacteria use the contaminant as the sole source of carbon and energy.

*Anaerobic:* These live in the absence of oxygen. Anaerobic bacteria are not as frequently used as aerobic bacteria. There is an increasing interest in anaerobic bacteria used for bioremediation of polychlorinated biphenyls (PCBs) in river sediments, dechlorination of the solvent trichloroethylene (TCE) and chloroform.

*Ligninolytic fungi*: Fungi such as the white rot fungus *Phanaerochaete chrysosporium* have the ability to degrade an extremely diverse range of persistent or toxic environmental pollutants. Common substrates used include straw, saw dust, or corn cobs.

*Methylotrophs*: Aerobic bacteria that grow utilizing methane for carbon and energy. The initial enzyme in the pathway for aerobic degradation, methane mono oxygenase, has a broad substrate range and is active against a wide range of compounds, including the chlorinated aliphatics trichloroethylene and 1,2-dichloroethane.

The objective of biotreatment is to accelerate the natural decomposition process by controlling oxygen, temperature, moisture, and nutrient parameters. Land application is a form of bioremediation that take place in more intensively managed programs, such as composting and bioreactors. Advantages of biological treatment are: it is relatively environmentally benign; it generates few emissions; wastes are converted into products; and it requires minimal, if any, transportation. Sometimes, bioremediation is used as an interim treatment or disposal step, which reduces the overall level of hydrocarbon and pesticide contamination prior to final disposal. Bioremediation can create a drier, more stable material for land filling, thereby reducing the potential to generate leachate. Depending on the composition of the hydrocarbon components, the bioremediation environment, and the type of treatment utilized, bioremediation may be a fairly slow process .

In composting, wastes are mixed with bulking agents such as wood chips, straw, rice hulls, or husks to increase porosity and aeration potential for biological degradation. The bulking agents provide adequate porosity to allow aeration even when moisture levels are high. To increase the water-holding capacity of the waste-media mixture, and to increase trace nutrients, manure or agricultural wastes may be added. Adding nitrogen- and phosphorus-based fertilizers and trace minerals can also enhance microbial activity and reduce the time required to achieve the desired level of biodegradation. Composting is similar to land treatment, but it can be more efficient. Also, with composting systems, treated waste is contained within the composting facility where its properties can be readily monitored. With composting, mixtures of the waste, soil and other additives may be placed in piles to be tilled for aeration, or placed in containers or on platforms to allow air to be forced through the composting mixture. To optimize moisture conditions for biodegradation, the compost mixture is maintained at 40 to 60% water by weight. Elevated temperatures (30-70°C) in compost mixtures increase microbial metabolism. Tilling the soil pile or forced aeration can help control temperature and oxygen levels. Composting in closed containers can control volatile emissions. Composted wastes that meet health-based criteria can be used to condition soil, cover landfills, and supply clean fill.

Bioreactors work according to the same aerobic biological reactions that occur in land treatment and composting, but the reactions occur in an open or closed vessel or impoundment. This environment accelerates the rate of biodegradation by allowing better control of the temperature and other conditions that affect the biodegradation rate. In a bioreactor, nutrients are added to a slurry of water and waste and air sparging or intensive mechanical mixing of the reactor contents provides oxygen. To accelerate system start-up, introduction of microbes capable of degrading the organic constituents of the waste may be useful. Many of the additives used for bioreactors are common agricultural products and plant or animal wastes. After the desired treatment level has been reached, and depending on the constituents, liquids may be reused, transported to wastewater treatment facilities, injected, or discharged.

Solids may be buried, applied to soils, used as fill, or treated further to stabilize components such as metals. In tank-based bioreactors, operating conditions (temperature, nutrient concentration, pH, oxygen transport and mixing) can be monitored and controlled easily. Optimized biological processes ensure the best rate of biodegradation and allow for reduced space requirements.

The conventional techniques used for remediation have been to dig up contaminated soil and remove it to a landfill, or to cap and contain the contaminated areas of a site. The methods have some drawbacks and a better approach is to completely destroy the pollutants if possible, or at least to transform them to innocuous substances. Bioremediation is an option that offers the possibility to destroy or render harmless various contaminants using natural biological activity. As such, it uses relatively low-cost, low-technology techniques, which generally have a high public acceptance and can often be carried out on site.

Different techniques *in situ* or *ex situ,* are employed depending on the degree of saturation and aeration of an area. *In situ* techniques are defined as those that are applied to soil and groundwater at the site with minimal disturbance. While *ex situ* techniques are those that are applied to soil and groundwater at the site which has been removed from the site via excavation (soil) or pumping (water).

There are many advantages of bioremediation. i) It is a natural process and is therefore perceived by the public as an acceptable waste treatment process for contaminated material such as soil. Microbes able to degrade the contaminant increase in numbers when the contaminant is present; when the contaminant is degraded, the biodegradative population declines. The residues left after the treatment are usually harmless products and include carbon dioxide, water, and cell biomass, ii) Theoretically, bioremediation is useful for the complete destruction of a wide variety of contaminants. Many compounds that are legally considered to be hazardous can be transformed to harmless products. This eliminates the chance of future liability associated with treatment and disposal of contaminated material, iii) Instead of transferring contaminants from one environmental medium to another, for example, from land to water or air, the complete destruction of target pollutants is possible, iv) Bioremediation can often be carried out on site, often without causing a major disruption of normal activities. This also eliminates the need to transport quantities of waste off site and the potential threats to human health and the environment that can arise during transportation, v) Bioremediation can prove less expensive than other technologies that are used for clean-up of hazardous waste.

## 7.2. Wastewater treatment

Wastewater treatment facilities simply comprises the organic decomposition processes which take place in nature. This is performed by a combination of physical, biological, and chemical treatment stages. Nature (receiving waters) can only accept small amounts of sewage before becoming polluted, that is, natural bacteria feed on the sewage organics and create an abnormal amount of dissolved oxygen uptake. Dissolved oxygen which exists in minute amounts, is required by all marine life for survival. One of the principle objectives of wastewater treatment is to prevent as much of this 'oxygen-demanding' organic material as possible from entering the receiving water. Other objectives of wastewater treatment include removal of objectionable items, nutrients and heavy metals.

Process of wastewater treatment involves several stages. i) Primary treatment involves the very first stage or pretreatment, for removal of larger materials and grit that if not removed could hinder subsequent treatment processes. It is accomplished through the use of equipment such as bar screens, macerators, comminutors, racks and grit removal systems, ii) Secondary treatment involving many biological processes in the treatment operation with the majority being variations of fixed film and mixed culture applications. The activated sludge process (mixed) is achieved by establishing large diversified cultures of bacteria. The bacteria metabolizes and provides the enzymatic breakdown of organic components, ie., liquids, carbohydrates, proteins and cellulose, in the wastewater. 90% removal rates of B.O.D. and suspended solids are typical of secondary treatment, iii) Tertiary treatment comprising a three stage process which further improves effluent quality after the primary and secondary treatment phases. Usually 95-98% removal rates are achieved by this final treatment.

## 7.3. Biofilteration

Biofiltration is a relatively new pollution control technology. It is an attractive technique for the elimination of malodorous gas emissions and of low concentrations of volatile organic compounds (VOCs).The most common style biofilter is just a big box. Some can be as big as a basketball court or as small as one cubic yard. A biofilter's main function is to bring microorganisms into contact with pollutants contained in an air stream. The box that makes up this biofilter contains a filter material, which is the breeding ground for the microorganisms. The microorganisms live in a thin layer of moisture, the "biofilm", which surrounds the particles that make up the filter media. During the biofiltration process, the polluted air stream is slowly pumped through the biofilter and the pollutants are absorbed into the filter media. The contaminated gas is diffused in the biofilter and adsorbed onto the biofilm. This gives microorganisms the opportunity to degrade the pollutants and to produce energy and metabolic byproducts in the form of $CO_2$ and $H_2O$. This biological degradation process occurs by oxidation, and can be written as follows:

$$\text{Organic Pollutant} + O_2 \rightarrow CO_2 + H_2O + \text{Heat} + \text{Biomass}$$

## 7.4. Biosorption and bioaccumulation

Biosorption refers to the passive uptake of metals by microbial cells. This sorption is passive in that no energy is required. Indeed, the biomass in biosorption is often non viable, some processes relying on nonviable or deliberately killed cells, others making use of a biomass that has a high percentage of viable cells. Such binding of inorganic pollutants is being used, or is being considered for use, with metals, metalloids, and radionuclides. Biosorption is characteristically rapid, and with the appropriate type of biomass, it can remove a high percentage of individual metallic cations from waste streams. The organisms that make up the biomass may be bacteria, filamentous fungi, yeasts, or algae. The choice of organisms is particularly important because of wide differences in their capacity for sorption or their affinity for the metal. Toxicity of the metal to microorganisms is not a disadvantage in biosorptive processes because a non viable biomass may then be used.The biosorption of U, Zn, Pb, Cd, Co, Ni, Cu, Hg, Th, Zn, Cs, Au, Ag, Sn, and Mn has been investigated. These studies have shown that extent of sorption varies markedly with the metal as well as with the microorganisms. A notable illustration of the use of biosorption is the treatment of the heavy metal-containing wastewater from an operating gold mine, which each day discharges 15 million liters. The system for removal of the toxic cations involves a series of rotating biological contactors or large disks on which a biofilm of *Pseudomanas sp.* has become attached. This biomass serves to reduce the concentration of toxic metallic cations, chiefly Cu and Fe, in solution by > 95% and to provide a discharge into a stream that following the biosorption, is capable of supporting a fishery. In some instances, the uptake of metallic cations is an energy-requiring process, which thus is associated with viable and actively metabolizing microorganisms. Such a process is termed bioaccumulation.

## 8. MICROORGANISMS AS BIOINDICATORS

Bioindicators are organisms, such as lichens and bacteria, that are used to monitor the health of the environment. The organisms are monitored for changes that may indicate a problem within their ecosystem. The changes can be chemical, physiological, or behavioural. Each organism within an ecosystem has the ability to report on the health of its environment. Bioindicators are used to i) detect changes in the natural environment, ii) monitor for the presence of pollution and its effect on the ecosystem in which the organism inhabit, iii) monitor the progress of environmental cleanup and iv) test substances, like drinking water, for the presence of contaminants. There are several types of bioindicators and their uses:

*Plant indicators :* The presence or absence of certain plant or other vegetative life in an ecosystem can provide important clues about the health of the environment. Lichens, often found on rocks and tree trunks, are organisms consisting of both fungi and algae. They respond to environmental changes in forests, including changes in forest structure, air quality, and climate. The disappearance of lichens in a forest may indicate environmental stresses, such as high levels of sulfur dioxide, sulfur-based pollutants, and nitrogen.

*Animal indicators :* An increase or decrease in an animal population may indicate damage to the ecosystem caused by pollution. For example, if pollution causes the depletion of important food sources, animal species dependent upon these food sources will also be reduced in number. In addition to monitoring the size and number of certain species, other mechanisms of animal indication include monitoring the concentration of toxins in animal tissues, or monitoring the rate at which deformities arise in animal populations.

*Microbial indicators :* Microorganisms can be used as indicators of aquatic or terrestrial ecosystem health. Found in large quantities, microorganisms are easier to sample than other organisms. Some microorganisms will produce new proteins, called stress proteins, when exposed to contaminants like cadmium and benzene. These stress proteins can be used as an early warning system to detect low levels of pollution.

Specific physiological and behavioural changes in bioindicators are used to detect changes in environmental health. The specific changes differ from organism to organism. The use of organisms as bioindicators encompasses many areas of science. Wildlife conservation genetics is an example of how traditional approaches can be combined with emerging. biotechnologies to improve accuracy, and to collect information not available through conventional methods. For example, bioluminescent bacteria are being used to test water for environmental toxins. If there are toxins present in the water, the cellular metabolism of the bacteria is inhibited or disrupted. This affects the quality or amount of light emitted by the bacteria. Unlike traditional tests, this one is very quick – taking from five to 30 minutes to complete. However, it only indicates the presence of a toxin and cannot identify the specific toxin causing the change in the organism. Wildlife conservation genetics combines traditional monitoring of wildlife populations, like racoons, with the scientific discipline of genetics, to gain information about the health of ecosystems. Behavioural and population changes in a species can be observed by scientists, but physiological changes must be detected using special tests. Bioassays require samples from organisms to detect changes in the environment. These tests may be used to ensure drinking water safety or to measure river health. In the future, as research identifies new ways to use microbes, these uses will expand to include testing of soil and air. Bioassays can be carried out in traditional ways and with new biotechnology-derived methods.

## 9. MICROORGANISMS IN TEXTILE INDUSTRY

Microorganisms have impacted the textiles industry through the development of more efficient and more environmentally friendly manufacturing processes, as well as through the design of improved textile materials. Some of their key roles have involved the implementation, production, and modification of enzymes for the improvement of textile manufacturing processes and these have also facilitated the production of novel and biodegradeable fibres from biomass feedstocks.

Microbial enzymes are used to treat and modify fibres, particularly during textile processing and in caring for textiles afterwards. For example, enzymes called catalases are used to treat cotton fibres and prepare them for the dyeing processes. Some bacterial enzymes are used to separate the tough stem of the flax plant from the flax fibres used in textiles. By degrading surface fibres, many enzymes, including some cellulases and xylanases, are used to finish fabrics, give jeans a stonewashed effect, or help in the tanning of leathers. A recombinant enzyme called laccase, made by certain fungi, may also be used to treat fabrics and even catalyse the synthesis of some synthetic fibres. There are even enzymes in regular laundry detergent to help break down dirt, clean clothes more effectively, and prevent the dulling of fabric colours. Enzymes are frequently used in laundry detergents. Without them, very high temperatures and mechanical shaking would be required to effectively clean clothes and other textiles. The main enzyme types used in laundry detergents are briefly described in Table 22.

The main applications of microbial enzymes in textile industry are discussed below:

*De-sizing of cotton :* Untreated cotton threads can break easily when being woven into fabrics. To prevent this breakage, they are coated with a jelly-like substance through a process called sizing. However, after the threads have been woven into fabrics, the agents

**Table 22** Applications of microbial enzymes in the textile industry

| Enzyme type | Sources | Brief description |
|---|---|---|
| Protease | *Bacillus licheniformis*<br>*B. amyloliquefaciens* | Breaks down protein-based dirt. May be made resistant to oxidation through protein engineering. |
| Amylase | *Bacillus licheniformis* | Breaks down starch-based dirt. May be made resistant to oxidation through protein engineering. |
| Lipase | *Aspergillus oryzae* | Breaks down fats and oils. Most are not sufficiently stable in washing machines to be very useful. Improved enzymes are being developed through genetic screening and modification. |
| Other | Various | Examples include:<br>Lipoxygenase and glucose oxidase enzymes, which produce hydrogen peroxide.<br>Cellulases, which degrade surface fibres to prevent colour fading and fuzzing. |

needed to further finish the material cannot adhere to the jelly-coated fabrics. Thus, the protective sizing agents must be removed by a process called de-sizing. Bacterial amylases from *Bacillus* spp. are widely used in de-sizing, as they do not weaken or affect cotton fibres, nor do they harm the environment.

*Retting of flax:* Flax plants are an important source of textile fibres. Useful flax fibres are separated from the plant's tough stems through a process called retting. Traditional retting methods consume large quantities of water and energy. Bacteria, which may be bred or genetically engineered to contain necessary enzymes, can be used to make this a more energy efficient process.

*Breakdown of hydrogen peroxide:* When cotton is bleached, a chemical called hydrogen peroxide, which can react with other dyes, remains on the fabric. Catalase enzymes specifically break down hydrogen peroxide and may be used to remove this reactive chemical before further dyeing.

*Biostoning and biopolishing :* Instead of using abrasive tools like pumice stones to create a stonewashed effect or to remove surface fuzz, microbial cellulase enzymes may be used to effectively stonewash and polish fabrics without abrasively damaging the fibres.

*Detergents:* Enzymes allows detergents to effectively clean clothes and remove stains. They can remove certain stains, such as those made by grass and sweat, more effectively than enzyme-free detergents. Without enzymes, a lot of energy would be required to create the high temperatures and vigorous shaking needed to clean clothes effectively. Enzymes used in laundry detergents must be inexpensive, stable, and safe to use. Currently, only protease and amylase enzymes are incorporated into detergents. Lipase enzymes break down too easily in washing machines to be very useful in detergents. However, their stability is being studied and further developed through methods such as genetic screening and modification.

*Novel Fibres:* Synthetic fibres made from renewable sources of biomass are environmentally sustainable, and are becoming increasingly economically sustainable. Biodegradable synthetic polymers include novel fibres such as polyglycolic acid and polylactic acid, which are made from natural starting materials. Not all novel fibres are synthetic; they may also be naturally derived. Some natural biological fibres come from basic materials found in nature, including chitin a type of sugar polymer found in crustaceans, collagen – a type of protein found in animal connective tissue and alginate – a type of sugar polymer found in certain bacteria. A prime example of a synthetic biomass fibre is polylactic acid (PLA), which is made by fermenting corn starch or glucose into lactic acid, and then chemically transforming it into a polymer fibre. With properties similar to other synthetic fibres, PLA based materials are durable with a silky feel, and may be blended with wool or cotton. PLA has potential applications in several areas, including the following: Textiles-clothing, fashions, upholstery, Agriculture-plant mats, tree nets, soil erosion control products, Sanitation – household wipes, diaper products, Medicine-disposable garments, medical textiles. PLA minimizes environmental waste, as it may be fully biodegraded by microorganisms under appropriate conditions into carbon dioxide and water. Unlike the non-renewable petroleum resources used to make traditional synthetic fibres, the supply of renewable corn biomass needed to make PLA is expected to surpass demand in the anticipated future. Biodegradable synthetic fibres and natural biological fibres may be used to make textiles for medical applications. Such textiles may be used in first aid, clinical, and hygienic practices (Table 23). Synthetic biomass fibres may also be used in drug delivery systems, which are designed to release drugs at a specified rate for a specified time.

**Table 23** Some important biopolymers and their uses

| Bio-Polymer | Use(s) |
|---|---|
| Polylactic acid and Polyglycolic acid | Used in sutures, absorbable wound closure products, orthopaedic repair absorbable pins, and fixation devices, as well as in tissue engineering structures |
| Chitin | Incorporated into wound dressings |
| Collagen | Uses in cell engineering structures, such as in artificial skin, or even as surgeon´s thread |
| Alginate | Used to protect and interact with wounds |

## 10. MICROORGANISMS IN OIL RECOVERY

Microbial enhanced oil recovery (MEOR) is the use of microorganisms to retrieve additional oil from existing wells, thereby enhancing the petroleum production of an oil reservoir. In this technique, selected natural microorganisms are introduced into oil wells to produce harmless by-products, such as slippery natural substances or gases, all of which help propel oil out of the well. Because these processes help to mobilize the oil and facilitate oil flow, they allow a greater amount to be recovered from the well. MEOR is used in the third phase of oil recovery from a well, known as tertiary oil recovery. Recovering oil usually requires two to three stages, which are briefly described as follows:

*Stage 1: Primary Recovery :* 12% to 15% of the oil in the well is recovered without the need to introduce other substances into the well.

*Stage 2: Secondary Recovery :* The oil well is flooded with water or other substances to drive out an additional 15% to 20% more oil from the well.

*Stage 3: Tertiary Recovery :* This stage may be accomplished through several different methods, including MEOR, to additionally recover up to 11% more oil from the well.

The microorganisms used in MEOR can be applied to a single oil well or to an entire oil reservoir. They need certain conditions to survive, so nutrients and oxygen are often introduced into the well at the same time. MEOR also requires that water be present. Microorganisms grow between the oil and the well's rock surface to enhance oil recovery by the following methods:

*Reduction of oil viscosity:* Oil is a thick fluid that is quite viscous, meaning that it does not flow easily. Microorganisms help break down the molecular structure of crude oil, making it more fluid and easier to recover from the well.

*Production of carbon dioxide gas:* As a by-product of metabolism, microorganisms produce carbon dioxide gas. Over time, this gas accumulates and displaces the oil in the well, driving it up and out of the ground.

*Production of biomass:* When microorganisms metabolize the nutrients they need for survival, they produce organic biomass as a by-product. This biomass accumulates between the oil and the rock surface of the well, physically displacing the oil and making it easier to recover from the well.

*Selective plugging:* Some microorganisms secrete slimy substances called exopolysaccharides to protect themselves from drying out or falling prey to other organisms. This substance helps bacteria plug the pores found in the rocks of the well so that oil may move past rock surfaces more easily. Blocking rock pores to facilitate the movement of oil is known as selective plugging.

*Production of biosurfactants:* Microorganisms produce slippery substances called surfactants as they breakdown oil. Because they are naturally produced by biological microorganisms, they are referred to as biosurfactants. Biosurfactants act like slippery detergents, helping the oil move more freely away from rocks and crevices so that it may travel more easily out of the well.

## 11. MICROORGANISMS IN BIOMINING

Biomining is the use of microorganisms to extract metals and minerals from ores in the mining process. Ores of high quality are rapidly being depleted and biomining allows environmentally friendly ways of extracting metals from low-grade ores (ores that have small amounts of valuable metals scattered throughout). Biomining includes two different chemical processes called bioleaching and biooxidation.

Bioleaching is the use of microorganisms to extract metals from an ore. Leaching is the process of removing a soluble substance from a solid structure by making it into a liquid form easy for extraction. In chemical leaching, this is done by treating the soluble substance with a chemical solution while in bioleaching, it is done through contact with bacteria. Bioleaching is used commercially in the extraction of copper.

Biooxidation uses microorganisms, not to extract metals, but to make the metals ready for extraction. Oxidation is the chemical reaction in which an element is changed by the addition of oxygen. Rust is an example of the oxidization of iron. Biooxidation is mainly used in gold mining. Gold is often found in ores with gold particles scattered throughout, called refractory ores, and the small particles of gold are covered by insoluble minerals. These minerals make the extraction difficult. Therefore, microorganisms that can "eat away" at the mineral coating are used to pre-treat the gold ores before they can be extracted. Bioleaching of copper, and biooxidation of refractory gold ores are the only well-established large scale processes that are commercially carried out today.

Biooxidation and bioleaching processes are usually done in heaps of ground ore. The low-grade ores are ground into powder and piled in an irrigated outdoor facility. The heaps are then treated with an acidic liquid that contains a fraction of the bacterial population required since some are naturally existing within the ore. The liquid with the metals extracted are then pumped into another section where the metal is recovered. Biomining in heaps has the advantage of being simple and cheap to perform. However, it has disadvantages such as the lack of complete control over the conditions of biomining, leading to unpredictable or inefficient extraction rates. For this reason, biomining is also done in reactors. Reactors are a series of tanks the ore mixture is treated in, and offers control of factors important in biomining, such as temperature and pH levels.

## 12. MICROORGANISMS IN GEOLOGICAL MICROBIOLOGY

The complex, large-scale investigations in the field of geological microbiology, which have been carried out for a long time, allowed a number of problems of practical importance to be solved. Methods were developed for the assessment of the rates of photosynthesis; bacterial chemosynthesis; and processes of the cycles of carbon, sulfur, and a number of other elements in various ecotopes; these methods find wide application in ecological studies, including those related to the production rates in water bodies, toxicant degradation, and monitoring of natural environments in connection with the progressive increase of the anthropogenic load, $H_2O_2$ and amino acids. The investigation of the physiology of microorganisms and their geochemical activity in the deposits of minerals promoted the establishment of biogeotechnology, a science dealing with microbial methods of the extraction of oil and processing of ores and concentrates. Thus, geomicrobiology has become biogeochemistry presently makes it possible not only to a large division of modern natural science but also to control the direction and rate of microbially driven processes under *in situ* conditions with the aim to develop new biotechnologies for bioremediation of the environment and for extraction and processing of minerals. At the modern stage of geomicrobiology development, attention is paid not only to the roles in the cycles of elements of separate functional groups of microorganisms but also to the cooperation observed in the natural multicomponent microbial communities and to their interactions with plants and animals. These investigations often employ not only microbiological and biogeochemical methods but also methods of molecular ecology. Further progress of geomicrobiology can be expected to occur along the following directions: i) Physiological–biochemical diversity of microorganisms : Detailed investigations of physiological–biochemical and molecular biological peculiarities of microorganisms involved in the key reactions of the global cycles of elements and isolation of new organisms active in these processes, ii) Microbial mineralogy : Experimental studies of the mechanisms by which microorganisms participate in the formation and destruction of rocks and their component minerals, iii) Microbial biogeochemistry : Assessment of the results of microbial activity in modern natural ecosystems of various levels of complexity, up to the biosphere level, iv) Microbial biogeotechnology. Development of a theoretical basis for the control of natural microbial processes of environment bioremediation and of formation and destruction of the deposits of minerals with the aim to minimize the negative effect of industrial and agricultural activities and develop more ecologically safe and less expensive methods for extraction and processing of mineral resources. It should be emphasized that recognition of the leading role of microorganisms in the global cycles of elements in the modern biosphere led to the origination of a new field of science at the interface of biology and geology, namely, of microbial paleontology, which studies the role of microorganisms in the formation and evolution of the biospheres of the past geological epochs of our planet.

## 13. MICROORGANISMS IN SPACE RESEARCH PROGRAMS

Microorganisms are natural constituents of the environment, existing in air, water, soil, and biotic systems. Microorganisms readily adapt to changes in environmental variables such as nutrient levels, temperature, oxygen concentration, atmospheric pressure, weightlessness, and light intensity, and exhibit and variety of physiological and morphological changes. Bacterial growth is defined as the "coordinated summation of a complex array of processes including chemical synthesis, assembly, polymerization, biosynthesis, fueling, and transport, the consequence of which is the production of new cells". It is possible that the conditions of space flight may alter one or more of these processes. Microorganisms can be introduced into the spacecraft through several avenues. Crewmembers will be a primary source of microorganisms in confined space habitats, as the healthy human body can host at least 50 microbial species. Biological payloads, resupply vehicles, hardware and supplies will provide additional sources of microorganisms. In the confined spacecraft environment, increases in interpersonal exchange of potential pathogens may facilitate infectious conditions. Microbial metabolism of metals, electronic components, fuels, or other spacecraft components may result in production of volatile organic compounds.

The presence of microorganisms and their potential effects has been a source of study since the advent of manned spaceflight. The unique aspects of the spaceflight environment, particularly the microgravity condition, have been shown to effect microbial and human physiology in many ways. Changes in microbial growth rates, survival, phage-induction, and antibiotic susceptibility have been observed. Effects of spaceflight on human physiology in such areas as cardiac, musculoskeletal, neurological, and immune function have been well documented. The unique microbial ecology of the spacecraft environment has also been the focus of many studies. Formation of an indigenous population within the closed environment of the spacecraft has been demonstrated. As we prepare for the longer duration spaceflights necessary to enter the era of manned planetary exploration, it is critical that we develop a better understanding of the changes that may be induced in the host-microbe relationship in the unique environment of spaceflight. Development of countermeasures to undesirable microbial interactions with the spacecraft and crewmembers is an important part of current research efforts. The study of microbial impacts on humans and spacecraft will continue to be a vital part of manned space exploration. Studies of spaceflight effects on microbial physiology, human physiology, and microbial ecology have generated many answers, as well as many more questions, about the nature of the host-microbe interaction in this unique environment.

## 14. MICROORGANISMS IN NATURAL TECHNOLOGY

Natural technology involves a combination of various naturally occurring microorganisms mostly used or found in foods, known as effective microorganisms (EM). EM technology can be considered a natural technology and has no known adverse effects on plants, animals, humans, or the environment after over a decade of application. EM is useful in a wide variety of fields. EM consists of a wide variety of effective, beneficial and nonpathogenic microorganisms of both aerobic and anaerobic types coexisting. It is produced through a natural process of fermentation and not chemically synthesized or genetically engineered. Professor Higa started his development of EM in 1968 with the first batch of what would eventually be called EM produced in 1982 and thereafter further developed and refined. His intention, at first, was to find an alternative to the use of agricultural chemicals. It has now spread into applications in the environment, industrial and health fields. However, and it is stressed that, it is not a synthetic chemical nor is it a medicine. In agriculture, EM has been used to enrich the soil and produce quality, healthy crops at a greater yield with decreases in pest, diseases, and the need for weeding and tilling and without the use of agricultural chemicals. In animal husbandry, EM has been used with noticeable decreases in foul smells, in the appearance of sickness and insect infestations, noticeable increase in fertility from artificial insemination, and increase of the quality of meat, dairy, and eggs. In the environment, EM has been used to clean up polluted waters in ponds, lakes, dams, and seashores, including in the cleanup of oil spills; make possible the recycling of water from sewage facilities into use in general cleaning; and the recycling of organic waste into quality fertilizer. In industrial uses, EM applied to cement mixing gave a measured rise to the cement's strength; and EM has been used in the plastics and metals waste separation facility to reduce the level of toxic fume emissions. If put in very simple words, EM lives off our waste while we off "their waste". Their waste simply translates to a healthy environment for us. Additionally, EM seems to induce strong antioxidation, that is, it encourages the presence of antioxidants and suppress the action or prevent the proliferation of active oxygen, also known as free radicals. Therefore, the presence of EM seems to help prevent the corrosion of inorganic materials, such as rusting, and help organic matter towards fermentation as opposed to putrefaction. A mutual existence can be enhanced between microorganisms and humanity by the utilization of EM to our environment from agricultural fields to households and in our polluted waters and soils. EM only creates the condition for best results, that is, the users should nurture the condition and provide the resources for EM to perform optimally. Microorganisms exists naturally throughout the environment from rock crevices to our internal organs. In our present day environment, putrefactive microorganisms, those types responsible for the rotting of organic matter to catalyzing the deterioration of inorganic matter, dominate much of this sphere. It has the potential to create an environment most suitable for the existence, propagation, and prosperity of life not in conflict with what we, humanity, consider healthy and hygienic. EM is available in a concentrated form. It is mainly diluted in water anywhere from 1:100 to 1: 10,000 depending on use, such as, irrigating farmland, in drinking water for animals, as a household air spray, in general cleaning, in treating waste water, in manufacturing processes, and so forth. EM is also used in mixing cement and in making ceramics, as well as, in treating and recycling plastics. EM is also mixed with organic matter or organic liquids for certain purposes as making fertilizer and treating organic waste. EM is just a liquid of effective microorganisms. It is not a fertilizer nor does it have mineral or nutritional value. It is used to create an advantageous condition, for instance, it is used to inoculate or treat soil to improve its microflora and make organic materials into fertilizers. All in all, EM is widely applicable product that can make a considerable change for the better, from the land onto which it is applied, to the waters that are influenced by runoffs, to the air affected by industrial emissions, and to human health affected by the foods consumed, the water drank, and air breathed.

## 15. CONCLUSIONS AND FUTURE OUTLOOK

Over millions of years of evolution, we humans have worked out a mutually beneficial partnership with the microbes that come to inhabit our guts. In return for their aid in digestion, we gave them a stable, protected home and plenty of nutrients via the food we eat. We need them as much as they need us. Microbes break down food molecules which our body's enzymes and acids can't dissolve, helping us squeeze all the nutrients out of our food. Some make valuable vitamins that our body needs. Many microbial species have proved to be consummate evolutionary wheelers and dealers, arranging collaborations, mergers and acquisitions that usually serve both partner well. In due course of time many new developments took place especially in the field of science and technology which enabled us to understand various microbial processes like their ecological diversities and interrelationships, physiology, biochemistry, metabolism, reproduction etc. more accurately and influentially. Consequently, this resulted in strengthening of our knowledge of various microbial machineries by the virtue of which could able to used them and exploit their potentialities for various industrial purposes as in food industry for production of food additives, proteins, enzymes for food proccessings; in pharmaceutical industry, for development of health care products like antibiotics, probiotics, prebiotics, vaccines, growth stimulating substances; in agriculture industry for obtaining high yield by employing them as microbial inoculants, biofertilizers and biocontrol agents; in environment monitoring by using them as natural bioremediators for waste management and as bioindicators of environmental pollution; in biogeochemical cycling for maintaining ecological balance and nurturing forest productivity; in energy research programs by using them as non conventional energy source by producing liquid and gaseous biofuels and generating microbial electricity; in biomining and oil recovery processes for obtaining high grade minerals and petroleum products respectively; in various space research programs for studying and evaluating vital processes related to human physiology by using them as research models; and as a tool for various biotechnological purposes for obtaining high quality, efficient and desired products for human use.

The 21st century has dawned with renewed hope for a better livelihood for the population of this earth and the science of microorganisms has also grown up and well established in the form of microbial biotechnology by its robust performance in the commercial and industrial sectors. This is primarily due to the blending in classical , basic microbiology with several new techniques that are likely to influence the development of the bioindustries of future with lot of applications. Therefore, for the long term harness of microbial potentialities, the first and foremost need is to explore, document and tap the natural existing microbial wealth for tommorows development. This requires the conservation of diversity of natural microbial germplasm and their patenting. Inspite of having a lot many discoveries, inventions and developments in microbial science, a serious attention is required to use microbial labour and capabilities more efficiently in various sectors. The first, important among them is health sector, which still need to develop newer, effective, products of health care for combating diseases like AIDS, tuberculosis, and polio; second, in food sector, there is need to put more efforts on encircling and developing microbial source of nutraceuticals (nutrient foods with health promoting substances); then in agriculture, stress should be given on development and production of safe genetically modified crops and other food commodities; attempts also required in bioenergy research programs especially towards the generation of microbial electricity. Above all, the indiscriminate use of microbial biotechnology should be avoided. It should be used only for development of sustainable products as its exposure/leakage into the environment may cause unexpected, undesired, harmful effects. Therefore, there should be a stringent legal control over malpracticing of microbial biotechnology. There should be the proper provision of law for regulation and monitoring of manufacturing, importing/exporting and marketing of various biotechnological products of microbial origin. Finally, the advantages of such technologies should reach to the common people so that they can use it for their sustainable development.

## 16. FURTHER READING

Andlaur, W. and Furst, P. (2002). Nutraceuticals : Piece of history present status and outlook. Food Research International 35 : 171-176.

Arora, D.K. , Mukerji, K. G. and Marth, E. H. (1991). Hand book of applied Mycology, Vol III Foods and Feeds. Marcel Dekker, Inc. New York

Banat, I.M., Makkar, R.S., and Cameotra, S.S. (2000). Potential commercial applications of microbial surfactants. Applied Microbiology and Biotechnology 53 : 495-508.

Bull, T.A. (2004). Microbial Diversity and Bioprospecting. ASM Press, Washington.

Chang, S.T. and Miles, P.G. (1991). Edible Mushrooms and their Cultivation. CRC Press,

Boca, Raton Fl.

Goldberg, I. and Williams, R. (1991). Biotechnology and Food Ingredients. Van Nostrand Reinhold, New York.

Huis In't Veld, J.H.J., Bosschaert, M.A.R. and Shortt, C. (1998). Health aspects of probiotics. Food science & Technology Today, 12 : 46-49.

Jjemba, P.K. (2004). Environmental Microbiology: Principles and Applications. Science Publishers, Enfield, New Hampshire, USA.

Jones, D.G. (1993). Exploitation of Microorganisms. Chapman and Hall, London

Koths, K. (1995). Recombinant proteins for medical use : the attractions and challenges. Current Opinion in Biotechnology 6 : 681-687.

Marks, T. and Sharp, R. (2000). Bacteriophages and Biotechnology : A Review. Journal of Chemical Technology & Biotechnology.

Moo-Young, M., Anderson, W.A. and Chakrabarty, A.M. (1996). Environmental Biotechnology : Principles and applications, Kluwer academic Press, Boston.

Ogawa, J. and Schimizu, S. (1999). Microbial enzymes : New industrial applications from traditional screening methods. Trends in Biotechnology 17 : 13-19.

Phillips, M.J. (2004). Pharmaceutical and Medicine Manufacturing. Indian Journal of Pharmacology. 44 : 232-255.

Rahi, D.K., Rajak, R.C. and Pandey, A.K. (2005). Mushroom Nutriceuticals : An Emerging Health Care Aid. In: Frontiers in Plant Sciences. (KG Mukerji, KVBR Tilak, S M Reddy, LV Gangwane, P Prakash, and I K Kunwar, Eds). I. K. International Pvt Ltd. New Delhi. pp. 481-496.

Soni. S.K. and Arora, J.K. (2000). Indian fermented foods: Biotehnological approaches. In: Food processing: Biotechnological approaches (S.S. Marwaha and J.K. Arora, Eds), Asiatech Publishers Inc., New Delhi. pp. 143-189.

Varma, A. and Behera, B. (2003). Green Energy: Biomass Processing and Technology. Capital Publishing Company, New Delhi

Walsh, G. (1998). Biopharmaceuticals : Biochemistry & Biotechnology. Wiley, Chichester.

□□□

# 3

# Microbial Ecology and Pollution Management

**S.K. SONI[1], D.K. RAHI[1] AND R. SONI[2]**

[1] *Department of Microbiology, Panjab University, Chandigarh-160 014*

[2] *Department of Biotechnology, D.A.V. College, Chandigarh - 160 011*

# 1. INTRODUCTION

Microorganisms have lived on Earth for billions of years and are essential to our very existence. They are ubiquitous and represent a huge biological diversity in soil, marine and aquatic ecosystems. A variety of microbes are found in common environments such as soil, water and air as well as exotic locations as diverse as deep sea hydrothermal vents and soda lime lakes. These are essentially bags of enzymes which along with minerals are some of the most reactive, catalytic surfaces in soils and sediments of the Earth's crust. Microbes crave energy, and while acquiring it, they chemically transform many important elements through oxidation and reduction reactions. Consequently, bacteria are responsible for most of the global decomposition of organic matter and the production of greenhouse gases. The biogeochemical cycles of the majority of biological elements are controlled over geologic time by microbially-mediated processes. A variety of microorganisms are found in the diverse environment and their number and types vary significantly and determined by various factors. Daubenmire (1947) defined the term ecology as "the study of the reciprocal relations between organisms and their environment". Thus Microbial Ecology is the study of microorganisms in their natural home or habitat including soil, water and air. The microorganisms in their natural habitats can be classified into various ecological groups based on the basis of oxygen requirement, use of carbon source for energy generation, temperature, habitat, extreme condition of the environment, and mode of nutrition. Following is the summary of various ecological groups of microorganisms :

- Based on oxygen requirement: Aerobes, anaerobes, facultative anaerobes.
- Based on carbon sources as energy: Autotrophs including cyanobacteria, algae and photosynthetic bacteria; Heterotrophs including majority of bacteria, all fungi, actinomycetes.
- Based on temperature: Psychrophiles including *Arthrobacter*, *Pseudomonas*, and *Flavobacterium;* Mesophiles including cyanobacteria (*Nostoc, Anabaena, Oscillatoria*), majority of spoilage bacteria, pathogenic bacteria, fungi, Amoebae; Thermophiles including *Bacillus stearothermophilus*; Hyperthermophiles including *Thermus aquaticus*, *Pyrococcus furiosus*.
- Based on habitat: Soil microorganisms, Aquatic microorganisms, Aeromicroflora.
- Microorganisms living in an extreme environment: Acidophiles including *Sulfolobus acidocaldarius;* Alkalophiles including *Natronobacterium gregoryi,* Halophiles including *Halobacterium halobium;* Osmophiles including *Schizosaccharomyces spp.*
- Based on the mode of nutrition: Saprophytes including *Aspergillus, Penicillium* and *Rhizopus;* Parasites including *Puccinia* spp., viruses; Symbionts including lichens, mycorrhiza, *Rhizobium.*

# 2. SOIL MICROBIOLOGY

Of the various natural homes of the microorganisms, soil is the richest source of microbes. The presence of many inorganic materials, organic matter in the form of plant and animal residues, water in the spaces between soil particles in which both organic and inorganic components of soil are dissolved in water and gases in the form of $CO_2$, $O_2$ and $N_2$ makes the soil as the most preferred habitat of the almost all the types of microorganisms. The number of types of microorganisms in soil differ from one place to other due to variation in soil composition. The main factors affecting microbial types and number in soil include the variation in amount and type of nutrients, moisture, degree of aeration, temperature, pH, existence of roots of plants which is known as rhizosphere and contains more amino acid and vitamin requiring bacteria than in root free zone. The average count of microorganisms in soil is billions/g. The various categories of microorganism generally found in soil include the following:

## 2.1. Bacteria

These are maximum of all types of microorganisms with their counts in the range of $1.5\times10^7 - 2.5\times10^9$/g. These include autotrophs, heterotrophs, mesophiles, thermophiles, psychrophiles, aerobes, anaerobes, cellulose digesters, sulphure oxidizers, nitrogen fixers and protein digesters. Several actinomycetes including *Nocardia, Streptomyces, Micromonospora* are also present in the range of $7\times10^5$/g. The most significant effect that procaryotes have on their environment is their underlying ability to recycle the essential elements that make up cells. The earth is a closed system with limited amounts of carbon, oxygen and nitrogen available to all forms of life. These essential elements must be converted from one form to another and shared among all living organisms. The total biomass of microbial cells in the biosphere, their metabolic diversity and their persistence in all habitats that support life, guarantee that microbes will play crucial roles in the transformations and recycling of these elements. Table 1 lists the major elements that make up a typical procaryotic cell. As expected, over 90 percent of the elemental analysis consists of carbon, hydrogen, oxygen, nitrogen, phosphorus and sulfur. These are the elements that become combined to form all the biochemicals that comprise living systems. C, H, O, N, P and S are the constituents of organic material (an organic compound is a chemical that contains a carbon to hydrogen bond). Organic compounds on earth are evidence of life. Organic compounds may be symolized as $CH_2O$, which is the emperical formula for a sugar such as glucose.) H and O are the constituents of water ($H_2O$), that makes up over 95 percent of the cell composition. Calcium ($Ca^{++}$), iron ($Fe^{++}$), magnesium ($Mg^{++}$) and potassium ($K^+$) are present as inorganic salts in the cytoplasm of cells.

The Table does not include the "trace elements" in cells. Trace elements are metal ions required in cellular nutrition in such small amounts that it is difficult to determine their presence in cells. The usual metals that qualify as trace elements are $Mn^{++}$, $Ca^{++}$, $Zn^{++}$, $Cu^{++}$ and $Mo^{++}$. Trace elements are usually built into vitamins and enzymes. For example, vitamin $B_{12}$ contains cobalt ($Co^{++}$) and the bacterial nitrogenase enzyme contains molybdenum ($Mo^{++}$). The structure and metabolism of any organism adapts it to its environment. Thus the various groups of microbes are adapted to certain environmental niches based on their predominant type of metabolism relevant to the elemental nutrients available.

**Table 1** Major elements, their sources and functions in cells of prokaryotes.

| Element | Source | % dry weight | Function |
|---|---|---|---|
| Carbon | Organic compounds or atmospheric $CO_2$ | 50 | Main constituent of cellular material |
| Nitrogen | Organic compounds or atmospheric $N_2$ | 14 | Constituent of amino acids, nucleic acids, nucleotides and coenzymes |
| Oxygen | Organic compounds, $H_2O$, $CO_2$ and $O_2$ | 20 | Constituent of cell material and cell water. Also act as ultimate electron acceptor in aerobic respiration |
| Hydrogen | Organic compounds, $H_2O$ and $H_2$ | 8 | Main constituent of organic compounds and cell water |
| Phosphorus | Inorganic phosphates | 3 | Constituent of nucleic acids, nucleotides, phospholipids, lipopolysaccharides, teichoic acids |
| Sulphur | Organic sulphur compounds, $SO_4$, $H_2S$, elemental sulphur, | 1 | Constituent of cysteine, methionine, glutathione, several coenzymes |
| Potassium | Potassium salts | 1 | Main cellular inorganic cation and cofactor for various enzymes |
| Magnesium | Magnesium salts | 0.5 | Inorganic cellular cation, cofactor for certain enzymes |
| Calcium | Calcium salts | 0.5 | Inorganic cellular cation, cofactor for certain enzymes and a component of endospores |
| Iron | Iron salts | 0.2 | Component of cytochromes and certain non-heme iron-proteins and a cofactor for certain enzymes |

The procaryotic bacteria and archaea, as a result of their diversity and unique types of metabolism, are involved in the cycles of virtually all essential elements. In two cases, methanogenesis (conversion of carbon dioxide into methane) and nitrogen fixation (conversion of nitrogen in the atmosphere into biological nitrogen) are unique to procaryotes and earns them their "essential role" in the carbon and nitrogen cycles. There are other metabolic processes that are unique, or nearly so, in the procaryotes that bear significantly on the cycles of elements. For example, procaryotes called lithotrophs use inorganic compounds like ammonia and hydrogen sulfide as a source of energy, and others called anaerobic respirers use nitrate ($NO_3$) or sulfate ($SO_4$) in the place of oxygen, so they can respire without air. Most of the archaea are lithotrophs that use hydrogen ($H_2$) or hydrogen sulfide ($H_2S$) as a source of energy, while many soil bacteria are anaerobic respirers that can use their efficient respiratory metabolism in the absence of $O_2$. The basic processes of heterotrophy are spread throughout the bacteria. Most of the bacteria in the soil and water, and in associations with animals and plants, are heterotrophs. Heterotrophy means living on dead organic matter, usually by some means of respiration (same as animals) or fermentation (same as yeast or lactic acid bacteria). Bacterial heterotrophs in the carbon chain are important in the processes of biodegradation and decomposition under aerobic and anaerobic conditions. In bacteria, there is a unique type of photosynthesis that does not use $H_2O$ or produce $O_2$ which impacts on the carbon and sulfur cycles. Meanwhile, the cyanobacteria (mentioned above) fix $CO_2$ and produce $O_2$ during photosynthesis and they make a very large contribution to the carbon and oxygen cycles. The list of examples of microbial involvement in the cycles of elements that make up living systems is endless, and probably every microbe in the web is involved in an intimate and unique way.

## 2.2. Fungi

Several hundred species of fungi are found in soil which are most abundant near the surface where aerobic conditions prevail, their average count is generally in the range of $4\times10^5$/g. The molds are aerobic organisms that utilize organic compounds for growth. They play an important role in decomposition or biodegradation of organic matter, particularly in soil. Yeasts can grow anaerobically (without oxygen) through the process of fermentation and play a role in fermentations in environments high in sugar. The prominant role of fungi in the environment is in the decomposition plant tissues including cellulose, lignin and pectin.

## 2.3. Algae and cyanobacteria

These are less than bacteria and fungi and are generally present on surface or just below the surface of soil. The common algae associated with soil include green algae and diatoms whose average count is generally in the range of $5\times10^4$/g. The algae are also an important part of the carbon cycle. They are the predominant photosynthetic organisms in many aquatic environments. The algae are autotrophs, which means they use carbon dioxide ($CO_2$) as a source of carbon for growth. Hence they convert atmospheric $CO_2$ into organic material (i.e., algal cells). Algae also play a role in the oxygen ($O_2$) cycle since their style of photosynthesis, similar to plants, produces $O_2$ in the atmosphere. The cyanobacteria are a group of procaryotic microbes, as prevalent as algae, that have this type of metabolism. Photosynthetic algae and cyanobacteria can be found in most environments where there is moisture and light.They are a major component of marine plankton which forms the basis of the food chain in the oceans. Due to photosynthetic activity they accumulate organic matter in soil - thus initiates the growth of bacteria and fungi.

## 2.4. Protozoa

Protozoans are heterotrophic organisms that have to catch or trap their own food. Therefore, they have developed elaborate mechanisms for movement and acquiring organic food which they can digest. Their food usually turns out to be bacterial cells, so one might argue that they are ecological predators that keep bacterial populations under control in soil, aquatic environments, intestinal tracts of animals, and many other environments. The average count of protozoa in soil is $3\times10^4$/g which are mostly flagellated and involved in ingestion of bacteria thus maintaining the equilibrium of microorganisms in soil.

### 2.5. Viruses

Various bacterial, plant and animal viruses are also associated with soil.

## 3. WATER MICROBIOLOGY

Water is another natural habitat of microorganisms including bacteria, algae, fungi, protozoa and viruses. Some organisms are indigenous in water while others are transient, entering from air, soil, industrial and domestic wastes. These are undesirable in water and many affect the human health and other animal's life. The number and type of microorganisms in water are also affected by certain factors including temperature, hydrostatic pressure which increases with depth with every 1 atmosphere per 10 meters, light, salinity, turbidity, pH, inorganic and organic constituents. Fresh water generally do not contain any harmful microflora but it becomes unfit for human consumption when it is mixed with wastes from domestic, industrial and agricultural or atmospheric sources. The microorganisms are distributed in water either as aggregation of floating microbial life on surface known as plankton, at bottom region as benthic microorganisms or in the gut of marine animals.

## 4. MICROBIAL ASSOCIATIONS

Microbial ecosystems include the sum of biotic and abiotic components of soil or water involving total microbial flora together with physical composition and characteristics of the natural habitats. Microbes living in natural habitats exhibit different types of associations/ interactions with other microorganisms or macroorganisms including plants and animals.

### 4.1. Neutral interactions

When two microorganisms live in the same environment without affecting each other . These utilize different nutrients and produce metabolites which are not inhibitory to other.

### 4.2. Positive interactions

When one or both members of the association are benefitted. These include commensalisms when one of the members of the association are benefitted and of mutualism where both are benefitted by the presence of the other. These include the association between some cellulose degrading fungi and non cellulolytic bacteria, association between nutritionally fastidious bacteria and less fastidious bacteria, association between *Thiobacillus ferrooxidans* ($CO_2$ fixer) & *Beijerinckia lacticogenes* ($N_2$ fixer) living in an association in an environment lacking C and N, the associations between rhizobia and roots of leguminous plants.

### 4.3. Negative interactions

When one of the members of the association is harmed by the presence of other. These include i) antagonism where one microbe makes the environment unfit for others by the production of some metabolites which are inhibitory to other, ii) competition which occurs among different microorganisms for essential nutrients and in which the best adapted ones predominate and others get eliminated due to limited nutrients, iii) parasitism where one organism lives in or on other. The one which lives in or on the other is called as a parasite while the other is known as a host. Parasite feeds on cells, tissues of host and harms it.

## 5. FUNCTIONS OF MICROORGANISMS IN SOIL & WATER

Microorganisms are essential for our existence and have specific jobs in the natural environments. They are responsible for recycling nutrients in our soils and purifying our waters. We also use microorganisms in constructed environments to serve our own functions. Some of the useful role of the microbes in the natural habitats are discussed below:

### 5.1. Nutrient recycling

When plants and animals take up nutrients, they are not available to other living organisms. When the plants and animals die, the nutrients remain in the carcass. If the dead material, or detritus, is not broken down by microbes, those nutrients will never become available to help sustain the life of other organisms. There are a finite amount of nutrients available in the environment and we cannot simply make more. The nutrients that are trapped in detritus must be released in order for life to continue.

### 5.2. Recycling of nutrients stored in organic matter to an inorganic form

Decomposition releases the mineral nutrients (e.g., N, P, K) bound up in dead organic matter in an inorganic form that is available for primary producers to use. Without this, recycling of inorganic nutrients, primary productivity on the globe would stop on land. Most of the decomposition (mineralization) of dead organic matter occurs at the soil surface, and the rate of decomposition is a function of moisture and temperature (too little or too much of either reduces the rate of decomposition). Fungi are important in terrestrial systems, but not in aquatic. They are present even before the leaves and twigs enter the soil and so decomposition starts in the living or senescent plant material. Fungi are the most important decomposers of structural plant compounds (cellulose and lignin — but note that lignin is not broken down when oxygen is absent). The fungi invade the organic matter in soils first and are then followed by bacteria.

In water, the decomposition of organic matter is mostly aerobic in streams as well as ocean and anaerobic in the bottoms of lakes or in swamps. Aerobic decomposition proceeds faster (produces higher energy yields for the bacteria) than decomposition in environments where there is no oxygen. In the open ocean, the water is so deep (average 3900 m) and contains so much oxygen, that most of the algal-formed organic matter at the surface decomposes aerobically before it reaches the bottom. For example, only 2% of the primary productivity in the upper ocean sinks to a depth of 3500 m. Most of the world is an ocean, and most of the ocean is deep, so most of the aquatic decomposition must be aerobic. But in shallow waters, coastal oceans, lakes and estuaries, 25-60% of the organic matter produced may settle out of the upper waters rapidly and be decomposed anaerobically.

Of course another important impact of decomposition besides generating inorganic nutrients is to produce $CO_2$ and $CH_4$ that is released to the atmosphere that is responsible for this nutrient recycling.

## 5.3. Nitrogen fixation from the atmosphere into a usable form

The only organisms capable of removing $N_2$ gas from the atmosphere and "fixing" it into a usable nitrogen form ($NH_3$) are bacteria. The specific bacteria that can perform N fixation are scattered throughout the groups including the cyanobacteria. All organisms that fix nitrogen use the same mechanisms and the same enzymes and this ability probably evolved only once and early in the history of life. Symbiotic $N_2$ fixation costs the plant photosynthate to support the fixation and the $NH_3$ assimilation; this cost could be from 15-30% of the total carbon assimilated by the plant. In fact, to fix one molecule of $N_2$ requires about 25 molecules of ATP, so it is expensive from the bacterial standpoint, and that means that the plant must support that energy requirement. In return the plant receives nitrogen, which may otherwise be a limiting nutrient.

Another difficulty for the bacteria is that one of the enzymes necessary for nitrogen fixation is destroyed by oxygen (which is necessary for efficient ATP formation). One solution to this problem is to form symbiotic relationships with other organisms that can provide carbohydrates: these include diatoms, the fungi of certain lichens, shipworms, termites, and certain plants especially in nodules of the roots.

## 5.4. Biodegradation

Microorganisms are responsible for getting rid of the waste generated by industry and households. They detoxify acid mine drainage and other toxins that we dump into the soil and water. The nutrients gained from the breakdown of these products then go to feed plants or algae, which in turn feed all animals.

## 5.5. Waste water treatment

When we flush things down the drain or the toilet, they go to a septic system or waste water treatment plant at the end of the line. After primary mechanical treatment and aeration microbes remove organic materials from the filthy waters that flow into these systems and, eventually, water can safely be returned to the rivers and streams. The methane produced during treatment can even be used to generate heat or electricity for the operation.

## 5.6. Generation of oxygen in the atmosphere

Almost all of the production of oxygen by bacteria on earth today occurs in the oceans by the cyanobacteria or 'blue-green algae'.

## 5.7. Deciding the fate of metal contaminants

Microbes are also involved in the weathering of rocks, the fate of metal-based contaminants in streams and rivers, and in the formation and disappearance of minerals. Sometimes bacteria must resort to scavenging the metals they need to survive. In order to cope, these bacteria send out iron-loving organic molecules called siderophores to pilfer iron from iron minerals when sufficient dissolved iron is not available. But siderophores also avidly bind other metals, including toxic ones like lead and cadmium. The potential geological consequences of indiscriminate metal binding by bacterially produced siderophores is enormous. Siderophores can release aluminum ions from the aluminum silicate mineral kaolinite.

This finding suggests that siderophores may mobilize toxic metals. But siderophores also may adsorb to mineral surfaces, trapping their toxic-metal cargo. It has been found that siderophore-mediated adsorption of metals to kaolinite can in fact occur, but the process is complex and pH-dependent. For instance, certain lead-loaded siderophores can enhance adsorption of lead to kaolinite.

### 5.8. Weathering of rocks

Bacteria and other organisms have a role in weathering of rocks and stones. Some bacteria produce a polysaccharide based gel which helps them to stick to the surface of rocks and depending upon the environment and their chemical structure, these polymers promote the dissolution of the rocks. One of the polymers, alginic acid, a linear copolymer consisting primarily of β-1,4-linked D-mannuronic acid and β-1,4-linked L-glucuronic acid is produced by many kinds of microorganisms which at concentrations as low as 0.1% dramatically increases calcite dissolution over a wide pH range.

## 6. ASSIMILATIVE AND DISSIMILATIVE PROCESSES IN ENERGY GENERATION FOR THE MICROORGANISMS IN NATURE

Microbes must acquire certain elements to grow and reproduce. These elements compose their protoplasm in the proportions listed in Table 1. In addition, they must produce ATP in order to use the stored energy in this molecule to operate various cellular processes. Assimilative processes are used to bring needed elements into the cell and to incorporate them into the cell protoplasm. Dissimilative processes do not incorporate elements into the cell, but instead they use the energy gained in the process to form ATP.

Microorganisms are classified as autotrophs or heterotrophs based on whether or not they require pre-formed organic matter. Autotrophs derive energy from either light absorption (photoautotrophs) or oxidation of inorganic molecules (chemoautotrophs). In most of the light reactions the bacteria are fixing carbon dioxide into organic carbon, just as green plants do. Some photosynthetic bacteria (photoheterotrophs) require pre-formed organic matter as reducing agents, but generate ATP from the absorption of light energy. Finally, some bacteria and fungi (heterotrophs) use pre-formed organic matter as both a source of energy to generate ATP and as a source of carbon for the cell, just as animals do. Table 2 summarizes the classification of the ways in which microbes process energy.

**Table 2** Classification of microorganisms on the basis of energy processes

| Classification | Energy source for generating ATP | Source of carbon for the synthesis of cellular mass | Organisms |
|---|---|---|---|
| Photoautotrophs | Light | $CO_2$ | Bacteria |
| Chemoautotrophs | Inorganic compounds | $CO_2$ | Bacteria |
| Photoheterotroph | Light | $CO_2$, organic matter | Bacteria |
| Heterotrophs | Organic matter | Organic matter | Bacteria, fungi |

Some of the common dissimilative reactions that bacteria perform to gain energy and to decompose organic matter are listed in detail in the Table 3. The energy yield listed in the last column is the amount of energy in kilocalories that is produced per mole of oxidant that is used. The formation of ATP requires about 7 kcal/mol of energy, so only reactions producing more than 7 kcal/mol can be used by bacteria for growth. As the most powerful oxidants (the electron donors that generate the greatest energy yields) are consumed, the major reaction that is performed by the bacteria shifts to the next most energy yielding process. For example, when oxygen is depleted from the environment, nitrate reduction occurs. If oxygen becomes available again, then nitrate reduction will stop, even if there is still $NO_3$ available in the environment, and aerobic respiration will continue. This shifting of electron donors continues until only $CO_2$ is left to serve as an oxidant, in which case the bacteria reduce the $CO_2$ to methane, $CH_4$. It is interesting to note that some bacteria can perform more than one of reactions, whereas other bacteria are quite specialized and can perform only a single, specific dissimilatory reaction. Remember that in all of these reactions the bacteria must have a source of carbon to incorporate into their cells in order to grow.

**Table 3** Common dissimilative reactions that bacteria perform to gain energy

| **Mechanism** | **Reactant** | **Oxidant** | **Reaction** | **Energy yield (kcal/mol)** |
|---|---|---|---|---|
| Aerobic respiration | CHO | $O_2$ | $C_6H_{12}O_6 + 6O_2 \rightarrow 6H_2O + 6CO_2$ | 686 |
| Nitrate reduction | CHO | $NO_3^-$ | $CHO + NO_3^- + H+ \rightarrow CO_2 + N_2 + H_2O$ | 649 |
| Sulphate reduction | CHO | $SO_4^{2-}$ | $2CHO + SO_4^{2-} + 2H+ \rightarrow 2CO_2 + HS^- + 2H_2O$ | 190 |
| Methanogenesis | $H_2$ | $CO_2$ | $4H_2 + CO_2 \rightarrow CH_4 + 2H_2O$ | 8.3 |

## 7. MICROORGANISMS AND COMPOSTING

Composting is nature's way of recycling which biodegrades organic waste including food waste, manure, leaves, grass trimmings, paper, wood, feathers, crop residue etc. and turns it into a valuable organic fertilizer. Compost can be used in a variety of applications. High quality compost can be used in agriculture, horticulture, landscaping and home gardening. Medium quality compost can be used in applications such as erosion control and roadside landscaping. Low quality compost can be used as a landfill cover or in land reclamation projects.

Composting is a natural biological process, carried out under controlled aerobic conditions (requires oxygen). In this process, various microorganisms, including bacteria and fungi, break down organic matter into simpler substances. The effectiveness of the composting process is dependent upon the environmental conditions present within the composting system i.e. oxygen, temperature, moisture, material disturbance, organic matter and the size and activity of microbial populations. Composting is not a mysterious or complicated process. Natural recycling (composting) occurs on a continuous basis in the natural environment.

Organic matter is metabolized by microorganisms and consumed by invertebrates. The resulting nutrients are returned to the soil to support plant growth.

Composting is relatively simple to manage and can be carried out on a wide range of scales in almost any indoor or outdoor environment and in almost any geographic location. It has the potential to manage most of the organic material in the waste stream including restaurant waste, leaves and yard wastes, farm waste, animal manure, animal carcasses, paper products, sewage sludge, wood etc. and can be easily incorporated into any waste management plan.

Since approximately 45-55% of the waste stream is organic matter, composting can play a significant role in diverting waste from landfills thereby conserving landfill space and reducing the production of leachate and methane gas. In addition, an effective composting program can produce a high quality soil amendment with a variety of end uses. The essential elements required by the composting microorganisms are carbon, nitrogen, oxygen and moisture. If any of these elements are lacking, or if they are not provided in the proper proportion, the microorganisms will not flourish and will not provide adequate heat. A composting process that operates at optimum performance will convert organic matter into stable compost that is odor and pathogen free, and a poor breeding substrate for flies and other insects. In addition, it will significantly reduce the volume and weight of organic waste as the composting process converts much of the biodegradable component to gaseous carbon dioxide. The composting process is carried out by three classes of microbes including low temperature microbes (psychrophiles), medium temperature microbes (mesophiles), high temperature microbes (thermophiles)Generally, composting begins at mesophilic temperatures and progresses into the thermophilic range. In later stages other organisms including actinomycetes, centipedes, millipedes, fungi, sowbugs, spiders and earthworms assist in the process. Composting process is affected by various factors some of which are discussed below:

## 7.1. Temperature

Temperature is directly proportional to the biological activity within the composting system. As the metabolic rate of the microbes accelerates, the temperature within the system increases. Conversely, as the metabolic rate of the microbes decreases, the system temperature decreases. Maintaining a temperature of 54°C or more for 3 to 4 days favors the destruction of weed seeds, fly larvae and plant pathogens. At a temperature of 68°C, organic matter will decompose about twice as fast as at 54°C. Temperatures above 68°C may result in the destruction of certain microbe populations. In this case temperature may rapidly decline. Temperature will slowly rise again as the microbe population regenerates.

Moisture content, oxygen availability, and microbial activity all influence temperature. When the pile temperature is increasing, it is operating at optimum performance and should be left alone. As the temperature peaks, and begins to decrease, the pile should be turned to incorporate oxygen into the compost. Subsequently ,the pile should respond to the turning and incorporation of oxygen, and temperature should again cycle upwards. The turning process should be continued until the pile fails to re-heat. This indicates that the compost material is biologically stable.

## 7.2. Moisture content

Composting microorganisms thrive in moist conditions. For optimum performance, moisture content within the composting environment should be maintained at 45 percent. Too much water can cause the compost pile to go anaerobic and emit obnoxious odors. Too little will prevent the microorganisms from propagating.

## 7.3. Particle size

The ideal particle size is around 2 to 3 inches. In some cases, such as in the composting of grass clippings, the raw material may be too dense to permit adequate air flow or may be too moist. A common solution to this problem is to add a bulking agent (straw, dry leaves, paper, cardboard) to allow for proper air flow. Mixing materials of different sizes and textures also helps aerate the compost pile.

## 7.4. Turning

During the composting process oxygen is used up quickly by the microbes as they metabolize the organic matter. As the oxygen becomes depleted the composting process slows and temperatures decline. Aerating the compost by turning should ensure an adequate supply of oxygen to the microbes.

## 7.5. The carbon to nitrogen ratio

The microbes in compost use carbon for energy and nitrogen for protein synthesis. The proportion of these two elements required by the microbes averages about 30 parts carbon to 1 part nitrogen. Accordingly, the ideal ratio of carbon to nitrogen (C:N) is 30 to 1 (measured on a dry weight basis). This ratio governs the speed at which the microbes decompose organic waste. Most organic materials do not have this ratio and, to accelerate the composting process, it may be necessary to balance the numbers. The C:N ratio of materials can be calculated by taking into consideration the total material. Example, if you have two bags of cow manure (C:N = 20:1) and one bag of corn stalks (C:N = 60:1) then combined you have a C:N ratio of (20:1 +20:1 +60:1)/3 = (100:1)/3 = 33:1

# 8. MICROORGANISMS AS BIOGEOCHEMICAL AGENTS

As the Earth is essentially a closed system with respect to matter, we can say that all matter on Earth cycles, it is neither created nor destroyed. The movement (or cycling) of matter through biological agents is termed as biogeochemical cycling. In general, the earth system is sub-divided into i) atmosphere ii) hydrosphere, iii) lithosphere and iv) biosphere, all of which take part in biogeochemical cycling of the matter. The cycling elements include the macronutrients in the form of nitrogen, carbon, oxygen, hydrogen, phosphorus and sulphur, required in large amounts by various organisms.

## 8.1. The nitrogen cycle

This involves the biogeochemical transformation of N and N compounds (Figure 1). The major source of nitrogen is air, which is about 78 percent $N_2$ by volume. Nitrogen is essential for many biogical processes; for example, it is included in all amino acids, is incorporated into proteins and is present in the four bases that make up nucleic acids, such as DNA. Processing is necessary to convert gaseous nitrogen into forms usable by living organisms. The nitrogen cycle is the most complex of the cycles of elements that make up biological systems. This is due to the importance and prevalence of N in cellular metabolism, the diversity of types of nitrogen metabolism, and the existence of the element in so many forms. Procaryotes (bacteria) are essentially involved in the biological nitrogen cycle in three unique processes including nitrogen fixation, nitrification and anaerobic respiration (denitrification) while eukaryotes (fungi) are also involved in the decomposition of organic nitrogen (proteins).

*Nitrogen fixation:* The first step in the nitrogen cycle starts with atmospheric Nitrogen fixation, a process which converts $N_2$ in the atmosphere into $NH_3$ (ammonia), which is assimilated into amino acids and proteins. Nitrogen fixation occurs in many free-living bacteria such as clostridia, azotobacters and cyanobacteria, and in symbiotic bacteria such as *Rhizobium* and *Frankia,* which associate with plant roots to form characteristic nodules. Biological nitrogen fixation is the most important way that $N_2$ from the air enters into biological systems.

$N_2 \rightarrow 2NH_3$ (nitrogen fixation)

*Organic N formation:* Fixed nitrogen in the soil is utilized by plants and converted into plant proteins which are consumed by animals and the proteins are thus converted into animal proteins. Nitrate present in the soil is also utilized by plants and ultimately get converted into proteins.

*Soil organic N:* Excretion products of animals, dead animals, plant tissues and microorganisms deposited in soil become soil's organic nitrogen. $NH_3$ present in the soil is also utilized by microorganisms and converted into cellular proteins.

*Organic N degradation:* The proteins are degraded by variety of microorganisms including bacteria like *Clostridium, Proteus, Pseudomonas*, many fungi and actinomycetes, through proteolysis and converted in to a variety of amino acids. The complete proteolysis involves two enzymatic steps with proteinases and peptidases.

$$\text{Proteins} \xrightarrow{\text{proteinases}} \text{Peptides} \xrightarrow{\text{peptidases}} \text{Amino acids}$$

Ammonia formation (ammonification): Amino acids deaminated by many microorganisms into ammonia.

Nitrification: Nitrification is a form of lithotrophic metabolism in which $NH_3$ is oxidized into $NO_3$ through two steps. Nitrifying bacteria such as *Nitrosomonas* utilize $NH_3$ as an energy source, oxidizing it to $NO_2$, while *Nitrobacter* oxidizes $NO_2$ to $NO_3$. Nitrifying bacteria generally occur in aquatic environments and their significance in soil fertility and the global nitrogen cycle is not well understood.

The overall process of nitrification involves:

$$NH_3 \longrightarrow NO_2 \text{ (\textit{Nitrosomonas})}$$

$$NO_2 \longrightarrow NO_3 \text{ (\textit{Nitrobacter})}$$

*Anaerobic respiration:* This relates to the use of oxidized forms of nitrogen ($NO_3$ and $NO_2$) as final electron acceptors for respiration. Anaerobic respirers such as *Bacillus, Pseudomonas, Achromobacter* and *Agrobacterium* are common soil inhabitants that will use nitrate ($NO_3$) as an electron acceptor. $NO_3$ is reduced to $NO_2$ (nitrite) and then to a gaseous form of nitrogen such as $N_2$ or $N_2O$ (nitrous oxide) or ammonia ($NH_3$). The process is called denitrification. Denitrifying bacteria are typical aerobes that respire whenever oxygen is available by aerobic respiration. If $O_2$ is unavailable for respiration, they will turn to the alternative anaerobic respiration which uses $NO_3$. Since $NO_3$ is a common and expensive form of fertilizer in soils, denitrification may not be so good for agriculture, and one rationale for tilling the soil is to keep it aerobic, thereby preserving nitrate fertilizer in the soil. The overall process of denitrification involves:

$$NO_3 \longrightarrow NO_2 \longrightarrow N_2$$

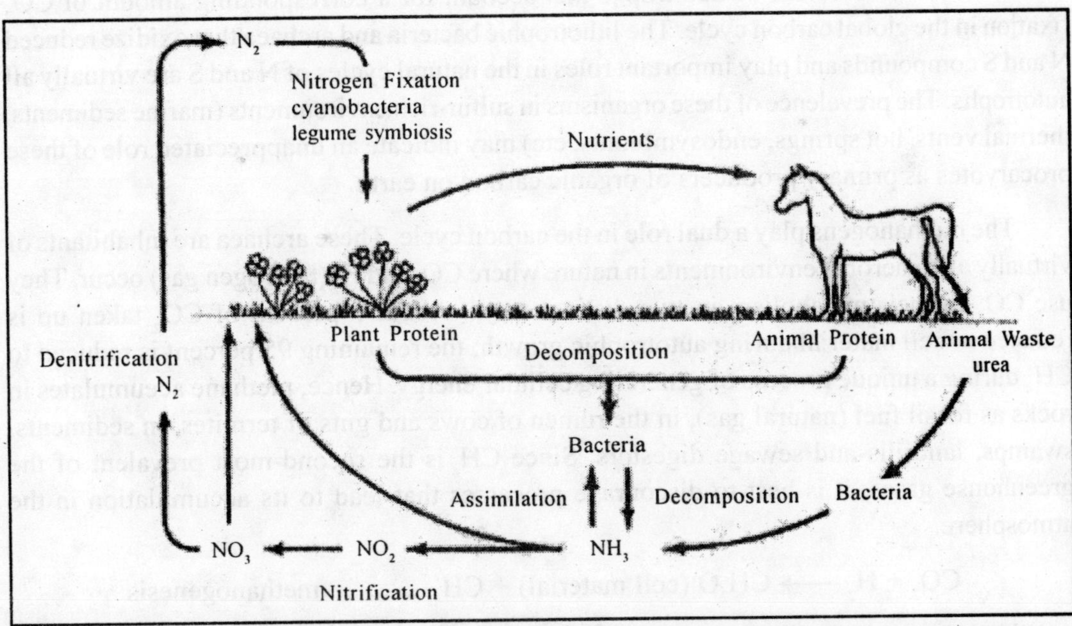


**Figure 1** Generalized Nitrogen cycle. The atmospheric $N_2$ is converted to $NH_3$ in soil by nitrogen fixation by prokaryotes and the same is then converted in to organic-N by plants which provide nutrition to heterotrophs. The proteins get deposited in soil after death of plants and animals and is decomposed by various bacteria and fungi in to $NH_3$ which is oxidized by various lithotrophic bacteria into $NO_3$ through nitrification which is used as ultimate electron acceptor in anaerobic respiration by various bacteria and gets ultimately reduced to $N_2$ which then again escapes into the atmosphere

## 8.2. The carbon cycle

This involves the biochemical transformations of C and C-compounds (Figure 2). Carbon is the backbone of all organic molecules and is the most prevalent element in cellular (organic) material. In its most oxidized form, $CO_2$, it can be viewed as an "inorganic" molecule.

*$CO_2$ fixation:* Autotrophs, which include plants, algae, photosynthetic bacteria, lithotrophs, and methanogens, use $CO_2$ as a sole source of carbon for growth, which reduces the molecule to organic cell material ($CH_2O$). Heterotrophs require organic carbon for growth, and ultimately convert it back to $CO_2$. Thus, a relationship between autotrophs and heterotrophs is established wherein autotrophs fix carbon needed by heterotrophs, and heterotrophs produce $CO_2$ used by the autotrophs.

$$CO_2 + H_2O \longrightarrow CH_2O \text{ (organic material) ......... autotrophy}$$

$$CH_2O + O_2 \longrightarrow CO_2 + H_2O \text{ ........... heterotrophy}$$

Since $CO_2$ is the most prevalent greenhouse gas in the atmosphere, it isn't good if these two equations get out of balance (i.e. heterotrophy predominating over autotrophy, as when rain forests are destroyed and replaced with cattle). Autotrophs are referred to as primary producers at the "bottom of the food chain" because they convert carbon to a form required by heterotrophs. Among procaryotes, the cyanobacteria, the lithotrophs and the methanogens are a formidable biomass of autotrophs that account for a corresponding amount of $CO_2$ fixation in the global carbon cycle. The lithotrophic bacteria and archaea that oxidize reduced N and S compounds and play important roles in the natural cycles of N and S are virtually all autotrophs. The prevalence of these organisms in sulfur-rich environments (marine sediments, thermal vents, hot springs, endosymbionts, etc) may indicate an unappreciated role of these procaryotes as primary producers of organic carbon on earth.

The methanogens play a dual role in the carbon cycle. These archaea are inhabitants of virtually all anaerobic environments in nature where $CO_2$ and $H_2$ (hydrogen gas) occur. They use $CO_2$ in their metabolism in two distinct ways. About 5 percent of $CO_2$ taken up is reduced to cell material during autotrophic growth; the remaining 95 percent is reduced to $CH_4$ during a unique process of generating cellular energy. Hence, methane accumulates in rocks as fossil fuel (natural gas), in the rumen of cows and guts of termites, in sediments, swamps, landfills and sewage digestors. Since $CH_4$ is the second-most prevalent of the greenhouse gases, it is best to discourage processes that lead to its accumulation in the atmosphere.

$$CO_2 + H_2 \longrightarrow CH_2O \text{ (cell material)} + CH_4 \text{ ........... methanogenesis}$$

Under aerobic conditions, methane and its derivatives (methanol, formaldehyde, etc.) can be oxidized as energy sources by bacteria called methylotrophs.

*Organic C in soil:* Organic carbon of primary producers is converted in to animal organic-C when the former are consumed by the latter. Upon death organic-C from both these forms gets deposited in soil.

*Organic C decomposition:* Biodegradation is the process in the carbon cycle for which microbes get most credit (or blame). Biodegradation is the decomposition of organic material ($CH_2O$) back to $CO_2$, $H_2O$ and $H_2$. In soil habitats, the fungi play a significant role in biodegradation, but the procaryotes are equally important. The typical decomposition scenario involves the initial degradation of biopolymers (cellulose, lignin, proteins, polysaccharides) by extracellular enzymes, followed by oxidation (fermentation or respiration) of the monomeric subunits. The ultimate end products are $CO_2$, $H_2O$ and $H_2$, perhaps some $NH_3$ (ammonia) and sulfide ($H_2S$), depending on how one views the overall process. These products are utilized by lithotrophs and autotrophs for recycling. Procaryotes which play an important role in biodegradation in nature include the actinomycetes, clostridia, bacilli, arthrobacters and pseudomonads.

Overall process of decomposition

Polymers (cellulose) ⟶ Monomers (glucose) ........ depolymerization

Monomers → fatty acids (lactic acid, acetic acid, propionic acid) + $CO_2$ + $H_2$ .... fermentation

Monomers + $O_2$ ⟶ $CO_2$ + $H_2O$ ............ aerobic respiration

The importance of microbes in biodegradation is embodied in the adage that "there is no known natural compound that cannot be degraded by some microorganism". The proof of the adage is that we aren't up to our ears in whatever it is that couldn't be degraded in the last 3.5 billion years. Actually, we are up to our ears in cellulose and lignin, which is better than concrete, and some places are getting up to their ears in teflon, plastic, styrofoam, insecticides, pesticides and poisons that are degraded slowly by microbes, or not at all.

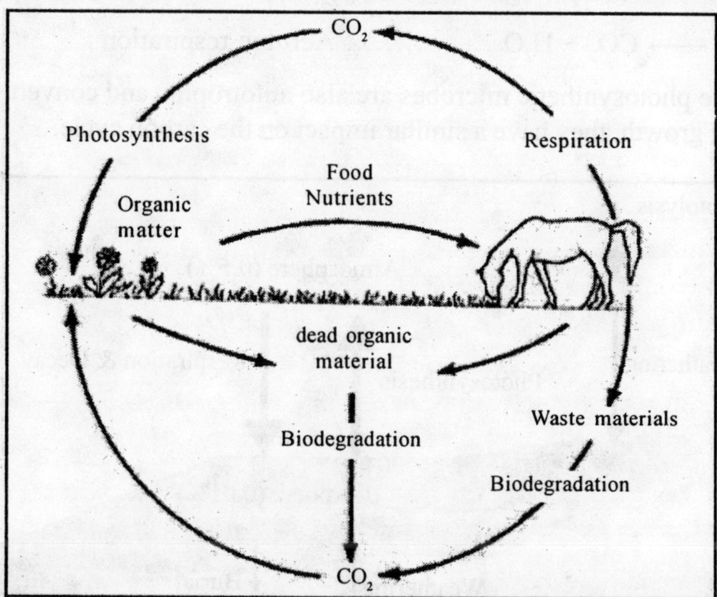


**Figure 2** The generalized carbon cycle. Organic matter ($CH_2O$) derived from photosynthesis (plants, algae and cyanobacteria) provides nutrition for heterotrophs (e.g. animals and associated bacteria), which convert it back to $CO_2$. Organic wastes, as well as dead organic matter in the soil and water, are ultimately broken down to $CO_2$ by microbial processes of biodegradation

## 8.3. The oxygen cycle

The oxygen cycle is the biogeochemical cycle that describes the movement of oxygen within and between its three main reservoirs: the atmosphere, the biosphere, and the lithosphere (Figure 3). The main driving factor of the oxygen cycle is photosynthesis, which is responsible for the modern Earth's atmosphere and life as we know it. If all photosynthesis were to cease, the Earth's atmosphere would be devoid all but trace amounts of oxygen within 5000 years. The oxygen cycle would no longer exist.

The vast majority of molecular oxygen is contained in rocks and minerals within the Earth (99.5%). Only a small fraction has been released as free oxygen to the biosphere (0.01%) and atmosphere (0.5%).

Basically, $O_2$ is derived from the photolysis of $H_2O$ during plant (oxygenic) photosynthesis. Two major groups of microorganisms are involved in this process, the eukaryotic algae, and the procaryotic cyanobacteria (blue-green algae). The cyanobacteria and algae are the source of much of the $O_2$ in the earth's atmosphere. Of course, plants account for some $O_2$ production as well, but the microbes predominate in marine habitats which cover the majority of the planet.

Since most aerobic organisms need the $O_2$ that results from plant photosynthesis, this establishes a relationship between plant photosyntheisis and aerobic respiration, the two prominant types of metabolism on earth. Photosynthesis produces $O_2$ needed for aerobic respiration. Respiration produces $CO_2$ needed for autotrophic growth.

$CO_2 + H_2O \longrightarrow CH_2O$ (organic material) + $O_2$ ..... Plant (oxygenic) photosynthesis

$CO_2 + O_2 \longrightarrow CO_2 + H_2O$ .............. Aerobic respiration

Since these photosynthetic microbes are also autotrophic and convert $CO_2$ to organic material during growth, they have a similar impact on the carbon cycle.

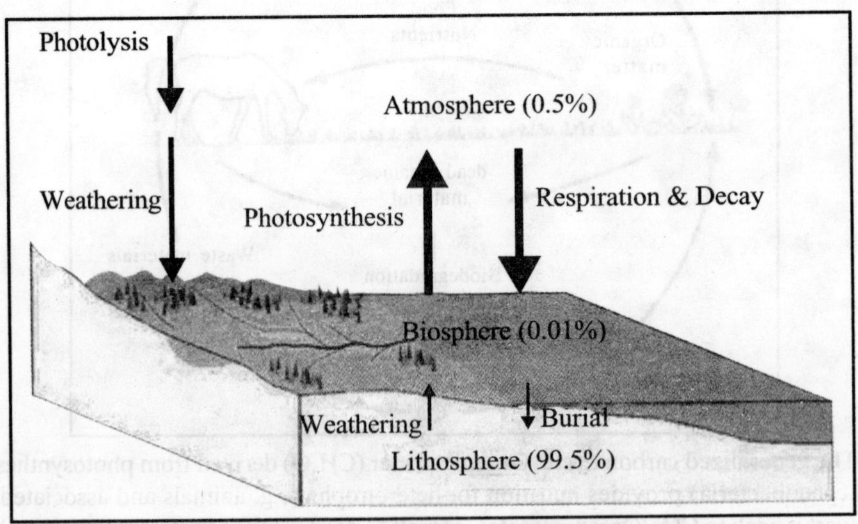


**Figure 3** The generalized oxygen cycle

An additional source of atmospheric oxygen comes from photolysis, whereby high energy ultraviolet radiation breaks down atmospheric water and nitrite into component molecules. The free H and N atoms escape into space leaving $O_2$ in the atmosphere:

$$2H_2O + energy \rightarrow 4H + O_2$$
$$2N_2O + energy \rightarrow 4N + O_2$$

The main way oxygen is lost from the atmosphere is via respiration and decay mechanisms in which animal life consumes oxygen and releases carbon dioxide. Because lithospheric minerals are reduced in oxygen, surface weathering of exposed rocks also consumes oxygen. An example of surface weathering chemistry is formation of iron-oxides (rust) such as that found in the red sands of Australia:

$$4FeO + 3O_2 \rightarrow 2Fe_2O_3$$

Oxygen is also cycled between the biosphere and lithosphere. Marine organisms in the biosphere create carbonate shell material ($CaCO_3$) that is rich in molecular oxygen. When the organism dies its shell is deposited on the shallow sea floor and buried over time to create limestone rock in the lithosphere. Weathering processes initiated by organisms can also free oxygen from the lithosphere. Plants and animals extract nutrient minerals from rocks and release oxygen in the process.

The presence of atmospheric oxygen has lead to the formation of the ozone layer within the stratosphere. The ozone layer is extremely important to modern life as it absorbs harmful ultraviolet radiation:

$$O_2 + uv\ energy \rightarrow 2O$$
$$O + O_2 + uv\ energy \rightarrow O_3$$

The absorbed ultraviolet energy also raises the temperature of the atmosphere within the ozone layer creating a thermal barrier that helps trap the atmosphere below (as opposed to bleeding out into space). An interesting theory is that phosphorus (P) in the ocean helps regulate the amount of atmospheric oxygen. Phosphorus disolved in the oceans is an essential nutrient to photosynthetic life and one of the key limiting factors. High oxygen levels in the oceans promote P removal by increasing the productivity of certain types of bacteria which uptake dissolved P to make their tissues. The decreasing phosphorus availability subsequently suppresses the population growth of the photosynthesizers and $O_2$ production diminishes. Low oxygen levels lead to increased mortality of oceanic organisms which, during decomposition, release P back into the ocean. This increasing phosphorus availability subsequently elevates the population growth of photosynthesizers and $O_2$ production increases.

## 8.4. The sulfur cycle

This involves the biochemical transformations of S and S-compounds (Figure 4). Microbial transformations of S-compounds have counterparts in microbial transformation of N-compounds. Sulfur is a component of a couple of vitamins and essential metabolites and it occurs in two amino acids, cysteine and methionine. In spite of its paucity in cells, it is an absolutely essential element for living systems. Like nitrogen and carbon, the microbes can transform sulfur from its most oxidized form (sulfate or $SO_4$) to its most reduced state (sulfide or $H_2S$). The sulfur cycle, in particular, involves some unique groups of procaryotes and procaryotic processes. Two unrelated groups of procaryotes oxidize $H_2S$ to S and S to $SO_4$. The first is the anoxygenic photosynthetic purple and green sulfur bacteria that oxidize $H_2S$ as a source of electrons for cyclic photophosphorylation. The second is the "colorless sulfur bacteria" (now a misnomer because the group contains many Archaea) which oxidize $H_2S$ and S as sources of energy. In either case, the organisms can usually mediate the complete oxidation of $H_2S$ to $SO_4$.

$$H_2S \longrightarrow S \longrightarrow SO_4 \text{ .......... litho or phototrophic sulfur oxidation}$$

Sulfur-oxidizing procaryotes are frequently thermophiles found in hot (volcanic) springs and near deep sea thermal vents that are rich in $H_2S$. They may be acidophiles, as well, since they acidify their own environment by the production of sulfuric acid. Since $SO_4$ and S may be used as electron acceptors for respiration, sulfate reducing bacteria produce $H_2S$ during a process of anaerobic respiration analogous to denitrification. The use of $SO_4$ as an electron acceptor is an obligatory process that takes place only in anaerobic environments. The process results in the distinctive odor of $H_2S$ in anaerobic bogs, soils and sediments where it occurs.

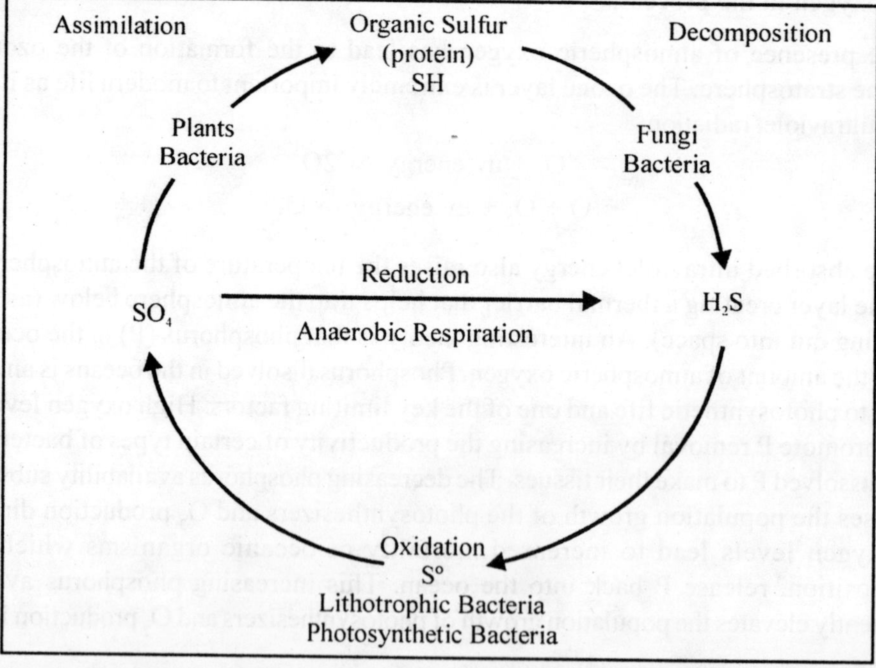


**Figure 4** Generalized sulfur cycle. Elemental S is not used by plants and animals but is oxidized by certain bacteria to sulphates which are used by plants and incorporated into S-containing amino acids and then into proteins which provide nutrition to animals and ultimately get deposited into soil upon death of plants and animals. The proteins upon decomposition by bacteria and fungi get converted into S-containing amino acids and ultimately to $H_2S$ which also comes from the reduction of sulphates by certain soil microorganisms. The oxidation of $H_2S$ by photosynthetic sulphur bacteria converts it into elemental S

Sulfur is assimilated by bacteria and plants as $SO_4$ for use and reduction to sulfide. Sulphate is used by plants and incorporated into S-containing amino acids and then into plant proteins → animal proteins → soil proteins. Sulphate may also be reduced to $H_2S$ by soil microorganisms.

Proteins as a result of proteolysis get converted into amino acids some of which contain sulphur. Many heterotrophic bacteria can remove the sulfide group from proteins as a source of S during decomposition.

Soil proteins → S-containing amino acids → $H_2S + NH_3$

$H_2S$ resulting from sulphate reduction and amino acid decomposition is oxidised to elemental sulphur.

$CO_2 + 2H_2S \xrightarrow{\text{light}} (CH_2O)_2 + H_2O + 2S$ .......... Photosynthetic sulphur bacteria

## 8.5. The phosphorus cycle

The phosphorus cycle is comparatively simple. Inorganic phosphate exists in only one form (Figure 5). It is interconverted from an inorganic to an organic form and back again, and there is no gaseous intermediate. Phosphorus is an essential element in biological systems because it is a constituent of nucleic acids, (DNA and RNA) and it occurs in the phospholipids of cell membranes. Phosphate is also a constituent of ADP and ATP which are universally involved in energy exchange in biological systems.

Dissolved phosphate ($PO_4$) inevitably ends up in the oceans. It is returned to land by shore animals and birds that feed on phosphorus containing sea creatures and then deposit their feces on land. Dissolved $PO_4$ is also returned to land by a geological process, the uplift of ocean floors to form land masses, but the process is very slow. However, the Figure below considers how $PO_4$ is recycled among land-based groups of organisms

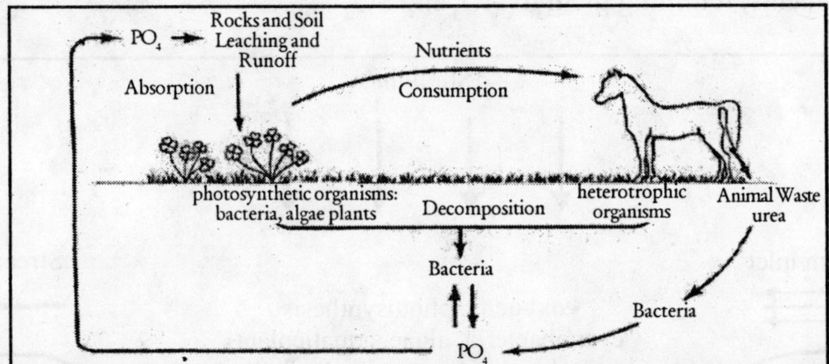


**Figure 5** Generalized phosphorus cycle. Plants, algae and photosynthetic bacteria can absorb phosphate ($PO_4$) dissolved in water, or if it washes out of rocks and soils. They incorporate the $PO_4$ into various organic forms, including such molecules as DNA, RNA, ATP, and phospholipid.The plants are consumed by animals wherein the organic phosphate in the plant becomes organic phosphate in the animal and in the bacteria that live with the animal. Animal waste returns inorganic $PO_4$ to the environment and also organic phosphate in the form of microbial cells. Dead plants and animals, as well as animal waste, are decomposed by microbes in the soil. The phosphate eventually is mineralized to the soluble $PO_4$ form in water and soil, to be taken up again by photosynthetic organisms

## 9. ECOLOGY OF A STRATIFIED LAKE

The role of microbes in the global cycle of elements (described above) can be visited on a smaller scale, in a lake, for example, like Lake Mendota, which may become stratified as illustrated in Figure 6. The surface of the lake is well-lighted by the sun and is aerobic. The bottom of the lake and its sediments are dark and anaerobic. Generally there is less $O_2$ and less light as the water column is penetrated from the surface. Assuming that the nutrient supply is stable and there is no mixing between layers of lake water, we should, for the time being, have a stable ecosystem with recycling of essential elements among the living systems. Here is how it would work.

At the surface, light and $O_2$ are plentiful, $CO_2$ is fixed and $O_2$ is produced. Photosynthetic plants, algae and cyanobacteria produce $O_2$, cyanobacteria can even fix $N_2$ aerobic bacteria, insects, animals and plants live here. At the bottom of the lake and in the sediments, conditions are dark and anaerobic. Fermentative bacteria produce fatty acids, $H_2$ and $CO_2$, which are used by methanogens to produce $CH_4$. Anaerobic respiring bacteria use $NO_3$ and $SO_4$ as electron acceptors, producing $NH_3$ and $H_2S$. Several soluble gases are in the water: $H_2$, $CO_2$, $CH_4$, $NH_3$ and $H_2S$.

The biological activity at the surface of the lake and at the bottom of the lake may have a lot to do with what will be going on in the middle of the water column, especially near the interface of the aerobic and anaerobic zones. This area, called the thermocline, is biologically very active. Bacterial photosynthesis, which is anaerobic, occurs here, using longer wave lenghts of light that will penetrate the water column and are not absorbed by all the plant chlorophyll above. The methanotrophs will stay just within the aerobic area taking up the $CH_4$ from the sediments as a carbon source, and returning it as $CO_2$. Lithotrophic nitrogen- and sulfur-utilizing bacteria do something analogous: they are aerobes that use $NH_3$ and $H_2S$ from the sediments, returning them to $NO_3$ and $SO_4$.

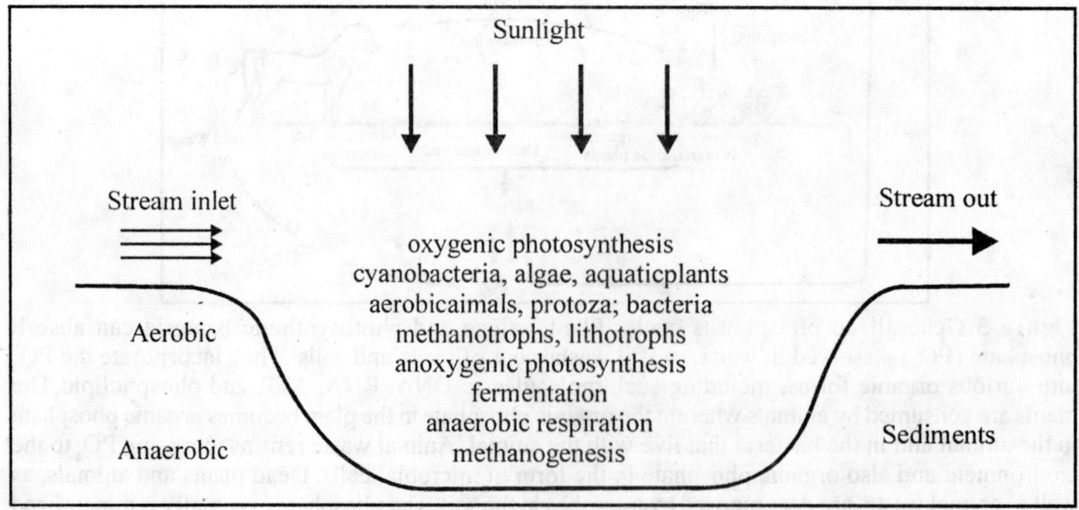


**Figure 6** Ecology of a stratified lake

## 10. POLLUTION MANAGEMENT

Pollution is the release of harmful environmental contaminants, or the substances so released. Generally the process needs to result from human activity to be regarded as pollution. Traditional forms of pollution include air pollution, water pollution, and radioactive contamination. Serious pollution sources include chemical plants, oil refineries, nuclear waste dumps, regular garbage dumps (many toxic substances are illegally dumped there), incinerators, PVC factories, car factories, plastics factories, and corporate animal farms creating huge amounts of animal waste (Figure 7). Some of the more common contaminants are chlorinated hydrocarbons (CFH), heavy metals like lead (in lead paint and until recently in gasoline), cadmium (in rechargeable batteries), chromium, zinc, arsenic and benzene. Pollutants are thought to play a part in a variety of maladies, including cancer, lupus, immune diseases, allergies, and asthma. Some illnesses are named in relation with certain pollutants: for example, Minamata disease, which is caused by mercury compounds. Since water is indispensible for the life, the pollution of surface waters by used waters of the community and industry is the most serious vectors of the environmental pollution. The used water of a community, also referred to as 'sewage' is generally a source of many pathogenic microbes, and/or toxic chemicals released by the industries in their effluents which may be detrimental for the environment (Figure 7). The domestic sewage is rich in organic matter and several pathogens while the nature of industrial waste depends upon the type of industry; food based industries produce an effluent rich in organic matter while metal based industries produce sewage rich in inorganic matter. These wastes generally has very high BOD levels and need to be treated before disposal with an aim to reduce BOD levels, to prevent oxygen imbalance in the water bodies, to prevent health hazards, to destroy pathogens and to prevent damage to concrete structures etc. There are many waste treatment activities generally required to ensure that waste has the least practicable impact on the environment. In many countries various forms of waste treatment are required by law. The principal waste treatment processes are:

- Sewage treatment – treatment and disposal of human waste. Sewage is produced by all human communities and is treated using processes that separate solid materials by settlement and then convert soluble contaminants into biological sludge and into gases such as carbon dioxide or methane.
- Waste management – re-use, recycling and disposal of solid waste from manufacturing, commerce and domestic sources.
- Industrial wastewater treatment – the treatment of wet wastes from manufacturing industry and commerce including mining, quarrying and heavy industries.
- Agricultural wastewater treatment – treatment and disposal of liquid animal waste, pesticide residues etc. from agriculture.
- Radioactive waste treatment – the treatment and containment of radioactive waste.

**Figure 7** Some common examples of water pollution

## 10.1. Sewage treatment

Sewage treatment is the process that removes the majority of the contaminants from waste-water or sewage and produces both a liquid effluent suitable for disposal to the natural environment and a sludge. To be effective, sewage must be conveyed to a treatment plant by appropriate pipes and infrastructure and the process itself must be subject to regulation and controls. Other wastewaters require often different and sometimes specialised treatment methods.

Sewage is the liquid waste from toilets, baths, showers, kitchens, etc. that is disposed of via sewers. In many areas sewage also includes some liquid waste from industry. The waste from toilets is termed foul waste, the waste from items such as basins, baths, kitchens is termed as sullage water, and the industrial and commercial waste is termed as trade waste.

The division of household water drains into "grey water"and "black water" is becoming more common in the developed world, with greywater being permitted to be used for watering plants or recycled for flushing toilets. Much sewage also includes some surface water from roofs or hard-standing areas. Municipal wastewater therefore includes residential, commercial, and industrial liquid waste discharges, and may include stormwater runoff.

In most of the countries including, India, the sewerage systems that transport liquid waste discharges and storm water together to a common treatment facility are called combined sewer systems. On the other hand, in some of the countries including U.S., Canada, UK and other European countries, liquid waste and storm waters are collected and conveyed in separate sewer systems, referred to as sanitary sewers and storm sewers or foul sewers and surface water sewers.

The conventional sewage treatment process typically involves the following three stages:

1. *Primary treatment :* To remove coarse and settleable solids by physical and mechanical means including sedimentation, filtration, aeration etc.
2. *Secondary treatment:* To remove the dissolved and emulsified organic components. It is biological in nature and may be aerobic or anaerobic. The aerobic process involves the oxidation of organic matter in which most of the free energy present in matter is

converted into $H_2O$ and $CO_2$. The microbes also conserve significant amount of energy per unit of substrate consumed. The solids, commonly termed as sludge, collected in the primary and in secondary treatment are subjected to anaerobic treatment in which most of the free energy is converted in to $CH_4$ (60-70%), also termed as biogas, fuel gas, digester gas, $CO_2$ (20-30%) with traces of $N_2$ and $H_2$.

3. *Tertiary treatment:* To make the effluent fit to be received in the environment involving disinfection and is chemical in nature.

### *10.1.1. Primary treatment*

Primary treatment is to reduce oils, grease, fats, sand, grit, and coarse (settleable) solids. This step is done entirely with machinery, hence the name mechanical treatment is also used for the primary treatment.

10.1.1.1. *Influx and removal of large objects:* In the mechanical treatment, the influx of sewage water is strained to remove all large objects that are deposited in the sewer system, such as condoms, sanitary towels (sanitary napkins) or tampons, cans, fruit, etc. This is most commonly done using a manual or automated mechanically raked screen. This type of waste is removed because it can damage the sensitive equipment in the sewage treatment plant.

10.1.1.2. *Sand and grit removal:* This stage typically includes a sand or grit channel where the velocity of the incoming wastewater is carefully controlled to allow sand grit and stones to settle but still maintain the majority of the organic material within the flow. This equipment is called a detritor or sand catcher. Sand grit and stones need to be removed early in the process to avoid damage to pumps and other equipment in the remaining treatment stages. Sometimes there is a sand washer followed by a conveyor that transports the sand to a container for disposal. The contents from the sand catcher may be fed into the incinerator in a sludge processing plant but in many cases the sand and grit is sent to a land-fill.

10.1.1.3. *Screening or maceration:* The grit free liquid is then passed through fixed or rotating screens to remove floating and larger material such as rags. Screenings are collected and may be returned to the sludge treatment plant or may be disposed of off site by landfilling or incineration. Maceration, in which solids are cut into small particles through the use of rotating knife edges mounted on a revolving cylinder, is used in plants that are able to process this particulate waste. Macerators are, however, more expensive to maintain and are less reliable than physical screens.

10.1.1.4. *Sedimentation:* In almost all plants there is a sedimentation stage where the sewage is allowed to pass through large circular or rectangular tanks. The tanks are large enough that faecal solids can settle and floating material such as grease and plastics can rise to the surface and be skimmed off. The main purpose of the primary stage is to produce a generally homogeneous liquid capable of being treated biologically and a sludge that can be separately treated or processed. Primary settlement tanks are usually equipped with mechanically driven scrapers that continually drive the collected sludge towards a hopper in the base of the tank from where it can be pumped to further sludge treatment stages.

### *10.1.2. Secondary treatment*

Secondary treatment is designed to substantially degrade the biological content of the sewage such as are derived from human waste, food waste, soaps and detergent. The majority of municipal and industrial plants treat the settled sewage liquor using aerobic biological processes. The generalized reaction carried out by microorganisms in aerobic treatment is as follows:

$$\text{Organic matter} + \text{microorganisms} + O_2 \longrightarrow \text{Inorganic matter} + \text{New cells} + \text{energy}$$

For the aerobic treatment to be effective, the biological agents require both oxygen and a substrate on which to live. There are number of ways in which this is done. In all these methods, the bacteria and protozoa consume biodegradable soluble organic contaminants (e.g. sugars, fats, organic short-chain carbon molecules, etc.) and bind much of the less soluble fractions into floc particles. The various forms of organic matter are oxidized into inorganic matter. Some of the reactions carried for such conversions are as follows:

$$\text{Organic C} + \text{microorganisms} \longrightarrow CO_2$$
$$\text{Organic N} + \text{microorganisms} \longrightarrow NO_3$$
$$\text{Organic S} + \text{microorganisms} \longrightarrow SO_4$$
$$\text{Organic H} + \text{microorganisms} \longrightarrow H_2O$$
$$\text{Organic P} + \text{microorganisms} \longrightarrow PO_4$$

Secondary treatment systems are classified as fixed film or suspended growth. In fixed film systems the biomass grows on media and the sewage passes over its surface. The sewage liquor is spread onto the surface of a deep bed made up of some inert materials to support the biofilms. Biofilm reactors include such processes as (1) trickling filters, (2) rotating biological contactors (RBC), (3) anaerobic filters, (4) submerged filters, (5) biological fluidized beds, (6) activated biofilters, and so on. The bulk substrate removal is thought to comprise three main processes as (a) substrate transport from the bulk to the biofilm, (b) molecular diffusion of the substrate within the biofilm, (c) metabolic reactions within the biofilm. In suspended growth systems - such as 'activated sludge' - the biomass is well mixed with the sewage. Typically, fixed film systems require smaller footprints than for an equivalent suspended growth system; however, suspended growth systems are more able to cope with shocks in biological loading.

10.1.2.1. *Trickling filters:* These have been popular for use in small plants, because of their ability to recover from shock loads and to perform well with a minimum of skilled technical supervision, and because of their economy in capital and operating costs. The design of the filters allows high hydraulic loading and a high flow-through of air. On larger installations, air is forced through the media using blowers. The resultant liquor is usually within the normal range for conventional treatment processes. In older plants and plants receiving more variable loads, trickling filter beds are used where the settled sewage liquor is spread onto the surface of a deep bed made up of coke (carbonised coal), limestone chips or specially fabricated plastic media. Such media must have high surface areas to support the biofilms that form. The liquor is distributed through perforated rotating arms radiating from a central pivot (Figure 8). The distributed liquor trickles through this bed and is collected in drains at the base. These drains also provide a source of air which percolates up through the bed, keeping it aerobic. Biological films of bacteria, protozoa and fungi form on the medias' surfaces and eat or otherwise reduce the organic content.

**Figure 8** A Trickling filter bed

10.1.2.2. *Rotating biological contactors (RBC):* In RBC, about 40 percents of the total surface area of disks are always submerged. During the rotation of disks, the fixed biofilm absorbs organics and oxygen. Although RBC is an efficient process removing organics, it would have a deficiency that the biomass detached from disks sometimes changes into fine dispersed flocks in the reactor and the color of effluent becomes white. Thus, a new system, in which the detached biofilm is drawn directly from the RBC reactor, has been under study.

10.1.2.3. *Activated sludge:* Activated sludge is a process in sewage treatment in which air or oxygen is forced into sewage liquor to develop a biological floc which reduces the organic content of the sewage. In all activated sludge plants, once the sewage has received sufficient treatment, excess mixed liquor is discharged into settling tanks and the supernatant is run off to undergo further treatment before discharge. Part of the settled material, the sludge, is returned to the head of the aeration system to re-seed the new sewage entering the tank. This fraction of the floc is called RAS - recycled activated sludge. The remaining sludge, also called WAS - waste activated sludge, is further treated prior to disposal. There are a variety of types of activated sludge plants. These include :

10.1.2.3.1. *Conventional activated sludge process:* The conventional activated sludge process is most commonly employed for a wastewater treatment process in developed countries. Figure 9 shows a typical flow diagram of the conventional activated sludge process. A key unit in this process is an aeration tank where organic and inorganic materials in the wastewaters are removed by activated sludge. An aeration tank has normally a depth of 4 to 6 meters, while a deep aeration tank has a depth of about 10 meters.

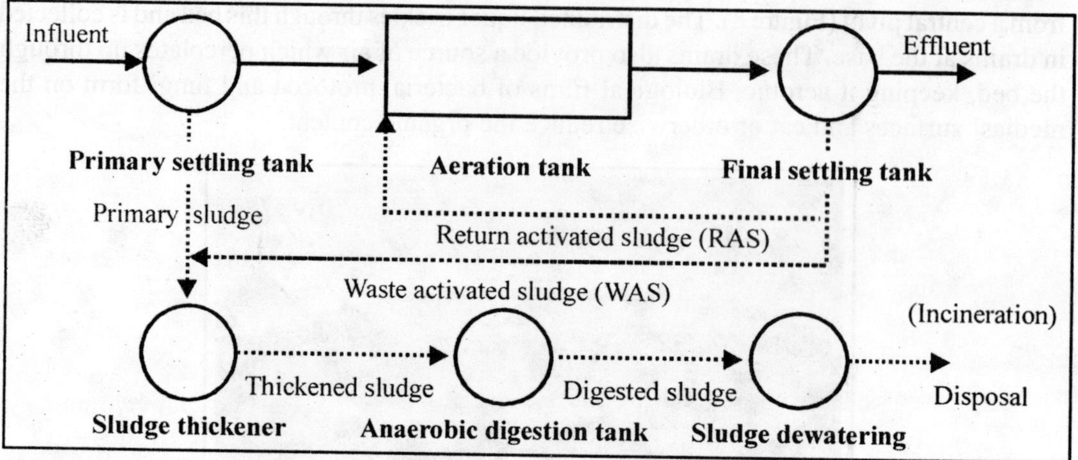


**Figure 9** Typical flow diagram of conventional Activated Sludge Process including sludge treatment processes.

Supply of air called aeration is done in an aeration tank through diffusers, though mechanical surface aeration is employed for small plants such as pre-treatment of industrial wastewaters. Aeration has mainly three purposes: first to supply oxygen to the mixture of wastewater and activated sludge called mixed liquor, second to move the mixed liquor fast enough to have good mixing of wastewater and activated sludge, and third to keep activated sludge in suspension. Oxygen transfer efficiency that expresses the percentage of oxygen actually dissolved in the water depends on types of diffusers since the bubble size, which may differ among diffusers, is an important factor for oxygen transfer efficiency.

Hydraulic retention time (HRT) in the aeration tank is 6 to 8 hours, and solids retention time (SRT) is 3 to 6 days. The concentration of solids of mixture of wastewater and activated sludge (mixed liquid suspended solids, MLSS) is 1,500 to 2,000 mg/l and food, and microorganism ratio (FM ratio) is 0.2 to 0.4 (kg BOD/kg SS/day). Although the ratio of returned activated sludge (RAS) to the influent flow is varied with the operational MLSS concentration, ratio of around 0.5 is usually adopted for MLSS in the range of 1,500 to 2,000 mg/l. As a reactor type, complete mixing is only employed for small plants and plug flow type is normally applied. Under these normal operational conditions, removal efficiencies of the conventional activated sludge process for BOD and suspended solids (SS) are expected more than 90 %. The process has, however, some deficiencies in that (1) the oxygen requirements at the head of aeration tank often exceed the aeration rate, and (2) the aeration tanks are not good approximation of stirred tanks or plug-flow processes, as there is a considerable amount of back mixing in long narrow tanks. To improve these deficiencies, a number of modifications of the conventional process were developed during the 1950s such as step aeration or tapered aeration.

In step aeration, influent of wastewater to an aeration tank is dividedly fed into several lower stream points of an aeration tank as well as the head of a tank. Such step feeding of wastewater into an aeration tank makes the oxygen consumption rate in an aeration tank relatively uniform and high oxygen consumption rate at the head of an aeration tank experienced in the conventional process can be avoided.

Since the oxygen consumption rate of activated sludge gradually decreases from the upper end to the lower end of an aeration tank, the rate of aeration can be reasonably varied according to the direction of flow in an aeration tank. This process is called tapered aeration. Supply of oxygen is matched to the oxygen demand of the activated sludge and the deficiency problems can be solved by the tapered aeration.

10.1.2.3.2. *Extended aeration process:* Since some of the other modification processes can be operated as extended aeration with relatively long solids retention time (SRT) and low food, and microorganism ratio (FM ratio), typical modification processes are explained. In the extended aeration process, production of the excess sludge can be reduced compared to the conventional process and nitrification proceeds relatively easily. Since ammonia is toxic to fish and other aquatic organisms, nitrification is favorable because the ammonia concentration of the effluent becomes relatively low. Primary settling process is usually neglected. HRT in the aeration tank of 16 to 24 hours and SRT of 13 to 50 days are generally applied. To keep the relatively long SRT, high concentration of MLSS in the range of 3,000 to 4,000 mg/l is maintained under the normal conditions. FM ratio is as low as 0.05 to 0.10 (kg BOD/kg SS/day). A rectangular aeration tank, an oxidation ditch (OD) type reactor, or a sequencing batch reactor (SBR) can carry out the extended aeration process. OD and SBR will be described as the processes of advanced wastewater treatment.

Recent advancement of material industries makes it possible to use membranes with very fine pore size. Membranes with nominal pore sizes of 0.1 to 0.5 microns are used as filters for the effluent to pass and for the activated sludge to remain in the reactor. Membrane separation or microfiltration (MF) activated sludge process is relatively a new process. There are several types of membranes such as plane membranes, string membranes, and tube membranes. MF activated sludge process is compacted in scale and produces high quality effluent. Since the activated sludge is retained in the reactor, there is no need of RAS and no need of either primary settling tanks or final settling tanks. MLSS is around 12,000 mg/l and SRT is 20 to 40 days, though HRT is usually around 6 hours. Removal efficiency of BOD and SS is as high as 98 % and almost 100%, respectively. Since the pore sizes are very fine, continuous cleaning of the surface of membranes is necessary to avoid clogging problems. For the cleaning purpose, gas to flow ratio (GQ ratio) of 40 to 50 is used in the membrane process, while GQ ratio of conventional activated sludge process is around 5. Therefore, in spite of its advantages of high quality effluent and less land requirements, high consumption of energy by membrane process discourages its adoption where price of electricity is high.

10.1.2.3.3. *Diffusers:* Sewage liquor is run into deep tanks with diffuser blocks attached to the floor. These are like the diffuser blocks used in tropical fish tanks but on a much larger scale. Air is pumped through the blocks and the curtain of bubbles formed both oxygenates the liquor and also provide the necessary stirring action. Where capacity is limited or the sewage is unusually strong or difficult to treat, oxygen may be used instead of air.

10.1.2.3.4. *Cones:* Vertically mounted tubes of up to 1 metre diameter extend from just above the base of a deep concrete tank to just below the surface of the sewage liquor. A typical shaft might be 10 metres high. At the surface end the tube is formed into a cone with helical vanes attached to the inner surface. When the tube is rotated, the vanes spin liquor up and out of the cones drawing new sewage liquor from the base of the tank. In many works each cone is located in a separate cell that can be isolated from the remaining cells if required for maintenance. Some works may have two cones to a cell and some large works may have 4 cones per cell.

10.1.2.3.5. *Package plants:* There are a wide range of other types of plants, often serving small communities or industrial plants that may use hybrid treatment processes often involving the use of aerobic sludge to treat the incoming sewage. In such plants the primary settlement stage of treatment may be omitted. In these plants, a biotic floc is created which provides the required substrate.

10.1.2.3.6. *Pasveer ditch:* In some areas where more land is available sewage is treated in large round or oval ditches with one or two rotating paddles called Kessener brush aerators which drive the liquor around the ditch and provide aeration. These are Pasveer ditches and have the advantage that they are relatively easy to maintain and are resilient to shock loads that often occur in smaller communities (at breakfast time and in the evening).

10.1.2.3.7. *Deep shaft:* Where land is in short supply sewage may be treated by injection of oxygen into a pressured return sludge stream which is injected into the base of a deep columnar tank buried in the ground. Such shafts may be up to 20 metres deep and are filled with settled sewage liquor. As the sewage rises the oxygen forced into solution by the pressure at the base of the shaft breaks out as molecular oxygen providing a highly efficient source of oxygen for the activated sludge biota. The rising oxygen and injected return sludge provide the physical mechanism for mixing of the sewage and sludge. Mixed sludge and sewage is decanted at the surface and separated into supernatant and sludge components. The efficiency of deep shaft treatment can be high but both installation costs and running costs are higher than most other methods and maintenance problems at many plants have given this method a poor rating.

10.1.2.3.8. *Rotating plates and spirals:* In some smaller plants slowly revolving plates or spirals are used which are partially submerged in the liquor. A biotic floc is created which provides the required substrate.

10.1.2.4. *Oxidation ponds (lagoons):* Lagooning provides the natural means of sewage treatment in large man-made ponds or lagoons. This involves the removal of organic matter by symbiotic action of aerobic bacteria and algae. Treatment of wastewaters can be simply done by introducing the wastewaters into ponds. The pond processes are classified as follows:

(a) *Aerobic ponds:* shallow ponds, less than 1 meter in depth, where dissolved oxygen is maintained throughout the entire depth mainly by the photosynthetic reactions.

(b) *Facultative ponds:* ponds with the depth of 1.0 to 2.5 meters, having an anaerobic lower zone, a facultative middle zone, and an aerobic upper zone maintained by photosynthesis and surface reaeration.

(c) *Anaerobic ponds:* deep ponds receiving high organic loadings such that anaerobic conditions prevail throughout entire pond depth.

(d) *Maturation or tertiary ponds:* ponds used for polishing effluents from other biological processes. Dissolved oxygen is furnished through photosynthesis and surface reaeration. This type of pond is also known as a polishing pond.

(e) *Aerated lagoons:* ponds oxygenated through the action of surface or diffused air aeration.

The processes (a), (b) and (e) are commonly used for secondary treatment of wastewater. Since sunlight penetrates into the bottom of aerobic ponds, the metabolic reactions of bacteria and algae complement each other. Since algae produce the oxygen in the photosynthetic reaction, aerobic bacteria can degrade organic compounds using the oxygen supplied by algae. In facultative ponds, not only aerobic zone but also anaerobic zone and facultative zone exist. Therefore, many kinds of biological reaction occur in the pond system.

10.1.2.5. *Secondary sedimentation:* The final step in the secondary treatment stage is to settle out the biological floc or filter material and produce an effluent with very low levels of organic material and suspended matter. Some systems use anaerobic digestion of sludge. These provide methane, which can be burned to produce electricity or process heat. If industrial toxin levels are low, the digested sludge is a mildly valuable organic amendment for farms.

### *10.1.3. Tertiary treatment*

Tertiary treatment provides a final stage to raise the effluent quality to the standard required before it is discharged to the receiving environment (sea, river, lake, ground, etc.) More than one tertiary treatment process may be used at any treatment plant. If disinfection is practiced, it is always the final process.

10.1.3.1. *Filtration:* Slow sand filtration removes much of the residual suspended matter. Filtration over activated carbon removes residual toxins.

10.1.3.2. *Nutrient removal:* Wastewater may also contain high levels of nutrients (nitrogen and phosphorus) that in certain forms may be toxic to fish and invertebrates at very low concentrations(e.g. ammonia) or that can create nuisance conditions in the receiving environment (e.g. weed or algal growth). Weeds and algae may seem to be an aesthetic issue, but algae can produce toxins, and their death and consumption by bacteria (decay) can deplete oxygen in the water and suffocate desirable fish. Where receiving rivers discharge to lakes or shallow seas, the added nutrients can cause severe eutrophication losing many sensitive clean water fish. The removal of nitrogen and/or phosphorus from wastewater can be achieved either biologically or by chemical precipitation.

10.1.3.3. *Nitrogen removal:* Nitrogen compounds have various forms in the water and transformations easily take place. Forms of nitrogen compounds in the municipal wastewaters are generally organic nitrogen or ammonia. When organic nitrogen compounds are degraded, ammonia is produced. Ammonia in the wastewaters can be oxidized to nitrite and nitrate as follows:

$$2NH_4^+ + 3O_2 \xrightarrow{Nitrosomonas} 2NO_2^- + 2H_2O + 4H^+$$
$$2NO_2^- + O_2 \xrightarrow{Nitrobactor} 2NO_3^-$$

Most of the heterotrophic bacteria can use nitrite and nitrate as electron acceptors when organic materials are available as electron donors. Then nitrite and nitrate will be reduced to molecular nitrogen and will be emitted to the air as nitrogen gas.

$$2NO_2^- + 3H_2 \rightarrow N_2 + 2OH^- + 2H_2O$$
$$2NO_3^- + 5H_2 \rightarrow N_2 + 2OH + 4H_2O$$

10.1.3.3.1. *Nitrification-denitrification activated sludge process:* Several types of nitrification-denitrification process have been developed. Figure 10 shows a flow diagram of one of the typical processes for nitrification-denitrification process. Biological reaction tanks are divided into two parts, one is anoxic and free from oxygen, and the other is oxic where oxygen is sufficient. Part of mixed liquor withdrawn from the end of aeration tanks is circulated into the head of anoxic tanks. Since the circulated mixed liquor contains nitrite and nitrate, and anoxic tanks receive fresh wastewater containing organic compounds, denitrification of nitrite and nitrate is easily taken place. Thus, the simultaneous removal of organic compounds and nitrogen can be achieved by using those processes. Although degradation of organic compounds also proceeds in oxic tanks, primary function of oxic tanks is in the nitrification.

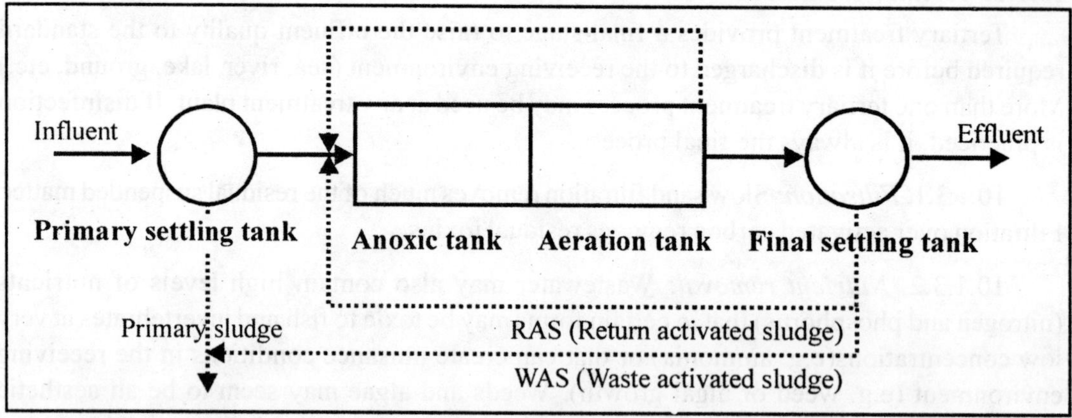


**Figure 10** Flow diagram of one of the typical Nitrification-Denitrification processes

In some treatment plants, aluminum salts or iron salts are dosed at the end of aeration tanks and the simultaneous removal of organic compounds, nitrogen and phosphorus by an activated sludge process can be achieved.

10.1.3.3.2. *Oxidation ditch (OD) process:* In OD process, primary sedimentation is usually neglected. A reactor is generally a continuous loop channel with 50 to 280 meters long, 4 to 10 meters wide, and 1 to 5 meters deep. Several kinds of aerators, such as a horizontal rotor-aerator, a jet aerator and a vertically mounted surface aerator, are used for oxygen supply. Schematic flow diagram of OD process is shown in Figure 11.

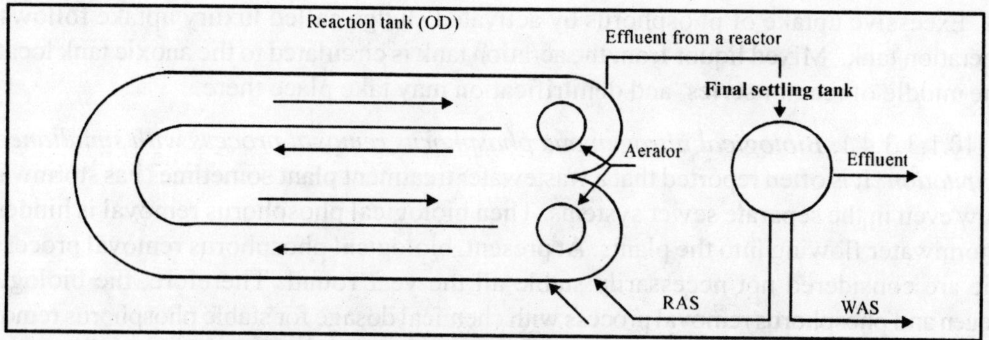


**Figure 11** Schematic flow diagram of OD process

The advantages of OD process are (1) ease of operation and maintenance, (2) low capital and operating costs, and (3) low production of excess sludge. It can also be designed to provide a stable and relatively high removal efficiency of organic and nitrogen compounds. However, because of relatively long HRT requirement and shallow reactor depth, OD process requires relatively wide space. Since OD can be operated under the extended aeration with long SRT and low FM ratio, nitrification is easily carried out. It is also possible to gradually decrease the concentration of dissolved oxygen along the flow direction by simply controlling the oxygen supply and the circulation flow rate. Under such conditions, denitrification can occur in the anoxic or partly anoxic zone. OD process with intermittent aeration is also useful to enhance denitrification. Dose of coagulants into the ditch is sometimes done to remove phosphorus. Then nitrogen and phosphorus are simultaneously removed along with organic compounds.

10.1.3.3.3. *Biofilm process:* It is possible in principle to remove organic compounds and nitrogen simultaneously by biofilm processes. In RBC, oxidation of organic compounds as well as ammonia is usually taking place at the same time. If RBC process is designed in series, denitrification can be carried out along with the oxidation of organic materials and ammonia, though the number of plants actually operating under such a condition is very small.

10.1.3.3.4. *Phosphorus removal:* Biological processes have remarkable phosphorus removal capability. The processes have considerable abilities to achieve simultaneous removal of organic compounds, nitrogen and phosphorus. Biological phosphorus removal depends on the principle that heterotrophic microorganisms that have a repeated experience of anaerobic and aerobic conditions in sequence, release phosphorus under the anaerobic condition and are able to take up excessive amount of phosphorus, on the other hand, under the subsequent aerobic condition. Therefore, if the activated sludge is withdrawn from an aerobic stage, phosphorus that is excessively accumulated in the sludge can be removed from the bulk of wastewater.

One of the activated sludge processes with biological nitrogen and phosphorus removal is known as $A_2O$ process. In $A_2O$ process, a biological reactor is operated under the conditions of anaerobic, anoxic and oxic in series. Since a head of the biological reactor is operated completely anaerobic condition, release of phosphorus from sludge will occur first. Excessive uptake of phosphorus by activated sludge called luxury uptake follows in the aeration tank. Mixed liquor from the aeration tank is circulated to the anoxic tank located in the middle of reactor series, and denitrification may take place there.

10.1.3.3.4.1. *Biological nitrogen and phosphorus removal process with simultaneous precipitation:* It is often reported that a wastewater treatment plant sometimes has stormwater inflow even in the separate sewer systems. Then biological phosphorus removal is hindered by stormwater flowing into the plants. At present, biological phosphorus removal processes alone are considered not necessarily stable all the year round. Therefore, the biological nitrogen and phosphorus removal process with chemical dosage for stable phosphorus removal becomes a practically reliable process. Figure 12 shows one of the typical processes of biological nitrogen and phosphorus removal process with chemical dosage.

As a coagulant to remove phosphorus, polyaluminum chloride (PAC) or polyiron salts are added at the end of an aeration tank. Since aluminum or iron ion will combine with phosphate, a salt is dosed around two times or less of stoichiometric requirement. Compared to the processes without biological phosphorus removal, dosage amount of salts can be reduced and effluent phosphorus concentration is stable.

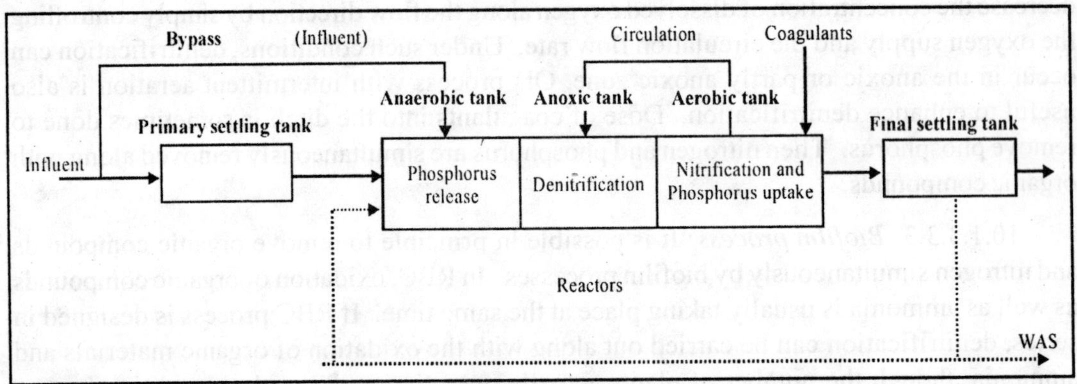


**Figure 12** A typical example of biological nitrogen and phosphorus removal process with chemical dosage

**10.1.3.3.4.2. Sequencing batch reactor (SBR):** Sequencing batch activated sludge process becomes in use for small-scale wastewater treatment. SBR is possible for the first time by the progress of reliable and inexpensive microprocessor-based automatic controls. SBR can be operated under various conditions such as anaerobic, anoxic, and oxic conditions. Then, SBR can remove nitrogen and phosphorus as well as organic materials simultaneously. Since SBR is a batch reactor, final solid-liquid separation is done relatively well.

SBR process is a method of wastewater treatment by the fill-and-draw activated sludge system composed of one or more tanks. In a typical SBR system, each tank containing activated sludge is filled with wastewater during a discrete period of time under aerated condition, and then operated in batch reaction mode. Figure 13 shows one of the typical operation modes of SBR.

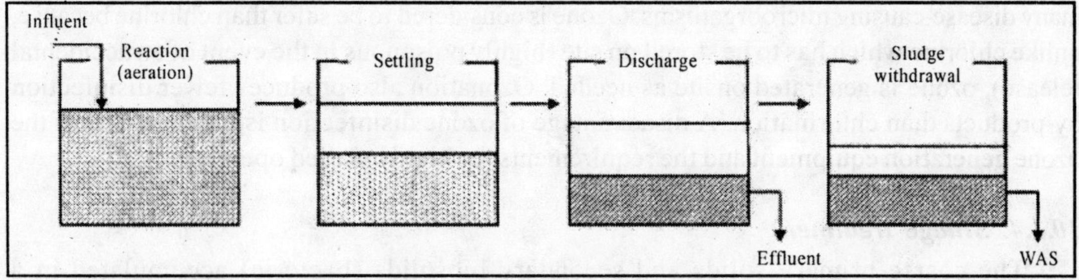


**Figure 13** An example SBR operation mode

After reaction, the mixed liquor is settled for some period of time and then the supernatant is withdrawn from the tank. After the filling, the wastewater flow is either directed to another tank in the system, as in a multiple tank configuration, or to a storage tank in a single tank configuration.

10.1.3.4. *Disinfection*: The purpose of disinfection in the treatment of wastewater is to substantially reduce the number of living organisms in the water to be discharged back into the environment. The effectiveness of disinfection depends on the quality of the water being treated (e.g., turbitidy, pH, etc.), the type of disinfection being used, the disinfectant dosage (concentration and time), and other environmental variables. Turbid water will be treated less successfully since solid matter can shield organisms, especially from ultraviolet light or if contact times are low. Generally, short contact times, low doses and high flows all militate against effective disinfection. Common methods of disinfection include ozone, chlorine, or UV light. Chloramine, which is used for drinking water, is not used in waste water treatment because of its persistence.

Chlorination remains the most common form of wastewater disinfection in North America due to its low cost and long-term history of effectiveness. One disadvantage is that chlorination of residual organic material can generate chlorinated-organic compounds that may be carcinogenic or harmful to the environment. Residual chlorine or chloramines may also be capable of chlorinating organic material in the natural aquatic environment. Further, because residual chlorine is toxic to aquatic species, the treated effluent must also be chemically dechlorinated, adding to the complexity and cost of treatment.

Ultraviolet (UV) light is becoming the most common means of disinfection in the UK because of the concerns about the impacts of chlorine in chlorinating residual organics in the wastewater and in chlorinating organics in the receiving water. UV radiation is used to damage the genetic structure of bacteria, viruses, and other pathogens, making them incapable of reproduction. The key disadvantages of UV disinfection are the need for frequent lamp maintenance and replacement and the need for a highly treated effluent to ensure that the target microorganisms are not shielded from the UV radiation (any solids present in the treated effluent may protect microorganisms from the UV light).

Ozone ($O_3$) is generated by passing oxygen $O_2$ through a high voltage potential resulting in a third oxygen atom becoming attached and forming $O_3$. Ozone is very unstable and reactive and oxidizes most organic material it comes in contact with, thereby destroying many disease-causing microorganisms. Ozone is considered to be safer than chlorine because, unlike chlorine which has to be stored on site (highly poisonous in the event of an accidental release), ozone is generated onsite as needed. Ozonation also produces fewer disinfection by-products than chlorination. A disadvantage of ozone disinfection is the high cost of the ozone generation equipment and the requirements for highly skilled operators.

### *10.1.4. Sludge treatment*

The coarse primary solids and secondary biosolids (bacteria) accumulated in a wastewater treatment process must be treated and disposed of in a safe and effective manner. This material is often inadvertently contaminated with toxic organic and inorganic compounds (e.g. heavy metals). The purpose of digestion is to reduce the amount of organic matter and the number of disease-causing microorganisms present in the solids. The most common treatment options include anaerobic digestion, aerobic digestion, and composting.

10.1.4.1. *Anaerobic digestion:* Anaerobic digestion is a bacterial process that is carried out in the absence of oxygen. The process can either be thermophilic digestion (in which sludge is fermented in tanks heated to about 38°C) or mesophilic digestion (cold digestion of sludge where sludge is maintained in large tanks for weeks to allow natural mineralisation of the sludge). Thermophilic digestion generates biogas with a high proportion of methane that may be used to both heat the tank and run engines or microturbines for other on-site processes. In large treatment plants sufficient energy can be generated in this way to produce more electricity than the machines require. The methane generation is a key advantage of the anaerobic process. Its key disadvantage is the long time required for the process (up to 30 days) and the high capital cost.

No treatment plants currently use the process, but under laboratory conditions it is possible to directly generate useful amounts of electricity from organic sludge using naturally occurring electrochemically active bacteria. Potentially, this technique could lead to positive ecological impact power generation, but in order to be effective such a microbial fuel cell must maximize the contact area between the effluent and the bacteria-coated anode surface, which could severely hamper throughput.

10.1.4.2. *Aerobic digestion:* Aerobic digestion is a bacterial process occurring in the presence of oxygen. Under aerobic conditions, bacteria rapidly consume organic matter and convert it into carbon dioxide. Once there is a lack of organic matter, bacteria die and are used as food by other bacteria. This stage of the process is known as endogenous respiration. Solids reduction occurs in this phase. Because the aerobic digestion occurs much faster than anaerobic digestion, the capital costs of aerobic digestion are lower. However, the operating costs are characteristically much greater for aerobic digestion because of energy costs for aeration needed to add oxygen to the process.

10.1.4.2.1. *Composting:* Composting is also an aerobic process that involves mixing the wastewater solids with sources of carbon such as sawdust or wood chips. In the presence of oxygen, bacteria digest both the wastewater solids and the added carbon source and, in doing so, produce a large amount of heat. Properly designed and controlled, the heat generated can be sufficient to significantly destroy a sufficient number of the disease-causing microorganisms to enable the resulting composted product to be safely used as a soil amendment material (with similar benefits to peat) for agricultural use.

Both anaerobic and aerobic digestion processes can result in the destruction of disease-causing microorganisms and parasites to a sufficient level to allow the resulting digested solids to be safely applied to land or used for agriculture as a fertilizer provided that levels of toxic constituents are sufficiently low.

The choice of a wastewater solid treatment method depends on the amount of solids generated and other site-specific conditions. However, in general, composting is most often applied to smaller-scale applications followed by aerobic digestion and then lastly anaerobic digestion for the larger-scale municipal applications.

10.1.4.2.2. *Thermal depolymerization:* Thermal depolymerization uses hydrous pyrolysis to convert reduced complex organics to oil. Basically, the premacerated, grit-reduced sludge is heated to 250°C and compressed to 40 MPa. The hydrogen in the water inserts itself between chemical bonds in natural polymers such as fats, proteins and cellulose. The oxygen of the water combines with carbon, hydrogen and metals. The result is oil, light combustible gases such as methane, propane and butane, water with soluble salts, carbon dioxide, and a small residue of inert insoluble material that resembles powdered rock and char.

All organisms and many organic toxins are destroyed. Inorganic salts such as nitrates and phosphates remain in the water after treatment at sufficiently high levels that further treatment is required. The energy from decompressing the material is recovered, and the process heat and pressure is usually powered from the light combustible gases. The oil is usually treated further to make a refined useful light grade of oil, such as no. 2 diesel and no. 4 heating oil, and then sold.

### *10.1.5. Sludge disposal*

When a liquid sludge is produced, further treatment may be required to make it suitable for final disposal. Typically, sludges are thickened (dewatered) to reduce the volumes transported off-site for disposal. Processes for reducing water content include lagooning in drying beds to produce a cake that can be applied to land or incinerated; pressing, where sludge is mechanically filtered, often through cloth screens to produce a firm cake; and centrifugation where the sludge is thickened by centrifugally separating the solid and liquid. Sludges can be disposed of by liquid injection to land or by disposal in a landfill. There are concerns about sludge incineration because of air pollutants in the emissions, along with the high cost of supplemental fuel, making this a less attractive and less commonly constructed means of sludge treatment and disposal. There is no process which completely eliminates the requirements for disposal of biosolids.

## 10.2. Waste management

Waste management is the collection, transport, processing or disposal of waste materials, usually ones produced by human activity, in an effort to reduce their effect on human health or local amenity. A subfocus in recent decades has been to reduce waste materials' effect on the environment and to recover resources from them. Waste management can involve solid, liquid or gaseous wastes, with different methods and fields of expertise for each. Waste management practices differ for developed and developing nations, for urban and rural areas, and for residential, industrial, and commercial producers. Waste management for non-hazardous residential and institutional waste in metropolitan areas is usually the responsibility of local government authorities, while management for non-hazardous commercial and industrial waste is usually the responsibility of the generator.

### *10.2.1. Waste management concepts*

Waste management has a number of different concepts, which vary in their usage between countries or regions. The waste hierarchy classifies waste management strategies according to their desirability. The term '3 Rs', or 'Reduce-Reuse-Recycle', has also been used for the same purpose. The waste hierarchy has taken many forms over the past decade, but the basic concept has remained the cornerstone of most waste minimisation strategies. The aim of the waste hierarchy is to extract the maximum practical benefits from products and to generate the minimum amount of waste. Some waste management experts have recently incorporated a 'fourth R': "Re-think", with the implied meaning that the present system may have fundamental flaws, and that a thoroughly effective system of waste management may need an entirely new way of looking at waste. Some "re-think" solutions may be counter-intuitive, such as cutting fabric patterns with slightly more "waste material" left - the now larger scraps are then used for cutting small parts of the pattern, resulting in a decrease in net waste. This type of solution is by no means limited to the clothing industry.

### *10.2.2. Waste management techniques*

Managing domestic, industrial and commercial waste has traditionally consisted of collection, followed by disposal. Depending upon the type of waste and the area, a level of processing may follow collection. This processing may be to reduce the hazard of the waste, recover material for recycling, produce energy from the waste, or reduce it in volume for more efficient disposal. Disposal methods also vary widely. In Australia, the most common method of disposal of solid waste is to landfills, because it is a large country with a low-density population. By contrast, in Japan it is more common for waste to be incinerated, because the country is smaller and land is scarce.

10.2.2.1. *Landfill:* Disposing of waste in a landfill is the most traditional method of waste disposal, and it remains a common practice in most countries. Historically, landfills were often established in disused quarries or mining voids. A well-run landfill can be a hygienic and relatively inexpensive method of disposing of waste materials. Older or poorly managed landfills can create number of adverse environmental impacts, including wind-blown litter, attraction of vermin and soluble contaminants (leachate) leaching into and polluting groundwater. Another product of landfills containing putrescible wastes is landfill gas (mostly composed of methane and carbon dioxide), which is produced as the waste breaks down.

Characteristics of a modern, well-run landfill should include methods to contain leachate, such as clay or plastic liners. Disposed waste should be compacted and covered to prevent vermin and wind-blown litter. Many landfills also have a landfill gas extraction system installed after they are closed to extract the gas generated by the decomposing waste materials. This gas is often burnt to generate power. Generally, even flaring the gas off is a better environmental outcome than allowing it to escape to the atmosphere, as this consumes the methane (a far more potent greenhouse gas than carbon dioxide).

Many local authorities (especially in urban areas) have found it difficult to establish new landfills, due to opposition from adjacent landowners. Few people want a landfill in their local neighbourhood. As a result, solid waste disposal in these areas has become more expensive as material must be transported further away for disposal. Some oppose the use of landfills in any way, anywhere, arguing that the logical end result of landfill operations is that it will eventually leave a drastically polluted planet with no canyons, and no wild space. Some futurists have stated that landfills will be the "mines of the future": as some resources become more scarce, they will become valuable enough that it would be necessary to 'mine' them from landfills where these materials were previously discarded as valueless. This fact, as well as growing concern about the impacts of excessive materials consumption, has given rise to efforts to minimise the amount of waste sent to landfill in many areas. These efforts include taxing or levying waste sent to landfill, recycling the materials, converting material to energy, designing products that require less material, etc. A related subject is that of industrial ecology, where the material flows between industries is studied. The by-products of one industry may be a useful commodity to another, leading to reduced waste materials.

10.2.2.2. *Incineration:* Incineration is the process of destroying waste material by burning it. Incineration is carried out both on a small scale by individuals, and on a large scale by industry. It is recognised as a practical method of disposing of hazardous waste materials (such as biological medical waste).

Though still widely used in many areas (especially developing countries), incineration as a waste management tool is becoming controversial for several reasons.

First, it may be a poor use of many waste materials because it destroys not only the raw material, but also all of the energy, water, and other natural resources used to produce it. Some energy can be reclaimed as electricity by using the combustion to create steam to drive an electrical generator, but even the best incinerator can only recover a fraction of the caloric value of fuel materials. Second, incineration creates toxic gas and ash, which can harm local populations and pollute groundwater. Modern, well-run incinerators take elaborate measures to reduce the amount of toxic products released in exhaust gas. But concern has increased in recent years about the levels of dioxins that are released when burning mixed waste.

Until recently, safe disposal of incinerator waste was a major problem. In the mid-1990s, experiments in France and Germany used electric plasma torches to melt incinerator waste into inert glassy pebbles, valuable in concrete production. Incinerator ash has also been chemically separated into lye and other useful chemicals.

10.2.2.3. *Resource recovery techniques:* A relatively recent idea in waste management has been to treat the waste material as a resource to be exploited, instead of simply a challenge to be managed and disposed off. There are a number of different methods by which resources may be extracted from waste: the materials may be extracted and recycled, or the calorific content of the waste may be converted to electricity. The process of extracting resources or value from waste is variously referred to as secondary resource recovery, recycling, and other terms. The practice of treating waste materials as a resource is becoming more common, especially in metropolitan areas where space for new landfills is becoming scarcer. There is also a growing acknowledgement that simply disposing of waste materials is unsustainable in the long term, as there is a finite supply of most raw materials. There are a number of methods of recovering resources from waste materials, with new technologies and methods being developed continuously.

10.2.2.4. *Recycling:* Recycling means to reuse a material that would otherwise be considered waste. The popular meaning of 'recycling' in most developed countries has come to refer to the widespread collection and reuse of single-use beverage containers. These containers are collected and sorted into common groups, so that the raw materials of the items can be used again (recycled).

In developed countries, the most common consumer items recycled include aluminium beverage cans, steel food and aerosol cans, HDPE and PET plastic bottles, glass bottles and jars, paperboard cartons, newspapers, magazines, and cardboard. Other types of plastic and PS are also recyclable, although not as commonly collected. These items are usually composed of a single type of material, making them relatively easy to recycle into new products. The recycling of obsolete computers and electronic equipment is important although more costly due to the separation and extraction problems. The recycling of junked automobiles also depends on the scrap metal market.

Recycled or used materials have to compete in the marketplace with new (virgin) materials. The cost of collecting and sorting the materials usually means that they are equally or more expensive than virgin materials. This is most often the case in developed countries where industries producing the raw materials are well-established. Practices such as trash picking can reduce this value further, as choice items are removed (such as aluminium cans). In some countries, recycling programs are subsidised by deposits paid on beverage containers (see container deposit legislation).

Not accounted for by most economic systems are the benefits to the environment of recycling these materials, compared with extracting virgin materials. It usually requires significantly less energy, water and other resources to recycle materials than to produce new materials. For example, recycling 1000 kg of aluminium cans saves approximately 5000 kg of bauxite ore being mined and 95% of the energy required to refine it.

In many areas, material for recycling is collected separately from general waste, with dedicated bins and collection vehicles. Other waste management processes recover these materials from general waste streams. This usually results in greater levels of recovery than separate collections of consumer-separated beverage containers, but are more complex and expensive.

10.2.2.5. *Composting and digestion:* Waste materials that are organic in nature, such as food scraps and paper products, are increasingly being recycled. These materials are put through a composting or artificial digestion process to decompose the organic matter and kill pathogens. The organic material is then recycled as mulch or compost for agricultural or landscaping purposes.

There are a large variety of composting methods and technologies, varying in complexity from simple window composting of shredded plant material, to automated enclosed-vessel digestion of mixed domestic waste. Composting methods can be broadly categorised into aerobic or anaerobic methods, although hybrids of the two methods also exist. Aerobic methods of composting seek to aerate the organic material continuously or frequently, in order to promote rapid and odourless decomposition. Anaerobic methods of composting seek to maximise the generation of gases such as methane during the process, in order to produce power from the waste materials.

10.2.2.6. *Incineration, pyrolysis and gasification:* Use of incinerators for waste management is controversial, and most Americans passionately oppose it. This controversy roots from the understandable conflict between short-term concerns and long-term ones, in this case between burning the wastes now, or postponing this problem by passing the waste burden to future generations. Whether any form of incineration or thermal treatment should be defined as "resource recovery" is a matter of dispute in policy-making circles.

Pyrolysis and gasification are two related forms of thermal treatment where materials are incinerated with limited oxygen. The process typically occurs in a sealed vessel, under high temperature and pressure. Converting material to energy this way is more efficient than direct incineration, with more energy able to be recovered and used.

Pyrolysis of solid waste converts the material into solid, liquid and gas products. The liquid oil and gas can be burnt to produce energy or refined into other products. The solid residue (char) can be further refined into products such as activated carbon.

Gasification is used to convert organic materials directly into a synthetic gas composed of carbon monoxide and hydrogen. The gas is then burnt to produce electricity and steam. Gasification is used in biomass power stations to produce renewable energy and heat.

## 10.3. Industrial wastewater treatment

Industrial wastewater treatment covers the mechanisms and processes used to treat waters that have been contaminated in some way by man's industrial or commercial activities prior to its release into the environment or its re-use. The different types of contamination of wastewater require a variety of strategies to remove the contamination.

### *10.3.1. Solids removal*

Most solids can be removed using simple sedimentation techniques with the solids recovered as slurry or sludge. Very fine solids and solids with densities close to one pose special problems. In such case filtration or ultra-filtration may be required. Alternatively, flocculation may be used using alum salts or the addition of poly-electrolytes.

### 10.3.2. Oils and greases removal

Many oils can be recovered from open water surfaces by skimming devices. However, hydraulic oils and the majority of oils that have degraded to any extent will also have a soluble or emulsified component that will require further treatment to eliminate. Dissolving or emulsifying oil using surfactants or solvents usually exacerbates the problem rather than solving it producing a very difficult to treat wastewater.

### 10.3.3. Soft organics removal

Organic material of plant or animal origin is usually possible to treat using extended conventional sewage treatment processes. Problems can arise if the wastewater is excessively diluted with washing water or is highly concentrated such as neat blood or milk. The presence of cleaning agents, disinfectants, pesticides, or anti-biotics can have detrimental impacts on treatment processes.

### 10.3.4. Hard organics removal

Synthetic organic materials including solvents, paints, pharmaceuticals, pesticides, coking products etc can be very difficult to treat . Treatment methods are often specific to the material being treated . Methods include distillation, adsorption, vitrification, incineration, chemical immobilisation or landfill disposal. Some materials such as some detergents may be capable of biological degradation and in such cases, a modified form of sewage treatment can be used.

### 10.3.5. Acids and alkalis removal

Acids and alkalis can usually be neutralised under controlled conditions. Neutralisation frequently produces a precipitate that will require treatment as a solid residue that may also be toxic. In some cases, gases may be evolved requiring treatment. Some other forms of treatment are usually required following neutralisation.

Waste streams rich in hardness ions as from de-ionisation processes can readliy loose the hardness ions in a buildup of preciptaed calcium and magnesium salts. This preciptation process can cause severe furring of pipes and can, in extreme cases, cause the blockage of disposal pipes. A 1 metre diameter industrial marine discharge pipe serving a major chemicals complex was blocked by such salts in the 1970s. Treament is by concentration of de-ionisation waste waters and disposal to landfill or by careful pH management of the released wastewater.

### 10.3.6. Toxic materials removal

Toxic materials including many organic materials, metals (such as zinc, silver, cadmium, thallium etc.) acids, alkalis, non-metallic elements (such as arsenic or selenium) are generally resistant to biological processes unless very dilute. Metals can often be precipitated out by changing the pH or by treatment with other chemicals. Many, however, are resistant to treatment or mitigation and may require concentration followed by landfilling or recycling.

## 10.4. Agricultural wastewater treatment

Agricultural wastewater treatment relates to the treatment of wastewaters produced in the course of agricultural activities. As agriculture is a highly intensified industry in many parts of the world, the range of wastewaters requiring treatment can be amongst the following:

### *10.4.1. Animal wastes*

The constituents of animal wastewater typically contain: i. Strong organic content-much stronger than human sewage, ii. High solids concentration, iii. High nitrate and phosphorus content, iv. Antibiotics, v. Synthetic hormones, vi. Often high concentrations of parasites and their eggs, vii. Spore of Cryptosporidum - a bacterium resistant to drinking water treatment processes, viii. Spore of *Giardia*, ix. Human pathogenic bacteria such as *Brucella* and *Salmonella*.

Animal wastes from cattle can be as produced as solid or semisolid manure or as a liquid slurry. The production of slurry is especially common in housed dairy cattle. Whilst solid manure heaps outdoors can give rise to polluting wastewaters from rain washing, this type of waste is usually relatively easy to treat by containment and/or covering of the heap.

Animal slurries require special handling and are usually treated by containment in lagoons before disposal by spray or trickle application to grassland. Constructed wetlands are sometimes used to facilitate treatment of animal wastes. Excessive application or application to sodden land or insufficient land area can result in direct runoff to watercourses with the potential for causing severe pollution. Application of slurries to land overlying aquifers can result in direct contamination or, more commonly, elevation of nitrogen levels as nitrite or nitrate.

The disposal of any wastewater containing animal waste upstream of a drinking water intake can pose serious health problems to those drinking the water because of the highly resistant spores present in many animals that are capable of causing disabling in humans. This risk exists even for very low level seepage via shallow surface drains or from rainfall run-off.

Some animal slurries are treated by mixing with straws and composted at high temperature to produce a bacteriologically sterile and friable manure for soil improvement

### *10.4.2. Piggery waste*

Piggery waste is comparable to other animal wastes except that many piggery wastes contain elevated levels of copper that can be toxic in the natural environment. Ascaris worms and their eggs are also common and can infect humans if wastewater treatment is ineffective. As for general animal waste although the liquid fraction of the waste is frequently separated off and re-used in the piggery to avoid the prohibitively expensive costs of disposing of a copper rich liquor.

### 10.4.3. Silage liquor

Fresh or wilted grass or other green crops can be made into the semi fermented product called silage which can be stored and used as winter forage for cattle and sheep. The production of silage often involves the use of an acid conditioner such as sulphuric acid or formic acid. The process of silage making frequently produces a yellow-brown strongly smelling liquid which is very rich in simple sugars, alcohol, short-chain organic acids and silage conditioner. This liquor is one of the most polluting organic substances known. The volume of silage liquor produced is generally in proportion to the moisture content of the ensiled material.

Silage liquor is best treated through prevention by wilting crops well before silage making. Any silage liquor that is produced can be used as part of the food for pigs. The most effective treatment is by containment in a slurry lagoon and subsequently spread on land following substantial dilution with slurry. Containment of silage liquor on its own can cause structural problems in concrete pits because of the acidic nature of silage liquor.

### 10.4.4. Pesticide runoff and surpluses

Inappropriate use of pesticides so that pesticide-containing wastewaters enter the environment can give rise to severe and long-lasting ecological damage. This is particularly true for insecticides used in sheep dips because of the volumes of pesticide-containing wastewater requiring disposal and because of the persistent and damaging nature of the pesticides. There are few safe ways of disposing of pesticide surpluses other than through containment in well managed landfills or by incineration. In some parts of the world, spraying on land is a permitted method of disposal.

### 10.4.5. Milking parlour wastes including milk

Although milk has a deserved reputation as an important and valuable food product, its presence in wastewaters is highly polluting because of its organic strength, which can lead to very rapid de-oxygenation of receiving waters. Milking parlour wastes also contain large volumes of wash-down water, some animal waste together with cleaning and disinfection chemicals. Milking parlour waste are often treated in admixture with human sewage in a local sewage treatment plant. This ensures that disinfectants and cleaning agents are sufficiently diluted and amenable to treatment. Running milking wastewaters into a farm slurry lagoon is a possible option although this tends to consume lagoon capacity very quickly. Land spreading is also a treatment option.

### 10.4.6. Slaughtering waste

Wastewater from slaughtering activities is similar to milking parlour waste although considerably stronger in its organic composition and therefore potentially much more polluting. The treatment of slaughtering waste is done as that of milking parlour waste.

### 10.4.7. Vegetable washing water

Washing of vegetables produces large volumes of water contaminated by soil and vegetable pieces. Low levels of pesticides used to treat the vegetables may also be present together with moderate levels of disinfectants such as chlorine. Most vegetable washing waters are extensively recycled with the solids removed by settlement and filtration. The recovered soil can be returned to the land.

### *10.4.8. Fire water*

Although few farms plan for fires, fires are nevertheless more common on farms than on many other industrial premises. Stores of pesticides, herbicides, fuel oil for farm machinery and fertilisers can all help promote fire and can all be present in environmentally lethal quantities in wastewater from firefighting at farms. All farm environmental management plans should allow for containment of substantial quantities of firewater and for its subsequent recovery and disposal by specialist disposal companies. The concentration and mixture of contaminants in fire-water make them unsuited to any treatment method available on the farm. Even land spreading has produced severe taste and odour problems for downstream water supply companies in the past.

## 11. CONCLUSIONS AND FUTURE OUTLOOK

Industry has become an essential part of modern society, and waste production is an inevitable outcome of the developmental activities. Currently in India even though hazardous wastes, emanations and effluents are regulated, solid wastes often are disposed off indiscriminately posing health and environmental risk. Therefore, management of hazardous wastes including their disposal in environment friendly and economically viable way is very important and needs R&D inputs for developing better strategies. Among the various categories of wastes, solid waste contributes a major share towards environmental degradation. Appraisal of the whole situation to the society and the development of better cost-effective microorganism based strategies for waste management is the need of the hour. Industrial and environmental biotechnology in collaboration are resulting in processes with "clean technologies", aimed at maximum production and least residues. Technologies of bioremediation are continuously being improved using naturally occurring or genetically modified microorganisms to clean the contaminated areas from toxic organics. Further, microbial enzymes are powerful tools being utilized for environmental cleaning and waste management in most of industries including agro-food, oil, animal feed, detergent, pulp and paper, textile, leather, petroleum etc. Recombinant DNA technology, protein engineering, and rational enzyme design are the emerging areas of research pertaining to environmental pollution management. The future will also see the employment of various technologies including gene shuffling, high throughput screening and nanotechnology.

The management of wastewater sludge from wastewater treatment plants, represents another major challenge in wastewater treatment, therefore, the focus in sludge minimization is steadily increasing. Various sludge disintegration technologies for sludge minimization including mechanical, physical, chemical and biological methods involving microorganisms or their enzymes are expected to play an important role in pollution management during the 21$^{st}$ century.

The quality of the potable water is also threatened by contamination with nutrients, primarily nitrogen and phosphorus. Animal manure can be a valuable resource for farmers, providing nutrients, improving soil structure, and increasing vegetative cover to decrease erosion potential. Apart from this, application of manure nutrients in excess of crop requirements can result in environmental contamination and pollution due to runoff of P. If agricultural practices continue as they have been in the past, continued damage to water resources and a loss of fishing and recreational activity are inevitable. Decreasing the P content of manure through nutrition is a powerful, cost-effective approach to reducing P losses from livestock farms and will help farmers meet increasingly stringent environmental regulations. To address the problem of solid waste, some countries have formulated a policy based on integrated waste management. The policy asks for reduction of waste at source, reuse, recycling (including composting), waste-to-energy technologies, and landfilling. Due to the implementation of this policy, all the large dumps have been closed, state-of-the art landfills have been built, and recovery rates have increased. Municipal solid waste is disposed and treated in an environmentally sound manner. Such a policy should also be implemented by other countries utilizing both enforcement and financial support so as to reduce the environmental pollution during the 21st century.

## 12. FURTHER READING


Adamse, A.D., Deinema, M.H. and Zehnder, A.J. (1984). Studies on bacterial activities in aerobic and anaerobic wastewater purification. Antonie Van Leeuwenhoek. 50: 665-82.

Focht , D.D. and Chang, A.C. (1975). Nitrification and denitrification processes related to wastewater treatment. Adv Appl Microbiol. 19:153-86.

Kadlec, R.H. (2005). Nitrogen farming for pollution control. J Environ Sci Health A Tox Hazard Subst Environ Eng. 40:1307-30.

Klimmek , O. (2005). The biological cycle of sulfur. Met Ions Biol Syst. 43:105-30.

Perez, J., Munoz-Dorado, J., de la Rubia, T. and Martinez, (2002). Biodegradation and biological treatments of cellulose, hemicellulose and lignin: an overview. J. Int Microbiol. 5: 53-63.

Rudolf, M. and Kroneck, P.M. (2005). The nitrogen cycle: its biology. Met Ions Biol Syst.;43: 75-103.

Street, J.H. and Paytan, A. (2005). Iron, phytoplankton growth, and the carbon cycle. Met Ions Biol Syst. 43:153-93.

Wagner, M. and Loy, A. (2002). Bacterial community composition and function in sewage treatment systems. Curr Opin Biotechnol. 13:218-27.

# 4

# Microbial Metabolism
# Energy Release and Conservation

**R. SONI[1] AND S.K. SONI[2]**

*[1]Department of Biotechnology, D.A.V. College, Chandigarh-160 011*

*[2]Department of Microbiology, Panjab University, Chandigarh-160 014*

## 1. INTRODUCTION

Every living organism needs energy which is the ability to do work. Energy is utilized for the construction of physical parts of the cell including wall, membrane, enzymes, nucleic acids, polysaccharides and other components. Many cells obtain energy by carrying out chemical reactions and others use light as a source of energy that too is first converted into chemical energy. A large number of organized chemical activities are going on in a cell which are a part of metabolism. The term metabolism refers to the sum of the biochemical reactions required for energy generation and the use of energy to synthesize cell material from small molecules in the environment. Hence, metabolism has an energy-generating component, called catabolism or dissimilatory metabolism, and an energy-consuming, biosynthetic component, called anabolism or assimilatory metabolism. Catabolic reactions produce energy, which can be utilized in anabolic reactions to buildup cell material. During catabolism, energy is transformed to another form according to the laws of thermodynamics. Energy transformations are never 100% because some energy is lost in the form of heat. The efficiency of a catabolic sequence of reactions is the amount of energy made available to the cell divided by the total amount of energy released during the reactions.

## 2. THE THERMODYNAMIC AND MECHANISTIC BASIS OF CELLULAR METABOLISM

The amount of energy available to drive useful work (Gibbs free energy) released in the course of the dissimilatory process performed is the most important bioenergetic parameter for any living organism. Therefore, the change in free energy ($\Delta G$) of any reaction or process determines whether it can (at least theoretically) be used for energy generation. Two types of reactions occur in the cells, exothermic and endothermic. Energy liberated or taken up during the reaction is called free energy change ($\Delta G$) of the reaction. If $\Delta G$ has a negative value the reaction releases energy (exergonic) and in case this has a positive value then the energy is taken up (endergonic). Exergonic and endergonic reactions are coupled with each other and energy released from exergonic reaction is used to drive endergonic reaction. The coupling of exothermic and endothermic reactions involve some common energy-rich or energy transfer compounds like adenosine triphosphate (ATP), guanosine triphosphate (GTP), uridine triphosphate (UTP) and cytidine triphosphate (CTP) etc., each of which release significant amount of energy upon breakdown.

Of the various energy-rich compounds, ATP is most common energy currency of the cell. This high energy compound has 3 phosphate groups attached to adenosine (Adenine-Ribose). Energy is released from ATP by hydrolysis during which 8.15 k. cal of energy is released leading to the generation of ADP. The change in free energy during any chemical reaction can be calculated according to the following equation:

$$\Delta G^{o} = \sum \Delta G^{o}_{f} \text{ (products)} - \sum \Delta G^{o}_{f} \text{ (reactants)}$$

$\Delta G_f^o$ represents the free energy (in kJ / $mol^{-1}$) required for the synthesis of the compounds involved from the elements of which they are composed. The sign "$o$" signifies that all calculations relate to standard conditions: all compounds that participate in the reaction are present at a concentration of 1 M (or in case of gases: 1 atmosphere), and the reaction temperature is 25°C.

In every living cell, there are two forms of energy available to the cell metabolism: (1) chemical energy in the form of ATP and a number of additional compounds possessing high-energy bonds, and (2), the proton electrochemical gradient (proton motive force, $\Delta\mu_H{}^+$), in rare cases accompanied by gradients of other ions such as $Na^+$, between the two sides of the cell membrane of prokaryotes, the inner membrane of mitochondria, or the thylakoids of chloroplasts. These two forms of energy are interchangeable by means of action of the ATP synthase located in the membrane. This enzyme enables the formation of ATP at the expense of the proton electrochemical gradient, and when acting in the opposite direction, it enables the build-up of the $\Delta\mu_H{}^+$ at the expense of high-energy bonds in ATP. In some of the modes in which organisms generate energy, ATP formation is the primary process (substrate-level phosphorylation, e.g., during bacterial and yeast fermentation). In other types of metabolism, such as electron transport during respiration or photosynthesis, the formation of the proton electrochemical gradient is the primary process, and ATP is produced as a secondary process in the course of the dissipation of the proton gradient ("electron transport phosphorylation").

## 2.1. Adenosine triphosphate

Adenosine triphosphate (ATP) is derived from the nucleotide adenosine monophosphate (AMP) or adenylic acid, to which two additional phosphate groups are attached (Figure 1) through pyrophosphate bonds (~P). These two bonds are energy rich in the sense that their hydrolysis yields a great deal more energy than a corresponding covalent bond. ATP acts as a coenzyme in energetic coupling reactions wherein one or both of the terminal phosphate groups is removed from the ATP molecule with the bond energy being used to transfer part of the ATP molecule to another molecule to activate its role in metabolism. For example, Glucose + ATP → Glucose-P + ADP or Amino Acid + ATP → AMP-Amino Acid + PPi. Because of the central role of ATP in energy-generating metabolism, it is involved as a coenzyme in most energy-producing processes in cells.

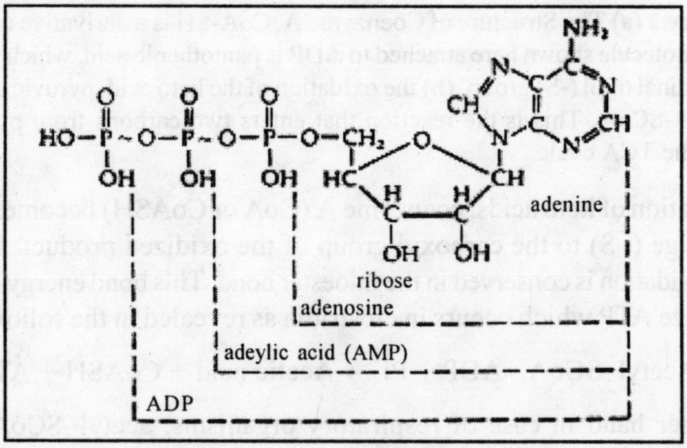

**Figure 1** The structure of ATP

Another coenzyme involved in energy-generating metabolism of prokaryotes and eukaryotes is Coenzyme A.

## 2.2. Coenzyme A

Coenzyme A (Figure 2) participates in ATP generating reactions in all respiratory organisms and some fermentative bacteria. The reaction occurs in association with the oxidation of keto acids such as pyruvic acid and α-ketoglutaric acid. These substrates are central to glycolysis and the TCA cycle, respectively, and they are direct or indirect precursors of several essential macromolecules in a cell. The oxidations of pyruvate and α-ketoglutarate, involving Coenzyme A, NAD, a dehydrogenation reaction and a decarboxylation reaction, are two of the most important, and complex, reactions in metabolism.

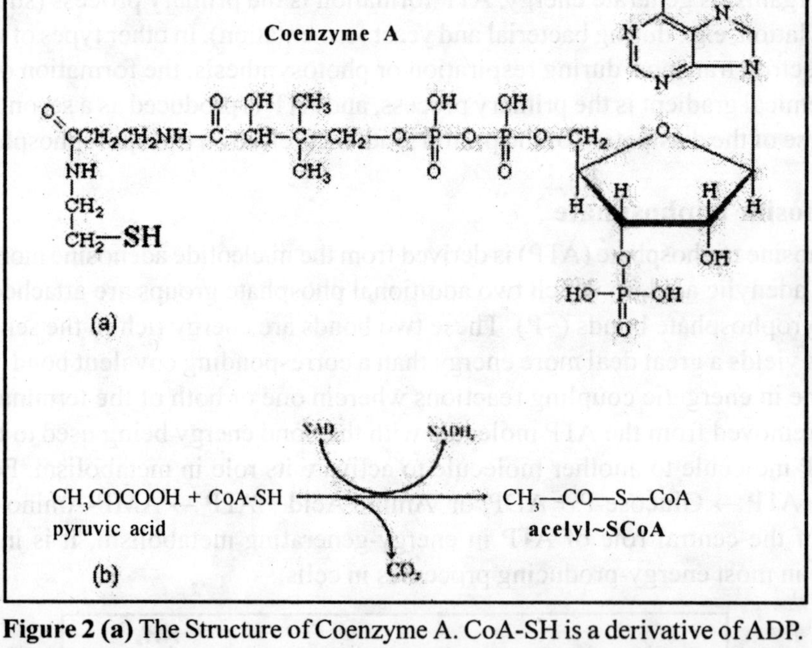

Coenzyme A

(a)

$$CH_3COCOOH + CoA\text{-}SH \xrightarrow{NAD \rightarrow NADH_2,\ -CO_2} CH_3\text{—}CO\text{—}S\text{—}CoA$$

pyruvic acid → acetyl~SCoA

(b)

**Figure 2 (a)** The Structure of Coenzyme A. CoA-SH is a derivative of ADP. The molecule shown here attached to ADP is pantothenic acid, which carries a terminal thiol (-S) group. **(b)** the oxidation of the keto acid, pyruvic acid, to acetyl~SCoA. This is the reaction that enters two carbons from pyruvate into the TCA cycle

In the oxidation of keto acids, coenzyme A (CoA or CoASH) becomes attached through a thioester linkage (~S) to the carboxyl group of the oxidized product. Part of the energy released in the oxidation is conserved in the thioester bond. This bond energy can be subseqently used to synthesize ATP which occurs in clostridia as revealed in the following equation:

$$\text{Acetyl\~SCoA} + \text{ADP} + \text{Pi} \rightarrow \text{Acetic acid} + \text{CoASH} + \text{ATP}.$$

On the other hand in case of respiratory organisms, acetyl~SCoA condenses with oxaloacetate in order to drive the TCA cycle into its oxidative branch.

## 2.3. Modes of ATP synthesis

In living organisms, ATP can be generated directly from the breakdown of organic compounds, accompanied by the synthesis of another organic compound, along with high energy phosphate bond in a process known as substrate level phosphorylation. In another mode, ATP molecules are liberated during the oxidation of reduced coenzymes, NADH, $FADH_2$ and the process is known as oxidative phosphorylation.

**2.3.1.** *Substrate level phosphorylation (SLP):* This is the simplest, oldest and least-evolved way to make ATP. It is based on the formation of compounds possessing phosphate groups bound with high-energy bonds (in most cases phosphate anhydride bonds), generated during the breakdown of organic compounds. Enzymatic reactions enable the substitution of this high-energy bond for another high-energy bond: the bond between the terminal phosphate and ADP in the ATP molecule (Figure 3). The number of different high-energy compounds used by the cell to produce ATP is limited. Table 1 presents the most important high-energy molecules involved in substrate-level phosphorylation.

**Table 1** The most common high-energy compounds involved in substrate-level phosphorylation

| High-energy compound | $\Delta G^{o\prime}$ of hydrolysis (k.cal / mol$^{-1}$) |
|---|---|
| 1,3-Diphosphoglycerate | - 12.44 |
| Phosphoenolpyruvate | -51.36 |
| Acetyl phosphate | -10.73 |
| Acetyl-CoA | - 8.55 |
| Carbamyl phosphate | -9.42 |
| Adenosine-5′-phosphosulfate | -21.09 |
| Succinyl-CoA | -35.10 |

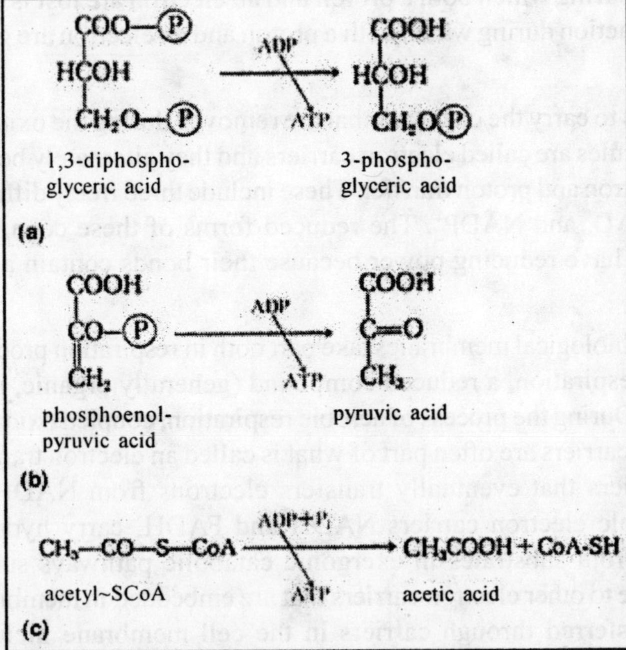


**Figure 3** Three examples of substrate level phos phorylation. **(a)** and **(b)** are the two substrate level phosphorylations that occur during the Embden-Meyerhof-Parnas pathway, but they occur in all other fermentation pathways which have an Embden-Meyerhof component. **(c)** is a substrate level phos-phorylation found in *Clostridium* and *Bifidobacterium*

**2.3.1.** *Oxidative phosphorylation:* This is also called electron transport phosphorylation (ETP) which is a complex pathway that evolved long after SLP. This involves the use of energy from electron transport through a chain of electron carriers such as flavines, quinones, and cytochromes, located in membranes, to make ATP. Thus as many as three ATP molecules may be synthesized from ADP and Pi when a pair of electrons pass from NADH to an atom of $O_2$. This is the same thing as saying that the phosphorus to oxygen (P/O) ratio is equal to 3. Because electrons from $FADH_2$ only pass two oxidative phosphorylation points, the maximum P/O ratio for $FADH_2$ is 2. The actual P/O ratios may be less than 3.0 and 2.0 in eucaryotic mitochondria.

The mechanism by which oxidative phosphorylation takes place has been studied intensively for years. Currently, the most widely accepted hypothesis about how oxidative phosphorylation occurs is the chemiosmotic hypothesis. According to the chemiosmotic hypothesis first formulated in 1961 by the British biochemist Peter Mitchell, the electron transport chain is organized so that protons move outward from the mitochondrial matrix and electrons are transported inward. To understand oxidative phosphorylation, it is important to first review the hydrogen atom and the process of oxidation and reduction. An atom of hydrogen contains only one proton ($H^+$) and one electron ($e^-$). Therefore, the term proton and the term hydrogen ion ($H^+$) are interchangable. The electrons have stored energy (potential energy), available to do work.

Oxidation-reduction reactions are coupled chemical reactions in which one atom or molecule loses one or more electrons (oxidation) while another atom or molecule gains those electrons (reduction). The compound that loses elecirons becomes oxidized, the compound that gains these electrons becomes reduced. In covalent compounds, however, it is usually easier to lose a whole hydrogen (H) atom - a proton and an electron - rather than just an electron. An oxidation reaction during which both a proton and an electron are lost is called dehydrogenation. A reduction reaction during which both a proton and an electron are gained is called hydrogenation.

Cells use specific molecules to carry the electrons that are removed during the oxidation of an energy source. These molecules are called electron carriers and they alternately become oxidized and reduced during electron and proton transfer. These include three freely diffusible coenzymes known as $NAD^+$, FAD, and $NADP^+$. The reduced forms of these coenzymes (NADH, $FADH_2$, and NADPH) have reducing power because their bonds contain a form of usable energy.

Electron transport chains in biological membranes take part both in respiration processes and in photosynthesis. During respiration, a reduced compound (generally organic, rarely inorganic) molecule is oxidized. During the process of aerobic respiration, coupled oxidation-reduction reactions and electron carriers are often part of what is called an electron transport chain, a series of electron carriers that eventually transfers electrons from NADH and $FADH_2$ to oxygen. The diffusible electron carriers NADH and $FADH_2$ carry hydrogen atoms (protons and electrons) from substrates in exergonic catabolic pathways such as glycolysis and the citric acid cycle to other electron carriers that are embedded in membranes. The electrons released are transferred through carriers in the cell membrane including flavoproteins, iron-sulfur proteins, quinones, and cytochromes, to a terminal electron acceptor. The final electron acceptors in respiration are oxygen (in the case of aerobic organisms) or other oxidized inorganic compounds, such as nitrate ($NO_3^-$) or sulfate ($SO_4^{2-}$) (in case of anaerobic respiration).

The chemiosmotic theory explains the functioning of electron transport chains. According to this theory, the tranfer of electrons down an electron transport system through a series of oxidation-reduction reactions releases energy (Figure 4). This energy allows certain carriers in the chain to transport hydrogen ions ($H^+$) across the membrane.

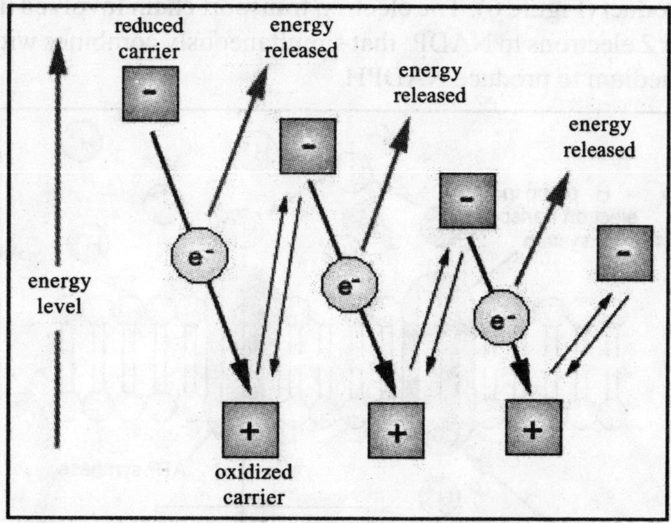


**Figure 4** Energy release from an electron transport system

Depending on the type of cell, the electron transport chain may be found in, the cytoplasmic membrane, the inner membranes of mitochondria and chloroplasts.

- In prokaryotic cells, the protons are transported from the cytoplasm of the bacterium across the cytoplasmic membrane to the periplasmic space located between the cytoplasmic membrane and the cell wall.
- In eukaryotic cells, protons are transported from the matrix of the mitochondria across the inner mitochondrial membrane to the intermembrane space located between the inner and outer mitochondrial membranes.
- In plant cells and the cells of algae, protons are transported from the stroma of the chloroplast across the thylakoid membrane into the interior space of the thylakoid.

As the hydrogen ions accumulate on one side of a membrane, the concentration of hydrogen ions creates an electrochemical gradient or potential difference (voltage) across the membrane. (The fluid on the side of the membrane where the protons accumulate acquires a positive charge; the fluid on the opposite side of the membrane is left with a negative charge.) The energized state of the membrane as a result of this charge separation is called proton motive force or PMF (Figure 5).

In an electron transport system, energy from electron transfer during oxidation-reduction reactions enables certain carriers to transport protons ($H^+$) across a membrane. As the $H^+$ concentration increases on one side of the membrane, an electrochemical gradient called proton motive force develops. Re-entry of the protons through an enzyme complex called ATP synthase provides the energy for the synthesis of ATP from ADP and phosphate.

At the end of the electron transport chain involved in aerobic respiration, the last electron carrier in the membrane transfers 2 electrons to half an oxygen molecule (an oxygen atom) that simultaneously combines with 2 protons from the surrounding medium to produce metabolic water as an end product (Figure 6). The electron transport chain involved in photosynthesis ultimately transfer 2 electrons to $NADP^+$ that simultaneously combines with 2 protons from the surrounding medium to produce NADPH.

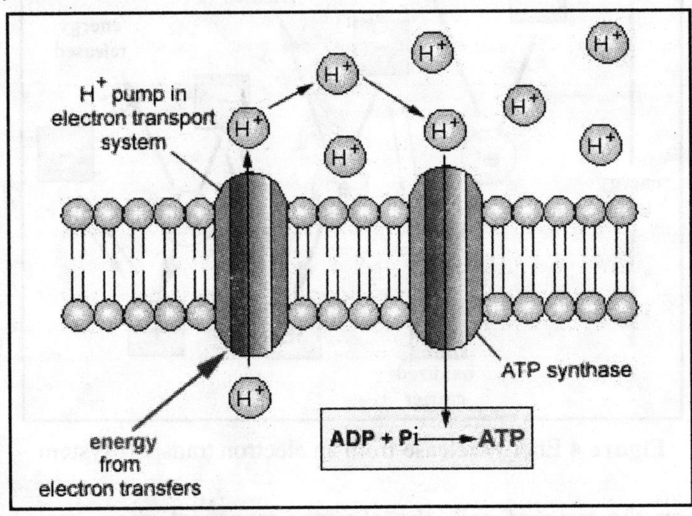


**Figure 5** Development of proton motive force from chemiosmosis and generation of ATP

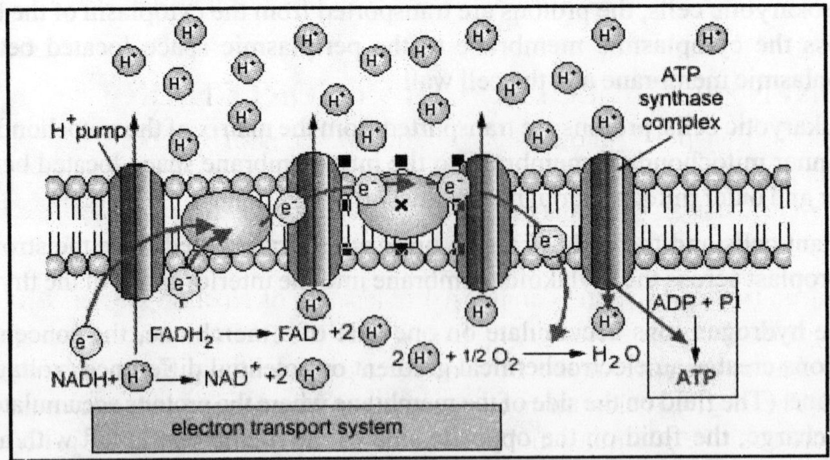


**Figure 6** ATP production during aerobic respiration by oxidative phosphorylation involving an electron transport system and chemiosmosis

When electron flow is from a low redox potential to a high one the process is exergonic, and part of the free energy released is conserved in the form of a proton electrochemical gradient. Photosynthesis is based on the use of light as the energy source. As a result of the absorption of photons by chlorophyll or bacteriochlorophyll molecules in the photosynthetic reaction centers, low-potential electrons are formed, and these are transferred though a chain of electron carriers in the photosynthetic membranes under formation of a primary proton gradient.

All processes mentioned above supply energy for the existence of the cell (dissimilatory processes). The energy that the cell gains in the form of ATP or in the form of a proton electrochemical gradient is utilized for all endergonic processes in the cell including assimilation of nutrients, biosynthesis of proteins, other macromolecules and for maintenance purposes. The energy available in the form of ATP or $\Delta\mu_H^+$ is the mediator between the dissimilatory and the assimilatory processes (Figure 7).

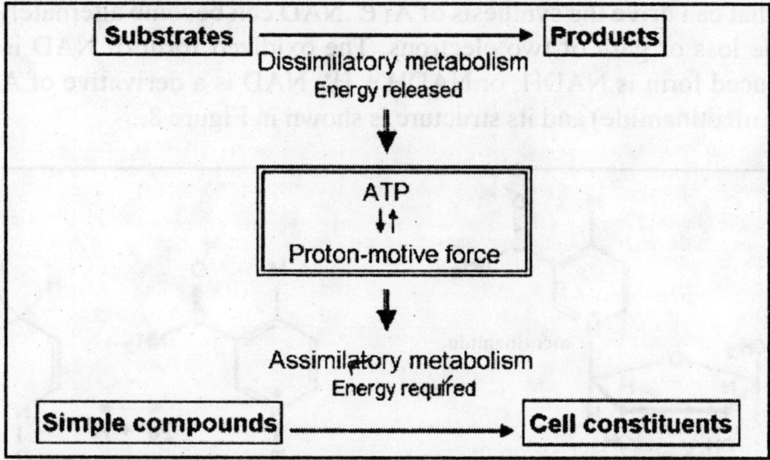


**Figure** 7 The function of ATP and the primary electrochemical gradient of protons ($\Delta\mu_H^+$) as a mediator between assimilatory and dissimilatory metabolism

## 3. COMPONENTS OF ELECTRON TRANSPORT CHAIN

The electron transport system is a coordinated series of reactions that operate in eukaryotic and prokaryotic microorganisms, which enables electrons to be passed from one carrier to another. The purpose of the electron transport system is to pump hydrogen ions to an enzyme that utilizes the energy from the ions to manufacture ATP molecules.

The electron transport chain is composed of a series of electron carriers that operate together to transfer electrons from donors, like NADH and $FADH_2$, to acceptors such as $O_2$ and proteins that participate in the flow of electrons including flavoproteins and the cytochromes. The electrons flow from carriers with more negative reduction potentials to those with more positive potentials and eventually combine with $O_2$ and $H^+$ to form water. The difference in reduction potentials between $O_2$ and NADH is large, about 1.14 volts, and makes possible the release of a great deal of energy. The potential changes at several points in the chain are large enough to provide sufficient energy for ATP production. The electron transport chain breaks up the large overall energy release into small steps. Some of the liberated energy is trapped in the form of ATP.

The ATP synthase attempts to restore the equilibrium of the hydrogen and hydronium ions by pumping a hydrogen ion back into the cell for each electron that is accepted. The energy supplied by the hydrogen ion is used to add a phosphate group to a molecule called adenine diphosphate (ADP), generating ATP.

## 3.1. Electron carriers in ETC

**$NAD^+$**, or Nicotinamide adenine dinucleotide, is a coenzyme that often works in conjunction with an enzyme called a dehydrogenase. The enzyme removes two hydrogen atoms ($2H^+$ and $2e^-$) from its substrate. Both electrons but only one proton are accepted by the $NAD^+$ to produce its reduced form, NADH, plus $H^+$. NADH is used to generate proton motive force that can drive the synthesis of ATP. NAD can become alternately oxidized or reduced by the loss or gain of two electrons. The oxidized form of NAD is symbolized NAD; the reduced form is $NADH_2$ or NADH + $H^+$. NAD is a derivative of ADP (NAD = ADP-ribose + nicotinamide) and its structure is shown in Figure 8.

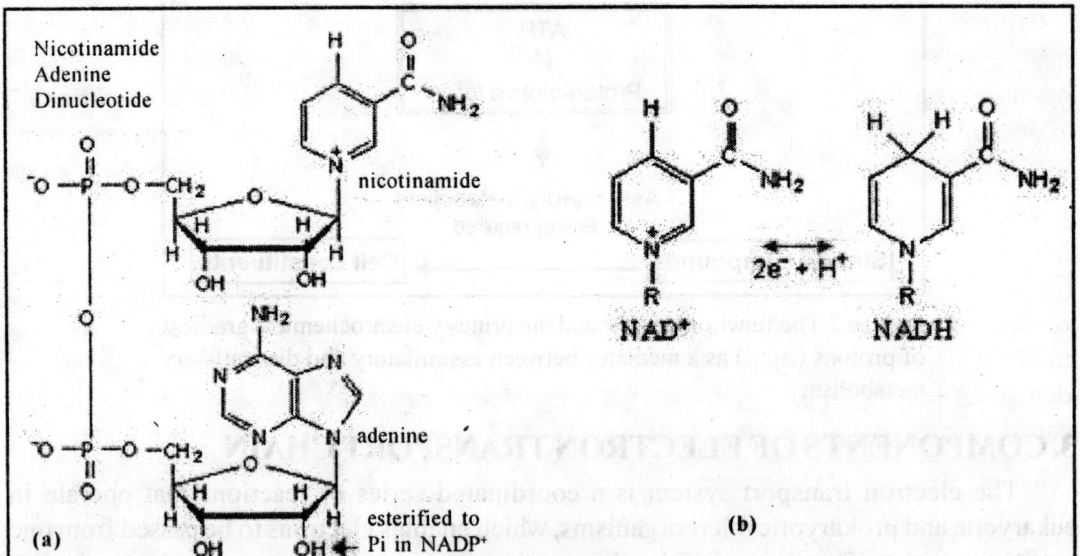

**Figure 8** The Structure of NAD. **(a)** Nicotinamide adenine dinucleotide is composed of two nucleotide molecules: Adenosine monophosphate (adenine plus ribose-phosphate) and nicotinamide ribotide (nicotinamide plus ribose-phosphate). NADP has an identical structure except that it contains an additional phosphate group attached to one of the ribose residues. **(b)** The oxidized and reduced forms of of the nicotinamide moiety of NAD. Nicotinamide is the active part of the molecule where the reversible oxidation and reduction takes place.The oxidized form of NAD has one hydrogen atom less than the reduced form and, in addition, has a positive charge on the nitrogen atom which allows it to accept a second electron upon reduction. Thus the correct way to symbolize the reaction is $NAD^+ + 2H \rightarrow NADH + H^+$. However, for convenience we will hereafter use the symbols NAD and $NADH_2$

**FAD**, or Flavin a denine dinucleotide, a coenzyme also works in conjunction with an enzyme called dehydrogenase. The enzyme removes two hydrogen atoms ($2H^+$ and $2e^-$) from its substrate (Figure 9). Both electrons and both protons are accepted by the FAD to produce its reduced form, $FADH_2$. $FADH_2$ is used to generate proton motive force that can drive the synthesis of ATP.

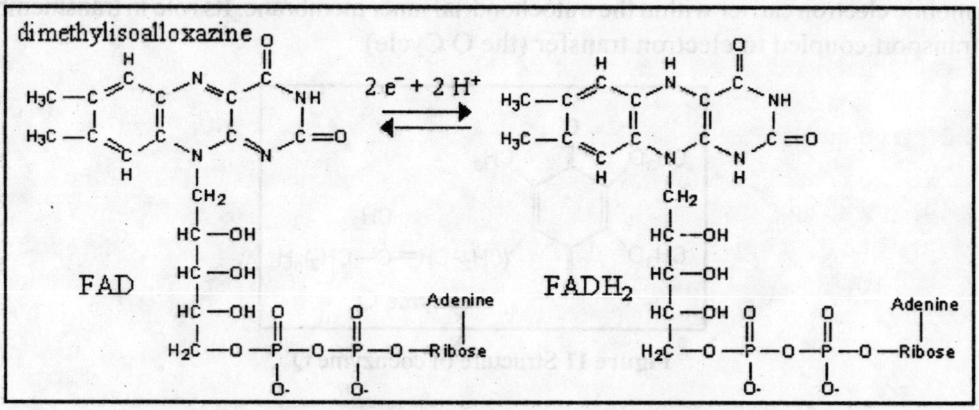

dimethylisoalloxazine

$2e^- + 2H^+$

FAD

$FADH_2$

Figure 9 Structure of Flavin adenine dinucleotide (FAD) and $FADH_2$

**NADP$^+$**, or Nicotinamide adenine dinucleotide phosphate, is a coenzyme that uses dehydrogenase to remove two hydrogen atoms ($2H^+$ and $2e^-$) from its substrate. Both electrons but only one proton are accepted by the $NADP^+$ to produce its reduced form, NADPH, plus $H^+$. NADPH is not used for ATP synthesis but its electrons provide the energy for certain biosynthesis reactions such as ones involved in photosynthesis.

**FMN** (Flavin mononucleotide) is a prosthetic group of some flavoproteins. It is similar in structure to FAD (Flavin adenine dinucleotide), but lacking the adenine nucleotide. FMN (like FAD) can accept $2e^- + 2H^+$ to yield $FMNH_2$ (Figure 10). When bound at the active site of some enzymes, FMN can accept 1 $e^-$, converting it to the half-reduced semiquinone radical. The semiquinone can accept a second $e^-$ to yield $FMNH_2$. Since it can accept/donate either 1 or 2 $e^-$, FMN has an important role in mediating electron transfer between carriers that transfer 2 $e^-$ (e.g., NADH) and carriers that can only accept 1 $e^-$ (e.g., $Fe^{+++}$).

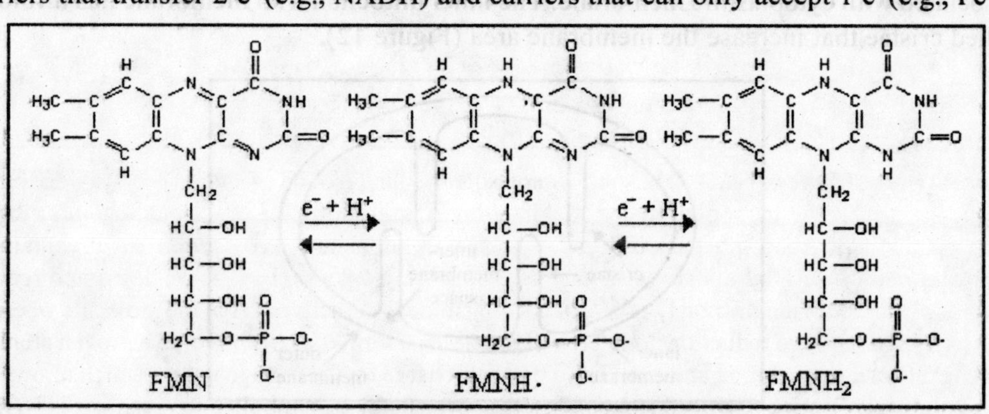

$e^- + H^+$

FMN

FMNH·

$FMNH_2$

Figure 10 Structure of FMN and $FMNH_2$

**Coenzyme Q** (also called CoQ, Q or ubiquinone) is very hydrophobic. It dissolves in the hydrocarbon core of a membrane. The structure of CoQ includes a long isoprenoid tail, with multiple units having a carbon skeleton comparable to that of the compound isoprene (Figure 11). Coenzyme Q is located within the lipid core of the inner membrane. There are also binding sites for coenzyme Q within protein complexes with which it interacts. It functions as a mobile electron carrier within the mitochondrial inner membrane. Its role in transmembrane $H^+$ transport coupled to electron transfer (the Q Cycle).

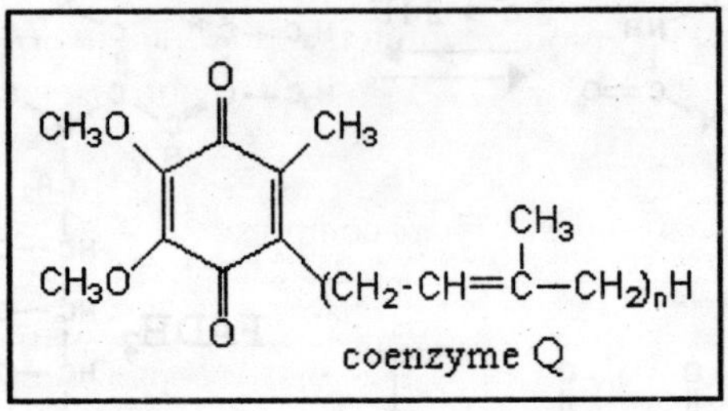

**Figure 11** Structure of coenzyme Q

**Cytochromes** are proteins with heme prosthetic groups. They absorb light at characteristic wavelengths. Changes in light absorbance upon oxidation or reduction of the heme iron provide a basis for monitoring the redox state of the heme. Some cytochromes are part of large integral membrane complexes, each consisting of several polypeptides and including multiple electron carriers. Individual heme prosthetic groups may be separately designated as cytochromes, even if associated with the same protein. For example, hemes a and $a_3$ that are part of the respiratory chain are often referred to as cytochromes a and $a_3$. Cytochrome c is instead a small, water-soluble protein, with a single heme group. Positively charged lysine residues surround the heme crevice on the surface of cytochrome c. These may interact with anionic residues on membrane complexes to which cytochrome c binds, when it is receiving or donating $e^-$.

The multicomponent electron-transport chains are associated with membranes. In eukaryotes, these are in mitochondrial or chloroplast membranes. In prokaryotes, they are associated with cytoplasmic membrane. The inner mitochondrial membrane has infoldings called cristae that increase the membrane area (Figure 12).

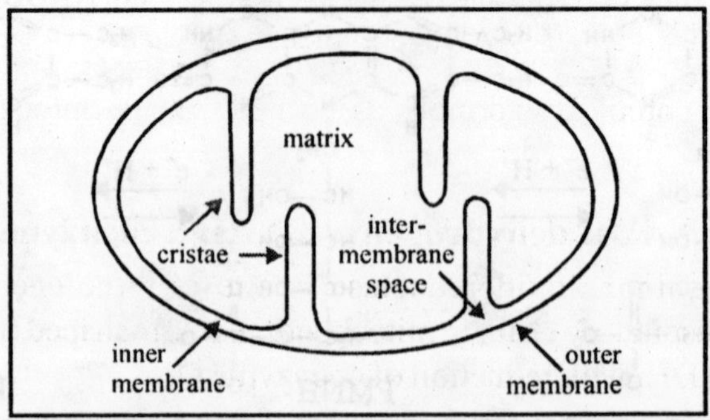


**Figure 12** Schematic diagram of a mitochondrion

The mitochondrial system is arranged into four complexes of carriers, each capable of transporting electrons part of the way to $O_2$. Coenzyme Q and cytochrome c connect the complexes with each other. The composition of each of the respiratory chain complexes is shown in Table 2. All of these are proteolipid complexes, with the first four containing either flavins, iron-sulfur clusters, copper centers, or heme moieties. Complexes I, III, and IV are proton pumps. Complex II is part of the Krebs cycle and does not pump protons, Complex IV is the terminus of the electron transfer chain, where oxygen from the lungs is reduced by electrons and hydrogen protons (provided by NADH and/or $FADH_2$) to make water. Sometimes an enzyme ATP synthase which is embedded in cristae of the inner mitochondrial membrane is considered as complex V of ETC and uses the electrochemical potential generated to create ATP.

**Table 2** The composition of respiratory chain complexes

| Complex | Name |
|---|---|
| Complex I | NADH Dehydrogenase |
| Complex II | Succinate-CoQ Reductase |
| Complex III | CoQ-cyt c Reductase |
| Complex IV | Cytochrome C Oxidase |

Electron transfer from NADH to $O_2$ involves multi-subunit inner membrane complexes I, III, & IV, plus coenzyme Q and cytochrome c (Figure 13) in which electrons pass sequentially through a series of electron carriers with the ejection of 10 protons from the mitochondrial matrix per 2 electrons transferred from NADH to $O_2$

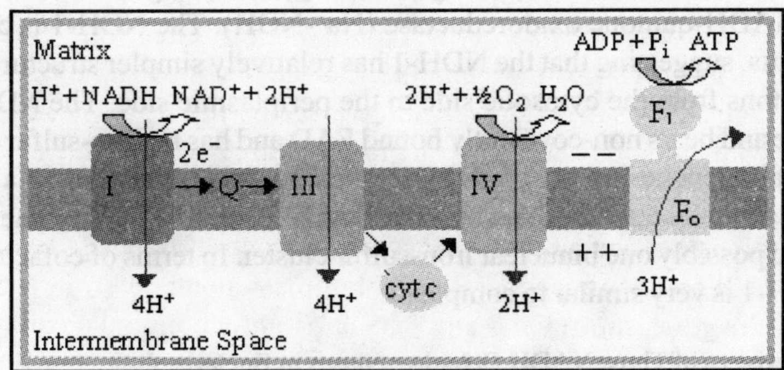


**Figure 13** Schematic transfer of electrons from NADH to $O_2$ and ATP generation in the chain

There are four complexes normally associated with the mitochondrial electron transport chain.

## 3.2. Complex I - NADH dehydrogenase (NADH coenzyme Q reductase)

The NADH-quinone oxidoreductase is one of three energy-transducing enzyme complexes of the respiratory chain in mitochondria. It is L-shaped (Figure 14) and is involved in oxidation of NADH, with reduction of coenzyme Q:

$$NADH + H^+ + Q \rightarrow NAD^+ + QH_2$$

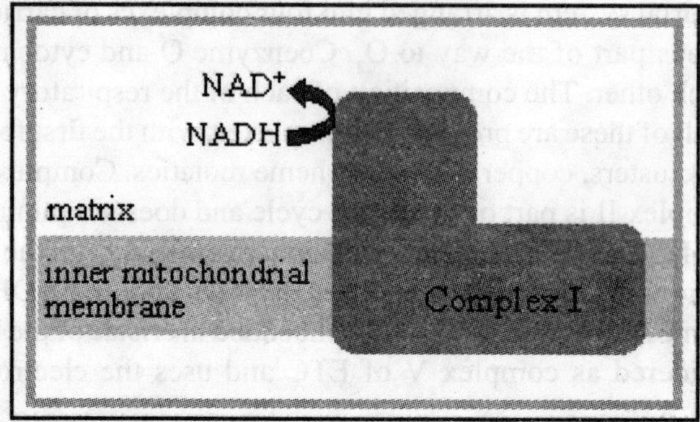


**Figure 14** Schematic structure of complex I

Electrons pass through a series of iron-sulfur centers in complex I, and are eventually transferred to coenzyme Q. Coenzyme Q accepts $2e^-$ and picks up $2H^+$ to yield the fully reduced $QH_2$. Complex I is the point of entry for the major fraction of electrons that traverse the respiratory chain eventually resulting in the reduction of oxygen. Coupled to the electron transfer, protons are pumped from the matrix side to the intermembrane space of mitochondria. It is the most intricate membrane-bound enzyme known to date, being composed of at least 46 unlike polypeptides. Of these, 7 are encoded by mitochondrial DNA. The NADH-quinone oxidoreductases of the bacterial respiratory chains can be divided into at least three groups (Table 3). These are the proton-translocating NADH-quinone oxidoreductase (NDH-1), the NADH-quinone oxidoreductase lacking an energy coupling site (NDH-2), and $Na^+$-translocating NADH-quinone oxidoreductase ($Na^+$-NDH). The NDH-1 is composed of 14 different subunits, suggesting that the NDH-1 has relatively simpler structure than complex I. It pumps protons from the cytosolic side to the periplasmic side. The NDH-2 is a single subunit enzyme and bears non-covalently bound FAD and has no iron-sulfur clusters similar to mitochondrial rotenone-insensitive NADH-quinone oxidoreductases. $Na^+$-NDH is made up of 6 unlike subunits and contains two covalently bound FMN and one noncovalently bound FAD and possibly one binuclear iron-sulfur cluster. In terms of cofactors and specific inhibitors, NDH-1 is very similar to complex I.

**Table 3** Characteristics of Three NADH Dehydrogenases in Bacteria

| | **NDH-1** | **NDH-2** | **$Na^+$-NDH** |
|---|---|---|---|
| Ion-pump | $H^+$ and $Na^+$ | not a pump | $Na^+$ |
| Cofactors | one noncovalent FMN<br>2 [2Fe-2S]<br>6 - 8 [4Fe-4S] | one noncovalent FAD | one noncovalent FAD<br>two covalent FMN<br>1 [2Fe-2S][a] |
| Subunit composition | 13-14 | 1 | 6 |
| Organisms | *Paracoccus denitrificans*<br>*Rhodobacter capsulatus*<br>*Escherichia coli*<br>*Synechocystis* PCC6803<br>*Thermus thermophilus*<br>*Klebsiella pneumoniae* | many gram-positive bacteria, many archaea<br>*Vibrio*<br>*Haemophilus influenzae*<br>*Escherichia coli* | *Vibrio*<br>*Haemophilus influenzae*<br>*Pseudomonas aeruginosa*<br>*Klebsiella* |

### 3.3. Complex II - Succinate - coenzyme Q reductase

Succinate dehydrogenase of the Kreb's cycle is also called complex II or Succinate-CoQ reductase. FAD is the initial electron receptor which gets reduced to $FADH_2$ during oxidation of succinate to fumarate. $FADH_2$ is then reoxidized by transfer of electrons through a series of three iron-sulfur centers to Coenzyme Q, yielding $QH_2$ (Figure 15). *E. coli* complex II indicates a linear arrangement of electron carriers within complex II, consistent with the predicted sequence of electron transfers:

$$FAD \rightarrow FeS_{center\ 1} \rightarrow FeS_{center\ 2} \rightarrow FeS_{center\ 3} \rightarrow CoQ$$

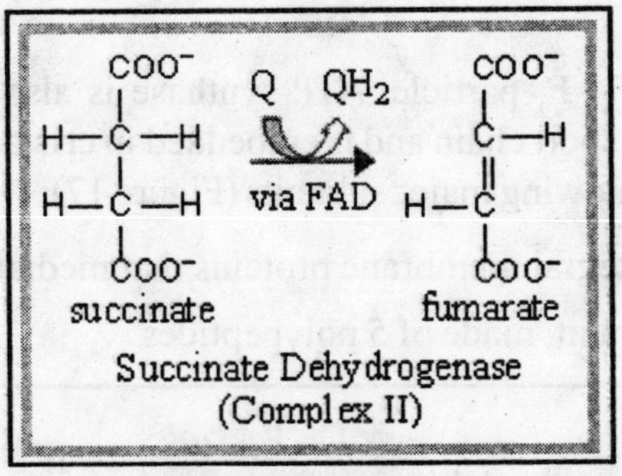

Q $QH_2$
via FAD
succinate
fumarate
Succinate Dehydrogenase
(Complex II)

**Figure 15** Role of complex II in oxidation of succinate to fumarate

### 3.4. Complex III - Coenzyme Q - cytochrome c reductase

Proton transport in complex III involves coenzyme Q (CoQ) and accepts electrons from coenzyme $QH_2$ that is generated by electron transfer in complexes I and II (Figure 16). The "Q cycle" depends on the mobility of CoQ within the lipid bilayer. According to some proposed models, the reaction cycle may involve one-electron transfer, with an intermediate semiquinone radical state.

Cytochrome $c_1$, a prosthetic group within complex III, reduces cytochrome c, which is the electron donor to complex IV.

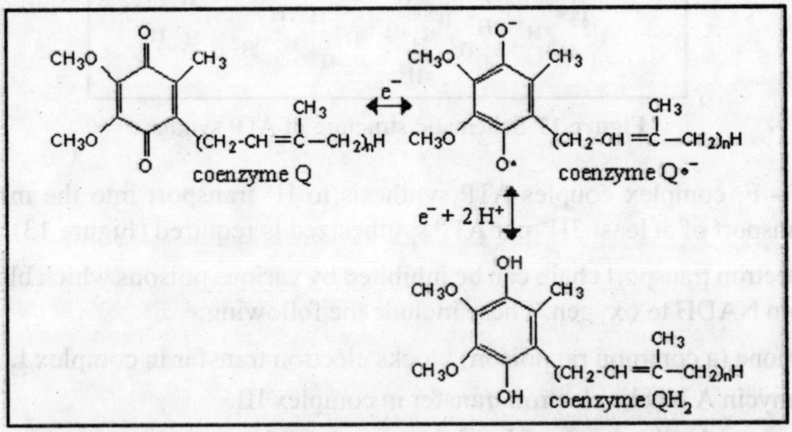

coenzyme Q
$e^-$
coenzyme $Q\bullet^-$
$e^- + 2H^+$
coenzyme $QH_2$

**Figure 16** Reactions involving complex III in electron transport chain

## 3.5. Complex IV - Cytochrome c oxidase

Cytochrome c donates electrons to complex IV which carries out the following irreversible reaction:

$$4e^- + 4H^+ + O_2 \rightarrow 2H_2O.$$

Protons utilized in this reaction are taken up from the matrix compartment.

In addition to the protons utilized in the reduction of $O_2$, there is electron transfer-linked transport of $2H^+$ per $2e^-$ ($4H^+$ per $4e^-$) from the matrix to the intermembrane space.

## 3.6. ATP synthase

Also known as the $F_o$ - $F_1$ particle, ATP synthase is also some times considered as complex V of electron transport chain and is embedded in cristae of the inner mitochondrial membrane, includes the following major subunits (Figure 17):

- $F_o$ - a complex of integral membrane proteins that mediates proton transport.
- $F_1$ - the catalytic subunit, made of 5 polypeptides

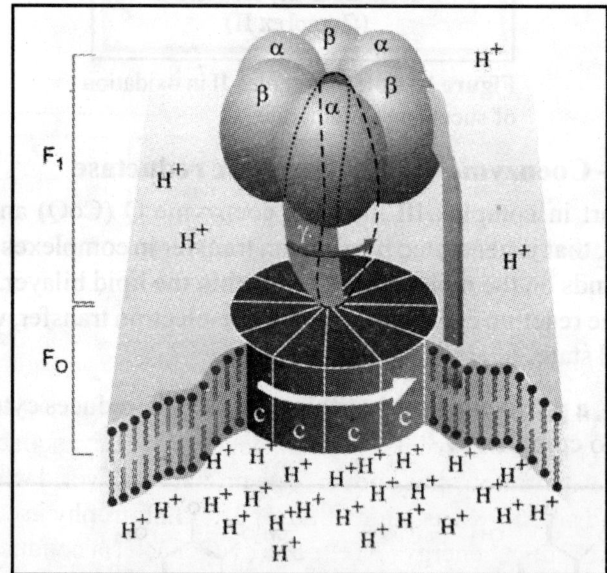


**Figure 17** Schematic structure of ATP synthase

The $F_o$- $F_1$ complex couples ATP synthesis to $H^+$ transport into the mitochondrial matrix. Transport of at least $3H^+$ per ATP synthesized is required (Figure 13).

The electron transport chain can be inhibited by various poisons which block electron transfer from NADH to oxygen. These include the following:

1. Rotenone (a common rat poison) blocks electron transfer in complex I.
2. Antimycin A blocks electron transfer in complex III.
3. Cyanide and carbon monoxide inhibit complex IV.

## 4. ENERGY GENERATING METABOLISMS

There are basically two different modes of obtaining energy in the living world, i) use of light energy, termed as the phototrophic way of life and ii) the use of chemical energy, also known as the chemotrophic way of life. Chemotrophs can be chemoheterotrophs or chemoorganotrophs that degrade organic compounds as their source of energy, and chemoautotrophs or chemolithotrophs that oxidize reduced inorganic compounds as their energy source (Figure 18). There is a great metabolic diversity among the representatives of the three domains of life - Bacteria, Archaea and Eucarya. Most eukaryotes live either as aerobic heterotrophs or as oxygenic photoautotrophs. Anaerobic respiration is virtually unknown among the eukaryotes, chemolithotrophic life has yet to be demonstrated in any representative of the domain, and fermentative growth is found only in a few fungi (e.g. yeasts involved in the making of bread and in beer and wine fermentation) and in a number of specialized anaerobic protozoa.

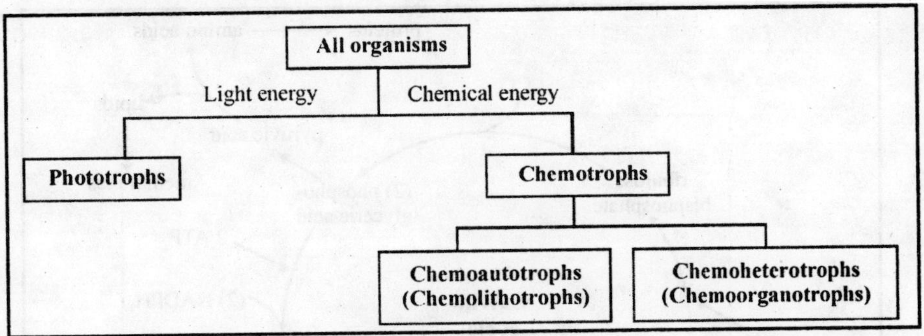


**Figure 18** Classification of the living organisms according to their mode of energy generation

### 4.1. Phototrophic way of life

Autotrophy involving the use of $CO_2$ as the sole source of assimilatory carbon is primarily found in the green plants and in eukaryotic algae. However, cyanobacteria, purple and green sulfur bacteria are also autotrophs known from prokaryotic world which obtain their energy from sun light and are known as photoautotrophs. Phototrophy involves the conversion of solar energy into chemical energy (ATP). The cyanobacteria conduct plant photosynthesis, called oxygenic photosynthesis; the purple and green bacteria conduct bacterial photosynthesis or anoxygenic photosynthesis; the halobacteria use a type of non photosynthetic photophosphorylation mediated by bacteriorhodopsin to transform light energy into ATP.

Photosynthesis is the conversion of light energy into chemical energy that can be used in the formation of cellular material from $CO_2$. Photosynthesis is a type of metabolism separable into a catabolic and anabolic component. The catabolic component of photosynthesis is the light reaction, wherein light energy is transformed into electrical energy. The anabolic component involves the fixation of $CO_2$ and its use as a carbon source for growth, usually called the dark reaction. In photosynthetic procaryotes there are two types of photosynthesis and two types of $CO_2$ fixation.

The most common mode of $CO_2$ fixation is the Calvin cycle (Figure 19), a biochemical cycle in which the central reaction is catalyzed by the enzyme Rubisco (ribulose bisphosphate carboxylase/oxygenase). This cycle mainly involves three major steps: a) carboxylation, b) glycolytic reversal, c) regeneration of RuBP. First of all $CO_2$ reacts with a five-carbon compound (ribulose-1,5-bisphosphate) to form two three-carbon molecules (3-phosphoglycerate). In the next stage, 3-phosphoglycerate is reduced to glyceraldehyde-3-phosphate using NADPH as the electron donor. In a series of additional reactions, ribulose-1,5-bisphosphate is regenerated to become available to accept additional $CO_2$ molecules. In addition to reducing power, energy is also required in the form of ATP, among other things in the reaction of the enzyme phosphoribulokinase, which produces ribulose-1,5-bisphosphate from ribulose-5-phosphate. This enzyme is the second enzyme that is specific for the Calvin cycle. The Calvin cycle can be summarized in the following overall equation:

$$3\ CO_2 + 9\ ATP + 6\ NAD(P)H + 6H^+ \rightarrow \text{Glyceraldehyde-3-P} + 9\ ADP + 8\ P_i + 6\ NAD^+$$

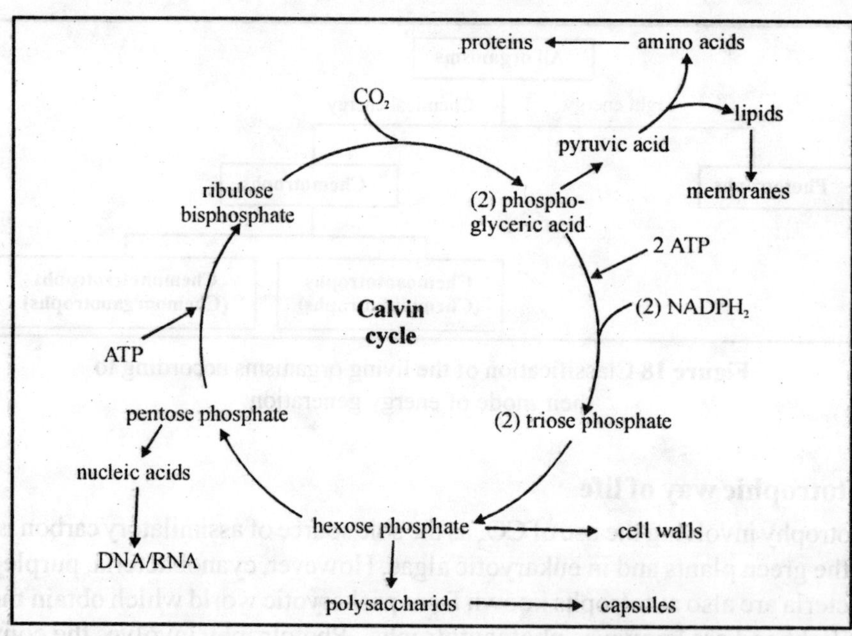


**Figure 19** The Calvin cycle and its relationship to the synthesis of cell materials

As glyceraldehyde-3-phosphate is also an intermediate in the pathways of glycolysis and gluconeogenesis, this compound is the starting point for additional biosynthetic conversions. The Calvin cycle is the basis for most of the $CO_2$ fixation by plants on Earth. Also many autotrophic bacteria, including the cyanobacteria and other prokaryotic oxygenic phototrophs, also use the reactions of the Calvin cycle as the central pathway in their metabolism. This is also the way in which aerobic chemoautotrophic bacteria, such as *Thiobacillus*, *Alcaligenes*, *Nitrosomonas*, and *Nitrobacter* fix $CO_2$. Many anoxygenic photosynthetic bacteria also use this pathway, including the purple sulfur bacteria (*Chromatium*, *Ectothiorhodospira* and others), and other purple bacteria such as *Rhodobacter* when growing autotrophically on hydrogen.

There are, however, alternative modes of $CO_2$ fixation that do not involve the Calvin cycle enzymes. A variety of different pathways of autotrophic $CO_2$ fixation have been identified in the green sulfur bacteria (*Chlorobium* and related organisms, *Chloroflexus*) and in those sulfate reducing bacteria and homoacetogenic bacteria that grow autotrophically on hydrogen and $CO_2$. Genes for Rubisco have been shown to occur in a number of Archaea, but a functional Calvin cycle has yet to be demonstrated in representatives of the Archaeal domain. Chemoautotrophic Archaea (many methanogenic bacteria, sulfur-oxidizing Archaea) all appear to use alternative $CO_2$ fixation pathways.

With the exception of those phototrophic processes based on light absorption by retinal pigments, all phototrophic life on Earth is based on magnesium tetrapyrrole derivatives - chlorophylls and bacteriochlorophylls - as the primary light-harvesting pigment in the membrane of photosynthetic organisms. Absorption of a quantum of light by a chlorophyll molecule causes the displacement of an electron at the reaction center. The displaced electron is an energy source that is moved through a membrane photosynthetic electron transport system, being successively passed from an iron-sulfur protein (X) to a quinone to a cytochrome and back to chlorophyll (Figure 20). As the electron is transported, a proton motive force is established on the membrane, and ATP is synthesized by an ATPase enzyme. This manner of converting light energy into chemical energy is called cyclic photophosphorylation. The functional components of the photochemical system are light harvesting pigments, a membrane electron transport system, and an ATPase. The photosynthetic electron transport system is fundamentally similar to a respiratory ETS, except that there is a low redox electron acceptor (e.g. ferredoxin) at the top of the electron transport chain, that is first reduced by the electron displaced from chlorophyll. Figure 21 presents the basic principles of phototrophic life. Most of the phototrophic procaryotes are obligate or facultative autotrophs, which means that they are able to fix $CO_2$ as a sole source of carbon for growth. Just as the oxidation of organic material yields energy, electrons and $CO_2$, in order to build up $CO_2$ to the level of cell material ($CH_2O$), energy (ATP) and electrons (reducing power) are required. The light reactions operate to produce ATP to provide energy for the dark reactions of $CO_2$ fixation. The dark reactions also need electrons. Usually, the provision of electrons is in some way connected to the light reactions. Coupling the light and dark reactions of photosynthesis is illustrated in Figure 22.

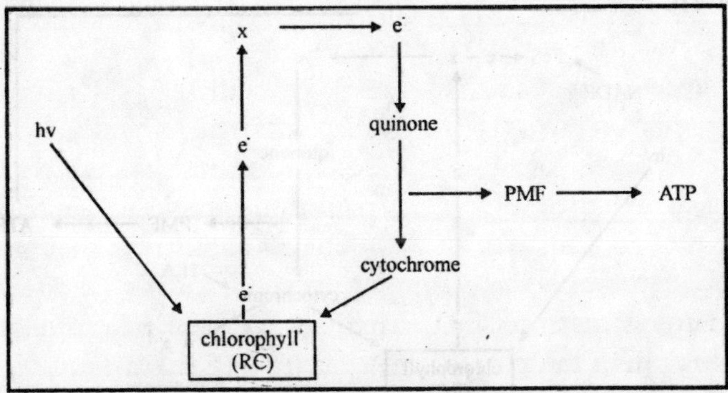


**Figure 20** Photosystem I: schematic representation of cyclical electron flow coupled to photophosphorylation

When water serves as the electron donor (oxygenic photosynthesis, with the release of molecular oxygen as waste product), two photons are required per electron to bring the electrons from water to a sufficiently low redox potential, and the photosynthetic machinery is therefore composed of two photosystems operating in series. Anoxygenic photosynthetic bacteria use a single photosystem, which suffices to reduce $NAD(P)^+$ while using relatively reduced electron donors.

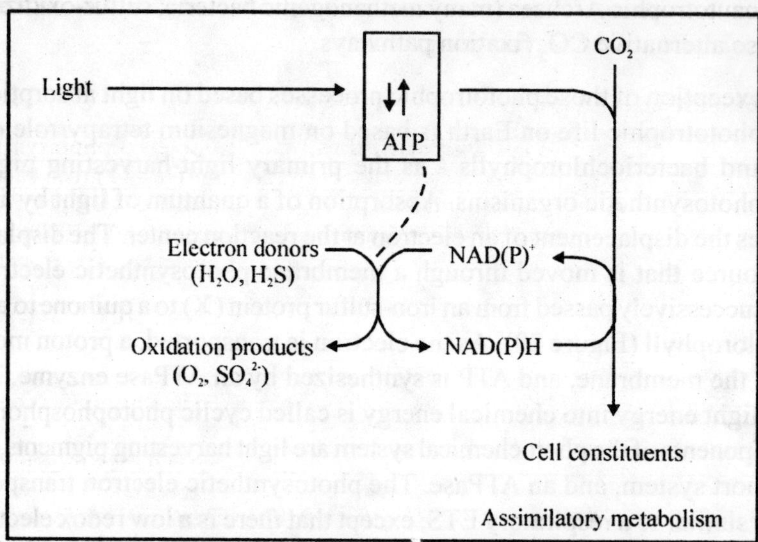


**Figure 21** Metabolism in phototrophic organisms

With respect to their assimilatory metabolism, most phototrophic organisms use $CO_2$ as their carbon source during photoautotrophic metabolism, as illustrated in Figure 18. However, certain phototrophic prokaryotes assimilate organic compounds as carbon sources while using light to drive the endergonic processes in the cell (photoheterotrophic metabolism).

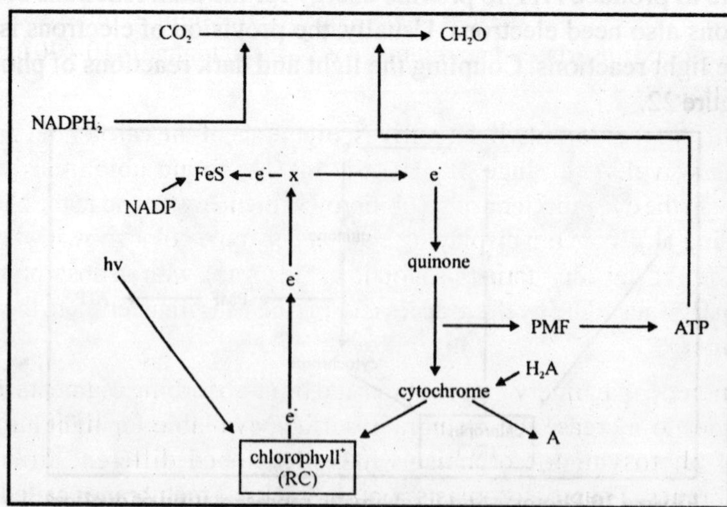


**Figure 22** Coupling the light and dark reactions of photosynthesis

The differences between plant and bacterial photosynthesis are summarized in Table 4. Bacterial photosynthesis is an anoxygenic process. The external electron acceptor for bacterial photosynthesis is never $H_2O$, and therefore, they never produce $O_2$ during photosynthesis. Furthermore, bacterial photosynthesis is usually inhibited by $O_2$ and takes place in microaerophilic and anaerobic environments. Bacterial chlorophylls use light at longer wave lengths not utilized in plant photosynthesis, and therefore they do not have to compete with oxygenic phototrophs for light. Bacteria use only cyclic photophosphorylation (Photosystem I) for ATP synthesis and lack a second photosystem.

**Table 4** Differences between plant and bacterial photosynthesis

| Characteristics | Plant photosynthesis | Bacterial photosynthesis |
|---|---|---|
| organisms | plants, algae, cyanobacteria | purple and green bacteria |
| type of chlorophyll | chlorophyll a<br>absorbs 650-750nm | bacteriochlorophyll<br>absorbs 800-1000nm |
| Photosystem I<br>(cyclic photophosphorylation) | present | present |
| Photosystem I<br>(noncyclic photophosphorylation) | present | absent |
| Produces $O_2$ | yes | no |
| Photosynthetic electron donor | $H_2O$ | $H_2S$, other sulfur compounds<br>or<br>certain organic compounds |

**4.1.1.** *Oxygenic photosynthesis of algae and cyanobacteria:* Oxygenic photosynthesis uses water as the electron donor with the release of molecular oxygen as a byproduct. We owe the presence of molecular oxygen in the Earth's atmosphere due to this process. The key pigment in the reaction centers of chloroplasts is chlorophyll *a*. Chlorophyll *a*, together with other chlorophylls (chlorophyll *b* in the chloroplast of higher plants, other accessory chlorophylls in different groups of algae) also serve as light-harvesting pigment. In the cyanobacteria we find only chlorophyll *a*; whereas in the Prochlorales, a branch of oxygenic prokaryotes related to the cyanobacteria, the divinyl derivatives of chlorophyll *a* and chlorophyll *b* are present. All these pigments absorb red light in the wavelength range 650-700 nm, with a second absorption maximum in the blue light range (410-470 nm, somewhat higher in the Prochlorales).

Apart from different chlorophyll derivatives, pigments of the carotenoid group, which absorb light in the wavelength range of 420-550 nm, are found ubiquitously within the phototrophic world. In the cyanobacteria, phycobiliprotein pigments bound to the photosynthetic membranes including phycocyanin displaying an intensely blue color showing an absorption maximum at 620 nm, and phycoerythrin (also found in red algae), with an absorption maximum at 540 nm. Light energy absorbed by these accessory pigments is transferred to the chlorophyll in the reaction center.

The photosynthetic machinery that contains the light-absorbing pigments is located on membranes. In order to increase the membrane surface available for light harvesting, the diverse groups of photosynthetic organisms have developed different arrangements of intracellular membranes. Thus, chloroplasts contain stacks of membranes called thylakoids, and cyanobacteria possess intracellular thylakoid-like membranes as well.

**All phototrophic bacteria are capable of performing cyclic photophosphorylation which is referred to as Photosystem I. Bacterial photosynthesis uses only Photosystem I (PSI), but the more evolved cyanobacteria, as well as algae and plants, have an additional light-harvesting system called Photosystem II (PSII). Photosystem II is used to reduce Photosystem I when electrons are withdrawn from PSI for $CO_2$ fixation. PSII transfers electrons from $H_2O$ and produces $O_2$, as shown in Figure 23.**

**Photosystem I and the mechanisms of cyclic photophosphorylation operate in plants, algae and cyanobacteria, as they do in bacterial photosynthesis. In plant photosynthesis, chlorophyll *a* is the major chlorophyll species at the reaction center and the exact nature of X and the components of the ETS are different than bacterial photosynthesis. But the fundamental mechanism of cyclic photophosphorylation is the same. However, when electrons must be withdrawn from photosystem I (shown by X→ferredoxin → NADP in upper left), those electrons are replenished by the operation of Photosystem II. In the Reaction Center of PSII, a reaction between light, chlorophyll and $H_2O$ removes electrons from $H_2O$ (leading to the formation of $O_2$) and transfers them to a component of the photosynthetic ETS (Q). The electrons are transferred from Q through a chain of electron carriers consisting of cytochromes and quinones until they reach plastocyanin in PSI. The movement of the electrons from Q to plastocyanin is a drop in redox potential that allows for the synthesis of ATP in a process called noncyclic photophosphorylation. The operation of photosystem II is what fundamentally differentiates plant photosynthesis from bacterial photosynthesis. Photosystem II accounts for the source of electrons for $CO_2$ fixation (provided by $H_2O$), the production of $O_2$, and ATP synthesis by noncyclic photophosphorylation.**

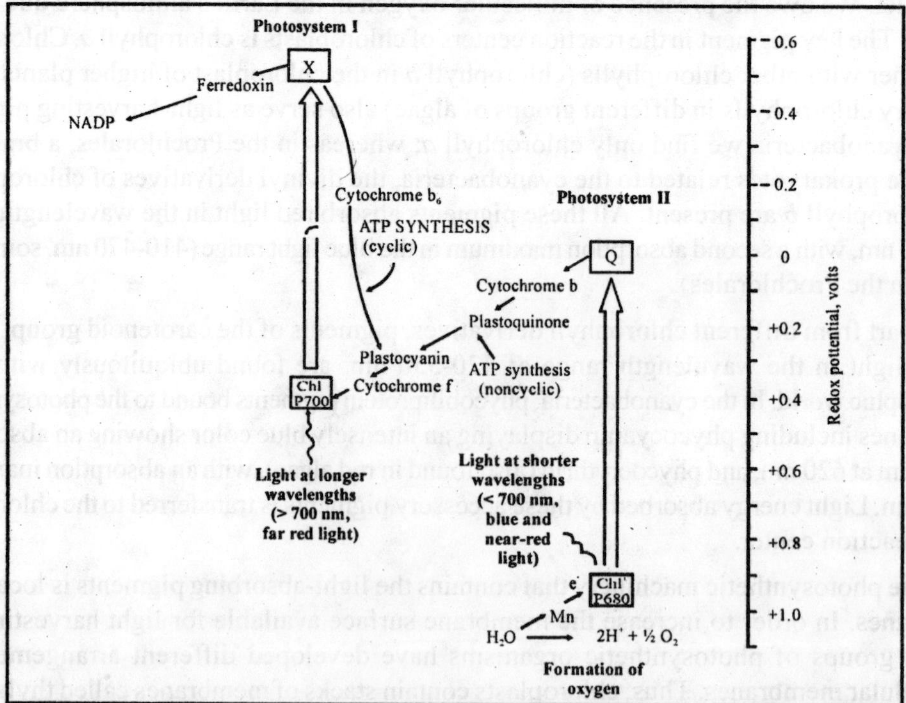


**Figure 23** Schematic representation of electron flow in plant (oxygenic) photosynthesis

The general scheme for finding electrons for $CO_2$ fixation is to open up Photosystem I and remove the electrons, eventually getting them to NADP which can donate them to the dark reaction. In bacterial photosynthesis the process may be quite complex. The electrons are removed from Photosystem I at the level of a cytochrome, then moved through an energy-consuming reverse electron transport system to an iron-sulfur protein, ferredoxin, which reduces NADP to $NADPH_2$ (Figure 23). The electrons that replenish Photosystem I come from the oxidation of an external photosynthetic electron donor, which may be $H_2S$, other sulfur compounds, $H_2$, or certain organic compounds. In plant photosynthesis, electron donor is $H_2O$, which is lysed by photosystem II, resulting in the production of $O_2$. Electrons removed from $H_2O$ travel through Photosystem II to Photosystem I as described in Figure 22. Electrons removed from Photosystem I reduce ferredoxin directly. Ferredoxin, in turn, passes the electrons to NADP.

**4.1.2.** *Anoxygenic photosynthesis in bacteria:* Anoxygenic phototrophs, abundant within the bacterial domain, use electron donors other than water to supply the electrons for autotrophic $CO_2$ fixation and other reductive processes. Suitable electron donors may be sulfide, other reduced sulfur compounds, hydrogen, and others. Such organisms are generally anaerobes, and the process of anoxygenic photosynthesis occurs only in specialized niches in which the presence of light is combined with the absence of oxygen and the availability of the necessary electron donors. The purple sulfur bacteria (*Chromatium*, *Thiocapsa*, *Ectothiorhodospira* etc., belonging to the Proteobacteria) use reduced sulfur compounds as electron donors for the fixation of $CO_2$, and these are oxidized to elemental sulfur which may be transiently accumulated intracellularly, or, is excreted from the cells as in the case of *Ectothiorhodospira*. The green sulfur bacteria (*Chlorobium* and related organisms) show a similar behavior. Another group of Proteobacteria is formed by the photoheterotrophic purple bacteria, such as *Rhodobacter* and *Rhodospirillum*. These organisms display a great degree of metabolic flexibility: there are species that are able to grow not only photoheterotrophically on a variety of carbon sources, but which may also develop photoautotrophically using hydrogen, sulfide, and in some cases even divalent iron as the electron donor, which is oxidized to trivalent iron through this process. In addition, they can shift to aerobic metabolism by respiration in the dark (heterotrophically on a variety of substrates, or chemolithotrophically on hydrogen), and some may even perform fermentation in the absence of oxygen and light.

The diverse groups of anoxygenic phototrophic bacteria contain different types of bacteriochlorophyll, molecules resembling chlorophyll in their structure, but all showing an absorption maximum in the infrared range. Thus, many purple bacteria have absorption maxima between 800 and 860 nm due to their content of bacteriochlorophyll *a*, and others containing bacteriochlorophyll *b* may even absorb light at wavelengths as high as 1020 nm. The green sulfur bacteria generally display an absorption maximum at 750-760 nm,due to the presence of bacteriochlorophyll *c*. The bacteriochlorophylls and the other components of the photosynthetic system are located on membranes, and also here we find different modes to increase the membrane surface in order to accommodate the large amounts of light harvesting and reaction center pigments.

The anoxygenic photosynthetic bacteria use a single photosystem (Figure 24). Some groups of photosynthetic bacteria, such as the green sulfur bacteria (*Chlorobium* and related organisms, that oxidize sulfide to sulfate with elemental sulfur as intermediate) reduce $NAD^+$ directly by the non-cyclic flow of electrons from the donor through the reaction center. In other groups, such as the purple sulfur bacteria (*Chromatium*, *Ectothiorhodospira*) the reduction potential of the excited electron in the reaction center is not sufficiently low to reduce $NAD^+$. In this case energy has to be spent to reduce $NAD^+$ in a process of reverse electron transport, and the photosynthetic machinery supplies energy only via cyclic electron flow.

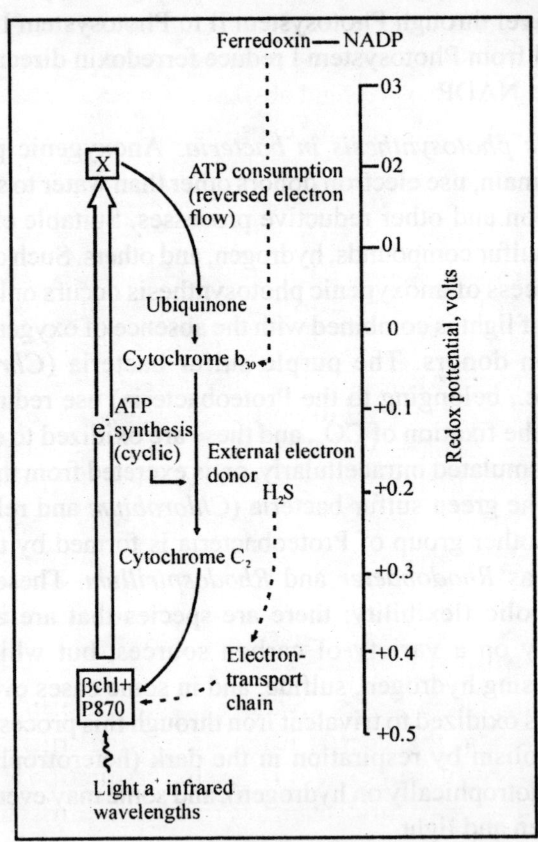


**Figure 24** Schematic representation of cyclic flow of electrons during bacterial (anoxygenic) photosynthesis

Certain species of (oxygenic) cyanobacteria are able to shift to anoxygenic photosynthesis under anaerobic conditions, then sulfide may be used as the electron donor and oxidized to elemental sulfur. Under these conditions, only Photosystem I participates, while Photosystem II remains inactive. Many anoxygenic photosynthetic bacteria use the Calvin cycle for $CO_2$ fixation. Alternative modes of $CO_2$ fixation exist in the two groups of green photosynthetic bacteria. Green sulfur bacteria of the *Chlorobium* group fix $CO_2$ using the reactions of the Kreb's cycle, which run in reverse direction under the expenditure of energy. Two-carbon units (acetyl-CoA) are carboxylated to three-carbon units (pyruvate), in a reaction in which reduced ferredoxin supplies the electrons. Whether this is the only mode of autotrophic $CO_2$ fixation in *Chlorobium* remains to be determined, especially in view of the fact that a Rubisco gene has been detected in the genome of *Chlorobium tepidum*. Yet another pathway of autotrophic $CO_2$ fixation has been demonstrated to occur in green bacteria of the *Chloroflexus* group.

**4.1.3.** *Retinal pigment-based phototrophy:* In the prokaryotic world, the bacteriorhodopsin system was originally identified in a number of representatives of the extremely halophilic archaea (genera *Halobacterium*, *Halorubrum*, and additional types). Bacteriorhodopsin is a membrane protein, carrying retinal as its prosthetic group. This protein resembles rhodopsin, the visual pigment of the eye, and absorbs light with a maximum absorbance at 570 nm. As a result of light absorption, a number of conformational changes occur within the molecule, resulting in the extrusion of a proton from the cytoplasmic side to the outer medium. The proton gradient thus generated is then used for the formation of ATP mediated by ATP synthase in the cell membrane. Bacteriorhodopsin-containing halophilic archaea do not grow photoautotrophically: *Halobacterium* and related organisms are heterotrophs, and the carbon source for their growth is therefore organic. However, the bacteriorhodopsin system enables them to obtain energy in situations in which the amount of available organic matter is limited. *Halobacterium* has additional ways to use light energy to drive bioenergetic processes in the cell: certain species contain an additional retinal pigment, halorhodopsin that serves as an inward-directed chloride pump. Bacteriorhodopsin genes occur in certain marine bacteria as well where the pigment "proteorhodopsin" functions as a proton pump. The demonstration of the presence of proteorhodopsin in marine surface waters worldwide shows that retinal pigments as light harvesting systems may be much more widespread than previously assumed, and are by no means restricted to specialized hypersaline environments in which the halophilic archaea thrive.

## 4.2. Chemoheterotrophic way of life

Organic compounds are a readily available source of energy. Nature has devised a variety of strategies to make this energy available to living organisms. The most common way of degradation of organic substances is aerobic respiration. This is the sole mode of energy generation in the animal world, but many microorganisms (fungi, most bacteria) also use a similar process to obtain energy. In aerobic respiration, organic carbon compounds are oxidized to $CO_2$, and the electrons released are transferred to molecular oxygen, which serves as the terminal electron acceptor. In most heterotrophs, the reactions of the glycolytic pathway (Embden-Meyerhof-Parnas pathway) and the Kreb's cycle (Tricarboxylic acid cycle) form the backbone of the metabolic pathways leading to the degradation of organic carbon compounds. Conversion of the different organic compounds available to intermediates of these central metabolic pathways allows their further degradation. A minor amount of ATP is also gained by substrate-level phosphorylation in the course of the degradation through the glycolytic pathway, but the major part of the energy gained is obtained by electron transport phosphorylation during transfer of the electrons to oxygen via cytochromes and other membrane-bound electron carriers, the energy being stored first as a proton electrochemical gradient across the cell membrane. Figure 25 summarizes the principle of the aerobic chemoorganotrophic life.

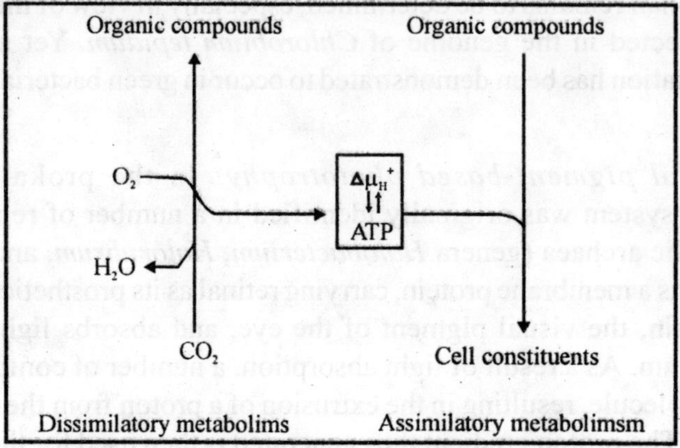


**Figure 25** Metabolism in aerobic organisms

When molecular oxygen is no longer available or the amount of oxygen available is insufficient for the breakdown of all organic material present, many prokaryotic microorganisms are still able to degrade the available material by respiration, this time by using terminal electron acceptors other than molecular oxygen (anaerobic respiration). The most important electron acceptors used by bacteria in anaerobic respiration processes are nitrate ($NO_3^-$) and nitrite ($NO_2^-$), oxidized forms of iron and manganese ($Fe^{3+}$, $Mn^{4+}$), and oxidized sulfur compounds such as sulfate ($SO_4^{2-}$), thiosulfate ($S_2O_3^{2-}$) and elemental sulfur (S°). Figure 26 summarizes the principle of the anaerobic chemoorganotrophic life.

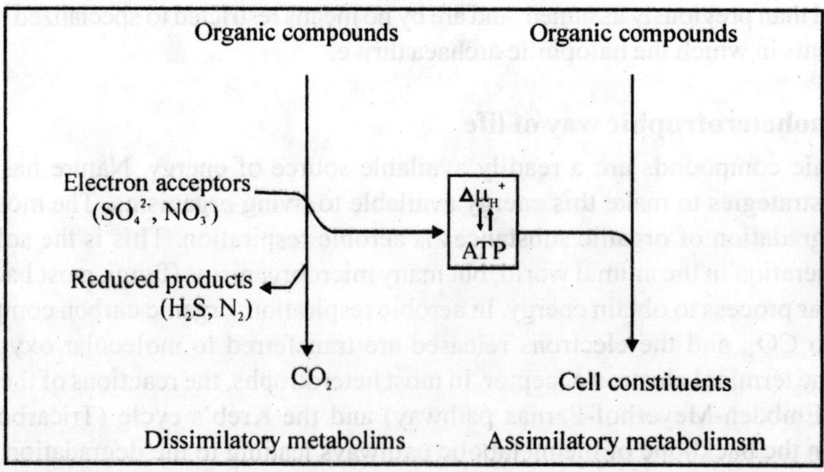


**Figure 26** The principle of the metabolism of organisms living by anaerobic respiration

A third way of obtaining energy by the degradation of organic compounds is that of fermentation. During fermentation, organic compounds are degraded to smaller products with the formation of ATP by substrate-level phosphorylation. In most cases, oxidation reactions are involved in the breakdown process. Intermediates of the degradation process itself then act as electron acceptors, so that no external electron acceptor is required. Figure 27 summarizes the principle of fermentative life.

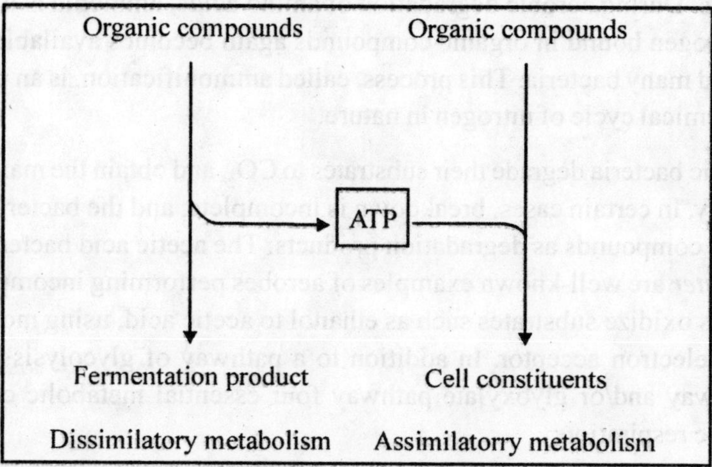


**Figure 27** The principle of the metabolism of organisms living by fermentation

A number of principles are common to all fermentations:

1. The reactions generally occur in the cytoplasm, and no membrane-bound reactions are involved.
2. In the course of the enzymatic reaction chain, high-energy compounds are generated, that lead to the formation of ATP by substrate-level phosphorylation. No primary electrochemical gradient of protons is formed.
3. In most cases stages are found in the chain of reactions in which intermediary products are oxidized, the electrons being transferred to the coenzyme $NAD^+$. A substantial part of the diversity that exists in fermentation processes is based on the various ways in which the cells reoxidize the reduced coenzymes.

### *4.2.1. Aerobic respiration*

Energetically, the most efficient way to degrade organic material is aerobic respiration, with molecular oxygen as the terminal electron acceptor. In most cases breakdown is complete, and the only products are carbon dioxide and water. Because of the high midpoint redox potential of the couple $O_2/H_2O$ (+ 0.82 V), much energy can be gained from the oxidation of organic compounds by transfer of the electrons to oxygen in respiration.

Aerobic dissimilation of organic compounds proceeds by transformation of the compounds to be degraded to one of the intermediates of the glycolytic pathway or the Kreb's cycle. In certain bacteria, the Embden-Meyerhof-Parnas pathway of sugar degradation, is replaced by alternative biochemical pathways such as the Entner-Doudoroff pathway or the oxidative pentose-phosphate cycle. Fatty acids are degraded stepwise to units of acetyl-

CoA which then enter the Kreb's cycle with additional involvement of the glyoxylic acid cycle enzymes, isocitrate synthase and malate lyase. Oil-degrading bacteria oxidize hydrocarbon chains to the corresponding fatty acids, which are then stepwise degraded to acetyl-CoA units. During aerobic degradation of amino acids, ammonia is liberated, and in this way the nitrogen bound in organic compounds again becomes available for uptake by plants, algae, and many bacteria. This process, called ammonification, is an important stage in the biogeochemical cycle of nitrogen in nature.

Most aerobic bacteria degrade their substrates to $CO_2$, and obtain the maximum possible amount of energy. In certain cases, breakdown is incomplete, and the bacteria excrete more reduced organic compounds as degradation products. The acetic acid bacteria *Acetobacter* and *Gluconobacter* are well-known examples of aerobes performing incomplete oxidation. These organisms oxidize substrates such as ethanol to acetic acid, using molecular oxygen as the terminal electron acceptor. In addition to a pathway of glycolysis and/or pentose phosphate pathway and/or glyoxylate pathway four essential metabolic components are needed in aerobic respiration:

1. The tricarboxylic acid cycle (TCA cycle or citric acid cycle or the Kreb's cycle): When an organic compound is utilized as a substrate, the TCA cycle is required for the complete oxidation of the substrate. The end product that always results from the complete oxidation of an organic compound is $CO_2$.

2. A membrane and an associated electron transport system (ETS): The ETS is a sequence of electron carriers in the plasma membrane that transports electrons taken from the substrate through the chain of carriers to a final electron acceptor. The operation of the ETS establishes a proton motive force (PMF) due to the formation of a proton gradient across the membrane.

3. An outside electron acceptor in the form of $O_2$: Molecular oxygen is reduced to $H_2O$ in the last step of the electron transport system.

4. A transmembranous ATPase enzyme (ATP synthetase): This enzyme utilizes the proton motive force established on the membrane to synthesize ATP in the process of electron transport phosphorylation.

The diagram below of aerobic respiration (Figure 28) integrates these metabolic processes into a scheme that represents the overall process of respiratory metabolism. A substrate such as glucose is completely oxidized to $CO_2$ by the combined pathways of glycolysis and the TCA cycle. Electrons removed from the glucose by NAD are fed into the ETS in the membrane. As the electrons traverse the ETS, a PMF becomes established across the membrane. The electrons eventually reduce an outside electron acceptor, $O_2$, and reduce it to $H_2O$. The PMF on the membrane is used by the ATPase enzyme to synthesize ATP by a process referred to as "oxidative phosphorylation".

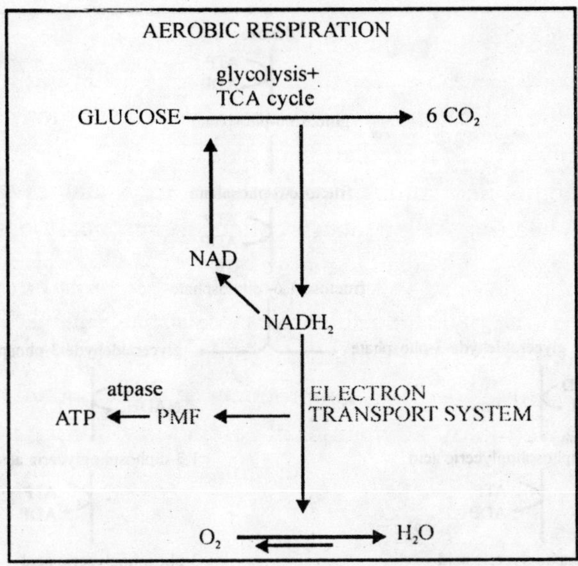


**Figure 28** Schematic representation of aerobic respiration

**4.2.1.1.** *Embden-Meyerhof-Parnas* pathway*:* Glycolysis as well as the TCA cycle are at the center of metabolism in nearly all organisms. Not only do these pathways dissimilate organic compounds and provide energy, they also provide the precursors for biosynthesis of macromolecules that make up living systems. These are rightfully called amphibolic pathways as they have both an anabolic and a catabolic function.

The first three steps of the pathway are prime (phos=phorylate) and rearrange the hexose for cleavage into 2 trioses by the enzyme fructose- 1,6-diphosphate aldolase, the key (cleavage) enzyme in the EMP pathway. Each triose molecule is oxidized and phosphorylated followed by two substrate level phosphorylations that yield 4ATP during the drive to pyruvate. Glucose ($C_6H_{12}O_6$) is split and oxidized through a ten step pathway to two molecules of pyruvic acid ($C_3H_4O_3$). For each initial glucose molecule; 2glyceraldehyde-3-phosphate oxidized to 2pyruvic acid, 4ATP produced, 2NADH produced (Figure 29). Net gain of 2ATP molecules, 4 from energy conserving phase (by substrate level phosphorylation) minus 2 from preparatory phase 2NADH molecules produced. Pyruvic acid can now undergo either fermentation or respiration.

The overall reaction is the oxidation of glucose to 2 pyruvic acid molecules. The two branches of the pathway after the cleavage are identical, drawn in this manner for comparison with other bacterial pathways of glycolysis.

**4.2.1.2.** *Pentose phosphate pathway or Hexose monophosphate shunt pathway:* A second pathway, the pentose phosphate or hexose monophosphate pathway may be used at the same time as the glycolytic pathway or the Entner-Doudoroff sequence (occurring in some fermentative bacteria) . It can operate either aerobically or anaerobically and is important in anabolism and in catabolism.

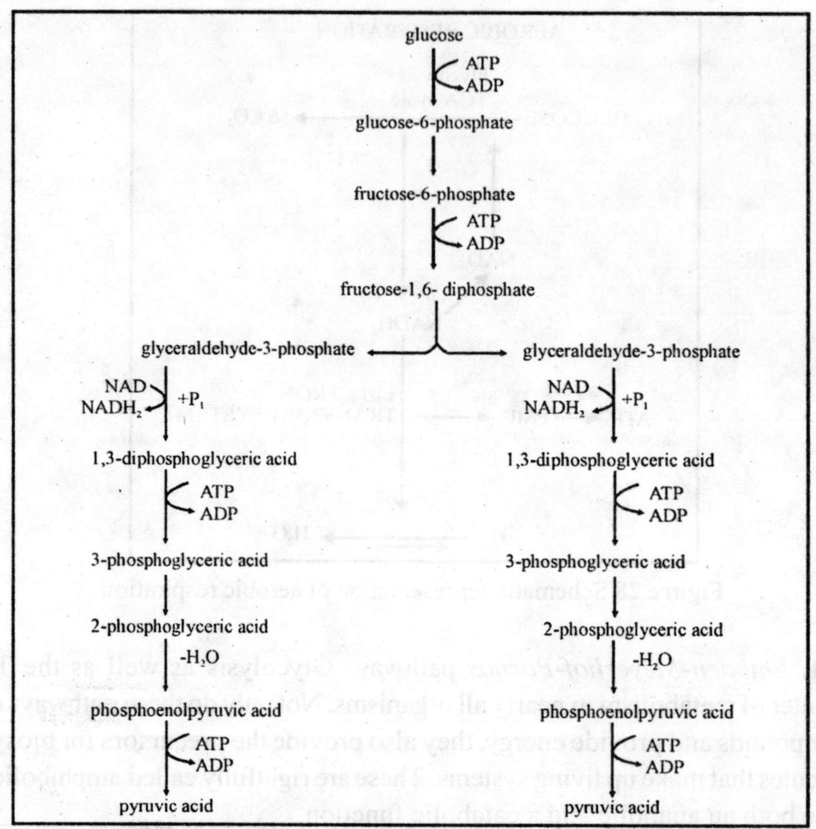


**Figure 29** The Embden-Meyerhof -Parnas pathway for glucose dissimilation

The pentose phosphate pathway begins with the oxidation of glucose-6-phosphate to 6-phosphogluconate followed by the oxidation of 6-phosphogluconate to the pentose, ribulose-5-phosphate and $CO_2$. NADPH is produced during these oxidations. Ribulose-5-phosphate is then converted to a mixture of three- through seven-carbon sugar phosphates (Figure 30). Two enzymes unique to this pathway play a central role in these transformations: (1) transketolase catalyzes the transfer of two-carbon ketol groups, and (2) transaldolase transfers a three-carbon group from sedoheptulose-7-phosphate to glyceraldehyde-3-phosphate (Figure 31). The overall result is that three glucose-6-phosphates are converted to two fructose-6-phosphates, glyceraldehyde-3-phosphate, and three $CO_2$ molecules. as shown in the following equation.

3 glucose-6-phosphate + 6NADP$^+$ + 3$H_2O$ $\rightarrow$ 2 fructose-6-phosphate + glyceraldehyde-3-phosphate + 3$CO_2$ + 6NADPH + 6H$^+$

These intermediates are used in two ways. The fructose 6-phosphate can be changed back to glucose 6-phosphate while glyceraldehyde 3-phosphate is converted to pyruvate by glycolytic enzymes. The glyceraldehyde 3-phosphate also may be returned to the pentose phosphate pathway through glucose 6-phosphate formation. This results in the complete degradation of glucose 6-phosphate to $CO_2$ and the production of a great deal of NADPH.

Glucose-6-phosphate + 12NADP$^+$ + 7$H_2O$ $\rightarrow$ 6$CO_2$ + 12NADPH + 12H$^+$ + $P_i$

The pentose phosphate pathway has several catabolic and anabolic functions that are summarized as follows:

1. NADPH from the pentose phosphate pathway serves as a source of electrons for the reduction of molecules during biosynthesis.
2. The pathway synthesizes four- and five-carbon sugars for a variety of purposes. The four-carbon sugar erythrose-4-phosphate is used to synthesize aromatic amino acids and vitamin $B_6$ (pyridoxal). The pentose ribose-5-phosphate is a major component of nucleic acids, and ribulose-1,5-bisphosphate is the primary $CO_2$ acceptor in photosynthesis. When a microorganism is growing on a pentose carbon source, the pathway also can supply carbon for hexose production (e.g. glucose is needed for peptidoglycan synthesis).
3. Intermediates in the pentose phosphate pathway may be used to produce ATP. Glyceraldehyde-3-phosphate from the pathway can enter the three-carbon stage of the glycolytic pathway and be converted to ATP and pyruvate. The latter may he oxidized in the tricarboxylic acid cycle to provide more energy. In addition, some NADPH can be converted to NADH, which yields ATP when it is oxidized by the electron transport chain. Because five-carbon sugars are intermediates in the pathway, the pentose phosphate pathway can be used to catabolize pentoses as well as hexoses.

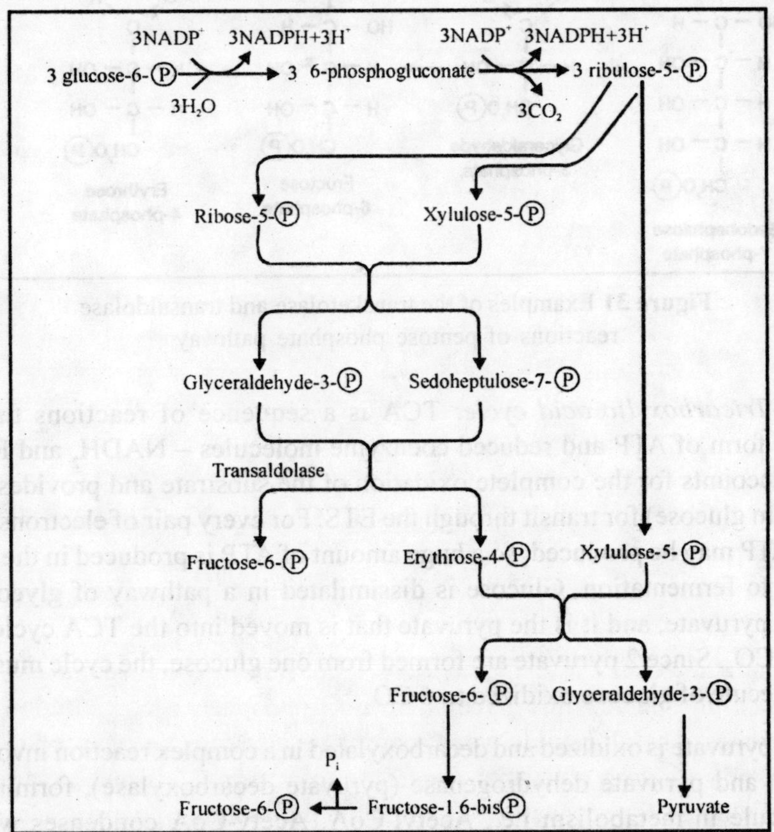


**Figure 30** Schematic representation of Pentose phosphate pathway

Although the pentose phosphate pathway may be a source of energy in many microorganisms, it is more often of greater importance in biosynthesis. The net equation of pentose phosphate pathway is:

$$\text{Glucose-6-phosphate} + 12NADP^+ \rightarrow 6\ CO_2 + 12\ NADPH + 12H^+ + Pi$$

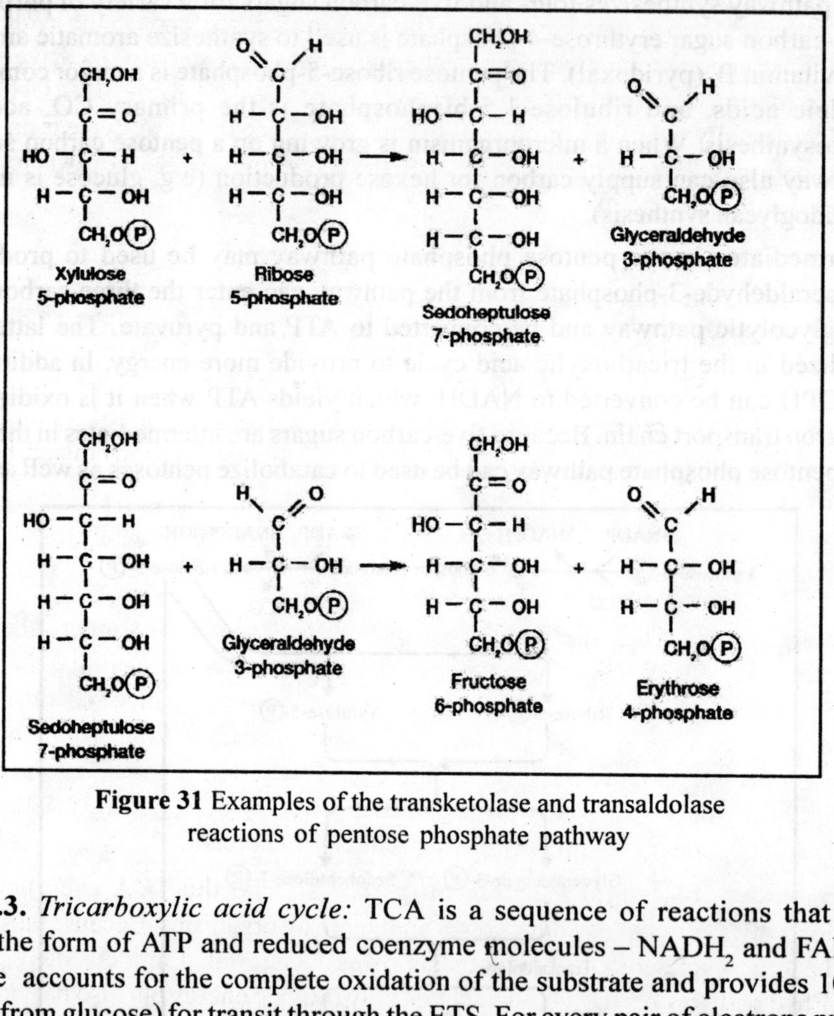

**Figure 31** Examples of the transketolase and transaldolase reactions of pentose phosphate pathway

**4.2.1.3.** *Tricarboxylic acid cycle:* TCA is a sequence of reactions that generate energy in the form of ATP and reduced coenzyme molecules – $NADH_2$ and $FADH_2$. The TCA cycle accounts for the complete oxidation of the substrate and provides 10 pairs of electrons (from glucose) for transit through the ETS. For every pair of electrons put into the ETS, 2 or 3 ATP may be produced, so a huge amount of ATP is produced in the respiration, as compared to fermentation. Glucose is dissimilated in a pathway of glycolysis to the intermediate, pyruvate, and it is the pyruvate that is moved into the TCA cycle to become oxidized to 3 $CO_2$. Since 2 pyruvate are formed from one glucose, the cycle must turn twice for every molecule of glucose oxidized to 6 $CO_2$.

Initially, pyruvate is oxidized and decarboxylated in a complex reaction involving NAD, Coenzyme A, and pyruvate dehydrogenase (pyruvate decarboxylase), forming the most central molecule in metabolism i.e., Acetyl-CoA. Acety-CoA condenses with the 4C-compound, oxaloacetate, to form the first stable intermediate of the TCA cycle, 6C-citric acid (citrate), a tricarboxylic acid. Citrate is isomerized to isocitrate, which is oxidized and decarboxylated forming α-ketoglutarate (αkg). α–ketoglutarate dehydrogenase uses CoA and NAD to oxidize αkg to succinyl-CoA in a reaction analogous to the pyruvate dehydrogenase reaction above. Succinyl-CoA is converted to succinate during a substrate level phosphorylation yielding high energy GTP (equivalent to ATP). This completes the decarboxylation of pyruvate forming $3CO_2$. The remaining three steps in the cycle complete the oxidation of succinate and regenerate the oxaloacetate necessary to drive the cycle. During the oxidation of pyruvate to $3CO_2$ by one turn of the TCA cycle, $4NADH_2$, $1FADH_2$ and one ATP (actually GTP) are produced (Figure 32). Since the TCA cycle is an important amphibolic pathway, several intermediates of the cycle may be withdrawn for anabolic (biosynthetic) pathways.

The overall reaction for the aerobic respiration of glucose is :

$$\text{Glucose} + 6\ O_2 \rightarrow 6\ CO_2 + 6\ H_2O + 688\ \text{kcal (total)}$$

which can be written as :

$$\text{Glucose} \rightarrow 6\ CO_2 + 10\ NADH_2 + 2\ FADH_2 + 4\ ATP$$

In *E. coli*, 2 ATP are produced for each pair of electrons that are introduced into the ETS by $NADH_2$. One ATP is produced from a pair of electrons introduced by $FADH_2$. Hence, the equation can be rewritten

$$\text{Glucose} + 6\ O_2 \rightarrow 6\ CO_2 + 6\ H_2O) + 20\ ATP\ (ETP) + 2\ ATP\ (ETP) +$$

$$4\ ATP\ (SLP) + 688\ \text{kcal (total)}$$

Since a total of 26 ATP is formed during the release of 688 kcal of energy, the efficiency of this respiration is 26×8/688 or about 30 percent. In *Pseudomonas* (or mitochondria), due to the exact nature of the ETS, 3ATP are produced for each pair of electrons that are introduced into the ETS by $NADH_2$ and 2ATP are produced from a pair of electrons introduced by $FADH_2$ (Figure 33). Hence, the overall reaction in *Pseudomonas*, using the same dissimilatory pathways as *E. coli*, is Glucose + $6O_2 \rightarrow 6CO_2 + 6\ H_2O$ + 38 ATP + 688 kcal (total) and the corresponding efficiency is about 45 percent.

The pathways of central metabolism (glycolysis and the TCA cycle), with a few modifications, always run in one direction or another in all organisms. Because these pathways provide the precursors for the biosynthesis of cell material. Embden-Meyerhof-Parnas pathway (EMP pathway) or the TCA cycle, not only provide energy but also provide chemical intermediates for the synthesis of cell material. Such pathways are known as amphibolic pathways. The main metabolic pathways, and their relationship to biosynthesis of cell material, are shown in Figure 34. The fundamental metabolic pathways of biosynthesis are similar in all organisms, in the same way that protein synthesis or DNA structure are similar in all organisms. When biosynthesis proceeds from central metabolism as drawn below, some of the main precursors for synthesis of procaryotic cell structures and components are as follows:

- Polysaccharide capsules or inclusions are polymers of glucose and other sugars.
- Cell wall peptidoglycan (N-acetyl glucosamine; NAG and N-acetyl muramic acid; NAM) is derived from glucose phosphate.

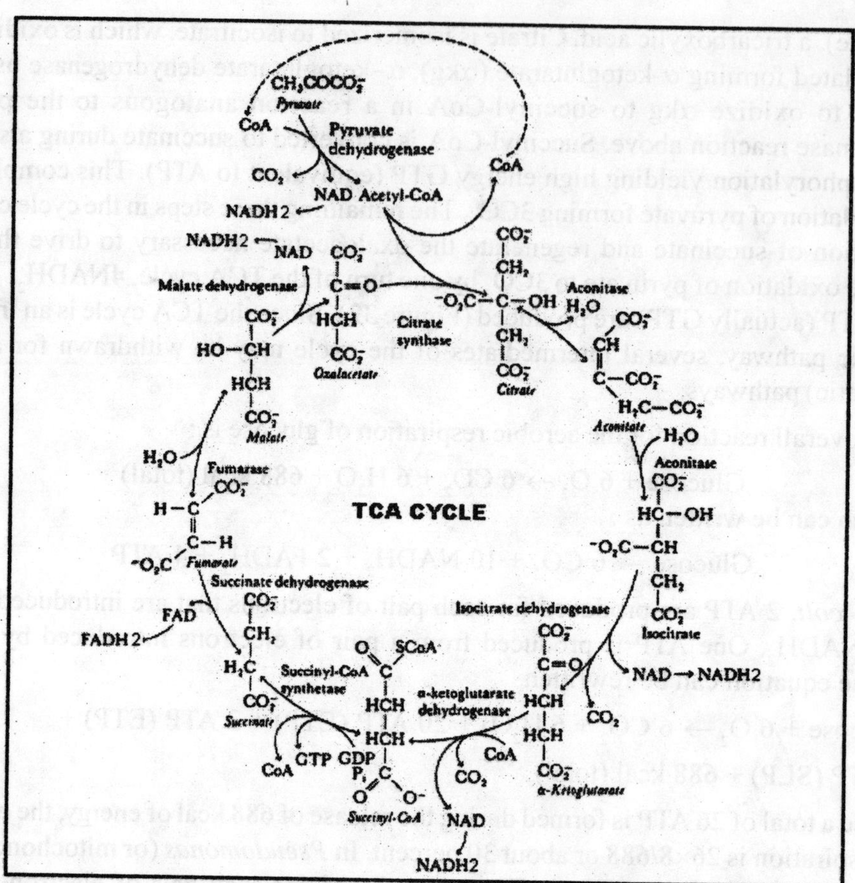


**Figure 32** Schematic representation of Tricarboxylic acid (TCA) or Kreb's cycle

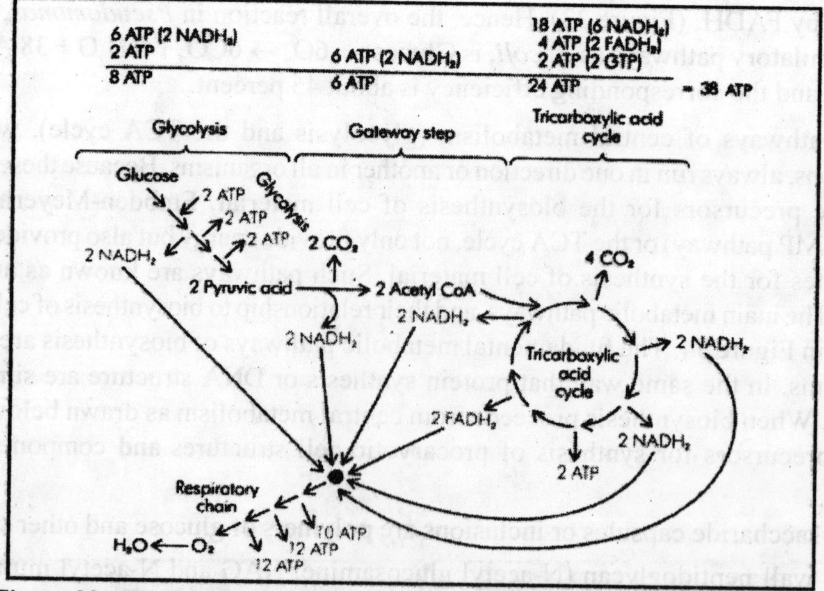


**Figure 33** ATP yield per glucose molecule broken down in aerobic respiration

- Amino acids have various sources.
- Nucleotides (DNA and RNA) are synthesized from ribose phosphate.
- Triose-phosphates are precursors of glycerol, and acetyl CoA is a main precursor of lipids for membranes.
- The main precursors of amino acids for manufacture of proteins are pyruvate, α-ketoglutarate and oxaloacetate.
- Vitamins and coenzymes are synthesized in various pathways that leave central metabolism.
- ATP and NAD are part of purine (nucleotide) metabolism.

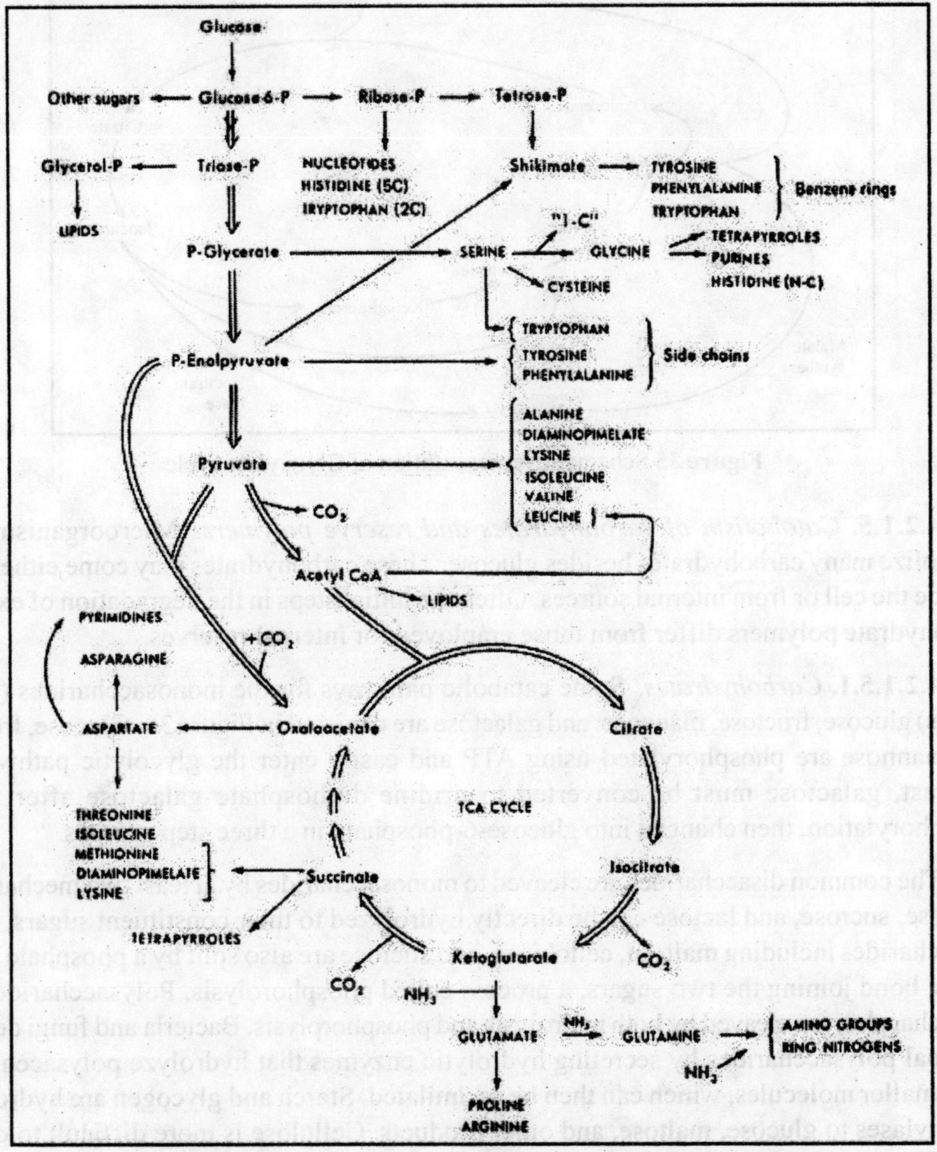


**Figure 34** The main pathways of biosynthesis in microorganisms

**4.2.1.4.** *Glyoxylate cycle:* In some organisms when acetate is the sole carbon compound or during the oxidation of primary substrates (higher fatty acids) that are cleaved to acetyl-CoA which enters the cycle at two places. It condenses with oxaloacetate to give citrate which is the entry point for TCA and further reaction leads to isocitrate which is split by isocitrate lyase to produce succinate and glyoxylate. Second acetyl-CoA molecule condenses with glyoxylate to give malate by malate synthetase as revealed in Figure 35.

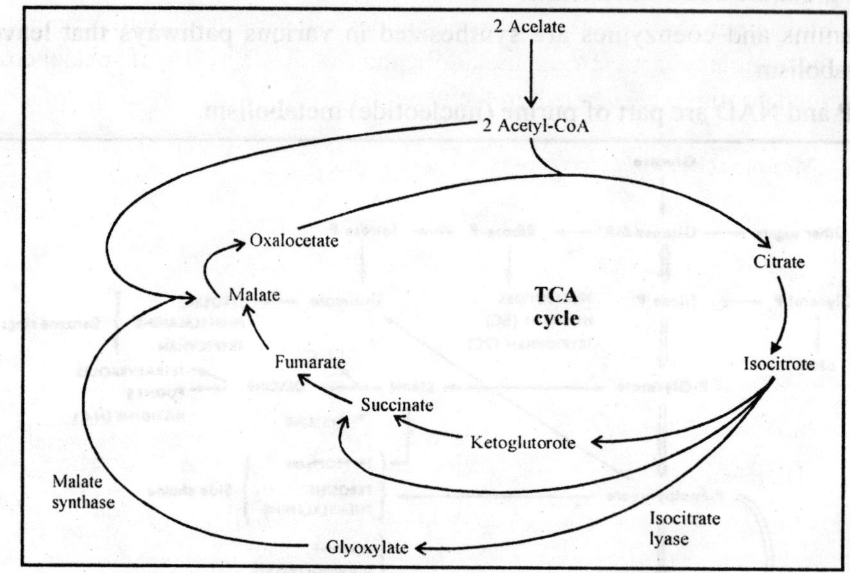


**Figure 35** Schematic representation of Glyoxylate cycle

**4.2.1.5.** *Catabolism of carbohydrates and reserve polymers:* Microorganisms can catabolize many carbohydrates besides glucose. These carbohydrates may come either from outside the cell or from internal sources. Often the initial steps in the degradation of external carbohydrate polymers differ from those employed for internal reserves.

**4.2.1.5.1.** *Carbohydrates:* Some catabolic pathways for the monosaccharides (single sugars) glucose, fructose, mannose, and galactose are depicted in Figure 36. Glucose, fructose and mannose are phosphorylated using ATP and easily enter the glycolytic pathway. In contrast, galactose must be converted to uridine diphosphate galactose after initial phosphorylation, then changed into glucose-6-phosphate in a three step process.

The common disaccharides are cleaved to monosaccharides by at least two mechanisms. Maltose, sucrose, and lactose can be directly hydrolyzed to their constituent sugars. Many disaccharides including maltose, cellobiose, and sucrose are also split by a phosphate attack on the bond joining the two sugars, a process called phosphorolysis. Polysaccharides like disaccharides, are cleaved by both hydrolysis and phosphorolysis. Bacteria and fungi degrade external polysaccharides by secreting hydrolytic enzymes that hydrolyze polysaccharides into smaller molecules, which can then be assimilated. Starch and glycogen are hydrolyzed by amylases to glucose, maltose, and other products. Cellulose is more difficult to digest; many fungi and a few bacteria (some gliding bacteria, clostridia, and actinomycetes) produce cellulases that hydrolyze cellulose to cellobiose and glucose. Some members of the bacterial genus *Cytophaga,* isolated from marine habitats, excrete an agarase that degrades agar. Many soil bacteria and bacterial plant pathogens degrade pectin, a polymer of galacturonic acid (a galactose derivative) that is an important constituent of plant cell walls and tissues.

In the context of compounds that are recalcitrant or difficult to digest, it should be noted that microorganisms also can degrade xenobiotic compounds (foreign substances not formed by natural biosynthetic processes) such as pesticides and various aromatic compounds. They transform these molecules to normal metabolic intermediates by use of special enzymes and pathways, then continue catabolism in the usual way.

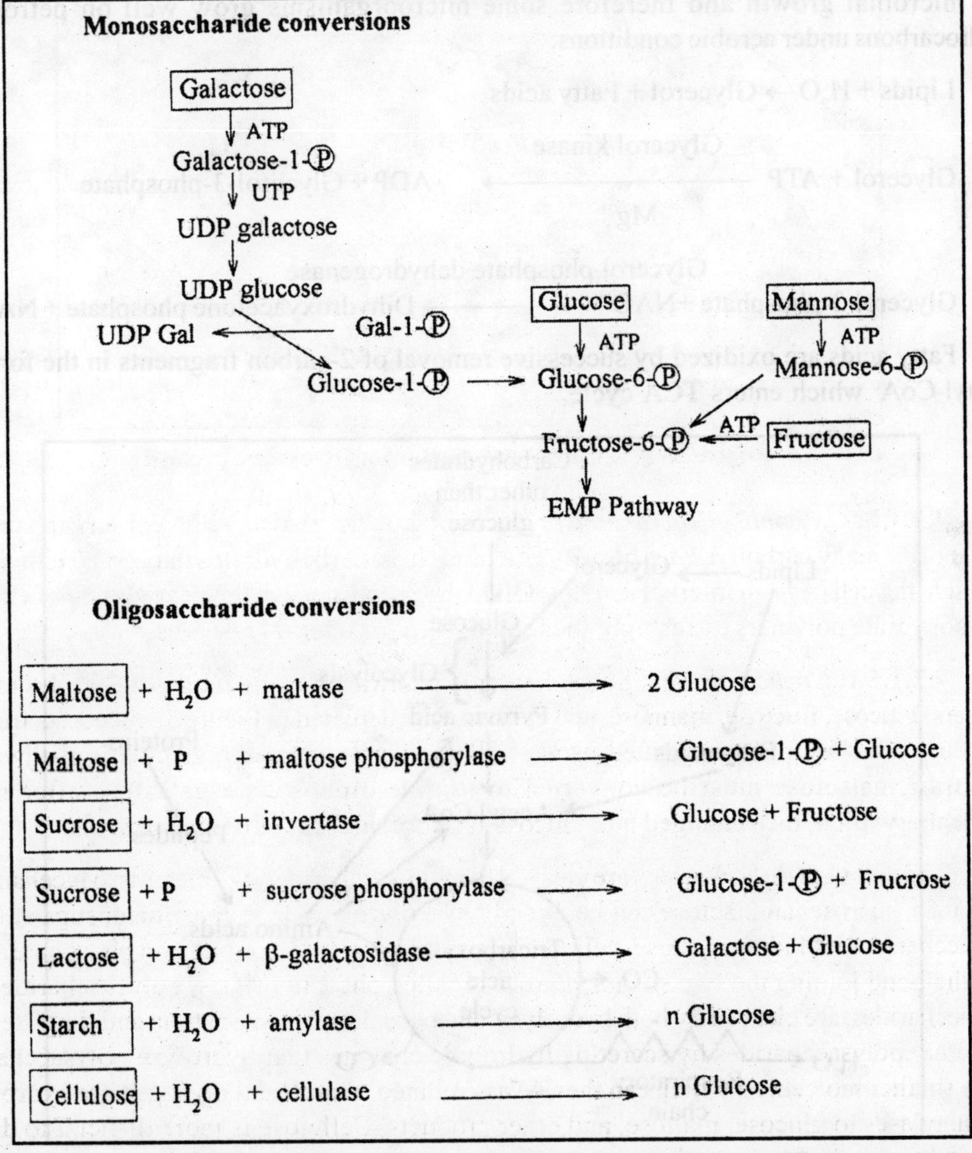


**Figure 36** Examples of enzymes and pathways used in the catabolism of various carbohydrates

**4.2.1.5.2.** *Lipids:* Microorganisms frequently use lipids as energy sources. Triglycerides or triacylglycerols, esters of glycerol and fatty acids, are common energy sources. They can be hydrolysed to glycerol and fatty acids by microbial lipases. The glycerol is then phosphorylated, oxidised to dihydroxyacetone phosphate and catabolized in the glycolytic pathway. Fatty acids from triacylglycerols and other lipids are often oxidized in the β-oxidation pathway after conversion to coenzyme A esters (Figure 37). In this cyclic pathway fatty acids are degraded to acetyl-CoA, which can be fed into the TCA cycle. One turn of the cycle produces acetyl-CoA, NADH and $FADH_2$: NADH and $FADH_2$ can be oxidized by the electron transport chain to provide more ATP. The fatty acyl-CoA, shortened by two carbons, is ready for another turn of the cycle. Lipid fatty acids are a rich source of energy for microbial growth and therefore some microorganisms grow well on petroleum hydrocarbons under aerobic conditions.

$$\text{Lipids} + H_2O \rightarrow \text{Glycerol} + \text{Fatty acids}$$

$$\text{Glycerol} + \text{ATP} \xrightarrow[Mg^{2+}]{\text{Glycerol kinase}} \text{ADP} + \text{Glycerol-3-phosphate}$$

$$\text{Glycerol-3-phosphate} + NAD^{+} \xrightarrow{\text{Glycerol phosphate dehydrogenase}} \text{Dihydroxyacetone phosphate} + NADH_2$$

Fatty acids are oxidized by successive removal of 2-carbon fragments in the form of acetyl-CoA which enters TCA cycle.

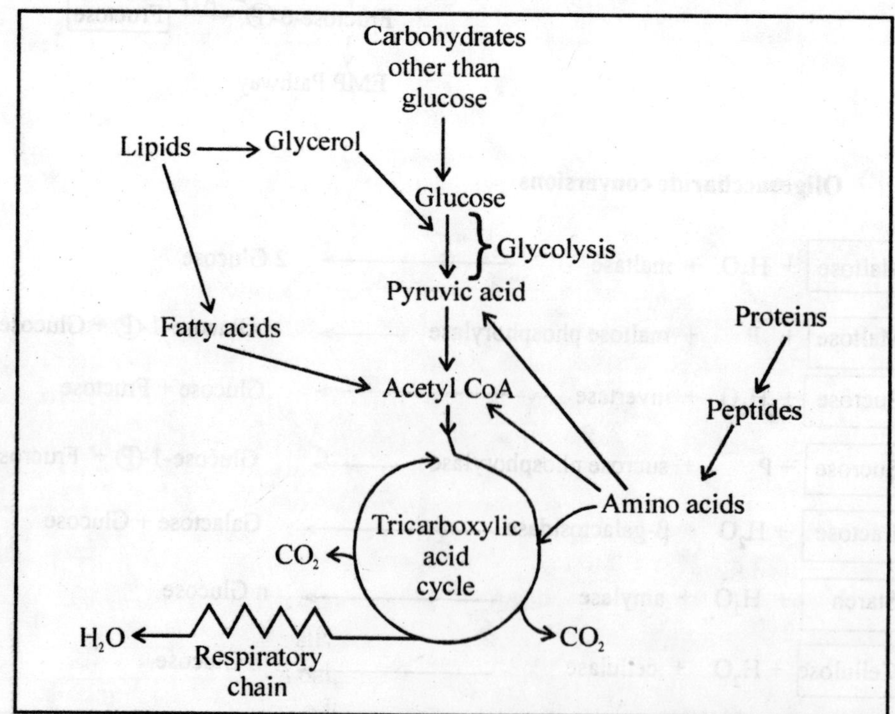


**Figure 37** Metabolism of carbohydrates, lipids, amino acids

**4.2.1.5.3.** *Proteins:* Some bacteria and fungi can use proteins as their source of carbon and energy. They secrete protease enzymes that hydrolyze proteins and polypeptides to amino acids, which are transported into the cell and catabolized. The first step in amino acid use is deamination, the removal of the amino group from an amino acid. This is often accomplished by transamination. The amino group is transferred from an amino acid to an α-keto acid acceptor (Figure 37). The organic acid resulting from deamination can be converted to pyruvate, acetyl-CoA or a TCA cycle intermediate and eventually oxidized in the TCA cycle to release energy.

Proteins → peptides → Amino acids

### *4.2.2. Anaerobic respiration*

Anaerobic respiration is a membrane bound biological process coupling the oxidation of electron donating substrates (e.g. sugars and other organic compounds, but also inorganic molecules like hydrogen, sulfide/sulfur, ammonia, metals or metal ions) to the reduction of suitable alternative electron acceptors other than molecular oxygen. During these redox processes, protons are translocated over the membrane from "inside" to "outside", establishing a concentration gradient over the membrane which temporarily stores the energy released in the chemical reactions. This energy is then converted into ATP by the same enzyme used during aerobic respiration, ATP synthase. Possible electron acceptors for anaerobic respiration are nitrate, nitrite, nitrous oxide, oxidised amines and nitro-compounds, fumarate, oxidised metal ions, sulfate, sulfur, sulfoxo-compounds, halogenated organic compounds, selenate, arsenate or carbon dioxide (in acetogenesis and methanogenesis). Some of the common electron acceptors used by procaryotes for respiration or methanogenesis (an analogous type of energy generation in archaea) are described in the Table 5. Like aerobic respiration, anaerobic respiration involves glycolysis, a transition reaction, the citric acid cycle, and an electron transport chain. The total energy yield per glucose oxidized is less than with aerobic respiration with a theoretical maximum yield of 36 ATP or less.

**Table 5** Electron acceptors for respiration and methanogenesis in procaryotes

| Electron acceptor | Reduced end product | Name of process | Organism |
|---|---|---|---|
| $O_2$ | $H_2O$ | aerobic respiration | *Escherichia*, *Streptomyces* |
| $NO_3$ | $NO_2$, $NH_3$ or $N_2$ | denitrification | *Bacillus*, *Pseudomonas* |
| $SO_4$ | S or $H_2S$ | sulfate reduction | *Desulfovibrio* |
| fumarate | succinate | anaerobic respiration | *Escherichia* |
| S° | $H_2S$ | sulfate reduction | *Desulfuromonas* and *Thermoproteus* |
| $Fe^{3+}$ | $Fe^{2+}$ | anaerobic respiration | *Bacillus*, *Geobacter* and *Pseudomonas* |
| $CO_2$ | $CH_4$ | methanogenesis | *Methanococcus* |

When nitrate and other oxidized forms of nitrogen, such as nitrite, are available in anoxic environments, these can be used as terminal electron acceptors by a variety of bacteria.

Nitrate is reduced stepwise via nitrite, nitric oxide, and nitrous oxide, the major product generally being gaseous nitrogen ($N_2$). In certain cases, especially in acidic environments, nitrous oxide ($N_2O$) can be formed as well. When the end products of the process are gases, the process is called denitrification. Not all bacteria that can reduce nitrate in their dissimilatory metabolism form gases, and many excrete nitrite as end product of the reduction. Some bacteria further reduce the nitrite formed to ammonia.

Of all possible terminal electron acceptors employed in anaerobic respiration, nitrate is preferentially used, as the amount of energy gained is relatively large. Because of the high midpoint redox potentials of the partial reactions, the free energy change during oxidation of organic compounds with nitrate as the electron acceptor being reduced to $N_2$ is not much lower than during aerobic respiration when oxygen serves as the terminal electron acceptor:

$$\text{Glucose} + 4.8\,NO_3^- + 4.8H^+ \rightarrow 6\,CO_2 + 2.4\,N_2 + 8.4H_2O\ \Delta G^{o\prime} = -\,2716\text{ kJ}$$

Some bacteria can use nitrate as the electron acceptor at the end of their electron transport chain and still produce ATP. Often this process is called dissimilatory nitrate reduction. Nitrate may be reduced to nitrite by nitrate reductase, which replaces cytochrome oxidase.

$$NO_3^- + 2e^- + 2H^+ \rightarrow NO_2^- + H_2O$$

However, reduction of nitrate to nitrite is not a particularly effective way of making ATP, because a large amount of nitrate is required for growth (a nitrate molecule will accept only two electrons).

$$2NO_3^- + 10e^- + 12H^+ \rightarrow N_2 + 6H_2O$$

Almost all carbon compounds that can be degraded aerobically can also be broken down under anaerobic conditions when nitrate serves as the terminal electron acceptor. A possible exception is formed by those compounds in which molecular oxygen is essential as a reagent in one of the enzymatic reactions involved in the decomposition e.g., hydrocarbons. All nitrate-reducing bacteria known are facultative aerobes that are able to use oxygen as terminal electron acceptor, but shift to denitrification when oxygen is depleted and nitrate is available.

There is considerable evidence that denitrification is a multistep process with four enzymes participating: nitrate reductase, nitrite reductase, nitric oxide reductase, and nitrous oxide reductase.

$$NO_3^- \rightarrow NO_2^- \rightarrow NO \rightarrow N_2O \rightarrow N_2$$

Interestingly, one of the intermediates is nitric oxide (NO). In mammals this molecule acts as a neurotransmitter and helps to regulate blood pressure and is used by macrophages to destroy bacteria and tumor cells. Two types of bacterial nitrite reductases catalyze the formation of NO in bacteria. One contains cytochromes *c* and $d_1$ e.g., *Paracoccus* and *Pseudomonas aeruginosa* and the other is a copper protein e.g., *Alcaligenes*. Nitrite reductase seems to be periplasmic in gram-negative bacteria. Nitric oxide reductase catalyzes the formation of nitrous oxide from NO and is a membrane-bound cytochrome *bc* complex. A well studied example of denitrification is gram-negative soil bacterium *Paracoccus denitrificans,* which reduces nitrate to $N_2$ anaerobically. Its chain contains membrane-bound nitrate reductase and nitric oxide reductase. whereas nitrite reductase and nitrous oxide reductase are periplasmic. The four enzymes use electrons from coenzyme Q and c-type cytochromes to reduce nitrate and generate PMF.

Denitrification is carried out by some members of the genera *Pseudomonas, Paracoccus* and *Bacillus.* They use this route as an altemative to normal aerobic respiration and may be considered facultative anaerobes. If $O_2$ is present, these bacteria use aerobic respiration (the synthesis of nitrate reductase is repressed by $O_2$). Denitrification in anaerobic soil results in the loss of soil nitrogen and adversely affects soil fertility.

Certain oxidized metal ions such as iron ($Fe^{3+}$) and manganese ($Mn^{4+}$) are used as terminal electron acceptors in anaerobic respiration by certain bacteria, the reduction products being $Fe^{2+}$ and $Mn^{2+}$, respectively. Some obligate anaerobes employing anaerobic respiration use $CO_2$ or carbonate as a terminal electron acceptor and are called methanogens because they reduce $CO_2$ to methane.

Use of sulfate as the terminal electron acceptor in anaerobic respiration is widespread in nature, and the process is quantitatively highly important. Sulfate is generally available in much higher concentrations than nitrate or oxidized iron, especially in marine environments. The sulfate concentration in sea water is around 25 mM. The final product of dissimilatory sulfate reduction is sulfide. The sulfate reducing bacteria are generally obligatory anaerobes, whose growth is inhibited by oxygen. In addition, the variety of organic substrates that each of the known sulfate reducing bacteria is able to oxidize is very limited (Table 6). Sulfate reducing bacteria belong to the Proteobacteria, with the exception of the endospore-forming types e.g., *Desulfotomaculum* which belong to the Gram-positive branch of the bacteria, and *Archaeoglobus*, which is a thermophilic archaeon that grows optimally at 83°C. The list of substrates that can be oxidized with concomitant reduction of sulfate includes hydrogen, alcohols with 2-5 carbon atoms, fatty acids from formate and acetate up to a chain length of 18 carbon atoms, branched-chain fatty acids, lactate, pyruvate, certain amino acids, aromatic compounds such as phenol, benzoate and aniline, and more.

**Table 6** A number of sulfate reducing bacteria and their electron donors

| Genus, species (partial list) | Substrates oxidized | Complete oxidation to $CO_2$ |
|---|---|---|
| *Desulfovibrio* | $H_2$, lactate, ethanol | No |
| *Desulfotomaculum* (most species) | $H_2$, lactate, ethanol | No |
| *Desulfotomaculum acetoxidans* | Acetate | Yes |
| *Desulfobulbus* | Propionate | No |
| *Desulfobacter* | Acetate | Yes |
| *Desulfobacterium* | Acetate | Yes |
| *Desulfococcus* | $C_1$-$C_{14}$ acids | Yes |
| *Desulfonema* | $C_2$-$C_{12}$ acids | Yes |
| *Desulfosarcina* | $C_2$-$C_{14}$ acids | Yes |
| *Archaeoglobus* | Lactate | Yes |

Sulfate is a much less favorable electron acceptor than nitrate: due to the relatively low midpoint redox potential of the couples APS/AMP + $HSO_3^-$ and $HSO_3^-/HS^-$, much less energy can be derived by reduction of sulfate than during reduction of oxygen or nitrate, as illustrated by the following examples:

Acetate$^-$ + $SO^{2-}_4$ → 2 $HCO^-_3$ + $HS^-$ ΔG°′ = - 47.7 kJ

Acetate$^-$ + 2$O_2$ → 2 $HCO^-_3$ + $H^+$ ΔG°′ = - 844.2 kJ

Acetate$^-$ + 1.6 $NO^-_3$ + 0.6 $H^+$ → 2 $HCO^-_3$ + 0.8 $N_2$ + 0.8 $H_2O$ ΔG°′ = - 792.2 kJ

4 $H_2$ + $SO^{2-}_4$ + $H^+$ → 4 $H_2O$ + $HS^-$ ΔG°′ = - 152.3 kJ

4 $H_2$ + 2 $O_2$ → 4 $H_2O$ ΔG°′ = - 948.8 kJ

4 $H_2$ + 1.6 $NO^-_3$ + 1.6 $H^+$ → 0.8 $N_2$ + 4.8 $H_2O$ ΔG°′ = - 896.8 kJ

In addition, before sulfate can be reduced, it has to be activated in a reaction that requires ATP. At the expense of the two high energy bonds in ATP, adenosine-5′-phosphate (APS) is formed, which is then reduced by APS reductase to sulfite and AMP. Part of the sulfate reducing bacteria do not perform a complete oxidation of their organic substrates to $CO_2$. These bacteria, which include the genera *Desulfovibrio*, *Desulfobulbus*, and certain species of *Desulfotomaculum* excrete acetate as the final product of their dissimilatory metabolism. The acetate formed may become available as carbon and is useful as an energy source to those sulfate reducing bacteria that are able to dissimilate acetate to $CO_2$. Some species of sulfate reducing bacteria can also generate energy by disproportionation of thiosulfate or sulfite, i.e., by oxidation of part of the substrate while reducing another part:

$S_2O^{2-}_3$ + $H_2O$ → + $SO^{2-}_4$ + $H_2S$ ΔG°′ = - 219.9 kJ

4 $SO^{2-}_3$ + 2$H^+$ → 3 $SO^{2-}_4$ + $H_2S$ ΔG°′ = - 235.7 kJ

Also, elemental sulfur can serve as terminal electron acceptor in anaerobic respiration. A number of sulfate reducing bacteria are also able to reduce elemental sulfur, but other bacteria were discovered that specialize in the reduction of sulfur, such as *Desulfuromonas*

Acetate$^-$ + $H^+$ + 2$H_2O$ + 4 S° → 2 $CO_2$ + 4 $H_2S$ ΔG°′ = - 16.8 kJ

Additional substrates exist that can serve as terminal electron acceptors in anaerobic respiration processes. Many aerobic bacteria are able to use dimethylsulfoxide, trimethylamine-*N*-oxide or fumarate as electron acceptors in the absence of oxygen, with the formation of dimethylsulfide, trimethylamine, and succinate, respectively.

Anaerobic respiration is not as efficient in ATP synthesis as aerobic respiration i.e. not as much ATP is produced by oxidative phosphorylation with nitrate, sulfate, or $CO_2$ as the terminal acceptors. Reduction in ATP yield arises from the fact that these altemate electron acceptors have less positive reduction potentials than $O_2$. The reduction potential difference between a donor like NADH and nitrate is smaller than the difference between NADH and $O_2$. Because energy yield is directly related to the magnitude of the reduction potential difference, less energy is available to make ATP in anaerobic respiration. Nevertheless, anaerobic respiration is useful because it is more efficient than fermentation and allows ATP synthesis by electron transport and oxidative phosphorylation in the absence of $O_2$. Anaerobic respiration is very prevalent in oxygen-depleted soils and sediments.

Often one will see a succession of microorganisms in an environment when several electron acceptors are present. For example, if $O_2$, $NO_3^-$, $Mn^{2+}$, $Fe^{3+}$, $SO_4^{2-}$ and $CO_2$ are available in a particular environment, a predictable sequence of oxidant use takes place when an oxidizable substrate is available to the microbial population. Oxygen is employed as an electron acceptor first because it inhibits nitrate use by microbes capable of respiration with either $O_2$ or nitrate. While $O_2$ is available, sulfate reducers and methanogens are inhibited because these groups are obligate anaerobes. Once the oxygen and nitrate are exhausted and fermentation products, including hydrogen, have accumulated, competition for use of other oxidants begins. Manganese and iron will be used first.

### *4.2.3. Fermentation*

When all possibilities to degrade organic material by aerobic respiration or by anaerobic respiration are exhausted (either because of lack of availability of electron acceptors, or in those cases in which the compounds cannot be broken down using the electron acceptors available), many organic compounds can be degraded further by fermentation. Fermentation is an ancient mode of metabolism, and it must have evolved with the appearance of organic material on the planet. In fermentation, energy is derived from the partial oxidation of an organic compound using organic intermediates as electron donors and electron acceptors. no outside electron acceptors are involved; no membrane or electron transport system is required; all ATP is produced by substrate level phosphorylation. By definition, fermentation may be as simple as two steps illustrated in a model described in Figure 38. Indeed, some amino acid fermentations by the clostridia are simple, but the pathways of fermentation are a bit more complex, usually involving several preliminary steps to prime the energy source for oxidation and substrate level phosphorylation.

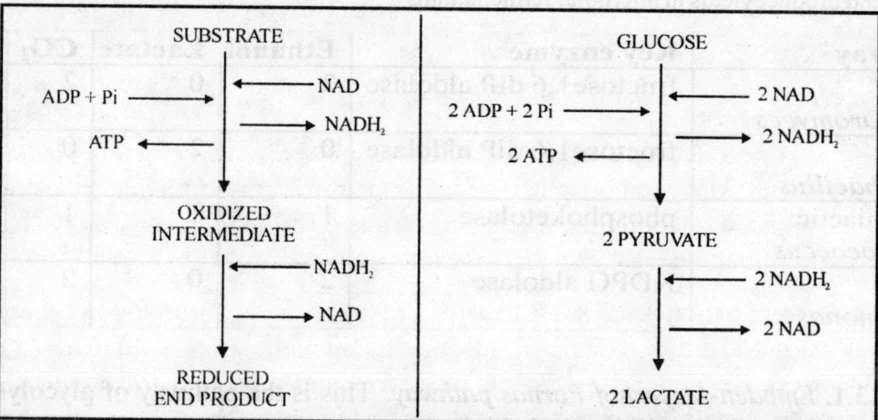


**Figure 38** Model fermentation: The substrate is oxidized to an organic intermediate; the usual oxidizing agent is NAD. Some of the energy released by the oxidation is conserved during the synthesis of ATP by the process of substrate level phosphorylation. Finally, the oxidized intermediate is reduced to end products. Note that $NADH_2$ is the reducing agent, thereby balancing its redox ability to drive the energy-producing reactions

In lactic acid fermentation by *Lactobacillus*, the substrate (glucose) is oxidized to pyruvate, and pyruvate becomes reduced to lactic acid.

In biochemistry, for the sake of convenience, fermentation pathways begin with glucose. This is because it is the simplest molecule, requiring the fewest catalytic steps, to enter into a pathway of glycolysis and central metabolism. In procaryotes their exist three major pathways of glycolysis (the dissimilation of sugars): the classic EMP pathway, which is also used by most eukaryotes, including yeast (*Saccharomyces*): the phosphoketolase or heterolactic pathway related to the hexose-pentose shunt; and the Entner-Doudoroff pathway. Whether or not a bacterium is a fermenter, it will likely dissimilate sugars through one or more of these pathways (Table 7) with various end products (Table 8).

**Table 7** Oxidative pathways of glycolysis employed by various bacteria

| **Bacterium** | **Embden-Meyerhof pathway** | **Phosphoketolase (heterolactic) pathway** | **Entner-Doudoroff pathway** |
|---|---|---|---|
| *Acetobacter aceti* | | + | - |
| *Agrobacterium tumefaciens* | | - | + |
| *Azotobacter vinelandii* | | - | + |
| *Bacillus subtilis* | major | minor | |
| *Escherichia coli* | + | - | - |
| *Lactobacillus acidophilus* | + | - | - |
| *Leuconostoc mesenteroides* | - | + | - |
| *Pseudomonas aeruginosa* | - | - | + |
| *Vibrio cholerae* | minor | - | major |
| *Zymomonas mobilis* | - | - | + |

**Table 8** End product yields in microbial fermentations

| **Pathway** | **Key enzyme** | **Ethanol** | **Lactate** | **$CO_2$** | **ATP** |
|---|---|---|---|---|---|
| EMP *Saccharomyces* | fructose1,6 diP aldolase | 2 | 0 | 2 | 2 |
| EMP *Lactobacillus* | fructose1,6 diP aldolase | 0 | 2 | 0 | 2 |
| Heterolactic *Streptococcus* | phosphoketolase | 1 | 1 | 1 | 1 |
| ED *Zymomonas* | KDPG aldolase | 2 | 0 | 2 | 1 |

**4.2.3.1.** *Embden-Meyerhof-Parnas pathway:* This is the pathway of glycolysis most familiar to biochemists and eukaryotic biologists, as well as to brewers, bread makers and cheese heads. The pathway is operated by *Saccharomyces* to produce ethanol and $CO_2$: it is operated by the (homo)lactic acid bacteria to produce lactic acid; and it is operated by other bacteria to produce a variety of fatty acids, alcohols and gases. Some end products of EMP fermentations are essential components of foods and beverages, and some are useful fuels and industrial solvents. Diagnostic microbiologists use bacterial fermentation profiles (e.g. testing an organism's ability to ferment certain sugars, or examining an organisms's end products) in order to identify them, down to the genus level.

The first three steps of the pathway are prime (phosphorylate) and rearrange the hexose for cleavage into 2 trioses by the enzyme Fructose-1,6-diphosphate aldolase, the key (cleavage) enzyme in the EMP pathway. Each triose molecule is oxidized and phosphorylated followed by two substrate level phosphorylations that yield 4 ATP during the drive to pyruvate. Yeast reduce the pyruvate to ethanol and $CO_2$, whereas lactic acid bacteria reduce the pyruvate to lactic acid.

The overall reaction is : Glucose → 2 lactate (or 2 ethanol +2 $CO_2$ in yeast) with a net gain of 2 ATP (Figure 39).

The oxidation of glucose to lactate yields a total of 56kcal per mole of glucose. Since the cells harvest 2 ATP (16 kcal) as useful energy, the efficiency of the lactate fermentation is about 29 percent (16/56).

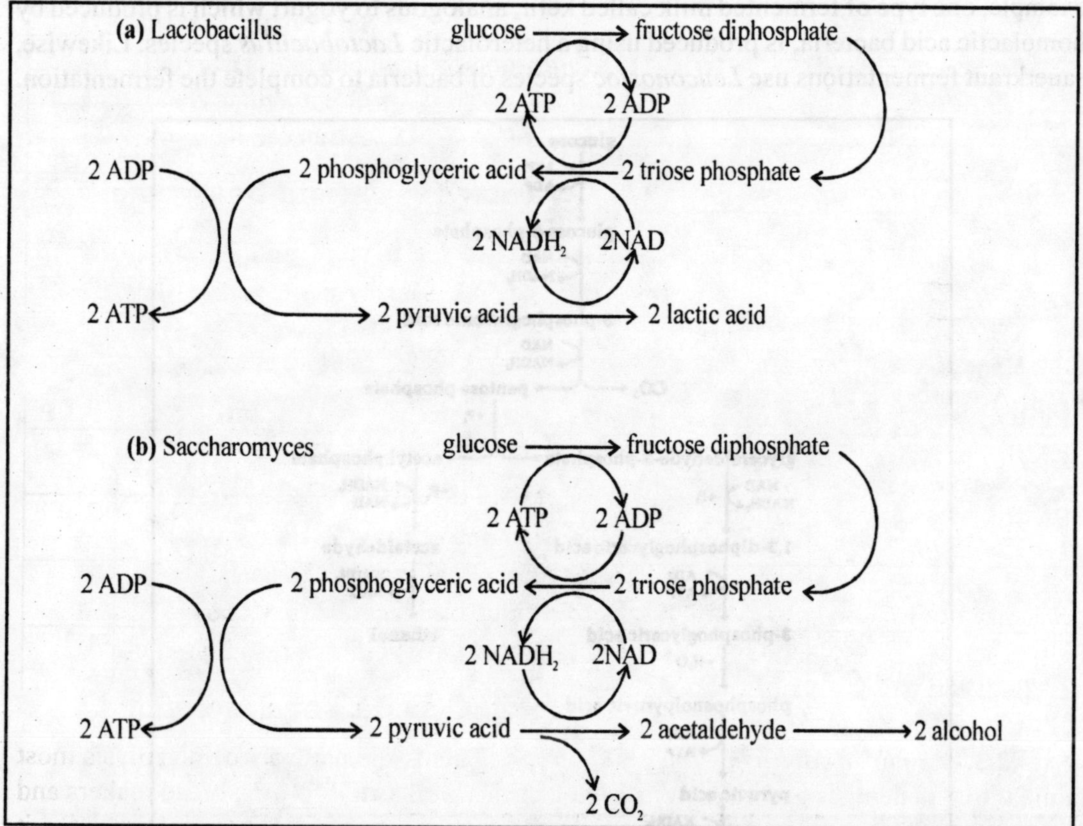


**Figure 39** The EMP pathway to lactic acid (in lactic acid bacteria) or ethanol and $CO_2$ (in yeast). The pathway yields two moles of end products and two moles of ATP per mole of glucose fermented. ATP is produced by substrate level phosphorylation at the steps indicated

**4.2.3.2.** *Heterolactic (Phosphoketolase) pathway:* The phosphoketolase pathway (Figure 40) is distinguished by the key cleavage enzyme, phosphoketolase, which cleaves pentose phosphate into glyceraldehyde-3-phosphate and acetyl phosphate. As a fermentation pathway, it is employed mainly by the heterolactic acid bacteria, which include some species of *Lactobacillus* and *Leuconostoc*. In this pathway, glucose-phosphate is oxidized 6-phosphogluconate, which becomes oxidized and decarboxylated to form pentose phosphate. Unlike the EMP pathway, NAD-mediated oxidations take place before the cleavage of the substrate being utilized. Pentose phosphate is subsequently cleaved to glyceraldehyde-3-phosphate (GAP) and acetyl phosphate. GAP is converted to lactic acid by the same enzymes as the EMP pathway. This branch of the pathway contains an oxidation coupled to a reduction as illustrated in Figure 40. Two ATP molecules are produced by substrate level phosphorylation. Acetyl phosphate is reduced in two steps to ethanol, which balances the two oxidations before the cleavage but does not yield ATP. The efficiency is about half that of the EMP pathway.

The overall reaction is: Glucose →1 lactate+1 ethanol +1 $CO_2$ with a net gain of 1 ATP.

Heterolactic species of bacteria are occasionally used in the fermentation industry. For example, one type of fermented milk called kefir, analogous to yogurt which is produced by homolactic acid bacteria, is produced using a heterolactic *Lactobacillus* species. Likewise, sauerkraut fermentations use *Leuconostoc* species of bacteria to complete the fermentation.

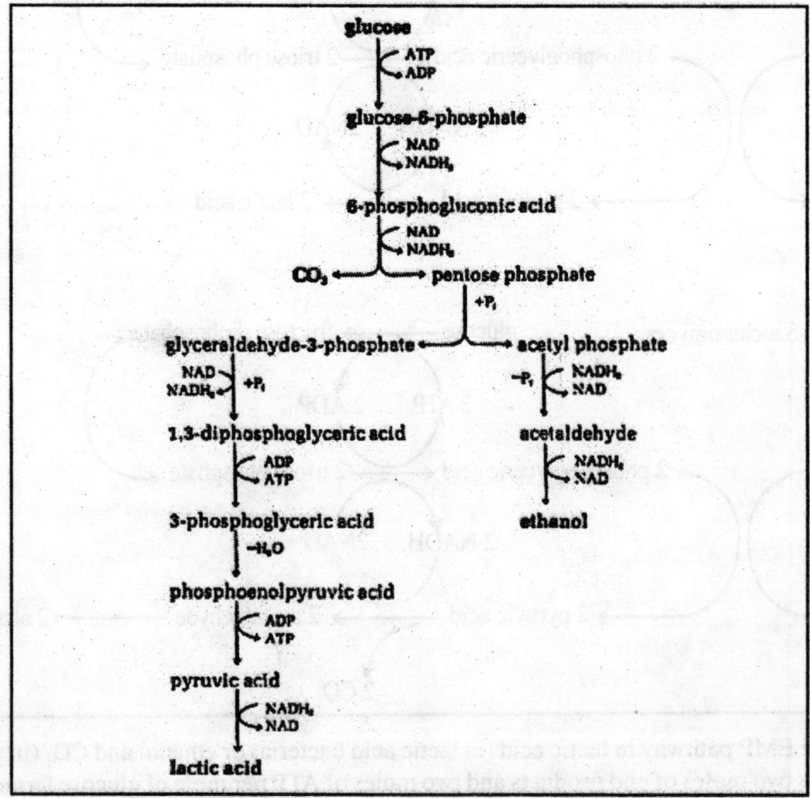


**Figure 40** The heterolactic (phosphoketolase) pathway of fermentation

**4.2.3.3.** *Entner-Doudoroff pathway:* Only a few bacteria, most notably *Zymomonas*, employ the Entner-Doudoroff pathway as a fermentation path. However, many bacteria, especially those grouped around the pseudomonads, use the pathway as a way to degrade carbohydrates for respiratory metabolism. The E-D pathway yields 2 pyruvate from glucose (same as the E-M pathway) but like the phosphoketolase pathway, oxidation occurs before the cleavage, and the net energy yield per mole of glucose utilized is one mole of ATP.

In the E-D pathway, glucose phosphate is oxidized to 2-keto-3-deoxy-6-phosphogluconate (KDPG) which is cleaved by KDPG aldolase to pyruvate and GAP. The latter is oxidized to pyruvate by E-M enzymes wherein 2 ATP are produced by substrate level phosphorylations. Pyruvate from either branch of the pathway is reduced to ethanol and $CO_2$, in the same manner as yeast, by the "yeast-like bacterium", *Zymomonas* (Figure 41).

Thus, the overall reaction is Glucose → 2 ethanol +2 $CO_2$, and a net gain of ATP.

*Zymomonas* is a bacterium that lives on the surfaces of plants, including the succulent maguey cactus which is indigenous to Mexico. Just as grapes are crushed and fermented by resident yeast to wine, so may the maguey flesh be crushed and allowed to ferment with *Zymomonas*, which gives rise to "cactus beer" or "pulque", as it is known in Mexico. Distilled pulque yields tequila in the state of Jalisco or mescal in the state of Oaxaca. Many cultures around the world prepare their native fermented brews with *Zymomonas* in deference to the yeast, *Saccharomyces*, although they may not have a choice in the matter. *Zymomonas* has potential advantageous over yeast for the industrial production of alcohol, but the industry is geared to do what it can do, and no change in organisms is forthcoming.

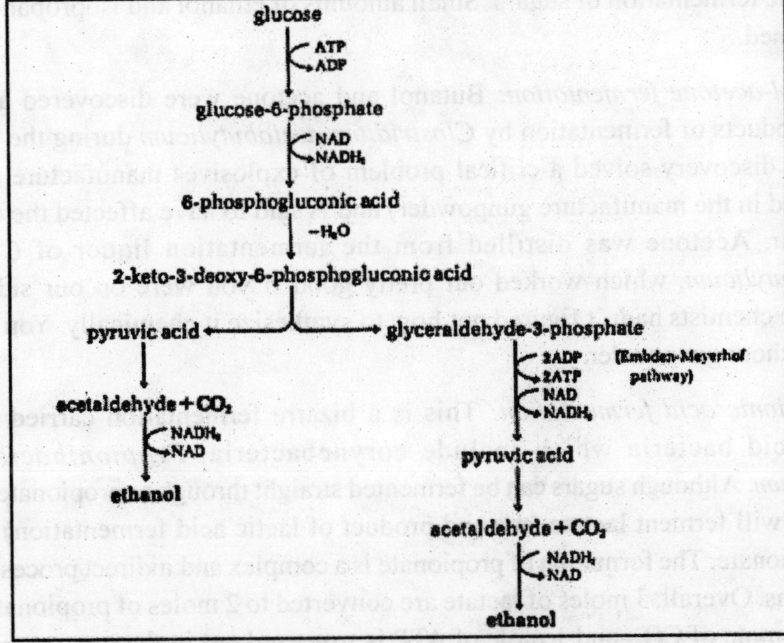


Figure 41 The Entner-Doudoroff pathway of fermentation

Embden-Meyerhof fermentations in bacteria can lead to a wide array of end products depending on the pathways taken in the reductive steps after the formation of pyruvate. Figure 42 shows some of the pathways proceeding from pyruvate in certain bacteria. Usually, bacterial fermentations are distinguished by their end products into the following groups:

1. *Homolactic fermentation:* Lactic acid is the sole end product and follows the pathway of the homolactic acid bacteria (*Lactobacillus* and most streptococci). The bacteria are used to ferment milk and milk products in the manufacture of yogurt, buttermilk, sour cream, cottage cheese, cheddar cheese, and most fermented dairy products.

2. *Mixed acid fermentations:* Mainly follows the pathway of the Enterobacteriaceae. End products are a mixture of lactate, acetate, formate, succinate and ethanol, with the possibility of gas formation ($CO_2$ and $H_2$) if the bacterium possesses the enzyme formate dehydrogenase, which cleaves formate to the gases.

2a. *Butanediol fermentation:* Forms mixed acids and gases as above, but, in addition, 2,3-butanediol from the condensation of 2 pyruvate. The use of the pathway decreases acid formation (butanediol is neutral) and causes the formation of a distinctive intermediate, acetoin. Water microbiologists have specific tests to detect low acid and acetoin in order to distinguish non fecal enteric bacteria (butanediol formers, such as *Klebsiella* and *Enterobacter*) from fecal enterics (mixed acid fermenters, such as *E. coli*, *Salmonella* and *Shigella*).

3. *Butyric acid fermentations:* These are run by the clostridia, the masters of fermentation. In addition to butyric acid, these clostridia form acetate, $CO_2$ and $H_2$ from the fermentation of sugars. Small amounts of ethanol and isopropanol may also be formed.

3a. *Butanol-acetone fermentation:* Butanol and acetone were discovered as the main end products of fermentation by *Clostridium acetobutylicum* during the World War I. This discovery solved a critical problem of explosives manufacture (acetone is required in the manufacture gunpowder) and is said to have affected the outcome of the war. Acetone was distilled from the fermentation liquor of *Clostridium acetobutylicum*, which worked out pretty good if you were on our side, because organic chemists hadn't figured out how to synthesize it chemically. You can't run a war without gunpowder.

4. *Propionic acid fermentation:* This is a bizarre fermentation carried out by the propionic acid bacteria which include corynebacteria, *Propionibacterium* and *Bifidobacterium*. Although sugars can be fermented straight through to propionate, propionic acid bacteria will ferment lactate (the end product of lactic acid fermentation) to acetate, $CO_2$ and propionate. The formation of propionate is a complex and indirect process involving 5 or 6 reactions. Overall, 3 moles of lactate are converted to 2 moles of propionate + 1 mole of acetate + 1mole of $CO_2$, and 1 mole of ATP is squeezed out in the process.

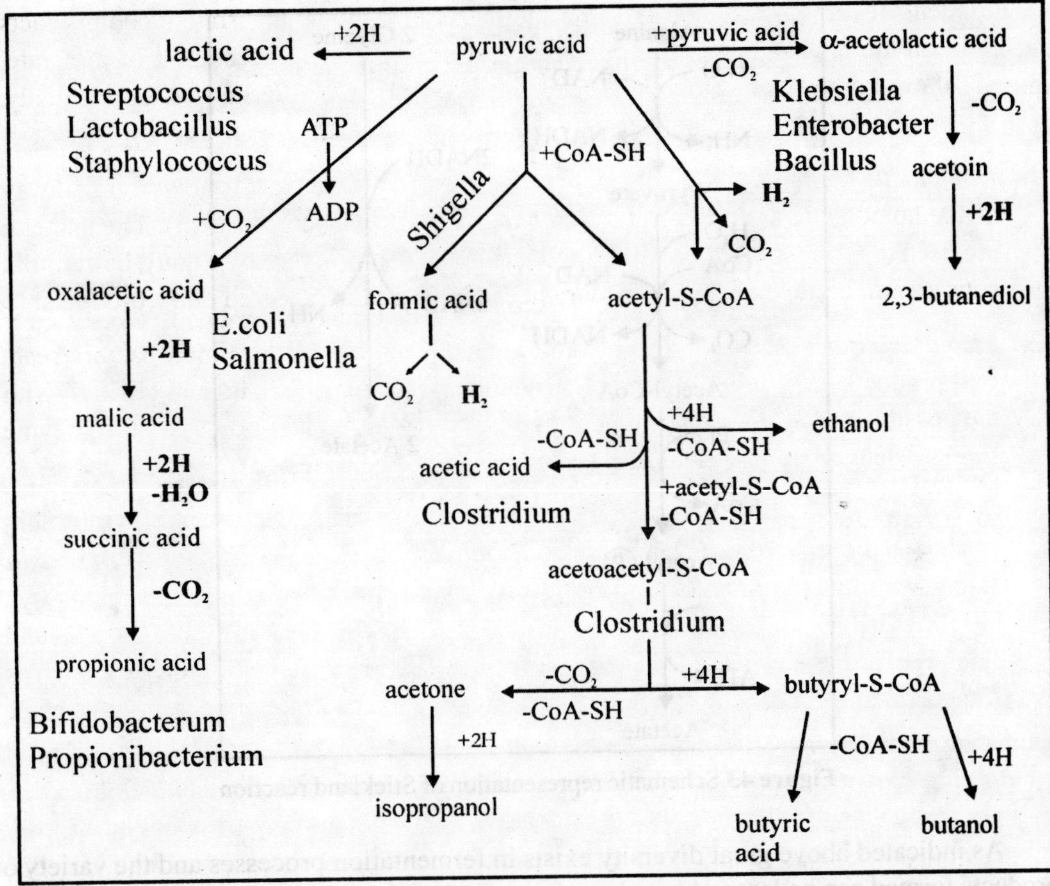


**Figure 42** Fermentations in bacteria that proceed through the Embden-Meyerhof pathway

The propionic acid bacteria are used in the manufacture of Swiss cheese, which is distinguished by the distinct flavor of propionate and acetate, and holes caused by entrapment of $CO_2$.

Substances other than sugars also are fermented by microorganisms. For example, some members of the genus *Clostridium* prefer to ferment mixtures of amino acids. Proteolytic clostridia such as the pathogens *Cl. sporogenes* and *Cl. botulinum* will carry out the Stickland reaction in which one amino acid is oxidized and a second amino acid acts as the electron acceptor. Figure 43 shows the way in which alanine is oxidized and glycine reduced to produce acetate, $CO_2$ and $NH_3$. Some ATP is formed from acetyl phosphate by substrate-level phosphorylation, and the fermentation is quite useful for growing in anaerobic, protein-rich environments. The Stickland reaction is used to oxidize several amino acids: alanine, leucine, isoleucine, valine, phenylalanine, tryptophan, and histidine. Bacteria also ferment amino acids (e.g., alanine, glycine, glutamate, threonine, and arginine) by other mechanisms. In addition to sugars and amino acids, organic acids such as acetate, lactate, propionate, and citrate are fermented. Some of these fermentations are of great practical importance e.g., citrate can be converted to diacetyl and give flavor to fermented milk.

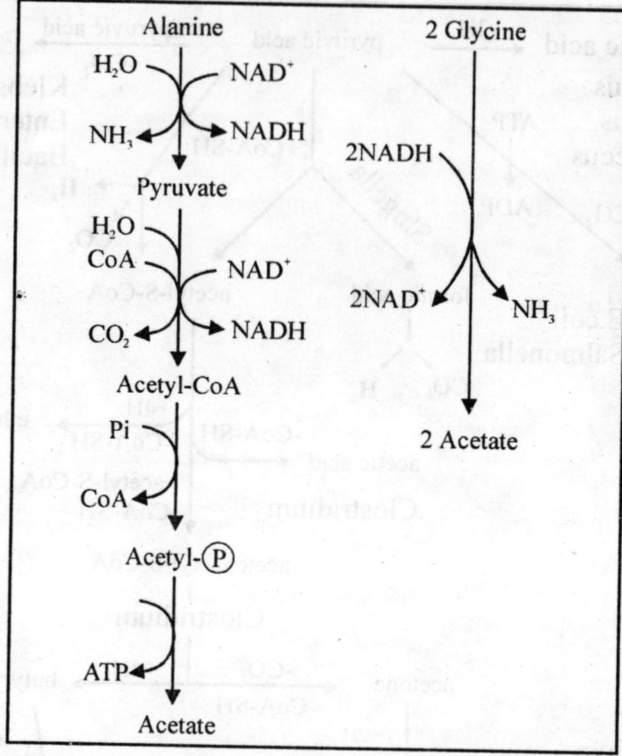


**Figure 43** Schematic representation of Stickland reaction

As indicated above great diversity exists in fermentation processes and the variety of products formed can be large. A number of representative examples with energy yields are as follows :

- The ethanol fermentation of yeasts:
  Glucose + $2H_2O \rightarrow$ 2 Ethanol + $2HCO_3^-$ + $2H^+$ $\Delta G^{o\prime} = -225.4$ kJ
- The homolactic fermentation, in which sugars such as glucose are fermented to lactic acid as the sole product
  Glucose $\rightarrow$ 2 Lactate$^-$ + $2H^+$ $\Delta G^{o\prime} = -198$ kJ
  This type of fermentation is characteristic of lactic acid bacteria such as *Lactobacillus* and *Streptococcus*. These bacteria play important role in industrial processes such as the production of cheeses and other milk products, pickling of vegetables, etc.
- The heterolactic fermentation of *Leuconostoc* and certain species of *Lactobacillus*:
  Glucose + $H_2O \rightarrow$ Lactate$^-$ + Ethanol + $HCO_3^-$ $2H^+$ $\Delta G^{o\prime} = -211.7$ kJ
- In many fermentative bacteria, formation of acids such as acetate and butyrate is accompanied by other products such as n-butanol, acetone, isopropanol, and generally also gases such as $CO_2$ and hydrogen. These fermentations are performed by a variety of Gram-positive bacteria such as *Clostridium* species, *Eubacterium* in the large intestine, and *Butyrivibrio* in the rumen of cows and other ruminant animals.

- Enteric bacteria, such as *Escherichia* and *Salmonella,* form a mixture of fermentation products under anaerobic conditions, including ethanol, lactate, acetate, hydrogen, succinate, $CO_2$ and formate. This type of fermentation is known as "mixed acid fermentation".

- The bacteria *Enterobacter* and *Serratia* excrete 2,3-butanediol and ethanol as the main products, together with hydrogen and formate.

- Certain bacteria (*Propionibacterium* and others) ferment sugars or lactate to propionate and acetate:

  1.5 Glucose + $3H_2O \rightarrow$ 2 Propionate$^-$ + Acetate$^-$ + $HCO_3^-$ + $4H^+$ $\Delta G^{o\prime}$ =- 249.7 kJ

  3 Lactate$^-$ $\rightarrow$ 2 Propionate$^-$ + Acetate$^-$ + $HCO_3^-$ + $H^+$ $\Delta G^{o\prime}$ = - 249.7 kJ

  Among the bacteria that form propionate, some are able to grow on succinate as well:

  Succinate$^{2-}$ + $H_2O \rightarrow$ Propionate$^-$ + $HCO_3^-$ $\Delta G^{o\prime}$ = - 20.6 kJ

- Also, many amino acids can be fermented e.g, certain *Clostridium* species are able to grow on glutamate, with the formation of acetate and butyrate as the main fermentation products:

  2 Glutamate$^-$ + $4H_2O \rightarrow$ 2 Acetate$^-$ + Butyrate$^-$ + $2HCO_3^-$ + $2H_2$ + $2NH_4^+$ H+ $\Delta G^{o\prime}$ = - 115.8 kJ

- Many bacteria can grown anaerobically on arginine:

  Arginine$^+$ + $3H_2O$ + $H^+$ $\rightarrow$ Ornithine+ + $HCO_3^-$ $2NH_4^+$ $\Delta G^{o\prime}$ = - 70 kJ

- Amino acids can also be fermented pair wise in a reaction in which one amino acid serves as the electron donor and the second as the electron acceptor as indicated in the Stickland reaction (Figure 38).

  Alanine + 2 Glycine + $3H_2O \rightarrow$ 3 Acetate$^-$ + $3NH_4^+$ + $HCO_3^-$ H + $\Delta G^{o\prime}$ = - 89.7 kJ

  Tryptophan, proline and arginine can in a similar way serve as electron donors, and histidine, valine, leucine, isoleucine, and serine are suitable electron acceptors.

## 4.3. Chemoautotrophic way of life

Many bacteria derive their energy from the oxidation of reduced inorganic compounds such as hydrogen, ammonia (both forms, non-ionized $NH_3$ and ionized ammonium $NH_4^+$), nitrite ($NO_2^-$), or reduced sulfur compounds such as sulfide ($S^{2-}$), elemental sulfur ($S^o$), or thiosulfate ($S_2O_3^{2-}$). These bacteria are called chemoautotrophic or chemolithotrophic. Oxygen generally serves as the terminal electron acceptor. Anaerobic chemoautotrophs also exist, e.g., those methanogenic and homoacetogenic bacteria that obtain their energy by reduction of $CO_2$ with hydrogen as the electron donor, while also using $CO_2$ as their sole or main source of assimilatory carbon. Other options for anaerobic chemoautotrophic life are, e.g., the oxidation of sulfide or hydrogen with nitrate as the electron acceptor. The exergonic nature of all these reactions becomes clear from the comparison of the midpoint redox potentials of the redox couples involved. Figure 44 shows the basic principles of the chemoautotrophic way of life.

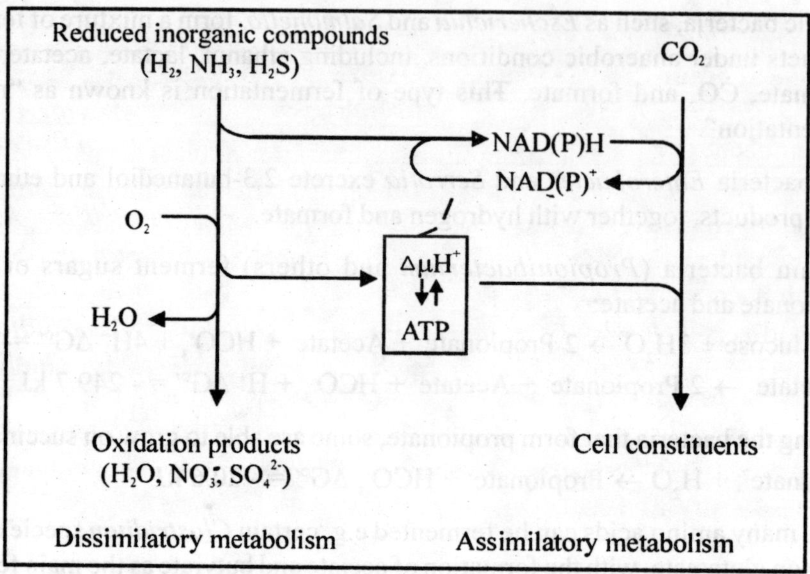


**Figure 44** The principle of the metabolism of aerobic chemoautotrophic microorganisms

During chemoautotrophic growth, reduced inorganic compounds are used both as an energy source and as electron donors for the fixation of $CO_2$. Oxidation of the inorganic compounds through an electron transport chain in the cell membrane, with the use of oxygen (or an alternative acceptor such as nitrate) as the terminal electron acceptor leads to the formation of a primary proton electrochemical gradient that can be exploited for the generation of ATP.

The idea that certain microorganisms may grow solely on inorganic compounds in the absence of light energy was first conceived by Sergei Winogradsky as early as 1887. He based his theory on observations of the filamentous sulfur bacterium *Beggiatoa*, which accumulated intracellular globules of elemental sulfur when fed with sulfide. Upon starvation of sulfide, these globules disappeared again, and the sulfur was found to be oxidized to sulfate. However, it took almost a hundred years to prove that certain *Beggiatoa* strains are indeed capable of autotrophic growth. In 1888, Winogradsky extended his studies to the iron oxidizers *Gallionella* and *Leptothrix*. When in 1890, he succeeded in growing nitrifying bacteria that oxidize ammonia to nitrate, and nitrite to nitrate, the concept of chemoautotrophic growth thus obtained a firm experimental basis.

Lithotrophy runs through the bacteria and the archaea. If one considers methanogen oxidation of $H_2$ a form of lithotrophy, then probably most of the Archaea are lithotrophs. Lithotrophs are usually organized into "physiological groups" based on their inorganic substrate for energy production and growth (Table 9) and their modes of energy generation are discussed below:

**Table 9** Physiological groups of lithotrophs

| Physiological group | Energy source | Oxidized end product | Organism |
|---|---|---|---|
| Hydrogen bacteria | $H_2$ | $H_2O$ | *Alcaligenes, Pseudomonas* |
| Methanogens | $H_2$ | $H_2O$ | *Methanobacterium* |
| Carboxydobacteria | CO | $CO_2$ | *Rhodospirillum, Azotobacter* |
| Nitrifying bacteria | $NH_3$ | $NO_2$ | *Nitrosomonas* |
| Nitrifying bacteria | $NO_2$ | $NO_3$ | *Nitrobacter* |
| Sulfur oxidizers | $H_2S$ or S | $SO_4$ | *Thiobacillus, Sulfolobus* |
| Iron bacteria | $Fe^{++}$ | $Fe^{+++}$ | *Gallionella, Thiobacillus* |

**4.3.1.** *Aerobic oxidation of hydrogen:* A variety of aerobic bacteria are able to generate energy by the oxidation of hydrogen. The amount of energy obtained from this reaction is large, due to the fact that the electron donor is highly reduced.

Among the organisms that grow on mixtures of hydrogen and oxygen (the so-called "Knallgas reaction"), there are soil bacteria and aquatic bacteria such as *Pseudomonas, Ralstonia,* and *Paracoccus* (all belonging to the *Proteobacteria*) and gram-positive bacteria such as *Bacillus schlegelii*. These bacteria can also grow chemoorganotrophically by the oxidation of organic compounds. Among those chemoautotrophic bacteria that grow on hydrogen, the thermophilic species *Hydrogenobacter* and *Aquifex* should also be mentioned, as they are species that represent separate phylogenetic lineages in the domain *'Bacteria'*.

**4.3.2.** *Anaerobic oxidation of hydrogen: Methanogens* and *Homoacetogens:* Under anaerobic conditions, hydrogen can serve as electron donor for chemolithotrophic growth in different ways:

- Some bacteria that oxidize hydrogen with oxygen as the electron acceptor are able to replace oxygen by nitrate and perform denitrification:
  $5\,H_2 + 2NO_3^- + 2H^+ \rightarrow N_2 + 6\,H_2O$ $\Delta G^{o\prime} = -1120$ kJ
- Some sulfate reducing bacteria can use hydrogen as the electron donor and use $CO_2$ as the assimilatory carbon source, thus displaying an autotrophic life style
  $4\,H_2 + 2O_2 \rightarrow 4\,H_2O$ $\Delta G^{o\prime} = -948.8$ kJ
- Most methanogenic Archaea (*Methanobacterium, Methanococcus*) obtain their energy by reducing $CO_2$ with hydrogen as the electron donor:
  $4\,H_2 + H^+ + HCO_3^- \rightarrow CH_4 + 3\,H_2O$ $\Delta G^{o\prime} = -135.7$ kJ
- Yet another group of anaerobes that makes a living by oxidizing hydrogen to support autotrophic growth is the homoacetogenic bacteria that reduce $CO_2$ to form acetate:
  $4\,H_2 + H^+ + 2\,HCO_3^- \rightarrow \text{Acetate}^- + 4\,H_2O$ $\Delta G^{o\prime} = -104.6$ kJ

During the process of acetogenesis, one molecule of $CO_2$ is reduced stepwise to a methyl group bound to a $B_{12}$ coenzyme. This methyl group then combines with a molecule of carbon monoxide formed by reduction of the second $CO_2$ molecule. A part of the energy released in these reactions is conserved in the form of primary proton and/or sodium ion gradients across the cell membrane.

**4.3.3.** *Aerobic oxidation of carbon monoxide:* An additional substrate that can serve as the electron donor for chemoautotrophic growth in certain bacteria is carbon monoxide (CO):

$$CO + 0.5\ O_2 \rightarrow CO_2 \qquad \Delta G^{o\prime} = -257.1\ kJ$$

CO is oxidized by the enzyme carbon monoxide dehydrogenase. Many of the bacteria that are able to use hydrogen as the electron donor for chemolithotrophic growth can also oxidize CO.

**4.3.4.** *Nitrification:* Aerobic oxidation of ammonia to nitrate (the process called nitrification) occurs in two stages, involving two distinct groups of bacteria:

$$NO^-_2 + 0.5\ O_2 \rightarrow NO^-_3 \qquad \Delta G^{o\prime} = -74.1\ kJ$$

$$NH^+_4 + 1.5\ O_2 \rightarrow NO^-_2 + 2H^+ + H_2O \qquad \Delta G^{o\prime} = -274.6\ kJ$$

The bacteria responsible for nitrification are Gram-negative and belong to the *Proteobacteria. Nitrosomonas*, *Nitrosolobus*, and a number of additional species oxidize ammonia to nitrite, and *Nitrobacter*, *Nitrococcus*, and related organisms produce nitrate from nitrite. Oxygen not only serves as the terminal electron acceptor in the electron transport chain, but molecular oxygen is also required in the first step of the oxidation of ammonia, catalyzed by the enzyme ammonia monooxygenase, with the formation of hydroxylamine.

$$NH^+_4 + O_2 + H^+ + 2e \rightarrow NH_2OH + H_2O$$

Chemolithotrophic organisms, such as the nitrifying bacteria, are bioenergetically severely stressed. First of all, the amount of energy that can be obtained from the oxidation of ammonia to nitrite or from oxidation of nitrite to nitrate is relatively small (for every 2 electrons 91.5 and 74.1 kJ, respectively, as compared to a yield of 273.2 kJ during the aerobic oxidation of hydrogen). Moreover, they also face the problem of obtaining reducing power in the form of NADH or NADPH for autotrophic $CO_2$ fixation in the Calvin cycle, as the electron donors available are not sufficiently reduced to enable the direct reduction of $NAD(P)^+$. The organisms have therefore to transport electrons "uphill" at the expense of ATP. As a result, the amount of substrate required for the growth of nitrifying bacteria is very large: *Nitrosomonas* needs to oxidize about 35 molecules of ammonia to fix one molecule of $CO_2$. In *Nitrobacter,* the ratio is even larger, where 100 molecules of nitrite are required in order to supply the energy and the electrons needed for the fixation of one molecule of $CO_2$.

The possible existence of anaerobic autotrophic oxidation of ammonia with the use of nitrite or nitrate as the electron acceptor, producing gaseous nitrogen as product, was already predicted on thermodynamic grounds:

$$NH^+_4 + NO^-_2 \rightarrow N_2 + 2\ H_2O \qquad \Delta G^{o\prime} = -357.8\ kJ$$

$$6\ NH^+_4 + 3\ NO^-_3 \rightarrow 4\ N_2 + 9\ H_2O + 2H^+ \qquad \Delta G^{o\prime} = -1483.5\ kJ$$

Almost twenty years, the process (termed "Anammox" for anaerobic ammonia oxidation) was indeed found to occur. The bacterium responsible belongs phylogenetically to the *Planctomyces* group. It is a true autotroph, which uses nitrite as the electron acceptor for the oxidation of ammonia, with the formation of gaseous nitrogen as a product. The discovery of the Anammox bacterium illustrates nicely that the possibilities presented in the framework of chemical thermodynamics are fully exploited by the microbial world. The biochemistry of anaerobic ammonia oxidation is completely different from the aerobic nitrification, and involves hydrazine as an intermediate.

**4.3.5.** *Oxidation of reduced sulfur compounds:* Chemoautotrophic sulfur-oxidizing bacteria are found both in the archaeal domain and in the bacterial domain. Within the *Proteobacteria* branch of the bacteria, rod-shaped or spiral-shaped bacteria such as *Thiobacillus* and *Thiomicrospira* are found, as well as long filamentous bacteria that move by gliding, such as the genus *Beggiatoa*. The Archaea that oxidize sulfide or sulfur include species such as *Sulfolobus* and *Acidianus*, which are able to grow at low pH values and at temperatures as high as 75-90°C. These are found worldwide in hot sulfur springs.

As sulfide is more reduced than ammonia or nitrate, the energy yield of aerobic sulfide oxidation is much higher than that of the nitrification reactions:

$H_2S + 2\ O_2 \rightarrow SO^{2-}_4 + 2H^+$ $\Delta G^{o\prime} = -\ 796.3$ kJ

Also elementary sulfur, thiosulfate, and other reduced sulfur compounds are suitable as electron donors and energy sources :

$S^o + 1.5\ O_2 + H_2O \rightarrow SO^{2-}_4 + 2H^+$ $\Delta G^{o\prime} = -\ 587.0$ kJ

$S_2O^{2-}_3 + H_2O + 2\ O_2 \rightarrow 2\ SO^{2-}_4 + 2\ H^+$ $\Delta G^{o\prime} = -\ 818.2$ kJ

In some cases nitrate can serve as alternative terminal electron acceptor:

$5\ H_2S + 8\ NO^-_3 \rightarrow 5\ SO^{2-}_4 + N_2 + 4\ H_2O + 2\ H^+$ $\Delta G^{o\prime} = -\ 3721.5$ kJ

As all these reactions are accompanied by the formation of protons, the sulfur-oxidizing bacteria cause a drop in pH of the medium in which they grow while producing sulfuric acid. Many chemoautotrophic bacteria that oxidize sulfur compounds are able to withstand very low pH values. Some are true acidophiles that grow optimally at pH 2-3, and a few can even grow below pH 1.

**4.3.6.** *Oxidation of reduced iron and manganese:* Oxidation of reduced iron and manganese ions can also supply energy for chemoautotrophic growth in certain bacteria. The amount of energy that can be derived from these processes is relatively small:

$2\ Fe^{2+} + 0.5\ O_2 + 2H^+ \rightarrow 2\ Fe^{3+} + H_2O$ $\Delta G^{o\prime} = -\ 9.0$ kJ

$2\ Fe^{2+} + 0.5\ O_2 + 2H^+ \rightarrow 2\ Fe^{3+} + H_2O$ $\Delta G^{o\prime} = -\ 65.8$ kJ at pH 2

$Mn^{2+} + 0.5\ O_2 + H_2O \rightarrow MnO_2 + 2H^+$ $\Delta G^{o\prime} = -\ 71.2$ kJ

Our understanding of these processes and of the bacteria that perform them is rather limited. This is mainly due to the fact that the reduced metal ions have a limited stability in the presence of oxygen, and chemical oxidation competes with the biological oxidation process. Divalent $Fe^{2+}$ are especially very unstable in oxidized environments at neutral or alkaline pH, and only in a limited number of cases have bacteria been shown to grow chemoautotrophically on iron under these conditions. The best example is *Gallionella*, a bacterium that forms a stalk of $Fe_2O_3$ deposited as a product of iron oxidation.

In acidic environments, however, divalent iron is chemically stable. Acidic environments containing high concentrations of iron ions are often found in the surroundings of mines, where coal and metal-containing ores from the depth of the Earth are brought to the surface. Sulfur, commonly found in these ores, is oxidized to sulfate by chemolithotrophic sulfur oxidizing bacteria such as *Thiobacillus*, with the formation of large amounts of sulfuric acid (see above). In the acidic environment, the divalent iron is oxidized to trivalent iron by *Thiobacillus ferrooxidans*, a bacterium able to grow chemoautotrophically using both reduced iron and reduced sulfur compounds as electron donor and energy source.

## 5. CONCLUSIONS

Microorganisms show a wide variation in metabolic diversity, especially in energy-generating metabolism and biosynthesis of secondary metabolites. The procaryotes, as a group, conduct all the same types of basic metabolism as eukaryotes, but, in addition, there are several types of energy-generating metabolism among the procaryotes that are non existent in eukaryotic cells or organisms. Even within a procaryotic species, there may be great versatility in metabolism. These can respire aerobically using $O_2$ as a final electron acceptor, some can respire under anaerobic conditions, using $NO_3$ or fumarate as a terminal electron acceptor, some can use glucose or lactose as a sole carbon source for growth, with the metabolic ability to transform the sugar into all the necessary amino acids, vitamins and nucleotides that make up cells, some having all the heterotrophic capabilities are also able to grow by photoautotrophic, photoheterotrophic or lithotrophic means. Fundamentally, most eukaryotes produce energy (ATP) through alcohol fermentation (e.g. yeast), aerobic respiration (e.g. molds, protozoa, animals) or oxygenic photosynthesis (e.g. algae, plants). These modes of energy-generating metabolism also exist among procaryotes, in addition to several additional means of energy production which are virtually non existent in eukaryotes. Because of metabolic diversity, the microorganisms are capable of supporting the recycling of matter and the functioning of earth as a sustainable ecosystem as well as producing a variety of metabolites including value-added products and biofuels.

## 6. FURTHER READING

Adams, MW. (1994). Biochemical diversity among sulfur-dependent, hyperthermophilic microorganisms. FEMS Microbiol Rev. 15:261-77.

Amend, J.P. and Shock, E.L. (2001). Energetics of overall metabolic reactions of thermophilic and hyperthermophilic Archaea and bacteria. FEMS Microbiol Rev. 25:175-243.

Jetten M.S.M., Strous M., van de Pas-Schoonen K.T., Schalk J., van Dongen U.G.J.M., van de Graaf A.A., Logemann S., Muyzer G., van Loosdrecht M.C.M., and Kuenen, J.G. (1998) The Anaerobic Oxidation of Ammonium. FEMS Microbiol. Rev. 22, 421-437.

Lehninger, A.L., Nelson, D.L. and Cox, M.M. (1993). Principles of Biochemistry. CBS Publishers and Distributors, New Delhi.

Madigan M.T. and Martinko J.M. (2006). Brock Biology of Microorganisms. 11th ed. Pearson Prentice Hall, Upper Saddle River.

Nath, S. (2002). The molecular mechanism of ATP synthesis by $F_0F_1$-ATP synthase: a scrutiny of the major possibilities. Adv Biochem Eng Biotechnol. 74: 65-98.

Pelczar (Jr), M.J., Chan, E.C.S., and Kreig, N.R. (1986). Microbial metabolism: energy production; Microbial metabolism: utilization of energy and biosynthesis. In: Microbiology. 5th ed. McGraw Hill Book company, Singapore. pp. 171-195; 196-226.

Prescott, L.M., Harley, J.P. and Klein, D.A. (2005). Microbiology. Microbial metabolism 6th ed. McGraw Hill Companies, Inc. New York. pp. 149-219.

Trumpower, BL. (1990). Cytochrome bc1 complexes of microorganisms. Microbiol Rev. 54:101-29.

# 5

# Biomass: A Source of Renewable Energy

**S.K. SONI[1], R. SONI[2] AND B. LUMBA[1]**
*[1] Department of Microbiology, Panjab University, Chandigarh-160 014*
*[2] Department of Biotechnology, D.A.V. College, Chandigarh - 160 011*

---



---

## 1. INTRODUCTION

Biomass is a general term used for living material derived from plants, animals, fungi and bacteria. Taken together, the Earth's biomass represents an enormous store of energy. It has been estimated that just one eighth of the total biomass produced annually would provide all of humanity's current demand for energy. And, since biomass can be regrown, it is a potentially renewable resource. The original source of the energy present in biomass is the sun. Small 'factories' in plant-leaves called chloroplasts use solar energy (in the form of light energy, or photons), together with carbon dioxide from the air and water from the soil, to manufacture a range of compounds. These compounds include sugars, starches and cellulose, collectively called carbohydrates. The original solar energy is now stored in the chemical bonds of these compounds. Some of this stored energy is passed on to animals when they eat plants (or eat other animals). So plants, animals and animal excretions, biomass, can be seen as storehouses of solar energy.

The photosynthesis process via solar energy builds plant material: polymers of carbon (C), hydrogen (H) and oxygen (O) to make cell walls and other plant parts. In addition to the main C, H, O constituents, biomass also contains minor concentration of phosphorous, nitrogen, and potassium as well as many other trace elements associated with cellular life processes. In the energy context biomass is a short hand form for phytomass which can be viewed as stored solar energy, and can be used as fuel or feedstock for processes that either produce energy - bioenergy, or a biofuel, which is an intermediate fuel such as charcoal, or ethanol. All applications of biomass involve a fuel chain with the following key steps:

Production of biomass
↓
Transportation of the biomass
↓
Conversion directly to energy or a biofuel
↓
Energy or biofuel transport
↓
Its final utilization

Prior to the industrialization of the 18th century, biomass was in fact the principal fuel for human existence, and most of the biomass used came from forests and woods as fuelwood. It was, and is, used for cooking, heating, and in industrial processes such as obtaining iron and steel from their ores after conversion of the fuelwood to charcoal as a reducing agent. Today biomass contributes about 12 % of the world's primary energy supply, though much of it is not commercially traded and is often missed from world energy statistics as a result.

## 2. PHOTOSYNTHESIS AND BIOMASS YIELDS

Biomass is basically derived from green plants, the primary producers, by the process of photosynthesis which can be summarized as the following chemical balance equation:

$$CO_2 + H_2O + \text{sunlight} \rightarrow CH_2O + O_2 \quad (1)$$

The carbon dioxide comes from the atmosphere and is absorbed by the leaf of the plant, while the plant's vascular system and roots supply the water. Light on the photosynthetic centers oxidizes water to produce oxygen, protons, and electrons. The leaves release oxygen back to the atmosphere and the protons react with carbon dioxide to produce a photosynthate identified in Eq. (1) as $CH_2O$. The $CH_2O$ unit is the basic building block of sugars, starch and cellulose and is used internally in plants as a "fuel" for metabolic processes and as the building block for the polymers that make up the cell walls of the plant's structure. While cellulose is the most common natural polymer on the earth, the plant also is able to synthesize compounds that are more chemically reduced and the second most common polymer is lignin that is composed of phenyl propane units. Energy is obtained from biomass by means of combustion with oxygen as in a fire, or through metabolic processes e.g. respiration to provide chemical energy for muscles and other processes:

$$CH_2O + O_2 \rightarrow CO_2 + H_2O + \text{Heat}$$

Either way, the net result is a closed cycle in which carbon dioxide and water vapor are returned to the atmosphere. This solar energy powered closed loop supports all life on earth, providing food for humanity, feed for animals, fiber for clothing and material for buildings in addition to fuel. The closed cycle has, in the last two centuries, been augmented by the use of fossil fuels such as coal, petroleum and natural gas, increasing the amount of carbon dioxide in the atmosphere to over 370 parts per millon by volume (ppmv) compared with the pre-industrial level of 280 ppmv. On a carbon basis the atmosphere contains approximately 720 gigaton (Gt), while the terrestrial biosphere contains about 2000 Gt of which about 50% is no longer living biomass. The aquatic biosphere is considered to contain about 1-2 Gt of living biomass, though the oceans also contain about 1000 Gt of organic material and almost 40 teraton (Tt) of inorganic carbon. Each year the "natural" carbon flux between the terrestrial biosphere and the atmosphere is about 60 Gt, while the addition of fossil carbon and the loss of terrestrial stored biomass due to changes in land use such as conversion of forest to agricultural land is about 8 Gt. It is this augmentation of the natural cycle that is responsible for the net increase in carbon dioxide, which is a green house gas that is implicated in climate change.

## 3. BIOMASS PROPERTIES

Biomass properties that affect the biofuels, and materials utilization for energy, include the chemical and the polymeric composition, and the physical traits such as density, toughness and so on.

### 3.1. Polymeric and chemical composition

An abundant raw material of renewable energy, biomass consists primarily of renewable polysaccharides and lignin (Figure 1). The traditional biomass used in energy applications has been fuelwood which is a fiber composed of lignin, cellulose and hemicellulose or in short hand, wood is a lignocellulosic material resource. Cellulose, hemicellulose and lignin are carbon-hydrogen-oxygen polymers that serve different structural purposes in the construction of the cell walls of woody plants (Tables 1 and 2). Lignocellulosic resources include trees, most woody plants, the straw and stalks of cereal crops, and are the most important biomass materials and energy resource as they represent much more than half of the above ground biomass produced by photosynthesis. Other major plant components containing polymers of carbon, hydrogen and oxygen that are also used for energy include starches that are the major part of the cereal grains, as well as the starch from tubers such as manioc and potatoes. Cellulose and starches can be hydrolyzed to simple sugars similar to those produced by sugar beet in temperate climates or sugar cane in the tropics and the sugars can be directly fermented to alcohols. Lipids are very low oxygen containing CHO polymers that are produced by oil seed bearing plants such as soya and rape or are in the fruits of oil palms and are possible diesel fuel substitutes when they are treated with simple alcohols such as methanol and ethanol.

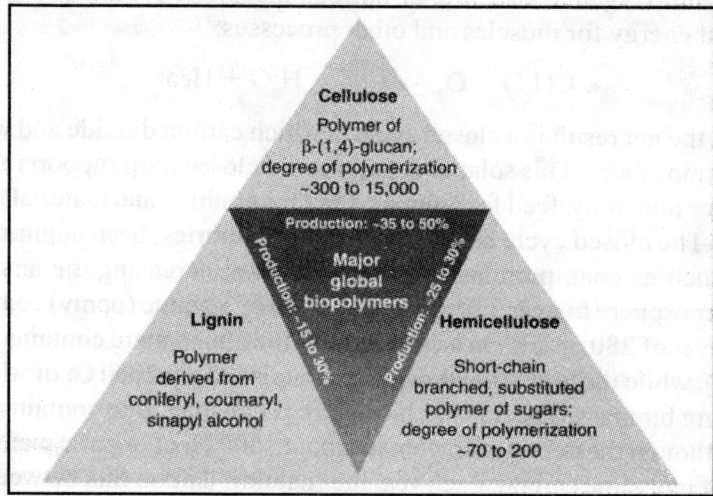


**Figure 1** Main components of fuelwood in the form of renewable polysaccharides and lignin which are generally fractionated and converted into a mixture of products through an array of processes

**Table 1** Typical levels of cellulose, hemicellulose and lignin in biomass

| Component | Percent dry weight |
|---|---|
| Cellulose | 40-60 |
| Hemicellulose | 20-40 |
| Lignin | 10-25 |

**Table 2** Sugar and ash composition of various biomass feedstocks (in percent)

| Material | Six-carbon sugars | Five-carbon sugars | Lignin | Ash |
|---|---|---|---|---|
| Hardwoods | 39-50 | 18-28 | 15-28 | 0.3-1.0 |
| Softwoods | 41-57 | 8-12 | 24-27 | 0.1-0.4 |
| Ag Residues | 30-42 | 12-39 | 11-29 | 2-18 |

In addition to the carbon-hydrogen-oxygen polymers, there are more complex polymers such as proteins (that can contain sulfur in addition to nitrogen), extractives, and inorganic materials. The inorganic materials range from anions such as chloride, sulfate and nitrates, and cations such as potassium, sodium, calcium and magnesium as major constituents and there are also many trace elements including manganese, and iron that are the metallic elements in key enzyme pathways involved in cell wall construction.

## 3.2. Fuels analysis

For combustion purposes there is not much need for information on the polymeric composition and the major analyses that are required are the proximate and ultimate analysis. The proximate analysis for fuels (Table 3) is usually of the material as received, includes the ash content (a reflection of the mineral content), moisture, the heating value and the fixed carbon.

**Table 3** Proximate analysis of representative solid fuels

| Fuel | Ash content (%) | Moisture (%) | Volatile matter (%) | HHV* (GJ/t) |
|---|---|---|---|---|
| Bituminous Coal | 2.72 | 2.18 | 33.4 | 34.5 |
| Sub-bituminous | 3.71 | 18.41 | 44.3 | 21.24 |
| Softwood | 1 | 20 | 85 | 18.6 |

*Higher heating value on a moisture and ash free basis (maf); GJ = gigajoules

The value of a proximate analysis is that it identifies the fuel value of the as-received biomass material, provides an estimate of the ash-handling requirement and describes something of the characteristics in burning. Generally fuels that are highly volatile such as biomass need to have specialized combustor designs to cope with the rapid evolution of gas when the fuel is heated, while those with low volatile matter or conversely a very high fixed carbon value need to be burnt on a grate as they take a long time to burn out or to be prior to combustion pulverized to a very small size. An ultimate analysis that is most often given on a moisture and ash free (maf) basis gives information on the elemental composition (Table 4).

**Table 4** Ultimate composition of biomass and selected solid fuels

| Material | C | H | N | S | O | Ash | HHV, (GJ / t) |
|---|---|---|---|---|---|---|---|
| Bituminous coal | 75.5 | 5.0 | 1.2 | 3.1 | 4.9 | 10.3 | 31.67 |
| Sub bituminous coal | 77.9 | 6.0 | 1.5 | 0.6 | 9.9 | 4.1 | 32.87 |
| Charcoal | 80.3 | 3.1 | 0.2 | 0.0 | 11.3 | 3.4 | 31.02 |
| Douglas fir | 52.3 | 6.3 | 0.1 | 0.0 | 40.5 | 0.8 | 21.00 |
| Douglas fir bark | 56.2 | 5.9 | 0.0 | 0.0 | 36.7 | 1.2 | 22.00 |
| Pine bark | 52.3 | 5.8 | 0.2 | 0.0 | 38.8 | 2.9 | 20.40 |
| Western hemlock | 50.4 | 5.8 | 0.1 | 0.1 | 41.4 | 2.2 | 20.00 |
| Eucalyptus grandis | 48.3 | 5.9 | 0.15 | 0.01 | 45.13 | 0.4 | 19.35 |
| Beech | 51.6 | 6.3 | 0.0 | 0.0 | 41.5 | 0.6 | 20.30 |
| Sugar cane bagasse | 44.8 | 5.4 | 0.4 | 0.01 | 39.6 | 9.8 | 17.33 |
| Wheat straw | 43.2 | 5.0 | 0.6 | 0.1 | 39.4 | 11.4 | 17.51 |
| Poplar | 51.6 | 6.3 | 0.0 | 0.0 | 41.5 | 0.6 | 20.70 |
| Rice hulls | 38.5 | 5.7 | 0.5 | 0.0 | 39.8 | 15.5 | 15.30 |
| Rice straw | 39.2 | 5.1 | 0.6 | 0.1 | 35.8 | 19.2 | 15.80 |

*HHV = Higher heating value ; GJ = gigajoules

The energy content of biomass is always reported for dry material; however, many of the lignocellulosic species are harvested fresh in a so-called green condition and contain as much as 50 % of their mass as water. There are also two different energy content reporting conventions in use. The term higher heating value HHV refers to the energy released in combustion when the water vapor resulting from the combustion is condensed thus realizing the latent heat of evaporation; much of the data from North America is reported in this way. The lower heating value or LHV reports the energy released when the water vapor remains in a gaseous state. For pure carbon, which only produces carbon dioxide when burned, the HHV and LHV are equal, while for methane when it is burned there are two molecules of water produced for each molecule of carbon dioxide, and the HHV is 11.1 % greater than the LHV.

## 3.3. Physical properties

In addition to the moisture content and chemical properties the physical properties are also important in the design of systems to handle biomass fuels and feed stocks. Compared with fossil fuels biomass in the raw state has a relatively low bulk energy density. Bituminous coal or crude oil for example, has a volumetric energy density of 30 $dm^3$ / GJ, solid wood has around 90 $dm^3$ / GJ, while in chip form the volume increases to 250 $dm^3$ / GJ for hardwood species and 350 $dm^3$ / GJ for coniferous species. Cereal straw has even less energy density and ranges from 450 $dm^3$ / GJ for large round bales to 1.2 $m^3$ / GJ for chopped straw, similar to that of sugar cane bagasse which is the fiber and pith left over after the extraction of sugar juice from green cane. Size reduction of biomass resources is often more difficult than with minerals as the materials are naturally strong fibers, and the production of uniform particle size feed stocks is correspondingly difficult. Traditionally the extraction of energy from biomass is split into 3 distinct categories:

*Solid biomass:* This includes the trees, crop residues, animal and human waste (although not strictly a solid biomass source, it is often included in this category for the sake of convenience), household or industrial residues for direct combustion to provide heat. Often the solid biomass will undergo physical processing such as cutting, chipping, briquetting, etc. but retains its solid form.

*Biogas:* This is obtained by anaerobically (in an air free environment) digesting organic material to produce a combustible gas known as methane. Animal waste and municipal waste are two common feedstocks for anaerobic digestion.

*Liquid biofuels:* These are obtained by subjecting organic materials to one of various chemical or physical processes to produce a usable, combustible, liquid fuel. Biofuels such as vegetable oils or ethanol are often processed from industrial or commercial residues such as bagasse (sugarcane residue remaining after the sugar is extracted) or from energy crops grown specifically for this purpose. Biofuels are often used in place of petroleum derived liquid fuels.

# 4. BIOMASS AND BIOENERGY SYSTEM

Biomass for bioenergy use comes either directly from the land or is a residue that is generated in primary processing of raw materials such as the production of pulp and paper from wood, or sugar from sugar cane. Another important contribution is from post consumer residue streams such as construction and demolition wood, pellets used in transportation and the clean fraction of MSW (municipal solid waste). In many respects the biomass and bioenergy system is part of the management of the solar generated flow of materials, food and fiber in society. These inter-relationships are illustrated in Figure 2, which shows the various resource types and their applications, and the cascade of their harvest, process and use residues to bioenergy applications. Not all biomass applications directly produce energy, but rather convert it into biofuels such as charcoal (a high energy density solid fuel) or ethanol (an automotive).

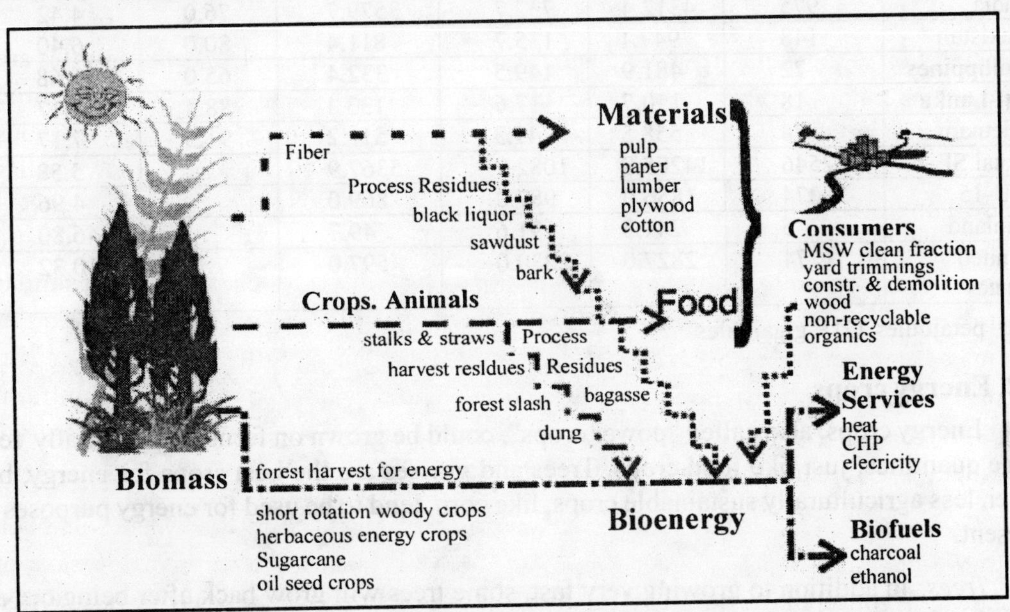


**Figure 2** The flow of biomass produced by photosynthesis and human activity

# 5. BIOMASS AS THE WORLD'S FOURTH FUEL

Biomass is fourth in rank after coal, oil, and natural gas in the world's primary energy supply. In 1995 the contribution of biomass to the world's primary energy supply was estimated to be about 11–14 % of the total primary energy supply (TPES). According to the International Energy Agency (IEA), combustible renewables and waste fuel shares in the TPES is estimated at 11.1% or 43.6 exajoule; $10^{18}$ Joules (EJ) of the total world consumption of 393 EJ.

## 5.1. Individual country usage

The biomass energy use is very dependent on the circumstances of individual countries (Table 5). For the six southeastern Asian countries (SE-6) which are home to over 40 % of the World's population the average biomass consumption is 5.6 GJ per capita, of which only

11 % goes to industry; and of the 89 % that goes to residential use, more than two thirds is used in cooking. This can be contrasted with the fifteen countries in the European Union (EU-15) whose industrial use is over 53 % of their biomass. The residential use is mainly for space heating in the northern countries, and very little being used for cooking. Overall biomass and bioenergy represents less than 10 % of the EU-15 primary energy supply. One country that stands out in the table is Finland that has one of the highest per capita consumption rates of biomass at 46.8 GJ per capita. Nearly 80 % of the consumption is in industry and 2/3 of that is from the combustion of black liquor in the pulp and paper industry.

**Table 5** Biomass shares of selected countries in total primary energy supply (TPES)

| | **Population (million)** | **Total (PJ)** | **Industry (PJ)** | **Residential (PJ)** | **Cooking (%)** | **Biomass (GJ per capita)** |
|---|---|---|---|---|---|---|
| China | 1255 | 7755.2 | 499.1 | 7256.1 | 89.0 | 6.18 |
| India | 975 | 4312.4 | 732.7 | 3579.7 | 76.0 | 4.42 |
| Pakistan | 148 | 947.1 | 135.7 | 811.4 | 80.0 | 6.40 |
| Philippines | 72 | 481.9 | 149.5 | 332.4 | 65.0 | 6.68 |
| Sri-Lanka | 18 | 150.7 | 17.6 | 133.1 | 88.4 | 8.17 |
| Vietnam | 78 | 558.5 | 47.3 | 511.2 | 91.5 | 7.17 |
| Total SE-6 | 2546 | 14205.8 | 1082.7 | 5367.9 | | 5.58 |
| EU-15 | 374 | 1856.0 | 987.0 | 869.0 | | 4.96 |
| Finland | 5 | 241.3 | 191.6 | 49.7 | | 46.80 |
| United States | 274 | 2827.0 | 2230.0 | 597.0 | | 10.32 |

*Pj = petajoules; GJ = gigajoules

## 5.2. Energy crops

Energy crops, also called "power crops", could be grown on farms in potentially very large quantities, just like food crops. Trees and grasses are the best crops for energy, but other, less agriculturally sustainable crops, like corn, tend to be used for energy purposes at present.

*Trees:* In addition to growing very fast, some trees will grow back after being cut off close to the ground, a feature called "coppicing". Coppicing allows trees to be harvested every three to eight years for 20 or 30 years before replanting. These "short-rotation woody crops" grow as much as 40 feet high in the years between harvests. In the cooler, wetter regions of the northern United States, varieties of poplar, maple, black locust, and willow are the best choice. In the warmer Southeast, sycamore and sweetgum are best, while in the warmest parts of Florida and California, eucalyptus is likely to grow well.

*Grasses:* Thin-stemmed perennial grasses used to blanket the prairies, before the settlers replaced them with corn and beans. Switchgrass, big bluestem, and other native varieties grow quickly in many parts of the country, and can be harvested for up to 10 years before replanting. Thick-stemmed perennials, like sugar cane and elephant grass, can be grown in hot and wet climates like those of Florida and Hawaii.

*Other crops:* A third type of grass includes annuals commonly grown for food, such as corn and sorghum. Since these must be replanted every year, they require much closer management and greater use of fertilizers, pesticides, and energy. While corn currently provides most of the liquid fuel from biomass in the United States, there are more sustainable ways to produce energy from plants.

A wide range of tree species are under consideration, some of which are multi-purpose trees in that they can be used for fiber and other applications including energy. Among these are *Eucalyptus*, *Populus* and *Leucaena* spp. Others, such as *Salix* spp, due to their growth habit, will likely only be useful for energy. Other crops include herbaceous perennials such as Switchgrass and Miscanthus. Annual crops have also been proposed, mainly because there would be no need for specialized production and harvesting equipment beyond what the farmer is currently using. With current energy prices, energy crops will require subsidy for establishment and use. In the United States there is a subsidy based on the concept of a closed loop, in which the energy crop is dedicated to a power plant and for each megawatt hours (mWh) generated there is a tax credit of $15. Wood based bioenergy crops are described in the United States as SRWC (short rotation woody crops) or in the UK and Europe as SRC (short rotation coppice). The energy crop for which the most experience has been gained is willow (*Salix* spp) in Sweden where there are 13,300 hectare (ha) of plantations that have been stimulated by subsidies for set-aside land and direct subsidies for willow culture, along with energy taxes which have favored the use of renewable fuels over fossil fuels, especially in the residential and district heating markets. With such a large land base in operation, it has been possible over the last two decades to evaluate many of the environmental and social factors that surround this new energy source. The crop has proven to be useful as a vegetation filter that can accept contaminated water from sewage plants, bio-solids in the form of sewage sludge and animal litter, and biofuel ashes. The nitrogen in these waste streams is taken up by the plants and offsets the use of inorganic fertilizer while avoiding non-point source releases of nitrogen to ground water. Despite being a plantation crop, there are in general, biodiversity benefits when it is used on former farmland that was in arable crops. The establishment of rules that will preserve and possibly improve biodiversity include: preservation of sensitive habitats (wetlands); enlargement of existing wooded areas by willow plantations that are close by, to improve continuity; careful planting to respect landscape aesthetics including multiple stands that are harvested in different years in rather large single blocks; use of several species and clones in a single stand to reduce vulnerability to pests and diseases; adoption of integrated pest management and avoiding excessive weed control in the establishment phase; and careful and appropriate application of fertilizers to withstand development requirements.

## 5.3. Biomass resources

Natural biomass resources vary in type and content, depending on geographical location. For convenience sake, we can split the world's biomass producing areas into three distinct geographical regions:

*Temperate regions:* These produce wood, crop residues, such as straw and vegetable leaves, and human and animal wastes. In Europe short rotation coppicing (SRC) has become popular as a means for supplying woodfuel for energy production on a sustainable basis.

Fast growing wood species, such as willow are cut every two to three years and the wood chipped to provide a boiler fuel. In the UK there is a functioning 12.6 Megawatt power plant which burns poultry litter (which has a relatively low moisture content) as fuel, and others which burn wheat straw. There are also many non-woody crops which can be grown for production of biofuels and biogas, and investigation of energy crops for direct combustion is underway. In western countries, where large quantities of municipal waste are generated, this is often processed to provide useful energy either from incineration or through recovery of methane gas from landfill sites.

*Arid and semi-arid regions:* These produce very little excess vegetation for fuel. People living in these areas are often the most affected by desertification and often have difficulty finding sufficient woodfuel.

*Humid tropical regions:* These produce abundant wood supplies, crop residues, animal and human waste, commercial, industrial and agro- and food-processing residues. Rice husks, cotton husks and groundnut shells are all widely used to provide process heat for power generation, particularly. Sugarcane bagasse is processed to provide ethanol as well as being burned directly; and many plants, such as sunflower and oil-palm are processed to provide oil for combustion. Many of the world's poorer countries are found in these regions and hence there is a high incidence of domestic biomass use. Tropical areas are currently the most seriously affected by deforestation, logging and land clearance for agriculture

## 5.4. Biomass residues

Biomass consists of organic substances produced by the activities of living creatures that are utilized as energy resources. It can be classified by its origin as plant-oriented biomass or animal-oriented biomass. Plant-oriented biomass includes wood, grasses, charcoal, and agricultural residues. Animal-oriented biomass includes excrements, waste meat, and waste fish. Another classification of biomass is based on its generation. Biomass can be produced as a crop, namely bioenergy crop, or can be obtained as by-products or waste of agricultural, industrial, and other human activities. Examples of biomass crop are corn and sugarcane cultivated in Brazil and in the United States for ethanol production. Sugar is obtained from the biomass crop, in order to produce ethanol by fermentation, and use the ethanol as a gasoline additive. By-product or waste biomass includes agroresidue (rice husk, straw, stalks), residues from city activity (sewage sludge, municipal waste), and wastes from wood industry (wood chips, sawdust). Table 6 shows biomass resources arranged according to this classification.

One of the important characteristics of biomass energy is its renewability. Unlike fossil fuel and uranium ore, biomass species can be grown with the aid of solar energy. Thus, biomass energy can be an energy source on which sustainable development is attained. Of the many kinds of primary energy forms, only biomass and solar energy itself can be considered as renewable energy sources to supply a large part of world energy consumption.

Another important characteristic of biomass energy is carbon neutrality. Because all living creature's activities are sustained by solar energy in the form of photosynthesized

Table 6 Classification of biomass resources

| Generation | Origin | |
|---|---|---|
| | **Plant-oriented** | **Animal -oriented** |
| Agroresidue | Agricultural residues including rice husk, straw, stalk etc.<br>Forestry residues including saw dust, wood chips, bark etc | Animal excreta from cows, buffaloes, pigs, chickens etc.<br>Fishery waste, meat waste |
| Bioenergy crops | Agricultural based including corn, sorghum, sugar cane, sugar beet, sweet sorghum etc<br>Wood based including eucalyptus, oak etc<br>Water based including water hyacinth, giant kelp, sea tangle etc. | |
| Waste | Municipal waste, food processing wa ste, pulp sludge, urban wastewood etc. | Sewage sludge |

carbohydrates, biomass is a renewable energy resource. It is noteworthy that hydrogen production from biomass is basically carbon neutral unlike hydrogen production from fossil fuel conversion. Carbon dioxide released in the process of hydrogen production is fixed again by photosynthesis of plants, and thus carbon dioxide is not accumulated in the atmosphere as long as biomass is reproduced. However, it is to be noted that if biomass is only harvested and not reproduced, utilization of biomass is also another source of carbon dioxide production. Slash-and-burn agriculture in Southeast Asia is a significant source of carbon dioxide emissions.

Biomass energy is also characterized by its storability. With direct use of solar energy there is always the difficulty of energy storage. Daily fluctuation of solar radiation causes fluctuation in energy production by solar devices. However, heat and electricity, which are the main products of solar devices, are difficult to store. Hydrogen production by solar energy improves the situation. However, hydrogen is not a good energy storage medium compared to other fuels because of its low volumetric energy density. Biomass energy is also developed by solar energy, but it is easily stored for long periods without energy loss because it takes the form of carbohydrates and other organic compounds. Storage in the form of biomass and conversion to hydrogen for utilization is a favorable strategy to meet requirements of storability and usability.

As biomass is basically solid, and sometimes contains large amount of moisture, conversion of biomass into a secondary energy form is necessary for efficient use and easy handling. Direct combustion of biomass usually results in efficiency as low as 15%. Fermentation and thermochemical conversion are widely accepted as useful biomass conversion methods. Ethanol production by fermentation is already being put into practice. Methane production by fermentation is called biomethanation, and is widely used in developing countries. To obtain solid fuel, biomass is pyrolyzed to produce char. By removing moisture and increasing carbon content, heating value and energy efficiency are improved.

## 5.5. Biomass waste

Biomass exists in the thin surface layer of the earth called the biosphere. It represents only a tiny fraction of the total mass of the earth but it is an enormous store of energy which is being replenished continuously. The sun is the main source of energy but only a very small fraction, about 0.5% of the solar energy striking the earth, is believed to be captured by plants through photosynthesis worldwide. Biomass includes mainly trees and plant wastes (e.g., wood, sawdust, leaves, and twigs), agricultural residues, animal waste, and coal.

### *5.5.1. Plant waste*

Any biodegradable material, whether of plant or animal origin can be used for the production of renewable energy (biogas or methane) through an anaerobic digestion process. Plant materials such as crop residues, weeds, and aquatic plants are also a source for methane production. Gas production is better if these materials are mixed with animal or human waste. *Eichornia crassipes,* commonly known as water hyacinth, is considered to be an obnoxious weed, yet this aquatic weed, which has almost the same C/N ratio as that of cow dung (i.e. 24), has proven to be an excellent plant material for methane production. Similarly, *Eupatorium adenophorum,* popularly called Banmara in Nepal, is another weed which has been devastating forest and pastoral land in Nepal. Experiments were carried out in Nepal using this weed for biogas production. Various agricultural residues such as rice straw, wheat straw, maize stalk, and leguminous plants have also been used to produce methane in conjunction with animal waste.

Cellulolytic plant materials can be easily degraded by the bacteria for methane generation while the ligneous materials are hard to digest by the bacteria and hence should be avoided. To facilitate digestion, plant materials have to be chopped into small pieces or crushed before feeding into the digester.

### *5.5.2. Animal waste*

Animal wastes are excellent raw materials for methane generation. In developing countries, where biogas technology is much advanced, it is customary to use cattle and buffalo dung to feed the digester. A homogenous mixture can be made while mixing the dung with water (slurry) facilitating the digestion process. Chicken or poultry manure is also a good source for biogas production. Raw materials from other animals include goat and sheep manure, horses and elephant dung and so on. It should be noted that while mixing with water, goat and sheep manure has the tendency of floating whereas horses and elephant dung contain fibrous materials.

### *5.5.3. Human waste*

Compared to animal waste, human faeces and latrine waste have been used for methane generation only on a limited scale in most of the developing countries due to social or religious reservations. The only exception is in China, where latrine waste is traditionally and socially acceptable and is used to produce biogas for cooking and lighting and bio–manure to enhance soil fertility. In recent years, the acceptability of latrine waste is increasing in Nepal, and about 40% of the installed biogas plants are found attached to latrines. Human faeces contain pathogens and has offensive odour. If not treated properly, it can cause diseases and can be detrimental to human health. Therefore, one of the best ways to dispose of human waste is to treat it in an anaerobic digester, producing biogas as energy and effluent as fertilizer.

## 6. CONVERSION OF BIOMASS TO ENERGY

The old way of converting biomass to energy, practiced for thousands of years, is simply to burn it to produce heat. This is still the use to which most biomass is put, in different parts of the world. The heat can be used directly, for heating, cooking, and industrial processes, or indirectly, to produce electricity. The problems with burning biomass are that much of the energy is wasted and that it can cause some pollution if it is not carefully controlled. A number of non combustion methods for converting biomass to energy have one thing in common - they convert raw biomass into a variety of gaseous, liquid, or solid fuels before using it. The carbohydrates in biomass, which are compounds of oxygen, carbon, and hydrogen, can be broken down into a variety of chemicals, some of which are useful fuels. There are several biomass energy technologies (Table 7) used these days for conversion of biomass to energy or fuel (Figure 3).

**Table 7** Biomass energy technologies

| **Technology** | **Conversion process type** | **Major biomass feedstock** | **Energy or fuel produced** |
|---|---|---|---|
| Direct combustion | Thermochemical | wood<br>agricultural waste<br>municipal solid waste<br>residential fuels | heat<br>steam<br>electricity |
| Gasification | Thermochemical | wood<br>agricultural waste<br>municipal solid waste | low or medium-Btu producer gas |
| Pyrolysis | Thermochemical | wood<br>agricultural waste<br>municipal solid waste | synthetic fuel oil (biocrude)<br>charcoal |
| Anaerobic digestion | Biochemical (anaerobic) | animal manure<br>agricultural waste<br>landfills<br>wastewater | medium Btu gas (methane) |
| Ethanol production | Biochemical (aerobic) | sugar or starch crops<br>wood waste<br>pulp sludge<br>grass straw | ethanol |
| Biodiesel production | Chemical | rapeseed<br>soy beans<br>waste vegetable oil<br>animal fats | biodiesel |
| Methanol production | Thermochemical | wood<br>agricultural waste<br>municipal solid waste | methanol |

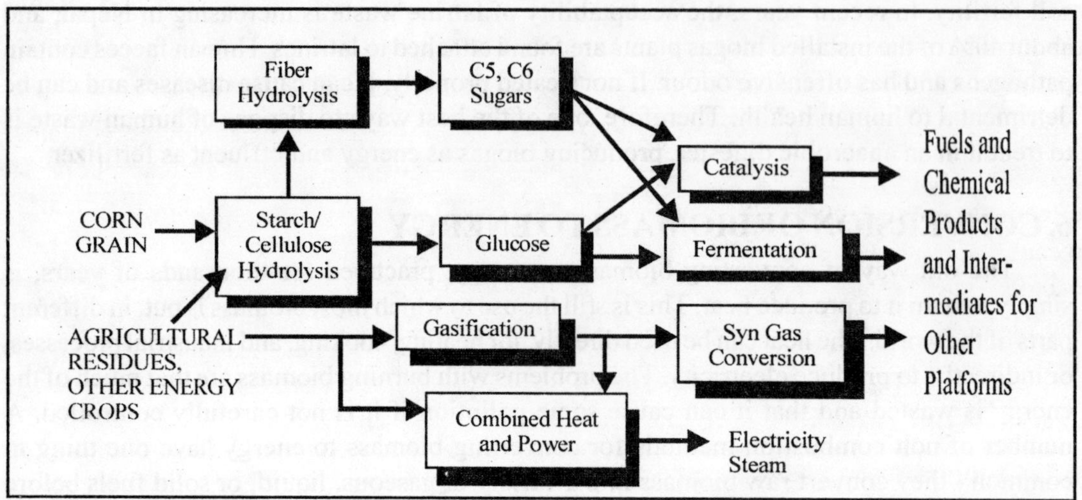


**Figure 3** A schematic representation indicating conversion of biomass to energy

The conversion of biomass to energy can be done in three ways:

*Thermochemical:* When plant matter is heated, but not burned, it breaks down into various gases, liquids, and solids. These products can then be processed into gas and liquid fuels like methane and alcohol. Biomass gasifiers capture methane released from the plants and burn it in a gas turbine to produce electricity. Another approach is to take these fuels and run them through fuel cells, converting the hydrogen-rich fuels into electricity and water, with few or no emissions.

*Biochemical:* Bacteria, yeasts, and enzymes also break down carbohydrates. Fermentation, the process used to make wine, changes biomass liquids into alcohol, which is inflammable. A similar process is used to turn corn into grain alcohol or ethanol, which is mixed with gasoline to make gasohol. Also, when bacteria break down biomass, methane and carbon dioxide are produced. This methane can be captured, in sewage treatment plants and landfills, for example, and burned for heat and power.

*Chemical:* Biomass oils, like soybean and canola oil, can be chemically converted into a liquid fuel similar to diesel fuel, and into gasoline additives. Cooking oil from restaurants, for example, has been used as a source to make "biodiesel" for trucks. (A better way to produce biodiesel is to use algae as a source of oils). Biomass is also used to make gas additives like ethyl tertiary-butyl ether (ETBE) and methyl tertiary-butyl ether (MTBE), which reduce air emissions from cars.

One persistent myth about biomass is that it takes more energy to produce fuels from biomass than the fuels themselves contain. In other words, that it is a net energy loser. In fact, current ethanol production uses corn, one of the most energy-intensive crops, and then uses just the kernels from the corn plant, and not even the entire kernel. Even so, this process yields 50 percent more energy than it takes to make the ethanol, so it is a net gainer. By making ethanol from energy crops, we can obtain between four and five times the energy that we put in, and by making electricity we could get perhaps 10 times or more. In the future, to make a truly sustainable biomass energy system, we would have to replace fossil fuels with biomass or other renewable fuels to plant and harvest the crops.

Another important consideration with biomass energy systems is that biomass contains less energy per pound than fossil fuels. This means that raw biomass typically can't be cost-effectively shipped more than about 50 miles before it is converted into fuel or energy. It also means that biomass energy systems are likely to be smaller than their fossil fuel counterparts, because it is hard to gather and process more than this quantity of fuel in one place. This has the advantage that local, rural communities- and perhaps even individual farms - will be able to design energy systems that are self-sufficient, sustainable, and perfectly adapted to their own needs.

## 6.1. Conversion of biomass directly into heat energy by combustion

Biomass combustion is not only the oldest form of combustion used by humanity, but it is also one of the most complex combustion systems to manage since it involves the use of solid fuels in a multi-phase reaction system with extensive interaction between the thermal and mass fluxes, processes that have only recently been properly analyzed and used in simulations to design efficient combustion systems. The key to understanding solid fuel combustion processes is to recognize that only fuel gases burn and release heat, that liquids and the solids do not burn themselves, but actually consume heat in the drying and volatilization processes needed for them to be chemically converted into fuel gases. The main fuel intermediates (Figure 4) are the volatile hydrocarbons and energy rich organic molecules, carbon monoxide (CO) and hydrogen ($H_2$).

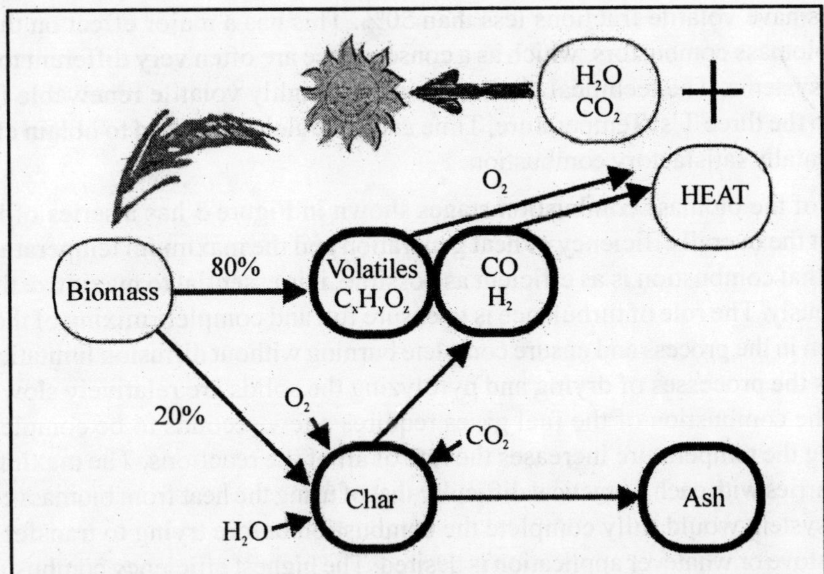


**Figure 4** Biomass combustion diagram and carbon cycle closed loop

The process of gas evolution due to heat is called pyrolysis, and progressively converts the biomass into gases, volatile liquids, and a carbon rich solid residue called char. In the pyrolysis stage, the rate of fuel gas evolution is a function of both the temperature and the intensity of the heat flux supplied to the solid surface and the pyrolysis liquids. When the char combustion stage is reached, after all of the volatiles have been removed, the char combustion rate is controlled by the velocity with which oxidizers namely oxygen, carbon dioxide and water vapor can reach the hot char surface to produce the fuel gases, hydrogen and carbon monoxide. The action of the hot char on water vapor and carbon dioxide is to convert the fully oxidized carbon and hydrogen forms back to fuel intermediates. This process of chemical reduction requires energy and is described as an endothermic reaction. The energy comes from the flame in which the carbon monoxide and hydrogen burn with oxygen from the air producing water vapor and carbon dioxide. This reaction is so endothermic that if the rate of heat release to the hot char surface is not sufficient the combustion reaction will come to a halt as the temperature at the surface will fall below 700°C. For this reason simple camp fires glow with only an red-orange color due to the limitation on the rate of diffusion of the gaseous reactants to the surface of the char. If the air flow is forced through the charcoal, the diffusion limitation is overcome and the temperatures become very high with a white incandescent appearance and temperatures greater than 1500°C sufficient to melt even stainless steels.

There are many different materials that are described as biomass ranging from solid wood, to oils, fats and proteins. While the majority of biomass used is a wood biomass or lignocellulosic, there is an increasing number of industrial and processing residues that have different material properties and combustion characteristics that are being used. However, a general and important biomass characteristic is the very high level of volatile material that is released during the pyrolysis stage. This is reported by the proximate analysis by the measurement of volatiles, which generally exceed 80% in most biomass materials, whereas many coals have volatile fractions less than 50%. This has a major effect on the technical design of biomass combustors, which as a consequence are often very different to traditional coal fired systems. The technical challenges of this highly volatile renewable fuel require attention to the three T's: Temperature, Time and Turbulence, needed to obtain efficient and environmentally satisfactory combustion.

Each of the biomass combustion stages shown in Figure 3 has a series of limitations, which limit the overall efficiency of heat generation and the maximum temperature reached. To ensure that combustion is as efficient as possible, it is essential to maximize the three T's simultaneously. The role of turbulence is to ensure full and complete mixing of the fuel gases with oxygen in the process and ensure complete burning without diffusion limitations. Time is required as the processes of drying and pyrolyzing the solids are relatively slow processes, and even the combustion of the fuel gases requires a few seconds to be complete. Finally, maximizing the temperature increases the rate of all of the reactions. The maximizing of all three T's carries with each a practical difficulty that of using the heat from biomass combustion. The ideal system would fully complete the combustion before trying to transfer the heat to the boiler, stove or whatever application is desired. The highest efficiency combustion systems do have a separate combustion zone from the heat transfer surfaces. The lowest efficiency systems, open fires, lose energy by radiation into space, and by drawing in a large amount of excess air by convection, lose energy in the hot air and smoke, in addition to large unburnt carbon losses due to the resulting low temperatures of combustion.

## 6.2. Conversion of biomass into gaseous fuels

One of the ways of using biomass is by conversion in to useful gaseous fuels. Three major processes for biomass gasification are available today: thermochemical gasification, biomethanation, and hydrothermal treatment (Table 8). Thermochemical gasification is treatment of biomass at temperatures as high as 627-927°C, where biomass is chemically decomposed into producer gas. This technology is well developed for coal gasification, but proper modification is desirable, in view of the differences between the properties of coal and biomass. For example, biomass is very reactive compared with coal. In order to raise the temperature of biomass feedstock, partial oxidation is often used. This technology is suitable for biomass feedstock whose moisture content is low. Otherwise, the biomass needs to be dried prior to gasification. Biomethanation is anaerobic fermentation of organic compounds to produce methane. Product gas contains carbon dioxide as much as methane. Hydrothermal treatment is a recent technology in which biomass is treated in water at temperature and pressure close to or higher than the critical temperature and pressure of water (374°C, 22.1 MPa). In this highly reactive water, with or without catalyst, biomass is easily decomposed at much lower temperatures than with conventional thermochemical gasifiers.

**Table 8** Biomass gasification technologies

| Technology for biomass gasification | | Temperature | Pressure | Product gas |
|---|---|---|---|---|
| Thermochemical gasification | Updraft gasifier | 627-927°C | Atmospheric | $H_2$, $CO_2$, CO, $CH_4$, other hydrocarbons $N_2$ (in case of air gasification) |
| | Downdraft gasifier | | | |
| | Fluidized bed gasifier | | | |
| | Circulating fluidized bed gasifier | | | |
| Biomethanation | | Room temperature 57°C | Atmospheric | $CH_4$, $CO_2$ |
| Hydrothermal gasification | Supercritical water gasification | 577-677°C | 25–35 MPa | $H_2$, $CO_2$, CO, $CH_4$, other hydrocarbons |
| | Thermochemical environmental energy system | 327-427°C | 20 MPa | |

The conversion of biomass into a gaseous fuel opens up modern applications in electric power generation, the manufacture of liquid fuels, and the production of chemicals from biomass (Figure 4). The process is feedstock flexible. The feedstock can contain organic rich materials from urban residue stream, industrial and commercial operations, agriculture/ food processing, black liquor, forest residues, and energy crops. At the contrary the biotechnology route to gas from biomass is based on anaerobic digestion, and this has substrate specificity that restricts it to biomass feedstock that can be easily hydrolyzed to sugars, amino acids, small lipids, alcohols and fatty acids. Figure 5 also shows the pathways for the use of gas from gasification processes. The low calorific value (LCV) gases produced from air gasification of biomass can be used to fire boilers (directly), as well as kilns, and will generate power in ICEs and gas turbines. The MCV gas has a wider range of applications: production of chemicals, liquid fuels, and hydrogen. After its purification, these are produced from the syngas via catalytic conversion. Carbon monoxide is converted into hydrogen using the water-gas shift reaction:

$$CO + H_2O \rightarrow CO_2 + H_2$$

Methane and other hydrocarbons are converted through reformation,

$$CH_4 + H_2O \rightarrow CO + 3H_2$$

Much of the interest in the last two decades has been in the production of electricity. Fischer-Tropsch chemistry is another approach for converting syn gas to valuable chemicals and fuels. The chemicals that can be produced include paraffins, monoolefins, aromatics, alcohols, aldehydes, ketones, and fatty acids. These molecules can contain from 1 to 35 carbons. Fischer-Tropsch chemistry utilizes either cobalt (fixed-bed) oriron (fixed- and fluid-bed) catalysts with high temperature and pressure to convert the syngas to chemicals and fuels. Yield, catalyst selectivity, and product composition depend on the catalyst, reaction conditions, and reactor type. The process is highly exothermic and the principal problem in

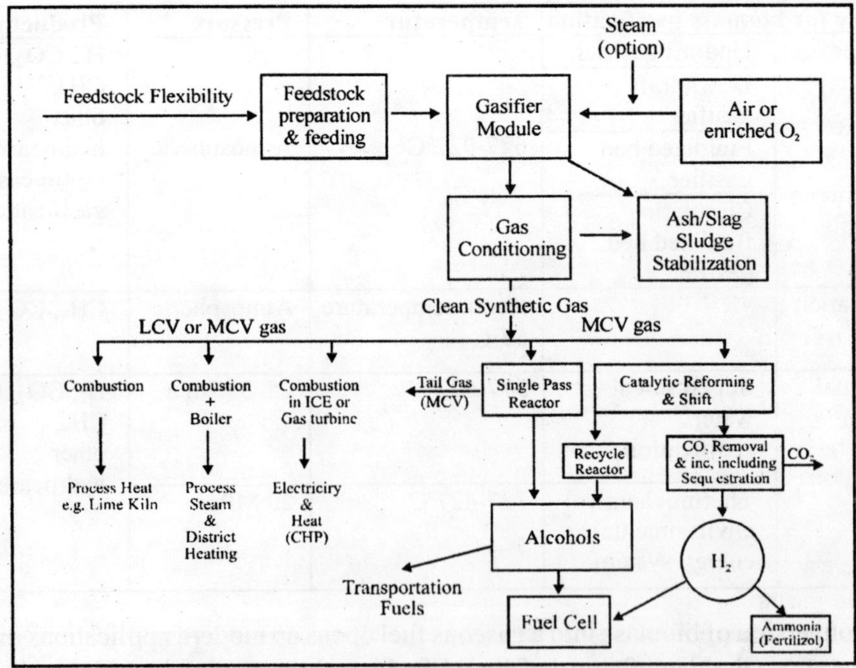


**Figure 5** Gasification process and product characterization

designing the reactor is heat removal. A schematic of the Fischer-Tropsch (F-T) process with product composition is shown in Figure 6. The lighter hydrocarbons, C1 and C2, can be used to generate the hydrogen used downstream to refine the heavier hydrocarbons. In the future, the hydrogen may instead be recovered and purified for use as a transportation fuel. The alcohols can be recovered, purified, and sold without further processing. The rest of the product stream is subjected to fractionation and processing to recover and upgrade the various products.

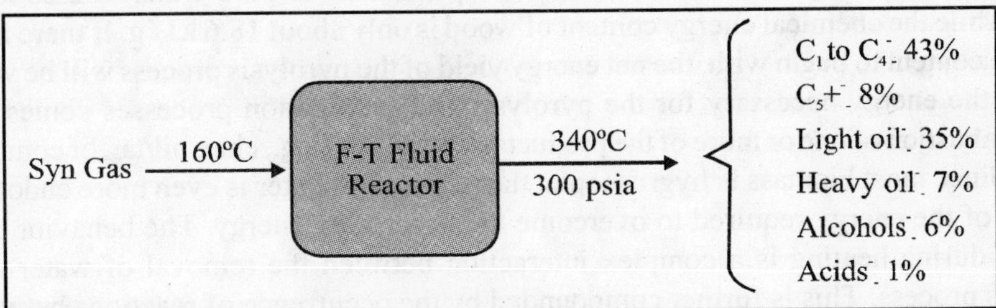


**Figure 6** Examples of Fischer product stream composition

### *6.2.1. Pyrolysis*

Pyrolysis is the fundamental chemical reaction process that is the precursor of both the gasification and combustion of solid fuels, and is simply defined as the chemical changes occurring when heat is applied to a material in the absence of oxygen. The products of biomass pyrolysis include water, charcoal (or more correctly a carbonaceous solid), oils or tars, and permanent gases including methane, hydrogen, carbon monoxide, and carbon dioxide. The nature of the changes in pyrolysis depend on the material being pyrolyzed, the final temperature of the pyrolysis process and the rate at which it is heated up. As typical lignocellulosic biomass materials such as wood, straws, and stalks are poor heat conductors, management of the rate of heating requires that the size of the particles being heated be quite small. Otherwise, in massive materials such as logs, the heating rate is very slow, and this determines the yield of pyrolysis products. Depending on the thermal environment and the final temperature, pyrolysis will yield mainly char at low temperatures, less than 450ºC, when the heating rate is quite slow, and mainly gases at high temperatures, greater than 800ºC, with rapid heating rates. At an intermediate temperature and under relatively high heating rates, the main product is a liquid bio-oil, a relatively recent discovery, which is just being turned to commercial applications. However, the bulk of commercial and technical pyrolysis processes are applied to the production of charcoal from biomass - a solid biofuel, which is then used as a reducing agent in metallurgy, as activated charcoal in absorption applications after chemical processing, and in domestic cooking in urban areas of the developing world.

Gasification is a complex thermal process that depends on the pyrolysis mechanism to generate gaseous precursors, which in the presence of reactive gases such as oxygen and steam convert the majority of the biomass into a fuel gas. The fuel gas product can in turn be further purified to syngas, which is mainly composed of carbon monoxide and hydrogen, and used to produce chemicals and liquid fuels over catalysts. The majority of fuel gas is used directly to fire processes such as kilns, as fuel in steam boilers, and increasingly as a gaseous fuel in internal combustion engines (ICEs) and gas turbines.

**6.2.1.1.** *Pyrolysis fundamentals:* The pyrolysis process consumes energy and, in chemists terms, is described as an endothermic reaction. It is, however, only mildly endothermic; and a factor of much more importance, in terms of energy demand in pyrolysis, is the water content of the starting biomass. The heat of vaporization of pure water is 2.26 kJ / g at 100°C, while the chemical energy content of wood is only about 18.6 kJ / g. If there is a high moisture content to begin with, the net energy yield of the pyrolysis process will be very low because the energy necessary for the pyrolysis and gasification processes comes mainly from combustion of one or more of the products of pyrolysis (e.g., char, oil/tar, or combustible gases). Since most biomass is hygroscopic, the removal of water is even more endothermic because of the energy required to overcome the absorption energy. The behavior of solid biomass during heating is a complex interaction between the removal of water and the pyrolysis process. This is further compounded by the occurrence of reactions between the pyrolysis products and the char.

**6.2.1.2.** *The chemical nature of pyrolysis in the solid and gaseous states:* Biomass is an organic material that is composed of polymers that have extensive chains of carbon atoms linked into macro-molecules. The polymer backbone chains consist of chemical bonds linking carbon with carbon, or carbon with oxygen, or sometimes other elements such as nitrogen or sulfur. Instead of describing polymers in terms of the atomic structure of the chain, most can be viewed as assemblies of some larger molecular unit. In the case of cellulose, that unit is the glucan moiety in the form of a molecule of glucose with one molecule of water missing ($C_6H_{10}O_5$). For hemicellulose, the unit is often a 5-carbon sugar, called xylose. However, hemicellulose polymers are not linear chains as in the cellulose polymer. Some are branched and other monomer units have side chains, with acetyl groups being very common. The lignin polymers are composed of phenyl propane subunits linked at various points on the monomer through C-C and C-O bonds. In addition, there are often side chain materials such as methoxy groups. As heat is applied, the chemical bonds become thermally activated and eventually some bonds break. In cellulose the bonds are broken at random locations along the chain composed of thousands of glucan moities. The average number of glucan subunits in a chain is described as the degree of polymerization (Dp). Thus, as heat continues to be applied, the polymer goes from a small number of large polymer molecules with a very high Dp, towards many smaller polymers with lower Dp values. As the value of Dp falls to less than 10 or so, the term, polymer, is replaced by the word, oligomer. Depending on the temperature and the applied pressure, eventually the oligomers ranging from Dp = 8 -> 1 are volatile (at typical pyrolysis temperatures between 400°C and 600°C) and evaporate from the solid mass. These small fragments are all anhydro sugars; when Dp = 2, it is called cellobiosan, and the monomeric version with Dp = 1 is called levoglucosan. If these volatile anhydro sugars are not quickly removed from the high temperatures of the pyrolysis, they will also undergo thermal fragmentation, producing highly reactive small intermediates. These in turn, if not removed from the original solid material (as in vacuum pyrolysis), will undergo chemical reactions with the remaining solid materials and may result in the creation of new polymers or accelerate the breakdown of the original chains. These reactions, unlike the endothermic chain-breaking reactions, can release heat (i.e., are exothermic), and in the pyrolysis of massive materials can result in the propagation of a thermal wave that accelerates the overall pyrolysis reaction. The removal of the products of pyrolysis and quenching them by sweeping them from the pyrolysis reactor into a cold zone (often a condenser), results in their capture for use as chemicals or fuels. Hemicellulose rapidly loses the side chains, which are often acetyl groups that are condensed as acetic acid. Likewise, lignin, which has methoxy substituents on the majority of the phenylpropane monomer units, leads to the production of methanol. Prior to the advent of petrochemical synthesis, wood pyrolysis was a major source of acetic acid, acetone, and methanol; the latter, as a result, is still known as wood alcohol. The volatile materials also undergo thermal rearrangements according to the temperature and the duration of exposure to that temperature.

**6.2.1.3.** *Thermal aspects of pyrolysis:* Most biomass has a cellular structure with extensive voids, such that while the density of the lignocellulosic cell wall material typically is between 1.5 - 2 g / $cm^3$, the density of wood ranges between 0.2 and 1.3 g / $cm^3$ for Balsa wood and Lignum Vitae, respectively. More typical softwoods, such as pine, have values around 0.4 g / $cm^3$ and typical hardwoods, such as oak, are about 0.6 g / $cm^3$. As a result, the heat transfer characteristics are very poor characterized by the Biot number, a dimensionless ratio of surface convective heat transfer to internal heat conductivity. When the Biot number is very small e.g. $< 10^{-3}$ then the material conducts heat rapidly to provide a uniform temperature throughout. However for biomass samples such as wood the Biot number often has values much greater than 0.2, and consequently there are large temperature gradients within the solid wood material. Thus, at high external heat fluxes with large particles of > 2 cm thick, the surface rapidly reaches the external temperature, while the center of the particle is still cold. For a 1 $cm^3$ cube of wood, a very slow heating rate of 0.01 K/min (Biot number @ $10^{-5}$) would result in an isothermal situation throughout the cube. In this case, the drying of the wood would take place independently of the pyrolysis process. Under the conditions that are often described as fast pyrolysis, the heating rates are on the order of 100 K / min. For the same 1 $cm^3$ cube, the Biot number 0.3 indicates a large thermal gradient, whereas a 1 $mm^3$ cube would more likely to be isothermal with a Biot number ~ 0.003.

For Biot numbers around unity, the passage of the thermal wave from the outside of the particle to the center takes a relatively long time and is opposed by the diffusion of the products of drying (water vapor) and pyrolysis (organic molecules and permanent gases) migrating to the surface. Physically this separates the drying process from the pyrolysis process such that each small volume of wood polymer is totally dry when it pyrolyzes. However, the net chemical result is that the primary products of pyrolysis have close contact with char and can react with the solid char matrix, modifying both it and the composition of the tars and gases. For thermally thick samples in high external heat flux regimes (Bi => 10), the processes of drying and pyrolysis travel together as a wave through the material. This pyrolysis wave is an exotherm and augments the rate of heat transfer through massive materials.

**6.2.1.4.** *Pyrolysis process technology:* Charcoal has been the major commercial product of biomass pyrolysis for a long time and is the largest single biofuel produced today. Both in historical times, and in developing countries today, relatively simple charcoal kilns have provided charcoal for use as a high quality fuel and as a reducing agent in the winning of ores. The Iron Age was characterized by the use of charcoal, and in Medieval times the increasing industrialization of society resulted first, in the over consumption of wood for charcoal production for use in iron production, and then charcoal was replaced with coke produced from coal.

The off gas and liquids represent as much as 40% of the original energy of the wood. In simple charcoal making, these are often not utilized, creating pollution of the soil, water and air. In the larger industrial systems, the recovery of byproducts for sale may not be economic compared to fossil-fuel derived products; however, the fuel value of both the gas and the tars (sometimes called pyroligneous liquids) may be utilized in the carbonization process to reduce energy loss, increase efficiency, and eliminate pollution. The destructive distillation of wood in retorts, prior to the coal and petroleum eras, provided a wide range of chemicals, in addition to the valuable charcoal, by condensing the liquids from the pyrolysis process. Before there was extensive chemical synthesis of methanol and acetic acid from fossil fuels, these were both byproducts of charcoal manufacture. Typical mass yields of charcoal and byproduct chemicals for Beech (*Fagus sylvatica*) are: charcoal - 32.5%; gases (with a heating value of 9 MJ / $Nm^3$) - 16%; tar - 14%; acetic acid - 7.7%, and methanol - 2.1% . Charcoal has value not only as fuel and metallurgical reducing agent, but is also the base for the production of special carbons in gas and liquid absorption applications. So-called activated charcoal is charcoal that has been chemically treated at high temperatures to produce an almost pure carbon product with a very high surface area. The absorptive capacity as a result of the large surface areas (e.g., $10^3$ $m^2$ / g) is extremely large and can trap a wide range of toxic substances and vapors.

**6.2.1.4.1.** *Slow prolysis:* Although pyrolysis has been used over a long period of time, the processes taking place during pyrolysis were not understood. This had to await the 20$^{th}$ Century evolution of polymer science and the application of rapid analysis techniques tools. Figure 7 is a summary of the extremes of pyrolysis behavior of lignocellulosic materials. The products of pyrolysis are the char, gases, and liquids, which are composed both of condensed water and the oils or tars from the pyrolysis process.

The triangles show the progression of traditional slow pyrolysis of a hardwood in a retort that collects the liquids and gases from the process. Starting with 100% solid material as unconverted hardwood at ambient temperature, by 250°C the solid mass is at 88%, less than 10% liquid, and only a few percent gas. Between 300°C and 350°C, the pyrolysis wave typically starts to move through the mass of the biomass, and the char rapidly decreases to less than 60% of the mass, the liquids are 20% to 30%, and the gases are between 15% and 25%. The char at this stage is not a pure carbon and contains both oxygen and hydrogen. Further heating of the char to 750°C will decrease the mass of the char, and while its composition moves closer to pure carbon, the yield of gas increases and that of liquids decreases. Typically the pyrolysis is conducted for hours to a maximum temperature of 400-500°C. The charcoal yield is 35% to 40% by weight. The central hexagon shows the theoretical yields of solid char, liquid as water, and gases (mainly methane and carbon dioxide) under the conditions of slow pyrolysis. Until recently it was thought that the theoretical yield was not attainable; however, studies by Professor Antal, at the University of Hawaii, and colleagues from around the world have demonstrated that high pressure carbonization conducted in a retort at 1.0 MPa pressure and a temperature of about 400°C will produce almost the theoretical yield; this process is now being commercialized. The high pressure restricts the vaporization and diffusion of the liquid products of pyrolysis out of the body of the pyrolyzing biomass, allowing them to react in the char matrix to produce more char, water and permanent gases.

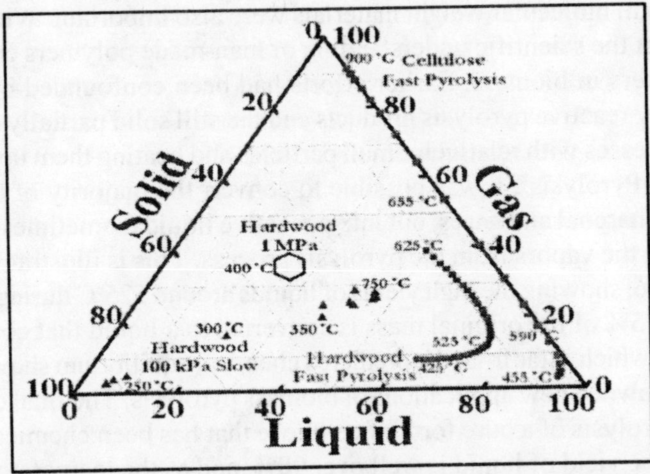


**Figure 7** Product gas, liquid and solid ratios as a function of pyrolysis process conditions

The slow pyrolysis process is an example of the pyrolysis wave phenomena in massive biomass, such that the only requirement is for the initiation of the process in which a small proportion of the wood charge is burnt to provide the initial heat in a well insulated reactor. Charcoal manufacture ranges from the artisanal methods of the earth pit or mound, to extremely high technology processes such as the Lambiotte continuous retort. The majority of industrial charcoal is produced in brick kilns, of which the Argentine half-orange and the Brazilian beehive are major examples. Portable steel kilns and permanent concrete kilns with rectangular steel doors are also used; these are sometimes called Missouri kilns. The capacities and throughput of the different systems are shown in Table 8.

**Table 8** Charcoal production systems

| **Charcoal Kilns** | | | | | |
|---|---|---|---|---|---|
| Technology | Production Cycle | | Performance Data | | |
| | Volume ($m^3$) | Days | Production (tons) | Efficiency tropical wood/tone of charcoal | Annual production (tons) |
| Earth Pit | 26 | 60 | 6 | 4.5 | 18 |
| Mound Cassemance | 100 | 24 | 12 | 5.5 | 144 |
| Brick - Argentine | 50 | 14 | 9 | 3.75 | 180 |
| Brick - Brazilian | 45 | 10 | 5 | 2.8 | 175 |
| Missouri Kiln | 180 | 25 | 20 | 3.0 | 240 |
| Lambiotte | 130 | Continuous | | 2.8 | 2500 |

**6.2.1.4.2.** *Fast pyrolysis:* Fast pyrolysis to produce bio-oils began only 25 years ago, following the development of analytical methods that made it possible to work with very small sample sizes in pyrolysis studies and ensure that heating of the biomass was uniform throughout the mass of the material. These techniques included: thermogravimetric analysis at very small scales, Fourier transform infra red spectroscopy, mass spectrometry, gas and liquid chromatography, as well as detailed simulation models. The rate of heating and the removal of the small molecular weight materials were also important. With these advances, it was realized that the scientific understanding of man-made polymers could be applied to the natural polymers in biomass. Earlier efforts had been confounded by the interactions between the highly reactive pyrolysis products and the still solid partially reacted materials. By designing processes with relatively small particles and heating them up rapidly by means of so called 'Fast Pyrolysis', it was possible to convert the majority of the input biomass material not into charcoal and gases, but into a reactive liquid (sometimes called bio-oil) by rapidly quenching the vapors from the pyrolysis process. This is illustrated with the square symbols in Figure 6, showing the high yields of liquids around 525°C during the pyrolysis of a hardwood. Over 75% of the original mass is converted to a liquid that contains about 10% water. This result, which is far from the predicted chemical equilibrium shown by the hexagon in Figure 6, has spawned new applications of biomass pyrolysis. The final curve in the figure is for the rapid pyrolysis of a pure form of cellulose that has been chemically isolated from wood. At 455°C the yield of liquid is well over 90%, and as the temperature is increased to 900°C , there is hardly any solid char and the liquid is converted into gas. At around 600°C a major component of the pyrolysis liquid from cellulose is levoglucosan, as would be expected from the normal thermal breakdown of a glucan chain.

### *6.2.2. Biomethanation*

This involves the conversion of organics into methane using the activity of microorganisms, which consume organic compounds for their life activity and releasing methane as the result. Because this activity is conducted in the absence of oxygen, it is called anaerobic digestion. When oxygen is present, aerobic digestion is preferred, and the product gas is basically carbon dioxide. Biomethanation is suitable for treatment of wet biomass feedstock. Thermochemical gasification of biomass is not applicable to biomass feedstock with high moisture content. It is necessary to reduce moisture content of the feedstock to less than 50 wt % to run a partial oxidation process because of the heat requirement for water evaporation. For wet biomass including sewage sludge, the cost and energy required for drying is too much, and partial oxidation is inappropriate. The main reaction is expressed as:

$$C_6H_{12}O_6 \rightarrow 3CH_4 + 3CO_2$$

Thus the resulting gas is composed of methane and carbon dioxide. This technology is cheap, and is employed in the developing as well as developed countries. In developing countries like India and China, there are many facilities to treat manure by this biomethanation process and product gas is utilized as fuel for daily life. In the developed countries, sewage sludge is treated by biomethanation to reduce its amount. It is also employed in some food industries to treat organic wastes and to recover some fuel gas. This technology has the disadvantage that the reaction takes time, usually 2–4 weeks. This makes it necessary to use large reactors. Another problem associated with this technology is the production of fermentation residue. Not all feedstocks can be gasified by anaerobic digestion. Lignin and compounds other than hydrocarbons cannot be decomposed, and are left in the sludge as well as wastewater. Because of the high biochemical oxygen demand (BOD) of this wastewater, it is not permitted to throw away the wastewater into sewage in developed countries. This is the reason why utilization of methane fermentation in developed countries is limited mainly to sewage sludge treatment where water from the reactor can be treated again in the activated sludge reactor.

Biogas is the common name for the gas produced either in specifically designed anaerobic digesters or in landfills by capturing the naturally produced methane. Biogas is typically about 60 percent methane and 40 percent carbon dioxide with a heating value of about 55 percent that of natural gas. Almost any biomass except lignin (a major component of wood) can be converted to biogas - animal and human wastes, sewage sludge, crop residues, carbon-laden industrial processing byproducts, and landfill material have all been widely used.

Anaerobic digesters generally consist of an inlet, where the organic residues and other wastes are fed into the digester tank; a tank, in which the biomass is typically heated to increase its decomposition rate and partially converted by bacteria into biogas; and an outlet where the biomass of the bacteria that carried out the process and non-digested material remains as sludge and can be removed. The biogas produced can be burned to provide energy for cooking and space heating or to generate electricity. Digestion has a low overall electrical efficiency (roughly 10–15 percent, strongly dependent on the feedstock) and is particularly suited for wet biomass materials. Direct non-energy benefits are especially significant in this process. The effluent sludge from the digester is a concentrated nitrogen fertilizer and the pathogens in the waste are reduced or eliminated by the warm temperatures in the digester tank.

## 6.3. Conversion of biomass in to liquid biofuels

Biofuels are produced in processes that convert biomass into more useful intermediate forms of energy. There is particular interest in converting solid biomass into liquids, which have the potential to replace petroleum-based fuels used in the transportation sector. However, adapting liquid biofuels to our present day fuel infrastructure and engine technology has proven to be non-trivial. Only oil producing plants, such as soybeans, palm oil trees and oilseeds like rapeseed can produce compounds similar to hydrocarbon petroleum products, and have been used to replace small amounts of diesel. This "biodiesel" has been marketed in Europe and to a lesser extent in the U.S.A., but it requires substantial subsidies to compete with diesel. Another family of petroleum-like liquid fuels that is produced from gasified biomass is a class of synthesized hydrocarbons called Fischer–Tropsch (F–T) liquids. The process synthesizes hydrocarbon fuels ($C_{10}$–$C_{12}$ hydrocarbons (kerosene) or $C_3$–$C_4$ hydrocarbons (LPG) from carbon monoxide and hydrogen gas over iron or cobalt catalysts. F–T liquids can be used as a sulfur-free diesel or blended with existing diesel to reduce emissions, an environmental advantage, but it has yet to be produced efficiently and economically on a large scale, and research and development (R&D) efforts are ongoing. In addition, to use as an automotive fuel F–T liquids can potentially be used as a more efficient, cleaner cooking fuel than traditional wood fuels from which it is synthesized. Other alternative biofuels to petroleum-based fuels are alcohols produced from biomass, which can replace gasoline or kerosene. The most widely produced today is ethanol from the fermentation of biomass. In industrialized countries ethanol is most commonly produced from food crops like corn, while in the developing world it is produced from sugarcane. Its most prevalent use is as a gasoline fuel additive to boost octane levels or to reduce dependence on imported fossil fuels. In the U.S.A. and Europe the ethanol production is still far from competitive when compared to gasoline and diesel prices, and the overall energy balance of such systems has not been very favorable. The Brazilian Proalcool ethanol program, initiated in 1975, has been successful due to the high productivity of sugarcane, although subsidies are still required. Two other potential transportation biofuels are methanol and hydrogen. They are both produced via biomass gasification and may be used in future fuel cells.

While ethanol production from maize and sugarcane, both agricultural crops, has become widespread and occasionally successful it can suffer from commodity price fluctuation relative to the fuels market. Consequently, the production of ethanol from lignocellulosic biomass (such as wood, straw and grasses) is being given serious attention. In particular, it is thought that enzymatic hydrolysis of lignocellulosic biomass will open the way to low cost and efficient production of ethanol. While the development of various hydrolysis techniques has gained attention in recent years, particularly in Sweden and the United States, cheap and efficient hydrolysis processes are still under development and some fundamental issues need to be resolved. Once such technical barriers are surmounted and ethanol production can be combined with efficient electricity production from unconverted wood fractions (like lignin), ethanol costs could come close to current gasoline prices and overall system efficiencies could go up to about 70 percent (low heating value). Though the technology to make this an economically viable option still does not exist, promising technologies are in the works and there are currently a number of pilot and demonstration projects starting up.

### *6.3.1. Types of biomass for liquid fuel development*

Agricultural produce in the form of grains, tubers and fruit surplus are mainly used for chemical manufacture and for bioconversion to motor fuel. The residues, the straws, sugar cane bagasse, Corn stover and the like, are prime sources of substrate for saccharification and fermentation to ethanol. Some of the examples of biomass used for conversion into liquid fuels are discussed below:

**6.3.1.1.** *Cane and beet sugar:* Until recently the bulk of food sugar was sucrose, and the term 'sugar' is ordinarily used referred to sucrose, a disaccharide of glucose and fructose. Sucrose is obtained from sugar cane and sugar beets, the former grown in tropical and semitropical countries such as South and Central America and the Caribbean Islands, as well as Africa and Asia, while the beets are grown in Europe and other temperate climatic regions. The molasses left in the purification and crystallization of sugar contains a very high proportion of fermentable sugars, and forms the basis of the Brazilian fuel ethanol programme. In the processing of these crops the residues, bagasse and beet pulp have been used for energy and for fodder, respectively. Both are lignocellulosics and potential feedstocks for fuel ethanol. A relatively new product which is made on a very large scale from corn and competes with sugar as a sweetener is high fructose corn syrup, a solution of glucose and fructose, used in such soft drinks as Coca Cola. This new sweetener helps keep the price of sugar down, making its conversion to ethanol more attractive.

**6.3.1.2.** *Grain:* Another major biomass product of photosynthesis is grain: wheat, corn (maize), rice, barley, sorghum, oats, millet and rye. These are primarily food crops, for human food directly and for animal feed. Total world production in 1982 was 1652 million tones, concentrated particularly in a few regions. Since grain is a food product, there is some hesitation about using it as a source of chemicals or fuel ethanol. However, present production is not a fair indication of production potential which is many times present production. The amount of straw produced when growing wheat varies with the variety from 1.5-4.0 tonnes/ tonne of useful grain. There are hundreds of millions of tonnes of straw produced annually. While much of this is ploughed back into farm operations, estimates have suggested that one-half to two-thirds could be diverted to other purposes. Like the straw, it is a lignocellulosic, and as such a potential source of fermentable sugar. On a worldwide basis, straw, chaff and stover have the greatest potential for motor fuel supply among present crops.

**6.3.1.3.** *Wood:* Cane sugar and corn (maize) are the sources of biomass which are presently converted to motor fuel, as well as man's other valuable products. on a large scale. These are also food crops and over the long haul development is expected to lead increasingly to the use of crop residues and wood as biomass feedstocks. Wood grows everywhere in the world from the equator north and south to near the arctic and the antarctic regions. Photosynthesis results in stored energy which in the form of biomass is about ten times the world's annual energy consumption. The present standing biomass primarily trees on the earth's surface equals the total amount of proven fuel reserves below the earth's surface. The annual forest biomass growth is estimated at 90 billion tonnes (dry matter). Wood has three main chemical components land (many 'minor' ones which lend it variety and interest), these being cellulose, hemicellulose and lignin.

In the operation of forest industries, the quantity of wood which ends up as pulp paper. Lumber and other commercial products is substantially less than what is harvested. Tops and trimmings of trees and damaged or undersized trees are normally by passed when harvesting. Saw mills produce substantial residues of sawdust, shaving blocks and trim. In the conversion industries making pulp from wood, a considerable amount of wood ends up as waste, generally lumped together as hog fuel. Much of this is burned with the plant it self for energy. and makes a significant contribution to the total energy requirement of the plant. Under proper conditions, hog fuel also could become available for ethanol production.

Another waste product of the forest industries, spent sulfite liquor was for many years an important source of ethanol, particularly in Scandinavia and North America. However the sulfite pulp mills have largely been converted to processes environmentally more benign and fermentation of spent sulfite liquor continues only in a few places. Switzerland has been a pioneer in this technology and their ethanol fermentation plant continues, as do also a few others elsewhere. This is a diminishing rather than expanding source of ethanol.

**6.3.1.4.** *'Surplus' fruit and other food:* A very large source of biomass presently untapped for fuel ethanol is the 'surplus' fruit and other food grown in Europe, and also on North American farms which cannot be marketed for food because of government price support policies. While the farmers involved would welcome the diversification which a fuel ethanol industry would represent so far no serious effort has been made to tap this potential feedstock literally lying there waiting to be picked up and meanwhile rotting. To achieve such development will require an act of political will which is so far slow in coming.

**6.3.1.5.** *Other crops:* There are several other crops which have potential for ethanol production as shown in Table 9. These include Jerusalem artichoke, cassava (or mandioca), sorghum, elephant grass and mangles (fodder beets).

**Table 9** World biomass potential for ethanol production

| Biomass | Million tons / per year | Million litres ethanol equivalent | Million barrels oil equivalent |
|---|---|---|---|
| Cane and beet molasses | 38 | 11000 | 69 |
| Cane and beet juice | | 6000 | 31 |
| Bagasse surplus to fuel | 24 | 7500 | 47 |
| Grain | 23 | 8000 | 50 |
| Grain low grade | 80 | 27000 | 170 |
| B starch | 116 | 52000 | 327 |
| Straw, chaff, stover | 3300 | 1000000 | 6290 |
| Cassava, cull | 2 | 1000 | 6 |
| Cassava, tops | 45 | 14400 | 90 |
| Potato, cull | 12 | 3800 | 24 |
| Jerusalem artichoke, tops | 3 | 1000 | 6 |
| Forest logging residues and non commercial harvest | 360 | 125000 | 786 |
| Plantation forests | 60 | 24000 | 152 |
| Municipal wastes | 250 | 37000 | 232 |
| World totals | 4313 | 1316700 | 8280 |

### *6.3.2. Bioethanol*

Bioethanol is an alcohol made by fermenting the sugar components of biomass. Today, it is made mostly from sugar and starch crops. With advanced technology being developed by the Biomass Program, celluosic biomass, like trees and grasses, are also used as feedstocks for ethanol production. Ethanol can be used as a fuel for cars in its pure form, but it is usually used as a gasoline additive to increase octane and improve vehicle emissions. Two reactions are key to understanding how biomass is converted to bioethanol:

*Hydrolysis* is the chemical reaction that converts the complex polysaccharides in the raw feedstock to simple sugars. In the biomass-to-bioethanol process, acids and enzymes are used to catalyze this reaction.

*Fermentation* is a series of chemical reactions that convert sugars to ethanol. The fermentation reaction is caused by yeast or bacteria, which feed on the sugars. Ethanol and carbon dioxide are produced as the sugar is consumed. The simplified fermentation reaction equation for the 6-carbon sugar, glucose, is:

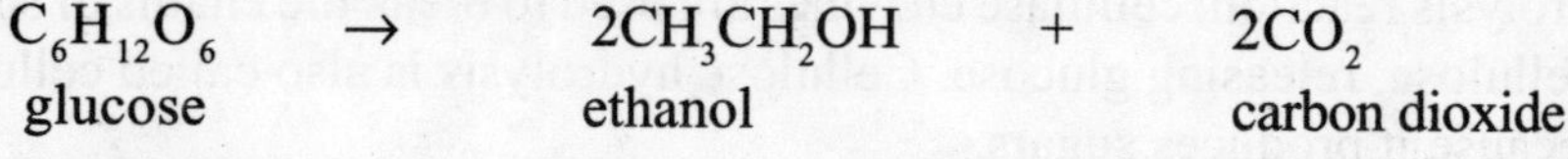

$$\underset{\text{glucose}}{C_6H_{12}O_6} \rightarrow \underset{\text{ethanol}}{2CH_3CH_2OH} + \underset{\text{carbon dioxide}}{2CO_2}$$

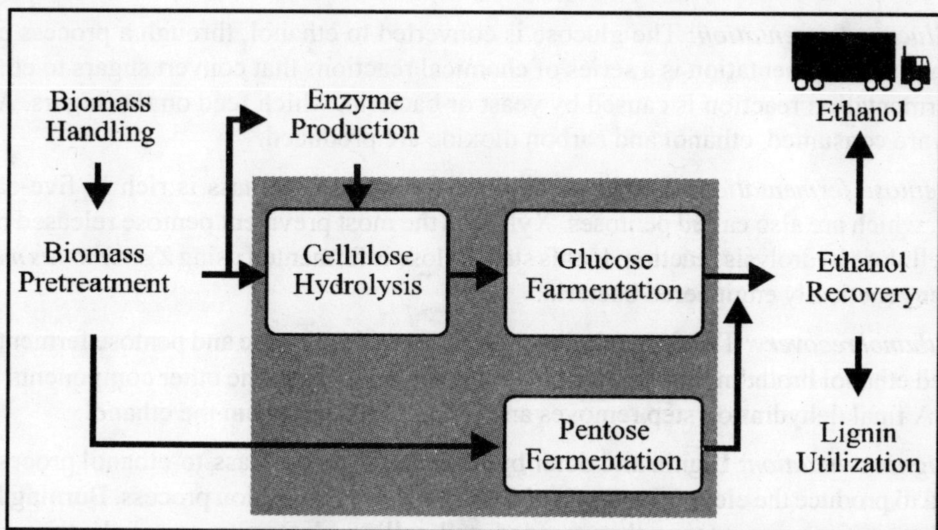


**Figure 8** Bioethanol production process diagram

**6.3.2.1.** *Process description:* The basic processes for converting sugar and starch crops are well-known and used commercially today. While these types of plants generally have a greater value as food sources than as fuel sources there are some exceptions to this. For example, Brazil uses its huge crops of sugar cane to produce fuel for its transportation needs. The current U.S. fuel ethanol industry is based primarily on the starch in the kernels of feed corn, America's largest agricultural crop. Bioethanol production process involves the following steps as shown in Figure 8.

*Biomass handling:* Biomass goes through a size-reduction step to make it easier to handle and to make the ethanol production process more efficient. For example, agricultural residues go through a grinding process and wood goes through a chipping process to achieve a uniform particle size.

*Biomass pretreatment:* In this step, the hemicellulose fraction of the biomass is broken down into simple sugars. A chemical reaction called hydrolysis occurs when dilute sulfuric acid is mixed with the biomass feedstock. In this hydrolysis reaction, the complex chains of sugars that make up the hemicellulose are broken, releasing simple sugars. The complex hemicellulose sugars are converted to a mix of soluble five-carbon sugars, xylose and arabinose, and soluble six-carbon sugars, mannose and galactose. A small portion of the cellulose is also converted to glucose in this step.

*Enzyme production:* The cellulase enzymes that are used to hydrolyze the cellulose fraction of the biomass are grown in this step. Alternatively the enzymes might be purchased from commercial enzyme companies.

*Cellulose hydrolysis:* In this step, the remaining cellulose is hydrolyzed to glucose. In this enzymatic hydrolysis reaction, cellulase enzymes are used to break the chains of sugars that make up the cellulose, releasing glucose. Cellulose hydrolysis is also called cellulose saccharification because it produces sugars.

*Glucose fermentation:* The glucose is converted to ethanol, through a process called fermentation. Fermentation is a series of chemical reactions that convert sugars to ethanol. The fermentation reaction is caused by yeast or bacteria, which feed on the sugars. As the sugars are consumed, ethanol and carbon dioxide are produced.

*Pentose fermentation:* The hemicellulose fraction of biomass is rich in five-carbon sugars, which are also called pentoses. Xylose is the most prevalent pentose released by the hemicellulose hydrolysis reaction. In this step, xylose is fermented using *Zymomonas mobilis* or other genetically engineered bacteria.

*Ethanol recovery:* The fermentation product from the glucose and pentose fermentation is called ethanol broth. In this step the ethanol is separated from the other components in the broth. A final dehydration step removes any remaining water from the ethanol.

*Lignin utilization:* Lignin and other byproducts of the biomass-to-ethanol process can be used to produce the electricity required for the ethanol production process. Burning lignin actually creates more energy than needed and selling electricity may help the process economics.

Converting cellulosic biomass to ethanol is currently too expensive to be used on a commercial scale. So researchers are working to improve the efficiency and economics of the ethanol production process by focusing their efforts on the two most challenging steps:

*Cellulose hydrolysis:* The crystalline structure of cellulose makes it difficult to hydrolyze to simple sugars, ready for fermentation. Researchers are developing enzymes that work together to efficiently break down cellulose.

*Pentose fermentation:* While there are a variety of yeast and bacteria that will ferment six-carbon sugars, most cannot easily ferment five-carbon sugars, which limits ethanol production from cellulosic biomass. Researchers are using genetic engineering to design microorganisms that can efficiently ferment both five- and six-carbon sugars to ethanol at the same time.

### *6.3.3. Biodiesel*

Biodiesel is a mixture of fatty acid alkyl esters made from vegetable oils, animal fats or recycled greases. Biodiesel can be used as a fuel for vehicles in its pure form, but it is usually used as a petroleum diesel additive to reduce levels of particulates, carbon monoxide, hydrocarbons and air toxics from diesel-powered vehicles. The main reaction for converting oil to biodiesel is called transesterification. The transesterification process reacts an alcohol (like methanol) with the triglyceride oils contained in vegetable oils, animal fats, or recycled greases, forming fatty acid alkyl esters (biodiesel) and glycerin. The reaction requires heat and a strong base catalyst, such as sodium hydroxide or potassium hydroxide. The simplified transesterification reaction is shown below:

$$\text{Triglycerides + Free Fatty Acids (<4\%) + Alcohol} \xrightarrow{\text{Base}} \text{Alkyl esters + Glycerin}$$

Some feedstocks must be pretreated before they can go through the transesterification process. Feedstocks with less than 4% free fatty acids, which include vegetable oils and some food-grade animal fats, do not require pretreatment. Feedstocks with more than 4% free fatty acids, which include inedible animal fats and recycled greases, must be pretreated in an acid esterification process. In this step, the feedstock is reacted with an alcohol (like methanol) in the presence of a strong acid catalyst (sulfuric acid), converting the free fatty acids into biodiesel. The remaining triglycerides are converted to biodiesel in the transesterification reaction.

$$\text{Triglycerides + Free Fatty Acids (>4\%) + Alcohol} \xrightarrow{\text{Acid}} \text{Alkyl esters + Triglycerides}$$

**6.3.3.1.** *Process description:* The basis technology involved in biodiesel production is indicated in Figure 9.

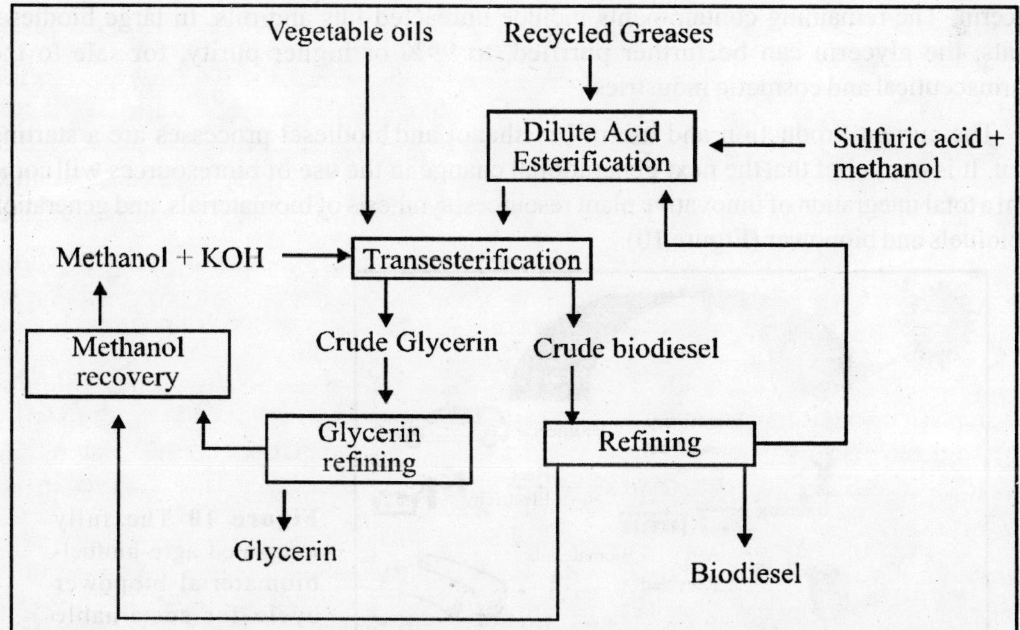


**Figure 9** A schematic representation of various steps involved in biodiesel production

*Acid esterification:* Oil feedstocks containing more than 4% free fatty acids go through an acid esterification process to increase the yield of biodiesel. These feedstocks are filtered and preprocessed to remove water and contaminants, and then fed to the acid esterification process. The catalyst, sulfuric acid, is dissolved in methanol and then mixed with the pretreated oil. The mixture is heated and stirred, and the free fatty acids are converted to biodiesel. Once the reaction is complete, it is dewatered and then fed to the transesterification process.

*Transesterification:* Oil feedstocks containing less than 4% free fatty acids are filtered and preprocessed to remove water and contaminants and then fed directly to the transesterification process along with any products of the acid esterification process. The catalyst, potassium hydroxide, is dissolved in methanol and then mixed with and the pretreated oil. If an acid esterification process is used, then extra base catalyst must be added to neutralize the acid added in that step. Once the reaction is complete, the major co-products, biodiesel and glycerin, are separated into two layers.

*Methanol recovery:* The methanol is typically removed after the biodiesel and glycerin have been separated, to prevent the reaction from reversing itself. The methanol is cleaned and recycled back to the beginning of the process.

*Biodiesel refining:* Once separated from the glycerin, the biodiesel goes through a clean-up or purification process to remove excess alcohol, residual catalyst and soaps. This consists of one or more washings with clean water. It is then dried and sent to storage. Sometimes the biodiesel goes through an additional distillation step to produce a colorless, odorless, zero-sulfur biodiesel.

*Glycerin refining:* The glycerin by-product contains unreacted catalyst and soaps that are neutralized with an acid. Water and alcohol are removed to produce 50%-80% crude glycerin. The remaining contaminants include unreacted fats and oils. In large biodiesel plants, the glycerin can be further purified, to 99% or higher purity, for sale to the pharmaceutical and cosmetic industries.

The current production and use of bioethanol and biodiesel processes are a starting point. It is our belief that the next generational change in the use of bioresources will come from a total integration of innovative plant resources, synthesis of biomaterials, and generation of biofuels and biopower (Figure 10).

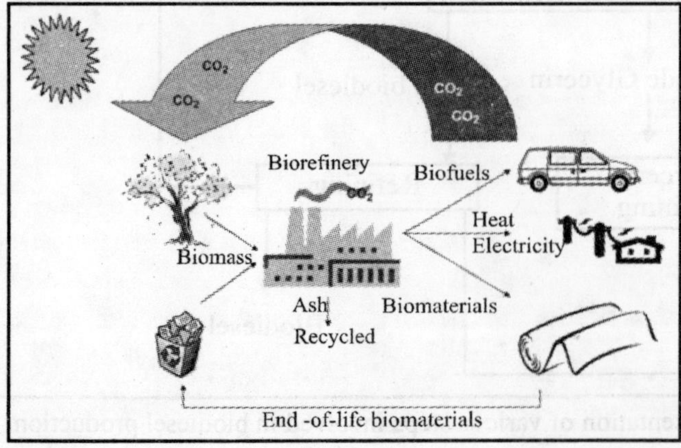


**Figure 10** The fully integrated agro-biofuel-biomaterial-biopower cycle for sustainable technologies

## 6.4. Organic commodities from biomass

Bioproducts are industrial and consumer goods, manufactured wholly or in part, from renewable biomass (plant-based resources). Today's industrial bioproducts are amazingly diverse, ranging from solvents and paints to pharmaceuticals, soaps, cosmetics and building materials (Table10). Industrial bioproducts are integral to our way of life - few sectors of the economy do not rely in some way on products made from biomass.

**Table 10** Common products from biomass

| Biomass Resource | Uses |
|---|---|
| Corn | Solvents, pharmaceuticals, adhesives, starch, resins, binders, polymers, cleaners, ethanol |
| Vegetable Oils | Surfactants in soaps and detergents. Pharmaceuticals (inactive ingredients), inks, paints, resins, cosmetics, fatty acids, lubricants, biodiesel |
| Wood | Paper, building materials, cellulose for fibers and polymers, resins, binders, adhesives, coatings, paints, inks, fatty acids, road and roofing pitch |

New bioproducts are continuously emerging, and the opportunities to use biomass in novel products are just beginning to be explored. In the following sections, key emerging and potential bioproducts are discussed according to the feedstock and technology platforms (biochemical, thermochemical) by which they are produced (Table 11).

### *6.4.1. Potential industrial bioproduct opportunities*

Potential markets for bioproducts are wide-ranging, and include polymers, lubricants, solvents, adhesives, herbicides, and pharmaceuticals. While bioproducts have already penetrated most of these markets to some degree, new products and technologies are emerging with the potential to further enhance performance, cost-competitiveness, and market share.

Organic chemicals represent the most direct and largest target for bioproducts. What makes a chemical organic is that it is primarily composed of carbon and hydrogen - the same primary constituents of biomass. Using novel chemistries, the carbons present biomass(carbohydrates) can be rearranged to yield products that are the equivalent of or superior to the products that are now produced from fossil energy (hydrocarbons). Many of the commodity organic chemicals serve as monomers for plastics (polymers), which represent a tremendous opportunity for bioproducts with an annual production of over100 billion pounds. Almost all of the polymers produced today are petroleum-based, with the only major exceptions being cellulose polymers and natural rubber.

Advances in biotechnology, chemicals processing, chemistry and separations are opening up new avenues for biobased polymers such as polylactide (now in production) and poly trimethylene terephthalate (soon to be partially bio based). Other major organic chemicals markets for bioproducts include organic acids, alcohols,and solvents. Biomass-based ethanol (used mostly for fuels, but some industrial uses) is an established industry. There are emerging markets for other alcohols and bio-derived acids.

**Table 11** Current industrial bioproduct production from domestic biomass

| Category | Principal technologies | Feedstock | Chemical | General product | Annual biobased production (Million lb) |
|---|---|---|---|---|---|
| Starch and sugars | Biochemical | Biomass Sugars derived from corn and sorghum | Lactic acid, citric acid, ethanol, starch, sorbitol, levulinic acid, itaconic acid | Polymers, solvents, cleaners, coatings, inks, detergents/ surfactants, pharmaceuticals, adhesives, paints, composites, laminates, toiletries, cosmetics | 5,413 |
| Oils/Lipids | Thermo-chemical | Oils/Lipids derived from soyabean, rapeseed | Glycerol/glycerine, alkyd resins, high erucic acid rapeseed, polyurethane, epoxialized soyabean oil, factice, sulfurized fatty oils, fatty acids, cyclo-pentadienized oils, lecithin, maleinized oils | Pharmaceuticals, personal care, urethanes, alkyd resins, plasticizers, lubricants, paints, resins, printing inks, industrial and textile finishes, semi-rigid foam, thermoplastic elastomers, cosmetics, coatings, surfactants, sealants, caulks, pesticides | 1,589 |
| Specialty crops | Thermo-chemical | Spearmint, peppermint, sweet almond | Spearmint oil, peppermint oil, sweet almond oil | Personal care, pharmaceuticals, epoxy and alkyd resins, paints, cosmetics and toiletries | 9 |
| Forest derivatives | Thermo-chemical | Pine, black, liquor and soft wood | Turpentine oil, rosin, toll oil, and cellulose derivatives (esters, acetates, etc.) | Solvents, soaps, detergents, toiletries, perfumes, rubber, adhesives, coatings, printing inks, phenolic resins, plastics, textiles | 5,326 |
| **Total** | | | | | **12,337** |

While lubricants and greases are now mostly petroleum-based, they were originally plant based because they could be made from vegetable oils with little modification. With energy prices on the rise, and growing environmental concerns over the impacts of petroleum-based products, vegetable oil-based lubricants and greases are making an entrance back in the market. With new chemistries and processes making them more economical, they could regain significant market share. The potential opportunities for new bioproducts, are summarized in Table 12. These are categorized according to the technology platform used to convert the biomass feedstock (e.g., fermentation, gasification). Biochemicals sector produces a wide range of bulk and specialty chemicals that find use as:

- ingredients for synthesis of other chemicals
- monomers for production of polymers and resins
- adhesives
- lubricants
- surfactants
- pest control agents
- paints and coatings
- pigments and inks
- ingredients for cosmetics, fragrances and food
- ingredients for pharmaceuticals and medical devices
- fertilizers.

**Table 12** Summary of industrial product opportunities

| Technology platform | Chemical | Applications | Current U.S. market size (million lb) | Potential 2020 biobased market size (million lb) |
|---|---|---|---|---|
| Sugars fermentation | Lactic acid | Acidulant (food, drink), electroplating bath additive, mordant, textile/leather auxiliary | <5 (Industrial uses) | Expect GDP-like growth |
| | Polylactide | Film and thermoformed packaging, fiber and fiberfill applications | Packaging: 21,289 Fiber/ fiberfill: 2,769 | 8,000 |
| | Ethyl lactate | Solvent (blending with methyl soyate), chemical intermediate | 8,000-10,000 | >1,000 |
| | 1,3-Propanediol | Apparel, upholstery, specialty resins, and other applications | Small | 500 |
| | Succinic acid | Surfactants/detergents, ion chelator (electroplating), food, pharmaceuticals, antibiotics, amino acids, and vitamin production | US: small World: 33 | GDP-like growth |

*Contd...*

| | | | | |
|---|---|---|---|---|
| | Succinic acid derivatives (tetrahydro-furan, 1,4-butanediol, -γ butyrolactone, N-methyl pyrrolidone, 2-pyrrolidone, succinate salts) | Solvents, adhesives, printing inks, magnetic tapes, coating resins, plasticizer/emulsifier, deicing compounds, herbicide ingredient, chemical and pharmaceutical ingredients. | 1,395 | >90 |
| | Bionolle 4,4 polyester | Thermoplastic polymer applications | 25,000-60,000 | >4,000 |
| | 3-Hydroxypropionic acid & derivatives (acrylic acid, acrylonitrile, acrolamide, 1,3-propanediol) | Acrylates, acrylic fibres, polymers, resins | 5,336 | Technology now just being developed |
| | N-Butanol | Solvent, plasticizer, polymers, resins | 1,850 | Could be significant |
| | Itaconic acid | <u>Current</u>: aluminium anodizing reagent, reactive comonomer. <u>Potential</u>: Methyl methacrylate applications, acrylic pressure sensitive adhesives | 1,808 (Itaconic Acid and its derivatives) | Significant if technology is successful |
| Sugars fermentation & thermochemical | Propylene glycol | Solvents, heat-transfer fluids, humectants, plasticizers, polyurethane chain extenders, antifreeze | 1,100 | >500 |
| Sugars thermochemical | Isosorbide (currently biobased) | <u>Current</u>: Active ingredient in diuretic and antianginal drugs <u>Potential</u>: Polymer additive | Small | If technology is successful, could be 100-300 |

*Contd...*

| | | | | |
|---|---|---|---|---|
| Oils and Lipids | Lubricants and hydraulic fluids | Lubricants and hydraulic fluids | 20,000 | Significant |
| | Solvents | Solvents | 8,000-10,000 | Significant |
| | Polymers (Polyurethanes are most significant) | Polyurethanes | 5,327 | Significant |
| | Proteins | Unknown | Unknown | Unknown |
| Biomass gasification | Fischer-Tropsch & gas -to-liquids products | Fuels (transportation, heating), solvent, aerosis, oxygenate, chemical intermediates | Very large markets | Will require more favorable economics |
| Biomass pyrolysis | Phenol-Formaldehyde resins | Plywood, oriented strand board (OSB), and other wood composites | 3,900 | Significant |
| Biocomposites | Biocomposites | Unknown | Unknown | Unknown |
| Plants and factories | Guayule | Natural rubber applications | 2,184 | Not projected |
| | Polyhydroxy-alkanoates (PHAs) | Thermoplastic polymer applications | 30,000 | If technology is successful, could be >4,000 |
| Photosynthesis, anaerobic digestion | Lignin, methane, carbon dioxide, other chemicals | Unknown | Unknown | Unknown |

# 7. CONCLUSIONS AND FUTURE OUTLOOK

In addition to the many environmental benefits, biomass offers many economic and energy security benefits. The use of biomass as a source of energy reduces methane emissions, which contributes 20 times more to the greenhouse effect than carbon dioxide. Biomass is a renewable source of energy and its use does not contribute to global warming. The use of biomass fuels produces low sulphur emissions and produces less ash than burning coal. Using organic waste for energy production is an effective use of waste products and helps reduce the need for waste disposal dumps. Biomass does not need to be imported and unlike oil and coal its price is not linked to world markets. Moreover, by growing our fuels at home, we reduce the need to import oil and reduce our exposure to disruptions in that supply. Farmers and rural areas gain a valuable new outlet for their products. Biomass already supports 60,000 jobs in the United States; if the Department of Energy's goal is realized, the industry would support three times as many jobs. However, the derivation of energy from biomass offers some disadvantages too as large volumes of biomass are needed to produce the same amount of energy as small volumes of fossil fuels. This can add to transport costs. Energy and food crops may compete for the same farmland. Biomass used for electricity production is still in its early stages of investigation and therefore is more expensive at this point.

Biomass remains a substantial source of energy in the modern age. New technologies and new approaches promise to bring biomass into the 21st century. Ultimately, the success of biomass as energy alternative will be determined by economics. Industries that use their waste biomass for energy simultaneously solve a waste disposal problem and save money on their energy needs and sometimes, earn money by selling excess electricity. As biomass technology becomes more efficient, the chances of biomass energy competing in the wider market place will increase. Most of the electricity will be provided by renewable energy sources by 2010. Many groups all over the world are working on various aspects of improving the energy generation from biomass. The renewable resources research labs at Hawaii Natural Energy Institute is working on synthesis of activated carbons from charcoal, ethylene from ethanol and microcrystalline cellulose from bagasse. The ongoing research has led to the discovery of flash carbonization. This quickly produces charcoal from biomass. The biomass research is also going on in Florida, where the Gas research institute of Chicago and institute of Food and Agricultural Science are carrying out the research on the genetic improvement of crops, development of crop management practices, harvesting and handling strategies and improvement of anaerobic digesters. The U.S. DOE, office of energy efficiency and renewable energy are focusing R and D in gasification and small modular biopower, Gasification R and D focuses on feasibility of producing $H_2$ from biomass, Small modular R and D focuses on producing commercially competitive small scale heat and power system.

## 8. FURTHER READING

Angenent, L.T., Karim, K., Al-Dahhan, M.H., Wrenn, B.A. and Domiguez-Espinosa, R. Production of bioenergy and biochemicals from industrial and agricultural wastewater. Trends Biotechnol. 22: 477-85.

Demain, A.L., Newcomb, M. and Wu, J.H. (2005). Cellulase, clostridia, and ethanol. Microbiol Mol Biol Rev. 69:124-154.

Ghoshm, P. and Ghose, T.K. (2003). Bioethanol in India: recent past and emerging future. Adv Biochem Eng Biotechnol. 85:1-27.

Moomaw, W.R. (2002). Energy, industry and nitrogen: strategies for decreasing reactive nitrogen emissions. Ambio. 31:184-189.

McKendry, P. (2002). Energy production from biomass (Part 1): Overview of biomass. Bioresour Technol. 83:37-46.

McKendry, P. (2002). Energy production from biomass (Part 2): Conversion technologies. Bioresour Technol. 83:47-54.

McKendry, P. (2002). Energy production from biomass (Part 3): Gasification technologies. Bioresour Technol. 83:55-63.

Ragauskas, A.J., Williams, C.K., Davison, B.H., Britovsek, G., Cairney, J., Eckert, C.A., Frederick, W.J. Jr, Hallett, J.P., Leak, D.J., Liotta, C.L., Mielenz, J.R., Murphy, R., Templer, R. and Tschaplinski, T. (2006). The path forward for biofuels and biomaterials. Science.311:484-489

Wayman, M. and Parekh, S.R. Biotechnology of biomass conversion, Fuels and chemicals from renewable resources, New Jersey, Prentice Hall. (1990).

# 6

# Microbes: Alternate Energy Sources

**S.K. SONI**
*Department of Microbiology, Panjab University, Chandigarh-160 014*

## 1. INTRODUCTION

In accordance with the prognoses of the World Energy Council in 1994, fossil fuels will continue to provide the bulk of energy demands for the next few decades. At current consumption rates, coal has some 250 years of reserves, oil over 40 years and natural gas over 65 years. The people have also become aware about pollution and side effects caused by these conventional non-renewable energy resources on health, global warming and expensive conversion technology in their production. Consequently, there is an urgent need to develop alternate energy sources. Most requirements are likely to be met from geothermal, nuclear, solar, water and wind sources. One of the alternative energy resources is renewable energy including biomass from which a variety of biofuels can be obtained. Biological fuel generation appears to become increasingly important, especially as it can provide both liquid and gaseous fuels. Importantly, these fuels are produced from renewable resources, primarily plant biomass, in the form of cultivated energy crops, natural vegetation, and agricultural, domestic and industrial wastes. The two main microbial fuel products currently derived from these resources, and currently in wide use, are methane and ethanol, but these are not the only fuels that can be formed from renewable resources. Other liquid and gaseous examples include methanol, butanol, biodiesel, hydrogen, propane and bioelectricity which can also be generated by microbiological systems are gaining worldwide attention these days. Numerous other important chemical compounds including organic acids, amino acids, industrial solvents, enzymes, a wide range of biopolymers are widely used as chemical feedstocks and functional ingredients in a wide range of industrial and food products. The use of alternative fuels and alternative fuel vehicles is increasing steadily as we try to find way to make use of home grown products and national resources and as we try to lessen our dependence on imported oil.

The knowledge of fermenting and distilling starch and sugar rich biomass for ethanol production has been practiced for thousands of years. Maximum efforts for the use of biomass energy came in 1906, after the world war and during energy crisis of 1972-73 and 1990-91. The Standard Oil Company of New Jersey, in 1922-23, started making a blend of 25% alcohol and 75% gasoline during the period of petroleum shortage. The Brazilian government established a National Alcohol Program (NAP) in 1975 to save foreign exchange by using alcohol as a supplement to nation's gasoline supply in which it set a target to produce and use (34 million litres) of biomass based fuel by 1980. First "gasohol", 1:9 blend of ethanol and gasoline, pump was opened in US in 1978. By 1979 there were outlets selling about 40 million gallon of ethanol fuel. By 1979, alcohol cars were made by Ford & Volkswagen of Brazil. The Indian government allocated Rs. 2.25 billion for non-conventional energy work in year 1994-95. The funds were used in biogas program to install over 200,000 biogas plants. India's Minister for Petroleum and Natural Gas gave his approval in December 2001 to a proposal to launch pilot projects to test the feasibility of blending ethanol with gasoline. In March 2002, the government decided to allow the sale of E-5 across the country. On 13 September, 2002, India's government mandated that nine states and four federally ruled areas will have to sell E-5 by law from 1 January 2003. Indian Railways also tested the use of diesel with 5%, 10%, 20% blend of Jatropha biodiesel. On March 12, 2005, Indian Union Minister of Petroleum and National Gas flagged off biodiesel bus service in Gandhinagar.

Although alternative fuels are seen as a substitute for gasoline and diesel fuel most of them still emit some carbon dioxide and other pollutants. The U.S. Department of Energy classifies the following fuels as "alternative fuels": biodiesel, biomass, ethanol, hydrogen, methanol, natural gas, propane, and p-series fuels. These biofuels are also gaining significance due to pollution and side effects caused by these conventional non-renewable energy resources on health, global warming and expensive conversion technology. Some of the alternative biofuels which can be produced from microbes are discussed below.

## 2. ALCOHOLS

There are three major subsets of alcohols - 'primary' (1°), 'secondary' (2°) and 'tertiary' (3°), based upon the number of carbons attached to the C-OH carbon is bonded to. Methanol is the simplest 'primary' alcohol (Figure 1). The simplest secondary alcohol is isopropanol (propan-2-ol), and a simple tertiary alcohol is *tert*-butanol (2-methylpropan-2-ol).

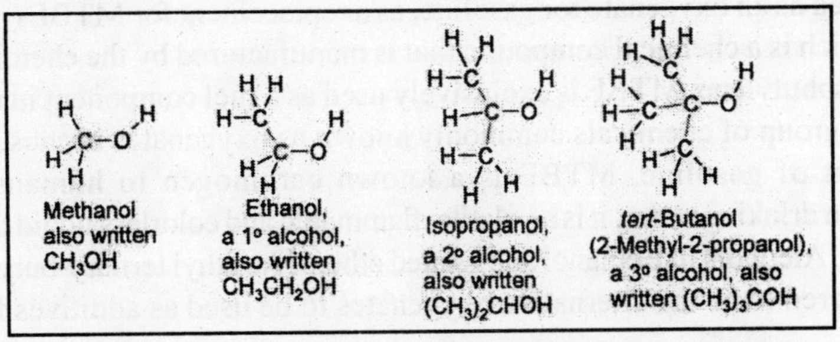

Methanol also written $CH_3OH$

Ethanol, a 1° alcohol, also written $CH_3CH_2OH$

Isopropanol, a 2° alcohol, also written $(CH_3)_2CHOH$

*tert*-Butanol (2-Methyl-2-propanol), a 3° alcohol, also written $(CH_3)_3COH$

**Figure 1** Structural classification of different alcohols

Alcohols are in wide use as industrial solvents, and also liquid fuels because of several properties that are favorable for motor fuel use, either alone or when blended with gasoline. These have good octane-enhancing properties and a relatively high heat of combustion (Table 1).

Ethanol and methanol can be made to burn more cleanly than gasoline or diesel. Because of its low toxicity and ability to dissolve non-polar substances, ethanol is also often used as a solvent in medical drugs, perfumes, and vegetable essences such as vanilla. In organic synthesis, alcohols frequently serve as versatile intermediates.

**Table 1** Characteristics of alcohols for use as co-solvents with motor fuels

| **Physical property** | **Methanol** | **Ethanol** | **Butanol** | **Gasoline** | **Diesel** |
|---|---|---|---|---|---|
| Heat of combustion (kJ/g) | 23.9 | 30.6 | 36.7 | 43.8 | 42.7 |
| Fuel values in terms of octane number with high speed, high temperature conditions | 91 | 90 | 87 | 83 | — |
| Miscibility with | | | | | |
| Gasoline | Poor | Fair | Good | — | — |
| Diesel | Poor | Poor | Good | — | |
| Water | High | High | Low | Low | Low |

Many alcohols can be produced by microorganisms by way of fermentation of fruits or grains but only ethanol is commercially produced chiefly this way. Other alcohols are generally produced by synthetic routes from natural gas, petroleum, or coal feed stocks, for example via acid catalyzed hydration of alkenes. Of all the alcohols, the use of ethanol as a fuel for internal combustion engines, either alone or in combination with other fuels, has been given much attention mostly because of its possible environmental and long-term economical advantages over fossil fuel. Both ethanol and methanol have been considered for this purpose. While both can be obtained from petroleum or natural gas, ethanol may be the most interesting because many believe it to be a renewable resource, easily obtained from sugar or starch in crops and other agricultural produce such as grain, sugarcane or even lactose. Since ethanol occurs in nature whenever yeast happens to find a sugar solution such as overripe fruit, most organisms have evolved some tolerance to ethanol, whereas methanol is toxic. When 10% alcohol fuel is mixed into gasoline, the result is known as gasohol or E10. Other experiments involve butanol, which can also be produced by fermentation of plants. Alcohol is also increasingly used as an oxygenate for gasoline, as a replacement for MTBE (methyl tertiary-butyl ether) which is a chemical compound that is manufactured by the chemical reaction of methanol and isobutylene. MTBE is exclusively used as a fuel component in motor gasoline and is one of a group of chemicals commonly known as oxygenates because they raise the oxygen content of gasoline. MTBE is a known carcinogen to humans even at low concentrations in drinking water; it is a volatile, flammable and colorless liquid that is relatively soluble in water. Alcohols like ethanol and related ethers like ethyl tertiary-butyl ether (ETBE) are also considered to be the alternative oxygenates to be used as additives for gasoline.

## 2.1. Methanol

Methanol is also known as methyl alcohol or wood alcohol, is a chemical compound with chemical formula $CH_3OH$. It is the simplest alcohol, and is a light, volatile, colorless, flammable, poisonous liquid that is used as an antifreeze, solvent, fuel, and as a denaturant for ethyl alcohol. It can be used as an alternative fuel in flexible fuel vehicles that run on M85 (a blend of 85% methanol and 15% gasoline). However, it is not commonly used because automakers are no longer supplying methanol-powered vehicles.

Methanol is produced naturally in the anaerobic metabolism of many varieties of bacteria. As a result, there is a small fraction of methanol vapor in the atmosphere. Over the course of several days, atmospheric methanol is oxidized by oxygen by the help of sunlight to carbon dioxide and water. Methanol has been considered as a fuel, mainly in combination with gasoline. It has received less attention than ethanol, however, because it has a number of problems of its own. Its main advantage is that it can be easily manufactured from methane. A methanol flame is almost colorless. Care should be exercised around burning methanol to avoid burning oneself on the almost invisible fire. Methanol burns in air forming carbon dioxide and water:

$$2CH_3OH + 3O_2 \rightarrow 2CO_2 + 4H_2O$$

Pure methanol, was first isolated in 1661 by Robert Boyle, who called it 'spirit of box', because he produced it via the distillation of boxwood. It was later named as 'pyroxylic spirit'. In 1923, the German chemist Matthias Pier, working for BASF, a chemical company, developed a means to convert synthesis gas (a mixture of carbon monoxide and hydrogen derived from coke and used as the source of hydrogen in synthetic ammonia production) into methanol. This process used a zinc chromate catalyst, and required extremely vigorous conditions - pressures ranging from 30–100 MPa (300–1000 atm), and temperatures of about 400°C. Modern methanol production has been made more efficient through the use of catalysts capable of operating at lower pressures. Today, synthesis gas is most commonly produced from the methane, a component in natural gas, or obtained by microbial based anaerobic methanogenesis, rather than from coal. At moderate pressures of 1 to 2 MPa (10–20 atm) and high temperatures (around 850°C), methane reacts with steam on a nickel catalyst to produce syngas according to the chemical equation:

$$CH_4 + H_2O \rightarrow CO + 3H_2$$

This reaction, commonly called steam-methane reforming or SMR, is endothermic and the heat transfer limitations place limits on the size of the catalytic reactors used.

Methane can also undergo partial oxidation with molecular oxygen to produce syngas, as the following equation shows:

$$CH_4 + 0.5O_2 \rightarrow CO + 2\ H_2$$

This reaction is exothermic and the heat given off can be used *in-situ* to drive the steam-methane reforming reaction. When the two processes are combined, it is referred to as autothermal reforming.

The carbon monoxide and hydrogen then react on a second catalyst to produce methanol. Today, the most widely used catalyst is a mixture of copper, zinc oxide, and alumina first used by ICI in 1966. At 5–10 MPa (50–100 atm) and 250°C, it can catalyze the production of methanol from carbon monoxide and hydrogen with high selectivity.

$$CO + 2H_2 \rightarrow CH_3OH$$

It is worth noting that the production of synthesis gas from methane produces 3 moles of hydrogen for every mole of carbon monoxide, while the methanol synthesis consumes only 2 moles of hydrogen for every mole of carbon monoxide. One way of dealing with the excess hydrogen is to inject carbon dioxide into the methanol synthesis reactor, where it, too, reacts to form methanol according to the chemical equation:

$$CO_2 + 3H_2 \rightarrow CH_3OH + H_2O$$

Although natural gas is the most economical and widely used feedstock for methanol production, other feedstocks can be used where natural gas is unavailable. The most economical means of forming methane is methanogenesis, which is an important and widespread form of microbial metabolism. In most environments, it is the final step in the decomposition of organic matter.

Organisms capable of methanogenesis are called methanogens which are considered to be a very old group of organisms, being members of the archaea. Methanogenesis is a form of anaerobic respiration in which methanogens do not use oxygen to breathe; in fact, oxygen inhibits the growth of methanogens. The terminal electron acceptor in methanogenesis is not oxygen, but carbon. The carbon can occur in a small number of organic compounds, all with low molecular weights. The two best described pathways involve the use of carbon dioxide and acetic acid as terminal electron acceptors:

$$CO_2 + 4H_2 \rightarrow CH_4 + 2H_2O$$

$$CH_3COOH \rightarrow CH_4 + CO_2$$

Methanogens cannot exist in the presence of oxygen, so they are only found in environments in which the oxygen has been depleted. Most commonly these are environments experiencing the rapid decay of organic matter, such as wetland soils, the digestive tracts of animals and aquatic sediments. Methanogenesis also occurs in areas where oxygen and decaying organic matter are both absent, such as the terrestrial deep subsurface, deep-sea hydrothermal vents and oil reservoirs.

Methanogenesis is the final step in the decay of organic matter. During the decay process, electron acceptors (such as oxygen, iron, sulfate, nitrate, and manganese) become depleted, while hydrogen and carbon dioxide accumulate. Light organics produced by fermentation also accumulate. During advanced stages of organic decay, all electron acceptors become depleted except carbon dioxide which is a product of most catabolic processes, so it is not depleted like other potential electron acceptors.

Only methanogenesis and fermentation can occur in the absence of electron acceptors other than carbon. Fermentation only allows the breakdown of larger organic compounds, and produces small organic compounds. Methanogenesis effectively removes the semi-final products of decay: hydrogen, small organics, and carbon dioxide. Without methanogenesis, a great deal of carbon (in the form of fermentation products) would accumulate in anaerobic environments.

Methanogenesis is useful to humanity. Through methanogenesis, organic waste can be converted to useful methane "biogas". Methanogenesis occurs in the guts of humans and animals. While methanogenesis is not believed to be necessary for human digestion, it is required for the nutrition of ruminant animals, such as cattle and goats. In the rumen (known incorrectly as the "second stomach" possessed by some animals), anaerobic organisms (including methanogens) digest cellulose into forms usable by the animal. Without the microbes of the rumen, cattle cannot survive without being fed a special diet.

Methanol's physical and chemical characteristics result in several inherent advantages as an automotive fuel. Some methanol benefits include lower emissions, higher performance, and lower risk of flammability than gasoline. In addition, methanol can be manufactured from a variety of carbon-based feedstocks such as natural gas; coal and biomass (e.g., wood) and the use of methanol would help reduce dependence on imported petroleum. On the down side, methanol produces a high amount of formaldehyde in emissions.

In addition, methanol can easily be converted into hydrogen and this is being worked upon. Some researchers are currently working to overcome the barriers to using methanol as a hydrogen fuel source. So methanol may potentially be used to create hydrogen for hydrogen fuel cell vehicles in the future. Methanol is used on a limited basis to fuel internal combustion engines, mainly by virtue of the fact that it is not nearly as flammable as gasoline. Methanol blends are the fuel of choice in various types of vehicles including racing cars and aeroplanes. When produced from wood or other organic materials, the resulting organic methanol (bioalcohol) has been suggested as renewable alternative to petroleum-based hydrocarbons. However, one cannot use BA100 (100% bioalcohol) in modern petroleum cars without modification.

Methanol is also used to produce a gasoline additive, methyl tertiary-butyl ether (MTBE) which has been used in gasoline at low levels since 1979 to replace tetra-ethyl lead as an octane enhancer and to help prevent engine knocking. Since 1992, MTBE has been used at higher concentrations in some gasoline to fulfill the oxygenate requirements of gasoline, however, since 1999, in California and other locations MBTE has begun to be phased out because of its health risk related to widespread groundwater contamination. Moreover, MTBE was found to be a carcinogen in animal studies. In the resulting backlash, several countries have banned the use of MTBE, and its future production remains uncertain.

Direct-methanol fuel cells are unique in their low temperature, atmospheric pressure operation, allowing them to be miniaturized to an unprecedented degree. This, combined with the relatively easy and safe storage and handling of methanol may open the possibility of fuel cell-powered consumer electronics.

## 2.2. Ethanol

The use of ethanol as a fuel for internal combustion engines, either alone or in combination with other fuels, has been given much attention mostly because of its possible environmental and long-term economical advantages over fossil fuel. Ethanol is mostly used to reduce emissions in gasoline. It is sometimes blended with gasoline to produce E10 fuel (10% ethanol and 90% gasoline), but it is also used in higher concentrations such as E85 or E95. Since ancient times, ethanol fuel was used for lamp oil and cooking, along with plant and animal oils. Before the American Civil War many farmers in the USA had an alcohol still to turn crop waste into free lamp oil and stove fuel for the farmers' family use. Since then, the use of alcohol as alternate fuel has come a long way as today, automakers have brought out multi-fuel vehicles that can run on any mixture of gasoline and ethanol. Some of the main milestones in the history of use of ethanol as fuel are as follows:

- In 1826, Samuel Morey used alcohol in the first American internal combustion engine prototype.
- By 1860, thousands of distilleries made 90 million gallons of alcohol or more per year for lighting, cooking and industry.
- In 1860, German inventor Nikolaus Otto used ethyl alcohol as a fuel in an early internal combustion engine.
- In the 1890s, alcohol fueled engines were used in farm machinery, train locomotives, and cars in the U.S. and Europe, making countries more fuel independent. Ethanol

was the first fuel used by American cars before gasoline. Henry Ford's first car, the Quadracycle, ran on ethanol.

- In 1902, the Paris alcohol fuel exposition exhibited alcohol powered cars, farm machinery, lamps, stoves, heaters, laundry irons, hair curlers, coffee roasters, and every conceivable household appliance and agricultural engine powered by alcohol.
- In 1908, the Ford Model T capable of running on ethanol or gasoline was introduced.
- In the 1920s and 1930s, Koolmotor, Benzalcool, Moltaco, Lattybentyl, Natelite, Alcool and Agrol were some of the ethanol-gasoline blends of fuels used in Britain, Italy, Hungary, Sweden, South Africa, Brazil and USA respectively.
- By the mid-1920s, ethyl alcohol was blended with gasoline in every industrialized nation except United States.
- In 1921, it was believed that world oil supplies would run out or be too rare and expensive in 25 years. Ethyl alcohol was considered to be the fuel that would eventually replace petroleum. About 100 million gallons of industrial alcohol supply was made available for use as fuel.
- In 1925, France, Germany, Brazil and other countries have a "mandatory blending" law. This law requires gasoline retailers to blend in large volumes of alcohol with all gasoline sold.
- In the 1930s, Ford Motors originally built cars that could be changed slightly to run on gasoline, alcohol or kerosene but Henry Ford, supported ethanol's use over gas.
- In 1942, more than 500 million gallons of alcohol is used for aviation fuel for World War II.
- In 1973, a worldwide energy crisis begins, causing ethanol to become cheaper than gasoline.
- By the mid-1980s, over a billion gallons of ethanol for fuel were sold per year.
- In 1984, the number of ethanol plants peaked at 163 in U.S., producing 595 million gallons of ethanol that year.
- In 1988, ethanol is first used as an oxygenate to lower pollution caused by burning gasoline.
- Between 1997 and 2002, three million U.S. cars and light trucks are produced which could run on E85, a blend of 85% ethanol with 15% gasoline. Almost no gas stations sell this fuel however.
- In the early 2000s, the invasion of Iraq made Americans aware of their dependence on foreign oil. This caused alternative energies like ethanol to expand 20 to 30% yearly.
- In 2003, California was the first state to start replacing the oxygenate, MTBE with ethanol. Several other states started switching soon afterward.

- In 2004, Crude oil prices rose by 80%, Gasoline prices rose by 30%, Diesel prices rose by almost 50% due to hurricane damage to oil rigs in the Gulf of Mexico and attacks on Iraqi oil pipelines. The ethanol industry in USA made 225,000 barrels per day in August, an all-time record.
- In July 2005, E85 was sold for 45 cents less than gasoline on average in U.S. More than 4 million flexible-fuel (runs on E85 and gasoline) vehicles came into existence in U.S. About 400 filling stations sold E85 fuel, mostly in the Midwest.

Ethanol is most interesting because it is a renewable resource, easily obtained from agricultural produce or even lactose. It is generally produced by fermentation from biomass and its products like sugarcane, sugar beet, molasses, and eight principal cereal grains: wheat, maize, rice, barley, sorghum, oats, millet, rye and root tubers. The substrates that have been classified into various categories are indicated in Table 2. Current economics dictate that the grains are the primary raw materials in various developed countries. Ethanol is produced by enzymatically converting the starch in grains to fermentable sugars, followed by metabolism of the sugars to ethanol by distiller's yeast, *Saccharomyces cerevisiae* as represented below:

$$\underset{\text{Starch}}{(C_6H_{10}O_5)_n} + \underset{\text{Water}}{nH_2O} \xrightarrow{\text{Amylolytic enzymes}} \underset{\text{Glucose}}{nC_6H_{12}O_6}$$

$$\underset{\text{Glucose}}{C_6H_{12}O_6} \xrightarrow{\textit{Saccharomyces cerevisiae}} \underset{\text{Ethanol}}{2C_2H_5OH} + \underset{\text{Carbon dioxide}}{2CO_2}$$

**Table 2** Various substrates used for ethanol production

| | |
|---|---|
| Directly Fermentable | Molasses, Sucrose, Sugarcane juice, Fruit juices, Sugar beet juice, Sorghum, Honey etc. |
| Easily hydrolysable substrates | Starchy materials-grains, tubers |
| Difficult to hydrolyse substrates | Lignocellulosics |

Although ethanolic fermentation is common to a wide variety of microorganisms, only yeasts have traditionally been used for industrial scale production of alcohol . The near-exclusive application of yeasts is due, in part, to their homofermentative mode (EMP pathway) which yields mostly a single product, ethanol, often in relatively high concentrations. *Saccharomyces cerevisiae* and *S. uvarum* (*S. carlbergensis*), for example, produce 12% ethanol, whereas in a slow fermentation *S. sake* produce up to 20% ethanol. Under anaerobic conditions, these yeasts normally produce small amounts of byproducts such as glycerol, methanol and fusel oil, which require that specific measures be taken for their separation during distillation of alcohol.

Most regions of the world have traditionally produced alcohol from locally available substrates. The annual world production of ethanol is over 37 billion liters, approximately 95% of which is produced by fermentation, the remainder being mostly manufactured by the catalytic hydration of ethylene. Almost 14% of the fermentation ethanol is beverage alcohol, 20% is for various industrial uses and the remaining 68% is fuel ethanol. While the America's account for over 65% of total alcohol production and Asia, a mere 20% (Figure 2). A major part of the production is tilted towards the Indian sub-continent and North East Asia. China is the largest producer of ethanol in the region; however, the industry remains fragmented, largely broken into small scale and state holdings across provinces. Brazil is the world leader in alcohol production, responsible for over 40% of annual world production, some 12 billion liters of ethanol.

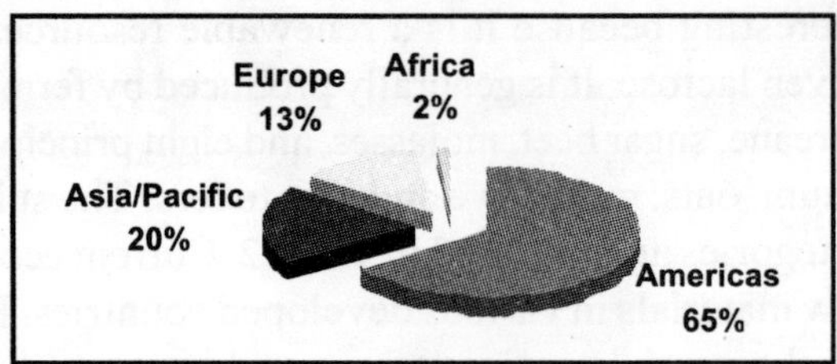


**Figure 2** World ethanol production by continent (2001)

The "fuel ethanolization" of the world alcohol industry is set to continue. If all recently announced ethanol projects are implemented, total fuel ethanol production worldwide could grow to 31 billion liters by 2006 against somewhat below 20 billion in 2001 (Figure 3). As a result, the share of fuel ethanol in 2005 could jump to almost 70 per cent compared with around 63 in 2001. Such a change is likely to have far reaching implications for the world alcohol sector. At a national level, such a growth can only be facilitated if plant sizes increase tremendously. This, in turn, would have repercussions on the feedstock market as well, where infrastructural adaptations would have to take place. At the same time, distribution systems would have to change. In countries where fuel ethanol receives financial assistance from the state (in the form of tax incentives or direct subsidies, etc.), fuel ethanol plants are likely to crowd out those producers that traditionally supply the industrial or beverage sector. This is so because fuel ethanol producers can fully utilize economies and are also able to cross-subsidise production for other markets. At an international level, the increased production and utilization of fuel ethanol will make necessary an international exchange mechanism, which can help to stabilize the market at times of regional production shortfalls. This will be a very difficult operation. After all, fuel ethanol production in most countries is being subsidized in order to support domestic farmers.

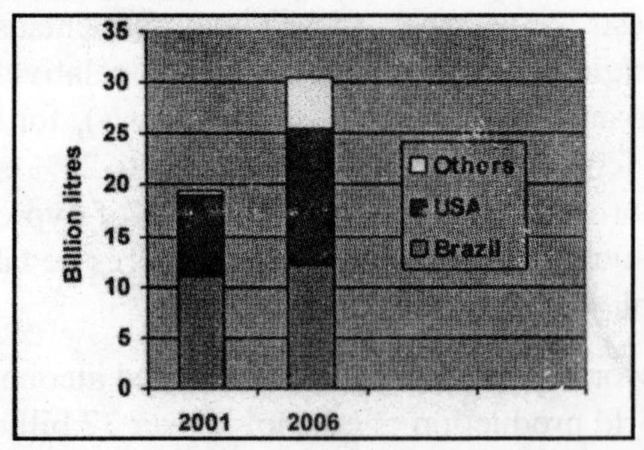


**Figure 3** World fuel ethanol production (2001 vs. 2006)

Ethyl alcohol has long been used as an automotive fuel in two ways: First, it replaces gasoline outright in a somewhat modified internal combustion engine; and secondly, it is an effective "octane booster" when mixed with gasoline in blends of 10 to 30 percent and requires no engine modification. These blends achieve the same octane boosting (or anti-knock) effects as petroleum-derived aromatics like benzene or metallic additives like tetra-ethyl lead. Ethanol used as part of the fuel, by blending with petrol, for a motor vehicle is called fuel-ethanol. The history of ethanol as a fuel dates back to the early days of the automobile. However, cheap petrol quickly replaced ethanol as the fuel of choice. Today, alcohol technology is again reviving and production of alcohol as a fuel is being given full attention. Ethanol could be blended in various proportions in petrol. Ethanol is usually added 5 to 10% by volume of petrol for such application. Both methanol and ethanol are probably superior to gasoline as a liquid fuel for automotive propulsion use. They are safer to handle, burn cleaner and significantly reduce pollution emissions from automotive exhaust systems. They have an excellent octane number, around 106, and should not require the addition of tetraethyl lead for better engine performance. Also the use of alcohol should permit greater improvements in engine design, because an explosive mixture of alcohol vapor and air can be used under far higher compression ratios without auto ignition than a similar mixture of gasoline and air. However, whereas both alcohols possess certain inherent basic advantages over gasoline, if a choice must be made, ethanol is superior to methanol.

Ethanol makes an excellent motor fuel. It has a research octane number of 109 and a motor octane number of 90, both of which exceed that of gasoline. Ethanol also has a lower vapour pressure than gasoline, which results, in lower evaporative emission (Table 3). The country has the potential to save nearly 800 million liters of petrol annually by blending up to 10 per cent ethanol in the gasoline used by the transport sector, according to the Federation of Indian Chambers of Commerce and Industry (FICCI). In the mid 1970's the Govt. of Brazil launched the National Fuel Alcohol Program, which aimed at increasing the share of domestically produced fuels in the country's fuel pool. The program proved to be spectacularly successful. By 1980s ethanol had a much larger market share than gasoline (Figure 4). India is initiating the use of ethanol as an automotive fuel. A move has been made by distilleries in India to use surplus alcohol as a blending agent or as an oxygenate in gasoline. Based on experiments by the Indian Institute of Petroleum, a 10 percent ethanol blend with gasoline and a 15 percent ethanol blend with diesel are being considered for use in vehicles in at least one state. It may be estimated that fuel ethanol accounted for roughly 70% of worlds ethyl alcohol production in 2003. At present, Brazil is the only country that uses ethanol as a 100% substitute for gasoline. Besides being a fuel, ethanol is also a very versatile chemical feedstock.

**Table 3** Fuel properties and characteristics of gasoline and ethanol

| | Gasoline | Ethanol |
|---|---|---|
| Specific calorific value (KJ/kg) | 43,900 | 26,700 |
| Octane number (RON/MON) | 91/80 | 101/98 |
| Latent heat of vaporization (KJ/kg) | 376-502 | 903 |
| Ignition temperature (°C) | 220 | 420 |

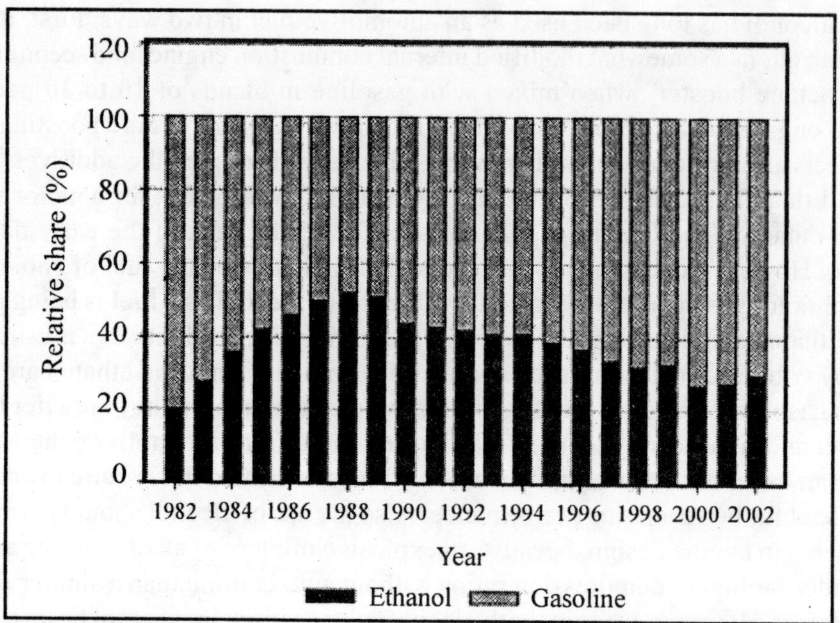


**Figure 4** Relative market share of ethanol vs. gasoline in Brazil

With the increasing shortage of petroleum, which leads to the shortage of petrochemicals as well, ethanol is expected to play a more significant role in the future. In Figure 5, the growth in fuel ethanol production, under a very optimistic assumption is forecasted. From this particular survey, it may be estimated that most of the growth will happen in U.S. Growth would also be strong in Brazil. The EU will be the third largest producer of fuel ethanol by 2005 and the growth rate will also be considerably above those in Brazil and the United States.

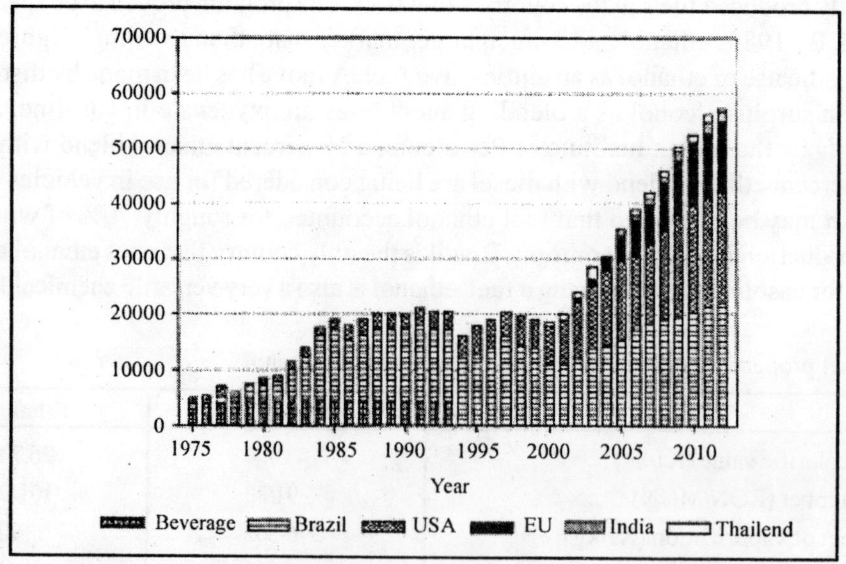


**Figure 5** An optimistic scenario of world fuel ethanol production

## 2.3. Butanol

Butanol is used as a solvent for a wide variety of chemical and textile processes, as a paint thinner, as well as a component of hydraulic and brake fluids. Butanol may also be used as a direct fuel in any standard internal combustion engine engineered for gasoline usage. Butanol is reported to yield 36,000 kJ/kg when burned.

Acetone, butanol, butyric acid and isopropanol, along with other organic acids and alcohols, may be obtained by clostridial fermentation. Members of the genus *Clostridium* are Gram-positive rods with peritrichous flagella and are accordingly very motile. They are characterized by their ability to form heat-resistant spores, a highly fermentative metabolism and their response to oxygen. They vary widely in the range of substrates that they can utilize including starch, molasses and hydrolysed cellulosic materials. The relative amount of each fermentation product is dependent upon the bacterial species and the specific strain used, and the environmental conditions of the fermentation. There are three main fermentation types:

1. Acetone-Butanol: This is brought about by *Clostridium acetobutylicum* with acetone and butanol as the main products along with butyric acid, acetic acid, acetoin, ethanol, $CO_2$ and $H_2$ as additional products.
2. Butanol-Isopropanol: This is brought about by *Clostridium butylicum* with butanol and isopropanol as the principal products along with butyric acid, acetic acid, $CO_2$ and $H_2$ as additional products.
3. Butyric acid-Acetic acid: This is brought about by *Clostridium butyricum* with butyric acid and acetic acid as main products along with $CO_2$ and $H_2$

The acetone-butanol fermentation has a long history a successful industrial fermentation process. It was Wiezmann in the UK, early in the 20th century, who conducted much of the basic research into the production of acetone, butanol and ethanol by *C. acetobutylicum.* This was invaluable during World War I, particularly for the production of acetone, which was needed in the manufacture of explosives. Following World War I, butanol became the main product of interest. It was used extensively as a feed-stock chemical in the production of lacquers, rayon, plasticizers, coatings, detergents, brake fluids and butadiene for synthetic rubber manufacture. Butanol was also used as a solvent for fats, waxes, resins, shellac and varnish, and as a valuable extractant and solvent in the food industry. The annual production of fermentation-derived butanol was over 20000 tones in 1945, but in the western world the process began to decline by the late 1940s due to changes in supply of fermentation raw materials like molasses, sugar cane, etc. and the increase in availability of inexpensive petrochemical feedstocks for its chemical synthesis.

Economically viable production of butanol requires a fermenter volume of atleast $1000m^3$. The fermenters were not stirred, as the evolution of gases provided sufficient mixing. These fermentations are operated as batch process, often using 5-7% starch or molasses as the carbon substrate. More recently, with the increasing demand for butanol, advanced processes based on corn, corn processing byproducts and other cellulosic wastes have been proposed. Prior to fermentation, the medium and fermenter are sterilized and purged with $CO_2$. The fermenter is then inoculated with relatively low level of inoculum, 0.03%. Over the first 18-24 hours the pH falls from an initial level of 5.8-6.0 to pH 5.2 due to the production of butyric acid during the rapid growth phase. Over the following 20-24 hours the pH rises back to 5.8-6.0, as these acids are metabolized to form the neutral solvents- acetone, butanol and ethanol. Product recovery is by fractional distillation.

Butanol is now predominantly manufactured from petroleum-based raw materials. In USA, for example, the current production of chemically synthesized butanol is more than 500000 tones/year, with annual growth of 3-4%. However, butanol fermentations are still operated in certain countries. In the former states of the Soviet Union some processes were based on beet molasses, whereas fermentations using sugar cane molasses continued until relatively recently in South Africa, and China still maintains some fermentation-based manufacturing plants. The future for fermentation-based production looks quite bright, particularly as worldwide consumption of butanol has surged over the past few years. This provides an opportunity for the introduction of new and more efficient fermentation process technology, especially as the supply of petrochemicals is declining.

Besides the existing role as a solvent and chemical feedstock, butanol has several properties that are favorable for motor fuel use, either alone or when blended with gasoline (Table 1). Butanol has good octane-enhancing properties, a relatively high heat of combustion and a much lower vapour pressure than both methanol and ethanol. These characteristics make butanol an even better liquid fuel extender than ethanol, which is currently used in the formulation of gasohol. Furthermore, butanol has low miscibility with water, but high miscibility with both diesel and gasoline. Owing to its high heat of combustion, butanol solutions containing as much as 20% (v/v) water have the same combustion value as anhydrous ethanol. This can indirectly reduce nitrogen oxide (NO) emissions by lowering the operating temperature of internal combustion engines. As fuel additives, alcohols such as butanol also have the potential for reducing carbon monoxide emissions.

## 3. BIODIESEL

It is a renewable fuel manufactured from vegetable oils or animal fats, including recycled oils from restaurants, and is therefore easy to produce domestically. Biodiesel is safe, completely biodegradable, and emits much less particulate matter, carbon monoxide, and other pollutants than regular diesel. Biodiesel includes variety of ester-based bio-fuels (fatty esters) generally defined as the monoalkyl esters (fatty acid alkyl esters) made from vegetable oils, or sometimes from animal fats through a simple transesterification process. It has a specific gravity 0.87-0.89 and is as efficient as petroleum diesel in powering unmodified diesel engine. Biodiesel was first made from vegetable oil in 1895 by Rudolf Diesel (1858-1913) who developed the first engine to run on peanut oil and demonstrated at the World Exhibition in Paris in 1900. Biodiesel could not be exploited after the death of Rudolf Diesel as the rapidly developing petroleum industry produced a cheap by-product "diesel fuel". Due to the depletion of petroleum reserves, biodiesel as fatty esters was considered as an alternative to petroleum diesel during mid 1970s. It is sold commercially in Europe, America and Australia. There are many advantages of biodiesel as blends of 20% biodiesel with 80% petroleum diesel can be used in unmodified diesel engines. Biodiesel can be used in its pure form but requires certain engine modifications to avoid maintenance and performance problems. Biodiesel is nontoxic and biodegradable. It reduces the emission of harmful pollutants (mainly particulates) from diesel engines (80% less $CO_2$ emissions, 100% less sulfur dioxide) but emissions of nitrogen oxides (precursor of ozone) are increased. Biodiesel has a high cetane number (above 100, compared to only 40 for diesel fuel), a measure of a fuel's ignition quality. The high cetane numbers of biodiesel contribute to easy cold starting and low idle noise. The use of biodiesel can extend the life of diesel engines because it is more lubricating and, furthermore, power output are relatively unaffected by biodiesel. It also replaces the exhaust odor of petroleum diesel with a more pleasant smell of popcorn or french fries. There are four routes to ester production from oils and fats:

1. Base catalyzed transesterification of oil with alcohol
2. Direct acid catalyzed esterification of oil with methanol
3. Conversion of oil to fatty acids and then to alkyl esters with acid catalysis.
4. Microbial or enzyme (lipase) catalyzed transesterification of oil with alcohol.

Base catalyzed transesterifications are preferred for commercial production because of low reaction time, high conversion rate (98%) with minimal side reactions and direct conversion to methyl ester with no intermediate steps. The transesterification process involves mixing methanol, ethanol, isopropyl alcohol, or butanol (50% excess) with a chemical catalyst in the form of NaOH/KOH or a biocatalyst which may be in the form of lipase or even the immobilized microbial cells capable of producing lipase at room temperature. The supernatant is biodiesel and contains a mixture of methylated/ ethylated/propylated/butylated fatty acids and methanol/ethanol/isopropanol/butanol, the catalyst remaining dissolved in the glycerol fraction. Industrially, the esters are subjected to purification process which consists of water washing, vacuum drying and filtration. Oil molecules (triglycerides) are broken apart and reformed into esters and glycerol which are then separated from each other and purified.

$CH_2COOR1$
|
$CHCOOR1 + 3CH_3OH \rightarrow (CH_2OH)_2CH\text{-}OH + 3\ CH_3COO\text{-}R1$
|
$CH_2COOR1$

Since any oil can be used, the alkyl groups of triglycerides may be different

$CH_2OC{=}OR1$
|
$CHOC{=}OR2 + 3CH_3OH \rightarrow (CH_2OH)_2CH\text{-}OH + CH_3COO\text{-}R1 +$
| $CH_3COO\text{-}R2 +$
$CH_2OC{=}OR3$ $CH_3COO\text{-}R3$

Triglyceride Methanol Glycerol Esters

6.25 g/l NaOH produces a very useful fuel. One uses about 6 g NaOH when the oil is light in colour and about 7 g NaOH when it is dark in colour. The reaction between the biolipid and alcohol is a reversible reaction so the alcohol must be added in excess to drive the reaction towards the right. Catalyst is dissolved in alcohol using an agitator or mixer. Alcohol/catalyst mixture is charged into a closed reaction vessel and the biolipid is added. The system is then totally closed to the atmosphere to prevent the loss of alcohol. The reaction mixture is kept just above the boiling point of alcohol (around 70°C) to speed up the reaction, some recommend the reaction to take place at room temperature. Reaction time varies from 1-8 hours.

$$\text{Triglycerides} + \text{Free Fatty acids } (<4\%) + \text{Alcohol} \xrightarrow{\text{Base}} \text{Alkyl esters} + \text{Glycerol}$$

$$\text{Triglycerides} + \text{Free Fatty acids } (>4\%) + \text{Alcohol} \xrightarrow{\text{Acid}} \text{Alkyl esters} + \text{Triglycerides}$$

$$\text{Triglycerides} + \text{Free Fatty acids} + \text{Alcohol} \xrightarrow{\text{Microbial Lipase}} \text{Alkyl esters} + \text{Glycerol}$$

The world biodiesel sources include rapeseed oil (84%), sunflower oil (13%), soybean oil (1%), palm oil (1%), and others (1%). If fats or solidified oil are used, it involves pre-heating up to 50°C prior to mixing with methanol and catalyst. By developing methods to use cheap and low quality lipids as feedstocks, it is hoped that a cheaper biodiesel can be produced, thus competing economically with petroleum resources. It can also be used in boilers or furnaces designed to use heating oils or in oil-fueled lighting equipment. It can be used neat (100% biodiesel), or it can be blended with petroleum diesel.

## 4. P-SERIES FUELS

P-Series fuel is a mixture of natural gas liquids (pentanes plus), ethanol, and methyl tetrahydrofuran (MTHF), a biomass-derived co-solvent. P-Series is predominantly derived from renewable resources and burns much cleaner than gasoline. It can be mixed with gasoline in any proportion and is used in multi-fuel vehicles. P-Series is a family of renewable, non-petroleum, liquid fuels that can substitute for gasoline. They are a blend of 25 or so domestically produced ingredients. About 35% of P-Series comes from liquid by-products, known as "C5+" or "pentanes-plus", which are left over when natural gas is processed for transport and marketing. Ethanol, fermented from corn, comprises about 45% and the remaining 20% is MTHF, an ether derived from lignocelullosic biomass including paper sludge, wastepaper, food waste, yard and wood waste, agricultural waste etc.

P-Series fuel addresses three problems including the need for non-petroleum energy sources, solid waste management, and affordability. Its production procedure solves the problem of the disposal of municipal wastes which are chemically digested, so there is no combustion with the accompanying toxic air emissions.

P-Series fuels were officially designated as an alternative fuel by the U.S. Department of Energy (DOE) in 1999. The Since P-Series is not derived from petroleum, these can effectively help to replace petroleum imports. These have environmental benefits because of the reductions in hydrocarbon and CO emissions, toxics, and greenhouse gases. Much like gasoline, P-Series fuels range from 89-93 octane (mid-grade to premium) and can be formulated specifically for winter or summer use. Refueling with P-Series is as quick and familiar as with gasoline. But P-Series is not gasoline and cannot be used in a regular gasoline car. The basic capability for utilizing P-Series in vehicles has already been incorporated into methanol/ethanol flexible-fuel vehicles (FFV's) which are designed to operate on alcohol, on gasoline, or on any mixture of the two. Nearly three million FFV's have been manufactured since 1996.

One of the attractions of FFV's is that they are very easy to use. There is no need for any special fuel management because gasoline and P-Series can be freely intermixed in any proportion with fuel that is already in the vehicle's fuel tank. So, even if P-Series is not available at a particular location, the vehicle can be filled up with gasoline.

All ingredients except for the MTHF are generally purchased as bulk commodities from natural gas processors and ethanol producers while MTHF is produced by hydrolysis of biomass. Several companies engineer hydrolysis plants. One, the Biofine process, is a commercialized technology that uses two-step dilute mineral acid hydrolysis to break down biomass containing lignocellulose into intermediate chemicals that can be further transformed into MTHF and other chemical byproducts.

## 5. ALKANES

Alkanes are aliphatic compounds which are in the form of saturated acyclic hydrocarbons in which the molecules have the maximum possible number of hydrogen atoms and so has no double bonds. Alkanes are also often known as paraffins, or collectively as the paraffin series. The first four members of the series (in terms of number of carbon atoms) are named as methane, $CH_4$; ethane, $C_2H_6$; propane, $C_3H_8$; butane; $C_4H_{10}$. Traces of methane (about 0.0001% or 1 ppm) occur in the Earth's atmosphere, produced primarily by forms of archaea. The content in the oceans is negligible due to the low solubility in water, however, at high pressures and low temperatures, methane can co-crystallize with water to form a solid methane hydrate. Although they cannot be commercially exploited at the present time, the calorific value of the known methane hydrate fields exceeds the energy content of all the natural gas and oil deposits put together. Methane extracted from methane hydrate is considered therefore a candidate for future fuels.

Alkanes are both important raw materials of the chemical industry and the most important fuels of the world economy. The starting materials for the processing are mainly natural gas and crude oil. The latter is separated in an oil refinery by fractional distillation and processed into many different products, for example gasoline. The different "fractions" of crude oil have different boiling points and can be isolated and separated quite easily and within the individual fractions the boiling points lie closely together.

The domain of usage of a certain alkane can be determined quite well according to the number of carbon atoms, although the following demarcation is idealized and not perfect. The first four alkanes are used mainly for heating and cooking purposes, and in some countries for electricity generation. Methane and ethane are the main componants of natural gas; they are normally stored as gases under pressure. It is however easier to transport them as liquids: this requires both compression and cooling of the gas. Propane and butane can be liquefied at fairly low pressures, and are well known as liquified petroleum gas (LPG). Propane, for example, is used in the propane gas burner, butane in disposable cigarette lighters (where the pressure is a mere 2 bar). The two alkanes are used as propellants in aerosol sprays. From pentane to octane the alkanes are highly volatile liquids. They are used as fuels in internal combustion engines, as they vaporise easily on entry into the combustion chamber without forming droplets which would impair the uniformity of the combustion. Branched-chain alkanes are preferred, as they are much less prone to premature ignition which causes knocking than their straight-chain homologues. This propensity to premature ignition is measured by the octane rating of the fuel, where 2,2,4-trimethylpentane (isooctane) has an arbitrary value of 100 and heptane has a value of zero. Apart from their use as fuels, the middle alkanes are also good solvents for nonpolar substances.

Alkanes also form nonane, the alkane hydrocarbon including, hexadecane (an alkane with sixteen carbon atoms) which is a liquid of higher viscosity and less suitable for use in gasoline and forms the major part of diesel and aviation fuel. Diesel fuels are charaterised by their cetane number, cetane being an old name for hexadecane. However the higher melting points of these alkanes can cause problems at low temperatures and in polar regions, where the fuel becomes too thick to flow correctly. Alkanes from hexadecane upwards form the most important components of fuel oil and lubricating oil. In latter function they work at the same time as anti-corrosive agents, as their hydrophobic nature means that water cannot reach the metal surface. Many solid alkanes find use as paraffin wax, for example in candles. This should not be confused however with true wax, which consists primarily of esters. Alkanes with a chain length of approximately 35 or more carbon atoms are found in bitumen, used for example in road surfacing. However the higher alkanes have little value and are usually split into lower alkanes by cracking.

Certain types of bacteria can metabolise alkanes: they prefer even-numbered carbon chains as they are easier to degrade than odd-numbered chains. On the other hand certain archaea, the methanogens, produce large quantites of methane by the metabolism of carbon dioxide or other oxidised organic compounds. The energy is released by the oxidation of hydrogen:

$$CO_2 + 4H_2 \rightarrow CH_4 + 2H_2O$$

Methanogens are also the producers of marsh gas in wetlands, and release about two billion tonnes of methane per year-the atmospheric content of this gas is produced nearly exclusively by them. The methane output of cattle and other herbivores, which can release up to 150 litres per day, and of termites, is also due to methanogens. They also produce this simplest of all alkanes in the intestines of humans. Methanogenic archaea are hence at the end of the carbon cycle, with carbon being released back into the atmosphere after having been fixed by photosynthesis. It is probable that our current deposits of natural gas were formed in a similar way. All alkanes react with oxygen in a combustion reaction, although they become increasing difficult to ignite as the number of carbon atoms increases. The general equation for complete combustion is:

$$2C_nH_{2n+2} + (3n+1)O_2 \rightarrow 2(n+1)H_2O + 2nCO_2$$

In the absence of sufficient oxygen, carbon monoxide or even soot can be formed from methane, as shown below:

$$2CH_4 + 3O_2 \rightarrow 2CO + 4H_2O$$

$$CH_4 + O_2 \rightarrow C + 2H_2O$$

Alkanes usually burn with a non-luminous flame with very little soot formation. The standard enthalpy change of combustion, $\Delta_c H^\circ$, for alkanes increases by about 650 kJ/mol per $CH_2$ group. Branched-chain alkanes have lower values of $\Delta_c H^\circ$ than straight-chain alkanes of the same number of carbon atoms.

## 5.1. Methane

Methane is used for both domestic and industrial fuel. At present, supplies mostly come from gas and oil fields or the gasification of coal. Consequently, methane production via fermentation is attractive only in limited small-scale local situations. However, this mode of production may become increasingly important later in the 21st century, when supplies from the non-renewable sources begin to deplete. Methane production by microorganisms is a very complex process, which involves a mixture of anaerobic microorganisms found naturally in marshes, organic sediments and in the stomach (rumen) of ruminant animals. Future microbial production of methane may be via the anaerobic digestion of agricultural, industrial and urban wastes. These wastes are predominantly plant biomass including lignocellulosic materials which have high collection costs. A potentially attractive, but costly mode of large-scale production is via landfilling with organic wastes, provided that stable long-term gas production can be developed. In contrast, biogas fermenters use low technology in the small-scale local production of methane. They often use animal excreta and are particularly valuable in locations where other fuels are not available. The biogas generated is primarily composed of 50-80% methane and 15-45% CO, along with some trace gases.

One technology that can successfully treat the organic fraction of waste is anaerobic digestion. When used in fully engineered system, anaerobic digestion not only provides pollution prevention, but also allows for sustainable energy, compost and nutrient recovery. Thus, anaerobic digestion can convert a disposal problem into a profit centre. As the technology continues to mature, anaerobic digestion is becoming a key method for both waste reduction and recovery of a renewable fuel and other valuable co-products. The basic points for this process are that:

i) Most of the important bacteria involved in biogas production process are anaerobes and slow growing.

ii) The greater degree of metabolic specialization is observed in these anaerobic microorganisms.

iii) Most of the free energy present in the substrate is found in the terminal product methane. Since less energy is available for the growth of organisms, less microbial biomass is produced and consequently, disposal of the sludge after the digestion may not be a major problem.

There are several advantages of using anaerobic digestion technology. These include:

i) Waste treatment benefits

a) Natural waste treatment process

b) It requires less land than aerobic composting or land filling

c) Reduces disposed waste volume and weight to be landfilled

d) Reduces concentrations of leachates

ii) Energy benefits

a) Net energy producing process

b) Generates a high quality renewable fuel

c) Biogas proven in numerous end-use applications

iii) Environmental benefits

a) Significantly reduces $CO_2$ and $CH_4$ emissions

b) Eliminates odour

c) Produces a sanitized compost and nutrient rich fertilizer

d) Maximizes recycling benefits

iv) Economic benefits

a) It is more cost-effective than other treatment options from a life cycle perspective

Mixed microbial populations associated with methane generation are highly versatile with regard to the range of substrates that they can utilize. Methane production from organic materials involves three specific phases. First, a group of microorganisms hydrolyse organic polymers, including fats, proteins and polysaccharides, to their respective soluble monomers. These compounds are then metabolized to organic acids by anaerobic acidogenic organisms. In the final phase, the organic acids are converted to alkanes and carbon dioxide.

$$\text{Polymers} \xrightarrow{\text{Hydrolysing microbes}} \text{Monomers} \xrightarrow{\text{Acidogenic bacteria}} \text{Organic acids}$$

Methanogenic bacteria

$$\text{Acetic acid} \longrightarrow \text{Methane} + CO_2$$

$$\text{Propionic acid} \longrightarrow \text{Ethane} + CO_2$$

$$\text{Butyric acid} \longrightarrow \text{Propane} + CO_2$$

Methanogenic bacteria also produce methane by acting on $H_2$ and coupling its oxidation to the reduction of $CO_2$.

$$4H_2 + CO_2 \longrightarrow CH_4 + 2H_2O$$

Some methanogens utilize CO for $CH_4$ biosynthesis

$$4CO + 4H_2O \longrightarrow 4CO_2 + 4H_2$$

$$CO_2 + 4H_2 \longrightarrow CH_4 + 2H_2O$$

Methanogenic bacteria produce methane from acetate, which is the major product. However, methane has a lower energy yield than the longer chain alkanes, ethane and propane (Table 4), which are derived from propionate and butyrate, respectively. Normally, only small amounts of these two organic acids are produced during acidogenesis, but the quantities generated depend upon the specific conditions. Therefore, there is potential for the future manipulation of such fermentations to produce a greater proportion of the more attractive fuels ethane and propane.

**Table 4** Comparative energy yields of short chain alkanes

| Alkane | Energy yield (kJ/m³) |
|---|---|
| Methane ($CH_4$) | 37 |
| Ethane ($C_2H_6$) | 64 |
| Propane ($C_3H_8$) | 94 |

## 5.2. Ethane

Ethane is an aliphatic hydrocarbon which is a colourless and odourless gas at ordinary pressure and temperature. It is the simplest saturated hydrocarbon containing more than one carbon atom and is refined on an industrial scale from natural gas as petrochemical feedstock.

Ethane was first prepared synthetically in 1834 by Michael Faraday, involving the electrolysis of a potassium acetate solution, but he mistook the hydrocarbon product of this reaction for methane and did not investigate it further. During the period 1847–1849, in an effort to vindicate the radical theory of organic chemistry, Hermann Kolbe and Edward Frankland produced ethane by the reductions of propionitrile (ethyl cyanide) and ethyl iodide with potassium metal, as done by Faraday, by the electrolysis of aqueous acetates. They also mistook the product of these reactions for methyl radical, rather than the dimer of methyl, ethane. This error, however, was corrected in 1864 by Carl Schorlemmer, who showed that the product of all these reactions was in fact ethane.

Ethane can be conveniently prepared by the electrolysis of an aqueous solution of an acetate salt in which acetate is oxidized at anode to produce carbon dioxide and methyl radicals, and the highly reactive methyl radicals combine to produce ethane:

$$CH_3COO^- \rightarrow CH_3\bullet + CO_2 + e^-$$

$$CH_3\bullet + \bullet CH_3 \rightarrow C_2H_6$$

After methane, ethane is the second-largest component of natural gas. Natural gas from different gas fields varies in ethane content from less than 1% to over 6% by volume. Prior to the 1960s, ethane was typically not separated from the methane component of natural gas, but simply burnt along with the methane as a fuel. Today, however, ethane is an important petrochemical feedstock, and it is separated from the other components of natural gas in most well-developed gas fields. Ethane can also be separated from petroleum gas, a mixture of gaseous hydrocarbons that arises as a byproduct of petroleum refining.

Ethane is most efficiently separated from methane by liquefying it at cryogenic temperatures. Various refrigeration strategies exist: the most economical process presently in wide use employs turbo-expansion, and can recover over 90% of the ethane in natural gas. In this process, chilled gas expands through a turbine; as it expands, its temperature drops to about -100°C. At this low temperature, gaseous methane can be separated from the liquefied ethane and heavier hydrocarbons by distillation. Further distillation then separates ethane from the propane and heavier hydrocarbons.

Ethane is also produced by mixed microbial populations associated with methane generation from organic materials as one of the byproducts. Ethane production from organic materials involves three specific phases. First, a group of microorganisms hydrolyse organic polymers to their respective soluble monomers. These compounds are then metabolized to organic acids including propionic acid by anaerobic acidogenic organisms. In the final phase, the butyric acid is converted to alkanes and carbon dioxide.

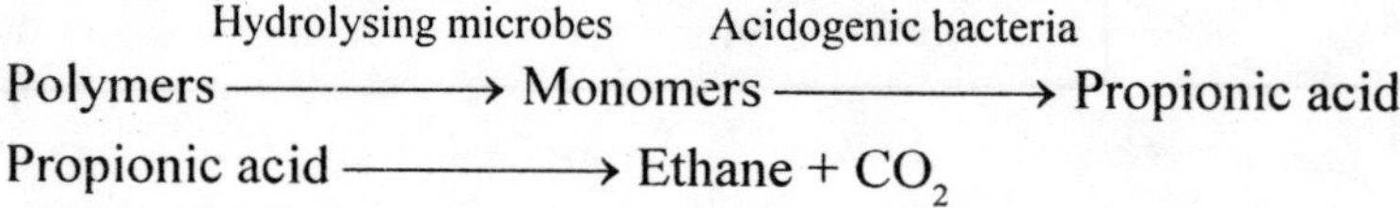


Ethane also reacts with oxygen in a combustion reaction and its energy yield is more than that of methane (Table 4), due to which it is also fast becoming a popular alternative fuel and is similar to natural gas in that it is already readily available to the public, and pollutes much less than gasoline.

## 5.3. Propane

Propane is a three-carbon alkane derived from other petroleum products during oil or natural gas processing. It is commonly used as a heat source for engines, barbecues, and homes. Its energy yield is more than that of methane and ethane (Table 4) and is fast becoming a popular alternative fuel and is similar to natural gas in that it is already readily available to the public, and pollutes much less than gasoline. When commonly sold as fuel it is also known as liquified petroleum gas (LPG or LP-gas) and is a mixture of propane with smaller amounts of propylene, butane and butylene, plus ethanethiol as an odorant to allow the normally odorless propane to be smelled. It is used as fuel in cooking on many barbecues and portable stoves and in motor vehicles. Propane powers some locomotives, buses, forklifts, and taxis and is used for heat and cooking in recreational vehicles and campers. In many rural areas of the US, propane is also used in furnaces, water heaters, laundry dryers, and other heat-producing appliances. Delivery trucks fill up large tanks that are permanently installed on the property (sometimes called pigs) or exchange bottles of propane.

Many industries involving glass making, brick kilns, poultry farms, and other industries that need portable heat use propane. Unlike natural gas, propane is heavier than air (1.5 times denser). In its raw state, propane sinks and pools at the floor. Liquid propane will flash to a vapor at atmospheric pressure and appears white due to moisture condensing from the air. Propane is the fastest growing fuel source in the Third World, especially in China and India. Its use frees up the huge rural populations from time-consuming ancient chores such as wood gathering and allows them more time to pursue other activities, such as increased farming or educational opportunties.

Propane is also being used increasingly more for vehicle fuels. In the U.S., 190,000 on-road vehicles use propane, and 450,000 forklifts use it for power. It is the third most popular vehicle fuel in America, behind gasoline and diesel. In other parts of the world, propane used in vehicles is known as autogas. About 9 million vehicles worldwide use autogas. Another use of propane is the application as propellant for aerosol sprays, especially after the ban of chlorofluorocarbons (CFCs). It is also used as a feedstock for the production of base petrochemicals in steam cracking. Propane is also instrumental in providing off-the-grid refrigeration, also called gas absorption refrigerators. Made popular by the 'Servel' company, propane-powered refrigerators are highly efficient, do not require electricity, and have no moving parts. Refrigerators built in the 1930s are still in regular use, with little or no maintenance. Today, the Unilever company is exploring the use of environmentally friendly propane as a refrigerant.

Propane is not produced for its own sake, but as a byproduct of two other processes: natural gas processing and petroleum refining. The processing of natural gas involves removal of propane and butane from the natural gas to prevent condensation of these liquids in natural gas pipelines. Additionally, oil refineries produce some propane as a by-product of production of gasoline or heating oil. Mixed microbial populations associated with methane generation from organic materials also leads to the generation of propane as one of the byproducts. Propane production from organic materials involves three specific phases. First, a group of microorganisms hydrolyse organic polymers to their respective soluble monomers. These compounds are then metabolized to organic acids including butyric acid by anaerobic acidogenic organisms. In the final phase, the butyric acid is converted to alkanes and carbon dioxide.

$$\text{Polymers} \xrightarrow{\text{Hydrolysing microbes}} \text{Monomers} \xrightarrow{\text{Acidogenic bacteria}} \text{Butyric acid}$$

$$\text{Butyric acid} \longrightarrow \text{Propane} + CO_2$$

## 6. HYDROGEN

Hydrogen has been an unrealized "fuel of the future" for over 30 years but has recently attracted world wide attention as it is a sustainable, non-polluting source of power that can be used in mobile and stationary applications. Several automobile manufacturers, including Ford and Toyota, will be introducing cars in the next few years that use hydrogen based fuel cells, Shell and BP have established core hydrogen divisions in their companies. Hydrogen is an obvious alternative to hydrocarbon fuels, such as gasoline because in its molecular form hydrogen can be used directly as a fuel to drive a vehicle, to heat water or indirectly to produce electricity for industrial, transport and domestic use. Some of the important advantages of hydrogen over other fuels indicate that it is non-polluting as on combustion it forms water as a byproduct, it is renewable source of energy, it does not evolve the "greenhouse gas" $CO_2$ in combustion, it liberates large amounts of energy per unit weight in combustion, and it is easily converted to electricity by hydrogen fuel cells which combines hydrogen and oxygen to produce electricity, heat, and water. Hydrogen is a very attractive fuel because of its high energy content (118.7kJ/g), which is about four-fold greater than ethanol and over two-fold higher than methane. The technology for its use has already been developed and the product of its combustion is water.

Hydrogen has many potential uses, is safe to manufacture, and above all it is environment friendly. All the means of transport we know today could be powered by hydrogen. There are two possibilities for doing so: Hydrogen can be burnt in conventional engines instead of gasoline or can be used in fuel cells which are generating electric power for an electric motor in the car. A hydrogen fuel cell generates electricity through an electrochemical reaction using hydrogen and oxygen. Hydrogen is sent into one side of a proton exchange membrane (PEM). The hydrogen proton travels through the membrane, while the electron enters an electrical circuit, creating a DC electrical current. On the other side of the membrane, the proton and electron are recombined and mixed with oxygen from room air, forming pure water. Because there is no combustion in the process, there are no other emissions, making fuel cells an extremely clean and renewable source of electricity. By combining the generating power of multiple PEM cartridges, our fuel cells can be built to meet specific loads from 500 Watts to 5 kilowatts. The use of fuel cells in cars has some decisive advantages: there is only water emitted from the exhaust, it operates without noise and without vibrations and it is more efficient than a combustion engine - so it saves energy. When a fuel cell car is waiting at a traffic light, there is no noise because the engine does not work. The noise from accelerating is significantly reduced as well. Our cities will become much quieter. World-wide, all the big motorcar producing companies are developing test cars with fuel cell drive systems. In Germany mainly Daimler Chrysler, Opel and Ford are the first to do so. BMW presented hydrogen powered cars very early. The Hydrogen Highway initiatives in both California and Florida have received added boosts with Chevron opening hydrogen refueling stations in both states. Instead of bringing in hydrogen, these stations will produce their own hydrogen, allowing them to be more cost-effective. These stations will provide hydrogen to fuel cell vehicles in California and passenger buses with internal combustion engines running on hydrogen in Florida great variety of other possible applications for hydrogen and fuel cells can be found in the energy supply of portable devices: mobile phones, laptops, walkman, camcorders and many other things could be powered by hydrogen and by fuel cells in the size of batteries.

Hydrogen, by far is the most plentiful element in the universe and one of the most abundant on the earth. Hydrogen is the simplest element also with one atom of hydrogen consisting of only one proton and one electron. Despite its simplicity and abundance, hydrogen doesn't occur naturally as a gas on the Earth - it's always combined with other elements such as oxygen and carbon. Once it has been separated, hydrogen is the ultimate clean energy carrier. Clean enough that the NASA Space Shuttle programs have been relying on hydrogen-powered fuel cells since 1970s to operate shuttle electrical systems; and the exhaust from the fuel cell – pure water – is used by the crew as drinking water. The car we drive in the future may well be powered by the breath of tiny living organisms. That's the promise, at least, of a new field of science known as biological energy production. Researchers in the field seek to find a cure for the world's dependence on fossil fuels by taking advantage of nature's ability to transform sunlight and simple sugars into usable fuels. Hydrogen is one such fuel that can be produced microbiologically and holds a great promise as the power fuel of the future but some risks are also equally high. The most important among these include high inflammability, difficulty in distribution and storage as the compression or liquefaction of the gas is expensive and the metallic containers absorb up to 1,000 times their own volume of hydrogen and become heavy and brittle after repeated use.

Most methods of producing hydrogen involve splitting water ($H_2O$) into its component parts of hydrogen ($H_2$) and oxygen (O). Currently, the most commonly used method for hydrogen production involves steam reforming of methane (from natural gas), although there are several other methods. Apart from these methods, biotechnologists are looking for microbial methods of hydrogen production that are less energy intensive and give a sustainable supply of hydrogen. The thermo-chemical and electrochemical methods for hydrogen generation from water are highly energy-intensive and not always environmentally friendly. Biological methods, on the other hand present a less energy-intensive means of hydrogen production. These occur at ambient temperatures and pressures and predominantly generate hydrogen and carbon dioxide. Low conversion efficiencies of biological systems can be compensated for, by low energy requirements and reduced initial investment costs. A wide range of microorganisms produce hydrogen as a part of mechanisms for disposing of electrons that are generated during metabolic reactions:

$$2e^- \quad + \quad 2H_3O^+ \quad \leftrightarrow \quad H_2 + H_2O$$

Hydrated hydrogen ion
(hydronium ion)

The generation of hydrogen using microorganisms is still very much in infancy. However there are three possible routes of hydrogen production through microbes: 1) Biophotolysis of water, 2) Photoreduction of organic compounds, 3) Fermentation of organic compounds.

1. Biophotolysis of water involves splitting water using light energy and does not require an exogenous substrate. This can be performed using photosynthetic systems, such as algal chloroplasts, which may be regarded as solar cells. In vivo, the energy generated is normally used to form reduced nicotinamide adenine dinucleotide phosphate (NADPH). However, in the presence of a bacterial hydrogenase and an appropriate electron carrier, molecular hydrogen can he generated (Figure 6).

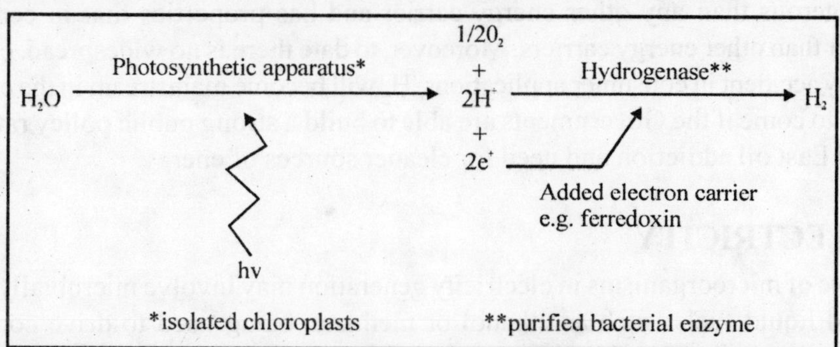


**Figure 6** Schematic representation of biophotolyis of water for $H_2$ production

2. Photoreduction, the light-dependent decomposition of organic compounds, is performed by photosynthetic bacteria. This is an anaerobic process requiring light and an exogenous organic substrate, which is inhibited by oxygen, dinitrogen and ammonium ions.

Formation of hydrogen is attributed to a nitrogenase that can reduce protons as well as dinitrogen. Members of the Chlorobiaceae, Chromatiaceae and Rhodospirillaceae carry out photoreduction. Those bacteria with most potential are probably the purple non-sulphur bacteria, such as *Rhodospirillium* species, which photometabolize organic acids.

3. Fermentation of organic compounds by many bacteria generates a small amount of hydrogen. For example, some enterobacteria produce hydrogen and $CO_2$ by cleaving formate, and in clostridia it is produced from reduced ferredoxin. Theoretically, 4 mol of hydrogen could be generated from each mole of glucose, which represents only a 33% energy yield. However, most organisms produce much less. Consequently, there is little possibility for the commercial production of hydrogen via this route in the near future.

$$C_6H_{12}O_6 + 2H_2O \rightarrow 2CH_3COO^- + 2H^+ + 2CO_2 + 4H_2$$

Biological hydrogen production is currently the most challenging area of biotechnology with respect to environmental problems but the yields are very low. The future of biological hydrogen production depends not only on research advances, i.e. improvement in efficiency through genetically engineering microorganisms and/or the development of bioreactors, but also on economic considerations (the cost of fossil fuels), social acceptance, and the development of hydrogen energy systems. There is a long way for this technology before we actually start driving microbe driven cars.

Bottom line in the roadmap for hydrogen policy of many countries indicates that $H_2$ has been produced extensively since World war - II with 9 million tons annual production. There are over 600 miles of pipelines and it is routinely transported by truck. There have been some accidents during transport, but not many. Hydrogen can be safer than current fuels with many favorable properties. There has been a solid track record in many countries. Extensive analysis by several research groups does not foresee any safety issues that would warrant cessation of hydrogen as a vehicle fuel. They also suggest that hydrogen is no more or less dangerous than any other energy carrier and has properties that in certain areas make it safer than other energy carriers. Moreover, to date there is no widespread, established record of any accident in consumer applications. $H_2$ will become mainstream in the community in the years to come if the Governments are able to build a strong public policy rationale for ending Mid East oil addiction and need for cleaner sources of energy.

## 7. BIO-ELECTRICITY

The role of microorganisms in electricity generation may involve microbially produced gaseous and liquid fuels, such as ethanol or methane, being used to drive conventional mechanical generators. The electricity generated by microbial based systems be used as a transportation fuel to power battery electric vehicles (BEVs), fuel cell vehicles (FCVs) and in hybrid-electric vehicles (HEVs). Alternatively, direct generation may be used in the form of microbial fuel cells, but this is still in the early stages of development. Possible routes involve microorganisms or microbial enzymes incorporated within fuel cells (Figure 7). A typical microbial fuel cell converts chemical energy to electrical energy by the catalytic reaction of microorganisms or their enzymes and consists of anode and cathode compartments separated by a cation specific membrane. In the anode compartment, fuel is oxidized by microorganisms or their enzymes, generating electrons and protons. Electrons are transferred to the cathode compartment through the circuit, and the protons through the membrane. Electrons and protons are consumed in the cathode compartment reducing oxygen to water. As most organic substrates undergo combustion with the evolution of energy, the biocatalyzed oxidation of organic substances by oxygen or other oxidizers at two-electrode interfaces provides a means for the conversion of chemical to electrical energy. Abundant organic raw materials such as methanol, organic acids or glucose can be used as substrates for the oxidation process, and molecular oxygen or $H_2O_2$ can act as the substrate being reduced. Intermediate formation of hydrogen as a potential fuel as possible as well.

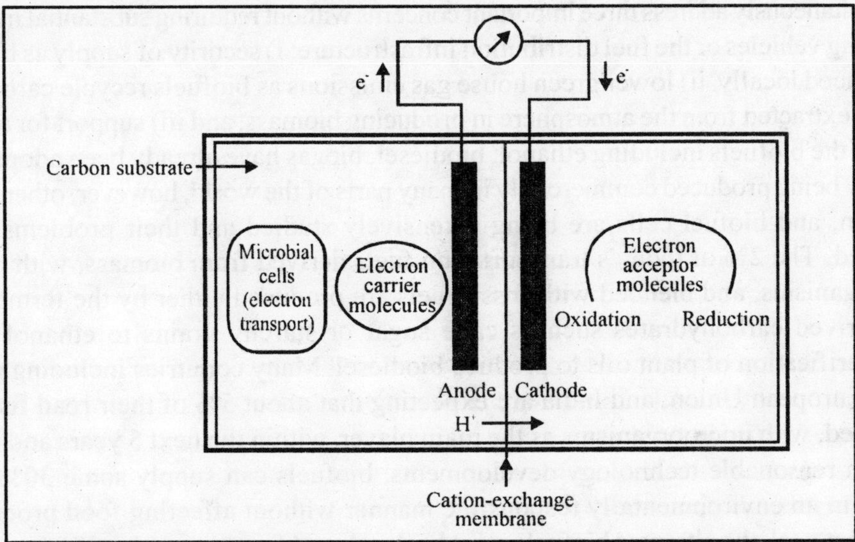


**Figure 7** Schematic representation of a typical microbial fuel cell

Microbial fuel cells can use biocatalysts, enzymes or even whole cell organisms in one of two ways. Either (i) the biocatalysts can generate the fuel substrates for the cell by biocatalytic transformations or metabolic processes, or (ii) the biocatalysts may participate in the electron transfer chain between the fuel substrates and the electrode surfaces. Unfortunately, most redox enzymes do not take part in direct electron transfer with conductive supports, and therefore a variety of electron mediators (electron relays) are used for the electrical contacting of the biocatalyst and the electrode. Recently, novel approaches have been developed for the functionalization of electrode surfaces with monolayers and multilayers consisting of redox enzymes, electrocatalysts and bioelectrocatalysts that stimulate electrochemical transformations at the electrode interfaces. The assembly of electrically contacted bioactive monolayer electrodes could be advantageous for biofuel cell applications as the biocatalyst and electrode support are integrated. Enzyme-based systems are preferred, as electron transfer between whole cells and electrodes is generally less efficient. In some cases, immobilized enzymes may be used. Possible candidates are microbial dehydrogenases coupled to electrode systems and catalysing the interconversion of hydrogen and electricity. Also, there is the possibility that phototrophic microorganisms, or their photoactive systems, could directly convert sunlight to electricity. For example, using artificial membranes incorporating bacteriorhodopsin-based systems from archaeans, e.g. *Halobacterium halobium.* Such systems facilitate the light-dependent translocation of protons and the resulting transmembrane electrochemical gradient created could be used to generate electricity.

## 8. CONCLUSIONS AND FUTURE OUTLOOK

Liquid hydrocarbons have been the main transport fuel during the last century because of their high energy density and handling convenience, however, the current reliance on petro-based fuels is not sustainable Although fossil fuels will be available for few decades, producing alternate fuels (biofuels) from biomass using microorganisms or their enzymes can simultaneously address three important concerns without requiring substantial modification of existing vehicles or the fuel distribution infrastructure: i) security of supply as biofuels can be produced locally, ii) lower green house gas emissions as biofuels recycle carbon dioxide that was extracted from the atmosphere in producing biomass, and iii) support for agriculture. Some of the biofuels including ethanol, biodiesel, biogas have already been adopted and are currently being produced commercially in many parts of the world, however, others including hydrogen, and biofuel cells are being extensively studied and their problems are being addressed. The 2% of today's transportation fuels derived from biomass, with the help of microorganisms, and blended with fossil fuels are produced either by the fermentation of food derived carbohydrates such as cane sugar or starchy grains to ethanol or by the transesterification of plant oils to produce biodiesel. Many countries including the United States, European Union, and India are expecting that about 5% of their road fuels will be bioderived, with microorganisms as the main player, within the next 5 years and it is hoped that with reasonable technology developments, biofuels can supply some 30% of global demand in an environmentally responsible manner without affecting food production. To realize that goal, the alternate biofuels must be developed from some identified energy crops, separately and distinctly from food. This is a multidisciplinary task in which microbiologists, biochemists, biotechnologists, agronomists, chemical engineers, fuel specialists, and social scientists must work to integrate and optimize several currently disjoint activities. In some cases, the yields of the alternate fuels from biomass are very less and demand the development of some novel microbial strains or enzymes for their production. Advances in genetics, biotechnology, process chemistry, and engineering are expected to lead to a new manufacturing concept for converting renewable biomass to valuable fuels and products, generally referred to as 'biorefinery'. There are major technological challenges in realizing these goals. It will be important to increase the yield and environmental range of energy crops while reducing agricultural inputs. Plant development, chemical composition, tolerance of biotic and abiotic stresses, and nutrient requirements are important traits to be manipulated. The combination of modern breeding and transgenic techniques should result in achievements greater than those of the Green Revolution in food crops, and in far less time. The cost of biomass transport determines the supply area of a biofuels processing facility and thus its scale and economics. Fuel production from the lignocellulosic biomass, available in abundance in every corner of the world will be a very important improvement. There is substantial technology "headroom" for alternate fuels to enhance energy security, reduce emissions, and provide economical transport. It exists largely because the world's scientific and engineering skills have not yet been focused coherently on the challenges involved. It is now time for becoming self sufficient in alternate fuels during the 21st century through a coordination of government, university, and industrial R&D efforts, facilitated by responsible public policies. The integration of agroenergy crops and biorefinery manufacturing technologies thus offers the potential for the development of sustainable biopower and biomaterials that will lead to a new manufacturing paradigm during the 21st century.

## 9. FURTHER READING

Koonin, S.E. (2006). Getting serious about biofuels. Science. 311: 435.

Dewulf, J., Van Langenhove, H. and Van De Velde, B. (2005). Exergy-based efficiency and renewability assessment of biofuel production. Environ Sci Technol. 39:3878-3882.

MacLean, H.L., Lave, L.B., Lankey, R. and Joshi, S. (2000). A life-cycle comparison of alternative automobile fuels. J Air Waste Manag Assoc. 50:1769-1779.

McLaren, J.S. (2005). Crop biotechnology provides an opportunity to develop a sustainable future. Trends Biotechnol. 23:339-342

Ragauskas. A.J., Williams, C.K., Davison, B.H., Britovsek, G., Cairney, J., Eckert, C.A., Frederick, W.J. Jr, Hallett, J.P., Leak, ,D.J., Liotta, C.L., Mielenz, J.R., Murphy, R., Templer, R., Tschaplinski, T. (2006). The path forward for biofuels and biomaterials. Science. 311:484-489.

Varma, A. and Behera, B. (2003). Green Energy : Biomass Processing and Technology. Capital Publishing Company, New Delhi

Waites, M.J., Morgan, N.L., Rockey, J.S. and Higton, G. (2002). Fuels and industrial chemicals. In: Industrial Microbiology: An introduction. Blackwell Science Ltd., USA. pp. 144-164.

□□□

corner of the world will be a very important improvement. There is substantial technology "headroom" for alternate fuels to enhance energy security, reduce emissions and provide economical transport fuels largely, because the world's scientific and engineering skills have not yet been focused coherently on the challenges involved. It is now time for becoming self sufficient in alternate fuels during the 21st century through a coordination of government, universities and industrial R&D efforts, facilitated by responsible public policies. The integration of agroenergy crops and biorefinery manufacturing technologies thus offers the potential for the development of sustainable biopower and biomaterials that will lead to a new manufacturing paradigm during the 21st century.

## 9. FURTHER READING

Koonin, S.E. (2006). Getting serious about biofuels. Science 311:435.

Dewulf, J., Van Langenhove, H. and Van De Velde, B. (2005). Exergy-based efficiency and renewability assessment of biofuel production. Environ Sci Technol. 39:3878-3882.

MacLean, H.L., Lave, L.B., Lankey, R. and Joshi, S. (2000). A life-cycle comparison of alternative automobile fuels. J Air Waste Manag Assoc. 50:1769-1779.

McLaren, J.S. (2005). Crop biotechnology provides an opportunity to develop a sustainable future. Trends Biotechnol. 23:339-342.

Ragauskas, A.J., Williams, C.K., Davison, B.H., Britovsek, G., Cairney, J., Eckert, C.A., Frederick, W.J. Jr, Hallett, J.P., Leak, D.J., Liotta, C.L., Mielenz, J.R., Murphy, R., Templer, R., Tschaplinski, T. (2006). The path forward for biofuels and biomaterials. Science 311:484-489.

Verma, Anand and Baheta, B. (2003). Green Energy : Biomass Processing and Technology. Capital Publishing Company, New Delhi.

Waites, M.J., Morgan, N.L., Rockey, J.S. and Higton, G. (2002). Fuels and industrial chemicals. In: Industrial Microbiology: An Introduction. Blackwell Science Ltd., USA. pp. 144-164.

# 7

# Microbes and Ethanol Production

**S.K. SONI**

*Department of Microbiology, Panjab University, Chandigarh-160 014*

## 1. INTRODUCTION

Alcohol production was the first fermentation known to mankind and today it is the largest volume industrial fermentation. Distilleries began to appear in Europe in the middle of the seventeenth century. There is semantic confusion with regard to the term ethanol. Very often the term is used as a synonym for alcoholic beverages. This is misleading, even though ethanol may be used as a raw material for the production of spirits. Ethanol is a clear, colourless, flammable oxygenated hydrocarbon, with the chemical formula $C_2H_5OH$. Even though the definition is fairly straight forward, there are various categories for describing a particular type of ethyl alcohol which can be classified by feedstock from which it is made, by composition and by end use.

The feedstocks and therefore the processes by which ethanol can be produced are diverse. Synthetic alcohol may be derived from crude oil or gas and coal. Agricultural alcohol may be distilled from grains, molasses, fruit, sugar cane juice, cellulose and numerous other sources. Both products, fermentation and synthetic alcohol are chemically identical. Synthetic alcohol is concentrated in the hands of a couple of mostly multi-national companies such as Sasol with operations in South Africa and Germany, SADAF of Saudi Arabia, a 50:50 joint venture between Shell of the UK and Netherlands, the Saudi Arabian Basic Industries Corporation, and BP of the UK as well as Equistar in the US. However, on a global scale synthetic feedstocks play a minor role. In 2003, less than 5% of overall output was accounted for by synthetic feedstocks. More than 95% came from agricultural crops and given the strong interest in fuel ethanol production world-wide this share can be expected to grow in the future.

Another distinction which is of importance in the field of ethanol is the one between anhydrous and hydrous alcohol. Anhydrous alcohol is free of water and at least 99% pure. This ethanol may be used in fuel blends. Hydrous alcohol on the other hand contains some water and usually has a purity of 96%. In Brazil, this ethanol is being used as a 100% gasoline substitute in cars with dedicated engines. The distinction between anhydrous and hydrous alcohol is of relevance not only in the fuel sector but may be regarded as the basic quality distinction in the ethanol market.

The final distinction which is necessary in order to understand the dynamics of the world ethanol market is by end-use. Certainly the oldest and largest form of use of alcohol is that of a beverage. The most important market for ethanol as an industrial application are solvents, primarily utilized in the production of paints and coatings, pharmaceuticals, adhesives inks and other products. Ethanol represents one of the most important oxygenated solvents in this category. Production and consumption is concentrated in the industrialized countries in Northern America, Europe and Asia. It is the only market where synthetic ethanol producers hold a significant market share. The last usage category is fuel alcohol. As mentioned before, fuel alcohol is either used in blends, for example in gasohol or diesohol, or in its pure form. However, at present Brazil is the only country that uses ethanol as a 100% substitute for gasoline. The history of ethanol as a fuel dates back to the early days of the automobile. However, cheap petrol quickly replaced ethanol as the fuel of choice and it was not until the early 1980s, when the Brazilian government launched the Proalcohol program, that ethanol made a come back to the market place. It may be estimated that fuel ethanol accounted for roughly 70% of world ethyl alcohol production in 2003. As can be seen from Figure 1, this share is forecast to rise to over 80% by the end of the decade. However, this projection only holds if the sometimes ambitious fuel ethanol programs which have been proposed in the last couple of years, come to fruition. Therefore, the figures presented here, represent more a potential than a hard forecast.

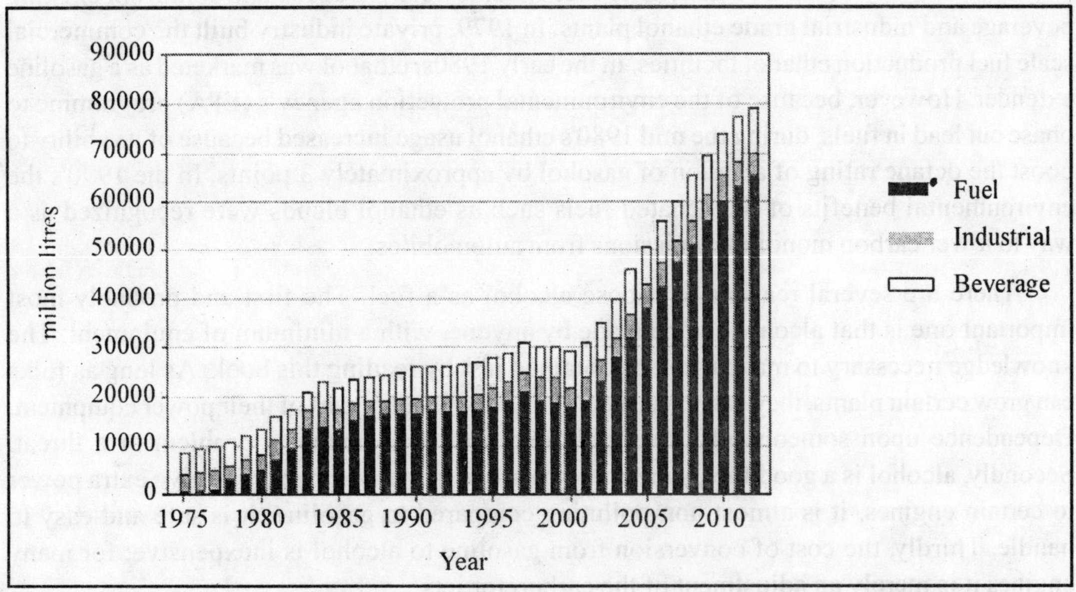


**Figure 1** Ethanol production by type in million litres. The industrial alcohol market is the smallest of the three which appears to remain constant. Demand for distilled spirits in most developed countries is stagnating or even declining, due to increased health awareness. But the demand for fuel alcohol is forecasted to show a significant increase in the future.

Ethyl alcohol as an automotive fuel can be used in two ways. It replaces gasoline outright in dedicated internal combustion engines and is an effective "octane booster" when mixed with gasoline in blends of 5 to 30%. In this case no engine modifications are required. These blends achieve the same octane boosting or anti-knock effect as petroleum derived aromatics like benzene or metallic additives like lead. Ethanol easily blends with gasoline but not with diesel. If the diesohol blend is to obtain more than 3% ethanol special emulsifiers are needed.

Ethanol, ethyl alcohol, alcohol, spirit, denatured spirit, rectified spirit, neutral spirit, extra-neutral spirit, absolute alcohol …., there are myriad descriptions for this agriculture-based product. A globally traded commodity, ethanol, slakes the thirst of many in Europe and finds its way in pharmaceutical and chemical industries, across the world. Large volumes of alcohol for use as chemical feed stock by fermentation have been produced commercially since World War II. After the war, the availability of low cost petroleum ushered in the growth era of the petrochemical industry. Ethylene and other petroleum-based compounds became building blocks for the chemical industry. It was not until the oil price stocks and energy shortages of 1973 and 1979 that the Carter administration (USA) established government policies to provide incentives for the establishment of alternative energy sources, including ethanol-gasoline blends (gasohol). The availability of surplus corn from increased yields and the Russian grain embargo also provided impetus for establishing a fuel ethanol programme.

In 1978, only 10 million gallons of fuel ethanol were produced and sold from existing beverage and industrial grade ethanol plants. In 1979, private industry built the commercial scale fuel production ethanol facilities. In the early 1980s, ethanol was marketed as a gasoline extender. However, because of the environmental protection agency's (EPA) programme to phase out lead in fuels, during the mid 1980's ethanol usage increased because of its ability to boost the octane rating of a gallon of gasohol by approximately 3 points. In the 1990's the environmental benefits of oxygenated fuels such as ethanol blends were recognized as a way to lower carbon monoxide emissions from automobiles.

There are several reasons to choose alcohol as a fuel. The first and probably most important one is that alcohol can be made by anyone, with a minimum of equipment. The knowledge necessary to make it can be obtained just by reading this book. As long as folks can grow certain plants, they can make alcohol fuel to run all or part of their power equipment. Dependence upon someone else to supply that fuel is no longer a problem or a threat. Secondly, alcohol is a good fuel, superior to gasoline in many ways: it can give extra power to certain engines, it is almost non-polluting compared to gasoline, it is safe and easy to handle. Thirdly, the cost of conversion from gasoline to alcohol is inexpensive: for many engines it is merely an adjustment of the carburetor jets.

There are several benefits of using ethanol as a fuel as it is a much cleaner fuel than petrol (gasoline). Some of these are as follows: i) It is a renewable fuel made from plants, ii) It is not a fossil-fuel: manufacturing it and burning it does not increase the greenhouse effect, iii) It provides high octane at low cost as an alternative to harmful fuel additives, iv) Its blends can be used in all petrol engines without modifications, v) It is biodegradable without harmful effects on the environment, vi) It significantly reduces emissions of hydrocarbons, a major contributor to the depletion of the ozone layer, vii) Its high oxygen content reduces carbon monoxide levels more than any other oxygenate by 25-30%, viii) Ethanol blends dramatically reduce nitrogen oxide emissions by up to 20%, ix) High-level ethanol blends can reduce emissions of volatile organic compounds (VOCs), the major sources of ground-level ozone formation, by 30% or more, x) As an octane enhancer, ethanol can cut emissions of cancer-causing benzene and butadiene by more than 50%, xi) It brings about a significant decrease in sulphur dioxide and particulate matter (PM) emissions.

## 2. RAW MATERIALS FOR ALCOHOL PRODUCTION

Feedstock suitable for use in ethanol production via fermentation must contain sugars, starches, or cellulose that may readily be convertible to fermentable sugars. The usual sources of raw material for alcohol production include i) sugar-producing plants such as sugar beets, sugarcane, fruits, and others, ii) starch-containing cereal grains such as corn, wheat, rye, barley, milo (sorghum grains), rice, potatoes of all kinds, other high-starch vegetables and iii) cellulose-containing agricultural residues, wood and waste sulphite liquor. Companies presently producing ethanol in the world vary by size, type of feed stock and technology. The substrates that have been classified into various categories and the characteristics of various feedstock are indicated in Tables 1 and 2. Current economics dictate that the sugar cane juice, molasses and cereal grains make the primary feedstock materials used in various parts of the world. Major raw materials used for ethanol production are discussed below:

**Table 1** Various substrates used for ethanol production

| | |
|---|---|
| Directly Fermentable | Sugarcane juice, fruit juices, sugar beet juice, molasses, sucrose, sweet sorghum, honey etc. |
| Easily hydrolysable substrates | Starchy materials-grains, tubers |
| Difficult to hydrolyse substrates | Lignoellulosics |

**Table 2** Characteristics of feedstock generally used for alcohol production

| Type of feedstock | Treatment needed before fermentation | Advantages | Disadvantages |
|---|---|---|---|
| Sugar crops (sugar cane, sugar beets, sweet sorghum, Jerusalem artichoke, fodder beets, fruits etc) | Squeezing, crushing to extract sugar | Preparation is simple<br>High yields of ethanol per acre<br>Crop co-products have value as fuel, livestock feed or soil amendment | Storage may result in loss of sugar<br>Cultivation practices are not wide-spread in non-conventional crops |
| Starch Crops:<br>Grains (corn, wheat, rice, sorghum, barley, oats, millet, rye etc)<br>Tubers: (potatoes, sweet potato, cassava etc) | Milling and enzymatic hydrolysis involving gelatinization, liquefaction and saccharification | Storage techniques are well developed and simple<br>Cultivation practices are widespread with grains and potato<br>Enzymatic hydrolysis is cheap and simple<br>Co-product, used as livestock feed is rich in protein | Preparation involves additional equipment, enzymes, labour and energy costs |
| Cellulosic:<br>Crop residues (sugarcane bagasse, corn stover, wheat straw etc)<br>Forages (alfalfa, Sudan grass, forage sorghum) | Milling and acid hydrolysis | Availability is widespread<br>Use involves no integration with livestock feed market | Preparation involves additional equipment, labour and energy costs<br>No commercially cost-effective process exists for hydrolysis |

## 2.1. Sugary substrates

These can be directly fermented without any pre-treatment, simply after squeezing from the crops or after dilution with water as in case of molasses.

**2.1.1.** *Sugarcane:* Sugar cane juice and the important co-products of sugar production, molasses are commonly used for alcohol production in various parts of the world. Molasses is thick brown fluid containing sugary residue which is left behind after primary sugar recovery. Of world's total sugar production of over 110 million tons, more than 65 million tons comes from cane. Cane sugar is produced in South and Central America, Africa, Asia and Oceania. Brazil has an area of 37,000 square kilometer under sugar cane plantation for ethanol production. Total annual sugar cane production in India is about 270 million tons. The juice from the crushed cane is weighed, adjusted for pH and clarified before evaporation and crystallization, then centrifuged to separate the molasses from the sugar crystals. Cane juice contains 15-16% dissolved solids of which 85% is sucrose. The juice is partially evaporated before fermentation is carried out.

**2.1.2.** *Molasses:* One of the cheapest sources of carbohydrates. In addition to sugar, molasses contains nitrogenous substances, vitamins and trace elements. Composition is variable depending upon the raw material used for sugar production. Quality of molasses varies depending upon the location, climatic conditions and production process of the sugar factory. In addition to conventional molasses, a residue from starch saccharification which accumulates after the crystallization of glucose is also used as a fermentation substrate, known as "Hydrol" molasses. The term 'molasses' is used for the viscous, sugary co-product of sugar refining, deriving from sugar cane, sugar beet and similar products. In refining cane juice for sugar production, 42 kg molasses (50% on sugar basis) are obtained from each ton of cane or 2.2 tons/hectare. Typical molasses compositions are given in Table 3. In addition to its use as a principal raw material for the production of alcohol, as used in various countries including India, molasses is also used in citric acid fermentation. Molasses has the advantage over the cane juice that it can be stored for longer periods, under optimal conditions for long enough to carry over one season to the next. After fermentation, the high ash content of molasses, which contains a substantial proportion of potassium, is of high value as fertilizer.

Molasses is the predominant raw material for alcohol production in the Asian region with the exception of China and Saudi Arabia, which produce alcohol from grains and synthetic route, respectively. Japan, which produces very small quantities of molasses, is the largest importer of alcohol ~450 million liters per year. It is apparent that molasses supply becomes a very potent force with the sugar industry which is likely to play it to its advantage. In Philippines, this factor was dealt with head-on by a distillery which faced shortage of molasses supply. They have set up a cassava/cane juice/molasses triple mode distillery to handle the vagaries of molasses supply and price. In India also, this trend has begun. Currently dual route plant operating on molasses as well as grain as a feedstock, are being set up.

**Table 3** Typical molasses composition

| Component | Range (%) | Average (%) |
|---|---|---|
| Fermentable sugars | 48-60 | 52 |
| a) Sucrose | 36 | |
| b) Fructose | 9 | |
| c) Glucose | 7 | |
| Other organics | 9-12 | 10.5 |
| Inorganic ash | 10-15 | 12.5 |
| Solids | 70-85 | 75 |
| Water | 15-30 | 25 |

Molasses production is showing gradual increase. The annual production levels in Asia are 46.5 million tons. India was more or less absent from the world's molasses market in first half of 1990's, when fortunes of the sugar industry turned with the steep increase in production which started in 1994/95. In 1996/97, India recorded as the world's largest supplier of molasses with exports of 1.5 million tons and production of 8.3 million tons with 42% share in the total asian production (Figure 2 )

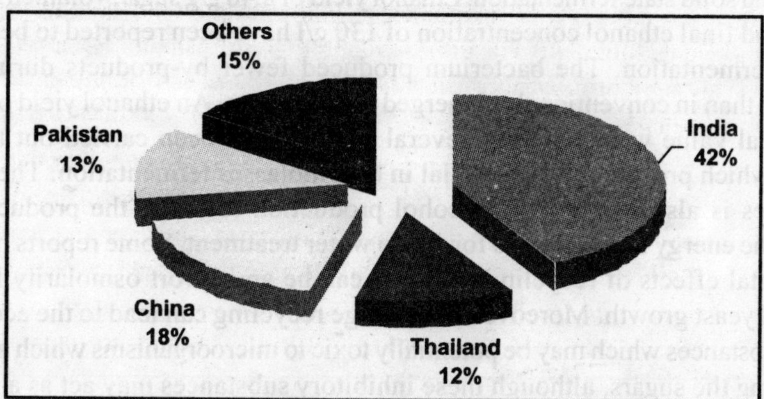


**Figure 2** The largest producers of molasses in Asia

The use of molasses is advantageous as it is a cheap source of fermentable sugars and is a byproduct; its prices are not politically controlled. The sugars in molasses can be easily utilized by most of the microorganisms and it also contains a number of other important nutrients such as vitamins, some elements such as N, P, Ca, Mg, K, Si, Al and Fe. However, there is a lack of consistency and batch-to-batch variation is there in the molasses composition. Some batches contain too much metal that pretreatment may be needed for some type of fermentation that are sensitive to specific metals. The quality of molasses is deteriorating, especially in India, due to decreased sugar content and increased solid content due to technological advancements in sugar industry aimed at recovering the last possible sugar crystal. The cost of molasses has also risen many a fold during the last decade due to Indian Government's decontrol policy. Some of the factors influencing the molasses composition and properties include soil composition in which the cane or beet is grown, harvesting practices, process of sugar manufacture, postproduction treatment to molasses and the storage conditions of molasses.

**2.1.3.** *Sugar beet:* Sugar beet is widely used in European countries for the production of sugar and alcohol. It is grown extensively in UK and Europe, however, tropical varieties are being developed for high yield of production in India and other tropical countries. Annual world wide production of sugar beet is 300 million tons. Sugar content of sugar beet varies from 14-19%. The sugar containing juice extracted from the crushed beet is crystallized resulting in the formation of molasses. Beet molasses has similar use as that of cane molasses and generally serving as a feed stock for alcohol fermentation, the yield of which is 90-100 l/ton. The short harvest season limits the use of sugar beet juice and its molasses for alcohol production. Two hundred and fifty kilograms of wet pulp is obtained from each ton of sugar beet. Sugar beet pulp or bagasse is a cellulosic fibrous residue. It is not as yet, commercially processed for alcohol, but it is expected that the hydrolysis and fermentation of pulp can produce 40 l alcohol/ton of beet. Several attempts have also been made to ferment the beet molasses by some alternative microorganism in place of the conventional distiller's yeast, *Saccharomyces cerevisiae*. The conversion of sugars present in sugar beet particles to ethanol by bacterium *Zymomonas mobilis* has also been reported in both conventional submerged and solid state fermentation. Ethanol yield of 0.48 g/g sugar, volumetric productivity of 12 g/l/h and final ethanol concentration of 130 g/l have been reported to be achieved in a solid state fermentation. The bacterium produced fewer by-products during solid state fermentation than in conventional submerged fermentation. An ethanol yield of upto 95% of the theoretical value was obtained. Several studies have been carried out to recycle the spent wash which proved to be beneficial in beet molasses fermentation. The stillage from beet molasses is also recycled for alcohol production to lower the production costs by decreasing the energy requirements for waste water treatment. Some reports have indicated the detrimental effects of recyclings as there can be an in wort osmolarity leading to the inhibition of yeast growth. Moreover, the stillage recycling can lead to the accumulation of inhibitory substances which may be potentially toxic to microorganisms which are responsible for fermenting the sugars, although these inhibitory substances may act as a shield against the contaminants. The production of alcohol from sugar beet molasses has also been reported without heat or filter sterilization wherein n-butyl p-hydroxy benzoate has been used as an antimicrobial substance in the sugar beet molasses medium. The yeast cells, immobilized in calcium alginate, completely utilized molasses containing upto 25% of sugar with an ethanol yield of 10.58%, equal to 83% of the theoretical value. The addition of EDTA, potassium ferrocyanide and zeolite-X into sugar beet molasses medium improves ethanol production. The effect is more pronounced when the three substances are added to fermentation medium rather than to the growth medium.

**2.1.4.** *Sweet sorghum:* Sweet sorghum is a name given to varieties of a species of sorghum: *Sorghum bicolor*. This crop has been cultivated on a small scale in the past for production of table syrup, but other varieties can be grown for production of sugar. The most common types of sorghum species are those used for production of grain. There are two advantages of sweet sorghum over sugar cane: its great tolerance to a wide range of climatic and soil conditions, and its relatively high yield of ethanol per acre. In addition, the plant can be harvested in three ways: (1) the whole plant can be harvested and stored as such; (2) it can be cut into short lengths (about 4 inches long) when juice extraction is carried out immediately; and (3) it can be harvested and chopped for ensilage. Since many varieties of sweet sorghum bear significant quantities of grain, the harvesting procedure will have to take this fact into account. The leaves and fibrous residue of sweet sorghum contain large quantities of protein, making the residue from the extraction of juice or from fermentation a valuable livestock feed and can also be used as boiler feed.

**2.1.5.** *Jerusalem artichokes:* The Jerusalem artichoke has shown excellent potential as an alternative sugar crop. A member of the sunflower family, this crop is native to North America and well-adapted to northern climates. Like the sugar beet, the Jerusalem artichoke produces sugar in the top growth and stores it in the roots and tuber. It can grow in a variety of soils, and it is not demanding of soil fertility. The Jerusalem artichoke is a perennial; small tubers left in the field will produce the next season's crop, so no plowing or seeding is necessary.

Although the Jerusalem artichoke traditionally has been grown for the tuber, an alternative to harvesting the tuber does exist. It has been noted that the majority of the sugar produced in the leaves does not enter the tuber until the plant has nearly reached the end of its productive life. Thus, it may be possible to harvest the Jerusalem artichoke when the sugar content in the stalk reaches a maximum, thereby avoiding harvesting the tuber. In this case, the harvesting equipment and procedures are essentially the same as for harvesting sweet sorghum or corn for ensilage.

**2.1.6.** *Fodder beets:* Another promising sugar crop which presently is being developed in New Zealand is the fodder beet. The fodder beet is a high yielding forage crop obtained by crossing two other beet species, sugar beets and mangolds. It is similar in most agronomic respects to sugar beets. The attraction of this crop lies in its higher yield of fermentable sugars per acre relative to sugar beets and its comparatively high resistance to loss of fermentable sugars during storage. Culture of fodder beets is also less demanding than sugar beets.

**2.1.7.** *Fruit crops:* Fruit crops including grapes, apricots, peaches, and pears are another type of feedstock in the sugar crop category. Typically, fruit crops such as grapes are used as the feedstock in wine production. These crops are not likely to be used as feedstocks for production of fuel-grade ethanol because of their high market value for direct human consumption. However, the co-products of processing fruit crops are likely to be used as feedstocks because fermentation is an economical method for reducing the potential environmental impact of untreated wastes containing fermentable sugars.

## 2.2. Starchy residues (cereal grains)

Annual world cereal production amounts to be more than two billion metric tons. Major cereal crops produced worldwide include wheat, rice, maize and barley. Other major cereal crops produced include sorghum, oats, millet and rye. Asia, America, and Europe produce more than 80 percent of the world's cereal grains. Wheat, rice, sorghum, and millet are produced in large quantities in Asia; corn and sorghum are principal crops in America, and barley, oats and rye are major crops in the former USSR and Europe. Apart from human and animal consumption, a major part of these cereal grains are used for the industrial production of starch and its derivatives, or for alcohol. The world wide production of major cereal crops from 1961 to 1996 are presented in Figure 3. Various common cereal grains employed for commercial production of ethanol in various parts of the world are discussed below:

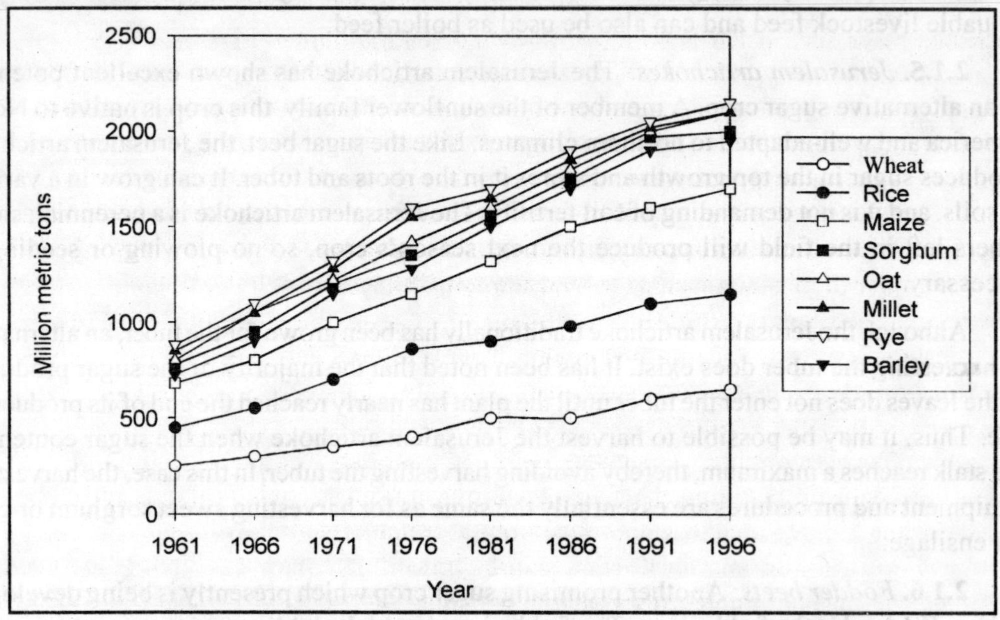


**Figure 3** Global production of major cereal crops from 1961-1996

**2.2.1.** *Maize:* Corn was originally cultivated in central America and became the staple of the Incas of Peru, the Mayas and Aztecs of Mexico and early cliff dwellers of the American Southwest. Columbus brought corn back to Europe where it became a popular crop in the south. Different types of maize are classified on the basis of their protein content and the hardness of the kernel. These include pop, flint, flour, Indian and sweet corns. Much of the niacin in corn is in a bound form and this led to pellagra in areas where corn became the food staple. It was not until the late 1920s that pellagra was identified as a vitamin deficiency. The traditional practice by early American natives, of boiling corn in 5 percent lime or ashes, releases bound niacin making it available as a nutrient. Maize is a major cereal crop in Canada and USA and a minor crop in UK. Hybrid corn varieties produce about 210 million tonnes of corn in north America every year which is used for human consumption, manufacture of starch, corn syrup (glucose) and high fructose corn syrup (HFCS). The annual production of maize in India is approximately 10 million tonnes. In a kernel of maize corn (Table 4), the various constituents are starch (72%); protein (10%); oil (4%); and fibre (14%). For fuel alcohol production, the separation into various components is unnecessary, and is not widely practicised. For most single product plants such as dry ground corn alcohol plants, the kernel corns are first steeped to swell and then ground, saccharified by amylases and the resulting sugar is fermented.

**Table 4** Chemical composition of corn (percent of dry matter)

| | Range | Average |
|---|---|---|
| Starch | 61.0 – 78.0 | 71.7 |
| Protein | 6.0 – 12.0 | 9.5 |
| Fat | 3.1 – 5.7 | 4.3 |
| Ash | 1.1 – 3.9 | 1.4 |
| Cellulose | 3.3 – 4.3 | 3.3 |
| Pentosans | 5.8 – 6.6 | 6.2 |
| Sugars | 1.0 – 3.0 | 2. |
| Other | 0.5-1.5 | 1.0 |

For alcohol production, whole maize corn kernels are steeped in warm water (50°C) for 24 h before grinding. The swollen ground corn is then liquefied by α-amylase at 95°C for 1 h at pH 6.5. The corn is then heated to 105°C by direct steam injection for 10 min followed by heating to 140°C for 2 min for gelatinization, then cooling to 90°C and held at that temperature for 1 h. The treated corn is then cooled to 50-53°C, pH is lowered to 4.5-5.0 and 0.5ml/l glucoamylase is added for saccharification. The addition of glucoamylase saccharifies the starch in the kernels more or less completely within 1-2 h at 50-53°C. The slurry is then cooled and yeast is added. The alcohol yield from such a process is 450 liter / ton of dry ground corn. The spent wash recycling in the fermentation of corn mashes increases the overall mashing and fermentation efficiencies and thus the alcohol yield.

**2.2.2.** *Wheat:* Wheat is one of the oldest of all cutivated plants. Today, there are more than 50000 cultivars of wheat in existence and as a result it can be grown in a relatively wide range of climatic conditions. Growing best in temperate climates, it is susceptible to disease in warm, humid regions and cannot be grown as far from the equator as can rye and oats. Wheat was brought to America early in the seventeenth century where it came to prominence in the Great Plains by 1855. Different types of wheat are classified based on planting season and endosperm composition. Wheat holds a special place amongst the cereals because upon mixing wheat flour with water, an elastic matrix called "gluten" required for the production of leavened breads is formed. "Hard wheats" contain relatively high levels of protein (10-14%), while the reverse is true for "soft wheats" (6-10%). High protein flours are best suited for pastas and breads, while flour from soft wheats is excellent for cakes and pastries, etc. Wheat is the second largest cereal crop of India mainly used as a major part human diet with annual productivity level of about 60 million tons. Wheat flour usually contains a starch content between 74-81%. About 5-10% of this is recovered as 'B' starch, and the remainder constitutes the 'A' starch or prime starch. The value of the prime starch is very high. For the production of alcohol the feed stock of choice is 'B' starch, which commands a lower price. For cereals, where the gluten does not have high value, dry milling, saccharification and fermentation of the whole grain is the preferable route to alcohol.

For the bioalcohol production, wheat starch 'B' is slurried in water at 15% dry solids and pH adjusted to 7.0. Then α-amylase is added and liquefaction is carried out at 85°C for 30 min under constant stirring. After liquefaction the slurry is cooled to 55°C, the pH lowered to 5.0 and glucoamylase is added. After 40 h at this temperature, the sugar solution contains about 130 g/l of glucose. To this solution, nutrients and yeast are added and after 24 h fermentation, alcohol concentration rises to 7.3% (v/v), the yield being 500 l ethanol/ton of starch 'B'. The wheat mashes contain growth limiting amounts of free amino nitrogen, so the fermentations by active dry yeast (*Saccharomyces cerevisiae*) is completed in 8 days at 20°C. Supplementing wheat mashes with yeast extract, casamino acids and glutamic acid stimulated the growth of the yeast and reduces the fermentation time. With 0.9% yeast extract as the supplement, the fermentation time is reduced from 8 days to 3 days and a final ethanol yield of 17.1% (v/v) is achieved. However, amino acids such as glycine reduces the cell yield and prolonged the fermentation time.

**2.2.3.** *Barley:* In the United States, barley is mostly used as feed, for brewing and alcohol production with only about 2 % used for human food. Barley flour is produced by abrasion dehulling, followed by milling of the "pearled" barley. In the UK, 2 million tons of barley is used each year in brewing and distilling, while another 6 million tons is used as animal fodder. In a dual enzyme process, whole barley grains are steeped in water for 18 h and ground. The ground barley is stirred in water at 20% solids, pH adjusted to 7.0, 0.1% α-amylase is added and the slurry is heated to 85°C for 45 min. The liquefied solution is then cooled to 50°C, pH adjusted to 5.0 and 0.1% glucoamylase added. The solution is held at 50°C for 48 h, pH adjusted to 5.5 and then cooled to 30°C and fermentation. Fermentation sugars after saccharification are around 147 g/l and the ethanol produced is 70 g/l or 9% (v/v), yield being 450 l/ton of barley. When saccharification and fermentation are combined, the liquefied starch is treated with 0.1% glucoamylase and fermented using yeast at 37°C, the results on the alcohol yield are the same. When saccharification and fermentation are separate operations, 60 h are required, whereas when the two steps were combined, alcohol production gets completed in 12 h.

**2.2.4.** *Malt extract:* An aqueous extract of malted barley is an excellent substrate for many fungi and actinomycetes. Dry malt extract contains 90-92% carbohydrates and is composed of hexoses (glucose, fructose), disaccharides (maltose, sucrose), trisaccharides (maltotriose) and dextrins (Table 5). Nitrogenous substances include proteins peptides, amino acids, purines, pyrimidines and vitamins. Amino acid composition varies according to grain used but proline always makes up about 50% of total amino acids.

Media containing malt extract must be carefully processed. Overheating in presence of low pH and high proportion of reducing sugars results in maillard reaction in which amino groups of amines, amino acids or proteins react with carbonyl groups of reducing sugars, aldehydes or ketones which lead to the formation of brown condensation products.

**Table 5** Typical composition of malt extract

| Component | % of dry weight |
|---|---|
| Maltose | 52.2 |
| Hexoses (glucose, fructose) | 19.1 |
| Sucrose | 1.8 |
| Dextrin | 15.0 |
| Other carbohydrates | 3.8 |
| Nitrogenous materials | 4.6 |
| Ash | 1.5 |
| Water content | 2.0 |

**2.2.5.** *Rice:* The second most abundant cereal crop originated in the Indian subcontinent and Africa. Today, 90 % of the world rice crop is grown in Asia. Its annual productivity level in India is over 80 million tons. Alexander the Great is credited with introducing rice to Europe around 300 BC. Growing rice requires more water than other cereal crops, although rice is a highly productive crop. There are several thousand rice cultivars which may differ in color, aroma and grain size. The main commercial distinction between rice types is the grain size, i.e. long, medium and short. Long grain rice, also called "Indian", tends to separate relatively easily on cooking and is dry and flakey. Short grain rice, also called "Japanese" is sticky, moist and firm when cooked. Unlike wheat, rice is most often consumed as grain rather than as a flour. Different grades of milling include brown rice (hull removed), unpolished rice (hull, bran and most of germ removed), and polished rice (aleurone layer removed from unpolished rice). Since polishing removes most of the lipid, the latter product is relatively stable during storage. The discovery that rice bran can alleviate beriberi led to the discovery of the vitamin thiamine. The traditional technique of parboiling rice in India and Pakistan (also called "converted rice") prior to milling improves the nutritional quality of the grain by allowing the B vitamins in the bran and germ to diffuse into the endosperm.

**2.2.6.** *Rye and mixed grains:* Rye is an impotant bread grain in Eastern Europe and Scandinavia. World production of rye is over 25 million tons. The main uses of rye is for bread and other foods, for animal feed, and/or rye whisky. The use of rye in not reported for industrial or fuel alcohol production. It can, however, be used as a component of mixed grains. While its starch content is lower than that of corn or barley, it is less expensive.

**2.2.7.** *Millet and sorghum:* Millet and sorghum are often grouped together because their growing conditions, processing and uses are similar. Millets are native to Africa or Asia and have been cultivated for more than 6,000 years. Millets grow well in regions with poor soils and are valued for their relatively high protein content among the cereals. Sorghum originated in East Africa and today is an important food crop in Africa, Asia, India and China where it is made into porridge, unleavened bread ("roti" in India) and beer. India is second largest producer of sorghum in the world with production of about 10–11 million tons from a total area of 12 million hectare. This crop is ideally suited for semi-arid agroclimatic regions of the country and, it gives reasonably good yield with minimal requirement of irrigation and fertilizers. On the other hand, cereals such as wheat and rice cannot withstand the harsh semi-arid climates. These crops also require fair amount of water and other inputs such as fertilizers and pesticides. Therefore, sorghum is one of the few cereals, which can be grown in semi-arid regions. However, demand for sorghum for human consumption is decreasing with enhanced socioeconomic status of population in general and easy availability of preferred cereals in sufficient quantities at affordable prices. Since sorghum must be cultivated in the semi-arid regions for fodder to feed the large cattle population of the country, industrial applications for this grain are needed so that sorghum cultivation becomes economically viable for farmers.

Whole grain sorghum flour has a relatively short shelf-life. The production of low-fat sorghum flour requires removal of about 20 % of the grain weight by abrasion. In North America, millet and sorghum are used primarily as livestock feed. *Sorghum vulgare* or *S.*

*bicolor* is a member of grass family. It bears starchy seeds upon its stalks and has ability to withstand adverse conditions of draught and water logging. The annual production of sorghum is about 75 million tons, grown on 48 million hectare. The leading producers are USA (31%), India (16%), China (10%), Argentina (10%), Mexico (9%) and Nigeria (5%). Over 95% of the total food use of sorghum occurs in Africa and Asia, where it serves as a staple food and constitutes a major source of energy and protein. About 97% of the total use of sorghum in developed countries is for animal feed. Nevertheless, sorghum is also a substrate for producing industrial alcohol, distilled spirits, starch, dextrose, syrups and edible oil. Traditional alcohol and non-alcoholic beverages are brewed using sorghum in parts of Africa and Asia. The use of sorghum, as an adjunct or as a substitute for barley in beer brewing has also been reported by several workers. The stalks of sorghum contain 14% sugars. It is very fast growing and two crops per year are cultivated in the most favourable areas. Yields of 35 tons/hectare, in addition to about 3 tons of seeds are recorded. The sorghum crop of 1 hectare allows production of 2800 l ethanol. Sweet sorghum, similar to the normal sorghum varieties, is also used in various parts of the world for alcohol production. However, in addition to the grains, the sweet sorghum variety contains sweet juice in the stalk materials. This juice is very similar to cane juice in characteristics and has between 12 to 16 % (w/w) sugar (sucrose). Sweet sorghum also contains a large amount of fibrous material in the stalks (similar to sugarcane), which generates equivalent bagasse for energy generation. Direct conversion of sorghum carbohydrates to ethanol by a mixed microbial culture has been reported by several workers who produced ethanol from sweet sorghum in submerged as well as solid state fermentation. One of the study involved simultaneous saccharification and fermentation of sweet sorghum carbohydrates to ethanol by *Fusarium oxysporum* alone or in mixed culture with *Saccharomyces cerevisiae* or *Zymomonas mobilis*. The process achieved its maximum value by the mixed culture of the fungus and the yeast. Under optimum conditions, ethanol yields and concentrations as high as 29.7 g of ethanol / 100 g of dry sorghum stalk and 7.5% (w/v), respectively, were obtained. Another study involved the production of bioethanol from sweet sorghum carbohydrates which were simultaneously saccharified and fermented to ethanol by a mixed culture of *Fusarium oxysporum* and *Saccharomyces cerevisiae*. The optimum yield of bioconversion and ethanol concentration was 5.2-8.4 g ethanol/100 g of fresh sorghum and 3.5-4.9 % (w/v) respectively. In all experiments, the ethanol yield exceeded the theoretical level, based on soluble sugars, by 20.0-32.1% due to the bioconversion of polysaccharides to ethanol.

**2.2.8.** *Raw cassava:* Cassava, perennial woody shrub is an annual crop. This is a low cost carbohydrate-producing crop, which finds applications in food as well as alcohol production. Brazil is the largest producer of cassava followed by, Thailand, Nigeria, Zaire and Indonesia. However, Thailand is a major exporter of cassava. The enlarged cassava roots contain maximum concentration of starch on dry weight basis among all food crops. The cassava is processed in different ways in different countries for food and for ethanol production.

## 2.3. Cellulosic materials (Crop residues)

The "backbone" of sugar and starch crops, the stalks and leaves, is composed mainly of cellulose. The individual six-carbon sugar units in cellulose are linked together in extremely long chains by a stronger chemical bond than exists in starch. As with starch, cellulose must be broken down into sugar units before it can be used by yeast to make ethanol. However, the breaking of the cellulose bonds is much more complex and costly than the breaking of the starch bonds. Breaking the cellulose into individual sugar units is complicated by the presence of lignin, a complex compound surrounding cellulose, which is even more resistant than cellulose to enzymatic or acidic pretreatment. Because of the high cost of converting liquefied cellulose into fermentable sugars, agricultural residues (as well as other crops having a high percentage of cellulose) are not yet a practical feedstock source for small ethanol plants. Current research may result in feasible cellulosic conversion processes in the future.

**2.3.1.** *Bagasse:* Bagasse is the fibrous residue of extraction of sugar from cane. About 270 Kg raw, wet bagasses or 130 Kg dry matter is produced from each ton of cane. Bagasse is a lignocellulosic residue with substantial carbohydrate content. It typically consists of cellulose, 35%; pentosans, 25%; lignin, 20%; other organics and ash. In most sugar mills, bagasse is used largely for its fuel value. India produces 50 million metric ton (MT) of bagasse a year. Whilst bagasse based paper production was considered to be a significant revenue-earner over two decades ago, the importance of bagasse as fuel has overridden this importance. In fact, co-generation of energy is the new wave which is being promoted actively in certain states to counter the deficit of electrical power as well as to improve financial health of the industry.

**2.3.2.** *Forage crops:* Forage crops (forage sorghum, Sudan grass) hold promise for ethanol production because, in their early stage of growth, there is very little lignin and the conversion of the cellulose to sugars is more efficient. In addition, the proportion of carbohydrates in the form of cellulose is less than in the mature plant. Since forage crops achieve maximum growth in a relatively short period, they can be harvested as many as four times in one growing season. For this reason, forage crops cut as green chop may have the highest yield of dry material of any storage crop. In addition to cellulose, forage crops contain significant quantities of starch and fermentable sugars which can also be converted to ethanol. The residues from fermentation containing nonfermentable sugars, protein, and other components may be used for livestock feed.

**2.3.3.** *Straw and chaff:* The annual production of straw and chaff is enormous. In addiiton to 1.5-4.0 tons of straw/tonne of grain, the chaff adds 100-150 kg/ton. Wheat straw contains 73% polysaccharides, with about equal quantities of cellulose and hemicellulose. Wheat chaff contains 68% polysaccharides of which 45% is cellulose. If half of all straw and chaff were hydrolysed and fermented, alcohol production would amount to 480,000 million l/year equivalent to 3000 million barrels of oil, about 15% of all crude oil production. Hemicellulose is made up of the 5 carbon sugar xylose arranged in chains with other minor 5 carbon sugars interspersed as side chains. Just as with cellulose, the hemicellulose can be extracted from the plant material and treated to release xylose which, in turn, can be fermented to produce ethanol (Figure 4). Xylose fermentation is not straight forward and depending upon the microorganism and conditions, a number of fermentations are possible. The array of products can include ethanol, carbon dioxide, and water.

The bioconversion of straw to fuel alcohol has been the subject of great deal of research in the past several years. There is a large pilot plant at Soustons in South Western France owned by "Institue Francais de Petrole, IFP" which is capable of processing 2-4 tons/h of straw to ethanol. After a steam pretreatment the fibres are enzymatically saccharified and the reducing sugars fermented. The fact that straws and chaffs being bulky materials, difficult to collect, transport and process. The use of straw and chaff as animal feed and farm yard manure has effectively hindered the development of industrial conversion of these to ethanol, though several reports have appeared indicating the production of alcohol from these, at laboratory scale. One study employed the fungus *Trichoderma viride* and *Pachysolen tannophilus* for the single batch bioconversion of wheat straw to ethanol. Another study reported production of ethanol from barley straw in a solid state fermentation using *Kluyveromyces marxianus*. After fermentation, the ethanol concentration increased to a maximum of 20 g ethanol/100 g of straw.

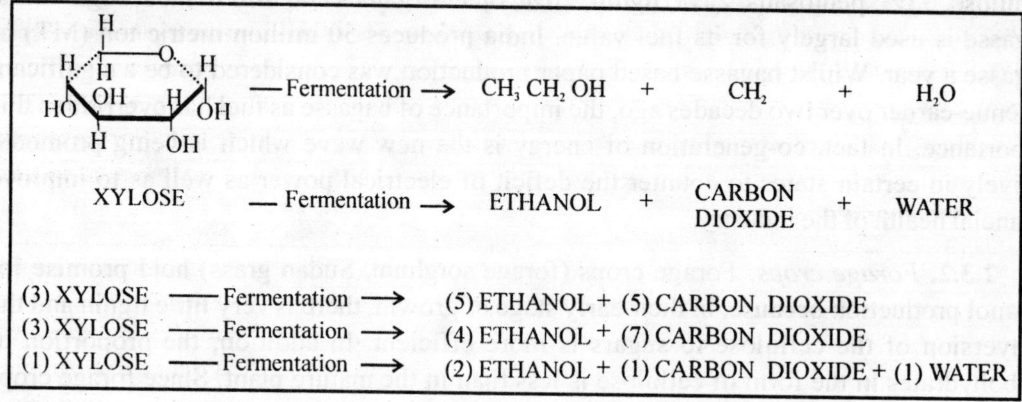


**Figure 4** Various routes of xylose fermentation

According to a recent report issued by the Energy Information Administration (EIA), USA, "Outlook for Biomass Ethanol Production and Demand," the advances in cellulose-ethanol technology over the next twenty years can reduce the cost of producing cellulose-ethanol.

## 3. MICROORGANISMS FOR ALCOHOL PRODUCTION

Efficient ethanol production depends upon the choice of microbes, which can adapt to the production procedures employed, and the composition of the feedstock. The microorganism selected should have i) large scale easy cultivation on cheap and simple substrates, ii) high fermentation rate, iii) high yield of product per unit substrate assimilated, iv) substantial tolerance for alcohol, v) stability under adverse environment conditions, including low and high pH, temperature etc, vi) inability to produce undesirable products.

Although ethanolic fermentation is common to a wide variety of microorganisms (Figure 5), with a few exceptions. Only yeasts have traditionally been used for industrial scale production of alcohol or alcoholic beverages. The near-exclusive application of yeasts is due, in part, to their homofermentative mode (EMP pathway) which yields mostly a single product, ethanol, often in relatively high concentrations. *Saccharomyces cerevisiae* and *S. uvarum* (*S.carlbergensis*), for example, produce 12% ethanol, whereas in a slow fermentation *S. sake* produce up to 20% ethanol. Under anaerobic conditions, these yeasts normally produce small amounts of byproducts such as glycerol, methanol and fusel oil, which require that specific measures be taken for their separation during distillation of alcohol.

One of the advantages of using yeasts has been their ability to ferment a wide spectrum of substrates. For example, when whey is used as substrate for ethanol production, the yeasts *Torula cremoris, Kluyveromyces fragilis, K. lactis* and *Candida pseudotropicalis* may be employed. Substrates containing low molecular weight oligomers of various sugars, dextrins, can also be fermented directly after partial hydrolysis. *K. marxianus* and *K. fragilis* can ferment inulin, a polyfructan directly, whereas *S. diastaticus* can produce ethanol from a partial digest of liquefied starch (maltodextrins) and *Torulopsis wickerhamii* has been shown to give good yields of ethanol from depolymerized cellulose (cellodextrins) while *Pichia stipitis* has been shown to be useful for the production of ethanol from sugars derived from the hydrolysis of lignocellulosic biomass.

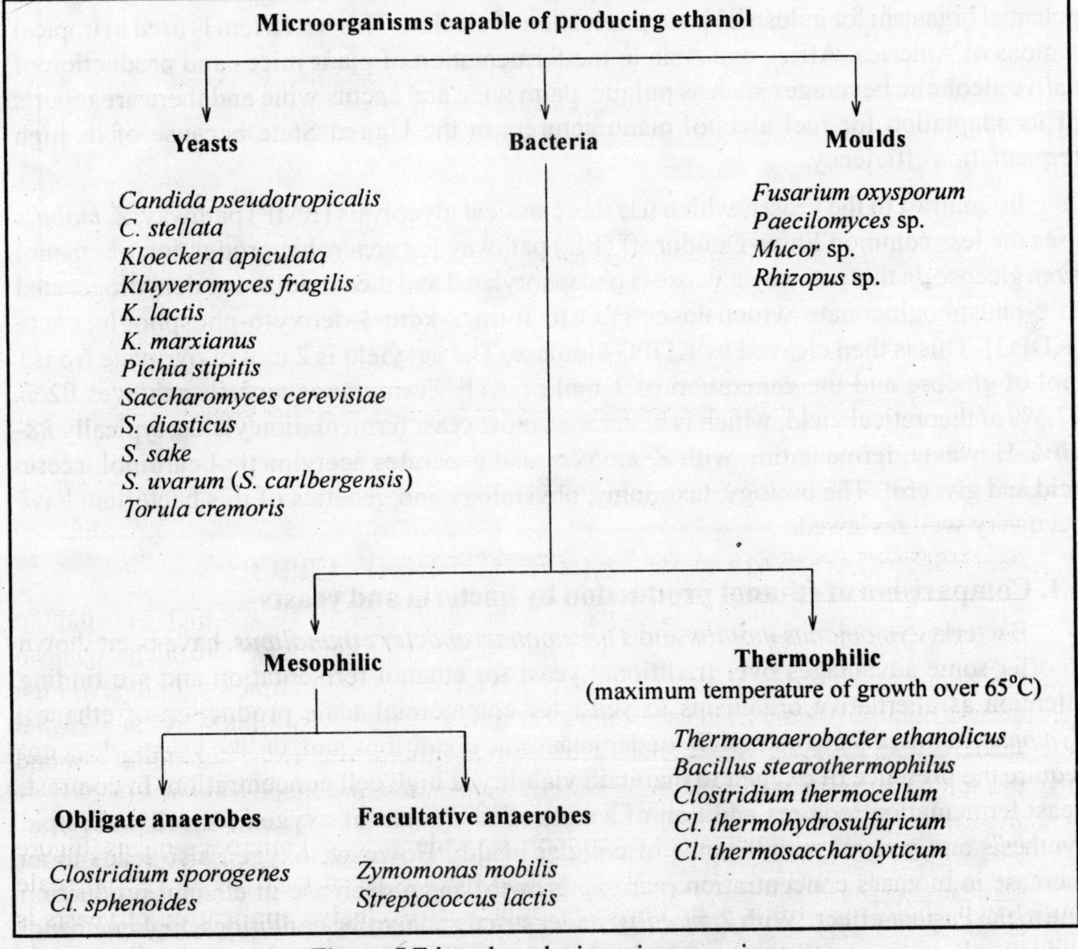


**Figure 5** Ethanol producing microorganisms

One disadvantage of ethanol-producing yeasts has been their inability to tolerate high temperatures as temperature above 40°C usually inhibits growth and fermentation. This means that during large-scale fermentation a cooling system is an absolute requirement. It has been reported that certain thermotolerant yeasts may overcome this handicap. The strains of *Pichia stipitis* have been reported to carry out fermentation well at 41°C, which results in reducing the cooling costs.

Some bacteria are capable of producing ethanol under both aerobic and anaerobic conditions as efficiently as yeasts, and therefore have attracted attention for their possible application as an alternative organism for alcohol production. However, in contrast to yeast, bacteria characteristically generate multiple end products in addition to ethanol. These include other alcohols (butanol, isopropanol), glycols (2-3-butanediol), organic acids (acetic, propionic, butyric, formic, lactic), polyols (arabitol, glycerol, xylitol), ketones (acetone), and gases (methane, $CO_2$, $H_2$). Of a number of bacteria capable of producing good yields of ethanol, only a few may be considered for a alcohol production at industrial scale. Within the mesophilic group only the gram-negative facultative anaerobic bacterium *Zymomonas mobilis* can be regarded as an efficient ethanol producer. It offers advantageous that makes it a potential organism for industrial scale production of alcohol. This bacterium is used in tropical regions of America, Africa and Asia in the fermentation of plant juices and production of native alcoholic beverages such as pulque, palm wine and cactus wine and there are reports of its adaptation for fuel alcohol manufacturers in the United State because of its high fermentation efficiency.

In contrast to the yeasts, which use the classical glycolysis (EMP) pathway, *Z. mobilis* uses the less common Enter-Doudoroff (ED) pathway for anaerobic production of ethanol from glucose. In this pathway glucose is phosphorylated and then oxidized or dehydrogenated to 6-phosphogluconate which loses $H_2O$ to form 2-keto-3-deoxy-6-phosphogluconate (KDPG). This is then cleaved by KDPG aldolase. The net yield is 2 mol of pyruvate from 1 mol of glucose and the generation of 1 mol of ATP. *Zymomonas mobilis* achieves 92.5-97.5% of theoretical yield, which is better than most yeast fermentation yields, typically 88-90%. However, fermentation with *Z. mobilis* also generates acetylmethyl carbinol, acetic acid and glycerol. The biology, taxonomy, physiology and genetics of this bacterium have been very well reviewed.

## 3.1. Comparision of ethanol production by bacteria and yeasts

Bacteria *Zymomonas mobilis* and *Thermoanaerobacter ethanolicus*, have been shown to offer some advantages over traditional yeast for ethanol fermentation and are finding attention as alternative organisms to yeast for commercial scale production of ethanol. *Zymomonas mobilis* grows rapidly under anaerobic conditions and, unlike yeasts, does not require the presence of oxygen to maintain viability at high cell concentration. In contrast, yeast fermentation requires addition of a controlled amount of oxygen for cell wall, lipid synthesis and general maintenance of cellular health. However, oxygen also leads to an increase in biomass concentration (aerobic growth) and a decrease in ethanol production due to the Pasteur effect. With *Z. mobilis*, under strictly anaerobic conditions, higher ethanol and lower biomass production are obtained. Lower cell production is also a consequence of the reduced energy available for growth (1 mol of ATP/mol of glucose consumed via the ED pathway versus 2 mol of ATP/mole of glucose via the EMP pathway with yeasts). Also *Z. mobilis* takes advantage of uncoupled growth metabolism in which glucose fermentation is continuous in the absence of cell growth. The fermentation rate of *Z. mobilis* is higher than that of yeasts. The key fermentation kinetic parameters indicating the superiority of this bacterium include higher specific growth rate, μ (2.4 times high than yeast), high specific ethanol producitivity Qp, (g/g/h) 2.9 times greater and high specific glucose uptake rate Qs, (g/g/h) 2.6 times greater than for *S. uvarum*.

*Zymomonas mobilis* is also reasonably ethanol (over 7%, v/v ethanol), glucose and thermotolerant. Another advantage of *Z. mobilis* is that it being a prokaryote, genetic manipulation can be carried out with greater ease than in yeasts. *Z. mobilis* was discovered in 1936 and despite its several favourable attributes and high production effectiveness, it has not yet worked its way into an industrial alcohol process for several reasons. *Zymomonas mobilis* being a temperamental, it requires some skill to work with; unless fermentation is closely controlled acetaldehyde tends to accumulate in the beer. *Z. mobilis* converts ethanol to acetate, in the presence of even traces of oxygen, through the Krebs respiration cycle, thus resulting in reduction in ethanol yield. Moreover, beer containing acetate can cause technical separation problems during distillation. *Z. mobilis* grows optimally at pH values above 5.0, which is more than the pH of most yeast fermentations. Higher pH increases the chances of contamination by other organisms, especially at the beginning of the fermentation when the ethanol concentration is low. *Z. mobilis* ferments a very limited range of substrates: glucose and fructose and only a few strains decompose sucrose. It cannot be utilized to ferment some classical substrates of the alcohol industry such as lactose, maltose, mannose, galactose or starch, or lignocellulosics.

The thermophilic bacteria, *Themonoanerobacter ethanolicus* and similar thermophiles have several advantages over yeast for ethanol production. These grow at high temperatures, exhibit rapid metabolic activity and a high fermentation efficiency with high product output. The ethanol yield by *T. ethanolicus* approaches that of yeasts. Fermentation at a high temperature would favour continuous operation and at the same time assist in ethanol recovery directly from the fermenter. Use of thermophiles minimize the chances of contamination and need for cooling. Viscosity of growth medium decreases at high temperature, thereby the energy requirement to maintain agitation is reduced. Reduced solubility of oxygen and other gases at high temperature assists maintenance of anaerobic conditions.

However, the thermophilic bacteria have the following disadvantages. Majority of thermophilic bacteria follow the EMP pathway, as do yeasts, but they also synthesize large amounts of acetic and lactic acids, thereby lowering yield of ethanol. The energy relation is 2.5-3.0 mol ATP/mol of glucose. High yields of organic acids and ATP diminish the amount of substrate available for the formation of ethanol e.g., *T. ethanolicus* produces 1.8 mol of ethanol, 0.1 mol of acetic acid and 0.1 mol of lactic acid per mol of glucose. *Clostridium thermohydrosulfuricum* produces only 0.5-0.6 mol of ethanol and 0.5 mol of each acid. For *Cl. thermocellum* and *Thermoanaerobium brockii*, the rates of ethanol production are below 1 mol alcohol/mol glucose. Technical problems are experienced in the separation of the acids from the alcohol. Thermophilic bacteria require specific substrates and growth conditions, for instance *Cl. thermohydrosulfuricum* produces large amounts of ethanol at a starting pH of 9.5 which decreases during the course of fermentation to 6.9 or even less. Also, the medium needs a significant amount of yeast extract as a growth factor for ethanol production. Thermophilic bacteria have low ethanol tolerance. Yeasts are known to tolerate 6-12% ethanol without a significant decrease in the growth rate, but, in contrast the thermophilic bacteria are able to tolerate only up to a maximum of 3-7% ethanol. Strains of *Cl. thermocellum* and *Cl. thermohydrosulfuricum* are reported to be inhibited by only 1% ethanol in the fermentation medium. Bacterial strains are susceptible to viral infection (bacteriophages), while the yeasts are not amenable to viral contamination.

It has become evident that the bacteria do not at present meet the criteria for industrial alcoholic fermentation as an alterative to yeasts. In order to use them as alternative to yeast for industrial alcohol production an improvement through genetic transformation is necessary. It is most unlikely that yeast will be replaced by other microbes for industrial scale production of ethanol in the near future, however, *Z. mobilis* may be adopted in some specific situations, especially where glucose is present as main sugar.

## 4. INDUSTRIAL PROCESSES FOR ETHANOL PRODUCTION

The present industrial fuel ethanol production is based on the use of yeast species of the genera *Saccharomyces*. Under anaerobic conditions, each g of glucose can theoretically give 0.51 g ethanol. In practise, however, the ethanol yield does not exceed 90-95 % and on industrial scale 86-90 % of the theoretical value. Some parts of the nutrients are required for biomass synthesis and cell maintenance reactions. About 4-5 % of the total substrate usually is utilised for glycerol and succinate formation. The medium for ethanol production requires small quantities of $O_2$, P, S, K, Mg, Mn, Co, Cu, Zn as well as organic growth factors such as amino acids, nucleic acids and vitamins which are generally present in the feedstock used in the fermentations. Yeasts are very susceptible to ethanol inhibition. For example, 1-2 % (w/v) ethanol decreases cell growth and 10 % (w/v) stops ethanol synthesis. During the fermentation process, fusel oils are formed from $\alpha$-keto acids, derived from deaminations of amino acids. The major fusel oil component is isoamyl alcohol (40-60 %) with optically active amylalcohol and iso-butylalcohol taking 15-20 percent each of the total. In general, the yield of fusel oil is 20 l / $m^{-3}$ of ethanol produced. Fuel oils can remain in the final product as a component of fuel ethanol.

### 4.1. Medium preparation from sugarcane

When sugarcane is used as a raw material, the raw cane is crushed and the juice extracted. After clarification by precipitating the inorganic fraction with milk of lime and $H_2SO_4$, the resulting cane juice is a green, sticky liquid more viscous than water and contains an average of 15% dissolved solids of which 85% (13% of the juice) is sucrose and less than 1% of the invert sugars glucose and fructose. The sugarcane juice forms the major raw material for the Brazilian ethanol industry and is directly used as a medium for fermentation after adjusting the pH to 4.5-5.0 and supplementing some nitrogenous compound in the form of urea, diammonium phosphate or ammonium sulphate at a concentration of 0.05% (w/v).

The fermentation is highly productive, producing 75 l of alcohol / ton of cane or 4800 l / hectare of land. This is six times the amount of alcohol obtainable from molasses based on the same land area. Several ethanol tolerant yeast strains have been isolated which are able to produce and tolerate above 20% (v/v) alcohol during the fermentation of sugar cane syrup. Because of low solid and salt contents the cane juice offers many advantages: i) higher efficiency of fermentation (up to 90%) , ii) savings in sugar due to yeast recycling, iii) reduction in water consumption by stillage recycling, iv) massive reduction in the spent wash generation to the tune of 3 l / l of ethanol produced, without using any additional energy, v) distillation efficiency of 98.5 % is achieved.

In sucrose producing industries, the sugarcane juice is concentrated to 80-82 % sucrose content to be able to crystallise raw sugar. The residual juice is referred to as molasses. Depending on the sugarmill efficiency, a maximum of three crystallisation steps can be carried out leading to an A-molasses (70-71 % sucrose), B-molasses (60-65 % sucrose) and/or C- or blackstrap molasses with a total carbohydrate (sucrose, glucose, fructose) concentration of 40-55 % (w/v). Molasses may be stored for long periods of time and diluted before fermentation, as either the high sugar content (A- and B-molasses) or the salt concentration (C-molasses) act as preservative. Molasses with 60% fermentable sugars or 1.32 tons/ha, upon fermentation, yields 11.5 l of alcohol/ton of cane or 730 l/hectare. Before fermentation, molasses may be clarified, refined and diluted with water to a solid concentration of about 20-25%. This may then be treated with $H_2SO_4$ (to precipitate calcium salts) and ferrocyanide (to remove excess levels of metals). After adjusting the pH to 4.5-5.0, some nitrogenous compound in the form of urea, diammonium phosphate or ammonium sulphate at a concentration of 0.05% (w/v). A similar technology can be applied for the medium preparation from sugar beet roots and sweet sorghum.

### 4.2. Medium preparation from cereal grains

In the case of grains such as corn (maize), wheat, rice, sorghum and barley as raw materials, the starch must first be enzymatically hydrolysed before fermentation can proceed. Amylase preparations are applied for liquefaction and saccharification of starchy substrates. The enzyme α-amylase (α -1,4-glucan 4-glucanohydrolase, EC 3.2.1.1) is an endoenzyme, which causes random hydrolysis of α -[1,4] linkages in the starch molecule resulting in the liquefaction of the medium thereby reducing the viscosity of starch. The β-amylase (α -1,4-glucan maltohydrolase, EC 3.2.1.2) is an exo-enzyme generating maltose and dextrins and is only available in plants. Finally, the enzyme glucoamylase (amyloglucosidase, α -1,4-glucane glucohydrolase, EC 3.2.1.3) is an exo-enzyme being able to hydrolyse α -(1,4) as well as α-(1,6) linkages and is able to produce glucose from the starch molecule. Two different grain processing technologies exist, dry- and wet-milling. In wet-milling, the major objective is to divide and convert the corn into a number of products, such as starch flour, fibre, gluten, germ, oil, meal residue, dextrose (glucose), fructose and other products. In this case, ethanol is produced from the lower grades of starch or from all of the starch. The dry-milling technology is much simpler as the whole grain is ground by hammer mill and slurred. The liquid starch slurry containing the liquefying enzyme α-amylase is cooked and thus liquefied. Saccharification using the glucoamylase can either be performed prior of or simultaneously during fermentation. The complete flow sheet involved in ethanol production from starchy grains is exhibited in Figure 6 and the various steps involved in this are discussed below:

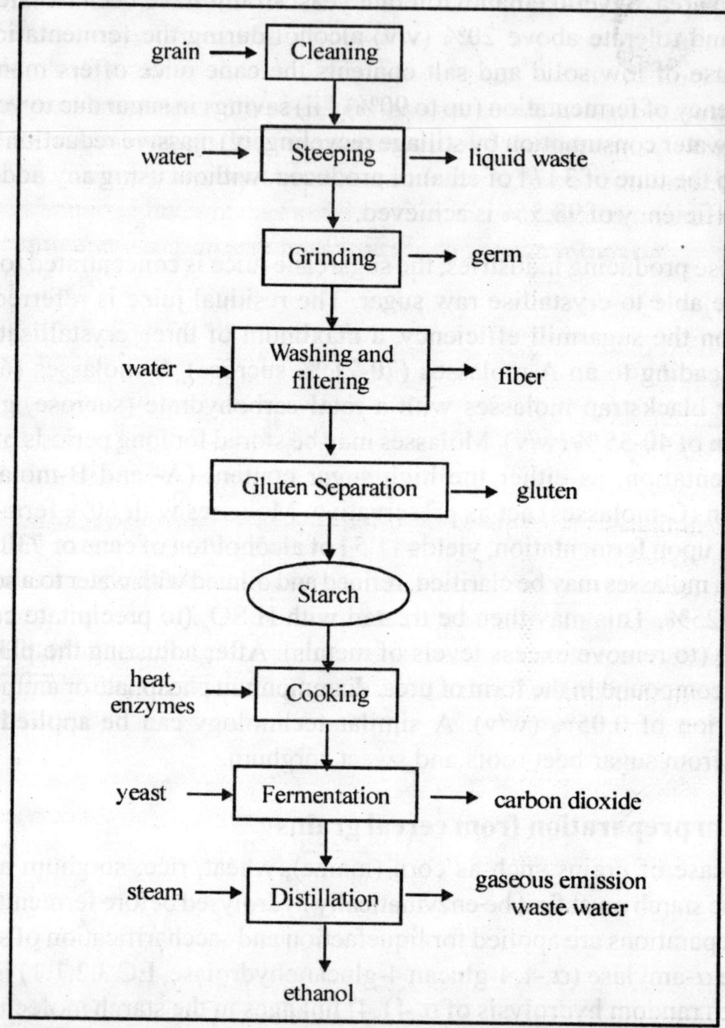


**Figure 6** A schematic representation of various steps in the conversion of starchy grains to alcohol involving wet mill process.

**4.2.1.** *Milling:* All grains must be ground before mashing to expose the starch granules and help them remain in suspension in a water solution. The grain should be ground into a meal, but not a flour, that will pass a 20-mesh screen.

Potatoes and similar high-moisture starch crops should be sliced or finely chopped. Since potato starch granules are large and easily ruptured, it isn't necessary to maintain the hard rapid boil which is required of the tougher, dryer "flinty" starches found in grains.

**4.2.2.** *Cooking:* Grain must be cooked to rupture the starch granules and to make the starch accessible to the hydrolysis agent. This process which increases the viscosity of the mash and converts it into a gel is also called gelatinization. Cooking time and temperature are related in an inverse ratio: high temperatures shorten cooking time. Industry practice is to heat the meal-water mixture by injecting steam directly rather than by heat transfer through the wall of the vessel. The latter procedure runs the risk of causing the meal to stick to the wall; the subsequent scorching or burning would necessitate a shutdown to clean the surface. The meal is added with water in the cooker slowly, to prevent lumps from forming. When, cooking with steam, or at higher temperatures, it is possible to save energy by using less water at the beginning. But for the "small batcher" with an ordinary cooking apparatus, the most complete conversion is obtained by using the full amount of water right from the start to encourage a rapid rolling boil.

The high temperature bacterial α-amylase is added next to the mixture and the temperature of the mash is raised to 77°C, the optimum working environment for the enzyme. The solution is then held at that temperature for 15 min while agitating it vigorously. At this point all the starch available at 77°C gets converted to dextrins. The temperature of the mash is then raised to the boiling point for cooking at which it is then held for 30 min to complete the liquefaction stage during which all the starches come in solution. Now the temperature is reduced back to 77°C, using the cooling coil, and more bacterial α-amylase is added. After 30 min of agitation at this temperature, all the previously released starches gets reduced to dextrins, thereby completing primary conversion. During secondary conversion the dextrins are further reduced to simple sugars (maltose and glucose) by the β-amylase or glucoamylase and the yeast necessary to carry out secondary conversion and proper fermentation simultaneously as soon as the temperature is brought down to 29°C using the cooling coils. Instead of using commercial enzymes, it is possible to affect conversion by employing barley malt, at the ratio of 15% by weight, in both the pre- and post-cooking. However, such a technique requires a more acidic medium (about pH 4-5) and lower temperatures, about 63°C.

Starch can also be hydrolysed by acids which is accomplished by directly contacting starch with dilute acid to break the polymer bonds. This process hydrolyzes the starch very rapidly at cooking temperatures and reduces the time needed for cooking. Since the resulting pH is lower than desired for fermentation, it may be increased after fermentation is complete by neutralizing some of the acid with either powdered limestone or ammonium hydroxide. It also may be desirable to add a small amount of glucoamylase enzyme after pH correction in order to convert the remaining dextrins.

Cooking can be accomplished with continuous or batch processes. Batch cooking can be done in the fermenter itself or in a separate vessel. When cooking is done in the fermenter, less pumping is needed and the fermenter is automatically sterilized before fermenting each batch. There is one less vessel, but the fermenters are slightly larger than those used when cooking is done in a separate vessel. It is necessary to have cooling coils and an agitator in each fermenter.

If cooking is done in a separate vessel, there are advantages to selecting a continuous cooker. The continuous cooker is smaller than the fermenter, and continuous cooking and hydrolysis lend themselves very well to automatic, unattended operation. Energy consumption is less because it is easier to use counterflow heat exchangers to heat the water for mixing the meal while cooling the cooked meal. The load on the boiler with a continuous cooker is constant. Constant boiler load can be achieved with a batch cooker by having a separate vessel for preheating the water, but this increases the cost when using enzymes.

Continuous cooking offers a high-speed, high-yield choice that does not require constant attention. Cooking at atmospheric pressure with a temperature a little over 93ºC yields a good conversion ratio of starch to sugar, and no high-pressure piping or pumps are required (Figure 7).

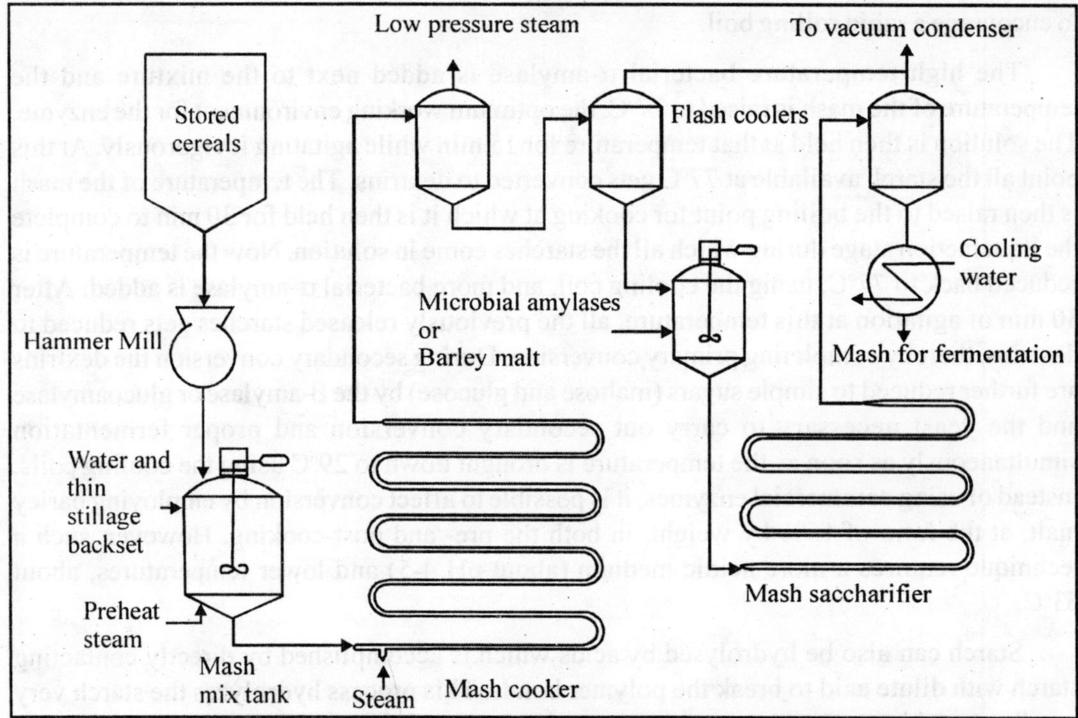


**Figure 7** Continuous flow grain cooking and saccharification

The hydrolyzed mash contains solids and dissolved proteins as well as sugar. There are some advantages to separating the solids before fermenting the mash, and such a step is necessary for continuous fermentation. Batch fermentation requires separation of the solids if the yeast is to be recycled. If the solids are separated at this point, the beer column will require cleaning much less frequently, thus increasing the feasibility of a packed beer column rather than plates. The sugars that cling to the solids are removed with the solids. If not recovered, the sugar contained on the solids would represent a loss of 20% of the ethanol. Washing the solids with the mash water is a way of recovering most of the sugar.

## 4.3. Medium preparation from lignocellulosics

Lignocellulosic materials are pretreated by milling and then chemically (by acid hydrolysis) or enzymatically hydrolysed before fermentation. The dilute sulfuric acid method for wood hydrolyses is basically a semicontinuous process with the following parameters: acid concentration 0.53 %, temperature of percolation 196°C, percolation time 145-190 min, percolation rate 8.69-14.44 l / min / $m^3$ and a total water: oven-dried wood ratio of 10. At the end of the percolation cycle the lignin-rich residues are discharged, recovered and used as fuel. Furfural is recovered by distillation. The hot acid hydrolysate is neutralised with a lime slurry and the precipitated calcium sulfate is separated. The neutralised hydrolysate is blended with yeast and fermented. Genetically modified producers are necessary to obtain ethanol from both hexoses and pentoses. The complete flow sheet involved in ethanol production from starchy grains is exhibited in Figure 8.

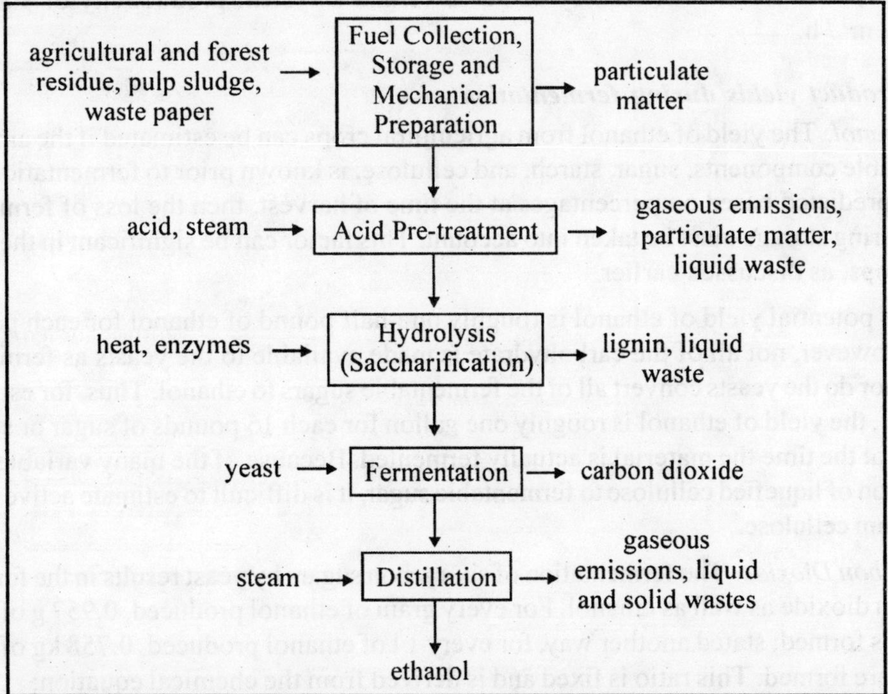


**Figure 8** A schematic representation of various steps in the conversion of cellulosic residues to alcohol involving acid hydrolysis

## 4.4. Fermentation

Yeast propagates in a solution containing free fermentable sugar feedstocks including sucrose from sugarcane, sugar beet and molasses as well as hydrolysed starch or lignocellulosics, with or without air. If the medium is continually agitated, the yeast will reproduce faster and make less carbon dioxide and alcohol. But if the solution becomes anaerobic (without air) the yeast slows down reproduction and makes more alcohol and carbon dioxide, so the wort is agitated only enough to saturate the wort with air and then let it stand still. Yeast also produces enzymes of its own to convert complex sugars. Since sugar conversion and alcohol conversion can take place simultaneously in case of starchy and cellulosic mashes, the enzymes and the yeast work in cooperation to convert the dextrins to glucose and fructose and then to alcohol and $CO_2$.

Fermentation is a biochemical process and produces heat. In concentrated or particularly large mashes, the temperature can actually rise to levels dangerous to yeast. Since the ideal temperature for yeast is around 30ºC, it's best to maintain that temperature by either utilizing cooling coils or spraying water on the outer surface of fermenters.

Conversion of sugars to alcohol and $CO_2$ gets completed in three to five days, depending on the temperature of the mixture and the type of yeast used. During fermentation, the rising $CO_2$ keeps the solids in constant motion, but when the bubbling stops, the solids fall to the bottom. At the end of the fermentation the solids get separated from the liquids and settle at the bottom. The final ethanol concentration is between 10-16 % (w/v) with a carbohydrate conversion efficiency of 90-95 % and a system productivity of 1.8-2.5 kg ethanol / $m^3$ / h.

### *4.4.1. Product yields during fermentation*

*Ethanol:* The yield of ethanol from agricultural crops can be estimated if the amount of fermentable components, sugar, starch, and cellulose, is known prior to fermentation. If the yield is predicted based on percentages at the time of harvest, then the loss of fermentable solids during storage must be taken into account. This factor can be significant in the case of sugar crops, as discussed earlier.

The potential yield of ethanol is roughly one-half pound of ethanol for each pound of sugar. However, not all of the carbohydrate is made available to the yeasts as fermentable sugars, nor do the yeasts convert all of the fermentable sugars to ethanol. Thus, for estimating purposes, the yield of ethanol is roughly one gallon for each 15 pounds of sugar or starch in the crop at the time the material is actually fermented. Because of the many variables in the conversion of liquefied cellulose to fermentable sugar, it is difficult to estimate active ethanol yields from cellulose.

*Carbon Dioxide:* The fermentation of six-carbon sugars by yeast results in the formation of carbon dioxide as well as ethanol. For every gram of ethanol produced, 0.957 g of carbon dioxide is formed; stated another way, for every 1 l of ethanol produced, 0.758 kg of carbon dioxide are formed. This ratio is fixed and is derived from the chemical equation:

$$C_6H_{12}O_6 \rightarrow 2C_2H_5OH + 2CO_2 + \text{Heat}$$

Glucose Ethanol Carbon dioxide

*Other co-products:* The conversion and fermentation of agricultural crops yield products in addition to ethanol and carbon dioxide. For example, even if pure glucose is fermented, some yeast will be grown, and they would represent a coproduct. These co-products have considerable economic value, but, since they are excellent cultures for microbial contaminants, they may represent a pollutant if dumped onto the land. Therefore, it becomes doubly important that these co-products be put to good use.

**4.4.2.** *Alcohol recovery:* Alcohol product recovery is energy intensive, typically accounting for more than 50% of the total fermentative ethanol plant energy consumption. When heat from burning of raw material residues (such as bagasse) is not available, this constitutes a significant operating cost. Depending on recovery system design, recovery equipment cost generally makes up 6-12% of the plant total capital investment.

Industrial alcohol is produced in various grades. The majority is 165ºP (94% v/v) alcohol used for solvent, pharmaceutical, cosmetic and chemical applications. Technical grade alcohol (containing up to 5% volatile organic aldehyde, esters and sometimes methanol) is used for industrial solvents and some chemical synthesis. A high-purity 175ºP anhydrous alcohol product is produced for specialized chemical applications. For fuel use in mixtures with gasoline (gasohol), a nearly anhydrous (99.2 %) alcohol, but with higher allowable levels of organic impurities, is used.

Today, distillation is used almost exclusively as the means for ethanol recovery and purification and various designs are used to produce the different product grades. Ethanol distillation technology was highly refined during the 1940's to reduce energy consumption to approximately 2.5 kg of steam / l of anhydrous ethanol produced. Recent further refinements in distillation technique make possible marginal improvements with increased capital investment. Alternatives to distillation are under study to further reduce costs.

## 4.5. Factors influencing ethanol production

Various process variables and conditions that influence the growth and efficiency of industrial ethanol production by yeasts are discussed below:

**4.5.1.** *pH:* Hydrogen ion concentration has a significant influence on industrial fermentation due to its importance in controlling bacterial contamination and has effect on yeast growth, fermentation rates, and by product formation. In an uncontrolled batch fermentation of a highly buffered medium, the best ethanol yields are generally obtained at pH 4.5-4.7. At higher pH more glycerol and organic acids are formed at the expense of ethanol.

In lightly buffered media, the optimum starting value is pH 5.5. At the completion of fermentation pH falls to about 3.5. If low pH is used initially in a lightly buffered medium, the final pH tends to fall enough to slow the fermentation rate. This difficulty can be overcome to a certain extent by the use of larger inocula. Yeasts survive in a pH range of pH 2.0-8.6.

**4.5.2.** *Choice of organism:* The ability to ferment rapidly and efficiently the sugars present in the fermentation medium containing high concentrations and sometimes unusual sugars, ethanol tolerance and ability to remain stable and viable under prevailing fermentation conditions determines the yeast to be used for industrial scale ethanol production. Presently, large-scale ethanol producing industry use *Saccharomyces* strains. The yeasts, however, have limitations in fermenting lignocellulosic biomass, due to the presence of cellobiose, xylose, arabinose and other wood sugars. Moreover, the yeasts are not able to ferment inulin from Jerusalem artichoke, an attractive renewable resource. However, a xylose-fermenting yeast like *P. stipitis* or an inulin-fermenting yeast such as *K. marxianus* can be used to ferment such feedstocks.

**4.5.3.** *Sugars fermented by yeasts:* Yeasts can ferment a wide variety of oligosaccharides and sugars in addition to glucose. Different yeasts have different capabilities in this regard due to the presence or absence of enzymes which are capable of converting these different sugars to substances which appear in the metabolic pathways of alcoholic fermentation.

With very few exceptions, L-sugars, including the common L-arabinose, are not fermented by yeasts Some yeasts are known to ferment other pentoses but not methyl pentoses although the latter may be utilized in the respiration of aerobic yeast. The substrates for alcoholic fermentation are, thus, primarily hexoses, xylose and oligosaccharides.

D-hexoses and oligosaccharides fermented most often by yeasts are glucose, mannose, fructose, galactose, maltose, sucrose, lactose, melibiose, trehalose and raffinose. The last, a trisaccharide, is partially fermented by some yeasts, but is assimilated completely by many more species during aerobic growth. This behaviour toward raffinose differentiates the bottom and top fermenting brewers' yeasts *S. cerevisiae* and *S. uvarum*.

Certain generalizations can be made regarding the fermentation characteristics of yeast on the basis of following rules or guidelines:

- Yeast unable to ferment D-glucose will not ferment other sugars.
- If D-glucose is fermented, so is D-mannose.
- If maltose is fermented, lactose is not, and vice versa (with a few exceptions, such as *Brettanomyces clausenii*, an ale yeast that ferments both sugars).
- If sucrose is fermented, so is raffionse.
- All sugars fermented, are also utilized aerobically.
- No L-sugars are fermented.

The advantage of using yeast strains *P. stipitis* is that in addition to sugars utilized by *S. cerevisiae*, this yeast can also ferment xylose and cellobiose. These sugars are normally present in lignocellulosic. Also the ability to ferment cellobiose and perhaps cellodextrins offers advantages in using partially hydrolysed lignocellulosic substrates.

**4.5.4.** *Sugar concentration:* The concentraion of sugar which can be fermented most efficiently depends to some extent on other components of the medium. Most industrial processes utilize a fermentation solution containing 12-20% sugars by weight. The advantages of a high concentration feed solution include reduced water requirement, suppression of osmosensitive contaminants and reduced distillation cost if ethanol inhibition can be limited.

**4.5.5.** *Temperature:* The rate of alcoholic fermentation increases with temperature and is optimum between 30 and 40°C. Both optimum and maximum temperature tolerance for growth and fermentation are strongly strain-dependent. However, for most ordinary yeasts, if the ethanol concentration reaches 8-9% by volume prior to completion of fermentation, the fermentation may stop if temperatures are held above 33°C. This effect has been attributed to intracellular ethanol being produced more rapidly than it can be transported through the cell membrane, causing inhibition of intracellular fermentation enzymes. Thus, in the course of fermentation, the temperature should not exceed 32°C, especially once the ethanol begins to accumulate. The temperature optimum for the growth of yeast is about 25°C. This temperature can be increased slightly by use of a rich nutrient medium. Since yeast growth and fermentation both produce heat, cooling may be necessary to maintain the desired temperature.

**4.5.6.** *Ethanol:* A limitation of ethanolic fermentation is the capacity of the microbe to tolerate this solvent, because ethanol inhibits alcoholic fermentation, which limits the concentration of ethanol which can be produced by a given strain of yeast. The maximum concentration of ethanol which can be produced by yeast varies with species and a maximum upto 20% ethanol by volume can be produced. The results of mating experiments with yeasts indicate that ethanol inhibition is a polygenic phenomenon, that it is a result of multiple enzyme and cell physiological functions. The degree of inhibition is also related to other environmental factors, in particular high sugar concentration and high temperature which cause the inhibition to be more severe. Brewers' yeast (*S. cerevisiae* var. *carlsbergensis*) ceases to ferment at about 6% ethanol by volume, whereas bakers' yeast (*S. cerevisiae* var. *diastaticus*) stops fermentation at about 12% ethanol by volume, and wine yeast (*S. cerevisiae* var. *ellipsoideus*) at about 15% ethanol by volume. Ethanol which is produced during fermentation (autogenous ethanol) is more inhibitory to cell growth than that from an exogenous source. Addition of ethanol to actively fermenting yeast cultures results in rapid reduction of fermentation and growth rates due to its influence on protein synthesis and irreversible denaturation of enzymes. Ethanol has also been shown to exhibit non-competitive inhibition of yeast growth, that is, independent of other inhibitors which may be present. The toxic effect of ethanol has also been attributed to damaging the cell membrane or changing its properties. The extent of ethanol tolerance of certain yeasts is highly strain dependent and appears to be related to the unsaturated fatty acids and the fatty acyl composition of the plasma membrane. The intolerance exhibited by some yeasts appears to be related to low cellular lipid content and the inability of the plasma membrane to maintain, during high rates of ethanol production, a high ethanol flux through the plasma membrane and out of the cell to avoid intracellular accumulation.

**4.5.7.** *Nutrients:* Most yeasts grow well on a variety of amino acids, purines, and pyrimidines as the sole source of nitrogen. Mixtures of amino acids are better that any single amino acid. About 15 mg of assimilable nitrogen / 100 ml of fermentation solution is sufficient to assure good fermentation rates. The nitrogen can also be supplied in the form of ammonium salts, urea, corn steep liquor or distillers' malt. The distillers' malt have the advantage of containing vitamins and minerals necessary for maximum yeast growth. Yeasts require trace amounts of biotin, thiamine, pyridoxine, calcium pantothenate, and inositol for maximum growth and fermentation rates. These vitamins also regulate yeast metabolism, vitamins being generally either coenzymes or precursors for fully active enzymes. Vitamin requirements are strain dependent. *Candida utilis* and *Hansenula anomala* are vitamin independent and synthesize all their own needs, whereas biotin and pantothenate are essential for all strains of *Saccharomyces*.

Phosphorus in the form of phosphate is an important ionic factor determining the rate of fermentation. It is required at a concentration of about 0.16 mM/g cells for optimum fermentation performance. Yeast growth also depends on compounds containing potassuim (K), sulphur (S), and traces of zinc (Zn), iron (Fe) and copper (Cu), required in concentrations of 0.1-1.0 mM. For anaerobic growth of yeasts, ergosterol and unsaturated fatty acids are beneficial additive while some thermophilic yeast strains have requirements for choline or leucine.

**4.5.8.** *Yeast inoculum:* Provided that temperature, pH, sugar concentration, assimilable nitrogen, vitamins and minerals are at their optimal values, a controlling factor is the amount of yeast added in the inoculum, or the "pitch" as it is often called in industrial operations. Using a larger inoculum of yeast will decrease the time required for growth, and alcoholic fermentation will set in sooner. For commerical production of alcohol, yeast is usually added to provide a starting population of 7-10 million cells/ml, about 0.2g of dry yeast/l of broth. At this inoculation level, fermentation is usually completed within a day or two, depending upon sugar concentration. Much higher inocula, up to 10 or more g cells/l, result in much more rapid fermentation, and is also beneficial in resisting inhibition and suppressing contamination.

**4.5.9.** *Dissolved oxygen:* Ethanol fermentation is not a purely anaerobic process. Dissolved oxygen concentration is an important variable for industrial alcohol production. In many instances, active starter cultures are grown under aerobic or semiaerobic conditions to improve yeast yields and growth rates. Yeasts are unable to multiply for more than four to five generations in solutions containing less than 1 ppm oxygen, even if sufficient nutrients, such as sugar, nitrogen, and vitamins, are present. It is the dissolved oxygen in the solution which limits the population of yeast in the fermentation medium. Once the dissolved oxygen is depleted during fermentation, the yeast growth slows and eventually ceases. Also during the time the metabolism of yeast switches from aerobic respiration to anaerobic alcoholic fermentation, and the synthesis of alcohol occurs until the sugar is depleted or until a limiting ethanol concentration is obtained. Oxygen is utilized by yeasts for at least two separate functions:

- as the ultimate electron acceptor in oxidative phosphorylation, that is for growth and respiration; and
- as an essential nutrilite in lipid, ergosterol and nicotinic acid biosynthesis.

During respiration yeasts have a high demand for dissolved oxygen which promotes complete oxidation of glucose to carbon dioxide and water via the tricarboxylic acid cycle at the expense of ethanol accumulation via fermentation (Pasteur effect). However, the extent of this respiration is tightly controlled by the concentration of fermentable carbohydrates in the medium. In the presence of high sugar concentrations, ethanol is formed during aerobic growth and the phenomenon is known as 'aerobic fermentation'. This happens due to the operation of the so-called reverse Pasteur, or Crabtree effect which results in the repression of both the synthesis and the activity of the respiratory enzymes, notably some Krebs cycle enzymes under fully aerobic conditions.

The Pasteur and Crabtree effect are illustrated in Figure 9. Conditions 3 and 4 are carried out exclusively for alcohol production, whereas, condition 2 is the mode for fodder yeast production, and condition 1 is practised during production of bakers' yeast. The Crabtree effect is also applied in brewery operation, where due to high initial sugar concentration (condition 1) a substantial degree of aeration of the wort prior to pitching, up to about 20% oxygen saturation, is practised to get maximum yeast yield, high viability, growth rate, and overall fermentation rate.

The effect of dissolved oxygen is critical in continuous yeast growth or single cell protein production, where it is usually the limiting factor. It is also important in continuous fermentation, in order to maintain yeast health as well as a satisfactory fermentation rate. The exact level of the minimal oxygen required is strain-dependent and varies with environment. At oxygen concentrations below 1 ppm in the broth, the addition of sterols (ergosterol) and unsaturated fatty acids (oleic acid) may be required during anaerobic growth, since oxygen is required for the synthesis of unsaturated fatty acids, essential components of yeast membranes. It is for these reasons that low concentrations of oxygen are generally maintained in industrial alcohol fermentation.

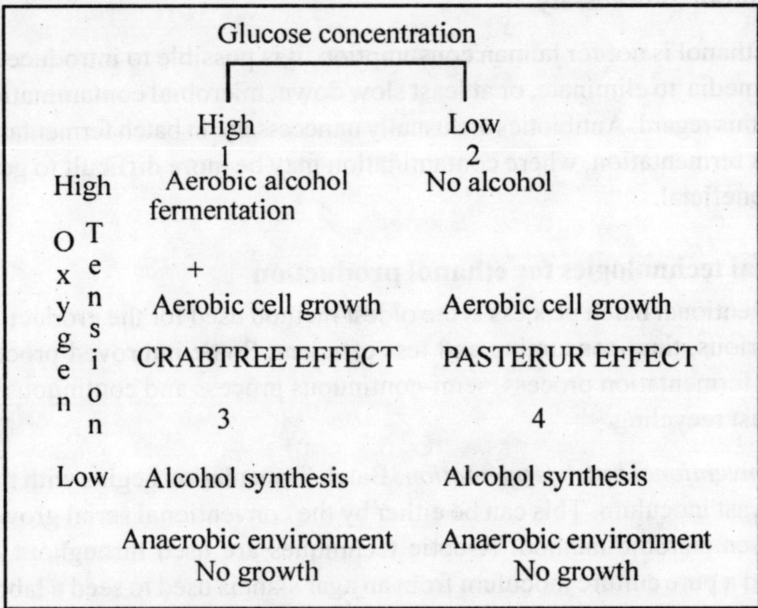


**Figure 9** The influence of oxygen on glucose metabolism

**4.5.10.** *Inhibitors:* Several trace elements including silver (Ag), arsenic (As) and mercury (Hg), with concentrations greater than 10-100 µM, are toxic to yeasts. In addition, anionic detergents, sorbic acid, diethyl pyrocarbonate (DEP) and the antibiotics cyclohexamide, antimycin A and nystatin completely inhibit the growth of most common yeasts. In fermentation of lignocellulosic hydrolysates, acetic acid, furfural and lignin-derived phenolics are inhibitory.

**4.5.11.** *Carbon dioxide:* At pressures greater than atmospheric, carbon dioxide inhibits both aerobic growth of yeast and anaerobic fermentation and this effect being enhanced markedly by low pH and high concentrations of ethanol. The inhibitory effect of $CO_2$ is due to its influence on the permeability and the composition of the cell membrane resulting in altered enzyme activity and changes in solute transport. A major inhibition due to dissolved $CO_2$ is to be expected in deep fermentations, for example in towers, where pressure may reach upto two atmospheres or more, and with fermenters having dissolved $CO_2$ levels as high as 0.02 M.

**4.5.12.** *Microbial contaminants:* The two most common types of contaminants encountered in alcoholic fermentation are acetic acid bacteria (*Acetobacter* spp.) and lactic acid bacteria, principally species of *Lactobacillus* and *Streptococcus*. Lactic acid bacteria are a more serious problem, since they metabolize and multiply under low pH and anaerobic conditions. Acetic acid bacteria metabolize ethanol in the presence of oxygen and thus appear as a gross infection in the early stages of yeast growth, when the fermentation liquid has been exposed to air for long periods. Strict precautions of cleanliness, sterilization of media and the addition of a high inoculum usually ensures that the yeast population will predominate in a batch fermentation. However, regular microscopic examination of the fermentation broth is necessary.

As fuel ethanol is not for human consumption, it is possible to introduce antibiotics to fermentation media to eliminate, or at least slow down, microbial contamination. Penicillin works well in this regard. Antibiotics are usually unnecessary in batch fermentation, however, in continuous fermentation, where contamination may be more difficult to get rid of, there use may be beneficial.

## 4.6. Industrial technologies for ethanol production

The conventional batch process is the oldest method used for the production of alcohol which is laborious, time-consuming and less efficient. Some improved processes include Melle-Boinot fermentation process, semi-continuous process and continuous process with or without yeast recycling.

**4.6.1.** *Conventional batch fermentation:* Batch fermentation begins with the production of an active yeast inoculum. This can be either by the conventional serial growth method or by the rapid semiaerobic method. Aseptic techniques are used throughout. In the serial growth method a pure culture inoculum from an agar slant is used to seed a laboratory shake flask. At the peak of growth (12-24 h) this culture is used to seed a succeeding culture 30-50 times larger. This is repeated, generally through three laboratory stages and two or three plant semiworks stages, to produce the final 2-5 vol% inoculum for the primary fermentation. The inoculum is grown on a medium similar to the final fermentation mash to minimize acclimatization time in the final fermenter, but higher levels of yeast growth nutrients may be used to produce a high cell density (typically 150 billion cells/l).

An inoculum 3-4 times more concentrated in yeast can be produced by the rapid semiaerobic method. Yeasts are grown in an aerated and agitated semiwork fermenter operated in fed batch mode. A large portion (20-25%) of the previous batch is retained to provide an inoculum. A high nutrient medium is added and pH and temperature are controlled. Sterile air is sparged at a rate of one-eighth volume of air per fermenter volume per minute. Aerobic metabolism is stimulated and. a cell density of 500 billion cells/l is reached in 5 h. This high cell density allows the use of a proportionately smaller inoculum to the final fermenter, and a smaller propagating fermenter can be used. However, aerobically grown yeast may require additional time to acclimatize to anaerobic fermentation conditions, resulting in an increased lag period between inoculation and rapid fermentation.

Cylindro-conical vessels are preferred for fermentation as these promote better circulation and allow thorough drainage for cleaning. For very large plants, sloped-bottom, cone-roof tanks of up to 1 million liters volume are used and these large vessels are often agitated only by carbon dioxide evolution during fermentation.

After emptying and rinsing from the previous batch, mash at 13-17 % sugar is pumped to the fermenters. Once 20% full, the inoculum is added to allow growth during the remainder of the filling cycle, which can last 4-6 h.

Fermentation temperature is regulated by circulating cooling water through submerged coils, circulating the mash through external heat exchanges, or simply spraying the vessel walls with cool water (adequate for small fermenters only). The feed is generally introduced at 25-30°C, and the temperature allowed to gradually rise as heat is evolved. The temperature thus varies from 30-35°C during the initial period (which is optimal for yeast growth). Cooling is then used to prevent the temperature from exceeding 35-38°C, which is optimal for ethanol production. These temperatures may be modified depending on the yeast strain used.

Stillage backset provides excellent buffering. The pH is set initially between 4.5-5.5 and decreases only slowly, generally holding at pH 4.0 or above. This is especially important for fermentation of grain mashes with simultaneous dextrin hydrolysis, as many amylase enzymes are rapidly denatured at lower pH.

The ethanol production rate is the product of specific (per cell) productivity and the concentration of cells. Initially, the rate of alcohol production is quite low, but as the number of yeast cells increases the overall rate increases, and with rapid carbon dioxide evolution the beer appears to boil. After 20 h, a maximum in ethanol productivity is reached. The effects of reduced sugar concentration and ethanol inhibition then become important. The fermentation continues at a decreasing rate until, at 36 h, 94% of the sugar is utilized and a final ethanol content of 69g/l is achieved. The average volumetric ethanol productivity over the course of the fermentation is 1.9g/l/h.

Fermentation time will vary depending on yeast strain and substrate. Molasses may require from 36 to 72h for fermentation. These times are reduced when molasses clarification is used. Grain fermentation requires 40-50 h to allow complete residual dextrin conversion. After fermentation, the beer is pumped to a beer well to provide a continuous feed to distillation. The fermenters are then cleaned and prepared for another cycle. The fermentation rate can be increased 30-40% by improved agitation and temperature regulation. Turbine impellers have been used in smaller fermenters (100000 l or less). For large fermenters, improved agitation and temperature control are achieved by rapidly circulating the beer through an external heat exchanger. The cooled beer is pumped back into the head of the fermenter tangentially and at a high velocity. The greatest portion of heat generation occurs during the 'boiling' fermentation period creating a high peak cooling demand. Investment in cooling equipment can be reduced by teaming three or four fermenters to share a single bank of exchangers. The fermentations are then carefully scheduled to give a uniform heat load. With improved agitation, continuous pH control can be used. Ammonium hydroxide or other base is added to the stirred fermenter or mixed with returning cycled beer. The pH can thus also be held at an optimum set point throughout the reaction period.

Contamination by lactic acid bacteria is occasionally a problem and alcohol yield can be reduced by as much as 20%.Such contamination is more likely when stillage backset is used. Allowing contaminating organisms to accumulate and acclimatize to the fermentation conditions. Aseptic operation, with complete sterilization of the very large mash volume, was considered impractical until recently. Growth of organisms other than the seed yeast is generally restricted by the adverse conditions of low pH and high sugar or alcohol concentration, and the rapid growth of the yeast compared with contaminants was relied upon in place of aseptic techniques. Highly efficient continuous media sterilization now makes aseptic operation quite practical. The medium is heated by steam injection to 135-140°C and held in plug flow for 1-2 min. resulting in essentially complete sterilization. Cooling is by flashing to regenerate steam or by heat exchange to preheat incoming feed. The steam requirement is only 3.5 kg steam per 100 kg mash. Small spherical head fermenter vessels can be sterilized by pressurizing with steam. For very large fermenters this is impractical and antiseptic solutions such as ammonium bifluoride, iodine, sodium hypochlorite or formalin are automatically sprayed to rinse the vessel walls before filling. Heat exchangers and transfer lines can be steamed.

**4.6.2.** *Melle-Boinot fermentation process:* This is characterized by reduced fermentation time and increased yield by recycling yeast. In the batch process the cell density (5-10 billion cells/l) is very low in the beginning and hence the initial growth phase is relatively unproductive upto 15 h. In Melle-Boinot process, yeast cells from the previous fermentation are recovered by centrifugation and upto 80% are recycled and hence the initial cell density is as high as 80 billion cells/l which initiates rapid fermentation almost immediately after inoculation (Figure 10). Contamination can be a problem with continuous recycling. This is taken care by adjusting the centrifugation conditions which favour the recovery of yeast but not bacteria. Centriguged yeast is held at pH 2.0 and in presence of $CO_2$ for 4 h.

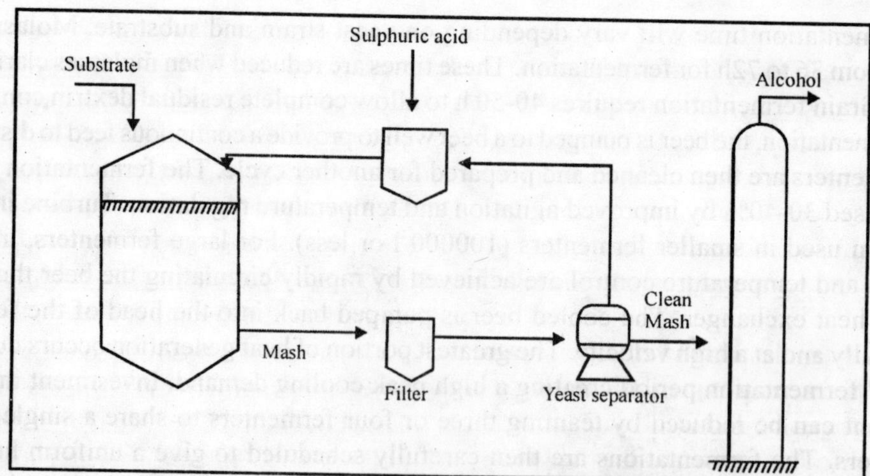


**Figure 10** Schematic representation of yeast recycling process (Melle Boinot)

**4.6.3.** *Semicontinuous fermentation process:* It is carried out in a conventional batch but after its completion 25% of the fermented broth is retained as seed for the next cycle. The fermenter is aerated for first few hours and a selective biocide is incorporated to kill bacterial contamination. Yeast is reused without any mechanical device causing reduction in the frequency of fresh yeast propagation (Figure 11). Fermentation efficiency and alcohol recovery are more than conventional batch process.

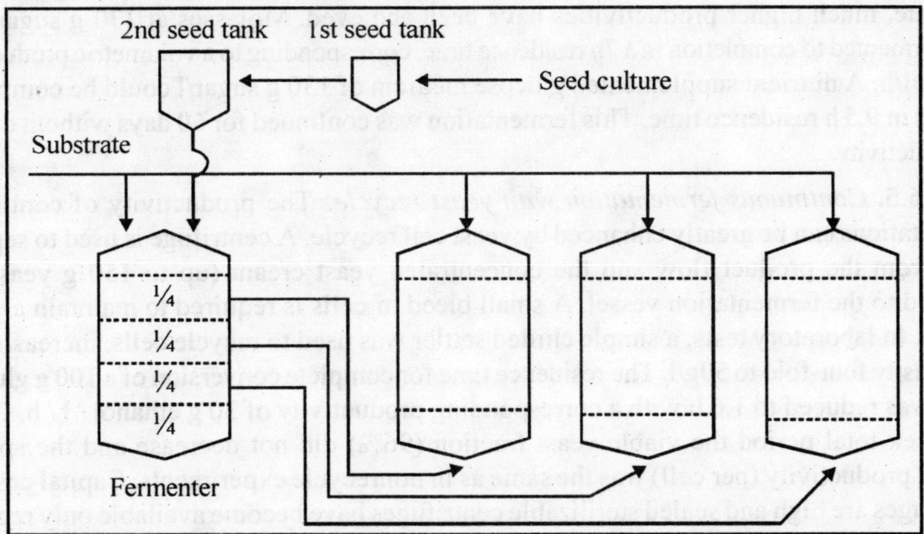


**Figure 11** Schematic representation of semi continuous fermentation process

**4.6.4.** *Simple continuous fermentation:* This includes a multi fermenter system consisting of several fermenters in series. First 1-2 fermenters are also aerated. Alcohol concentration goes on increasing in the subsequent fermenters. Continuous flow fermentation processes have been used in industrial sulfite waste liquor fermentation since the 1930s. The antiseptic qualities of sulfite liquor minimize the possibility of adverse contamination and allow long continuous runs without shutdowns for cleaning. Early attempts at continuous fermentation of molasses and grain hydrolysates on an industrial scale were unsuccessful due to contamination problems and these plants were retrofit for batch operation. With continuous media sterilization and aseptic plant techniques, the contamination problem has been overcome, as is illustrated by the success of continuous molasses fermentation plants in Europe and Japan, and many continuous beer brewing plants especially in New Zealand and Britain.Continuous culture techniques have been reviewed extensively. Detailed results of pilot scale (1800 l fermenter) tests of molasses fermentation in a simple continuous stirred flow reactor have been published. Feed is pumped continuously into the fermenter, displacing beer which then overflows from the vessel. The product beer composition is the same as the composition in the fermenter vessel and thus, if a high ethanol product concentration is desired, the fermentation will be relatively slow as the entire course of the fermentation must take place under inhibitory high ethanol concentration conditions. The throughput rate must be adjusted to be slow enough to allow growth of new yeast in the fermenter to replace yeast washed out in the overflow and to allow essentially complete utilization of the sugar.

Agitation by stirring or gas sparging was found to be especially important for successful continuous flow fermentation. With agitation, blackstrap molasses at 140g/l concentration could be 95% utilized in a 21 h residence time compared with the batch fermentation time of 40 h. Without agitation, the continuous flow fermentation residence time for complete sugar utilization was 55 h.

In laboratory tests with carefully optimized conditions of temperature, pH, agitation and flow rate, much higher productivities have been achieved. Molasses at 130 g sugar/l has been fermented to completion in a 7h residence time, corresponding to a volumetric productivity of 8.3g/l/h. A nutrient supplemented glucose medium of 130 g sugar/l could be completely utilized in 9.5 h residence time. This fermentation was continued for 60 days without decline in productivity.

**4.6.5.** *Continuous fermentation with yeast recycle:* The productivity of continuous fermentations can be greatly enhanced by yeast cell recycle. A centrifuge is used to separate yeast from the product flow and the concentrated yeast cream (up to 150 g yeast/l) is recycled to the fermentation vessel. A small bleed of cells is required to maintain a viable culture. In laboratory tests, a simple chilled settler was used to recycle cells, increasing the cell density four-fold to 50g/l. The residence time for complete conversion of a 100 g glucose/l feed was reduced to 1.6 h with a corresponding productivity of 30 g ethanol / l / h. Over a two-week total period the viable yeast fraction (96%) did not decrease and the specific ethanol productivity (per cell) was the same as in nonrecycle experiments. Capital costs for centrifuges are high and sealed sterilizable centrifuges have become available only recently. This has slowed the application of cell recycling in industrial fermentation. Attempts have been made to develop cheaper alternative cell recycle methods. Simple cell settling systems have been developed wherein the cells are thermally shocked (to temporarily halt $CO_2$ evolution) and allowed to gravity settle. Very large settling vessels are required, however, whirlpool separators have been used industrially in beer manufacture. The fermenter broth is pumped tangentially into a vertical cylindrical vessel and flocculent yeast cells are deposited in a central cone for recycle. A very simple partial recycle fermenter has been developed and tested at pilot scale. The overflow is taken from a vertical pipe rising through the fermenter base and jacketed by a baffled sleeve. The region between overflow pipe and sleeve escapes agitation, allowing yeast to settle back and separate from rising beer. The cell density in the fermenter can be increased 2.5 times using flocculent rapid settling yeast strains with essentially no added equipment cost.

**4.6.6.** *Series arranged, continuous flow fermentation:* Fermentation throughput can be increased when very complete sugar utilization or high ethanol product concentration is required by using continuous flow fermenters arranged in series. In laboratory tests the residence time for complete sugar utilization (upto 0.3% residual sugar) in a single fermenter increased from 9.5 h for a 100 g/l glucose feed to 25 h for a 160 g /l feed.

With two fermenters arranged in series the residence time can be chosen so that the sugar is only partially utilized in the first with fermentation completed in the second. Ethanol inhibition is reduced in the first vessel, allowing a faster throughput. The second, lower productivity fermenter must now convert less sugar than if operated alone and it too can be operated at an increased throughput.

For a high product concentration (80 g ethanol/l) the productivity of a two-stage system has been as much as 2.3 times higher than a single stage. In other tests, by adding equal volume fermentation stages at constant molasses feed rate, feed sugar concentration could be increased. Alcohol product concentration at essentially complete sugar utilization was increased from 52g/l for one stage to 73 g/l for two and 99 g/l for three stages.

When multiple vessels are used in series, conditions in each can be controlled independently for optimal performance. The first vessel is kept at 27°C and aerated conditions to optimize cell growth to start the fermentation. The remaining vessels are not aerated and are kept at 30°C for higher specific ethanol productivity.

For fermentation of grain, dextrin conversion would ordinarily be limiting and no advantage would be found in the series continuous flow process. It has been found economical to hydrolyze the grain completely and then conduct a continuous fermentation. Two fermenters in series are generally used with the first aerated to promote yeast growth. Slow fermenting brewer's yeast and low temperatures (15°C) are used so that the fermentation time is typically 28 h. This is a considerable improvement, however, over the typical 2-4 day batch beer fermentation. Alternatively, acid hydrolyzed grain has been fermented in a two-stage continuous flow system at pilot plant (4000 l fermenter) scale to produce palatable beer in a 12 h residence time. Stable operation was maintained for 6 months. Another industrial process combines simultaneous *Aspergillus* enzymatic dextrin conversion and yeast fermentation in a four-vessel continuous series flow system with yeast recycle. Conversion of glucose to ethanol apparently increases the dextrin conversion rate and the fermentation time is reduced from 68 h to only 17 h. Series flow arrangements are especially advantageous in sulfite liquor processing, where the sugar concentration is low and a high degree of utilization results in yeast starvation effects in a single vessel.

**4.6.7.** *Biostil process:* Modification of continuous fermentation process with yeast recycle whereby the fermentation and distillation are closely coupled and a very high stillage backsetting rate is used. The fermenter medium is continuously cycled (through a centrifuge for yeast recycle) to a small rectifying column where ethanol is removed (Figure 12). The yeast cell density is maintained at 500 billion cells/l where a special osmophilic yeast strain – *Schizosaccharomyces pombe* is used.

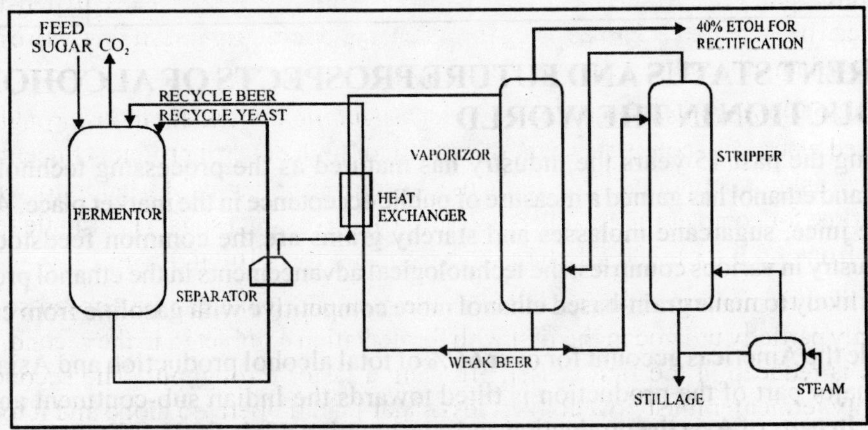


**Figure 12** Schematic representation of Biostil process

### 4.7. Comparison of industrial fermentation processes

The table 6 compares fermentation cycle times and alcohol product concentrations for Melle-Boinot batch and series vessel continuous molasses alcohol plants as well as the conventional batch fermentation. The Melle-Boinot process results in a 66% reduction in required fermenter capacity as compared with batch fermentation. Two 80000 l fermenter vessels are eliminated while one added 250 l/min capacity centrifuge is required. A net reduction in purchased equipment cost results. The continuous process gives an 80% reduction in required fermenter capacity for an even greater reduction in capital investment. Additional advantages cited for the continuous process are: (1) stable steady state fermentation characteristics; (2) improved automation and better control based on real time sampling of the product; (3) a slight increase in ethanol yield; (4) reduced labor; (5) reduced distillation and stillage drying energy requirement for the higher concentration beer and (6) reduced cleaning expenses.

Continuous fermentation is also preferred for waste sulfite liquor fermentation. The series continuous process can be used in the simultaneous microbial enzyme saccharification and yeast fermentation of starch with a 70% reduction in fermentation time. Fermentation savings comparable to those achieved with molasses are possible with continuous fermentation of starch which has been completely prehydrolyzed with acid or high activity glucoamylase. The further adoption of continuous fermentation for starches should be expected.

The Biostil process allows substantial savings in stillage processing costs and should become more widely accepted as further industrial experience is accumulated.

**Table 6** Comparison of industrial fermentation processes

| Process | Cycle Time (h) | | | Product concentration (g/l) | Productivity (g/l/h) |
|---|---|---|---|---|---|
| | Fermentation | Emptying | Total | | |
| Batch | 32 | 4 | 36 | 79 | 2.2 |
| Melle-Boinot | 8 | 4 | 12 | 79 | 6.6 |
| Series Continuous | 8 | 0 | 8 | 86 | 10.8 |

## 5. CURRENT STATUS AND FUTURE PROSPECTS OF ALCOHOL PRODUCTION IN THE WORLD

During the past 15 years the industry has matured as the processing technology has improved and ethanol has gained a measure of public acceptance in the market place. Although sugarcane juice, sugarcane molasses and starchy grains are the common feedstocks used by the industry in various countries, the technological advancements in the ethanol production process is likely to make grain-based ethanol more competitive with gasoline from crude oil.

While the Americas account for over 66% of total alcohol production and Asia, a mere 19%. A major part of the production is tilted towards the Indian sub-continent and North East Asia. In parts of Asia the production and consumption of alcohol is affected by religious considerations, like in West Asia. Pakistan, which has a substantial molasses production, mainly trades in molasses instead of processing it. China is the largest producer of ethanol in the region; however, the industry remains fragmented, largely broken into small scale and state holdings across provinces. At the last estimate, there were about 431 units producing alcohol at 3,000 million l / year. India is the second largest producer of Ethanol with an organized industry producing 1800 million l / year. Japan, although the highest per capita consumer of alcohol, is the largest importer of alcohol/ raw spirit which it then redistills locally. Thailand and Indonesia are the other two countries in the region, which have seen a steady growth of alcohol producing industry in the past five years. Going by the production figures, Indonesia has achieved the highest rate of growth followed by Thailand. The reason for this growth lies in its export potential. Both Thailand and Indonesia export alcohol to Japan. Thailand is the largest exporter in the region with over 60 million liters exported in 1997. Indonesia and Thailand have a fairly organized sugar-alcohol link that makes it a very steady business with assured raw material supply.

The world ethanol production in 2001 got a strong rebound after several years of decline. Overall output was 31.4 billion liters as compared to 29.9 billion liters in 2000 and 31.1 billion liters in 1999. Nevertheless, the world total is still below the all-time high reached in 1997, when a total of 33.0 billion liters were produced. Even though the bulk of production still comes from Brazil and the USA, two countries with an elaborate fuel ethanol program, there are interesting developments in other countries as well. There are some quarters that seek to establish fuel ethanol as an energy commodity, just like gasoline or diesel. There is some likelihood that this plan will be successfully implemented, once all these ambitious fuel ethanol programmes, which have been announced over the last two years, are put in place. In this case, the world market would have enough liquidity to facilitate a transparent price formation process and, as a result, attract the interest of the players involved. In fact, this instrument could have positive repercussions on use and production world-wide. If the burden of fuel ethanol production is shared by a sufficient number of producers, crop shortfalls in a specific country need not necessarily disrupt supplies, as these could be sourced from elsewhere.

Brazil led world ethanol production in 2004, distilling 15 billion liters from sugarcane juice followed by United States with the annual production of 13 billion liters almost exclusively from corn. China's wheat- and corn-rich provinces produced nearly 1 billion gallons of ethanol, and India turned out 1.8 billion litres made from sugarcane. France, the front-runner in the European Union's attempt to boost ethanol use, produced over 0.7 billion litres from sugar beets and wheat. In all, the world produced enough ethanol to displace roughly two percent of total gasoline consumption.

India produces only about 1.4 billion liters from sugarcane molasses in about 300 distilleries largely by conventional batch fermentation process. This production in India is very low in terms of its size and per capita production. India uses this alcohol in the ratio of 40:60 for potable and industrial purposes. As per reports at present actual production of alcohol is around 1.8 billion liters per annum against the installed capacity of 3.2 billion liters in the country. India consumes about 1.2 billion liters of alcohol. This leaves a surplus stock of 0.4 billion liters which can be exported. Efforts are on at Government of India's level to utilize this surplus as fuel-ethanol additive to gasoline. Government of India is in process of framing a National Policy for alcohol production similar to the alcohol policy already exercised by Brazil and some advanced countries. In July-August 2000, there was sharp rise in crude oil prices. While there has been a drop in consumption of fuel world-over to counter the increase in prices, in India we saw a renewed interest in the promotion of ethanol production for use as fuel. Ministry of Petroleum announced its intention to introduce fuel ethanol programme. In October 2000 pilot projects were announced in the state of Maharashtra and Uttar Pradesh. Five percent addition of ethanol to gasoline would require approximately 350 million liters of ethanol annually at the current rate of consumption of motor fuel. An addition of 10% would mean 700 million liters which constitutes about 40% of the annual ethanol production. As a sharp departure from other, for fuel ethanol programmes, no subsidies are planned - neither for the producers nor for the oil companies. Full scale programme has been chalked out on the success of pilot projects firm schedule for the full scale programme has not been drawn yet, but, is in different stages of framing up for the implementation of the ambitious project on oil saving by substituting with fuel alcohol.

The recent figures of world ethanol production and a summary of ethanol production examples with recent developments worldwide are depicted in Tables 7 and 8. If ethanol is to become a major part of the world fuel supply without competing with food and forests, it's primary source will not be grains or even sugar crops; it will be more-abundant and land-efficient cellulosic feedstocks, such as agricultural and forest residues, grasses, and fast-growing trees. Promising new technologies are being developed that use enzymes to break down cellulose and release the plants' sugars for fermentation into ethanol. A demonstration plant using this technology opened in Canada last year, and large-scale production is expected to be commercially viable by 2015.

**Table 7** World Ethanol Production, 2004

| Country | Million liters | Country | Million liters |
|---|---|---|---|
| Brazil | 15098,365 | Australia | 124.905 |
| United States | 13379.975 | Japan | 117.335 |
| China | 3648.74 | Pakistan | 98.41 |
| India | 1748.67 | Sweden | 98.41 |
| France | 828.915 | Philippines | 83.27 |
| Russia | 749.43 | South Korea | 83.27 |
| South Africa | 416.35 | Guatemala | 64.345 |
| United Kingdom | 401.21 | Cuba | 60.56 |
| Saudi Arabia | 299.015 | Ecuador | 45.42 |
| Spain | 299.015 | Mexico | 34.065 |
| Thailand | 280.09 | Nicaragua | 30.28 |
| Germany | 268.735 | Mauritius | 22.71 |
| Ukraine | 249.81 | Zimbabwe | 22.71 |
| Canada | 230.885 | Kenya | 11.355 |
| Poland | 200.605 | Swaziland | 11.335 |
| Indonesia | 166.54 | Others | 1279.33 |
| Argentina | 158.97 | **Total** | **40764.45** |
| Italy | 151.4 | | |

*Contd...*

**Table 8** Ethanol production examples worldwide

| Location | Feedstock | Ethanol Production |
|---|---|---|
| **South America** | | |
| **Brazil** | Sugarcane | The Brazilian government launched its National Fuel Alcohol Program in the mid-1970s, and by 1980 ethanol use had overtook gasoline. Since price liberalization in 1999, ethanol has maintained prices a third lower than gasoline. Brazil is the world's leading ethanol producer and exporter, distilling nearly 15 billion liters in 2004. |
| **Peru** | Sugarcane | In 2002 Peru announced the "Mega-Project," a plan to build up to 20 distilleries and an ethanol pipeline from the interior to the port of Bajovar. Up to 600,000 acres of sugarcane will be planted in forest areas now used for coca leaf production. The government hopes to export 1136 million liters of ethanol by 2010. |
| **Colombia** | Sugarcane | From 2006 the use of 10 percent ethanol in fuel will be mandated in cities with populations over 500,000, requiring the cultivation of an additional 370,000 acres of sugar cane and nine new ethanol plants to produce the necessary 984 million liters a year. |
| **Central America** | | Sugarcane El Salvador, Guatemala, Honduras, Nicaragua and Costa Rica project total output to reach 500 million liters by 2010, enough to allow for a 10 percent ethanol blend in gasoline. Costa Rica, Jamaica and El Salvador export ethanol fuel blends to the United States duty-free under the Caribbean Basin Economic Recovery Act, and are looking to increase exports. |
| **North America** | | |
| **United States** | Corn | In 2004, 35 million tons of corn (12 percent of the U.S. corn crop) were used to produce 13 billion liters of ethanol. Capacity is expected to top 16.7 billion by late 2005 as 16 new plants come on line. Currently there are 81 plants ranging in size from 4 to 1136 million liters annual capacity, half of which are farmer-owned. |
| **Canada** | Corn, wheat, barley | Canada produced 231 million liters of ethanol in 2004. To meet Kyoto Protocol commitments, the country aims to replace 35 percent of its gasoline use with E10 blends, requiring production of 1325 million liters of ethanol. 7 new plants with total capacity of 757 million liters are planned under the Ethanol Expansion Program. Ontario, Saskatchewan, and Manitoba are already promoting ethanol through production subsidies, tax breaks, and blending requirements. |
| **Asia / Oceania** | | |
| **China** | Corn, wheat | China is the third largest ethanol producer, with 3649 million liters in 2004. Since 2001, China has promoted ethanol-based fuel on a pilot basis in five cities in its central and northeastern regions (Zhengzhou, Luoyang and Nanyang in Henan and Harbin and Zhaodong in Heilongjiang province). The Jilin Tianhe Ethanol Distillery, the largest in the world, is producing 908 million liters per year, and has a potential final capacity of 1211 million liters per year. |

*Contd...*

| | | |
|---|---|---|
| **India** | Sugarcane | Since 2003, India's government has mandated use of E5 in nine states and enacted an excise duty exemption for ethanol. Sugar producers are planning to build 20 new ethanol plants in addition to 10 existing plants, with most located in Uttar Pradesh, Maharashtra and Tamil Nadu. Estimated annual ethanol needs for an E5 blend is 371 million liters, against actual production of 1749 million liters in 2004 and total capacity of 2699 million liters per year. |
| **Thailand** | Cassava, sugarcane, rice | Thailand has mandated a 10 percent ethanol mix starting in 2007, which would boost production from 280 million liters in 2004 to 1499 million liters. 18 new ethanol plants are being developed, and producers will enjoy several tax breaks. |
| **Australia** | Grains, sugarcane, sweet sorghum | The Australian government has supported ethanol since 2000 with a range of tax exemptions & production subsidies, aiming to produce 348.22 million liters of biofuel by 2010, enough to replace one percent of total fuel supply. 2004 production stood at 125 million liters. |
| **Europe** | | |
| **EU** | Grains, sugar beet | A non-binding directive by the European Commission asks EU countries to meet 2 percent of vehicle fuel demand with biofuels by the end of 2005 and 5.75 percent by 2010. In practice, it is only likely to reach 1.5 percent by the end of 2005, up from 0.2 percent in 2000 and 0.6 percent in 2003. Assuming 12.9 billion liters of ethanol were produced, some 12.6 million acres of grain and 1.5 million acres of sugar beet would be required. Member states may also exempt biofuels from the tax on petroleum products. |
| **France** | Sugarbeet, wheat, corn | France leads the European Union, with ethanol production jumping from 88.6 million liters in 2003 to 829 million liters in 2004, and plans to add another 367 million liters by 2008. |
| **Germany** | Rye, wheat | 2004 production stood at 269 million literss. Three new distilleries should bring domestic capacity to nearly 568 million liters annually, requiring an additional 1.4 million tons of rye and wheat in 2005, 3 percent of Germany's 2004 grain crop. |
| **Spain** | Wheat, barley | In 2004, ethanol production stood at 299 million liters, up from 208 million liters in 2003. Total Spanish capacity has been estimated at about 500 million liters per year. |
| **Sweden** | Wheat | E5 blends have been in wide use since 2003 and E85 is now available at some fifty service stations. Ethanol production reached 98.4 million liters in 2004, up from 60 million liters in 2003, and Sweden now imports ethanol from Brazil and Spain. |
| **Africa** | | |
| **South Africa** | Corn, sugarcane | South Africa's state gas company is a leading producer of synthetic ethanol from coal, but the country is now moving toward crop-based ethanol production. Total ethanol production in 2004 was 416.3 million liters. 8 new plants are may be built by the end of 2006, enough to produce 1211 million liters of ethanol. |

*Contd...*

| Zimbabwe | Sugarcane | Most gasoline sold in Zimbabwe for the past 20 years has contained 12-15 percent ethanol. Production capacity has exceeded 37.85 million liters since 1983, though actual production stood at only 22.7 million liters in 2004. |
|---|---|---|

# 6. SUCCESS FACTORS OF INCREASED WORLD ETHANOL PRODUCTION

Fuel ethanol production and use is expected to rise strongly and it will go along with an ever wider geographical spread. Ten years ago, there were only a handful of countries producing ethanol. The largest was Brazil, where ethanol is produced from molasses and sugar cane juice. The US produces mostly corn alcohol and in France, sugar beets are being used. In some African countries, sugar cane was processed into fuel alcohol.

In 2003, there were some 13 countries spread over all five continents which actually use ethyl alcohol as a fuel component. Looking into the future, the world fuel ethanol map may look like this in ten years time: the Americas are likely to be almost completely covered by fuel ethanol programs. Moreover, the green fuel will likely be established in the European Union as well as in India, Thailand, China, Australia and possibly Japan to name the largest nations.

What are the reasons for the overwhelming success of fuel ethanol? As fuel ethanol is competing with gasoline, a direct comparison between the two products is possible. Because ethanol is invariably more expensive to produce than gasoline, if actual market prices are taken account of, political objectives come into play. Ethanol has been promoted because it has a positive net energy balance, that means that the energy contained in a tonne of ethanol is greater than the energy required to produce this tonne. Moreover, it has been demonstrated that it has a less severe impact on the environment than conventional gasoline or other petroleum derived additives. As such it is also less dangerous to health. From a macro-economic point of view, it is thought to be good for the development of disadvantaged rural areas by promoting an industry which creates jobs. Furthermore it can help to reduce the dependence on oil imports and, finally, it may be regarded as a means to promote advances in biotechnology, particularly if one thinks of all the research that is going on in the biomass-to-ethanol sector.

If we look at the biofuel programs that are already in existence, there are three key success factors which must be considered:

- Firstly, the abundance and cheapness of feedstocks used for their production.
- Secondly the technology involved.
- Thirdly a supportive political framework.

## 6.1. The feedstock issue

Let's look at the feedstocks issue first. According to our 2003 survey, around 61% of world ethanol production is being produced from sugar crops, be it sugar beet, sugar cane or molasses, while the remainder is being produced from grains and here maize or corn is the dominating feedstock. Feedstocks crucially determine the profitability of fuel ethanol production. There are various ways to look at the issue. In Figure 13, the theoretical per ha ethanol yields of the three major feedstocks currently in use are plotted.

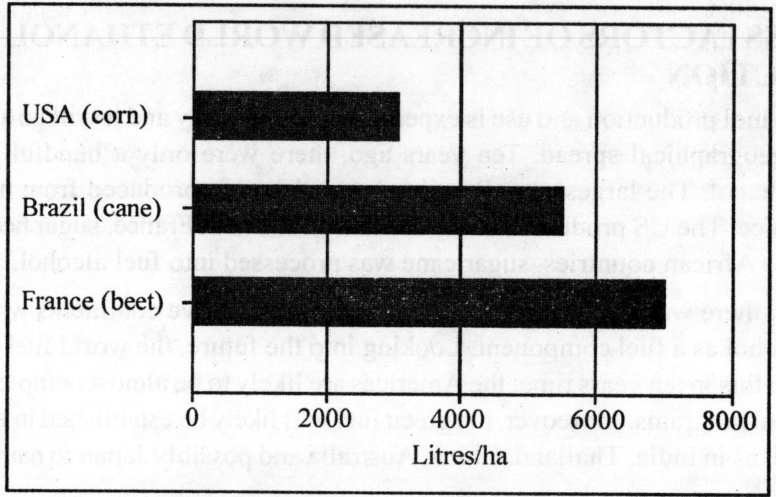


**Figure 13** Ethanol yields (per hectare) of three major feedstocks

In the USA, corn (maize) is the predominant raw material for fuel ethanol production. (Figure 13). The ethanol yields per ha are the lowest by comparison. A middle position is held by sugar cane in Brazil but the highest ethanol yields per ha may be realized with sugar beets, particularly if calculations are based on the rather high yields that may be achieved in the EU's leading producer, France.

However, if we look at the factor productivity of the various ingredients (Figure 14), we can see that corn clearly takes the top spot with almost 400 liters of ethanol produced per tonne of feedstock. Sugar cane has an even lower factor productivity than sugar beet.

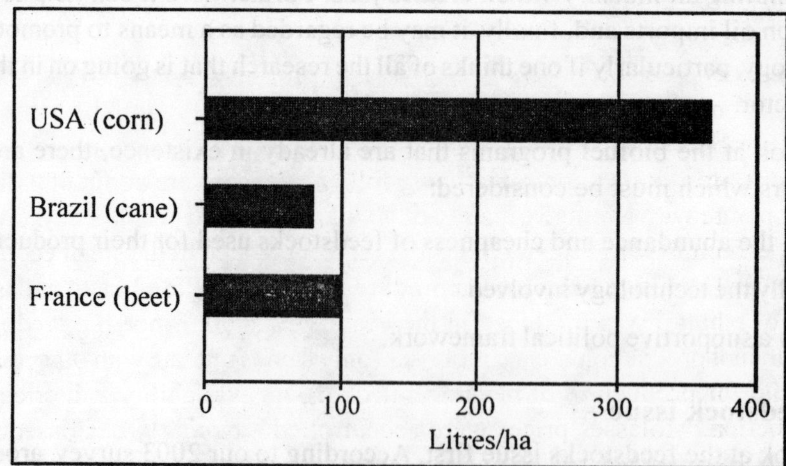


**Figure 14** Ethanol yields (per ton) of three major seed stocks

If we look at the gross feedstock costs per gallon of fuel ethanol produced, it is sugar cane grown in the Centre/South of Brazil which clearly leaves the rest of the competition behind (Figure 14). We may arrive at a first conclusion concerning the role of feedstocks in biofuel production. Leaving aside biomass as a feedstock, the raw material accounts for around 70 to 80% of the overall costs of fuel ethanol. Therefore, their relative abundance plays a crucial role in getting the fuel alcohol industry started in a particular country. The highly regulated price of sugar beet in the case of the European Union may have acted as an obstacle to the emergence of a viable large-scale ethanol industry there. This may change in the future, not least because of developments in the political sphere.

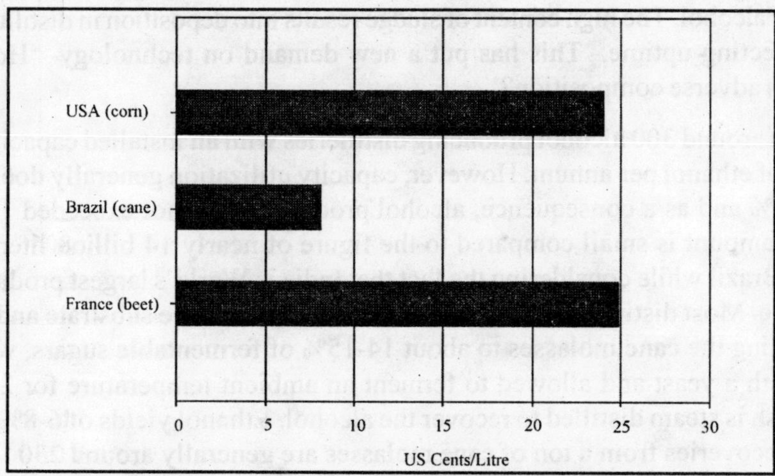


**Figure 15** Gross feedstock cost per liter of ethanol

## 7. STATUS OF ALCOHOL PRODUCTION IN INDIA

The ethanol industry in India has been a traditional industry, however, winds of change are blowing in India. This is very evident from the fact that molasses, which has been the most controlled commodity, has been decontrolled across all states with only a few states still exercising partial control. To predict the future trends in the ethanol industry we need to study some of the key trends in the market over the past decade. India is 28 states in the context of one. Each state has different laws for controlling taxes and movement of molasses and ethanol. In some states, like in the northern state of Uttar Pradesh, use of molasses for industrial and potable use -local liquor and for superior grade liquor category - is still controlled, whilst in other states total decontrol is being followed. There are about 300 distilleries in India which operate with average capacity of 40,000 to 50,000 liters per day, producing 1.8 billion liters per annum. The ethanol industry is closely related to the sugar economy and its cycles. With over 18 million MT of sugar produced, about 8 million MT of molasses and 1.5 billion liters of ethanol was produced in the year 1999-2000. Since the predominant raw material for ethanol production is cane molasses (only 8 units operate with other than molasses as raw material for production of beverage alcohol), its availability and price affects the alcohol production. Molasses prices were decontrolled around 1992. Consequently price shot up from US $ 6 per MT to US $ 60 per MT. This, of course, was an initial reaction as a result of which, the price settled down to around US $ 25-30 within a year. However, due to the increasing interest in the use of alcohol as biofuel the price of molasses is again experiencing an upward trend.

In terms of quality of molasses, there has been a significant change in the composition of molasses. As the sugar mills are improving their extraction efficiency, composition of molasses is adversely affecting the distillery industry. Content of fermentable sugar has dropped from 47-50% to as low as 35% by weight. This, of course, has notable impact on the yield and efficiency. Content of volatile organic acids, caramel and sludge has been on the increase as the extraction efficiency has been increasing. Volatile organic acids and caramel pose severe problems to activity of yeast, thereby affecting not only efficiency but also quality of alcohol. The high content of sludge results into deposition in distillation column adversely affecting uptime. This has put a new demand on technology- 'How to tackle molasses with adverse composition?'

India has around 300 alcohol producing distilleries with an installed capacity nearly 3.2 billion liters of ethanol per annum. However, capacity utilization generally does not exceed more than 60% and as a consequence, alcohol production has not exceeded 1.8 billion l / annum. This amount is small compared to the figure of nearly 14 billion liters of alcohol produced by Brazil while considering the fact that India is World's largest producer of sugar and sugar cane. Most distilleries in India use cane molasses as the substrate and the process involves diluting the cane molasses to about 14-15% of fermentable sugars, which is then inoculated with a yeast and allowed to ferment an ambient temperature for 24-48 h after which the wash is steam distilled to recover the alcohol. Ethanol yields of 6-8% are average and alcohol recoveries from a ton of cane molasses are generally around 230 l / ton. There is no control over the quality of molasses and as a consequence the recoveries are largely variable. The major substrate for ethanol production in India is cane molasses. Generally about 50-60% of the cane produced in India is processed in sugar mills and the yield of cane molasses i.e. around 30% of the amount of the sugar produced annually. With the price of sugar on the steady rise, there is tendency to recover the last crystal of sugar from the cane and as a consequence, harsh methods of processing are used. As a consequence, the quality of molasses produced is extremely poor. Table 9 shows the quality of Indian cane molasses vis a vis the cane molasses from other countries. Indian cane molasses contain the lowest amount of fermentable sugars and the highest amount of ash material. Further, the quality of molasses in India also varies from factory to factory.

**Table 9** Average composition of Indian cane molasses vis a vis other cane producing countries

| Component | India | Brazil | Philipines | Thailand |
|---|---|---|---|---|
| pH | 4.5-5.0 | - | 5.3 | 5.2 |
| Total solids g% | 75 | - | 73 | 72 |
| Total ash g% | 12 | 7 | 6 | 7 |
| Total sugars % | 42-45 | 56 | 57 | 53 |
| Invert sugars % | 15 | - | 20 | 19 |
| Unfermentable sugars % | 12 | 5 | 6 | 7 |
| CaO % | 2 | 0.5 | 0.7 | 0.5 |
| Total Nitrogen % | 0.1 | 0.3 | 0.2 | 0.6 |

Table 10 shows the composition of molasses obtained from different distilleries in India. The composition not only varies with regard to the fermentable sugars but also with the nutrients available. As a consequence, each batch of molasses needs to be examined for nutrient supplementation particularly nitrogen and phosphate, to ensure satisfactory recoveries. For efficient production of ethanol, a number of factors are involved. These include the substrate, the yeast strain, the fermentation conditions, distillation conditions and finally the effluent treatment process. All these factors individually and collectively determine the final efficiency of the fermentation process. In recent years, a number of limitations have been faced by the Indian alcohol distilleries at each of these stages.

**Table 10** Composition of cane molasses from different Indian distilleries

| Distillery No. | % Total sugars | % Fermentable sugars | pH | % Ash |
|---|---|---|---|---|
| 1 | 51.8 | 47.2 | 5.5 | 13.0 |
| 2 | 54.2 | 49.6 | 4.8 | 9.0 |
| 3 | 50.6 | 44.2 | 5.0 | 13.6 |
| 4 | 48.5 | 43.9 | 5.3 | 11.6 |
| 5 | 50.6 | 45.6 | 5.0 | 18.2 |
| 6 | 49.3 | 44.2 | 4.8 | 13.8 |
| 7 | 48.5 | 4.5 | 5.1 | 13.0 |
| 8 | 42.4 | 38.4 | 4.8 | 12.0 |

Ethylene, being cheaper, was used earlier (World War II) for ethanol production by chemical synthesis. As now, the ethylene cost is high, production of ethanol from agricultural renewable sources is preferred. Due to short supply of sugarcane molasses, its rising cost, deteriorating quality and technological advancement in Indian sugar industry, the industries have shown keen enthusiasm to utilize other alternate substrates for ethanol production. However, the cost of production of ethanol from such substrates is always more than that of cost of ethanol production from molasses (Table 11).

**Table 11** Approximate cost of production of ethanol from different substrates by fermentation

| Substrate | Cost/ton Rs. | EtOH yield l/ton | Appoximate Cost / l EtOH (Rs.) |
|---|---|---|---|
| Sucrose | 14,000 | 520 | 27 |
| Jaggery | 10,000 | 400 | 25 |
| Molasses | 3000 | 225 | 13 |
| Wheat and other starchy materials | 5,000 | 350 | 15 |

Other than cane molasses there are a number of additional substrates which can be employed for ethanol production (Table 11) in India, but, due to some limitations, many have not been exploited yet on commercial scale. In the directly fermentable category of substrates, the substrates other than cane molasses are not available in adequate quantities in India for use in alcohol production. However, considering the recent increase in cost of cane molasses as a result of price decontrol, sugarcane juice appears to be an attractive substrate for ethanol production. Use of this substrate does not need any change in the fermentation conditions and further will create no problems of effluent disposal. Average yields of 50-60 l of ethanol / ton of cane can be expected. At the current production of nearly 250 million tons of cane annually, a part of this can be diverted for the production of ethanol by direct fermentation. Production of ethanol by direct fermentation of cane juice has also been examined in India and yields of 7-8% ethanol in the fermented wash have been reported (Table 12). Economical production of ethanol from direct fermentation of sugarcane is now possible mainly because of the changed price structure of cane molasses.

**Table 12** Ethanol production from cane juice

| Fermentation time (h) | %(v/v) Ethanol in Farm cane juice | % (v/v) Ethanol in sugar mill juice |
|---|---|---|
| 6 | 1.7 | 1.6 |
| 24 | 5.9 | 4.9 |
| 30 | 7.8 | 5.0 |
| 36 | 8.3 | 5.1 |

The other directly fermentable substrates available are sweet sorghum juice, jaggery, sugar beet etc. The use of these substrates have some limitations. Sweet sorghum appears to be an excellent substrate since the stems in addition to yielding the juice, can also yield valuable grain and fodder. However, difficulties in its successful cultivation in India, has placed restrictions on its use. Should easy cultivation be possible, sweet sorghum appears to be an interesting crop for ethanol production. The fermentation of the juice does not need any changes in the fermentation conditions and as discussed later, the grain can also be used for ethanol production after hydrolysis.

Substrate such as Jaggery is being used for ethanol production, especially when molasses is in short supply. Fermentation of this does not need any changes in fermentation procedure. Infact, use of jaggery or even concentrated cane juice appears to be best alternative for producing ethanol from cane molasses. These substrates are directly fermentable and moreover cause no serious effluent disposal problems.

Another substrate that has been examined in India for the production of ethanol is sugar beet. This crop is next only to sugarcane in using the solar energy efficiently. However, it is not grown at large scale in India except in Ganganagar region of Rajasthan and in the Sunderbans of West Bengal. With slight modification of fermentation conditions, both the sugar beet as well as beet juice can be efficiently used for the production of ethanol.

The second group of substrates available for ethanol fermentation in India include the starchy substrates. These substrates need to be first hydrolysed before they can be fermented by yeast to produce ethanol. The four major starchy substrates available in adequate quantities in country are rice, wheat, corn and sorghum. Since the first three substrates are a part of staple diet of majority of people of India and are not available in surplus the only potential substrate that can be exploited for commercial production of alcohol in India, in future, appears to be sorghum. This is already in use by a one of the largest alcohol producing companies of India, but, the yields are not sufficient. The procedures used by this company are kept secret which needs to be looked into to improve the ethanol prodution.

The technology of producing alcohol from starchy substrates is well documented. Briefly, the process involves gelatinisation of starchy material using bacterial α-amylases and fungal glucoamylases. The hydrolysate is supplemented with nutrients and fermented by strains of *Saccharomyces cerevisiae*. Recently, high gravity wheat hydrolysate fermentations have been reported and alcohol yields as high as 18% have been recorded. The main problem in using starchy substrates as alternate substrates for alcohol production is that the thermostable microbial enzymes for the hydrolysis of starch are not produced in India in sufficient quantities, though, Biocon (India) Limited, Bangalore is the main producer of such enzymes, which are generally used in the hydrolysis of barley malt, carried out in the temperarture range of 60-65°C. However, the high-temperature amylases, required for the liquefaction of wheat, rice, corn and sorghum, are generally procured from Novo Nordisk, Denmark, the leader in the world enzyme market. According to a recent survey, a largest alcohol industry of Northern India, (the third largest alcohol group of India) imports the thermostable amylases worth rupees 1.0 crore annually for the hydrolysis of various cereal grains for ethanol production. The main advantage of this category of substrates is that they does not lead to the production of an effluent that is difficult to handle. Infact, the beer after the alcohol is removed, which can be processed to a high value low alcohol beverage rich in yeast nutrients. While in the western world, the fermented grains are the main high value products of the grain based distilleries. In India, the effluent can be the main revenue earner with ethanol becoming a by product. Microbiological techniques to quantitatively convert the starch present in the cereal grains to alcohol need to be only standardized and the development of some efficient, in-house amylase preparations is the requirement of the present scenario, if the cost of ethanol production from starchy biomass is to be made comparable to that of molasses. At the moment, the cost of sorghum and other cereal grains including wheat in India is comparable or even less than that of molasses in some parts the country, but the cost of ethanol production from starchy biomass is very high as compared to that from molasses (Table 11).

At present the hydrolysis of starch is required prior to its fermentation since yeasts that can ferment starch directly to yield high levels of ethanol are not available at the moment. Strains capable of fermenting starch substrates directly to ethanol have been derived by genetic manipulation, but, these do not produce significantly high levels of ethanol for use in distilleries. Also currently available technologies for the production of ethanol from starchy substrates need high temperature gelatinization of the substrate before enzymatic hydrolysis. Fungi producing enzymes that hydrolyze uncooked starch have been reported and their use in raw starch hydrolysis would substantially reduce the cost of alcohol production from starchy materials.

While India is the second largest producer of sorghum in the world, the yield of 840 kg / ha is the lowest amongst the major sorghum-producing countries in the world. The world average was 1435 kg / ha in 1994. Although yield of sorghum in India is much lower than the world average, it has been consistently increasing during the recent past. The major applications of sorghum in India are animal feed, ethanol production, and production of starch/starch derivatives. Sorghum grain fetches lower price when used for animal feed and the quantity used for starch production is very small. Therefore, use of sorghum for production of alcohol has a lot of potential and this will also help the poor farmers because they will get a good price for their produce. Sorghum also offers several advantages over maize for alcohol production. Firstly, it has higher starch compared to maize (Table 13). Secondly, sorghum is grown in both kharif and rabi seasons and the kharif crop is mostly F1 hybrids, which have good fodder and grain yield. Lastly, out of 10–11 million tons of sorghum produced annually in India, about 2–3 million tons is wasted due to grain blackening following unseasonal rains. This grain is not suitable for human or animal consumption. Hence it is sold at a low price and thus gives low returns to the farmer. Thousands of marginal farmers will be benefited if such grains are used for alcohol production.

Although there are few distilleries in India already using sorghum, their alcohol yields are low. Sorghum can potentially give good yield of alcohol of about 380 to 390 l absolute alcohol / ton of grain provided the process is optimized.

**Table 13** Comparison of composition of sorghum and maize

| Component | Content (%) | |
|---|---|---|
| | **Sorghum** | **Maize** |
| Starch | 63–68 | 60–64 |
| Moisture | 9–13 | 8–11 |
| Proteins | 9–11 | 9–11 |
| Fats and oils | 1–1.5 | 3–5 |
| Crude fiber | 1.5–2 | 1.5–2 |
| Ash | 1–2 | 1–2 |
| Other organics | 8–12 | 7–9 |

The third group of substrates includes the difficult to hydrolyze lignocellulosics. Although available in plenty in India, and is of renewable type, there are still difficulties in quantitatively converting this substrate to fermentable sugars. Also, hydrolysis of this material leads to the production of a hydrolysate containing both hexoses and pentoses and as yet there is no organisms that utilize both substrates efficiently to produce alcohol. Technology for acid hydrolysis of the cellulosic wastes and to produce alcohol is well documented. However, acid hydrolysates contain products that are inhibitory to yeast growth. Enzymatic hydrolysis of lignocellulosics has been investigated for a long time, all over the world, but as yet no efficient method is reported and commercialized. Attempts are also being made to derive yeasts that can ferment the cellulose present in lignocellulosics directly but as yet, not much success is reported. Recently much research has been focussed on the utilization of lower value substrate such as lignocellulose and agrowaste for fuel ethanol production because cost of the substrate constitutes 55-70% of the total process costs. Thus the low value lignocellulose offers a great potential for reducing the production cost and increasing the use of ethanol as a fuel additive. The biological process of ethanol fuel production utilizing lignocellulose requires:

- Delignification to liberate cellulose and hemicellulose from their complex with lignin.
- Depolymerization of the carbohydrate polymers (cellulose and hemicellulose) to produce free sugars.
- Fermentation of mixed hexose and pentose sugars to produce ethanol.

The development of the feasible biological delignification process is possible if lignin degrading microorganisms, their ecophysiological requirements and optimal biorector design are effectively coordinated. Some thermophilic anaerobic and recombinant bacteria have advantageous features for direct microbial conversion of cellulose to ethanol i.e. the simultaneous depolymerisation of cellulosic carbohydrate polymers with ethanol production. Few workers have used a verified mathematical model to examine the most critical biochemical engineering aspects of ethanol production from cellulose. Extensive simulations of the simultaneous saccharification and fermentation of cellulose were conducted to identify the effect of operating conditions, pretreatment effectiveness, microorganism parameters and enzyme characteristics on ethanol production. The simulation studies have shown that the biomass-enzyme interaction play a dominant role in determining the performance of solid state fermentation in batch and continuous operating modes. The digestibility of the substrate and the cellulase enzyme dosage, specific activity and composition had a profound effect on ethanol yield. Some researchers have developed a low cost fermentation medium for ethanol production from biomass. The low cost medium containing 0.3% corn steep liquor and 2.5 mM $MgSO_4.7H_2O$ alongwith pretreated poplar was similar in performance to a nutrient rich medium. Besides its low cost, the medium consisted of components that are avilable on a commercial scale. Bioconversion of pentose sugars (xylose etc.) and rice straw hydrolysate has also been studied using free and immobilized cells of *Candida shaetae* NCL 3501 in batch, fed batch and continuous culture conditions. *C. shaetae* NCL 3501 effectively utilized the sugars present in rice straw hydrolysate. Ethanol yields were higher with immobilized cells in all reactor types, i.e. batch, fed batch and continuous processes. Few workers have reported an increase in ethanol production by metabolic modulation of cellulose fermentation in *Clostridium thermocellum.* Significant quantitative difference in ethanol yield alongwith repression in acetic acid production were observed in *Clostridium thermocellum* SS21 and SS22 in the presence of $H_2$, acetone and sodium azide. The use of a thermotolerant strain of *Kluyveromyces maxianus* IMB-3 has been studied in solid state fermentation of cellulose to ethanol. At maximum achievable cellulose concentration, ethanol production increased to 6.6 g/l at 45°C, representing 21% of the theoretical yield.

## 7.1. The Indian ethanol industry - in transition

Although China is the largest producer of alcohol in the region, it is India which has the largest number of significantly sized plants in Asia that produce ethanol of 95-96% v/v concentration. The fate of the ethanol industry in India is linked to the sugar economy. Almost 50% of the distilleries are attached to sugar factories, with as many as 70% of them in the co-operative sector. In India, the distillery industry has shown remarkable resilience and progressiveness by adopting modern technology and trying to meet the environmental norms. The fact that almost 90% of the continuous fermentation plants in Asia are operating in India, is in itself a testimony to the aspiration of the industry to move to an international arena.

Cane molasses is the single largest raw material. India produces almost 25% of the world's molasses. State levies and control factors often affect the availability and the price of molasses thereby creating a lopsided demand-supply situation. The cost of production is driven higher by these factors. Marketers around the world have understood that India is in fact 28 countries in the context of one. A number of controls are exercised by the individual states (provinces) on cross-border movement of molasses and ethanol. The tax on ethanol is determined by the state and not the Federal Govt., leading to vastly different tax levels in even neighboring states. This creates some regional imbalances within the country. For example, the states of Maharashtra and Uttar Pradesh have surplus molasses which is trading within the states at US$ 10 to 20 per MT. On the other hand, in the states of Madhya Pradesh and Kerala which are deficit in molasses, the price of molasses is as high as US$ 70 to 90. This made it prudent for the distillers in these states to put up ethanol plants based on grain and cassava.

Production methods across the region are traditional. Fermentation is done in batch mode. India took a lead in successful introduction of continuous ethanol ferementation a decade ago. About 100 continuous plants have been set up all over during the last decade. The major reasons for this shift have been:

- Easier operation of the plant
- Higher efficiency of conversion
- Consistent quality of product
- Lower consumption of steam due to higher concentration of ethanol in the mash.

Lower steam consumption and consistently high qualities are the issues in distillation. The Asian region makes its liquor largely from molasses-based ethanol. With competition in the liquor industry, the requirement of a better quality of ethanol as base spirit is going up day by day. This has necessitated the installation of better distillation systems. Earlier, the level of instrumentation and controls in a distillation plant was fairly low. However, this has undergone a change and plants with better controls to achieve better and consistent quality are being put up. Several refining or re-distillation plants have been put in the last decade.

Out of the 900 million liters being consumed as beverage alcohol, 60% goes as cheap liquor and the balance for fine liquor production. Tax structure in each state shapes the supply of alcohol within the state and the price thereof. Beverage alcohol commands a premium in price of 20%-40% over industrial alcohol. Thus, the value-addition is higher for sugar mills that go in for beverage-grade alcohol. Considering the downstream blending and bottling unit, the return on investment is higher than the alcohol-based chemical manufacturing industry. However, State control over production of liquor restricts the growth of this alternative for sugar mills.

The size of the alcohol-based chemical industry is estimated to be around US$ 1 billion. Through licensing and price controls on feedstock, Central and State governments encouraged the alcohol-based chemicals industry prior to 1993. In June 1993, the Govt. of India decontrolled movement and pricing of molasses to encourage this industry.

At present, there are over 200 alcohol-based products manufactured in India, the single largest alcohol-based chemical being acetic acid. Acetic acid itself is the building block for the manufacture of several other alcohol-chemicals like ethyl acetate (EA), butyl acetate (BA), acetic anhydride, etc. Ethyl alcohol is also used to produce ethylene and its downstream products. The Table 14 lists out the capacities of major alcohol based chemicals in the country:

**Table 14** Alcohol based chemicals production in India

| Alcohol chemicals | Capacities (MT per annum) |
|---|---|
| Acetic acid | 290,000 |
| Acetic anhydride | 78,000 |
| Vinyl acetate monomer | 52,000 |
| Pyridines, picolines and other derivatives | 15,000 |
| Ethyl acetate | 65,000 |
| Mono ethylene glycol | 25,000 |
| Amines, ethoxylates and others | 50,000 |
| Pentaerythritol | 15,000 |
| Ethyl/butyl acrylates | 12,000 |
| Glyoxal | 8,000 |

Fuel ethanol, power alcohol are buzz words today. There are many reasons to justify the use of alcohol as a blend in petrol for automobiles. Environmental causes will prevail in utilizing one of the best sources of oxygenating agents in gasoline. Countries that are short of hard currency and/or have to import crude petroleum, the oxygenating agents see more reasons in doing so. In India, there are missions working on this aspect. The Ministry of Non-conventional Energy Sources, the regional Pollution Control Boards in India are following the leads in this regard. Ethanol can have a brighter future if the fuel-ethanol develops in the region. More stringent norms for environment could increase the cost of production and could restrict volumes of production. The issue needs to be addressed today. Molasses as a raw material, although available in plenty, could be leveraged for bringing down the cost of production and, hence, there could be a stronger linkage between sugar producing mills and distilleries for consistent molasses supply. Other raw materials are being looked at. As the trend for alcoholic beverages across the region is shifting to 'white liquors', the need to switch to 'grain-spirit' is greater. This will also balance out the use of molasses for industrial purposes. A review of local levies and taxes on this commodity could alter the scenario significantly. Technological upgradation could take a driver's seat in bringing down the cost of production and in ensuring improved and consistent quality of the product. The growth has been consistent at 10% (Table 15). As the movement of molasses is restricted by various constraints and barriers, almost 60% of production of ethanol is limited to 4 states, that of Uttar Pradesh, Maharashtra, Tamil Nadu and Karnataka. Another state which is fast catching up is Andhra Pradesh.

**Table15** Ethanol production in India

| Year | Production, million liters |
|---|---|
| 1999 | 1569 |
| 2000 | 1652 |
| 2001 | 1739 |
| 2002 | 1832 |
| 2003 | 1929 |

Almost 50% of the ethanol produced is consumed by the industrial sector for production of ethanol-based chemicals and the remaining 40-50% is utilized by the potable sector. It is expected that the pattern of utilization of ethanol may change only slightly in the coming years. However, with the Government removing control on price and allocation of molasses, the ethanol-based chemicals industry is facing problems as the procurement price of alcohol has increased substantially in the last two years. One probable reason for the drop in industrial consumption is realisation of lower price for chemicals as a result of heightened competition. Also, due to depressed oil prices in the last decade, the petroleum route had been quite attractive. With crude prices rallying around US$ 30 per barrel, it is now competitive to use ethanol as feedstock. A spurt is already seen in this area. Consumption of ethanol is expected to rise significantly (Table 16), provided the growth in sugar production keeps pace and fuel ethanol programme takes shape.

**Table 16** Ethanol consumption in India

| Year | Consumption, million liters |
|---|---|
| 2000 | 1240 |
| 2001 | 1290 |
| 2002 | 1342 |
| 2003 | 1396 |
| 2004 | 1451 |

Some concerns which need to be kept in perspective are: vagaries of nature could seriously influence the availability of ethanol impacting the prices adversely. This, in turn, would influence the downstream applications. The use of fuel ethanol would require production competitiveness amongst producers of ethanol.

India's transport sector is growing rapidly and presently accounts for over half of the country's oil consumption whilst the country has to import a large part of its oil needs. Hastening interest in an ethanol program was the country's sugar glut (part of which the industry is now exporting to the world market) and burgeoning supplies of molasses. The sugar industry lobbied the government to embrace a bio-ethanol programme for several years. The industry emphasised that producing fuel ethanol would absorb the sugar surplus and help the country's distillery sector, which is presently burdened with huge overcapacity, and also allow value adding to by-products, particularly molasses.

India's Minister for Petroleum and Natural Gas gave his approval in December 2001 to a proposal to launch pilot projects to test the feasibility of blending ethanol with gasoline.

Mid-March 2002 the government decided to allow the sale of E-5 across the country. On 13 September, 2002, India's government mandated that nine states and four federally ruled areas will have to sell E-5 by law from 1 January, 2003. In response India's sugar producers reportedly planned to build 20 ethanol plants before the end of the year in addition to 10 plants already constructed. Most of the plants were being constructed in Uttar Pradesh, Maharashtra and Tamil Nadu, the key sugar producing states and will chiefly use cane sugar molasses as a feedstock.

Estimated annual ethanol needs for a E-5 blend is 0.37 billion liters. A 10% blend increases the need to 0.72 billion liters. This is against installed annual production capacity of 2.7 billion liters/year and annual consumption of 1.5 billion liters. These figures have to be treated with some caution. The chemical industry, fearing higher ethanol prices as a result of the fuel alcohol programme, usually estimates the surplus to be much lower or even non-existent. The sugar industry, on the other hand, estimates capacity at 3.2 billion liters inflating the surplus.

The success of ethanol in India will depend to a significant degree on pricing. The sugar industry originally claimed that it could provide ethanol at 19 rupees / l ($0.38/liter), which is at a lower cost than the product it would substitute, MTBE, which costs 24-26 rupees per liter ($0.49-0.53/liter). The oil industry however is seeking parity between ethanol and the price of gasoline on an ex-refinery or import basis. In April 2002 the government announced a Rs 0.75 excise duty exemption. Implementation of the excise duty for ethanol which, however, was delayed however until February 2003, because the chemical industry opposed it, fearing higher prices and shortages of alcohol.

However, pricing appears to becoming a stumbling block and in June 2003 India's Petroleum Ministry announced that it would appoint a Tariff Commission to fix an appropriate price for ethanol sourced from sugar mills. Ethanol pricing in India is also complicated by differences in excise duty and sales tax across states and the central government is trying to rationalize ethanol sales tax across the country. More significantly perhaps, there are still substantial differences in the profitability of potable alcohol as against fuel alcohol and in several states. Consequently, insufficient fuel alcohol is being produced to meet demand. Other states have yet to set up sufficient production capacity. Analysts expect that there is a deficit of around 150 mln liters under the current geographic base to the fuel ethanol program; a deficit that will grow once the mandated blending requirement is extended to all states in India. Consequently, there may be a short-term market for imported Brazilian ethanol.

## 8. CONCLUSIONS

Fuel ethanol will not go away in the foreseeable future. On the contrary, world production is set to continue to grow vigorously at least up to 2012. There are various fuel ethanol projects in the pipelines around the world and, even though their implementation may be delayed, there is enough momentum in the political arena to push them through. Political support is there and in many instances the industry and the authorities are very close to reaching an agreement over a viable framework of support for fuel ethanol.

World trade is likely to grow as well but the rate of growth will depend on several factors. First of all, the sugar and alcohol economics as has been illustrated in the case of Brazil. Unless the strong link between sugar and alcohol production can be severed an additional element of volatility will be present in the equation. The same applies to the corn and corn products market in the United States, even though this relationship is not very obvious at present because of the depressed state of the corn sweeteners market. Before significant increases in ethanol exports can be expected, new investments in the origins will have to be made. It cannot be expected that the sugar and alcohol industries in the origins will be able to make these investments all by themselves. Instead, a new partnership between the producers and the importers will have to be created in order to provide the significant funds which are required to facilitate this growth. Moreover, a viable trading system would have to be established. A futures market in particular would be required in order to provide the possibility to hedge against price fluctuations. There cannot be any doubt that the big futures markets in London or New York would be willing to create such a contract as long as it can be assured that there would be sufficient liquidity in the market to make it sustainable.

Finally, the problem of subsidized production and exports would have to be resolved. At the moment, the fact that fuel ethanol is being subsidized almost anywhere in the world provides a powerful justification for high import tariffs in order to neutralize these subsidies. In fact, potential producers in the European Union argue strongly in favour of high import tariffs so that the fledgling industry in the Community can establish itself. However, if this notion forms the basis for future policy making there is every reason to be pessimistic about the prospective development of world trade. Without an effective system of international exchange fuel ethanol supplies are bound to be volatile resulting in fluctuating prices and consumer uncertainty. Despite these controversies the outlook for fuel ethanol is bright and strong rates of growth in both production and trade can be expected for the next several years.

## 9. FUTURE OUTLOOK

With a view to give boost to agriculture sector and reduce environmental pollution, Government of India have been examining for quite some time supply of ethanol-blended-petrol in the country. In order to ascertain financial and operational aspects of blending 5% ethanol with petrol as allowed in the specifications of Bureau of Indian Standards for petrol. Government had launched three pilot projects; two in Maharashtra and one in Uttar Pradesh during April and June 2001 and these pilot projects have been supplying 5% ethanol-doped-petrol only to the retail outlets under their respective supply areas since than. Apart from the aforesaid field through pilot projects, R&D studies also were undertaken simultaneously. Both pilot projects and R&D studies have been successful and established blending of ethanol up to 5% with petrol and usage of ethanol-doped-petrol in vehicles.

The Society for Indian Automobile Manufacturers (SIAM) has confirmed the acceptance for use of 5% ethanol-doped-petrol in vehicles. State Governments of major sugar producing States and the representatives of sugar/distillery industries have confirmed availability / capacity to produce ethanol. Government have set up an Expert Group headed by the Executive

Director of the Centre for High Technology for examining various options of blending ethanol with petrol including use of ETBE in refineries. Considering the logistical and financial advantages, this Group has recommended blending of ethanol with petrol at supply locations (terminals / depots) of oil companies. In view of the above, Government of India resolved to supply 5% ethanol-doped-petrol in the following nine States and four Union Territories including Andhra Pradesh, Goa, Gujrat, Haryana, Karnataka, Maharashtra, Punjab, Tamilnadu, Uttar Pradesh, Chandigarh, Daman and Diu, Dadra and Nagar Haveli, Pondicherry with effect from 1st January, 2003 and has assured to supply this ethanol-blended petrol in all states by November, 2006. Today, fuel ethanol producers also dominate the beverage and industrial markets. However, the structure of the distilling industries in other countries is different and therefore the development path needs not necessarily be the same. However, the forces at work will be the same and something similar can therefore be expected. At an international level, the increased production and utilisation of fuel ethanol will make necessary an international exchange mechanism, which can help to stabilise the market at times of regional production shortfalls. This will be a very difficult operation. After all, fuel ethanol production in most countries is being subsidised in order to support domestic farmers. If large scale imports occurred, foreign farmers and ethanol industries would benefit from these support mechanisms. Nevertheless, an international exchange will be the only way to firmly establish large-scale fuel ethanol programmes in any one country. Biofuels are produced from renewable resources, supplies of which are dependent on the weather. This is a fundamental difference to non-renewables, such as oil and gas. As a result, the level of production cannot be fully controlled by mankind. On the other hand, being produced from agricultural crops, fuel ethanol production need not be concentrated in a handful of countries, as is the case in the oil sector. Therefore, it is unlikely that a typical importer/exporter relationship will develop. Instead, international trade will occur in order to compensate for temporary production shortfalls and, as such, will remain at a rather limited level relative to world production. There could be some countries that will start to regularly import more environment friendly fuel ethanol to be sold in the domestic market. However, these countries will remain the exception. Fuel ethanol, besides its environmental value, is and will remain first and foremost instrument to support farmers. It is they, who in the first place, will profit from fuel ethanol programs and it is they who primarily lobby for the production and the utilisation of fuel ethanol. Roughly two-thirds of the world's ethanol production is used as fuel. As against this, despite being the world's fourth largest ethanol producer, India has not been able to use even one third of this production as transport fuel. This is unlike Brazil, which during 1998-99 produced an estimated 15,000 million liters of ethanol fuel from sugarcane, saving the country about 2,50,000 barrels per day of petrol imports. Currently 41 per cent of Brazil's transportation fuel demand is met by ethanol and there are 25,000 outlets where ethanol can be purchased. According to the President of the Indian Sugar Mills Association (ISMA), any molasses-based distillery is capable of producing anhydrous ethanol with the addition of two columns and associated equipment in a fairly short period of time. In fact, 90 distilleries have already indicated their willingness to set up plants for a total production of approximately 1,400 kiloliters / day of anhydrous alcohol. However, they are awaiting a firm policy decision from the Government. India has the potential to save nearly 800 million liters of petrol annually by blending up to 10 % ethanol in the gasoline used by the transport sector and thus biofuel production from agro-residues has an enormous potential in future.

## 10. FURTHER READING

Dien, B.S., Cotta, M.A. and Jeffries, T.W. (2003). Bacteria engineered for fuel ethanol production: current status. Appl. Microbiol. Biotechnol. 63:258-66.

Ghosh, P. and Ghose, T.K. (2003). Bioethanol in India: recent past and emerging future. Adv. Biochem. Eng. Biotechnol. 85:1-27.

Kurtzman, C.P. In: Yeast Biotechnology and Biocatalysis. (Eds. H. Verachter and R. de Mot) New York, Marcel Dekker. (1990) 1-34.

Martini, A. and Martini, A.V. In: Yeast Technology, (Eds. J.F.T. Spencer and D.M. Spencer) Berlin, Springer-Verlag. (1990) 105-123.

Singh, A. and Kumar, P.K. (1991). Fusarium oxysporum: status in bioethanol production. Crit. Rev. Biotechnol. 11:129-47.

Soni, S.K. and Marwaha, S.S. (2003). Biotechnological approaches for ethanol production from agro-residues. In: Biotechnological strategies in agro-processing (Eds. S.S. Marwaha and J.K. Arora). Asiatech Publishers Inc., New Delhi. pp.151-189.

Soni, S.K. and Sandhu D.K. (1999). Microbiology of Fermentation's In: Biotechnology: Food Fermentation Microbiology, Biochemistry and Technology. (Eds. V.K. Joshi and A. Pandey) Volume I: Basic, Educational Publishers and Distributors, New Delhi, India. pp. 25-85.

Ward, O.P. and Singh, A. (2002). Bioethanol technology: developments and perspectives. Adv. Appl. Microbiol. 51:53-80.

Wayman, M. and Parekh, S.R. Biotechnology of biomass conversion, Fuels and chemicals from renewable resources, New Jersey, Prentice Hall. (1990).

Wood, W.A. Trends in the biology of fermentation for fuels and chemicals, New York, Plenum Press. (1981).

Zaldivar, J., Nielsen, J. and Olsson, L. (2001). Fuel ethanol production from lignocellulose: a challenge for metabolic engineering and process integration. Appl Microbiol Biotechnol. 56:17-34.

# 8

# Microbes and Biodiesel Production

**S.K. SONI**

*Department of Microbiology, Panjab University, Chandigarh-160 014*

## 1. INTRODUCTION

Biodiesel is an alternative fuel for diesel engines that is gaining attention in various parts of the world including India after reaching a considerable level of success in Europe. Its primary advantages are that it is one of the most renewable fuels currently available and it is also non-toxic and biodegradable. It can also be used directly in most diesel engines without requiring extensive engine modifications. Chemically, it comprises a mixture of mono esters of long chain fatty acids. A lipid transesterification production process is used to convert the base oil to the desired esters and remove free fatty acids. The most common form uses methanol to produce methyl esters, though ethanol can be used to produce an ethyl ester biodiesel. A byproduct of the transesterification process is the production of glycerol. After this processing, unlike straight vegetable oil, biodiesel has combustion properties very similar to those of petroleum diesel, and can replace it in most current uses. However, it is at present most often used as an additive to petroleum diesel, improving the otherwise low lubricity of pure ultra low sulfur petrodiesel fuel. It is one of the possible candidates to replace fossil fuels as the world's primary transport energy source, because it is a renewable fuel that can replace petrodiesel in current engines and can be transported and sold using today's infrastructure.

Biodiesel, with a flash point of 150°C, is not as readily ignited as petroleum diesel whose flash point is 64°C. Indeed, it is classified as a non-flammable liquid by the Occupational Safety and Health Administration (OSHA), the main federal agency charged with the enforcement of safety and health legislation, although it will of course burn if heated to a high enough temperature. This property makes a vehicle fueled by pure biodiesel far safer in an accident. Biodiesel gels at higher temperatures (around 0 °C) than petroleum diesel, which limits its use in pure form in cold climates.

Unlike petrodiesel, biodiesel significantly reduces toxic and other emissions when burned as a fuel. Advantages of biodiesel over petroleum–derived diesel fuel include sulfur and aromatic contents so low as to be considered absent. Furthermore, diesel engines need not be modified to run on biodiesel, and emissions testing shows lower production of carbon monoxide, unburned hydrocarbons, and particulate matter than in the case of diesel produced from petroleum. These environmental considerations are of major importance to urban areas suffering from poor air quality. Two other advantages exist: the vegetable matter remaining after oil extraction can be used as feed for livestock; and a valuable by product of the transesterification reaction, glycerol, can be isolated and sold separately. Once the transesterified fatty acids have been made, another previous option becomes a possibility again, their blending with conventional diesel fuel. This allows some of the favorable environmental aspects to be maintained while keeping production costs low. Biodiesel can be used as B 100 (neat) or in a blend with petroleum diesel. A blend of 20 % biodiesel with 80 % petrodiesel, by volume, is termed "B 20".

Biodiesel has physical properties very similar to conventional diesel (Table 1). Due to the depletion of petroleum reserves, biodiesel as fatty esters was considered as an alternative to petroleum diesel during mid 1970s. Biodiesel has become more attractive recently because of its environmental benefits, the depleting reserves of petroleum and the fact that it is made from renewable resources. The cost of biodiesel, however, is the main hurdle to commercialization of the product in India but it is already being sold commercially in Europe, America and Australia. There are two aspects of cost of biodiesel, the cost of raw material and the cost of processing. The cost of raw materials accounts for 60-75% of the total cost of biodiesel fuel. Numerous studies have been conducted on biodiesel production and emission testing in the past two decades. Most of the current challenges are targeted to reduce its production cost, as the cost of biodiesel is still higher than its petro-diesel counterpart. This opens a golden opportunity for the use of waste or recycled oils as its production feedstock. Everywhere in the world, there is an enormous amount of waste lipids generated from restaurants, food processing industries and fast food shops everyday. Reusing of these waste greases can not only reduce the burden of the government in disposing the waste, maintaining public sewers and treating the oily wastewater, but also lower the production cost of biodiesel significantly. Table 2 reveals the pattern of global biodiesel production during the last 15 years.

**Table 1** Biodiesel's physical characteristics

| | |
|---|---|
| Specific gravity | 0.87 to 0.89 |
| Kinematic viscosity @ 40°C | 3.7 to 5.8 |
| Cetane number | 46 to 70 |
| Higher heating value (btu/lb) | 16,928 to 17,996 |
| Sulfur, wt% | 0.0 to 0.0024 |
| Cloud point °C | -11 to 16 |
| Pour point °C | -15 to 13 |
| Iodine number | 60 to 135 |
| Lower heating value (btu/lb) | 15,700 to 16,735 |

**Table 2** World biodiesel production, 1991-2003

| Year | Biodiesel (Million Litres) |
|---|---|
| 1991 | 11.355 |
| 1992 | 87.055 |
| 1993 | 143.830 |
| 1994 | 283.875 |
| 1995 | 401.210 |
| 1996 | 541.255 |
| 1997 | 548.825 |
| 1998 | 541.255 |
| 1999 | 681.300 |
| 2000 | 950.035 |
| 2001 | 1230.125 |
| 2002 | 1502.645 |
| 2003 | 1767.595 |

## 2. FEEDSTOCKS FOR BIODIESEL PRODUCTION

The primary raw materials used in the production of biodiesel are vegetable oils, animal fats, and recycled greases. These materials contain triglycerides, free fatty acids, and other contaminants depending on the degree of pretreatment they have received prior to delivery.

Since biodiesel is a mono-alkyl fatty acid ester, the primary alcohol used to form the ester is the other major feedstock.

Most processes for making biodiesel use a catalyst to initiate the esterification reaction. The catalyst is required because the alcohol is sparingly soluble in the oil phase. The catalyst promotes an increase in solubility to allow the reaction to proceed at a reasonable rate. The most common catalysts used are strong mineral bases such as sodium hydroxide and potassium hydroxide. After the reaction, the base catalyst must be neutralized with a strong mineral acid. Typical proportions for the chemicals used to make biodiesel are:

| | |
|---|---|
| Reactants | • Fat or oil (100 kg) |
| | • Primary alcohol (10 kg methanol) |
| Catalyst | • Mineral base (0.3 kg sodium hydroxide) |
| Neutralizer | • Mineral acid (0.25 kg sulfuric acid) |

## 2.1. Fats and oils

Fats and oils are primarily water-insoluble substances in the plant and animal kingdom that are made up of one mole of glycerol, three moles of fatty acids and are commonly referred to as triglycerides. Fatty acids vary in carbon chain length and in the number of unsaturated double bonds. Table3 shows typical fatty acid compositions of common oil sources, while Tables 4 and 5 list the most common fatty acids and their corresponding methyl esters.

**Table 3.** Typical fatty acid composition-common oil source

| Fatty acid | Soybean | Cottonseed | Palm | Lard | Tallow | Coconut |
|---|---|---|---|---|---|---|
| Lauric | 0.1 | 0.1 | 0.1 | 0.1 | 0.1 | 46.5 |
| Myristic | 0.1 | 0.7 | 1.0 | 1.4 | 2.8 | 19.2 |
| Palmitic | 10.2 | 20.1 | 42.8 | 23.6 | 23.3 | 9.8 |
| Stearic | 3.7 | 2.6 | 4.5 | 14.2 | 19.4 | 3.0 |
| Oleic | 22.8 | 19.2 | 40.5 | 44.2 | 42.4 | 6.9 |
| Linoleic | 53.7 | 55.2 | 10.1 | 10.7 | 2.9 | 2.2 |
| Linolenic | 8.6 | 0.6 | 0.2 | 0.4 | 0.9 | 0.0 |

**Table 4.** Chemical structure of common fatty acids and their methyl esters

| Fatty acid | Structure[a] | Common acronym | Methyl ester |
|---|---|---|---|
| Palmitic acid/ Hexadecanoic acid | R-$(CH_2)_{14}$-$CH_3$ | C16:0 | Methyl palmitate/ Methyl hexadecanoate |
| Stearic acid / Octadecanoic acid | R-$(CH_2)_{16}$-$CH_3$ | C18:0 | Methyl stearete/ Methyl octadecanoate |
| Oleic acid / 9(Z)-octadecanoic acid | R-$(CH_2)_7$-CH=CH-$(CH_2)_7$-$CH_3$ | C18:1 | Methyl oleate/ Methyl 9(Z)-octadecanoate |
| Linoleic acid / 9(Z), 12(Z)-octadecadienoic acid | R-$(CH_2)_7$-CH=CH-$(CH_2)$-CH=CH-$(CH_2)_4$-$CH_3$ | C18:2 | Methyl linoleate/ Methyl 9(Z), 12(Z)-octadecadienoate |
| Linolenic acid / 9(Z), 12(Z), 15(Z)-octadecatrienoic acid | R-$(CH_2)_7$-(CH=CH-$CH_2)_3$-$CH_3$ | C18:3 | Methyl linolenate/ Methyl 9(Z), 12(Z), 15(Z)-octadecatrienoate |

a) R=COOH ($CO_2H$) or $COOCH_3$ ($CO_2CH_3$); $(CH_2)_7$=$CH_2$-$CH_2$-$CH_2$-$CH_2$-$CH_2$-$CH_2$-$CH_2$, etc

**Table 5** Characteristics of common fatty acids and their methyl esters

| Fatty acid Methyl ester | Formula | Molecular weight | Melting point (ºC) |
|---|---|---|---|
| Palmitic acid | $C_{16}H_{32}O_2$ | 256.428 | 63-64 |
| Methyl palmitate | $C_{17}H_{34}O_2$ | 270.457 | 30.5 |
| Stearic acid | $C_{18}H_{36}O_2$ | 284.481 | 70 |
| Methyl stearate | $C_{19}H_{38}O_2$ | 298.511 | 39 |
| Oleic acid | $C_{18}H_{34}O_2$ | 282.465 | 16 |
| Methyl oleate | $C_{19}H_{36}O_2$ | 296.495 | -20 |
| Linoleic acid | $C_{18}H_{32}O_2$ | 280.450 | -5 |
| Methyl linoleate | $C_{19}H_{34}O_2$ | 294.479 | -35 |
| Linolenic acid | $C_{18}H_{30}O_2$ | 278.434 | -11 |
| Methyl linolenate | $C_{19}H_{32}O_2$ | 292.463 | -52/-57 |

Methanol is the preferred alcohol for obtaining biodiesel because it is the cheapest and most available alcohol. However, for the reaction to occur in a reasonable time, a substance called a 'catalyst' must be added to the mixture of the vegetable oil and methanol. The transesterification reaction for biodiesel production and the relationship for prediction of biodiesel from fats and oils is provided in Figure 1.

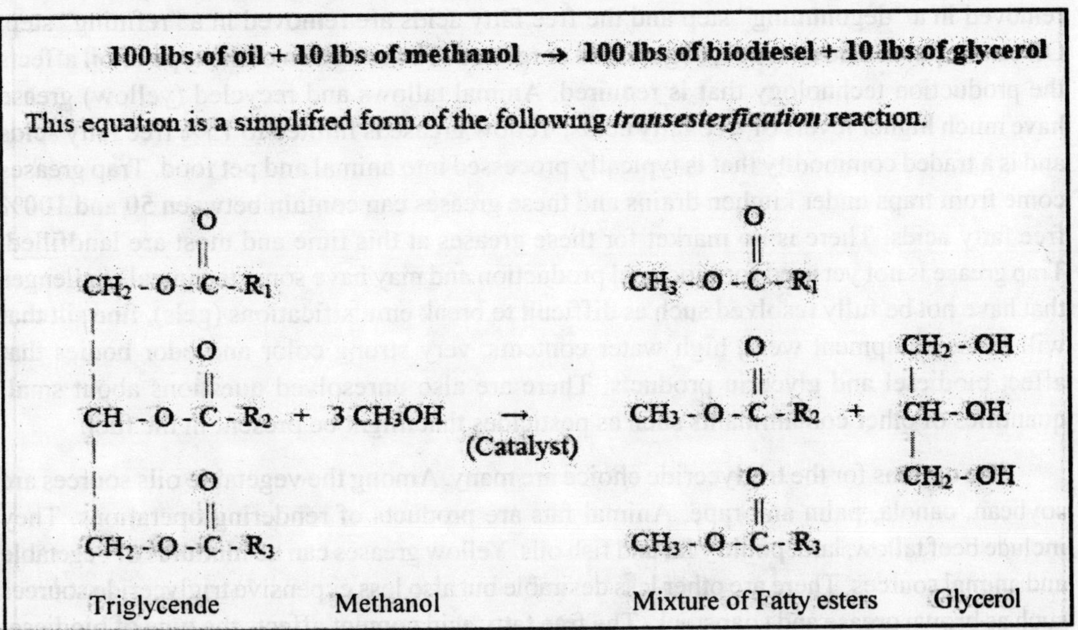


**Figure 1** Transesterification reaction

**where $R_1$, $R_2$, and $R_3$ are long chains of carbons and hydrogen atoms, sometimes called fatty acid chains. There are five types of chains that are common in soybean oil and animal fats (others are present in small amounts):**

| | | |
|---|---|---|
| Palmitic: | $R = -(CH_2)_{14}-CH_3$ | **16 carbons, (including the one that R is attached to.) (16:0)** |
| Stearic: | $R = -(CH_2)_{16}-CH_3$ | **18 carbons, 0 double bonds (18:0)** |
| Oleic: | $R = -(CH_2)_7\ CH{=}CH(CH_2)_7CH_3$ | **18 carbons, 1 double bond (18:1)** |
| Linoleic: | $R = -(CH_2)_7\ CH{=}CH\text{-}CH_2\text{-}CH{=}CH(CH_2)_4CH_3$ | **18 carbons, 2 double bonds (18:2)** |
| Linolenic: | $R = -(CH_2)_7\ CH{=}CH\text{-}CH_2\text{-}CH{=}CH\text{-}CH_2\text{-}CH{=}CH\text{-}CH_2\text{-}CH_3$ | **18 carbons, 3 double bonds (18:3)** |

Choice of the fats or oils to be used in producing biodiesel is both a process chemistry decision and an economic decision. With respect to process chemistry, the greatest difference among the choices of fats and oils is the amount of free fatty acids that are associated with the triglycerides. Other contaminants, such as color and odor bodies can reduce the value of the glycerin produced and reduce the public acceptance of the fuel if the color and odor persist in the fuel. Most vegetable oils have a low percentage of associated free fatty acids. Crude vegetable oils contain some free fatty acids and phospholipids. The phospholipids are removed in a "degumming" step and the free fatty acids are removed in a "refining" step. Oil can be purchased as crude, degummed or refined. The selection of the type of oil affects the production technology that is required. Animal tallows and recycled (yellow) grease have much higher levels of free fatty acids. Yellow grease is limited to 15% free fatty acids and is a traded commodity that is typically processed into animal and pet food. Trap greases come from traps under kitchen drains and these greases can contain between 50 and 100% free fatty acids. There is no market for these greases at this time and most are landfilled. Trap grease is not yet used for biodiesel production and may have some technical challenges that have not be fully resolved such as difficult to break emulsifications (gels), fine silt that will cause equipment wear, high water contents, very strong color and odor bodies that affect biodiesel and glycerin products. There are also unresolved questions about small quantities of other contaminants such as pesticides that might be present in the fuel.

The options for the triglyceride choice are many. Among the vegetable oils sources are soybean, canola, palm and rape. Animal fats are products of rendering operations. They include beef tallow, lard, poultry fat and fish oils. Yellow greases can be mixtures of vegetable and animal sources. There are other less desirable but also less expensive triglyceride sources such as brown grease and soapstock. The free fatty acid content affects the type of biodiesel process used and the yield of fuel from that process. The other contaminants present can affect the extent of feedstock preparation necessary to use a given reaction chemistry.

In accordance with the names of fatty acids and their esters (Tables 4 and 5), the methyl ester of soybean oil is often called methyl soyate. The term soybean oil methyl ester (SME) is also very common. The same holds for the esters of other vegetable oils. Another common abbreviation that is commonly used is FAME (fatty acid methyl ester). Besides triglycerides, mono- and diglycerides can also exist. They are formed as intermediates during the transesterification reaction. Other materials that can contaminate biodiesel are residual methanol (or other alcohol), glycerol and catalyst. During the transesterification reaction, all materials don't mix with each other and there exist two phases. At the end of the reaction, there are two layers (phases), one consisting mainly of glycerol and the other of the methyl esters. Obviously, glycerol and methyl esters do not mix readily. How readily one compound will dissolve in another depends on the structural features of the compounds, for example the existence of OH groups. Thus compounds containing OH groups and those not containing OH groups often will not readily mix.

## 2.2. Alcohol

The most commonly used primary alcohol used in biodiesel production is methanol, although other alcohols such as ethanol, isopropanol and butyl can be used. A key quality factor for the primary alcohol is the water content. Water interferes with transesterification reactions resulting in poor yields and high levels of soap, free fatty acids and triglycerides in the final fuel. Unfortunately, all the lower alcohols are hygroscopic and are capable of absorbing water from the air.

Many alcohols have been used to make biodiesel. As long as the product esters meet ASTM D-6751, a biodiesel standard laid down by American Society for Testing and Materials, it does not make any chemical difference which alcohol is used in the process. Other issues such as cost of the alcohol, the amount of alcohol needed for the reaction, the ease of recovering and recycling the alcohol, fuel tax credits and global warming issues influence the choice of alcohol. Some alcohols also require slight technical modifications to the production process such as higher operating temperatures, longer or slower mixing times, or lower mixing speeds.

Since the reaction to form the esters is on a molar basis and we purchase alcohol on a volume basis, their properties make a significant difference in raw material price. It takes three moles of alcohol to react completely with one mole of triglyceride. In addition, a base catalyzed process typically uses an operating mole ratio of 6:1 mole of alcohol rather than the 3:1 ratio required by the reaction. The reason for using extra alcohol is that it "drives" the reaction closer to the 99.7% yield we need to meet the total glycerol standard for fuel grade biodiesel. The unused alcohol must be recovered and recycled back into the process to minimize operating costs and environmental impacts. Methanol is considerably easier to recover than the ethanol. Ethanol forms an azeotrope with water so it is expensive to purify the ethanol during recovery. If the water is not removed it will interfere with the reactions. Methanol recycles easier because it doesn't form an azeotrope. These two factors are the reason that even though methanol is more toxic, it is the preferred alcohol for producing biodiesel. Methanol has a flash point of 10°C, while the flash point of ethanol is 8°C, so both are considered highly flammable. You should never let methanol come into contact with your skin or eyes as it can be readily absorbed. Excessive exposure to methanol can cause blindness and other health effects. For student demonstrations, ethanol may be safer to use. For simplicity, consider an oil such as soybean oil to consist of pure triolein. Triolein is a triglyceride in which all three fatty acid chains are oleic acid. This is near the actual number of carbons and hydrogens and gives a molecular weight that is near the value for soybean oil. If triolein is reacted with methanol, the reaction will be that shown in Figure 2. Note that weights for each of the compounds in the reaction are given. These are based on the fact that one molecule of triolein reacts with 3 molecules of methanol to produce 3 molecules of methyl oleate, the biodiesel product and one mole of glycerol. Chemists typically multiply all the terms of this equation by a large number that corresponds to the number of molecules in a quantity equal to the molecular weight of the substance. This quantity is called a mole of the substance. To calculate the molecular weight of triolein, we count the number of carbons in the molecule and multiply this by 12.011, the molecular weight of carbon. Doing the same thing for hydrogen and oxygen gives:

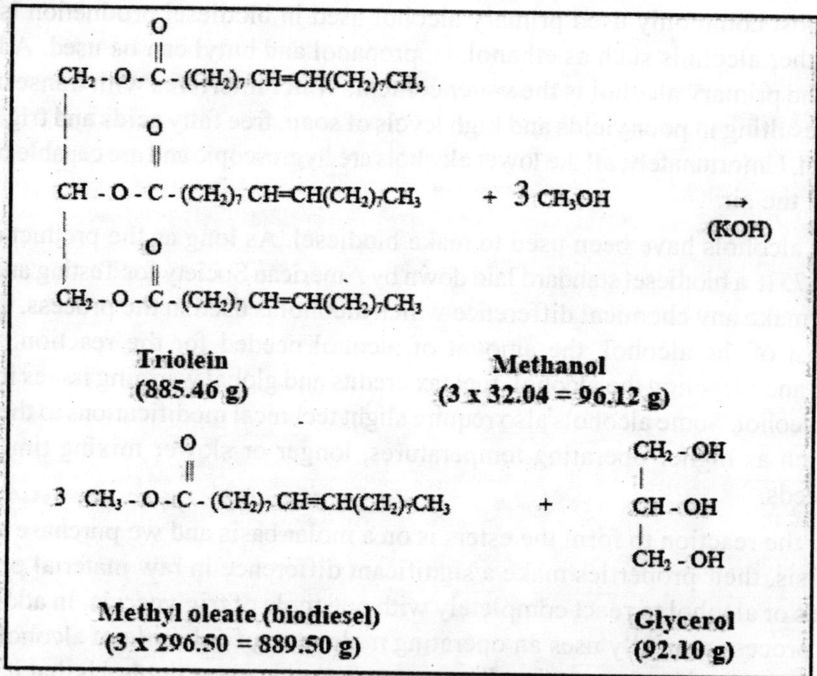


Figure 2 Transesterification of Triolein

| | | | |
|---|---|---|---|
| 57 × 12.0111 | = | 684.63 | |
| 104 × 1.00797 | = | 104.83 | |
| 6 × 16.000 | = | 96.00 | |
| Total | = | 885.46 | grams per mole |

Therefore, the molecular weight of triolein is 885.46 and one mole of triolein weighs 885.46 grams. Three moles of methanol weigh 96.12 g, 3 moles of methyl oleate weigh 889.50 g, and 1 mole of glycerol weighs 92.10 g. We do not actually conduct the reaction this way. We usually add 60% to 100% excess methanol to ensure that the reaction goes to completion. In general, reactions can be encouraged to progress by adding an excess of one of the reactants or by removing one of the products. The reaction of triolein with 100% excess (XS) methanol is shown in Figure 3.

| Triolein (885.46 g) | + | 2X Methanol (6 x 32.04 = 192.24 g) | | | | |
|---|---|---|---|---|---|---|
| ⟶ (Catalyst) | | Methyl oleate (3 x 296.50 = 889.50 g) | + | Glycerol (92.10 g) | + | XS Methanol (96.12 g) |

**Figure 3** Transesterification of Triolein with 100% Excess Methanol

On the basis of 100 kg of oil, the reaction mass balance with 100% XS methanol becomes:

100 kg oil + 21.71 kg methanol → 100.45 kg biodiesel + 10.40 kg glycerol + 10.86 kg XS methanol

The reaction also requires about 1% (based on the weight of oil) of sodium hydroxide or a similar catalyst that mostly ends up in the glycerol. These quantities can be converted to volumes by including the densities of the reactants and products given in Table 6.

**Table 6** Densities of biodiesel reactants (kg/liter)

| | |
|---|---|
| Triolein | 0.8988 |
| Methanol | 0.7914 |
| Methyl Oleate | 0.8739 |
| Glycerol | 1.2613 |

On a volume basis, the reaction becomes:

100 liters of oil + 24.65 liters of methanol

→ 103.3 liters of oleate + 7.42 liters glycerol + 12.33 liters XS methanol

## 2.3. Catalysts and neutralizers

The catalyst used for carrying out the transesterification is usually sodium hydroxide (NaOH) or potassium hydroxide (KOH). These compounds belong to a class of materials known as bases and also are inorganic compounds (inorganic compounds are often used in organic chemistry for carrying out or catalyzing reactions). Other bases are also suitable for the transesterification reaction. The counterparts of bases are known as acids. Many acids can also be used as catalysts in the transesterification reaction. However, the base-catalyzed reaction has advantages such as a higher reaction rate. In the soap formation reaction, a fatty acid and base reacted to form a new compound, which was called soap and water.

Compounds such as soap, in which the hydrogen (proton) of an acid has been replaced with a metal ion, are often called salts. The reason that such compounds exist is that materials such as NaOH (or KOH) can split apart (dissociate) in a fashion that gives $Na^+$ and $OH^-$ (or $K^+$ and $OH^-$) in which the protons and electrons are not evenly distributed, leading to charged particles. Thus, having the same charge, $Na^+$ or $K^+$ can replace $H^+$ here. Another important aspect of the chemistry of fatty acids and their esters is that the more unsaturated fatty acids and their esters (linoleic, linolenic) can relatively easily react with air (more specifically, the oxygen in air) and form degradation products with time (the time depends also on other factors such as temperature). This is due to the existence of $CH_2$ between two carbons double-bonded to other carbons.

Catalysts may either be base, acid or enzyme materials. The most commonly used catalyst materials for converting triglycerides to biodiesel are sodium hydroxide, potassium hydroxide and sodium methoxide. Most base catalyst systems use vegetable oils as a feedstock. If the vegetable oil is crude, it contains small amounts (<2%) of free fatty acids that will form soaps that will end up in the crude glycerin. Refined feedstocks, such as refined soy oil can also be used with base catalysts.

The base catalysts are highly hygroscopic and they form chemical water when dissolved in the alcohol reactant. They also absorb water from the air during storage. If too much water has been adsorbed the catalyst will perform poorly and the biodiesel may not meet the total glycerin standard.

Although acid catalysts can be used for transesterification they are generally considered to be too slow for industrial processing. Acid catalysts are more eommonly used for the esterification of free fatty acids. Acid catalysts include sulfuric acid and phosphoric acid. Solid calcium carbonate is used as an acid catalyst in one experimental homogeneous catalyst process. The acid catalyst is mixed with methanol and then this mixture is added to the free fatty acids or a feedstock that contains high levels of free fatty acids. The free fatty acids convert into biodiesel. The acids will need neutralization when this process is complete, but this can be done as base catalyst is added to convert any remaining triglycerides. There is continuing interest in using lipases as enzymatic catalysts for the production of alkyl fatty acid esters.

Lipases (triacylglycerol acylhydrolases E.C. 3.1.1.3) are ubiquitous enzymes of considerable physiological significance as well as industrial potential. According to a classical definition lipases are the enzymes which, in contrast to esterases, become active only when adsorbed on to an oil-water interface and donot hydrolyse soluble substrates. The unique property is known as interfacial activation. These catalyze the hydrolysis of triacyl glycerols to glycerol and free fatty acids. A true lipase is one which splits emulsified esters of glycerine and fatty acids with long chains such as triolein and tripalmitin. Lipases are serine hydrolases and unlike esterases, which show a normal Michaelis-Menten activity, lipases display little activity in aqueous solutions with soluble substrates. Lipases with industrial potential are usually obtained from microorganisms which produce a wide spectrum of extracellular lipases. Many are active in organic solvents where they catalyze a number of useful reactions including transesterification, regioselective acylation of glycols and menthols, even synthesis of peptides.

Microorganisms producing lipases are widespread and include fungi, yeast, bacteria and actinomycetes. The important molds belonging to the genera *Rhizopus, Aspergillus* & *Mucor*, yeasts belonging to *Candida, Rhodotorula* & *Saccharomyces* and bacteria belonging to *Bacillus* and *Pseudomonas* are reported to produce significant levels of extracellular lipases.

The commercial use of lipases in biodiesel production is currently limited to countries like Japan, where energy costs are high or for the production of specialty chemicals from specific types of fatty acids. The commercial use of enzymes is limited because costs are high, the rate of reaction is slow and yields to methyl esters are typically less than the 99.7% required for fuel-grade biodiesel. Enzymes are being considered for fatty acid conversion to biodiesel as a pretreatment step, but this system is not commercial at this time.

Neutralizers are used to remove the base or acid catalyst from the product biodiesel and glycerol. If one is using a base catalyst, the neutralizer is typically an acid and vice versa. If the biodiesel is being washed, the neutralizer can be added to the wash water. While hydrochloric acid is a common choice to neutralize base catalysts, as mentioned earlier, if phosphoric acid is used, the resulting salt has value as a chemical fertilizer.

***Catalyst selection***

Two of the most commonly used catalysts for transesterification are NaOH and KOH. These catalysts operate by reacting with the alcohol according to the reaction given below (written using methanol and NaOH but other alcohols and catalysts could be substituted).

$$CH_3OH + NaOH \rightarrow CH_3O\text{-}Na + H_2O \qquad \text{(Eq. 1)}$$

**Similar to $H_2O$ "consisting" of $H^+$ and $OH^-$, $CH_3O$-Na can be seen as consisting of $CH_3O^-$ (alkoxide; alkylate) and $Na^+$. $CH_3O^-$ is the species that attacks the ester moieties in the glycerol molecule in the following fashion:**

$$\underset{\textbf{Alkoxide}}{CH_3\text{-}O^-} + \underset{\textbf{Triglyceride}}{\begin{array}{l} CH_2\text{-}O\text{-}\overset{O}{\overset{\|}{C}}\text{-}R \\ | \\ CH\text{-}O\text{-}\overset{O}{\overset{\|}{C}}\text{-}R \\ | \\ CH_2\text{-}O\text{-}\overset{O}{\overset{\|}{C}}\text{-}R \end{array}} \rightarrow \underset{\textbf{Methyl ester}}{CH_3\text{-}O\text{-}\overset{O}{\overset{\|}{C}}\text{-}R} + \underset{\textbf{Triglyceride anion}}{\begin{array}{l} CH_2\text{-}O\text{-}\overset{O}{\overset{\|}{C}}\text{-}R \\ | \\ CH\text{-}O^- \\ | \\ CH_2\text{-}O\text{-}\overset{O}{\overset{\|}{C}}\text{-}R \end{array}} \qquad \text{(Eq. 2)}$$

While the ester molecule is complete after this reaction, the anion of the triglyceride needs to pick up a proton to give a stable product (a diglyceride in this case). If this proton is taken from methanol, then the alkoxide catalyst will be recovered as shown in Eq. 3.

$$\begin{array}{l} CH_2\text{-}O\text{-}\overset{O}{\overset{\|}{C}}\text{-}R \\ | \\ \quad + \; CH_3OH \;\rightarrow \\ | \\ CH_2\text{-}O\text{-}\overset{O}{\overset{\|}{C}}\text{-}R \end{array} \qquad \begin{array}{l} CH_2\text{-}O\text{-}C\text{-}R \\ | \\ CH\text{-}OH \; - \; CH_3\text{-}O^- \\ | \\ CH_2\text{-}O\text{-}\overset{O}{\overset{\|}{C}}\text{-}R \end{array} \qquad \text{(Eq. 3)}$$

Several other reactions could be written that would allow the anion of the triglyceride to pick up a proton, such as reaction with free fatty acids and water. The presence of water affects the transesterification negatively because the triglyceride anion in Eq. 3 will react with the water to form $OH^-$, which can behave in a fashion similar to $CH3\text{-}O^-$ but a free fatty acid will result instead of a methyl ester.

$$\underset{\text{Hydroxide}}{OH^-} + \underset{\text{Triglyceride}}{\begin{array}{l} CH_2\text{-}O\text{-}\overset{O}{\overset{\|}{C}}\text{-}R \\ | \\ CH\text{-}O\text{-}\overset{O}{\overset{\|}{C}}\text{-}R \\ | \\ CH_2\text{-}O\text{-}\overset{O}{\overset{\|}{C}}\text{-}R \end{array}} \rightarrow \underset{\text{Free fatty acid}}{HO\text{-}\overset{O}{\overset{\|}{C}}\text{-}R} + \underset{\text{Triglyceride anion}}{\begin{array}{l} CH_2\text{-}O\text{-}\overset{O}{\overset{\|}{C}}\text{-}R \\ | \\ CH\text{-}O^- \\ | \\ CH_2\text{-}O\text{-}\overset{O}{\overset{\|}{C}}\text{-}R \end{array}} \qquad \text{(Eq. 4)}$$

The free fatty acid can react with the $Na^+$ to form soap. Some water in the system can be tolerated, because $R\text{-}O^-$ is a stronger base than $OH^-$ so the transesterification reaction occurs at a higher rate than the "saponification" of glycerol leading to free fatty acids. The equilibrium of the following reaction (formation of free fatty acids from the ester) is far on the left side of the reaction equation:

$$OH^- + \underset{\text{alkyl ester}}{R\text{-}O\text{-}\overset{O}{\overset{\|}{C}}\text{-}R} \rightarrow R\text{-}O^- + \underset{\text{free fatty acid}}{HO\text{-}\overset{O}{\overset{\|}{C}}\text{-}R} \qquad \text{(Eq. 5)}$$

As a result of the slow reaction rate, only very minor amounts of free fatty acids are formed during transesterification if the reaction is free of water at the beginning. Another aspect of the weaker basicity of $OH^-$ *vs.* $R\text{-}O^-$ is that the ester moieties in the triglyceride molecules will react less with $OH^-$ than $R\text{-}O^-$. However, when too much catalyst (or water) is present, Eq. 5 becomes more prevalent and causes enhanced formation of mono- and diglyceride molecules instead of reactions with all positions in the glycerol backbone. Thus, the effect of too much catalyst or too much water leads to the same result, mainely, enhanced formation of undesirable mono- and diglycerides as well as free fatty acids. Too much catalyst can lead to soap formation when conditions encourage free fatty acid production. Because of the possibility of the reactions described above leading to free fatty acids and mono and diglycerides, direct use of sodium or potassium alkylate (R-ONa or R-OK; the alkylate moiety must correspond to the R- moiety in the alcohol) as catalysts is becoming of greater interest. The reaction of alcohol and XOH given above (Eq. 1) cannot occur in this case (assuming, of course, that the reaction system is free of water). Instead, the transesterification according to Eq. 2 can occur directly.

Base catalysts are used for essentially all vegetable oil processing plants. The initial free fatty acid content and the water content are generally low. Tallows and greases with free fatty acid contents greater than about 1% must be pretreated to either remove the FFA or convert the FFA to esters before beginning the base catalyzed reaction. Otherwise, the base catalyst will react with the free fatty acids to form soap and water. The soap formation reaction is very fast and goes to completion before any esterification begins. Essentially all of the current commercial biodiesel producers use base catalyzed reactions. Base catalyzed reactions are relatively fast, with residence times from about 5 minutes to about 1 hour, depending on temperature, concentration, mixing and alcohol:triglyceride ratio. Most use NaOH or KOH as catalysts, although glycerol refiners prefer NaOH. KOH has a higher cost but the potassium can be precipitated as $K_3PO_4$, a fertilizer, when the products are neutralized using phosphoric acid. This can make meeting water effluent standards a bit more difficult because of limits on phosphate effluents.

Sodium methoxide, usually as a 25 % solution in methanol, is a more powerful catalyst on a weight basis than the mixture of NaOH and methanol. This appears to be, in part, the result of the negative effect of the chemical water produced *in situ* when NaOH and methanol react to form sodium methoxide. Acid catalyst systems are characterized by slow reaction rates and high alcohol: TG (triglyceride) requirements (20:1 and more). Generally, acid catalyzed reactions are used to convert FFAs to esters, or soaps to esters as a pretreatment step for high FFA feedstocks. Residence times from 10 minutes to about 2 hours are reported.

Counter current acid esterification systems have been used for decades to convert pure streams of fatty acids into methyl esters at yields above 99%. These systems tend to force yields to 100% and wash water out of the system at the same time because the feedstock and the sulfuric acid/methanol mix are moving in opposite directions. Acid esterification systems produce a byproduct of water. In batch systems, the water tends to accumulate in the vessel to the point where it can shut the reaction down prematurely. The sulfuric acid tends to migrate into the water, out of the methanol, rendering it unavailable for the reaction. All acid esterification systems need to have a water management strategy. Good water management can minimize the amount of methanol required for the reaction. Excess methanol (such as the 20:1 ratio) is generally necessary in batch reactors where water accumulates. Another approach is to approach the reaction in two stages: fresh methanol and sulfuric acid is reacted, removed and replaced with more fresh reactant. Much of the water is removed in the first round and the fresh reactant in the second round drives the reaction closer to completion.

Chemical transesterification is efficient in terms of reaction time; however, the chemical approach to synthesize biodiesel from triglyceride has drawbacks, such as difficulty in the recovery of glycerol and the energy-intensive nature of the process. In contrast, biocatalysts allow synthesis of specific alkyl esters, easy recovery of glycerol and transesterification of glycerides with high free fatty acid content. Much work has been done on the lipase catalyzed transesterification of triglyceride. One common drawback with the use of enzyme-based processes is the high cost of the enzyme. Immobilization of enzymes has generally been used to obtain reusable enzyme derivatives. This enables recycling of the biocatalyst and hence lowers the cost. In the case of biocatalysts in nonaqueous media, immobilization is also reported to result in better activity. Thus, many transesterification processes employing lipases have used an immobilized form of the enzyme. Lipase catalyzed reactions have the advantage of reacting at room temperature without producing spent catalysts. The enzymes can be recycled for use again or immobilized onto a substrate. If immobilized, the substrate will require replacement when yields begin to decline. The enzyme reactions are highly specific. Because the alcohol can be inhibitory to some enzymes, a typical strategy is to feed the alcohol into the reactor in three steps of 1:1 mole ratio each. The reactions are very slow, with a three step sequence requiring from 4 to 40 hours or more. The reaction conditions are modest, ranging from 35 to 45 °C.

Much work has been done on the lipase-catalyzed transesterification of triglyceride but most of the studies have made use of commercial lipases active at ambient temperature. Many studies have been demonstrated that temperature, methanol/oil molar ratio and glycerol have a great influence on lipase-catalysed transesterification during non-continuous and continuous batch operation for biodiesel production. Within the optimized conditions, activity of lipase remains relatively high during continuous batch operation which demonstrates that enzymic transesterification of renewable oil is very promising for large-scale production of biodiesel. The enzymic activity and the transesterification process is affected by various factors:

***a) Effect of glycerol on enzymic activity***

Enzymic transesterification of renewable oil with short-chain alcohols as the acyl acceptor for biodiesel production has been studied extensively in recent years. However, with these short-chain alcohols as the acyl acceptor, glycerol, as one of the major by-products, has been demonstrated to have some serious negative effects on enzymic activity. During repeated use, lipase loses its activity dramatically (Fig 4).

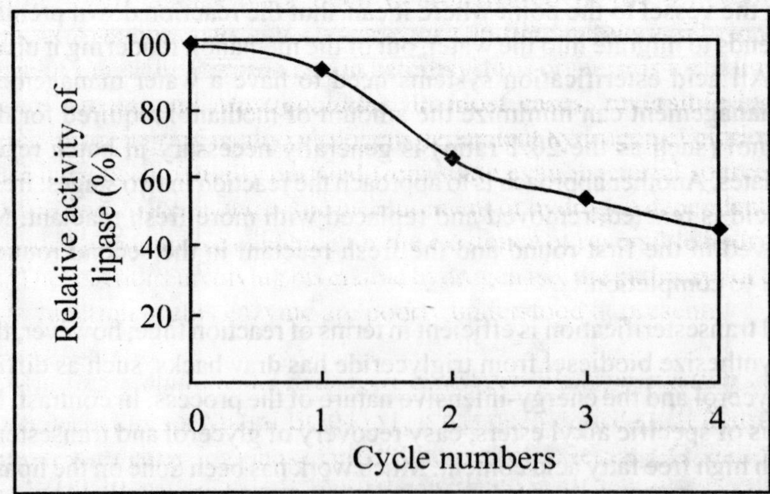


**Figure 4** Operational stability of lipase without glycerol removal

Some efforts have been made to remove glycerol from lipase during enzymic transesterification for biodiesel production and water has been found to be the ideal solvent to desorb glycerol from the lipase. However, water significantly influences the enzymic catalysis for biodiesel production in non-aqueous media, so some extra procedures (such as lyophilization) are usually needed to remove excess water from the lipase and this manipulation is complicated for biodiesel production, especially in large-scale operation. One of the alternatives involves the use of some hydrophilic organic compounds (propanol, isopropanol, butanol, t-butanol) which desorb glycerol from reaction mixture.

## Operational stability with glycerol removal by isopropanol

The enzyme loses its activity rapidly, even after one cycle of operation. Therefore glycerol removal should be carried out after each cycle. The operational stability of lipase with glycerol removal by isopropanol (Figure 5) shows good operational stability and there is negligible loss of lipase activity after 10 batches of biodiesel production.

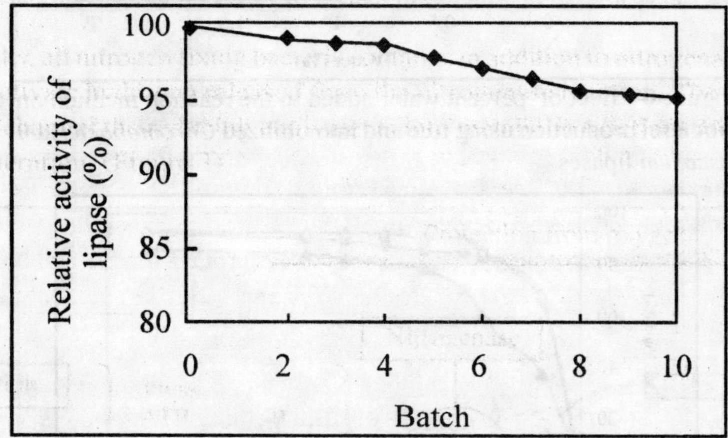


**Figure 5** Operational stability of lipase with glycerol removal by isopropanol during continuous batch operation

***b) Effect of percent water in reaction media***

Amount of water present in the reaction media is another critical parameter, which is known to influence biotransformations in nonaqueous media. The general picture available is that less than a mono-layer of water is required for an enzyme to show biological activity. As the water level increases, it increases the enzyme flexibility and the expressed activity. After an optimum level of water, hydrolytic reactions become significant and transesterification yield is expected to go down. Figure 6 shows the effect of different amounts of water present in the solvent-free system on biodiesel yield.

***c) Effect of molar ratio of methanol/oil***

Excessive short-chain alcohols such as methanol inactivate lipase seriously. However, at least three molar equivalents of methanol are required for the complete conversion of the oil to its corresponding methyl ester. As shown in Figure 7, the highest methyl ester yield

(92%) can be obtained at a methanol/oil molar ratio of 4:1. Either a higher (5:1) or lower (3:1) methanol concentration decreases the methyl ester yield to some degree. On the other hand high alcohol concentration leads to a heavy loss of enzymic activity during long-term operation (Figure 8).

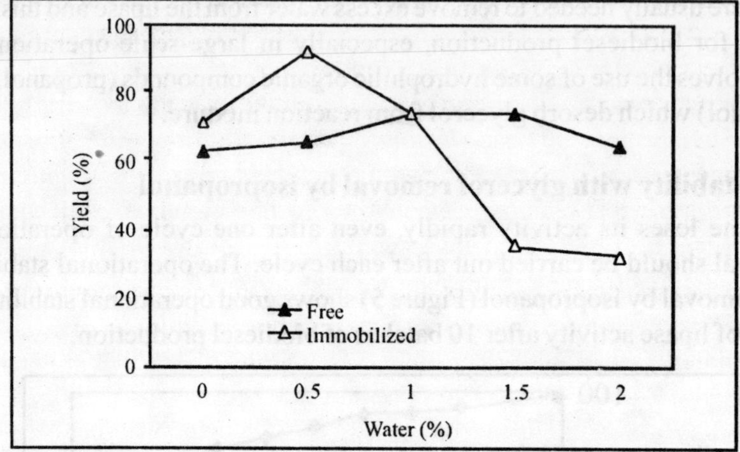


**Figure 6** Effect of percent water added to the reaction medium on biodiesel production using free and immobilized *Chromobacterium viscosum* lipases

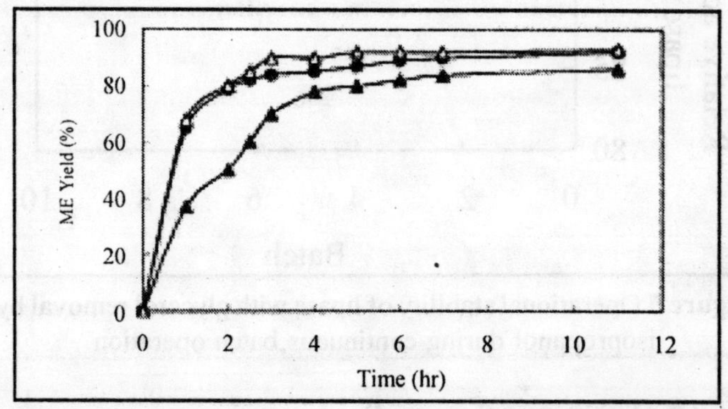


**Figure 7** Stability of lipase with different methanol/oil molar ratios

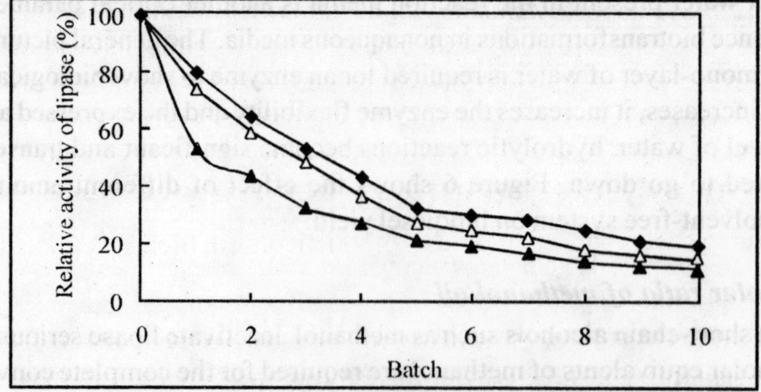


**Figure 8** Stability of lipase with different methanol/oil molar ratios

***d) Effect of temperature***

Lipase can express relatively high activity during short-term operation in non-continuous reactions, but during long-term operation it loses its activity rapidly at the relatively high temperature (Figures 9 and 10).

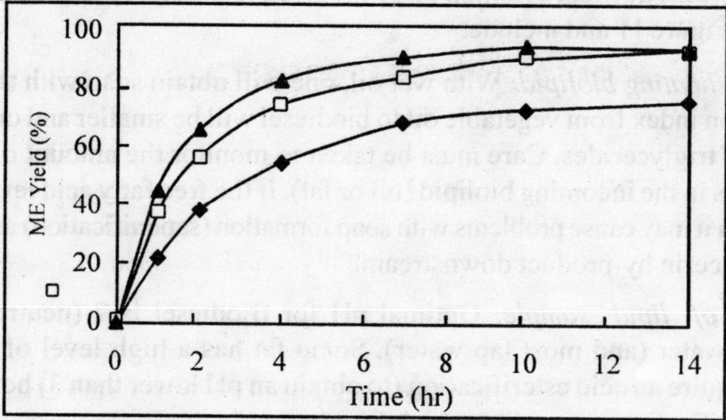


**Figure 9** Stability of lipase during continuous batch operation at different temperatures

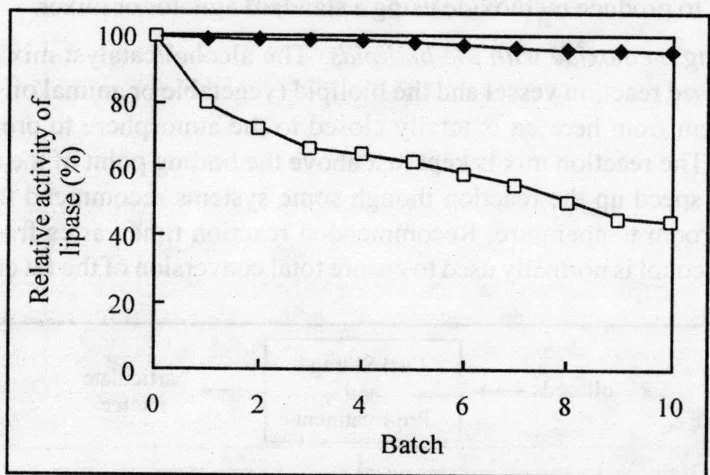


**Figure 10** Stability of lipase during continuous batch operation at different temperatures

## 3. TYPES OF BIODIESEL PRODUCTION PROCESS

There are three basic routes to biodiesel production from biolipids (biological oils and fats):

- Base catalyzed transesterification of the biolipid.
- Direct acid catalyzed transesterification of the biolipid.
- Conversion of the biolipid to its fatty acids and then to biodiesel.

Almost all biodiesel is produced using base catalyzed transesterification as it is the most economical process requiring only low temperatures and pressures and producing a 98% conversion yield. Transestrification is crucial for producing biodiesel from biolipids. The transesterification process is the reaction of a triglyceride (fat/oil) with an bioalcohol to form esters and glycerol. The most common steps involved in biodisel production from oil seeds are depicted in Figure 11 and include:

1. *Cleaning/heating biolipid:* With wet oil, one will obtain soap with the biodiesel, the conversion index from vegetable oil to biodiesel will be smaller and one will obtain an excess of triglycerides. Care must be taken to monitor the amount of water and free fatty acids in the incoming biolipid (oil or fat). If the free fatty acid level or water level is too high it may cause problems with soap formation (saponification) and the separation of the glycerin by-product downstream.
2. *Titration of lipid sample:* Optimal pH for Biodiesel is 7 (neutral), the same as distilled water (and most tap water). Some fat has a high level of free fatty acids which require an acid esterification (to obtain an pH lower than 3) before the alkaline transesterification.
3. *Mixing* the bioalcohol (methanol or ethanol) and catalyst (sodium hydroxide) in exact amounts, to produce methoxide using a standard agitator or mixer.
4. *Combining methoxide with the biolipids:* The alcohol/catalyst mix is then charged into a closed reaction vessel and the biolipid (vegetable or animal oil or fat) is added. The system from here on is totally closed to the atmosphere to prevent the loss of alcohol. The reaction mix is kept just above the boiling point of the alcohol (around 70°C) to speed up the reaction though some systems recommend the reaction take place at room temperature. Recommended reaction time varies from 1 to 8 hours. Excess alcohol is normally used to ensure total conversion of the fat or oil to its esters.

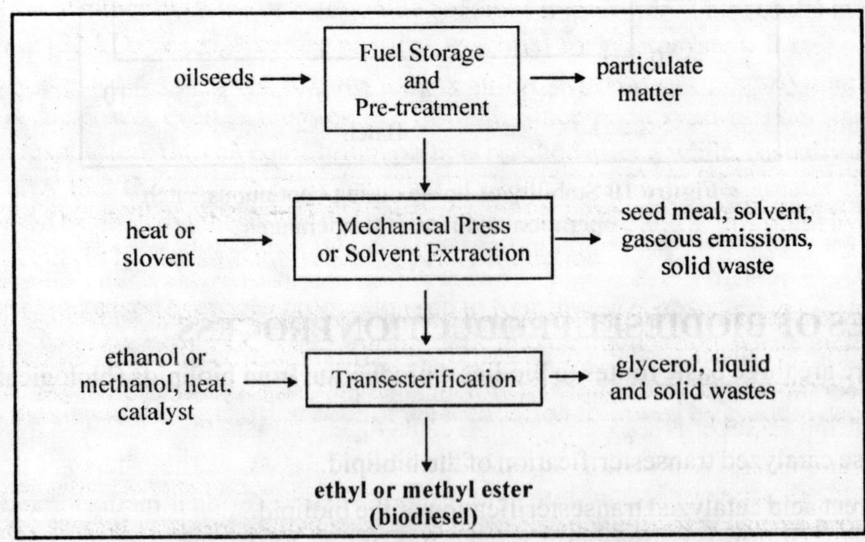


**Figure 11** Flow sheet of biodiesel production from oilseeds

5. *Separation of biodiesel and glycerol :* The glycerin phase is much more dense than biodiesel phase and the two can be gravity separated with glycerin simply drawing off the bottom of the settling vessel. In some cases, a centrifuge is used to separate the two materials faster. The glycerin by-product contains unused catalyst and soaps that are neutralized with an acid and sent to storage as crude glycerin (water and alcohol are removed later, chiefly using evaporation, to produce 80-88% pure glycerin).
6. *Removal of alcohol (by distillation):* Once the glycerin and biodiesel phases have been separated, the excess alcohol in each phase is removed with a flash evaporation process or by distillation. In other systems, the alcohol is removed and the mixture neutralized before the glycerin and esters have been separated. In either case, the alcohol is recovered using distillation equipment and is re-used. Care must be taken to ensure no water accumulates in the recovered alcohol stream.
7. *Biodiesel purification:* The separation of the wastes (catalyst and soap) is achieved by washing and drying the biodiesel. Once separated from the glycerin, the biodiesel is sometimes purified by washing gently with warm water to remove residual catalyst or soaps, dried, and sent to storage.
8. *Disposal* of the waste material.

## 3.1. Batch processing

The simplest method for producing alcohol esters is to use a batch, stirred tank reactor. Alcohol to triglyceride ratios from 4:1 to 20:1 (mole:mole) have been reported, with a 6:1 ratio most common. The reactor may be sealed or equipped with a reflux condenser. The operating temperature is usually about 65°C, although temperatures from 25°C to 85°C have been reported. The most commonly used catalyst is sodium hydroxide, with potassium hydroxide also used. Typical catalyst loadings range from 0.3 % to about 1.5%. Thorough mixing is necessary at the beginning of the reaction to bring the oil, catalyst and alcohol into intimate contact. Towards the end of the reaction, less mixing can help increase the extent of reaction by allowing the inhibitory product, glycerol, to phase separate from the ester – oil phase. Completions of 85% to 94 % are reported. Some groups use a two-step reaction, with glycerol removal between steps, to increase the final reaction extent to >95%. Higher temperatures and higher alcohol:oil ratios also can enhance the percent completion. Typical reaction times range from 20 minutes to more than one hour. Figure 12 shows a process flow diagram for a typical batch system. The oil is first charged to the system, followed by the catalyst and methanol. The system is agitated during the reaction time. Then agitation is stopped. In some processes, the reaction mixture is allowed to settle in the reactor to give an initial separation of the esters and glycerol. In other processes the reaction mixture is pumped into a settling vessel or is separated using a centrifuge. The alcohol is removed from both the glycerol and ester stream using an evaporator or a flash unit. The esters are neutralized, washed gently using warm, slightly acid water to remove residual methanol and salts and then dried. The finished biodiesel is then transferred to storage. The glycerol stream is neutralized and washed with soft water. The glycerol is than sent to the glycerol refining section.

For yellow grease and animal fats, the system is slightly modified with the addition of an acid esterification vessel and storage for the acid catalyst. The feedstock is sometimes dried (down to 0.4% water) and filtered before loading the acid esterification tank. The sulfuric acid and methanol mixture is added and the system is agitated. Similar temperatures to transesterification are used and sometimes the system is pressurized or a cosolvent is added. Glycerol is not produced. If a two-step acid treatment is used, the stirring is suspended until the methanol phase separates and is removed. Fresh methanol and sulfuric acid is added and the stirring resumes. Once the conversion of the fatty acids to methyl esters has reached equilibrium, the methanol/water/acid mixture is removed by settling or with a centrifuge. The remaining mixture is neutralized or sent straight into transesterification where it will be neutralized using excess base catalysts. Any remaining free fatty acids will be converted into soaps in the transesterification stage. The transesterification batch stage processes as described above

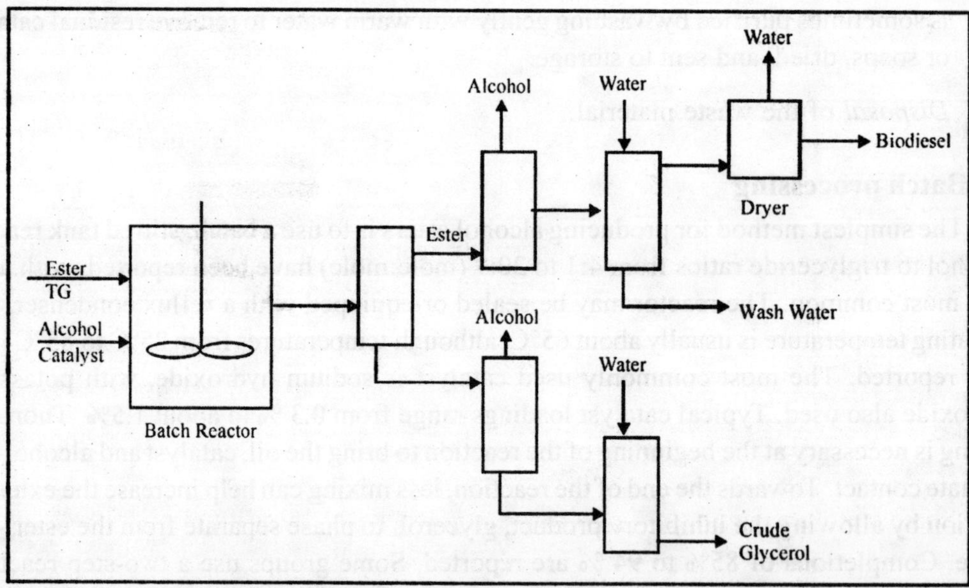


**Figure 12** A schematic representation of Batch reaction process for biodiesel production.

## 3.2. Continuous process systems

A popular variation of the batch process is the use of continuous stirred tank reactors (CSTRs) in series. The CSTRs can be varied in volume to allow for a longer residence time in CSTR 1 to achieve a greater extent of reaction. After the initial product glycerol is decanted, the reaction in CSTR 2 is rather rapid, with >98% completion not uncommon. An essential element in the design of a CSTR is sufficient mixing input to ensure that the composition throughout the reactor is essentially constant. This has the effect of increasing the dispersion of the glycerol product in the ester phase. The result is that the time required for phase separation is extended. There are several processes that use intense mixing, either from pumps or motionless mixers, to initiate the esterification reaction. Instead of allowing time for the reaction in an agitated tank, the reactor is tubular. The reaction mixture moves through this type of reactor in a continuous plug, with little mixing in the axial direction. This type of reactor, called a plug-flow reactor (PFR) behaves as if it were a series of small CSTRs chained together (Figure 13). The result is a continuous system that requires rather short residence times, as low as 6 to 10 minutes, for near completion of the reaction. The PFRs can be staged, as shown, to allow decanting of glycerol. Often this type of reactor is operated at an elevated temperature and pressure to increase reaction rate.

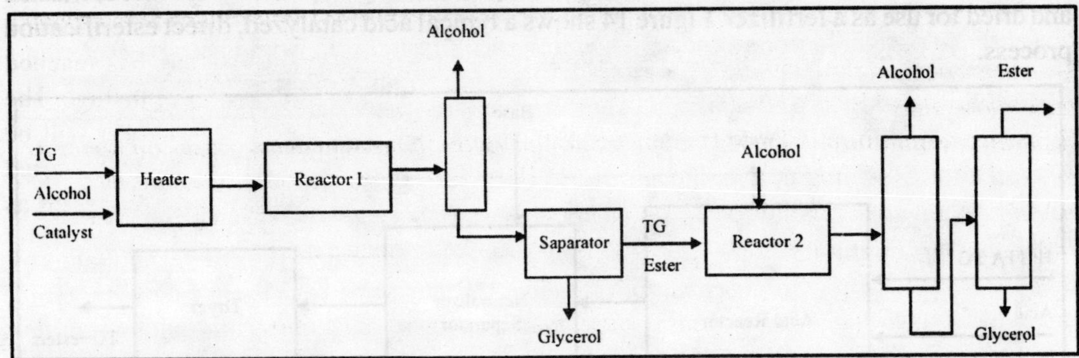


**Figure 13** A schematic representation of Plug flow reaction system for the biodiesel production

## 3.3. High free fatty acid systems

High free fatty acid feedstocks will react with the catalyst and form soaps if they are fed to a base catalyzed system. The maximum amount of free fatty acids acceptable in a base catalyzed system is less than 2 percent, and preferably less than 1 percent. Some approaches to using high free fatty acid feedstocks use this concept to "refine" the free fatty acids out of the feed for disposal or separate treatment in an acid esterification unit. The caustic is added to the feedstock and the resulting soaps are stripped out using a centrifuge. This is called caustic stripping.

Some triglycerides are lost with the soaps during caustic stripping. The soap mixture can be acidulated to recover the fatty acids and lost oils in a separate reaction tank. The refined oils are dried and sent to the transesterification unit for further processing. Rather than waste the free fatty acids removed in this manner, they can be transformed into methyl esters using an acid esterification process. As described earlier, acid catalyzed processes can be used for the direct esterification of free fatty acids in a high FFA feedstock. Less expensive feedstocks, such as tallow or yellow grease, are characteristically high in free fatty acids (FFA). The standard for tallow and yellow grease is = 15 % FFA. Some lots may exceed this standard.

Direct acid esterification of a high free fatty acid feed requires water removal during the reaction, or the reaction will be quenched prematurely. Also, a high alcohol to FFA ratio required, usually between 20:1 and 40:1. Direct esterification may also require rather large amounts of the acid catalyst depending on the process used.

The esterification reaction of FFAs with methanol produces byproduct water that must be removed, but the resulting mixture of esters and triglyceride, can be used directly in a conventional base catalyzed system. The water can be removed by vaporization, settling or centrifugation as a methanol-water mixture. Counter-current continuous-flow systems will wash out the water with the exiting stream of acidic methanol.

One approach to the acid catalyst system has been to use phosphoric acid as the initial catalyst, neutralize with an excess of KOH for the base step, and then neutralize with phosphoric acid upon completion. The insoluble potassium phosphate is recovered, washed and dried for use as a fertilizer. Figure 14 shows a typical acid catalyzed, direct esterification process.

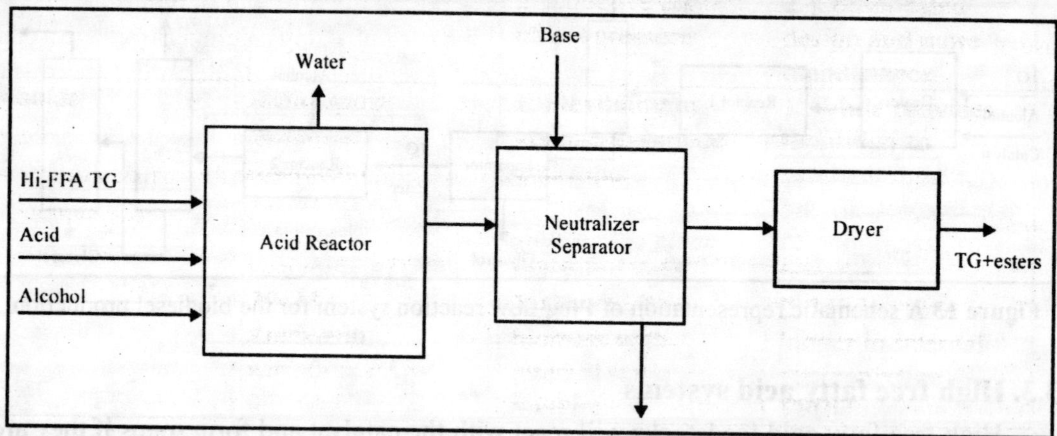


**Figure 14** A schematic representation of acid catalyzed direct esterification process for biodiesel production

An alternative approach to utilization of high FFA feedstocks is to use a base catalyst to deliberately form soap from the FFAs. The soap is recovered, the oil dried and then used in a conventional base catalyzed system. This strategy can lead to a false sense of economy. If the soapstock is discarded, the effective price of the feedstock is increased in inverse proportion to the percentage of remaining oil. The soapstock can, however, be converted into esters by using an acid catalyzed reaction. The problem with this strategy is that the soapstock system contains a large amount of water that must be removed before the product esters can meet the biodiesel standard. The soapstock process is shown in Figure 15.

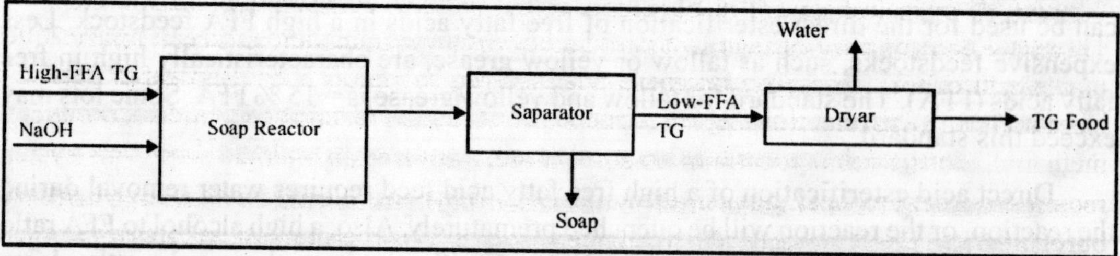


**Figure 15** A schematic representation of preparation of soapstock from a high FFA feed for biodiesel production

An alternative procedure for processing high FFA feeds is to hydrolyze the feedstock into pure FFA and glycerine. Typically this is done in a counter current reactor using sulfuric/ sulfonic acids and steam. The output is pure free fatty acids and glycerin. Any contaminants in the feedstock move mostly into the glycerin and some may leave with the steam/water effluent. Some contaminants continue with the FFA and can be removed or left in, depending on the processes and product specifications. The pure FFA are then acid esterified in another counter current reactor to transform them into methyl esters. The methyl esters are then neutralized and dried. Yields can exceed 99%. Equipment needs to be acid resistant but generally feedstock costs are extremely low.

A variation of the base-catalyzed system that avoids the problem of high FFAs is the use of a fixed bed, insoluble base. An example of this system, using calcium carbonate as the catalyst, has been demonstrated at the bench-scale. This process is depicted in Figure 16.

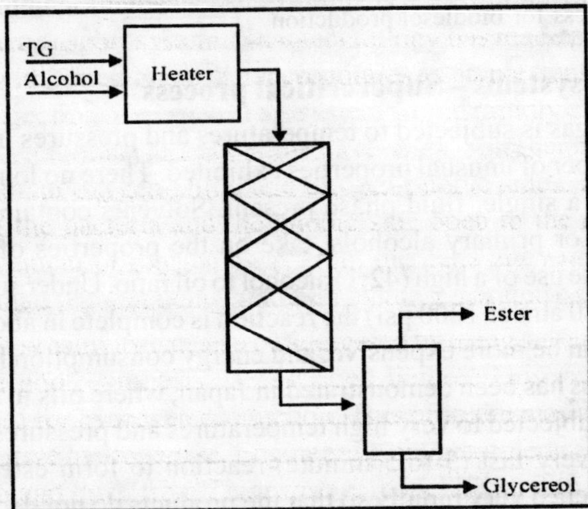


**Figure 16** A schematic fixed-bed, base catalyzed reactor system for biodiesel production

## 3.4. Non-catalyzed systems – Biox process

Co-solvent options are designed to overcome slow reaction time caused by the extremely low solubility of the alcohol in the TG phase. One approach that is nearing commercialization is the Biox Process. This process uses a co-solvent, tetrahydrofuran to solubilize the methanol. The result is a fast reaction, of the order of 5 to 10 minutes and there is no catalyst residues in either the ester or the glycerol phase. The THF co-solvent is chosen, in part, because it has a boiling point very close to that of methanol. After the reaction is complete, the excess methanol and the tetrahydrofuran co-solvent are recovered in a single step. This system requires a rather low operating temperature, 30°C. Other cosolvents, such as methyl tertiary-butyl ether (MTBE), have been investigated. The ester-glycerol phase separation is clean and the final products are catalyst- and water-free. The equipment volume has to be larger for the same quantity of final product because of the additional volume of the co-solvent.

The Biox Process is depicted in Figure 17. Co-solvents that are subject to the hazardous require special "leak proof" equipment. Fugitive emissions are tightly controlled. The co-solvent must be completely removed from the glycerine as well as the biodiesel.

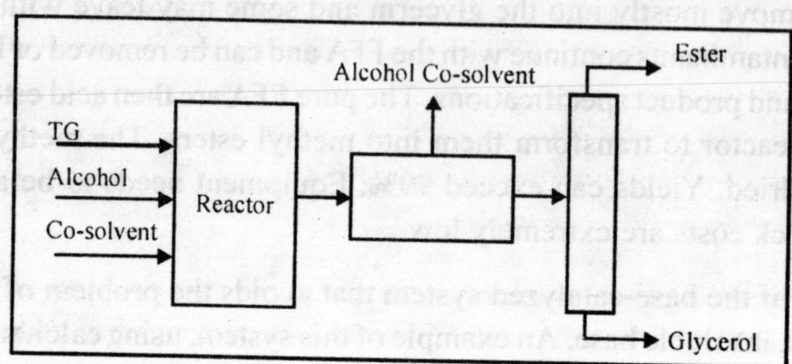


**Figure 17** A schematic representation of Biox co-solvent process for biodiesel production

## 3.5. Non-catalyzed systems – Supercritical process

When a fluid or gas is subjected to temperatures and pressures in excess of its critical point, there are a number of unusual properties exhibited. There no longer is a distinct liquid and vapor phase, but a single, fluid phase present. Solvents containing a hydroxyl (OH) group, such as water or primary alcohols, take on the properties of super-acids. A non-catalytic approach is the use of a high (42:1) alcohol to oil ratio. Under supercritical conditions (350 to 400 °C and > 80 atm or 1200 psi) the reaction is complete in about 4 minutes. Capital and operating costs can be more expensive, and energy consumption higher. An intriguing example of this process has been demonstrated in Japan, where oils in a very large excess of methanol have been subjected to very high temperatures and pressures for a short period of time. The result is a very fast (3 to 5 minute) reaction to form esters and glycerol. The reaction must be quenched very rapidly so that the products do not decompose. The reactor used in the work to date is a 5 ml cylinder that is dropped into a bath of molten metal, and then quenched in water. Figure 18 depicts one conception of a configuration for a supercritical esterification process.

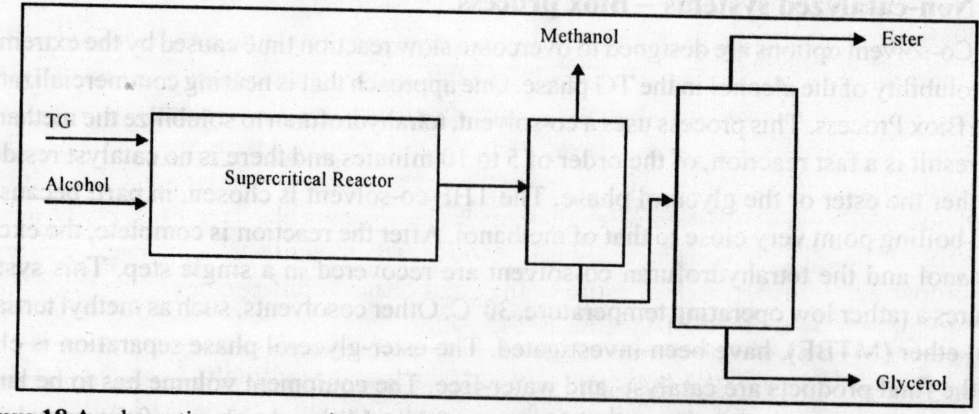


**Figure 18** A schematic representation of supercritical esterification process for biodiesel production

The refining of glycerol is discussed in a separate chapter, although it is a topic of great importance to the choice of biodiesel production process. Generally the quality of the glycerol produced and its value as a co-product is an important economic variable. The typical glycerol produced by a biodiesel plant is 50% glycerol or less and contains water, salts, methanol, methyl esters and unreacted glycerides, color and odor bodies, dimers, timers, and other minor compounds. This is commonly referred to as "biodiesel crude" and is generally worth less than 5 cents per pound. Removing the water and methanol and bringing the glycerol content up to 88% can generate a credit based on the value of crude glycerol. Salts and particularly sulfur salts or potassium salts reduces the value of the crude glycerol produced.

## 3.6. The Washing process

Washing the ester product is necessary in order to improve its fuel properties, largely by removing residual free glycerol and small amounts of potassium remaining from the catalyst. The best method so far devised was previously described. It is a combination of: i) Mixing the glycerol layer into the ester after the initial settling has occurred; adding 15% water; stirring and settling, ii) A water wash with agitation and aeration after the glycerol/water layer has been drained off.

***Wash Methodology***

**1.** *Agitation:* During washing, many of the impurities in the ester have a greater affinity for the water and they are transferred by diffusion across the phase boundary into the water. This process is greatly hastened by agitation, which can increase the area of phase contact by emulsion formation or can promote transfer by maintaining the most effective concentration gradient for transfer across the interface.

**2.** *Mechanical stirring:* Best results have been obtained using a mechanical stirrer whose rotation can be strictly controlled. The best speed for the equipment used has been about 50 to 70 RPM. The stirrer shaft should have two blades with one in the water phase and one in the ester phase rotating to lift the solution upward. This orientation, along with aeration develops maximum contact between ester and water.

**3.** *Aeration mixing:* A unique method of aeration mixing was discovered. If air is introduced deep into the water layer through a sandstone, glass or stainless steel gas diffusion disk, numerous air bubbles are formed in the water phase. These numerous water coated bubbles rise through the liquid interface into the ester, carrying large amounts of water in the film and accomplishing washing as they rise up through the ester. Upon reaching the surface, the bubbles burst and form droplets of water which fall back down through the ester, further washing it. These bubbles and droplets seem to be of such size and nature that the droplets formed do not remain emulsified when they reach the aqueous phase, but quickly coalesce and disappear into the aqueous layer. This method greatly magnifies the interface area, and at the proper aeration rate, half or more of the ester phase volume seems to be filled with quite rapidly settling droplets of aqueous phase.

**4.** *Combination aeration and mechanical stirring:* By combining (2) and (3) above very efficient and rapid washing can be achieved using minimum amounts of water.

## 3.7. Product completeness

1. *Bench wash test:* Throughout the washing, a rough idea of the completeness of the washing may be obtained by washing (in a 100 ml beaker with magnetic stirring) 50 ml ester with 25 ml water for about 1/2 hour, and then determining the pH of the wash water. If the ester is sufficiently washed, the pH should be around pH 6 to 7. There is also a good way to determine washing completeness by noting the emulsion-forming tendency during the wash-test. If the wash has been satisfactory, it is possible to stir a sample rather vigorously in this test and form an emulsion of large, clear, shiny droplets which will quickly separate, settle, and disappear upon cessation of stirring.

2. *Turbidity:* Occasionally a batch of washed ester may end up with turbidity caused by traces of condensed moisture. This moisture may be conveniently removed (evaporated) by aeration with dry air, using the gas diffusion disks from the washing step, to increase the surface contact between the air and the ester. Slight warming along with the aeration also hastens the removal of this trace of moisture.

## 3.8. Product quality

The standard for biodiesel allows 0.24% total glycerol in the final product. What does this actually mean? It is clear that a molecule of a triglyceride can be considered to contain a molecule of glycerol, sometimes called the glycerol backbone. In the case of triolein, the mole of glycerol would weigh 92.10 g and the mole of triolein weighs 885.46 g. Therefore, triolein can be considered to consist of 92.10/885.45 = 0.104, or 10.4% glycerol. This glycerol is called 'bound glycerol' because it is chemically bound to the triolein molecule. Bound glycerol can also be associated with monoglycerides and diglycerides, the partial reaction products of the conversion of triglycerides to alkyl esters. The structures of these molecules are shown below:

Diglyceride:

$CH_2 - OH$
|
$CH - O - C(=O) - R_2$
|
$CH_2 - O - C(=O) - R_3$

**Diglyceride**

Monoglyceride:

$CH_3 - OH$
|
$CH_3 - OH$
|
$CH_3 - O - C(=O) - R_3$

**Monoglyceride**

Bound glycerol is added to any fully reacted glycerol, or free glycerol, that may still be in the biodiesel, to get the total glycerol. If the original oil contains 10.4% glycerol and the final biodiesel can only contain a total glycerol level of 0.24% then the transesterification reaction must be

$$\frac{10.4 - 0.24}{10.4} \times 100 = 97.7\% \quad \text{or 97.7\% complete.}$$

***Competing reactions***

It is common for oils and fats to contain small amounts of water and free fatty acids. Free fatty acids consist of the long carbon chains described that are disconnected from the glycerol backbone. They are sometimes called carboxylic acids

$$\mathrm{HO-\overset{\overset{\displaystyle O}{\|}}{C}-R}$$

Carboxylic acid (R is a carbon chain)

The oleic group used, gives oleic acid, one of the free fatty acids that can be found in unrefined vegetable oils and animal fats.

$$\mathrm{HO-\overset{\overset{\displaystyle O}{\|}}{C}-(CH_2)_7\,CH{=}CH(CH_2)_7CH_3}$$

Oleic acid

If an oil or fat containing a free fatty acid such as oleic acid is used to produce biodiesel, the alkali catalyst typically used to encourage the reactiỏn will react with this acid to form soap when the catalyst is potassium hydroxide (KOH):

$$\underset{\textbf{Oleic Acid}}{\mathrm{HO-\overset{\overset{\displaystyle O}{\|}}{C}-(CH_2)_7\,CH{=}CH(CH_2)_7CH_3}} \quad + \quad \underset{\textbf{Potassium Hydroxide}}{\mathrm{KOH}}$$

$$\rightarrow \quad \underset{\textbf{Potassium oleate (soap)}}{\mathrm{K^+\,O-\overset{\overset{\displaystyle O}{\|}}{C}-(CH_2)_7\,CH{=}CH(CH_2)_7CH_3}} \quad + \quad \underset{\textbf{Water}}{\mathrm{H_2O}}$$

Formation of soap

This reaction is undesirable because it binds the catalyst into a form that does not contribute to accelerating the reaction. Excessive soap in the products can inhibit later processing of the biodiesel, including glycerol separation and water washing. Water in the oil or fat can also be problem. When water is present, particularly at high temperatures, it can hydrolyze the triglycerides to diglycerides and form a free fatty acid. When an alkali catalyst is present, the free fatty acid will react to form soap following the reaction given earlier. When water is present in the reaction it generally manifests itself through excessive soap production. The soaps of saturated fatty acids tend to solidify at ambient temperatures so a reaction mixture with excessive soap may gel and form a semi-solid mass that is very difficult to recover.

$$
\begin{array}{lcccccc}
CH_2 - O - \overset{\displaystyle O}{\overset{\|}{C}} - R_1 & & & & CH_3 - OH & & \\
| & & & & | & & \\
CH - O - \overset{\displaystyle O}{\overset{\|}{C}} - R_2 & + & H_2O & \rightarrow & CH_3 - O - \overset{\displaystyle O}{\overset{\|}{C}} - R_2 & + & HO - \overset{\displaystyle O}{\overset{\|}{C}} - R_1 \\
| & & & & | & & \\
CH_2 - O - \overset{\displaystyle O}{\overset{\|}{C}} - R_3 & & & & CH_3 - O - \overset{\displaystyle O}{\overset{\|}{C}} - R_3 & & \\
\textbf{Triglyceride} & & \textbf{Water} & & \textbf{Diglyceride} & & \textbf{Fatty acid}
\end{array}
$$

Hydrolysis of a triglyceride to form free fatty acids

**3.8.1.** *Quality testing:* This is the most useful all-round test, and it's very simple: Put 150 ml of unwashed biodiesel (settled for 12 hours or more, with the glycerine layer removed) in a half-litre glass jar. Add 150 ml of water, screw the lid on tight and shake it up and down violently for 10 seconds or more. Then let it settle. The biodiesel should separate from the water in half an hour or less, with amber biodiesel on top and milky water below (Figure 19). This is quality fuel, a completed product with minimal contaminants. Wash it and then use it with confidence.

**Figure 19** Wash-test with unwashed biodiesel — left, after a violent 10-second shaking; right, biodiesel and water separated cleanly within minutes

## 4. BIO-DIESEL IN INDIA

The main commodity sources for Bio-diesel in India can be non-edible oils obtained from plant species such as *Jatropha curcas* (Ratanjyot), *Pongamia pinnata* (Karanj), *Calophyllum inophyllum* (Nagchampa), *Hevcca brasiliensis* (Rubber) etc. Of these, Jatropha appears to be the feedstock of choice because of its ability to thrive under a variety of geo-climatic conditions, its low gestation period and its high seed yield. *Jatropha* (Euphorbiaceae) is a genus comprising species growing in tropical and subtropical countries. Nine species are reported to occur in India. The seed kernel contains 40-60% (w/w) oil. Saturated fatty acids constitute 20% of this, whereas those remaining are unsaturated ones. Oleic acid is the most abundant (44.8%) followed by linoleic acid (34%), palmitic acid (12.8%), and stearic acid (7.3%). While composition of the oil is similar to other oils, which are used for edible purposes, the presence of some antinutritional factors such as toxic phorbol esters renders this oil unsuitable for use in cooking. Thus, it is a good choice as the starting oil for production of biodiesel. In fact, the seed oil of Jatropha was used as a diesel fuel substitute during World War II. Later, its blends with diesel fuel were tested. It is interesting to note that Jatropha seeds themselves are reported to contain lipase activity which could also catalyze transesterification reactions.

India has nearly 60 million hectares of wasteland, of which 30 million hectares are available for energy plantations like "Jatropha". Once grown, the crop has a life of 50 years (Figure 20). Each acre will produce about 2 tonnes of bio-diesel at about Rs. 20 per litre. India has a potential to produce nearly 60 million tones of bio-fuel annually, thus making a significant and important contribution to the goal of Energy Independence. Indian Railways has already taken a significant step of running two passenger locomotives (Thanjavur to Nagore section) and six trains of diesel multiple units (Tiruchirapalli to Lalgudi, Dindigul and Karur sections) with a 5% blend of bio-fuel sourced from its in-house esterification plants. In addition, they have planted 75 lakh Jatropha saplings in Railway land which is expected to give yields from the current year onwards. This is a pioneering example for many other organisations to follow. Similarly many States in our country have energy plantations. What is needed is a full economic chain from farming, harvesting, extraction to esterification, blending and marketing. Apart from employment generation, bio-diesel has a significant potential to lead our country towards energy independence.

**Figure 20** Plantations of Jatropha

*India Planning Comprehensive National Biodiesel Policy:* Indian government has announced a comprehensive policy for the use of B20 biodiesel (20% biodiesel) in the country on its Independence Day, August, $15^{th}$, 2005. The Government is planning to produce 13 million tonnes of alternate fuel every year. However, this will require 11 million hectares of land. India currently imports more than 70% of its crude oil. The government is not only looking to stabilize its energy sources and reduce its outbound payments; it hopes to become a global sourcing hub for both feedstock and processed bio-diesel. The industry is expected to be a US$2-billion revenue earner within the next three years. "Bio-diesel processing cost in India is almost one-third of that in European countries".

*Buses in Haranya, India, Shifting to Biodiesel:* Started in June 2005, all buses running out of Gurgaon in the state of Haryana, India (next to New Delhi) are running on B5 biodiesel (5% biodiesel). The decision from the Haryana government came after a year-long pilot test. IOC (Indian Oil Corporation) is providing the biodiesel, processed from jatropha, at no extra cost. "Smoke levels in buses running on fuel mixed with bio-diesel have certainly gone down and this home-grown fuel like this can go a long way in providing energy-security to the country". Although only B5 was used in the test, the participants are looking to increase the blend to a B20 over the next few months.

*Gujarat, India Begins First Biodiesel Bus Service:* Gujarat is the first State in India to run commercial buses using biodiesel fuel. Initially, four buses using a B5 (5% biodiesel) blend will service a route between Gandhinagar (the capital city of the State) and Ahmedabad (India's worst-polluted city). "Trial runs had been carried out in Haryana and Maharashtra. Even the Indian Railways had run a Jan Shatabdi on bio-diesel. But this is the first time that a regular commercial service is being started with bio-diesel," the Gujarat Chief Minister, Mr Narendra Modi, said. Although the Gujarat State Transport Corporation will begin with the B5 blend, it will work toward B20. Gujarat Oleo Chem Ltd, which has set up a biodiesel blending and extraction plant in South Gujarat, is providing the biodiesel.

*Indian State Chief Minister's Official Car Uses Biodiesel; All State Vehicles to Follow:* Chief Minister of India's state of Chhattisgarh, is the first head of a Indian state government to use biodiesel to power his official car, a Tata Safari. "This is just the start of the bio-fuel energy revolution in Chhattisgarh. Jatropha diesel will power all government vehicles within three months" Singh told reporters. "The state is tipped to cultivate 80 million jatropha seedlings in rural areas to make Chhattisgarh a bio-fuel self-reliant state by 2015."

*State Bank of India Funds Jatropha Cultivation for Biodiesel:* UK-based D1 Oils, through its 50:50 Indian joint venture, D1 Mohan Bio Oils Ltd., signed a Memorandum of Understanding (MOU) with the State Bank of India for Rupees 1.3 billion (approximately $30 million, or €23 million) in funding for farmers in Tamil Nadu to plant up to 40,000 hectares of jatropha. The anticipated yield is between 100,000 and 120,000 metric tons of crude jatropha oil per annum (assuming the full 40,000 hectares are planted). D1 Mohan has sought to plant up to 100,000 hectares of jatrop1ha across India in 2005. D1 Oils has the option to export 25% of the raw oil to its international customers, with the bulk of the oil being retained for domestic (Indian) biodiesel production.

## 5. FUTURE OUTLOOK

Currently, biodiesel is generally somewhat more expensive to produce than petroleum diesel, which is often stated as the primary factor keeping it from being in more widespread usage. Economies of scale in biodiesel production, however, as well as the rising cost of petroleum, may reduce, eliminate, or even reverse this cost differential in the future. Current worldwide production of vegetable oil and animal fat, however, is not enough to replace liquid fossil fuel use. Some environmental groups, notably the NRDC (Natural Resources Defense Council), object to the vast amount of farming and the resulting over-fertilization, pesticide use and land use conversion that would be needed to produce the additional vegetable oil. Another alternative for meeting the increased demand of oils appears to be microalgae, which like higher plants, produce storage lipids in the form of triacyglycerols (TAGs). This property makes microalgae well suited for use in a commercial-scale biodiesel production facility. Many species exhibit rapid growth and high productivity and many microalgal species can be induced to accumulate substantial quantities of lipids, often greater than 60% of their biomass. Microalgae can also grow in saline waters that are not suitable for agricultural irrigation or consumption by humans or animals. The growth requirements are very simple, primarily carbon dioxide ($CO_2$) and water, although the growth rates can be accelerated by sufficient aeration and the addition of nutrients. Many microalgae can be induced to accumulate lipids under conditions of nutrient deprivation. If this process could be understood, it might be possible to manipulate either the culture conditions, or to manipulate the organisms themselves, to increase lipid accumulation in a particular strain.

## 6. FURTHER READING

Du, W., Wu, Y. and Liu, D. (2003). Lipase-catalysed transesterification of soya bean oil for biodiesel production during continuous batch operation. Biotechnol. Appl. Biochem. 38: 103-106.

Isigigür A., Karaosmanoglu F., Aksoy H. A., Hamdullahpur F. and Gülder Ö. L. (1993) Safflower seed oil of Turkish origin as a diesel fuel alternative. Applied Biochemistry and Biotechnology 39: 89-105.

Krawczyk, T. (1996). Biodiesel – alternative fuel makes inroads but hurdles remain. INFORM 7: 801-829.

Ma, F and Hanna, M.A. (1999). Biodiesel production: a review. Bioreseource Technology. 70: 1-15.

Rao P. S. and Gopalakrishnan K. V. (1991). Vegetable oils and their methyl esters as fuels for diesel engines. Indian Journal of Technology 29: 292-297

Reed T. B., Graboski M. S. and Gaur S. (1992). Development and commercialization of oxygenated diesel fuels from waste vegetable oils. Biomass and Bioenergy 3: 111-115.

Shah, S., Sharma, S and Gupta, M.N. (2004). Biodiesel Preparation by ;ipase-catalyzed transesterification of Jatropha oil. Energy & Fuels. 18: 154-159

Stern, R. Hillion, G. and Rouxel, J.J. (1995). Improved process for the production of esters from fatty substances having a natural origin. US Patent. 5: 424-466.

Tickell J. and Tickell K. (2000). From the Fryer to the Fuel Tank: the complete guide to using vegetable oil as an alternative fuel. Green Tech Publishing.

Wu, H., Zong, M-H., Luo, Q. and Wu, H-C. (2003). Enzymatic conversion of waste oil to bio-diesel in a solvent-free system. Prepr. Pap. Am. Chem. Soc.Div. fuel Chem. 48: 533-534.

# 9

# Microbes and Methane Production

**S.K. SONI**
*Department of Microbiology, Panjab University, Chandigarh-160 014*

## 1. INTRODUCTION

The simplest hydrocarbon, methane, is a gas with a chemical formula of $CH_4$. Pure methane is odorless, but when used commercially is usually mixed with small quantities of strongly-smelling sulfur compounds such as ethyl mercaptan to enable the detection of leaks. A principal component of natural gas, methane is a significant fuel. Burning one molecule of methane in the presence of $O_2$ (oxygen) releases one molecule of $CO_2$ (carbon dioxide) and two molecules of $H_2O$ (water):

$$CH_4 + 2O_2 \rightarrow CO_2 + 2H_2O$$

Methane can be used in any appliance or utility that uses natural gas. The natural gas requirements of an average person with a U.S. standard of living is about 60 $ft^3$/day. This is equivalent to 4.5 kg of chicken or pig manure per day (7 pigs and 100 chickens) or 9.0 kg of horse manure (about 2 horses). Methane is a greenhouse gas with a global warming potential of 21. Principal methane sources are decomposition of organic wastes, natural sources (marshes): 23 % and mineral fuel extraction: 20 %. Methane is extracted from geological deposits as a mineral fuel which is associated with other hydrocarbon fuels and is also produced by the anaerobic degradation of organic matter by microorganisms. 60% of the world emissions are from sources affected by humans. They come primarily from agricultural and other human activities. During the past 200 years, the concentration of this gas in the atmosphere doubled, passing from 0.8 to 1.7 ppm. Methane is also classified as a biogas because it can be created by the anaerobic decomposition of certain organic matters.

Methane, the lightest organic gas, has two fundamental drawbacks to its use in heat engines: it has a relatively low fuel value and it takes nearly 5,000 psi to liquefy it for easy storage (87.7 $ft^3$ methane gas = 3.78 l of liquid methane or 1 $ft^3$ methane gas = 9 tablespoons of liquid methane). So a great deal of storage is required of methane for a given amount of work, in comparison to propane which liquefies around 250 psi. Methane has been used in tractors and automobiles. The gas bottles carried by such vehicles are often about 5 ft long by 9 in diameter (1.9 $ft^3$) charged to 2800 psi so that about 420 $ft^3$ of methane is carried, which is comparable to 13.25 l of gasoline fuel. However, it seems that the most efficient use of methane would be in stationary heat engines located near the digester (e.g., compressors and generators). There are two reasons for this:

1. The engine's waste heat can be recirculated in digester coils instead of dissipating in the open.

2. Gas can be used directly as it is produced, without the need of compressors. For example, bio-gas produced from pig manure was used at ordinary pressures by John Fry to power a Crossley Diesel Engine. The diesel ran an electric generator and the waste heat was recirculated directly back into the digesters. It is likely that bio-gas produced from mixed wastes would have to be "scrubbed" of corrosive hydrogen sulfide (by passing through iron filings), and possibly $CO_2$ (by passing through lime water).

In nature, anaerobic decay is probably one of the earth's oldest processes for decomposing wastes and is one of the few natural processes that hasn't been fully exploited until recent times. Pasteur once discussed the possibilities of methane production from farmyard manure and the Chinese have used "covered lagoons" to supply methane fuel to communes and factories for decades. But the first attempt to build a digester to produce methane gas from organic wastes (cow dung) appears to have been in Bombay, India in 1900. At about this time, sewage plants started digesting sewage sludge in order to improve its quality. This started a mass of laboratory and small-scale experiments during the 1920's and 1930's.

During World War II, the shortage of fuel in Germany led to the development of methane plants in rural areas, where the gas was used to power tractors. The idea spread into Western Europe, until fossil fuels once again became available (although, today, many farmers in France and Germany continue to use home digesters to produce their own methane fuel gas). Currently, the focus of organic digester/biogas research is in India. India's impetus has been the overwhelming need of a developing country to raise the standard of living of rural population. Cows in India produce over 800 million tons of manure per year, over half of which is burned for fuel and thus lost as a much needed crop fertilizer. The problem of how to obtain cheap fuel and fertilizer at a local level led to several studies by the Indian Agricultural Research Institute, New Delhi in the 1940's to determine the basic chemistry of anaerobic decay. In the 1950's, simple digester models were developed which were suitable for village homes. These early models established clearly that bio-gas plants could:

1. provide light and heat in rural villages, eliminating the need to import fuel, to burn cow dung, or to deforest land;
2. provide a rich fertilizer from the digested wastes; and
3. improve health conditions by providing air-tight digester containers, thus reducing diseases borne by exposed dung.

More ambitious designs were tested by the Planning Research and Action Institute in the late 1950's. Successes led to the start of the Gobar Gas Research Station at Ajitmal in Northern India with practical experience from the Khadi and Village Industries Commission (KV / C). Methane, when artificially produced by anaerobic decomposition of organic matter is recovered as a Biogas and when naturally produced from organic waste dumped in land is recovered as Landfill gas.

## 2. LANDFILL GAS

A large proportion of ordinary domestic refuse, municipal solid wastes, is biological material and its disposal in landfills creates suitable conditions for anaerobic digestion. The landfill sites producing methane has been known for decades and recognition of the potential hazard led to the fitting of systems for burning it off. However, it was only in the 1970s that serious attention was paid to the idea of using this undesirable product. The waste matter is more miscellaneous in a landfill than in a biogas digester and the conditions are neither warm nor wet. So the process is much slower, taking place over years rather than weeks. The end product known as landfill gas, is again a mixture consisting mainly of $CH_4$ and $CO_2$. In theory, the lifetime yield of a good site should lay in the range 150-300 $m^3$ of gas/ton of wastes, with between 50% and 60% by volume of methane. This suggests a total energy of 5-6 GJ per ton of refuse, but in practice yields are much less. In developing a site, each area is covered with a layer of impervious clay or similar material, after it is tilled producing an environment which encourages anaerobic digestion. The gas is collected by an array of interconnected perforated pipes buried at depths up to 20 metres in the refuse. In new sites this pipe system is constructed before the wastes start to arrive and in a large well-established landfill there can be several miles of pipes with as much as 1000 $m^3$ an hour of gas being pumped out.

## 3. BIOGAS

Biogas is a colourless, odourless, inflammable gas, produced by organic waste and biomass decomposition (fermentation). Biogas can be produced from animal, human and plant (crop) wastes, weeds, grasses, vines, leaves, aquatic plants and crop residues etc. The composition of different gases in biogas is depicted in Table 1.

**Table 1** Composition of Biogas

| Sr. No. | Substances | Percentage (%) |
|---|---|---|
| 1 | Methane ($CH_4$) | 55-70 |
| 2 | Carbon dioxide ($CO_2$) | 30-45 |
| 3 | Hydrogen sulphide ($H_2S$) | 1-2 |
| 4 | Nitrogen ($N_2$) | 0-1 |
| 5 | Hydrogen ($H_2$) | 0-1 |
| 6 | Oxygen ($O_2$) | Traces |

Biogas is a valuable fuel which in many countries is produced in purpose built digesters filled with the feedstock like dung or sewage. Digesters range in size from 1$m^3$ for a small 'household' unit to more than 1000 $m^3$ used in large commercial installation or farm plants. The input may be continuous or in batches, and digestion is allowed to continue for a period of ten days to a few weeks. The bacterial action itself generates heat, but in cold climates additional heat is normally required to maintain the ideal process temperature of at least 35°C and this must be provided from the biogas. A well-run digester will produce 200-400 $m^3$ of biogas with a methane content of 50 to 75% for each dry tonne or input.

### 3.1. Benefits of Biogas

Biogas technology makes optimal utilization of the valuable natural resource of dung; it provides nearly three times more useful energy than directly burnt dung, and also produces nutrient-rich manure.

As a cooking fuel, it is cheap and extremely convenient. Based on the effective heat produced, a 2$m^3$ biogas plant could replace, in a month, fuel equivalent to 26 kg of LPG (nearly two standard cylinders), or 37 liters of kerosene, or 88kg of charcoal, or 210 kg of fuelwood, or 740 kg of animal dung. In terms of cost, biogas is cheaper, on a life cycle basis, than conventional biomass fuels (dung, fuelwood, crop wastes, etc.) as well as LPG, and is only fractionally more expensive than kerosene; the commercial fuels like kerosene and LPG, however, have severe supply constraints in the rural areas. To a housewife, biogas is easy to use and saves time in the kitchen; biogas stove has an efficiency of about 55% which is comparable to that of a LPG stove. Cooking on biogas is free from smoke and soot, and can substantially reduce the health problems, which are otherwise quite common in most rural areas in India where biomass is the chief source of fuel. The use of biogas is helpful to improve the quality of life in household. However, the use of biogas is by no means confined to cooking alone. It can be used, through a specially designed mantle, for lightning, too. Further, biogas can partially replace diesel to run IC (internal combustion) engines for water pumping, small industries like flour mill, saw mill, oil mill etc. This would not only reduce dependence on diesel, but also help in reducing carbon pollutants which adversely affect the atmosphere. Dual fuel engines (80% biogas and 20% diesel) are now commercially manufactured in India. Biogas can be similarly used to produce electricity, though this has not been attempted on a large scale in the country so far. Nevertheless, the versatility of biogas is its greatest advantage as a source of energy for the rural areas. While biogas has multiple benefits at the individual family level, it also has several qualitative and quantitative benefits at the societal level. Firstly, a shift to biogas from traditional biomass fuels results in less dependence on natural resources such as forests, checking their indiscriminate and unsustainable exploitation. Since dung is collected systematically when used in biogas, environment can be kept clean and hygienic.

Biogas can be used for lighting in rural areas which lack electricity supply. Special types of gauze mantle lamps, consuming 0.07 to 0.14 $m^3$ of gas per hour, are used for household lighting. Several companies in India manufacture a great variety of lamps which have single or double mantles. Generally, 1 mantle lamp is used for indoor purposes, and 2 mantle lamps for outdoors. Such lamps emit clear and bright light equivalent to 40 to 100 candle powers. These are generally efficient, easy to adjust and to maintain but they are unfortunately costly. The other advantage is that, unlike centralized systems such as thermal power plants and fertilizer factories, which entail huge capital investments and need elaborate distribution networks, biogas plants are decentralized systems which can be installed even in remote areas with very low investments.

## 3.2. Potential of biogas

In India, the dissemination of large scale biogas plants has began in the mid-seventies and the process has become consolidated with the advent of the National Project on Biogas Development (NPBD) in 1981, which has been continuing since. Against the estimated potential of 12 million biogas plants, 2.9 million family type and 2700 community, institutional and nightsoil-based plants have been set up till December 1999. This is estimated to have helped in a saving of 3 million tonnes of fuelwood per year and manure containing nitrogen equivalent to 0.7 million tonnes of urea.

However, in terms of total dung that is available in the country, the potential is much more. The bovine population in India is 260 millions and an adult bovine produces an average of 10 kg of dung per day. Since grazing is a common practice in India, all the dung produced cannot be collected. If it is assumed that 75% of the dung is collected, nearly 2 millions tonnes of dung would be available everyday. At 25 kg /m$^3$, this dung can feed as many as 40 millions biogas plants of 2 m$^3$ capacity, which can be considered the ultimate potential for biogas technology. But even this high potential of biogas is based on animal dung only. However, all organic matter can technically be used to generate methane; if the scientific experiments that are going on in the country under the patronage of Ministry of Non-conventional Energy Sources (MNES) to develop alternative feedstocks (such as water hyacinth, kitchen waste, and poultry waste) are successful, potential for biogas generation could be virtually unlimited. It can be mentioned in this context that human waste is an excellent source of biogas which would enhance the potential; substantially. With such high potential, which can be routed to hitherto unemphasized applications of shaft power and electricity generation, biogas can make a significant contribution to the development of small industries and agriculture, and thus to the overall advancement of the rural areas.

## 3.3. Biogas production system

The biogas (mainly mixture of methane and carbon dioxide) is produced/generated under both, natural and artificial conditions. However, for techno-economically-viable production of biogas for wider application, the artificial system is the best and most convenient method. The production of biogas is a biological process which takes place in absence of air (oxygen), through which the organic material is converted into, methane ($CH_4$) and carbon dioxide ($CO_2$) and in the process gives excellent organic fertilizer and humus as the second by-product. The one essential requirement in producing biogas is an airtight container. Biogas is generated only when the decomposition of biomass takes place under the anaerobic conditions, as the anaerobic bacteria (microbes) that live without oxygen are responsible for the production of this gas through the destruction of organic matter. The airtight container used for the biogas production under artificial condition is known as digester or reactor.

### *3.3.1. Raw materials*

*Plant wastes:* The primary advantage to plant wastes is their availability. Their disadvantage for a small farm operation is that plant wastes can often be put to better use as livestock feed or compost. Also, plants tend to be bulky and to accumulate lignin and other indigestible materials that must be regularly removed from digesters. This severely limits the use of plant wastes in continuous-feeding digesters. There seem to be three possible ways to take advantage of plant wastes in continuous digesters:

1. Press plant fluids out of succulent plants (cacti, iceplant, etc.) and digest juices directly, or use them as a diluter for swill.
2. Culture algae for digestion.
3. Digest plants not containing lignin (seaweed).

*Animal manures:* The main advantage to animal manure, with respect to continuous digesters, is that it is easy to collect (with proper design of livestock shelters) and easy to mix as slurry and load into digesters. Successful continuous digesters have been set up using pig manure, cow dung and chicken manure. The general consensus seems to be that, among animal manures, chicken manure is easily digested, produces large quantities of gas and makes a fertilizer very high in nitrogen.

*Human waste:* Human waste or "night soil" has long been used as a fertilizer, especially in the Orient. However, there seems to be little information on using human wastes as raw materials for anaerobic digesters. A well-designed privy digester which paid special attention to the transmission of diseases peculiar to humans would be a real asset to homestead technology. A solution to this problem would be welcomed. One suggestion is a seat with a clip-on plastic bag. When filled it could be dropped into a digester intact. The plastic would have to be a material which would decompose only in the presence of methane bacteria, or liquids generally after so many hours.

*Garbage and municipal solid waste:* Basically the municipal solid waste is composed of the above categories of the waste. In recent years, the technology of production of energy from refuse has been well developed in many countries. It has been recognized that if city wastes are not managed properly but are left to rot in the streets, the putrefying smell produced from these wastes is obnoxious to humans with a possible risk of occurrence of fatal diseases. One of the best solutions to tackle the growing problem of pollution due to these municipal wastes is to treat them through anaerobic digesters in order to produce biogas and bio-fertilizer.

*3.3.1.1. Concept and options for treating the municipal solid waste*

A plan for treating the municipal waste for production of compost and bio-energy has been schematically presented in Figure 1; this also shows that there are several options for the treatment of MSW including the biogas production.

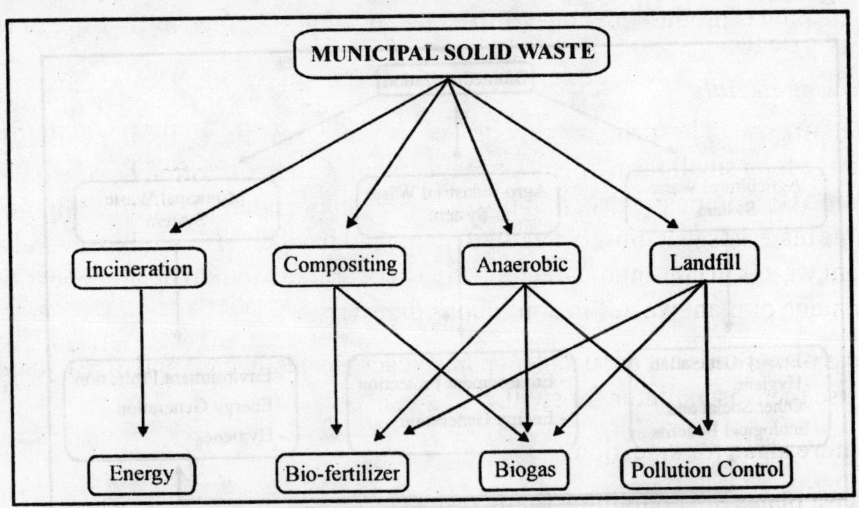


**Figure 1** Energy and fertilizer production from municipal solid waste

One of the options is to treat the waste in an incinerator for the production of energy. The other option is to treat the waste to prepare compost for use as organic fertilizer. The best option would be to process MSW in an anaerobic reactor with the aim of producing both biogas and bio-fertilizer with simultaneous improvement in the environmental pollution. If these options are not available, the city waste can be collected and directly transferred to the landfill site. The waste conversion process with the aim of producing biogas and high quality compost for cogeneration of electricity and heat or conversion of biogas to vehicle fuel has been conceptualized in Figure 2.

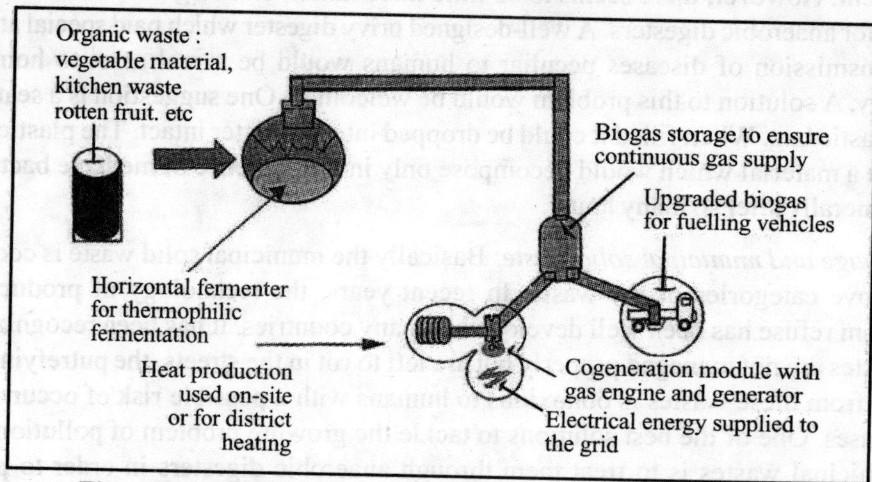


**Figure 2** A schematic representation of waste conversion process

### 3.3.2. *Biogas technology*

Biogas technology is best suited to convert organic waste from agriculture, livestock, industries, municipalities, and other human activities into energy and manure. The use of energy and manure can lead to environmental, health, and other socio economic gain as shown in Figure 3.

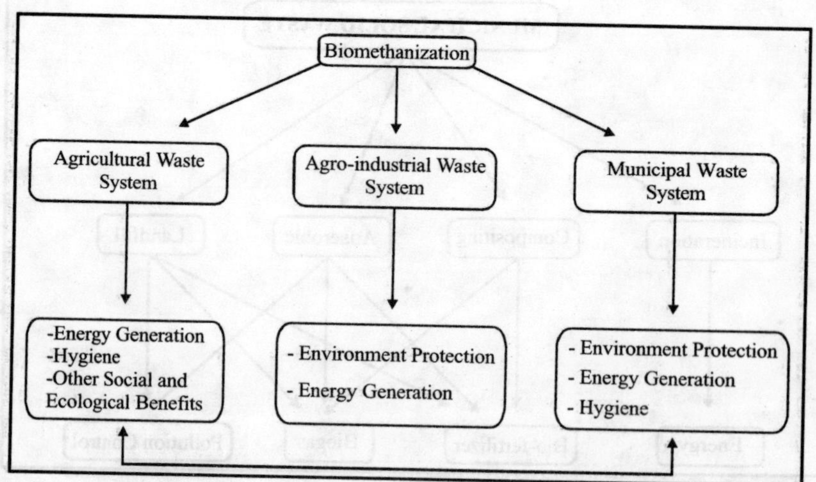


**Figure 3** Biomethane implementation and its effects

Methane is produced by the bacterial decomposition of organic materials in the absence of oxygen. Anaerobic digestion is of note because it can effectively extract the carbon from human, animal and farm residues for use as fuel while leaving the nitrogen in a sludge which can be used as a farm fertilizer. The gas is used for cooking, lighting and to run electric motors, irrigation pumps, refrigerators and compressors. As a source of renewable energy, biogas or anaerobic digestion technology has become quite popular in developing countries including India Methane fermentation is a versatile biotechnology capable of converting almost all types of polymeric materials to methane and carbon dioxide under anaerobic conditions. This is achieved as a result of the consecutive biochemical breakdown of polymers to methane and carbon dioxide in an environment in which a variety of microorganisms which include fermentative microbes (acidogens), hydrogen-producing, acetate-forming microbes (acetogens) and methane-producing microbes (methanogens) harmoniously grow and produce reduced end-products. Anaerobes play important role in establishing a stable environment at various stages of methane fermentation.

Methane fermentation offers an effective means of pollution reduction, superior to that achieved via conventional aerobic processes. Although practiced for decades, interest in anaerobic fermentation has only recently focused on its use in the economic recovery of fuel gas from industrial and agricultural surpluses. Methane fermentation is the consequence of a series of metabolic interactions among various groups of microorganisms. The first group of microorganisms, commonly known as non-methanogens, secrete enzymes which hydrolyze polymeric materials to monomers such as glucose and amino acids, which are subsequently converted to higher volatile fatty acids, $H_2$ and acetic acid. In the second stage, hydrogen-producing acetogenic bacteria, also called as non-methanogens, convert the higher volatile fatty acids including propionic and butyric acids, produced, to $H_2$, $CO_2$, and acetic acid. Finally, the third group, methanogenic bacteria convert $H_2$, $CO_2$, and acetate, to $CH_4$ and $CO_2$.

Sludge from the sewage treatment plants or the sewage itself or any source of organic matter is taken to an anaerobic digester "Imhoff tank" which combines a settling basin and sludge digestion tank. The process of anaerobic digestion of organic residues is as follows and is schematically presented in Figures 4 and 5.

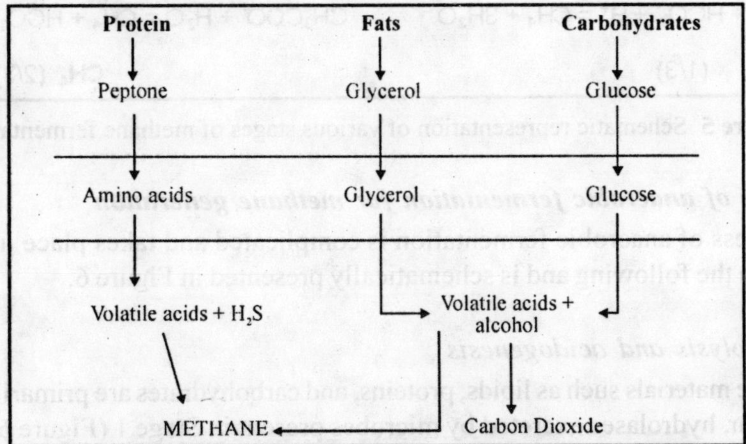


**Figure 4** Anaerobic digestion process of organic residues

Non-methanogens resemble the bacteria of human intestinal tract and may include *E. coli, Micrococcus varians, Pseudomonas reptilivora, Microcollus luteus, Clostridium, Peptococcus, Eubacterium, Lactobacillus, Actinomyces, Bacillus, Vibrio* and *Spirillium*. On the other hand the methanogens belong to *Methanococcus vennelli, M. omelianskii, M. sarcina, Methanobacterium ruminantum, M. mobilis, Methanospirillum* spp. The complete digestion takes place in 30-40 days of incubation at 30°C and pH 6.4-7.5. Compounds containing sodium, magnesium or calcium are preferred to control the pH of the digester. Presence of acid forming bacterial population enhances $CH_4$ formation by producing a combination of compounds acting as substrates – methanol, acetate, formate, $H_2$ and other acids for methanogenic bacteria. Addition of Fe and S compounds in traces is beneficial. C, N and P ratio required for optimal gas production is 100:15:1.

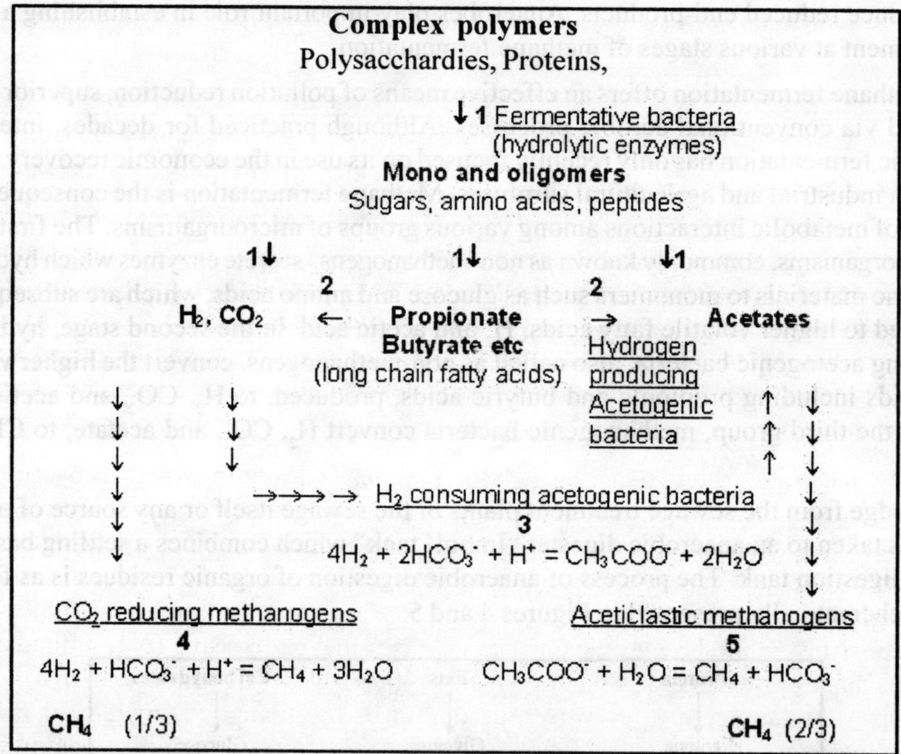


**Figure 5** Schematic representation of various stages of methane fermentation

### *3.3.3. Process of anaerobic fermentation for methane generation*

The process of anaerobic fermentation is complicated and takes place in three stages as described in the following and is schematically presented in Figure 6.

#### *3.3.3.1. Hydrolysis and acidogenesis*

Polymeric materials such as lipids, proteins, and carbohydrates are primarily hydrolyzed by extracellular, hydrolases excreted by microbes present in Stage 1 (Figure 6). Hydrolytic enzymes, (lipases, proteases, cellulases, amylases, etc.) hydrolyze their respective polymers into smaller molecules, primarily monomeric units, which are then consumed by microbes. In methane fermentation of waste waters containing high concentrations of organic polymers, the hydrolytic activity relevant to each polymer is of paramount significance, in that polymer hydrolysis may become a rate-limiting step for the production of simpler bacterial substrates to be used in subsequent degradation steps.

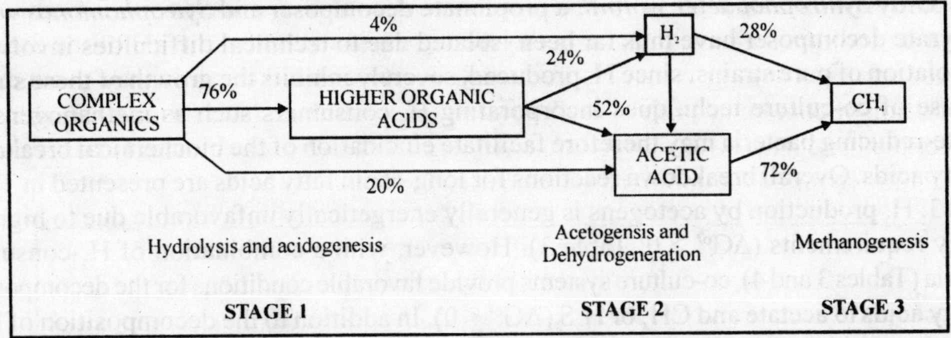


**Figure 6** Schematic representation of various stages of methane fermentation

Lipases convert lipids to long-chain fatty acids. A population density of $10^4$-$10^5$ lipolytic bacteria per ml of digester fluid has been reported. Clostridia and the micrococci appear to be responsible for most of the extracellular lipase producers. The long-chain fatty acids produced are further degraded by β-oxidation to produce acetyl CoA. Proteins are generally hydrolyzed to amino acids by proteases, secreted by *Bacteroides, Butyrivibrio, Clostridium, Fusobacterium, Selenomonas,* and *Streptococcus.* The amino acids produced are then degraded to fatty acids such as acetate, propionate, and butyrate, and to ammonia as found in *Clostridium, Peptococcus, Selenomonas, Campylobacter,* and *Bacteroides* species.

Polysaccharides such as cellulose, starch, and pectin are hydrolyzed by cellulases, amylases, and pectinases. The majority of microbial cellulases are composed of three species: (a) endo-β-l, 4-glucanases; (b) exo-β-l, 4-glucanases; (c) cellobiase or β-glucosidase. These three enzymes act synergistically on cellulose effectively hydrolyzing its crystal structure, to produce glucose. Microbial hydrolysis of raw starch to glucose requires amylolytic activity, which consist of 5 amylase species: (a) α-amylases that endocleave α–1, 4 bonds; (b) β-amylases that exocleave α–1, 4 bonds; (c) amyloglucosidases that exocleave α–l, 4 and α–l, 6 bonds; (d) debranching enzymes that act on α–l, 6 bonds; (e) maltase that acts on maltose liberating glucose. Pectins are degraded by pectinases, including pectinesterases and depolymerases. Xylans are degraded with endo-xylanase and β-xylosidase to produce xylose. Hexoses and pentoses are generally converted to $C_2$ and $C_3$ intermediates and to reduced electron carrier like NADH via common pathways. Most anaerobic bacteria undergo hexose metabolism via the Emden-Meyerhof-Parnas pathway (EMP) which produces pyruvate as an intermediate along with NADH. The pyruvate and NADH thus generated, are transformed into fermentation endo-products such as lactate, propionate, acetate, and ethanol by other enzymatic activities which vary tremendously with microbial species. Thus, in hydrolysis and acidogenesis (Figure 6; Stage 1) sugars, amino acids, and fatty acids produced by microbial degradation of biopolymers are successively metabolised by groups of bacteria & are primarily fermented to acetate, propionate, butyrate, lactate, ethanol, carbon dioxide, and hydrogen.

*3.3.3.2. Acetogenesis and dehydrogenation*

Although some acetate (20%) and $H_2$ (4%) are directly produced by acidogenic fermentation of sugars, and amino acids, both products are primarily derived from the acetogenesis and dehydrogenation of higher volatile fatty acids (Figure 6 ; Stage 2). Obligate $H_2$-producing acetogenic bacteria are capable of producing acetate and $H_2$ from higher fatty acids. Only *Syntrophobacter wolinii*, a propionate decomposer and *Sytrophomonos wolfei*, a butyrate decomposer have thus far been isolated due to technical difficulties involved in the isolation of pure strains, since $H_2$ produced, severely inhibits the growth of these strains. The use of co-culture techniques incorporating $H_2$ consumers such as methanogens and sulfate-reducing bacteria may therefore facilitate elucidation of the biochemical breakdown of fatty acids. Overall breakdown reactions for long-chain fatty acids are presented in Tables 2 and 3. $H_2$ production by acetogens is generally energetically unfavorable due to high free energy requirements ($\Delta G^{o\prime} > 0$; Table 3). However, with a combination of $H_2$-consuming bacteria (Tables 3 and 4), co-culture systems provide favorable conditions for the decomposition of fatty acids to acetate and $CH_4$ or $H_2S$ ($\Delta G^{o\prime} < 0$). In addition to the decomposition of long-chain fatty acids, ethanol and lactate are also converted to acetate and $H_2$ by an acetogen and *Clostridium formicoaceticum*, respectively.

The effect of the partial pressure of $H_2$ on the free energy associated with the conversion of ethanol, propionate, acetate, and $H_2/CO_2$ during methane fermentation has been studied by several workers who have found that an extremely low partial pressure of $H_2$ ($10^{-5}$ atm) appears to be a significant factor in propionate degradation to $CH_4$. Such a low partial pressure can be achieved in a co-culture with $H_2$-consuming bacteria (Tables 3 and 4).

*3.3.3.3. Methanogenesis*

Methanogens are physiologically united as methane producers in anaerobic digestion (Figure 6; Stage 3). Although acetate and $H_2/CO_2$ are the main substrates available in the natural environment, formate, methanol, methylamines, and CO are also converted to $CH_4$ (Table 4).

**Table 2** Proposed reactions involved in fatty acid catabolism by *Syntrophomonas wolfei*

| Fatty Acids | Reaction |
|---|---|
| Even-numbered | |
| $CH_3CH_2CH_2COO^-$ | $+ 2H_2O \rightleftharpoons 2CH_3COO^- + 2H_2 + H^+$ |
| $CH_3CH_2CH_2CH_2CH_2COO^-$ | $+ 4H_2O \rightleftharpoons 3CH_3COO^- + 4H_2 + 2H^+$ |
| $CH_3CH_2CH_2CH_2CH_2CH_2CH_2COO^-$ | $+ 6H_2O \rightleftharpoons 4CH_3COO^- + 6H_2 + 3H^+$ |
| Odd-numbered | |
| $CH_3CH_2CH_2CH_2COO^-$ | $+ 1H_2O \rightleftharpoons CH_3CH_2COO^- + CH_3COO^- + 2H_2 + H^+$ |
| $CH_3CH_2CH_2CH_2CH_2CH_2COO^-$ | $+ 4H_2O \rightleftharpoons CH_3CH_2COO^- + 2CH_3COO^- + 4H_2 + 2H^+$ |
| Branched-chained | |
| $CH_3CHCH_2CH_2CH_2COO^-$<br>\|<br>$CH_3$ | $+ 2H_2O \rightleftharpoons CH_3CHCH_2COO^- + CH_3COO^- + 2H_2 + H^+$<br>\|<br>$CH_3$ |

**Table 3** Free-energy changes for reactions involving anaerobic oxidation in pure cultures or in co-cultures with $H_2$-utilizing methanogens or *Desulfovibrio* spp.

| Equations | $\Delta G^0$ (kJ/reaction) |
|---|---|
| 1. Proton-reducing ($H_2$-producing) acetogenic bacteria | |
| A. $CH_3CH_2CH_2COO^- + 2H_2O \rightleftharpoons 2CH_3COO^- + 2H_2 + H^+$ | +48.1 |
| B. $CH_3CH_2COO^- + 3H_2O \rightleftharpoons CH_3COO^- + HCO_3^- + H^+ + 3H_2$ | +76.1 |
| 2. $H_2$-using methanogens and desulfovibrios | |
| C. $4H_2 + HCO_3^- + H^+ \rightleftharpoons CH_4 + 3H_2O$ | -135.6 |
| D. $4H_2 + SO_4^{2-} + H^+ \rightleftharpoons HS^- + 4H_2O$ | -151.9 |
| 3. Co-culture of 1 and 2 | |
| A + C $2CH_3CH_2CH_2COO^- + HCO_3^- + H_2O \rightleftharpoons 4CH_3COO^- + H^+ + CH_4$ | -39.4 |
| A + D $2CH_3CH_2CH_2COO^- + SO_4^{2-} \rightleftharpoons 4CH_3COO^- + H^+ + HS^-$ | -55.7 |
| B + C $4CH_3CH_2COO^- + 12H_2 \rightleftharpoons 4CH_3COO^- + HCO_3^- + H^+ + 3CH_4$ | -102.4 |
| B + D $4CH_3CH_2COO^- + 3SO_4^{2-} \rightleftharpoons 4CH_3COO^- + 4HCO_3^- + H^+ + 3HS^-$ | -151.3 |

**Table 4** Energy-yielding reactions of methanogens

| S.No. | Reaction | $-\Delta G^{0'}$ (kJ/mol substrate) |
|---|---|---|
| 1. | $CO_2 + 4H_2 \rightarrow CH_4 + 2H_2O$ | -130.7 |
| | $HCO_3^- + 4H_2 + H^+ \rightarrow CH_4 + 3H_2O$ | -135.5 |
| 2. | $CH_3COO^- + H^+ \rightarrow H_2 + CO_2$ | -37.0 |
| | $CH_3COO^- + H_2O \rightarrow CH_4 + HCO_3^-$ | -32.3 |
| 3. | $HCOO^- + H^+ \rightarrow 0.25\,CH_4 + 0.75\,CO_2 + 0.5\,H_2O$ | -36.1 |
| 4. | $CO + 0.5\,H_2O \rightarrow 0.25\,CH_4 + 0.75\,CO_2$ | -52.7 |
| 5. | $CH_3OH \rightarrow 0.75\,CH_4 + 0.25\,CO_2 + 0.5\,H_2O$ | -79.9 |
| 6. | $CH_3NH_3^+ + 0.5\,H_2O \rightarrow 0.75\,CH_4 + 0.25\,CO_2 + NH_4^+$ | -57.4 |
| 7. | $(CH_3)_2NH_2^+ + H_2O \rightarrow 1.5\,CH_4 + 0.5\,CO_2 + NH_4^+$ | -112.2 |
| 8. | $(CH_3)_2NCH_2CH_3H^+ + H_2O \rightarrow 1.5\,CH_4 + 0.5\,CO_2 + {}^+H_3NCH_2CH_3$ | -105.0 |
| 9. | $(CH_3)_3NH + 1.5H_2O \rightarrow 2.25\,CH_4 + 0.75\,CO_2 + NH_4^+$ | -170.8 |

Since methanogens, as obligate anaerobes, require a redox potential of less than -300 mV for growth, their isolation and cultivation was somewhat elusive due to technical difficulties encountered in handling them under completely $O_2$-free conditions. However, as a result of a greatly improved methanogen isolation techniques developed by Hungate , more than 40 strains of pure methanogens have now been isolated. Methanogens can be divided into two groups: $H_2/CO_2$ and acetate-consumers. Although some of the $H_2/CO_2$ consumers are capable of utilizing formate, acetate is consumed by a limited number of strains, such as *Methanosarcina* spp. and *Methanothrix* spp. (now, *Methanosaeta),* which are incapable of using formate. Since a large quantity of acetate is produced in the natural environment (Figure 6), *Methanosarcina* and *Methanothrix* play an important role in completion of anaerobic digestion and in accumulating $H_2$, which inhibits acetogens and methanogens. $H_2$ consuming methanogens are also important in maintaining low levels of atmospheric $H_2$.

$H_2/CO_2$ consuming methanogens reduce $CO_2$ as an electron acceptor via the formyl, methenyl, and methyl levels through association with unusual coenzymes, to finally produce $CH_4$. The overall acetoclastic reaction can be expressed as:

$$CH_3COOH \rightleftharpoons CH_4 + CO_2$$

Since a small part of the $CO_2$ is also formed from carbon derived from the methyl group, it is suspected that the reduced potential produced from the methyl group may reduce $CO_2$ to $CH_4$.

### *3.3.4. Conditions for anaerobic fermentation*

Necessary conditions for the anaerobic digestion of organic materials are described in the following paragraphs:

#### *3.3.4.1. Temperature*

Microbial growth rates are governed by temperature. Psychrophilic bacteria, which are relatively few in number, grow at low temperature (0 to 5°C). Their function for methane production is not clear. Optimum temperature for mesophilic bacteria lies in the range of 30 to 40°C, whereas thermophilic bacteria work optimally at 50 to 60°C.

The production of methane gas depends upon temperature. A temperature of 35°C is considered ideal for gas production. Rates of gas production can be increased up to 50 to 55 °C beyond which it diminishes due to the destruction of enzymes by heat. It is quite difficult to maintain this temperature in the digester in temperate climates. It has been observed that gas plants work with difficulty between 12 to 18°C. Gas production can be significantly reduced or even stopped below 10°C as methanogenic bacteria do not thrive in the psychrophilic range. Methane bacteria are sensitive to a sudden change of temperature, and as such, a more or less stabilized working temperature is needed for continuous and efficient gas production.

#### *3.3.4.2. pH*

The pH plays an important role in methane production. Both mesophilic and thermophilic bacteria require an optimum pH in the range of 6 to 7. During the initial phase of activity when organic acids are produced, the pH of the media may fall below 6.0. At this point, a greater amount of $CO_2$ is liberated. Acid producing bacteria thrive at a pH range of 6.5 and they are still active below 5.5. As fermentation proceeds, volatile acids and nitrogen are digested and ammonia is formed and there is an increment in pH. As pH rises to 7.2, acidity is decreased in the medium. The quantity of $CO_2$ is diminished and that of methane is increased. As methane production is stabilized, the pH range remains buffered between 7.2 to 8.2. Methane bacteria are very sensitive to pH and they do not thrive below pH 6.5.

#### *3.3.4.3. Dilution*

A certain amount of water must be present in the digester for the fermentation process. Generally, the quantity of water to be added should be about 10% of the total weight of the materials. For efficient gas production, the total solids in the slurry should be 7 to 9%. This can be verified by measuring the consistency of the slurry using a hydrometer. If the specific gravity is 1.030, 1.100 and 1.190, the solids in the mixture should be 7, 8, and 9 respectively.

It is customary to add one part of water to one part of dung for slurry making. If the consistency is too thick, gas can not penetrate easily through the slurry; if it is too thin, solid particles will fall out at the bottom. However, correct consistency can be obtained by using a hydrometer. Experience shows that many users enjoy trouble free operation with a proportion of 4 parts dung to 5 parts water.

*3.3.4.4. Necessary elements and C/N ratio*

Necessary elements such as carbon, hydrogen, nitrogen, phosphorus and many other micro-elements must be present in adequate quantities for the normal growth of the micro-organisms. It has been recognized that all living organisms need nitrogen for the synthesis of protein. In the absence of sufficient nitrogen, the bacteria are not be able to utilize all the carbon present and the process will be less efficient. In general, a C/N ratio of around 20-30:1 is considered best for anaerobic digestion. The C/N ratio should never be more than 35, with an optimum of 30. If the C/N ratio is very high, nitrogen will be consumed rapidly and the rate of reaction will decrease. On the other hand, if the C/N ratio is very low, nitrogen will be liberated and accumulated in the form of ammonia, producing toxicity under certain conditions.

Animal waste particularly cattle dung. has an average C/N ratio of about 24. The plant materials such as straw and sawdust contain a higher percentage of carbon. C/N ratio of some of the commonly used materials are presented below:

| Raw Materials | C/N Ratio |
|---|---|
| Duck dune | 8 |
| Human excreta | 8 |
| Chicken dung | 10 |
| Goat dung | 12 |
| Pig dung | 18 |
| Sheep dung | 19 |
| Cow dung / Buffalo dung | 24 |
| Water hyacinth | 25 |
| Elephant dung | 43 |
| Straw (maize) | 60 |
| Straw (rice) | 70 |
| Straw (wheat) | 90 |
| Saw dust | above 200 |

Materials with high C / N ratio could be mixed with those of low C/ N ratio to bring the average ratio of the composite input to a desirable level. In China, as a means to balance C / N ratio, it is customary to load rice straw at the bottom of the digester upon which latrine waste is discharged.

*3.3.4.5. Loading rate*

The loading rate is the amount of raw material (expressed in volatile solids) fed to the digester per day per unit volume of digester capacity. The correct rate of loading is essential for efficient gas production. If the plant is overfed, acidity will accumulate and methane production will be inhibited and if the loading rate is lower, the gas production will not be sufficient. A daily loading rate of 0.19 kg of volatile solids/ft$^3$ (16.7 kg/ m$^3$) is recommended for a cow dung plant. Municipal plants are run at a loading rate of 0.03-0.07 kg of volatile solids/ft$^3$.

*3.3.4.6. Detention time*

Detention time (also known as retention time) is the average duration of time a sample remains in the digester. In a cow dung plant, the detention time is calculated by dividing the total volume of the digester by the volume of slurry added daily. Usually, for a cow dung plant a detention time of 30 to 40 days is required. Thus, the fermenting pit should have a volume of from 30 to 40 times the slurry added daily. But for a night soil digester, a longer detention time (60 to 90 days) is needed.

*3.3.4.7. Inputs and their characteristics*

Any biodegradable organic material can be used as inputs for processing inside the biodigester. However, for economic and technical reasons, some materials are more preferred as inputs than others. Easily available biodegradable wastes are used as inputs, then the benefits could be of two folds:

(a) economic value of biogas and its slurry

(b) environmental cost avoided in dealing with the biodegradable waste in some other ways such as disposal in landfill.

One of the main reactions of biogas technology is its ability to generate biogas out of organic wastes that are abundant and freely available. In case of India, it is the cattle dung that is most commonly used as an input mainly because of its availability.

***Animal dung***

Gas production potential of various types of dungs is given below:

| Types of Dung | Gas Production / kg dung (m$^3$) |
|---|---|
| Cattle (cows and buffaloes) | 0.023 - 0.040 |
| Pig | 0.040 - 0.059 |
| Poultry (Chickens) | 0.065 - 0.116 |
| Human | 0.020 - 0.028 |

***Plant materials***

Plant materials can also be used to produce biogas and bio-manure. For example, one kg of pre-treated crop waste and water hyacinth have the potential of producing 0.037 and 0.045 m$^3$ of biogas, respectively. Since different organic materials have different biochemical characteristics, their potential for gas production also varies. Two or more of such materials can be used together provided that some basic requirements for gas production or for normal growth of methanogens are met.

Typical operating conditions for anaerobic digestion processes are indicated below:

| | |
|---|---|
| Temperature | |
| Mesophilic | 35ºC |
| Thermophilic | 54ºC |
| pH | 7-8 |
| Alkalinity | 2500 mg/l minimum |
| Retention time | 10-30 days |
| Loading rate | 0.07-0.16 kg VS/ft$^3$/d |
| Biogas yield | 6.6-17.6 ft$^3$/lb VS |
| Methane content | 60-70% |

### *3.3.5. Sludge-algae-methane system*

Another system of biogas production involves the use of sludge and algae in which the green algae is grown in diluted sludge, then harvested, dried and digested to produce methane for power and sludge for recycling. This procedure of transforming solar energy and sludge nutrients into the chemical energy of methane is potentially a very efficient and rapid biological process: (1) It is a closed nutritional system and (2) the rate of turnover is extremely high; organic matter is decomposed relatively quickly by anaerobic bacteria in the pond while it is most rapidly made by green algae. The complete sludge-algae-methane system involves a series of processes. The principle features of the system are integration of the algae culture with the gas in such a manner that nutrients and water are recycled from one process to the other (Figure 7). Most of the information concerning this system has been developed by researchers at Berkeley in a manner that has real potential for the homestead or small farm.

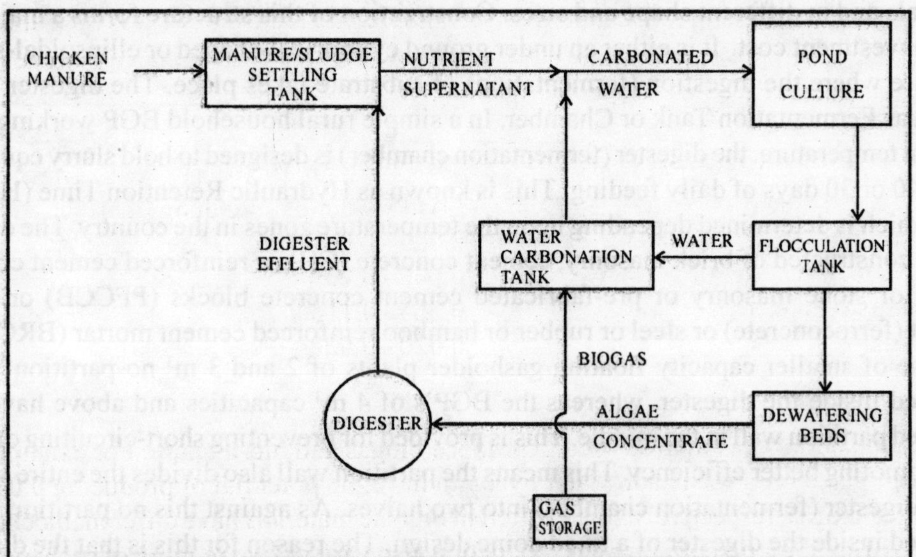


Figure 7 A schematic representation of sludge-algae-methane system

### *3.3.6. Biogas plant*

It is an air tight container that facilitates fermentation of material under anaerobic condition. The other names given to this device are "biogas digester", "biogas reactor", "methane generator" and "methane reactor". The recycling and treatment of organic waste (biodegradable material) through anaerobic digestion (fermentation) technology not only provides biogas as a clean and convenient fuel but also as an excellent and enriched bio-manure. Thus the biogas plant also acts as a miniature bio-fertilizer factory, hence, some people prefer to refer it as a "biogas-fertilizer plant" or "bio-manure plant". Then fresh organic material, generally in a homogenous slurry form is fed into the digester of the plant from one end, known as inlet pipe or inlet tank. The decomposition (fermentation) takes place inside the digester due to bacterial (microbial) action, which produces biogas and organic fertilizer (manure) rich in humus and other nutrients. There is a provision for storing biogas on the upper portion of the biogas plant. There are some biogas plant designs that have floating gas holders and others have fixed gas storage chamber. On the other hand of the digester outlet pipe or outlet tank is provided for the automatic discharge of the liquid digested manure.

#### *3.3.6.1. Components of biogas plant*

The major components of BGP include (i) Digester. (ii) Gasholder or Gas Storage Chamber, (iii) Inlet. (iv) Outlet. (v) Mixing Tank and (vi) Gas Outlet Pipe.

##### *3.3.6.1.1. Digester*

The biodigester is a physical structure, commonly known as a biogas plant. Since various chemical and microbiological reactions take place in the biodigester, it is also known as a bioreactor or anaerobic reactor. The main function of this structure is to provide anaerobic conditions. The chamber should be air and watertight. It can be made of various construction materials and in different shape and sizes. Construction of this structure forms a major part of the investment cost. It is either an under ground cylindrical-shaped or ellipsoidal-shaped structure where the digestion (fermentation) of substrate takes place. The digester is also known as Fermentation Tank or Chamber. In a simple rural household BGP working under ambient temperature. the digester (fermentation chamber) is designed to hold slurry equivalent to 55, 40 or 30 days of daily feeding. This is known as Hydraulic Retention Time (HRT) of BGP which is determined depending upon the temperature zones in the country. The digester can be constructed of brick masonry, cement concrete (CC) or reinforced cement concrete (RCC) or stone masonry or pre-fabricated cement concrete blocks (PFCCB) or Ferro-cement (ferroconcrete) or steel or rubber or bamboo reinforced cement mortar (BRCM). In the case of smaller capacity floating gasholder plants of 2 and 3 m$^3$ no partition wall is provided inside the digester, whereas the BGP's of 4 m$^3$ capacities and above have been provided partition wall in the middle. This is provided for preventing short-circuiting of slurry and promoting better efficiency. This means the partition wall also divides the entire volume of the digester (fermentation chamber) into two halves. As against this no partition wall is provided inside the digester of a fixed dome design. The reason for this is that the diameter of the digesters in all the fixed dome models are comparatively much bigger than the floating drum BGPs, which takes care of the short-circuiting problems to a satisfactory level, without adding to additional cost of providing a partition wall. There are many basic digester designs and some of the commonly used designs are discussed below:

*3.3.6.1.1.1. Floating drum digester*

Experimentation with biogas technology in India began in 1937. In 1956, Jashu Bhai J. Patel developed a design of floating drum biogas plant popularly known as the Gobar Gas Plant. In 1962, Patel's design was approved by the Khadi and Village Industries Commission (KVIC) of India, and this design soon became popular in India and around the world. The design of such a KVIC plant is shown in Figure 8.

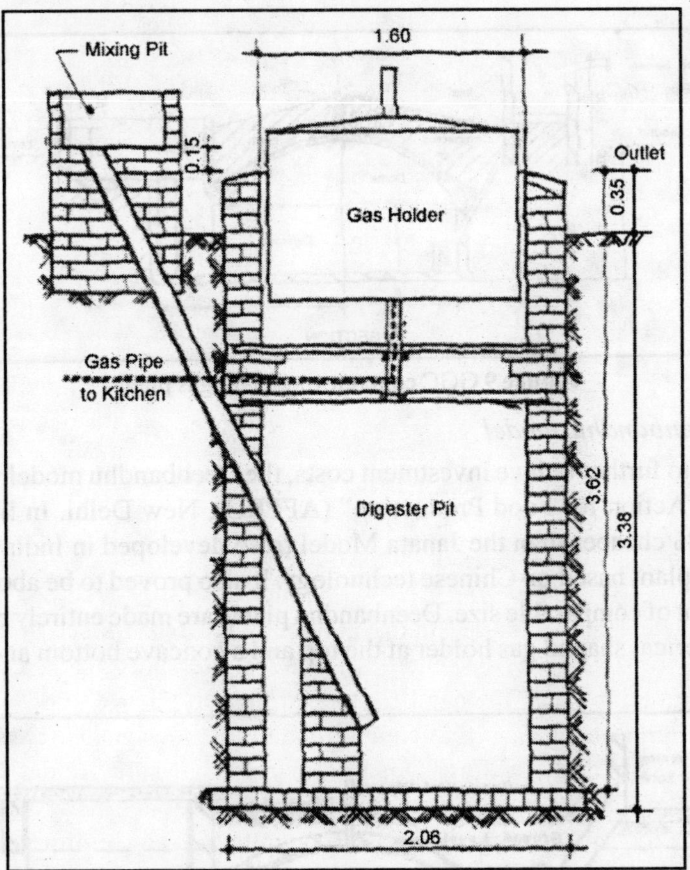


**Figure 8** KVIC Floating gas holder system

In this design, the digester chamber is made of brick masonry in cement mortar. A mild steel drum is placed on top of the digester to collect the biogas produced from the digester. Thus, there are two separate structures for gas production and collection. With the introduction of the fixed dome Chinese model plant, the floating drum plants became obsolete because of comparatively high investment and maintenance cost as well as other design weaknesses. In Nepal, KVIC design plants have not been constructed since 1986.

*3.3.6.1.1.2. Fixed dome digester*

Fixed dome Chinese model biogas plants (also called a drum-less digester) were built in China as early as 1936. They consist of an underground brick masonry compartment (fermentation chamber) with a dome on the top for gas storage. In this design, the fermentation chamber and gas holder are combined as one unit. This design eliminates the use of the costlier mild steel gas holder which is susceptible to corrosion. The life of the fixed dome type plant is longer as compared to the KVIC plant. Based on the principles of the fixed dome model from China, the Gobar Gas and Agricultural Equipment Development Company (GGC) of Nepal has developed a new design and has been popularizing it over the last 17 years. The concrete dome is the main characteristic of the GGC design and is shown in Figure 9.

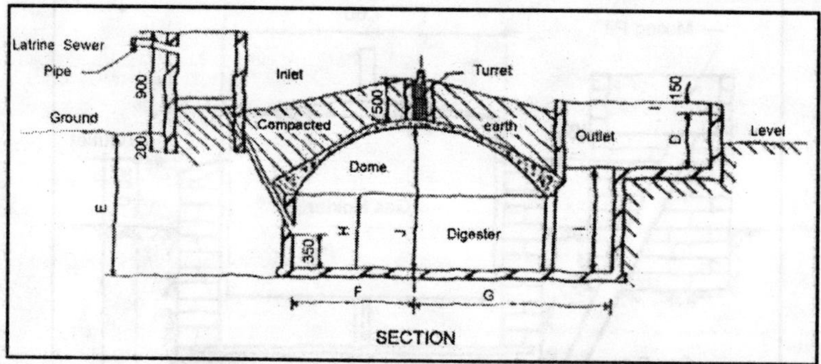


**Figure 9** GGC concrete model biogas plant

*3.3.6.1.1.3. Deenbandhu model*

In an effort to further reduce investment costs, the Deenbandhu model was put forward in 1984 by the "Action for Food Production" (AFPRO), New Delhi. In India, this model proved to be 30% cheaper than the Janata Model (also developed in India) which was the first fixed dome plant based on Chinese technology. It also proved to be about 45% cheaper than a KVIC plant of comparable size. Deenbandhu plants are made entirely of brick masonry work with a spherical shaped gas holder at the top and a concave bottom and is exhibited in Figure 10.

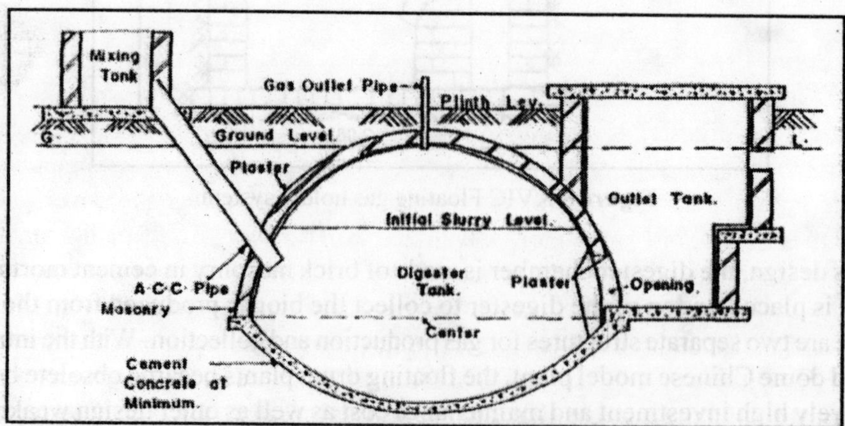


**Figure 10** Deenbandhu biogas plant

The South Asian Partnership/Nepal (SAP/N), an International Non Governmental Organization (INGO) working in Nepal, has introduced the Deenbandhu model in the Bardiya district of Nepal. About 100 plants were constructed by the South Asian Partnership/Nepal (SAP/N) in the villages of Bardiya district in 1994. Preliminary studies carried out at that time by the Biogas Support Programme (BSP) of the Netherlands Development Organization (SNV) did not find any significant difference in the investment costs of GGC and the Deenbandhu design plants.

In addition to the above developed plants, there are other designs suitable for adoption under other specific conditions, particularly for household use in developing countries. Although they are not of much relevance to present conditions in Nepal, they could prove useful in the future. These designs are briefly described below for reference.

*3.3.6.1.1.4. Bag digester*

The bag digester design was developed in the 1960s in Taiwan. The bag digester was developed to solve problems experienced with brick and metal digesters and consists of a long cylinder made of PVC or red mud plastic (Figure 11).

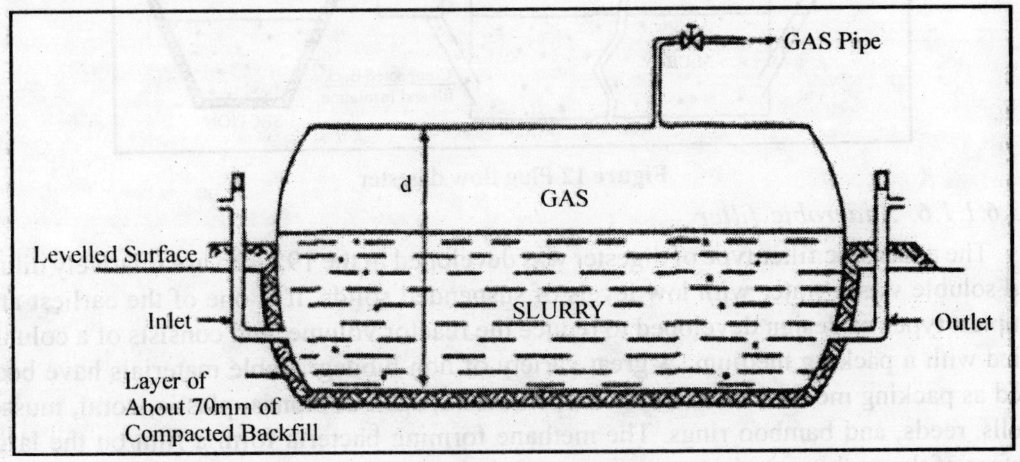


**Figure 11** Bag digester

*3.3.6.1.1.5. Plug flow digester*

The plug flow digester is similar to the bag digester. It consists of a trench (whose length has to be considerably greater than the width and depth) lined with concrete or an impermeable membrane. The reactor is covered with either a flexible cover gas holder anchored to the ground, or concrete or Galvanized Iron (GI) top. The first documented use of this type of design was in South Africa in 1957. Figure 12 shows a sketch of such a reactor. Plug-flow digesters are suitable for ruminant animal manure that has a solid concentration of 11-13%. A typical design for a plug-flow system includes a manure collection system, a mixing pit and the digester itself. In the mixing pit, the addition of water adjusts the proportion of solids in the manure slurry to the optimal consistency. The digester is a long, rectangular container, usually built below-grade, with an airtight, expandable cover.

New material added to the tank at one end pushes older material to the opposite end. Coarse solids in ruminant manure form a viscous material as they are digested, limiting solids separation in the digester tank. As a result, the material flows through the tank in a "plug". Average retention time (the time a manure "plug" remains in the digester) is 20 to 30 days. Anaerobic digestion of the manure slurry releases biogas as the material flows through the digester. A flexible, impermeable cover on the digester traps the gas. Pipes beneath the cover carry the biogas from the digester to an engine-generator set.

A plug-flow digester requires minimal maintenance. Waste heat from the engine-generator can be used to heat the digester. Inside the digester, suspended heating pipes allow hot water to circulate. The hot water heats the digester to keep the slurry at 25-40°C, a temperature range suitable for methane-producing bacteria. The hot water can come from recovered waste heat from an engine generator fueled with digester gas or from burning digester gas directly in a boiler.

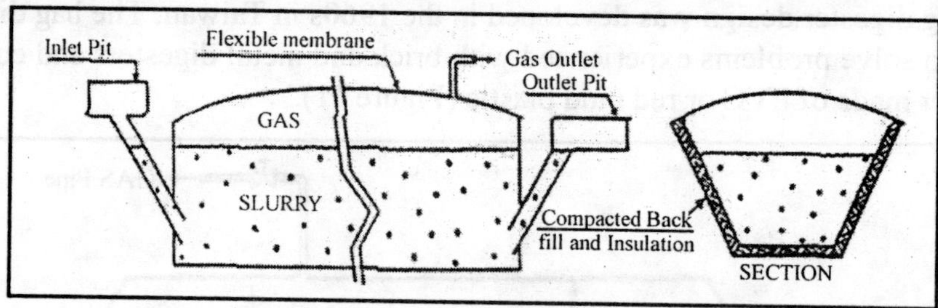


**Figure 12** Plug flow digester

*3.3.6.1.1.6. Anaerobic filter*

The anaerobic filter type of digester was developed in the 1950s to use relatively dilute and soluble waste water with low levels of suspended solids. It is one of the earliest and simplest types of design developed to reduce the reactor volume, and consists of a column filled with a packing medium. A great variety of non-biodegradable materials have been used as packing media for anaerobic filter reactors, such as stones, plastic, coral, mussel shells, reeds, and bamboo rings. The methane forming bacteria form a film on the large surface of the packing medium and are not carried out of the digester with the effluent. For this reason, these reactors are also known as "fixed film" or "retained film" digesters. Figure 13 presents a sketch of the anaerobic filter. This design is best suited for treating industrial, chemical, and brewery wastes.

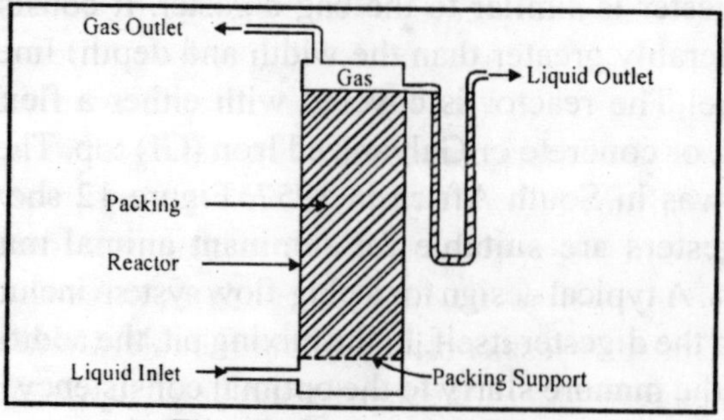


**Figure 13** Anaerobic filter

*3.3.6.1.1.7. Upflow anaerobic sludge blanket (UASB)*

The Up-flow Anaerobic Sludge Blanket (UASB) digester design was developed in 1980 in the Netherlands. It is similar to the anaerobic filter which involves a high concentration of immobilized bacteria in the reactor. However, the UASB reactors contain no packing medium, instead, the methane forming bacteria are concentrated in the dense granules of the sludge blanket which covers the lower part of the reactor. The feed liquid enters from the bottom of the reactor and biogas is produced while liquid flows up through the sludge blanket (Figure 14). Many full-scale UASB plants are in operation in Europe using waste water from sugar beet processing and other dilute wastes that contain mainly soluble carbohydrates.

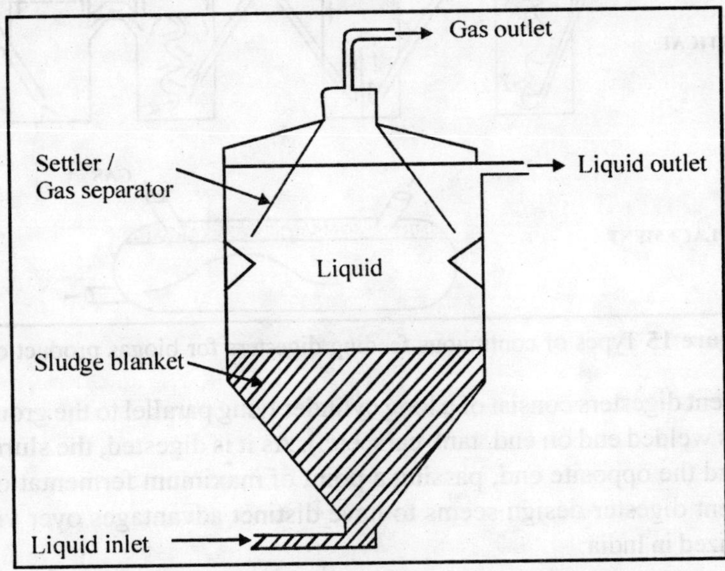


**Figure 14** Upflow anaerobic sludge blanket [UASB] digester

Digesters can be designed for batch-feeding or for continuous feeding. With batch digesters a full charge of raw material is placed into the digester which is then sealed off and left to ferment as long as gas is produced. When gas production has ceased, the digester is emptied and refilled with a new batch of raw materials. Batch digesters have advantages where the availability of raw materials is sporadic or limited to coarse plant wastes (which contain undigestible materials that can be conveniently removed when batch digesters are reloaded). Also, batch digesters require little daily attention. Batch digesters have disadvantages, however, in that a great deal of energy is required to empty and load them; also gas and sludge production tend to be quite sporadic. You can get around this problem by constructing multiple batch digesters connected to the same gas storage. In this way individual digesters can be refilled in staggered sequence to ensure a relatively constant supply of gas. Most early digesters were of the batch type. With continuous-load digesters, a small quantity of raw material is added to the digester every day or so. In this way the rate of production of both gas and sludge is more or less continuous and reliable. Continuous-load digesters are especially efficient when raw materials consist of a regular supply of easily digestible wastes from nearby sources such as livestock manures, seaweed, river or lake flotsam or algae from production sludge-ponds. Continuous-feeding digesters can be of two basic designs: vertical-mixing or displacement (Figure 15). Vertical-mixing digesters consist of vertical chambers into which raw materials are added. The slurry rises through the digester and overflows at the top. In single-chamber designs the digested or "spent" slurry can be withdrawn directly from effluent pipes. In double-chamber designs the spent slurry, as it overflows the top, flows into a second chamber where digestion continues to a greater degree of completion.

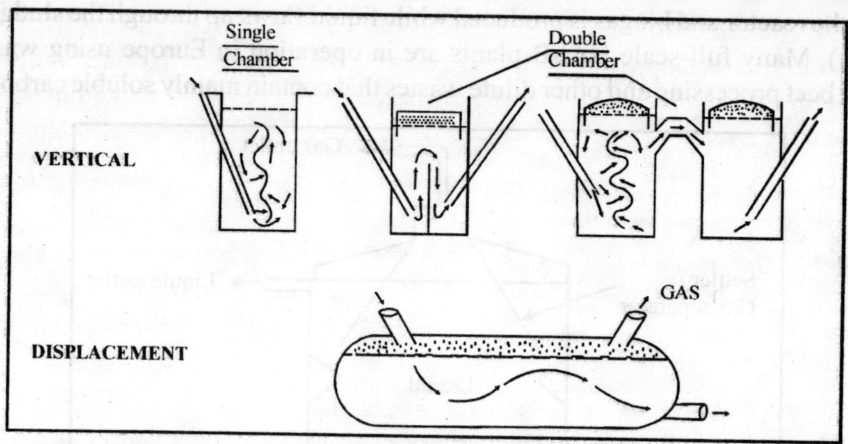


**Figure 15** Types of continuous feeding digesters for biogas production

Displacement digesters consist of a long cylinder lying parallel to the ground (e.g., inner tubes, oil drums welded end on end, tank cars, etc.). As it is digested, the slurry is gradually displaced toward the opposite end, passing a point of maximum fermentation on the way. The displacement digester design seems to have distinct advantages over vertical-mixing designs popularized in India:

1. In vertical-mixing digesters raw material is subject to a vertical pumping motion and often escapes the localized action of digesting bacteria. Slurry introduced at one time can easily be withdrawn soon afterwards as incompletely digested material. In displacement digesters, slurry must pass an area of maximum fermentation activity so that all raw materials are effectively digested (much like the intestines of an animal).
2. From a practical point of view, displacement digesters are easier to operate. If digester contents begin to sour for one reason or another, strongly buffered material at the far end can be recirculated efficiently by simply reversing the flow of material along the line of the cylinder. In addition, raw materials can be digested to any desired degree without the need for constructing additional chambers or digesters.
3. The problem of scum accumulation is reduced in displacement digesters. Since scum forms evenly on the surface of the digesting slurry, the larger the surface area, the longer it takes to accumulate to the point where it inhibits digestion. A prone cylinder has a larger surface area than an upright one.
4. Any continuous-load digester will eventually accumulate enough scum and undigested solid particles so that it will have to be cleaned. The periodical washing out of displacement digesters is considerably easier than vertical-mixing digesters.

*3.3.6.1.1.8. Maintaining temperature of digesters*

For the most efficient operation, especially in temperate climates, digesters should be supplied with an external supply of heat to keep them around 35°C; there are several ways to do this. Methods which heat the outside of digesters (e.g., compost piles, light bulbs, and water jackets) could be more effectively used as insulation since much of their heat dissipates to the surroundings. (Since digesters should be constantly warmed rather than sporadically heated, compost "blankets" are not very practical unless you coordinate a regular program of composting with digestion.) Similarly, green houses built over digesters tend to overheat the digester during the day and cool it down at night. The most effective method of keeping digesters warm is to circulate heated water through pipes or coils placed within the digester. The water can be heated by solar collectors or by water boilers heated with methane.

Gas-heated water boilers are a good idea since they allow the digestion process to feed back on itself, thus increasing efficiency. The thermostat in the water boiler is set at 60°C because slurry will cake on surfaces (the water coils) warmer than this. The digester thermostat is set at the optimum temperature of 35°C. Until the digester begins producing methane, propane can be used as a fuel source for the water boiler. For optimum heat exchange within the digester, a ratio of 1 $ft^2$ coil area / 100 $ft^3$ of digester volume is recommended.

Sometimes the digesters are insulated for maintaining temperature but some precautions are needed in that. Methane is not only combustible but highly explosive when it makes up more than 9% of the surrounding air in confined spaces. If one uses synthetic insulation, avoid porous materials such as spun glass which can trap gas mixtures. It's easy to scrounge styrofoam sheets since they are so commonly used as packing material and regularly discarded. Styrofoam is one of the best insulating materials, although it is slightly flammable.

*3.3.6.1.2. Gas holder or gas storage chamber*

In the case of floating gas holder BGPs, the gas holder is a drum like structure, fabricated either of mild steel sheets or ferro-cement (ferroconcrete) or high density plastic (HDP) or fibre glass reinforced plastic (FRP). It fits like a cap on the mouth of digester where it is submerged in the slurry and rests on tho ledge, constructed inside the digester for this purpose. The drum collects gas, which is produced from the slurry inside the digester as it gets decomposed, and rises upwards, being lighter than air. To ensure that there is enough pressure on the stored gas so that it flows on its own to the point of utilisation through pipeline when the gate valve is open, the gas is stored inside the gasholder at a constant pressure of 8-10 cm of water column. This pressure is achieved by making the weight of biogas holder as 80-100 kg/$cm^3$. In its up and down movement the drum is guided by a central guide pipe. The gas formed is otherwise sealed from all sides except at the bottom. The scum of the semidried mat formed on the surface of the slurry is broken (disturbed) by rotating the biogas holder, which has scum-breaking arrangement inside it. The gas storage capacity of a family size floating biogas holder BGP is kept as 50% of the rate capacity (daily gas production in 24 hours). This storage capacity comes to approximately 12 h of biogas produced every day. In the case of fixed dome designs, the biogas holder is commonly known as gas storage chamber (GSC). The GSC is the integral and fixed part of the main unit of the plant (MUP) in the case of fixed dome BGPs. Therefore the GSC of the fixed dome BGP is made of the same building material as that of the MUP. The gas storage capacity of a family size fixed dome BGP is kept as 33% of the rate capacity (daily gas production in 24 hours). This storage capacity comes to approximately 8 hours of biogas produced during the night when it is not in use.

*3.3.6.1.3. Inlet*

In the case of floating biogas holder pipe, the inlet is made of cement concrete (CC) pipe. The Inlet Pipe reaches the bottom of the digester well on one side of the partition wall. The top end of this pipe is connected to the Mixing Tank. In the case of the first approved fixed dome models, the inlet is like a chamber or tank-it is a bell mouth shaped brick masonry construction and its outer wall is sloppy. The top end of the outer wall of the inlet chamber has an opening connecting the mixing tank, whereas the bottom portion joins the inlet gate. The top (mouth) of the inlet chamber is kept covered with heavy slab. The Inlet of the other fixed dome models (Deenbandhu and Shramik Bandhu) has asbestos cement concrete (ACC) pipes of appropriate diameters.

*3.3.6.1.4. Outlet*

In the case of floating gas holder pipe the outlet is made of cement concrete (CC) pipe standing at an angle, which reaches the bottom of the digester on the opposite side of the partition wall. In smaller plants (2 & 3 $M^3$ capacity BGPs) which has no partition walls, the 'outlet is made of small (approx. 2 ft. length) cement concrete (CC) pipe inserted on top most portion of the digester, submerged in the slurry. In the two Fixed Dome (Janata & Deenbandhu models) plants, the outlet is made in the form of rectangular tank. However, in the case of Shramik Bandhu model the upper portion of the outlet (known as outlet displacement chamber) is made hemi-spherical in shape, designed to save in the material and labour cost. In all the three-fixed dome models (Janata, Deenbandhu & Shramik Bandhu models), the bottom end of the outlet tank is connected to the outlet gate. There is a small opening provided on the outer wall of the outlet chamber for the automatic discharge of the digested slurry outside the BGP, equal to approximately 80-90% of the daily feed. The top mouth of the outlet chamber is kept covered with heavy slab.

*3.3.6.1.5. Mixing tank*

This is a cylindrical tank used for making homogenous slurry by mixing the manure from domestic farm animals with appropriate quantity of water. Thoroughly mixing of slurry before releasing it inside the digester, through the inlet, helps in increasing the efficiency of digestion. Normally a feeder fan is fixed inside the mixing tank for facilitating easy and faster mixing of manure with water for making homogenous slurry.

*3.3.6.1.6. Gas outlet pipe*

The gas outlet pipe is made of GI pipe and fixed on top of the drum at the centre in case of floating biogas holder BGP and on the crown of the fixed dome BGP. From this pipe the connection to gas pipeline is made for conveying the gas to the point of utilisation. A gate valve is fixed on the gas outlet pipe to close and check the flow of biogas from plant to the pipeline.

*3.3.6.2. Functioning of rural household biogas plants (BGPs)*

The fresh organic material (generally in a homogenous slurry form) is fed into the digester of the plant from one end, known as Inlet. Fixed quantity of fresh material fed each day (normally in one lot at a predetermine time) goes down at the bottom of the digester and forms the 'bottom-most active layer', being heavier then the previous day and older material. The decomposition (fermentation) takes place inside the digester due to bacterial (microbial) action, which produces biogas and digested or semi-digested organic material. As the organic material ferments, biogas is formed which rises to the top and gets accumulated (collected) in the gas holder (in case of floating gas holder BGPs) or gas storage chamber (in case of fixed dome BGPs). A gas outlet pipe is provided on the top most portion of the gas holder (gas storage chamber) of the BGP. Alternatively, the biogas produced can be taken to another place through pipe connected on top of the gas outlet pipe and stored separately. The slurry (semi-digested and digested) occupies the major portion of the digester and the sludge (almost fully digested) occupies the bottom most portion of the digester. The digested slurry (also known as effluent) is automatically discharged from the other opening, known as outlet, is an excellent bio-fertilizer, rich in humus. The anaerobic fermentation increases the ammonia content by 120% and quick acting phosphorous by 150%. Similarly, the percentage of potash and several micro-nutrients useful to the healthy growth of the crops also increases. The nitrogen is transformed into ammonia that is easier for plant to absorb. This digested slurry can either be taken directly to the farmer's field along with irrigation water or stored in a slurry pits (attached to the BGP) for drying or directed to the compost pit for making compost along with other waste biomass. The slurry and also the sludge contain a higher percentage of nitrogen and phosphorous than the same quantity of raw organic material fed inside the digester of the BGP. The digestion of organic materials in simple rural household biogas plants can be classified under three broad categories. They are:

(i) Batch-fed digestion.
(ii) Semi-continuous digestion.
(iii) Semi-batch-fed digestion.

*Batch-fed digestion:* In batch-fed digestion process, material to be digested is loaded with seed material or inoculum into the digester at the start of the process. The digester is then sealed and the contents left to digest (ferment). At completion of the digestion cycle, the digester is opened and sludge (manure) removed (emptied). The digester is cleaned and once again loaded with fresh organic material, available in the season.

*Semi-continuous digestion:* This involves feeding of organic matter in homogenous slurry form inside the digester of the BGP once in a day, normally at a fixed time. Each day digested slurry; equivalent to about 85-95% of the daily input slurry is automatically discharged from the outlet side. The digester is designed in such a way that the fresh material fed comes out after completing a HRT cycle (either 55, 40 or 30 days), in the form of digested slurry. In a Semi-continuous digestion system, once the process is stabilized in a few days of the initial loading of the BGP, the biogas production follows a uniform pattern.

*Semi-batch fed digestion:* A combination of batch and semi-continuous digestion is known as Semi-batch fed Digestion. Such a digestion process is used where the manure from domestic farm animals is not sufficient to operate a plant and at the same time organic waste like crop residues, agricultural wastes (paddy and weed straw), water hyacinths and weeds etc, are available during the season. In as semi-batch fed digestion, the initial loading is done with green or semi-dry or dry biomass (that can not be reduced in to slurry form) mixed with starter and the digester is sealed. This plant also has an inlet pipe for daily feeding of manure slurry from animals. The semi-batch fed digester will have much longer digestion cycle of gas production as compared to the batch-fed digester. It is ideally suited for the poor peasants having 1-2 cattle or 3-4 goats to meet the major cooking requirement and at the end of the cycle (6-9 months) will give enriched manure in the form of digested sludge.

### *3.3.6.3. Stratification (layering) of digester due to anaerobic fermentation*

In the process of digestion of feedstock in a BGP many by-products are formed. They are biogas, scum, supernatant, digested slurry, digested sludge and inorganic solids. If the content of biogas digester is not stirred or disturbed for a few hours then these byproducts get formed into different layers inside the digester. The heaviest by-product, inorganic solids will be at the bottom most portion, followed by digested sludge, and so on and so forth.

*Scum:* Mixture of coarse fibrous and lighter material that separates from the manure slurry and floats on the top most layer of the slurry is called Scum. The accumulation and removal of scum is sometimes a serious problem. In moderate amount scum can't do any harm and can be easily broken by gentle stirring, but in large quantity can lead to slowing down biogas production and even shutting down the BGPs.

*Supernatant:* The spent liquid of the slurry (mixture of manure and water) layering just above the sludge, in case of batch-fed and semi batch-fed digester, is known as supernatant. Since supernatant has dissolved solids, the fertilizer value of this liquid (supernatant) is as great as that of effluent (digested slurry). Supernatant is a biologically active by-product; therefore must be sun dried before using it in agricultural fields.

*Digested slurry (effluent):* The effluent of the digested slurry is in liquid form and has its solid content (total solid TS) induced to approximately 10-20% by volume of the original (Influent) manure (fresh) slurry, after going through the anaerobic digestion cycle. Out of the three types of digestion processes mentioned above, the digested slurry in effluent-form comes out only in semi-continuous BGP. The digested slurry effluent, either in liquid-form or after sun drying in slurry pits makes excellent bio-fertilizer for agricultural and horticultural crops or aquaculture.

*Sludge:* In the batch-fed or semi batch-fed digester where the plant wastes and other solid organic materials are added, the digested material contains less of effluent and more of sludge. The sludge precipitates at the bottom of the digester and is formed mostly of the solids substances of plant wastes. The sludge is usually composted with chemical fertilizers as it may contain higher percentage of parasites and pathogens and hookworm eggs etc., especially if the semi-batch digesters are either connected to the pigsty or latrines. Depending upon the raw materials used and the conditions of the digestion, the sludge contains many elements essential to the plant life including nitrogen, phosphorous, potassium plus a small quantity of salts (trace elements), indispensable to the plant growth- the trace elements such as boron, calcium, copper, iron, magnesium, sulphur and zinc etc. The fresh digested sludge, especially if the night soil is used, has high ammonia content and in this state may act like a chemical fertilizer by forcing a large dose of nitrogen than required by the plant and thus increasing the accumulation of toxic nitrogen compounds.

For this reason, it is probably best to let the sludge age for about two weeks in open place. The fresher the sludge the more it needs to be diluted with water before application to the crops, otherwise very high concentration of nitrogen may kill the plants.

*Inorganic solids:* In village situation the floor of the animals shelters are full of dirt, which gets mixed with the manure. Added to this the collected manure is kept on the unlined surface which has plenty of mud and dirt. Due to all this the feed stock for the BGP always has some inorganic solids, which goes inside the digester along with the organic materials. The bacteria can not digest the inorganic solids, and therefore settles down as a part of the bottom most layer inside the digester. The Inorganic Solids contains mud, ash, sand, gravel and other inorganic materials. The presence of too much inorganic solids in the digester can adversely affect the efficiency of the BGP. Therefore to improve the efficiency and enhance the life of a semi-continuous BGP, it is advisable to evenly empty it in a period of 5-10 years for thorough cleaning and washing it from inside and then reloading it with fresh slurry.

### *3.3.6.4. Efficiency of digestion*

The efficiency of anaerobic digestion can be estimated by comparing the energy available in a specific amount of raw material to the energy of the methane produced from that material. Four such estimates are given in Figure 16.

It seems fair to conclude that anaerobic digestion is about 60-70% "efficient" in converting organic waste to methane. However, it would probably be more accurate to call this a conversion rate since, like all biological processes, a great deal of energy is required to maintain the system, and most of this extra energy is not included in the conversion.

### *3.3.6.5. Fuel value*

The fuel value of bio-gas is directly proportional to the amount of methane it contains (the more methane, the more combustible the bio-gas). This is because the gases, other than methane, are either non-combustible, or occur in quantities so small that they are insignificant. Since tables of "fuel values of bio-gas" may not show how much combustible methane is in the gas, different tables show a wide variety of fuel values for the same kind of gas, depending on the amount of methane in the gas of each individual (Table 5).

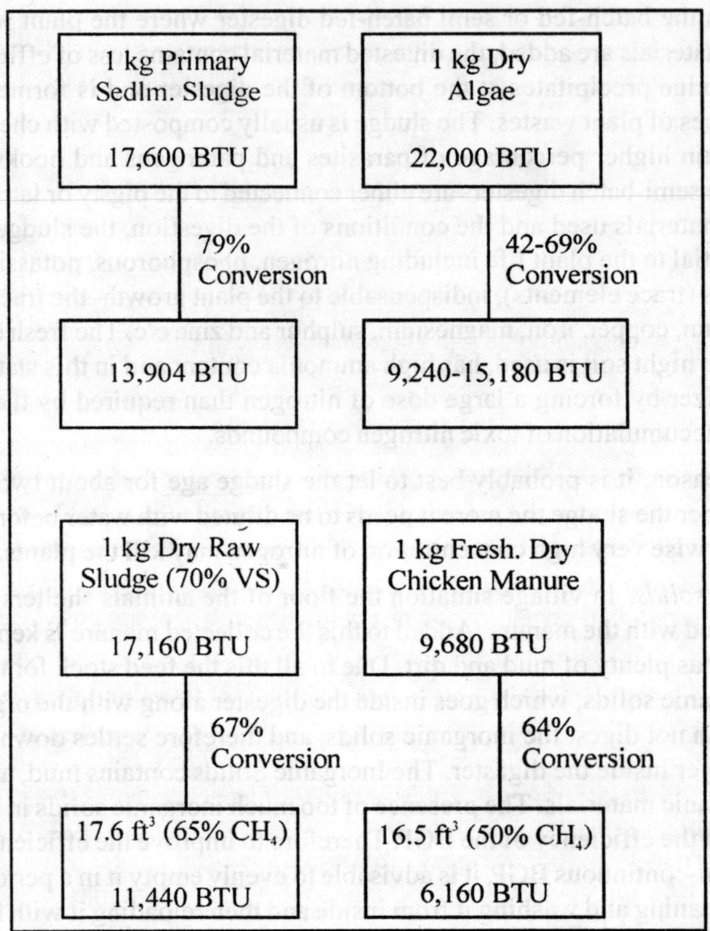


**Figure 16** Efficiency of methane production from different materials

**Table 5** Fuel value of biogas and other major fuel gases

| Fuel gas | Fuel value (BTU/ft³) |
|---|---|
| Coal (town) gas | 450-500 |
| Bio-gas | 540-700 |
| Methane | 896-1069 |
| Natural gas (methane or propane-based) | 1050-2200 |
| Propane | 2200-2600 |
| Butane | 2900-3400 |

**The composition and fuel value of bio-gas from different kinds of organic wastes depend on several things:**

1. **The temperature at which digestion takes place. This has already been discussed.**
2. **The nature of the raw material.**

*3.3.6.6. Amount of gas from different wastes*

The actual amount of gas produced from different raw materials is extremely variable depending upon the properties of the raw material, the temperature, the amount of material added regularly, etc. (Table 6).

**Table 6** Cubic feet of gas produced by volatile solids of combined wastes

| Material | Proportion | $Ft^3$ Gas/kg VS Added | $CH_4$ content of gas (%) |
|---|---|---|---|
| Chicken manure | 100% | 5.0 | 59.8 |
| Chicken manure & paper pulp | 31% 69% | 7.8 | 60.0 |
| Chicken manure & newspaper | 50% 50% | 4.1 | 66.1 |
| Chicken manure & grass clippings | 50% 50% | 5.9 | 68.1 |
| Steer manure | 100% | 1.4 | 65.2 |
| Steer manure & grass clippings | 50% 50% | 4.3 | 51.1 |

# 4. FUTURE OUTLOOK

At the start of the new millennium waste management has become a political priority in many countries. One of the main problems today is to cope with an increasing amount of primary waste in an environmentally acceptable way. Biowastes, i.e., municipal, agricultural or industrial organic waste, as well as contaminated soils etc., have traditionally been deposited in landfills or even dumped into the sea or lakes without much environmental concern. In recent times, environmental standards of waste incineration and controlled land filling have gradually improved, and new methods of waste sorting and resource/energy recovery have been developed. Treatment of biowastes by anaerobic digestion processes is in many cases the optimal way to convert organic waste into useful products such as energy (in the form of biogas) and a fertilizer product. Other waste management options, such as land filling and incineration of organic waste has become less desirable, and legislation, both in Europe and elsewhere, tends to favor biological treatment as a way of recycling minerals and nutrients of organic wastes from society back to the food production and supply chain. Removing the relatively wet organic waste from the general waste streams also results in a better calorific value of the remainder for incineration, and a more stable fraction for land filling.

Biogas, a clean and renewable form of energy could very well substitute (especially in the rural sector) for conventional sources of energy (fossil fuels, oil, etc.) which are causing ecological-environmental problems and at the same time depleting at a faster rate. Despite its numerous advantages, the potential of biogas technology could not be fully harnessed or tapped as certain constraints are also associated with it. Most common among these are: the large hydraulic retention time of 30-50 days, low gas production in winter, etc. Therefore, efforts are needed to remove its various limitations so as to popularize this technology in the rural areas. Researchers have tried different techniques to enhance gas production.

The production of biogas for reducing fossil $CO_2$ emissions is one of the key strategic issues of the German government and has resulted in the development of new process techniques and new technologies for the energetic use of biogas. Progress has been made in cultivating energy crops for biogas production, in using new reactor systems for anaerobic digestion, and in applying more efficient technologies for combined heat and power production. Recently, integration of fuel cells within the anaerobic digestion process was started, and new technologies for biogas upgrading and conversion to hydrogen were tested.

## 5. FURTHER READING


Angelidaki, I., Ellegaard, L. and Ahring, B.K. (2003). Applications of the anaerobic digestion process. Adv Biochem Eng Biotechnol.82:1-33

Angelidaki, I., and Ellegaard, L. (2003). Codigestion of manure and organic wastes in centralized biogas plants: status and future trends. Appl Biochem Biotechnol. 109:95-105.

Bal, A.S. and Dhagat, N.N. (2001). Upflow anaerobic sludge blanket reactor-a review. Indian J Environ Health.43:1-82.

Lissens, G., Vandevivere, P., De Baere, L., Biey, E.M. and Verstrae, W. (2001). Solid waste digestors: process performance and practice for municipal solid waste digestion. Water Sci Technol. 44:91-102.

Nishio, N. and Nakashimada, Y. (2004). High rate production of hydrogen/methane from various substrates and wastes. Adv Biochem Eng Biotechnol. 90:63-87

Oremland RS (1988) Biogeochemistry of methanogenic bacteria. In: Biology of Anaerobic Microorganisms (ed. AJB Zehnder). John Wiley & Sons, Inc, pp. 641-705.

Oremland RS, King GM (1989) Methanogenesis in hypersaline environments. In: Microbial Mats: the Physiological Ecology of Benthic Microbial Communities (eds Y Cohen, E Rosenberg). American. Society for Microbiology, Washington, D.C.USA, pp.180-190.

Oremland RS, Marsh L, Des Marais DJ (1982) Methanogenesis in Big Soda Lake, Nevada: an alkaline, moderately hypersaline desert lake. Limnology and Oceanography 43, 462468.

Weiland, P. (2003). Production and energetic use of biogas from energy crops and wastes in Germany. Appl Biochem Biotechnol.109:263-74.

Yadvika, S., Sreekrishnan, T.R., Kohli, S., Rana, V. (2004). Enhancement of biogas production from solid substrates using different techniques-a review. Bioresour Technol. 95:1-10.


□□□

# 10

# Microbes and Hydrogen Production

**S.K. SONI**

*Department of Microbiology, Panjab University, Chandigarh-160 014*

# 1. INTRODUCTION

Hydrogen is a colorless, odorless and completely non-poisonous gas. It has a specific gravity of 0.0899g/l and it condenses at -252.77°C. Liquid-hydrogen has a specific gravity of 70.99 g/l. Because of this, hydrogen has the highest energy density in relation to mass of all the fuels and energy carriers: 1 kg hydrogen contains as much energy as 2.1 kg natural gas or 2.8 kg petrol. The energy density referring to volume of liquid hydrogen is a quarter of petrol's and a third of natural gases. When burnt or combined, with pure oxygen, the only by-products are heat and water. When burnt with air, which has about 68% nitrogen, some oxides of nitrogen (Nitrogen oxides or NOx) are formed. Even then, burning hydrogen produces less air pollutants relative to fossil fuels. Hydrogen is an obvious alternative to hydrocarbon fuels, such as gasoline because in its molecular form hydrogen can be used directly as a fuel to drive a vehicle, to heat water or indirectly to produce electricity for industrial, transport and domestic use. It is potentially available wherever there is water and is a clean source of power. The data presented in the Table 1 shows that when 1g of Hydrogen burns, it yields more energy than the same amount of conventional fuels.

Hydrogen has many potential uses, is safe to manufacture, and above all it is environment friendly. All the means of transport we know today could be powered by hydrogen. There are two possibilities for doing so, hydrogen is burnt in conventional engines instead of gasoline. The other option is the use of fuel cells which are generating electric power for an electric motor in the car. A hydrogen fuel cell generates electricity through an electrochemical reaction using hydrogen and oxygen. Hydrogen is sent into one side of a proton exchange membrane. The hydrogen proton travels through the membrane, while the electron enters an electrical circuit, creating a DC electrical current. On the other side of the membrane, the proton and electron are recombined and mixed with oxygen from the room air, forming pure water. Because there is no combustion in this process, there are no other emissions, making fuel cells an extremely clean and renewable source of electricity. By combining the generating power of multiple proton exchange membrane (PEM) cartridges, fuel cells can be built to meet specific loads from 500 Watts to 5 kilowatts. The use of fuel cells in cars has some decisive advantages as only water is emitted from the exhaust, it operates without noise and without vibrations and it is more efficient than a combustion engine - so it saves energy. When a fuel cell car is waiting at a traffic light there is no noise because the engine does not work. The noise from accelerating is much reduced as well. World-wide all the big motorcar producing companies are developing test cars with fuel cell drive systems. In Germany mainly Daimler Chrysler, Opel and Ford are the first to do so. BMW presented hydrogen powered cars very early. The Hydrogen Highway initiatives in both California and Florida have received added boosts with Chevron opening hydrogen refueling stations in both states. Instead of bringing in hydrogen, these stations will produce their own hydrogen, making it more cost-effective. These stations will provide hydrogen to fuel cell vehicles in California and passenger buses with internal combustion engines running on hydrogen in Florida. A great variety of other possible applications of hydrogen and fuel cells can be found in the energy supply of portable devices including mobile phones, laptops, walkmen, camcorders and many other things could be powered by hydrogen and by fuel cells in the size of batteries.

Hydrogen, by far is the most plentiful element in the universe and one of the most abundant on the earth. Hydrogen is the simplest element with one atom of hydrogen consisting of only one proton and one electron. Despite its simplicity and abundance, hydrogen doesn't occur naturally as a gas on the Earth - it's always combined with other elements such as oxygen and carbon. Once it has been separated, hydrogen is the ultimate clean energy carrier. Clean enough that the NASA Space Shuttle programs have been relying on hydrogen-powered fuel cells since 1970s to operate shuttle electrical systems; and the exhaust from the fuel cell – pure water – is used by the crew as drinking water. The car we drive in the future may be well powered by the breath of tiny living organisms. That's the promise, at least, of a new field of science known as 'biological energy production'. Researchers in the field seek to find a cure for the world's dependence on fossil fuels by taking advantage of nature's ability to transform sunlight and simple sugars into usable fuels. Hydrogen is one such fuel that can be produced microbiologically and holds a great promise as the power fuel of the future but some risks are also equally high. Hydrogen gas is highly inflammable and compressing or liquefying the gas is expensive, thus distribution of the gas will be a problem. Metal alloys can absorb up to 1,000 times their own volume of hydrogen but they are heavy and some become brittle after repeated use. Carbon nanotubes, tiny cylinders of carbon atoms, have an even greater storage capacity, but this wonder technology is years from finding its way into a fuel tank.

**Table 1** Energy value of some common fuels

| **Fuel** | **Energy produced during combustion (KJ / g)** |
|---|---|
| Hydrogen gas ($H_2$) | 143 |
| Methane gas ($CH_4$) | 56 |
| Petrol (Octane, $C_8H_{18}$) | 48 |
| Coal (Carbon, C) | 33 |
| Ethanol ($C_2H_5OH$) | 30 |
| Methanol ($CH_3OH$) | 23 |
| Carbohydrates (e.g. $C_6H_{12}O_6$) | 16 |
| Carbon Monoxide gas (CO) | 10 |

The huge advantages that hydrogen has over other fuels suggest that it is non-polluting and a renewable source of energy which does not evolve the "greenhouse gas", $CO_2$ in combustion. It liberates large amounts of energy per unit weight in combustion and is easily converted to electricity by fuel cells which combine hydrogen and oxygen to produce electricity, heat, and water. Hydrogen gas is thought to be the ideal fuel for a world in which air pollution has been alleviated, global warming has been arrested, and the environment has been protected in an economically sustainable manner. Hydrogen and electricity could team up to provide attractive options in transportation and power generation. Interconversion between these two forms of energy suggests on-site utilization of hydrogen to generate electricity, with the electrical power grid serving in energy transportation, distribution utilization, and hydrogen regeneration as needed. A challenging problem in establishing hydrogen as a source of energy for the future is the renewable and environment friendly generation of large quantities of hydrogen gas.

## 2. CONVENTIONAL METHODS OF HYDROGEN PRODUCTION

Hydrogen production is a large industry. Globally, about 50 million metric tons of hydrogen are produced each year. However, for a hydrogen economy to emerge, current hydrogen production will not be able to satisfy a fraction of the demand. There are several processes which can yield hydrogen from several sources at different efficiencies and costs. 48% of current hydrogen production is from natural gas, 30% comes from oil, 18% is from coal, and electrolysis accounts for about 4%.

Hydrogen is usually commercially produced by the steam reforming of natural gas which uses high temperatures (700–1100 °C) for reacting steam ($H_2O$) with methane ($CH_4$) to yield syngas ($CO + H_2$).

$$CH_4 + H_2O \rightarrow CO + 3H_2$$

Additional hydrogen can be recovered from the carbon monoxide (CO) through the water-gas shift reaction:

$$CO + H_2O \rightarrow CO_2 + H_2$$

Essentially, the oxygen (O) atom is stripped from the water (steam) to oxidize the carbon (C), liberating the hydrogen formerly bound to the carbon and oxygen. The byproduct carbon dioxide ($CO_2$), which is a greenhouse gas, is usually released into the atmosphere, but there is some research into interning it underground or undersea.

Coal can also be converted into syn gas and methane, also known as 'town gas', via coal gasification.

Electrolysis is an alternative to using fossil fuels to generate hydrogen. The only requirements are electricity and water. However, electricity is much more expensive per unit of energy than methane, and hence the process is not economical for large scale production. Research into high-temperature electrolysis may eventually lead to a viable process that is cost-competitive with natural gas steam reforming. The only additional requirement is a high temperature heat source, which may be provided by a thermal power plant. There has also been a research in to other high-efficiency thermochemical processes such as the sulfur-iodine process. Again, a high temperature heat source would be needed. An example of a $CO_2$ emission-free system, would be where concentrated solar thermal power is used to produce hydrogen from water, using either the sulfur-iodine process or high-temperature electrolysis, and then hydrogen fuel cells would be used to produce electricity for mobile applications.

The cleanest way to produce hydrogen is by using sunlight to directly split water into hydrogen and oxygen. Photo-electrochemical systems use semi-conducting materials (like photovoltaics) to split water using only sunlight. It uses two types of electro-chemical systems to produce hydrogen. One uses soluble metal complexes as a catalyst, while the other uses semi-conductor surfaces. When the soluble metal complexes dissolve, the complex absorbs solar energy and produces an electrical charge that drives the water splitting reaction. This process mimics photosynthesis. The other method uses semiconducting electrodes in a photochemical cell to convert optical energy into chemical energy. The semi-conductor surface serves two functions, to absorb solar energy and to act as an electrode. Light induced corrosion, limits the use of light by the semi-conductor. Researchers at the University of Tennessee and US Department of Energy's (DOE) Oak Ridge National Laboratory are finding ways to use photosynthesis to produce hydrogen from sunlight. The researchers extracted two photosynthetic complexes from spinach plants: called Photosystem I and Photosystem II which work together to produce carbohydrates for the plant. By attaching platinum atoms to the Photosystem I complex, the researchers were able to produce hydrogen from visible light. Unfortunately, the process required the use of an added chemical that makes the overall process impractical. The researchers are working to combine the platinum Photosystem I complex with Photosystem II complexes forming a molecular system that can convert light and water directly into hydrogen, without the help from an added chemical.

A photoelectrochemical (PEC) system combines the harvesting of solar energy with the electrolysis of water. When a semiconductor of proper characteristics is immersed in an aqueous electrolyte and irradiated with sunlight, the energy can be sufficient to split water into hydrogen and oxygen. Depending on the type of semiconductor material and the solar intensity, this current density is 10-30 $mA/cm^2$. At these current densities, the voltage required for electrolysis is much lower, and therefore, the corresponding electrolysis efficiency is much higher. One of the major advantages of a direct conversion photoelectrochemical system is that it not only eliminates most of the costs of the electrolyzer, but it also has the possibility of increasing the overall efficiency of the process leading to a further decrease in cost.

Hydrogen can also be produced via pyrolysis or gasification of biomass resources such as agricultural residues like peanut shells, consumer wastes including plastics and waste grease, or biomass specifically grown for energy uses. Biomass pyrolysis produces a liquid product (bio-oil) that contains a wide spectrum of components that can be separated into valuable chemicals and fuels, including hydrogen. Technologies for the generation of hydrogen from biomass are not commercially available so far.

Gasification uses heat to break down biomass or coal into a gas from which pure hydrogen can be generated. Along with the commercial methods of biomass utilization, this can be used to produce hydrogen via pyrolysis and gasification. Coke, methanol and primary gases are obtained in the first stage. In the second stage, the reaction with (air) oxygen and/ or steam results in a mixture of 20% $H_2$, 20% CO, 10% $CO_2$, almost 5% $CH_4$ and 45% $N_2$. Using pure oxygen or steam only eliminates the nitrogen component. The transformation of this gas mixture into a hydrogen rich gas is named, depending on the feedstock, as gasification (solids) or reforming (gas). Endothermic reactions of hydrocarbons with steam create synthetic gases with high hydrogen content, whereby the so called shift reaction ($CO + H_2O \rightarrow CO_2 + H_2$) can be used to alter the molar $CO/H_2$ ratio. The hydrogen content of the gas is determined by the process parameters pressure and temperature. Before the actual gasification, the organic substance breaks up under the application of heat into coke, condensates and gases. This preliminary stage is known as thermal decomposition or pyrolysis. The presence of oxygen in the reactor leads to partial oxidation of the intermediate products rather than reformation (Figure 1).

The development of this process began in 1993. The original concept was that the pyrolysis of oil could be fractionated into two fractions based on water solubility. The water-soluble fraction is to be used for hydrogen production and the water insoluble fraction could be used in adhesive formulations. The bio-oil can be stored and shipped to a centralized facility where it is converted to hydrogen via catalytic steam reforming and shift conversion. Catalytic steam reforming of bio-oil at 750-850ºC over a nickel-based catalyst is a two-step process that includes the shift reaction:

$$\text{Bio-oil} + H_2O \rightarrow CO + H_2$$

$$CO + H_2O \rightarrow CO_2 + H_2$$

The overall stoichiometry gives a maximum yield of 17.2 g $H_2$/100 g bio-oil (11.2 wt.% based on wood).

$$CH_{1.9}O_{0.7} + 1.26H_2O \rightarrow CO_2 + 2.21H_2$$

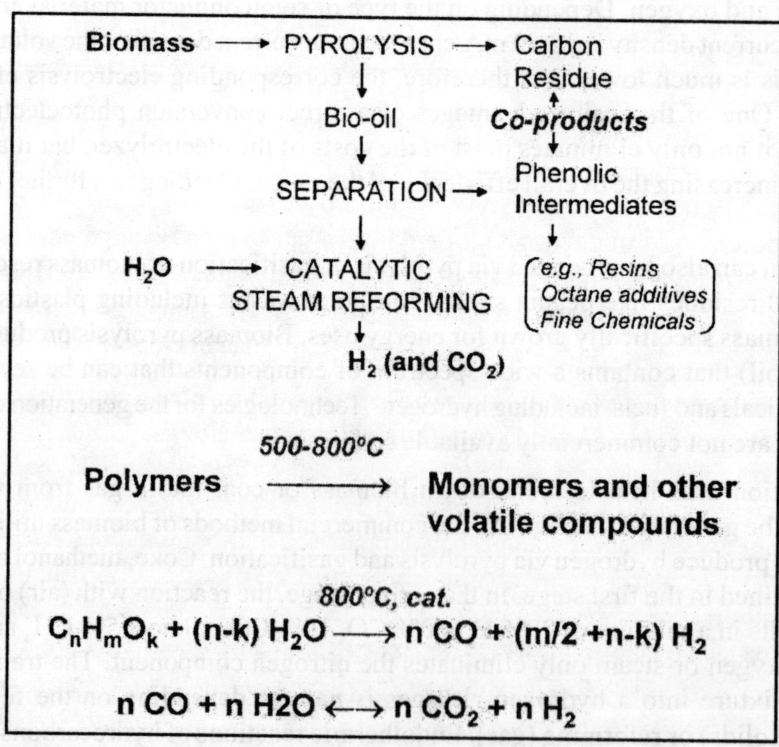


**Figure 1** Schematic representation of pyrolysis process concept and reactions for hydrogen production from biomass

## 3. BIOLOGICAL METHODS OF HYDROGEN PRODUCTION

Biological and photo-biological processes can use algae and bacteria to produce hydrogen. Under specific conditions, the pigments in certain types of algae absorb solar energy. The enzyme in the cell acts as a catalyst to split the water molecules. Some bacteria are also capable of producing hydrogen, but unlike algae, they require a substrate to grow on. The organisms not only produce hydrogen, but can clean up pollution as well. Apart from these methods, biotechnologists are looking for microbial methods of hydrogen production that are less energy intensive and give a sustainable supply of hydrogen. The thermo-chemical and electrochemical methods for hydrogen generation are highly energy-intensive and not always environmentally friendly. Biological methods, on the other hand present a less energy-intensive means of hydrogen production. These occur at ambient temperature and pressure and predominantly generate hydrogen and carbon dioxide. Low conversion efficiencies of biological systems can be compensated for, by low energy requirements and reduced initial investment costs.

The production of hydrogen in photosynthetic bacteria, under anaerobic conditions, was discovered by Roelofson in 1934 and Nakamura in 1937. The rate of hydrogen photoproduction is dependent on experimental methods and conditions. Periods of peak productivity ranged from few hours to a few days. In algae, it was first reported by Gaffron and Rubin and in photosynthetic bacteria by Gest and Kamen. Since then, many scientists have worked with various photosynthetic microorganisms. (Table 2).

**Table 2** Hydrogen producing fermenting bacteria, photosynthetic bacteria and algae

| **Fermenting bacteria** | **Photosynthetic bacteria** |
|---|---|
| *Escherichia coli* | *Rhodospirillum rubrum* |
| *Citrobacter intermedius* | *Rhodopseudomonas* |
| *Aerobacter* sp. | *capsulate* |
| *Serratia* sp. | *R. palustris* |
| *Clostridium acetobutylicum* | *R. spheroides* |
| *C. butylicum* | *Chromatium* sp. |
| *C. kluyveri* | |
| *C. botulinum* | **Green algae** |
| *C. hitolyticum* | *Ankistrodesmus braunii* |
| *C. sticklandii* | *Chlorella protothecoides* |
| *C. sporogenes* | *C. fusca* |
| *C. tetanomorphum* | *C. vulgaris* |
| *C. butyricum* | *Coelastrum proboscideum* |
| *C. tyrobutyricum* | *Selesatrum* sp. |
| *Lactic acid bacteria* | *Scenedesmus obliquus* |
| *Spirulina maxima* | *Kirchneriella lunaris* |
| *Bacillus polymyxa* | *Chlamydomonas reinhardtii* |
| *Ruminococcus* sp. | |
| *Vibrio succinogenes* | **Blue green algae** |
| | *Anabaena flosaquae* |
| | *A. cylindrical* |
| | *Nostoc muscorum* |
| | *Oscillatoria brevis* |
| | *Synechococcus* sp. |

Some species of clostridia and *Escherchia* under anaerobic conditions convert pyruvate to acetic acid and carbon dioxide with the formation of hydrogen gas. While one mole of ATP synthesized, one mole of proton is formed along with one mole of hydrogen as a result of substrate dehydrogenation. Cells of *Clostridium thermocellum* degrade cellulose and cellobiose to produce primarily $H_2$, $CO_2$, ethanol and acetic acid. The biological production of hydrogen in intact cells is too slow for commercial exploitation. To overcome this, many workers have used pure enzyme for the same. However, the major problem of a cell free system is the cost involved to procure the pure enzyme. The generation of hydrogen using microorganisms is still very much in infancy. However there are three possible routes of hydrogen production through microbes: 1) Biophotolysis of water, 2) Photoreduction of organic compounds, 3) Fermentation of organic compounds.

## 3.1. Hydrogen production by biophotolysis of water

Certain photosynthetic microbes produce hydrogen from water in their metabolic activities using light energy. This involves splitting of water using light energy and does not require an exogenous substrate. Cyanobacteria and microalgae are the organisms with the photosynthetic systems that can carry out this photolysis according to the following reaction:

$$CO_2 + H_2O \rightarrow 6\,[CH_2O] + O_2$$

Organic compounds.

Cyanobacteria, formerly known as blue-green algae, are now recognized as bacteria since the anatomical characteristics of their cells are prokaryotic. Microalgae are primitive microscopic plants living in aqueous environments. Photosynthesis consists of two processes involving light energy conversion to biochemical energy by a photochemical reaction, and $CO_2$ reduction to organic compounds such as sugar phosphates, through the use of this biochemical energy by Calvin-cycle enzymes. Under certain conditions, instead of reducing $CO_2$, a few groups of microalgae and cyanobacteria consume biochemical energy to produce molecular hydrogen.

### *3.1.1. Photolysis by cyanobacteria (Blue green algae)*

Cyanobacteria form a large and diverse group of oxygenic photoautotrophic prokaryotes, many of which have the ability to produce hydrogen (Table 3). The group reveals wide differences in morphology, physiology and reproduction. Cyanobacteria possess phycobiliproteins and chlorophyll, and have the capacity to perform oxygenic photosynthesis. Generally cyanobacteria fix $CO_2$ but there are some cyanobacteria which lead facultative heterotrophic life. Many are noticed to be obligate photoautotrouphs. The hydrogen photoproduction by cyanobacteria has been investigated for both heterocystous and non-heterocystous forms. Hydrogen production occurs within at least 14 cyanobacteria genera, under a vast range of culture conditions. Unicellular non-diazotrophic cyanobacteria *Gloeocapsa alpicola* under sulphur starvation shows increased hydrogen production. *Arthrospira* (*Spirulina platensis*) can produce hydrogen (1 µmole $H_2$/12 hr/mg cell dry weight) in complete anaerobic and dark condition. Another nitrogen-fixing cyanobacterium, *Anabaena cylindrica*, produces hydrogen and oxygen gas simultaneously in an argon atmosphere after 30 days in light limited condition. Symbiotic cyanobacteria within coralloid roots of the cycads, *Cycas revoluta* (king Sago palm) and *Zamia furfuracea* show a significant *in vivo* hydrogen uptake. *Anabaena* sp. is able to produce significant amount of hydrogen. Among them, nitrogen-starved cells of *Anabaena cylindrica* produces highest amount of hydrogen (30 ml of $H_2$/l culture/h). Hydrogenase-deficient cyanobacteria *Nostoc punctiforme* NHM5 when incubated under high light for a long time, until the culture was depleted of $CO_2$ shows increase in hydrogen production.

**Table 3** Cyanobacterial organisms that produce hydrogen

| **Organism** | **Characteristics** | **Maximum Hydrogen evolution** | **Growth condition** |
|---|---|---|---|
| *Anabaena cylindrical* B-629 | Marine cyanobacteria | 0.103 µmol/mg dry wt/h | Air contained 5% $CO_2$; 7000 lx at the surface of culture vessels |
| *Oscillatoria brevis B-1567* | Marine cyanobacteria | 0.168 µmol/mg dry wt/h | Air contained 5% $CO_2$; 7000 lx at the surface of culture vessels |
| *Calothrix scopulorum 1410/5* | Marine cyanobacteria | 0.128 µmol/mg dry wt/h | Air contained 5% $CO_2$; 7000 lx at the surface of culture vessels |
| *Calothrix membrnacea B-379* | Marine cyanobacteria | 0.108 µmol/mg dry wt/h | Air contained 5% $CO_2$; 7000 lx at the surface of culture vessels |
| *Oscillatoria* sp. strain Miami BG7 | Marine cyanobacteria | 0.250 µmol/mg dry wt/h | Air; 100 µE/m$^2$/s; $NH_4Cl$ (25 mg/l) used as combined nitrogen source |
| *Oscillatoria limosa* | Marine cyanobacteria | 0.83 µmol/mg chl a/h | Air; incubation in 16 h light, 8 h darkness cycles |
| *Cyanothece* 7822 | Marine unicellular cyanobacteria | 0.92 µmol/mg chl a/h | $N_2$ with 5% $CO_2$ |
| *Anabaena* sp. PCC 7120 | Heterocystous cyanobacteria | 2.6 µmol/mg chl a/h | Air; 20 µE/m$^2$/s |
| *Anabaena cylindrica* IAMM-1 | Heterocystous cyanobacteria | 2.1 µmol/mg chl a/h | Air; 20 µE/m$^2$/s |
| *Anabaena variabilis* IAMM-58 | Heterocystous cyanobacteria | 4.2 µmol/mg chl a/h | Air; 20 µE/m$^2$/s |

*Contd...*

*Table 3 Contd...*

| | | | |
|---|---|---|---|
| *Anabaena cylindrica* UTEX B 629 | Heterocystous cyanobacteria | 0.91 μmol/mg chl a/h | Air; 20 μE/$m^2$/s |
| *Anabaena flos-aquae* UTEX 1444 | Heterocystous cyanobacteria | 1.7 μmol/mg chl a/h | Air; 20 μE/$m^2$/s |
| *Anabaena flos-aquae* UTEX LB 2558 | Heterocystous cyanobacteria | 3.2 μmol/mg chl a/h | Air; 20 μE/$m^2$/s |
| *Anabaenopsis circularis* IAM M-13 | Heterocystous cyanobacteria | 0.31 μmol/mg chl a/h | Air; 20 μE/$m^2$/s |
| *Nostoc muscorum* IAM M-14 | Heterocystous cyanobacteria | 0.60 μmol/mg chl a/h | Air; 20 μE/$m^2$/s |
| *Nostoc linckia* IAM M-30 | Heterocystous cyanobacteria | 0.17 μmol/mg chl a/h | Air; 20 μE/$m^2$/s |
| *Nostoc commune* IAM M-13 | Heterocystous cyanobacteria | 0.25 μmol/mg chl a/h | Air; 20 μE/$m^2$/s |
| *Anabaena variabilis* AVM13 | Heterocyst filamentous | 68 μmol/mg chl a/h | Air and 1% $CO_2$; 100 μE/$m^2$/s |
| *Anabaena variabilis* PK84 | Heterocyst filamentous | 32.3 μmol/mg chl a/h | Air and 2% $CO_2$; continuous tarbidostat mode; 113 μE/$m^2$/s |
| *Anabaena variabilis* PK84 | Heterocyst filamentous | 167.6 μmol/mg chl a/h | 73%Ar, 25%$N_2$, 2% $CO_2$; 90 μE/$m^2$/s |
| *Anabaena variabilis* PK84 | Heterocyst filamentous | 0.11 μmol/mg chl a/h | Air and 2% $CO_2$; outdoor condition |
| *Anabaena variabilis* ATCC 29413 | Heterocyst filamentous | 45.16 μmol/mg chl a/h | 73%Ar, 25%$N_2$, 2% $CO_2$; 90 μE/$m^2$/s |
| *Anabaena variabilis* ATCC 29413 | Heterocyst filamentous | 0.05 μmol/mg dry wt/h | 5000 lx at the surface of culture vessels. |
| *Anabaena variabilis* ATCC 29413 | Heterocyst filamentous | 39.4 μmol/mg chl a/h | Air and 2% $CO_2$; continuous tarbidostat mode; 113 μE/$m^2$/s |
| *Anabaena variabilis* 1403/4B | Heterocyst filamentous | 20 μmol/mg chl a/h | Air; 15 μE/$m^2$/s |
| *Anabaena azollae* | Heterocyst filamentous | 38.5 μmol/mg chl a/h | Air and 1% $CO_2$; 100 μE/$m^2$/s |
| *Anabaena variabilis* PK17R | Heterocyst filamentous | 59.18 μmol/mg chl a/h | 73%Ar, 25%$N_2$, 2% $CO_2$; 90 μE/$m^2$/s |

*Contd...*

*Table 3 Contd...*

| | | | |
|---|---|---|---|
| *Anabaena variabilis* SPU 003 | Heterocyst filamentous cyanobacteria | 5.58 nmol/mg dry wt/h | Air; incubation in 16 h light (3000 lx light intensity), 8 h darkness cycles; mannose used as carbon source. |
| *Synechococcus* PCC 6830 | Non-nitrogen-fixing unicellular cyanobacteria | 0.26 μmol/mg chl a/h | Air; photon fluence rate 20 $\mu E/m^2/s$ |
| *Synechococcus* PCC 602 | Non-nitrogen-fixing unicellular cyanobacteria | 0.66 μmol/mg chl a/h | Air; photon fluence rate 20 $\mu E/m^2/s$ |
| *Synechococcus* PCC 6307 | Non-nitrogen-fixing unicellular cyanobacteria | 0.02 μmol/mg chl a/h | Air; photon fluence rate 20 $\mu E/m^2/s$ |
| *Synechoccus* PCC 6301 | Non-nitrogen-fixing unicellular cyanobacteria | 0.09 μmol/mg chl a/h | Air; photon fluence rate 20 $\mu E/m^2/s$ |
| *Microcystis* PCC 7820 | non-nitrogen-fixing unicellular cyanobacteria | 0.16 μmol/mg chl a/h | Air; photon fluence rate 20 $\mu E/m^2/s$ |
| *Gloebacter* PCC 7421 | Non-nitrogen-fixing unicellular cyanobacteria | 1.38 μmol/mg chl a/h | Air; photon fluence rate 20 $\mu E/m^2/s$ |
| *Synechocystis* PCC 6308 | Non-nitrogen-fixing unicellular cyanobacteria | 0.13 μmol/mg chl a/h | Air; photon fluence rate 20 $\mu E/m^2/s$ |
| *Synechocystis* PCC 6714 | Non-nitrogen-fixing unicellular cyanobacteria | 0.07 μmol/mg chl a/h | Air; photon fluence rate 20 $\mu E/m^2/s$ |
| *Aphanocapsa montana* | Non-nitrogen-fixing unicellular cyanobacteria | 0.40 μmol/mg chl a/h | Air; photon fluence rate 20 $\mu E/m^2/s$ |
| *Gloeocapsa alpicola* CALU 743 | Unicellular non-diazotrophic cyanobacteria | 0.58 μmol/mg protein | Sulphur free 4% $CO_2$; 25 μmol photons/$m^2$/s |
| *Chroococcidiopsis thermalis* CALU 758 | Unicellular non-nitrogen-fixing | 0.7 μmol/mg chl a/h | Ar and1% $CO_2$; during 20 h. |
| *Mycrocystis* PCC 7806 | Unicellular/colony embedded in matrix | 11.3 nmol/mg prot/h | Air; incubation in 16 h light, 8 h darkness cycles |
| *Microcoleus chthonoplasts* | Mat-building cyanobacteria | 1.7 nmol/mg prot/h | Air; ferric ammonium citrate added (46 μmol) 30 $\mu E/m^2/s$ |

The cyanobacteria have two photosystems (PSI and PSII) and are capable of generating reducing potential from water. The photoproduction of hydrogen in cyanobacteria mainly depends on the type of species (Table 3). Besides this, other physical and physiological factors are also responsible for regulating the production of hydrogen. Cyanobacteria use two sets of enzymes to generate hydrogen gas. The first one is 'nitrogenase' and it is found in the heterocysts of filamentous cyanobacteria when they grow under nitrogen limiting conditions. Hydrogen is produced as a byproduct of fixation of nitrogen into ammonia. The other hydrogen-metabolizing/producing enzymes in cyanobacteria are 'hydrogenases'; they occur as two distinct types in different cyanobacterial species. One type of them, 'uptake hydrogenase' , has the ability to oxidize hydrogen and the other type of hydrogenase is reversible or 'bidirectional hydrogenase' and it can either take up or produce hydrogen. Uptake hydrogenase enzymes are found in the thylakoid membrane of heterocysts from filamentous cyanobacteria, where it transfers the electrons from hydrogen for the reduction of oxygen via the respiratory chain in a reaction known as oxyhydrogenation or Knallgas reaction.

#### *3.1.1.1. Nitrogenase dependent hydrogen production*

Nitrogenase is distributed mainly among prokaryotes, including cyanobacteria, but does not occur in eukaryotes, under which microalgae are classified. Molecular nitrogen is reduced to ammonium with consumption of reducing power ($e^-$ mediated by ferredoxin) and ATP. The reaction is substantially irreversible and produces ammonia. Even in the presence of nitrogen, part of the reducing equivalents are consumed for proton reduction and hydrogen production and three quarters of the reducing equivalents are used for the reduction of nitrogen. This can be expressed as

$$N_2 + 8H^+ + 8e^- + 16\ ATP \rightarrow 2NH_3 + H_2 + 16\ ADP + 16\ Pi$$

However, nitrogenase catalyzes proton reduction in the absence of nitrogen gas (i.e. in an argon atmosphere).

$$2H^+ + 2e^- \rightarrow H_2$$

$$4ATP \rightarrow 4ADP + 4Pi$$

Hydrogen production catalyzed by nitrogenase occurs as a side reaction at a rate of one-third to one-fourth that of nitrogen-fixation, even in a 100% nitrogen gas atmosphere. In the absence of molecular nitrogen, the nitrogenase system catalyses the reduction of protons to molecular hydrogen. This reaction is ATP dependent and also activated by hydrogenase enzyme.

Nitrogenase itself is extremely oxygen-labile, however, cyanobacteria have developed mechanisms for protecting nitrogenase from oxygen and supplying it with energy (ATP) and reducing power. The most successful mechanism is the localization of nitrogenase in the heterocysts of filamentous cyanobacteria (Figure 2). Vegetative cells (ordinary cells) in filamentous cyanobacteria carry out oxygenic photosynthesis. Organic compounds produced by $CO_2$ reduction are transferred into heterocysts and are decomposed to provide nitrogenase with reducing power. ATP can be provided by PSI-dependent and anoxygenic photosynthesis within heterocysts. It has been found out that the hydrogen-producing activity of cyanobacteria is stimulated by nitrogen starvation.

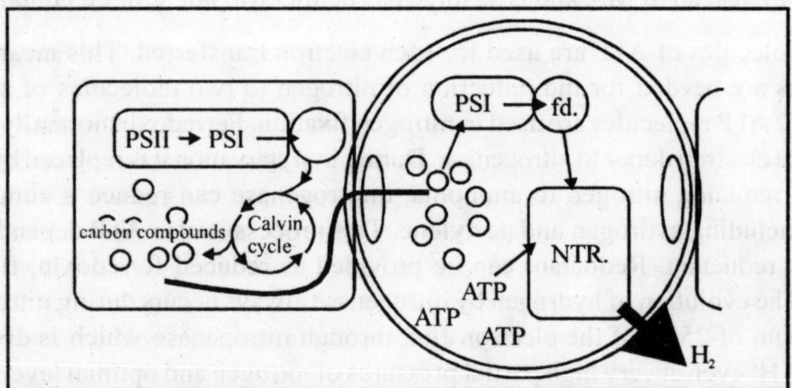


**Figure 2** Nitrogenase-mediated hydrogen production in heterocystous cyanobacteria

Generally, all nitrogen fixing bacteria contains, in addition to nitrogenase, a hydrogenase, which can activate hydrogen released from the nitrogenase reaction. The hydrogenase thus serves to re-channel the reducing equivalents into metabolic reactions and conserves them for ATP generation. (Figure 3).

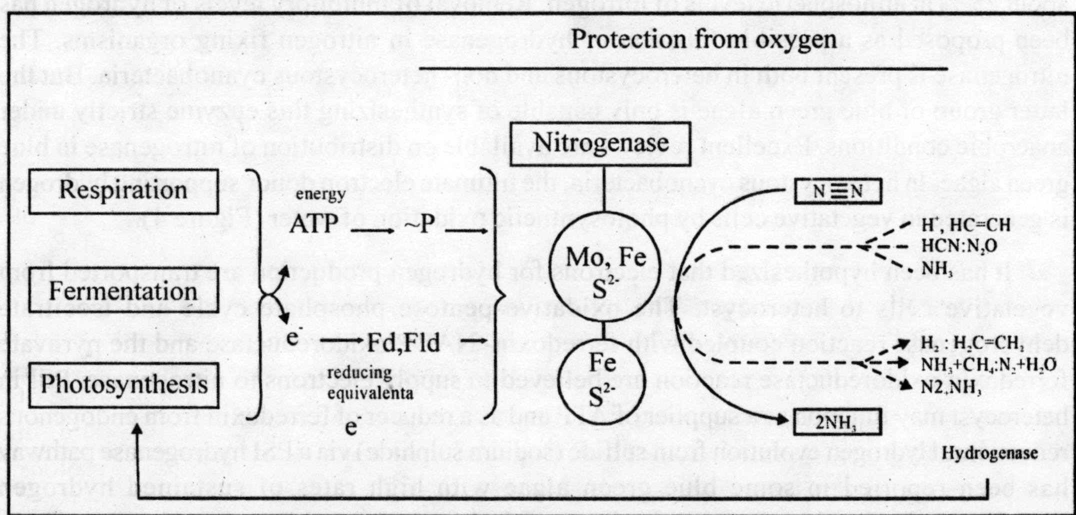


**Figure 3** A schematic diagram of relationship of nitrogenase and hydrogenase of cyanobacteria during utilization of molecular nitrogen and production of hydrogen

A nitrogenase enzyme consists of two parts, one is dinitrogenase (MoFe Protein, encoded by the genes *nifD* and *nifK*, $\alpha$ and $\beta$ respectively) and the other is dinitrogenase reductase (Fe Protein, encoded by *nifH*). Dinitrogenase is a $\alpha_2\beta_2$ heterotetramer, having molecular weight of about 220 to 240 kDa respectively, breaks apart the atoms of nitrogen. Dinitrogenase reductase is a homodimer of about 60 to 70 kDa and plays the specific role of mediating the transfer of electrons from the external electron donor (a ferredoxin or a flavodoxin) to the dinitrogenase. There are three types of dinitrogenase found in nitrogenase, which vary depending on the metal content. Type one contains molybdenum (Mo), type two contains vanadium (V) instead of Mo, and type three has neither Mo nor V but it contains iron (Fe).

Two molecules of ATP are used for each electron transferred. This means altogether six electrons are needed for the reduction of nitrogen to two molecules of ammonia. So altogether 12 ATP molecules are used in nitrogen fixation. Ferredoxin normally functions as an immediate electron donor to nitrogenase. During iron starvation it is replaced by flavodoxin. Apart from reducing nitrogen to ammonia, dinitrogenase can reduce a number of other substrates including hydrogen and acetylene. This process is also ATP dependent and also need strong reductant. Reductant can be provided as reduced ferredoxin, flavodoxin or dithionite. The evolution of hydrogen by nitrogenase always occurs during nitrogen fixation on a minimum of 25 % of the electron flux through nitrogenase which is diverted to the reduction of $H^+$ even at very high partial pressures of nitrogen and optimal levels of ATP and reductant. If the rate of dinitogenase turn over is retarded owing to limitation in the supply of ATP or reductant, the relative proportion of electrons diverted to $H^+$ reduction increases. The evolution of hydrogen by nitrogenase, therefore consumes a substantial amount of the reductant and ATP. The biological significance of hydrogen evolution by nitrogenase is still not clearly understood. However, it has been understood that in a crude preparation of nitrogenase from *Azotobacter vinelandii*, 0.3 atm hydrogen inhibited nitrogen reduction by about 25 % at atmospheric levels of nitrogen. Removal of inhibitory levels of hydrogen has been proposed as a possible function of hydrogenase in nitrogen fixing organisms. The nitrogenase is present both in heterocystous and non- heterocystous cyanobacteria. But the latter group of blue green algae is only capable of synthesizing this enzyme strictly under anaerobic conditions. Excellent reviews are available on distribution of nitrogenase in blue green algae. In heterocystous cyanobacteria, the ultimate electron donor supporting hydrogen is generated in vegetative cells by photosynthetic oxidation of water (Figure 4).

It has been hypothesized that electrons for hydrogen production are transported from vegetative cells to heterocyst. The oxidative pentose phosphate cycle and isocitrate dehydrogenase reaction coupled with ferredoxin- NADP oxidoreductase and the pyruvate ferredoxin oxidoreductase reaction are believed to supply electrons to nitrogenase. PSI in heterocyst may function as a supplier of ATP and as a reducer of ferredoxin from endogenous reductions. Hydrogen evolution from sulfide (sodium sulphide) via a PSI hydrogenase pathway has been reported in some blue green algae with high rates of sustained hydrogen photoproduction by nitrogen starved cultures of the heterocystous blue green algae *Anabaena cylindrica* have been reported . However, a few others have been reported with conflicting results. The organism, therefore, constitutes a biophotolytic system in which hydrogen and oxygen are generated from water using light energy with hydrogen and oxygen evolution occurring in separate cells. The observed rate of hydrogen evolution , however, is not solely dependent on the level of activity of nitrogenase , since hydrogen consuming enzymes occur in a number of cyanobacteria and, under appropriate conditions will recycle most of the hydrogen evolved by nitrogenase. The hydrogen recaptured by hydrogenase can support ATP synthesis or provide electrons for biosynthetic reductions, thereby allowing the organism to recover some of the energy that would otherwise be lost upon diffusion of hydrogen from the cell.

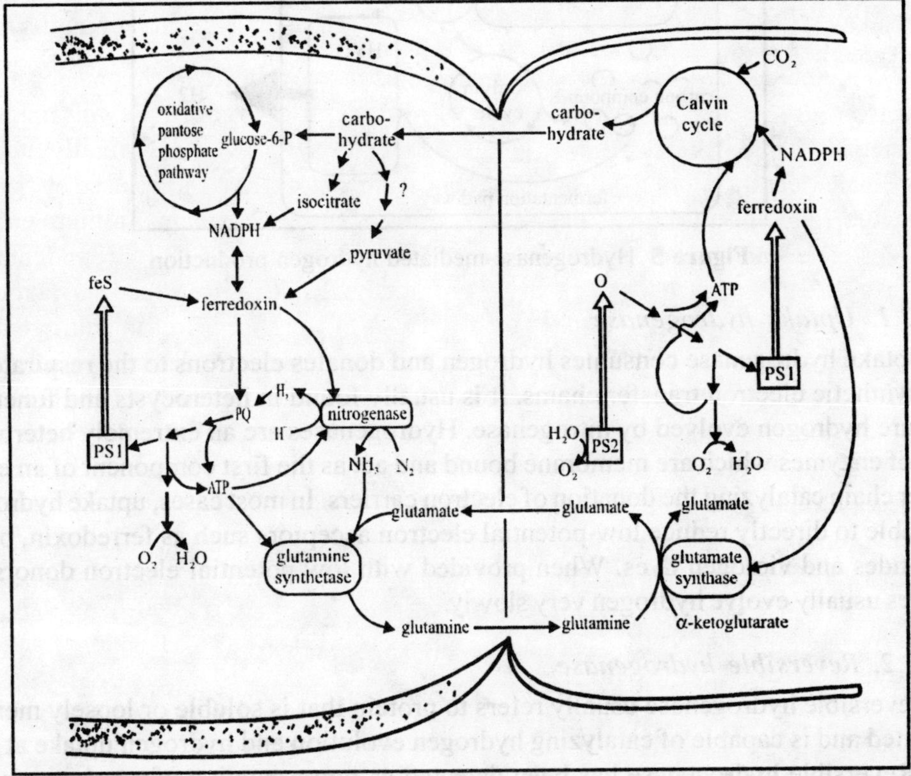


**Figure 4** Hydrogen production in blue green algae and metabolic relation between vegetative cells and heterocysts

*3.1.1.2. Hydrogenase-dependent hydrogen production*

Hydrogenase, the enzyme responsible for this hydrogen production, catalyses the following reaction:

$$2H^+ + 2X \text{ (reduced)} \leftrightarrow 6\ H_2 + 2X \text{ (oxidized)}$$

The electron carrier, X is thought to be ferredoxin. Since ferredoxin is reduced with water as an electron donor by the photochemical reaction, cyanobacteria and green alga are theoretically water-splitting microorganisms (Figure 5). It has been determined that the reducing power (electron donation) of hydrogenase does not always come from water, but may also originate intracellularly from organic compounds such as starch. The contribution of the decomposition of organic compounds to hydrogen production is dependent on the algal species concerned, and on culture conditions. Even when organic compounds are involved in hydrogen production, an electron source can be derived from water, since organic compounds are synthesized by oxygenic photosynthesis. Hydrogenase is inactive during the normal photosynthetic growth conditions of algae and the reason for this is unclear and becomes active in order to excrete excess reducing power under anaerobic conditions.

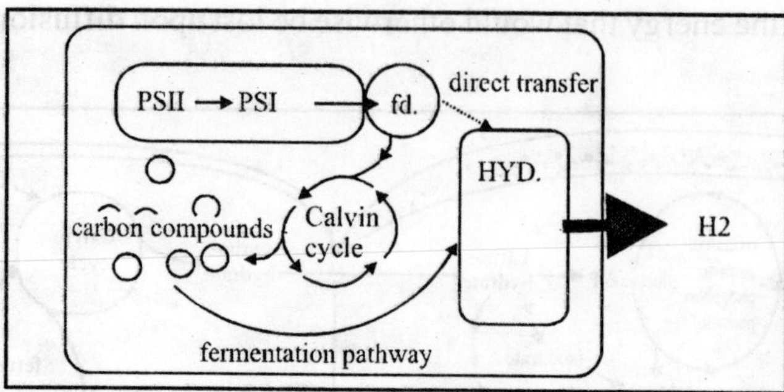


**Figure 5** Hydrogenase-mediated hydrogen production

*3.1.1.2.1. Uptake hydrogenase*

Uptake hydrogenase consumes hydrogen and donates electrons to the respiratory and photosynthetic electron transfer chains. It is usually found in heterocysts and functions to recapture hydrogen evolved by nitrogenase. Hydrogenases are an extremely heterogenous group of enzymes which are membrane bound and act as the first component of an electron transfer chain catalyzing the donation of electron carriers. In most cases, uptake hydrogenase are unable to directly reduce low-potential electron acceptors such as ferredoxin, pyridine nucleotides and viologen dyes. When provided with low potential electron donors, these enzymes usually evolve hydrogen very slowly.

*3.1.1.2.2. Reversible hydrogenase*

Reversible hydrogenase usually refers to protein that is soluble or loosely membrane associated and is capable of catalyzing hydrogen evolution and hydrogen uptake at similar rates. Reversible hydrogenase has been detected as representative of nearly every major cyanobacteria. It is tetrameric, $NAD^+$ - dependent enzyme, consisting of the two subunits hoxY and hoxH which show homology to other dimeric (NiFe) – hydrogenase and two subunits, F and hoxU constituting the diaphorase moiety of the enzyme. Reactions catalyzed by reversible hydrogenase have been demonstrated in the unicellular organisms *Synechococcus, Synechocystis and Aphanocapsa,* in filamentous non heterocystous strains such as *Spirulina* and *Oscillatoria,* and in heterocystous organisms including several strains of *Anabaena, Nostoc* and *Mastigocladus*. In most cases, reversible hydrogenase was demonstrated by measuring methyl viologen- dependent hydrogen evolution and reversible hydrogenase have been partially purified from some cyanobacterial sources. However, for *Synechococcus and Synechocystis,* the measurement of hydrogen dependent photoreduction of $CO_2$ provided presumptive evidence for the existence of reversible hydrogenase in these organisms. The reactions involving reversible hydrogenase, the pathways of electron transfer and *in vivo* function of this enzyme are poorly understood at present.

*3.1.1.2.3. One step simultaneous hydrogen & oxygen photoproduction in cyanobacteria*

In *Synechococcus* sp. Miami BG04351, it has been noticed that hydrogen production was maximum with early log phase cells. Removal of external $CO_2$ and $N_2$ from the gas phase enhanced hydrogen production significantly, as these gases were found to be competitive inhibitors of hydrogen production. Besides this, it has been understood that light saturated conditions were also essential for stoichiometric hydrogen and oxygen production from water. This strain grew exceptionally well even under vigorous aeration without any combination of nitrogen source and ammonia. Molecular nitrogenase is known to be extremely sensitive to oxygen since oxygen is produced photosynthetically inside cells during growth. These strains probably have unique mechanism to protect nitrogenase from the inhibitory effect of oxygen.

As stated earlier, the simultaneous evolution of hydrogen and oxygen continued for a short period of five days at a molar ratio of 2:1. However, this problem can be solved by special treatment such as inert gas sparging by continuous flow through the system to remove accumulated hydrogen and oxygen, or applying chemicals to trap oxygen. During this process there was no net change in cellular carbohydrate content. This sort of simultaneous hydrogen and oxygen production, with no carbon dioxide production, indicates that the water is ultimate source of hydrogen as illustrated in Figure 6.

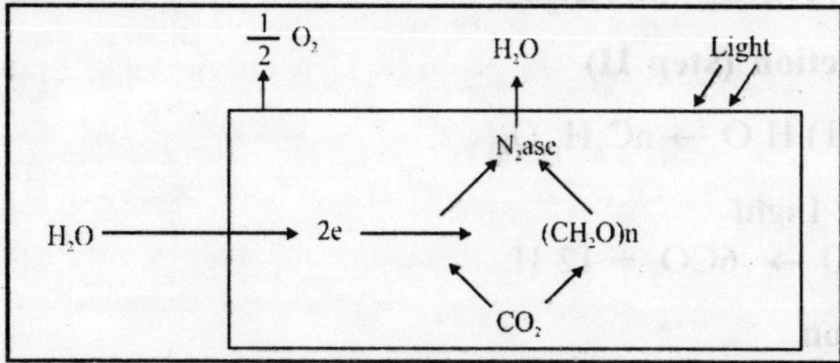


**Figure 6** One-step stoichiometric hydrogen and oxygen photoproduction from water by *Synecliococcus* sp.

*3.1.1.2.4. Two step hydrogen photoproduction*

A two step hydrogen photoproduction technique was developed with a strain of *Oscillatoria* sp. Miami BG7 (Figure 7). In the first step, cells were grown in a combined nitrogen limited medium. While culturing the level of intracellular hydrogen donor substance, i.e. glycogen, increased remarkably. This causes a high cellular C/N ratio resulting in the enhancement of nitrogenase synthesis. Meanwhile, the concentration of photosynthetic pigments like chlorophyll and phycobilin, responsible for oxygen photoproduction, decreased significantly and, thus, minimizing oxygen inhibition of nitrogenase.

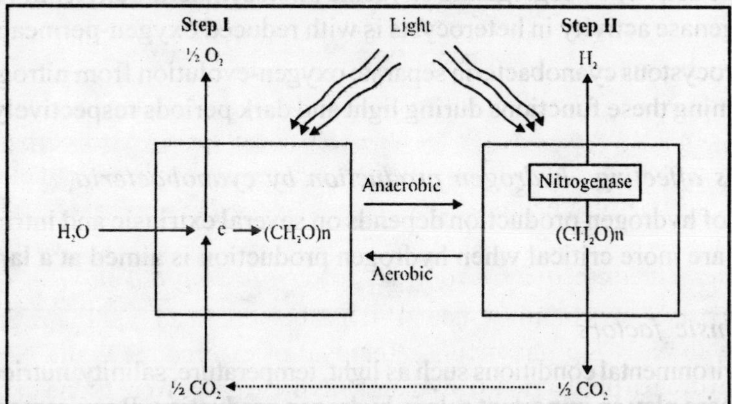


**Figure 7** Two step hydrogen production from filamentous blue green algae

In the second step, the cultures were grown under anaerobic conditions (argon atmosphere). Upon illumination, the accumulated intracellular glycogen was hydrolyzed to glucose and this was subsequently converted to hydrogen and carbon dioxide in a stoichimetric ratio of 2. Experiment yielded a value of 9.8 as the molar ratio of hydrogen production per ton glucose consumption. In other words, 82% of the combustible energy used in the form of photosynthetically produced glycogen is converted to hydrogen. The overall reaction of the steps, stated above are as follows:

**Aerobic reaction (step I)**

$$12H_2O + 6CO_2 \xrightarrow{\text{light}} C_2H_{12}O_6 + 6H_2O + 6CO_2$$

$$nC_2H_{12}O_6 \rightarrow \text{glycogen} + (n-1)H_2O$$

**Anaerobic reaction (step II)**

$$\text{Glycogen} + (n-1)\,H_2O \rightarrow nC_6H_{12}O_6$$

$$C_6H_{12}O_6 + 6H_2O \xrightarrow{\text{Light}} 6CO_2 + 12\,H_2$$

**Overall reaction**

$$H_2O \rightarrow H_2 + 1/2O_2$$

Overall inputs in these two steps are water and light whereas overall outputs are hydrogen and oxygen.

Hydrogen photo evolution catalyzed by nitrogenases or hydrogenases can only function under anaerobic conditions due to their extreme sensitivity to oxygen. Since oxygen is a byproduct of photosynthesis, organisms have developed the following spatial and temporal strategies to protect the enzyme from inactivation by oxygen, these are:

(a) Heterocyst-containing cyanobacteria physically separate oxygen evolution from nitrogenase activity by segregating oxygenic photosynthetic activity in vegetative cells and nitrogenase activity in heterocysts is with reduced oxygen-permeability and

(b) Non-heterocystous cyanobacteria separate oxygen-evolution from nitrogenase activity by performing these functions during light and dark periods respectively.

### *3.1.1.3. Factors affecting hydrogen production by cyanobacteria*

Efficiency of hydrogen production depends on several extrinsic and intrinsic factors or parameters that are more critical when hydrogen production is aimed at a large scale.

#### *3.1.1.3.1. Extrinsic factors*

Several environmental conditions such as light, temperature, salinity, nutrient availability, gaseous atmosphere play an important role in hydrogen production. Requirements of different cyanobacterial species are different for optimum hydrogen production.

*Light:* Although most cyanobacterial species preferentially absorb red light near 680 nm, light requirement for hydrogen production varies among different species of cyanobacteria. While *Spirulina (Arthrospira platensis)* produces hydrogen under anaerobic conditions, both in the dark and in the light but several other species produces hydrogen only in the presence of light. Hydrogen production mediated by native hydrogenases in *Synechococcus* PCC7942 occurs in the dark under anaerobic condition. *Spirulina platensis* can produce hydrogen optimally at 32°C in complete anaerobic and dark condition. The highest volumetric hydrogen production was found in *Anabaena variabilis* ATCC 29413 and in its mutant PK84. Hydrogen evolution by *Anabaena variabilis* PK84 with air containing 2% $CO_2$ was stimulated by light. Hydrogen production in *Nostoc muscorum* is catalyzed by nitrogenase, more hydrogen is produced in this strain in the light than in the dark. *Anabaena cylindrica,* produces hydrogen under an argon atmosphere for 30 days in limited light (luminous intensity 6.0 W/m$^{-2}$) and 18 days under elevated light (luminous intensity 32 W/m$^{-2}$). Continuous hydrogen production by *Anabaena cylindrica* for a prolonged period under light limited condition occurs in the absence of exogenous nitrogen. The effect of light on nitrogenase, mediated hydrogen production by most cyanobacteria is well studied. Nitrogenase function is saturated only at much higher light intensities than required for optimal growth. Thus hydrogen production rates can be doubled if the luminous intensity exposure to cultures is changed from 20 W/m$^2$ to 60 W/m$^2$.

*Temperature:* The optimum temperature for hydrogen production for most cyanobacterial species is between 30–40°C and varies from species to species of cyanobacteria. For example, *Nostoc* cultured at 22°C showed higher rates of hydrogen production than at 32°C, while *Nostoc muscorum* SPU004 showed optimum hydrogen production at 40°C. *Anabaena variabilis* SPU 003 on the other hand show optimum hydrogen production at 30°C.

*Salinity:* Salinity does effect hydrogen production by cyanobacteria. In general fresh water cyanobacteria shows lower rate of hydrogen production with increase in salinity. This occurs likely because of diversion of energy and reductants for extrusion of $Na^+$ ions from within the cells or prevention of $Na^+$ influx.

*Micronutrients:* Trace elements such as cobalt (Co), copper (Cu), molybdenum (Mo), zinc (Zn), and nickel (Ni) affects hydrogen production. Many of these metals have shown pronounced enhancement of hydrogen production and thought to be due to their involvement in the nitrogenase enzyme. For example, *Anabaena variabilis* SPU003 is highly sensitive to Co, Cu, Mn, Zn, Ni, Fe ions and shows no hydrogen production at concentrations below 10 mM for these ions. A culture of *Anabaena cylindrica* grown with 5.0 mg of Ferric ions per liter produce hydrogen at a rate about twice that of culture with 0.5 mg of Ferric ions per litre.

*Carbon source:* Carbon sources are also known to influence the hydrogen production considerably by influencing nitrogenase activity. Presence of different carbon sources cause variation in electron donation capabilities by the cofactor compounds to nitrogenase thus influencing hydrogen production.

*Nitrogen source:* Several inorganic nitrogenous compounds influence hydrogen production rates in many ways. Nitrite, nitrate and ammonia have been reported to inhibit nitrogenase in *Anabaena variabilis* SPU003 and *Anabaena cylindrica*. Generally all exogenously added nitrogen sources inhibit nitrogenase synthesis. Although in some *Anabaena cylindrica* strains, ammonium addition (0.2 mM $NH_4^+$) at a given time point eventually suppresses hydrogen production, but periodic addition of lower amounts (0.1 mM ammonium chloride) do not inhibit hydrogen evolution. However, influence of nitrogen source does not always part pronounced effects, and interpretation is not straightforward. While some studies showed that there are significant differences in hydrogen production depending on the nitrogen content of the media while other studies show the reverse. In *Anabaena cylindrica* culture, oxygen production gets dominated with the incremental addition of ammonium chloride (from 0.1 mM to 0.5 mM). The hydrogen to oxygen production ratio (4:1) in totally nitrogen starved condition decrease (1.7:1) when ammonium ions are added.

*Molecular nitrogen:* Molecular nitrogen is a competitive inhibitor for hydrogen production and removal of molecular nitrogen is often very necessary for hydrogen production. Hydrogen production may be considerably inhibited in presence of molecular nitrogen.

*Oxygen:* Due to their extreme sensitivity to oxygen, hydrogen photo evolution catalyzed by nitrogenases or hydrogenases can function only under anaerobic conditions. As oxygen is a byproduct of photosynthesis, nitrogenase-containing organisms have developed several spatial and temporal separation/compartmentalization strategies to protect the enzyme from inactivation by oxygenation.

*Sulfur:* Sulfur starvation enhances the rate of hydrogen production in several cyanobacterial species (*Gloeocapsa alpicola* and *Synechocystis* PCC 6803). It is possible to inhibit the oxygenic photosynthesis and enhance hydrogen production by incubating the cynobacteria in nutrients that lack sulfur. Sulfur is a very important component in the photosystem II repair cycle and without sulfur the protein biosynthesis is heavily impaired and production of either cysteine or methionine becomes impossible. This results in lack of the D1 protein (32-kDa reaction center protein), essential for photosystem II and needs to be constantly replaced. For these reasons during sulfur deprivation, photosynthesis and respiration is decreased, even in the presence of light. Since photosynthesis decline much quicker than respiration, thus an equilibrium point is reached after a while, (usually after 22 hours) and after that the amount of oxygen that is used in respiration is greater than the oxygen released by photosynthesis and the cell becomes anaerobic and at this point hydrogen production occurs in higher amounts reaching peak production.

*Methane:* Increased hydrogen production (up to four times) is observed in *Gloeocapsa alpicola* and *Synechocystis* PCC 6803 during dark anoxic incubation when methane is present and the medium pH is between 5.0–5.5. The effect of methane on the hydrogen evolution is maximum during the first hour of the incubation followed by gradual decline.

### *3.1.1.3.2. Intrinsic factors*

There are several intrinsic factors such as genetic components or sensitive proteins in cyanobacteria that may affect hydrogen production.

*Presence of uptake hydrogenase and decreased hydrogen yield:* The net hydrogen yield is affected in strains containing uptake hydrogenase. Much of the produced hydrogen is lost due to the activity of the uptake hydrogenase. Knocking out the genes coding uptake hydrogenase is therefore contemplated to result in higher hydrogen production in the species of cyanobacteria that harbor uptake hydrogenase. It is, however, also critical to over express the genes for bidirectional hydrogenase and this may be achieved by transfecting cyanobacteria with plasmids containing particular genes.

*Sensitivity of hydrogenase and nitrogenase to molecular oxygen:* The molecular oxygen acts as an inhibitor for hydrogenase and nitrogenase. However, innovative technical interdisciplinary solutions, as described below, are now available to reduce or eliminate presence of molecular oxygen and increase yield of hydrogen.

a) With the advances in nanotechnology it has become possible to build semi-permeable membranes around the organisms. An example is a membrane that discriminates by size with an active transport system. Nanotechnology has made possible the creation of cyanobacteria with a membrane that incorporates an active transport protein system for the facilitated ejection of oxygen. This system would rapidly expel the oxygen that is created during metabolism and not allow its re-entry. Another membrane is the one that possesses both an oxygen-philic and oxygen-phobic sides. Having the oxygen-philic side facing the bacteria and the phobic side open to the environment would facilitate the movement of oxygen away from the cells allowing the production of hydrogen to continue without being hindered by the generated oxygen.

b) Problem to oxygen sensitivity can also be addressed by using sulfur stress phenomenon to down-regulate photosynthesis as described before. This creates the anaerobic conditions required for hydrogen production. This problem may also be addressed by engineering the native hydrogenase. Engineering oxygen-tolerant hydrogenase genes, for example, *hydS* and *hydL* from *Thiocapsa roseopersicina* into sensitive organisms may help reducing the oxygen sensitivity. An expression vector pEX-Tran used for *Synechococcus* sp. PCC7942 transformation is readily available and with minimal modification should be suitable for other cyanobacterial systems as well.

*Heterocysts:* Heterocystsous cyanobacteria are more efficient to produce hydrogen than cyanobacteria with vegetative cells. These types of cyanobacteria do engage in simultaneous oxygen and hydrogen production coupled with $CO_2$ fixation. Problems associated with these types of cyanobacteria are, however, high-energy requirement and separation of hydrogen and oxygen. The over expression of the genes responsible for changing vegetative state into heterocyst cell type is *hetR* (coding for the HetR protein involved in heterocyst frequency regulation, for example, in *Anabaena sp.* PCC 7120). Recombinant strains can be created with HetR protein harboring vectors that may allow variable and controllable heterocyst frequency.

### *3.1.1.4. Bioreactors for cyanobacterial hydrogen production*

Bioreactors are essential for large-scale production of hydrogen. Since light is an essential parameter for cyanobacterial growth so all such bioreactors must be transparent and hence are called photobioreactors. All photobioreactors require adequate entry of light, which usually is sunlight but in some photobioreactors other artificial sources of light is also used for providing controlled light. Inside photobioreactor there should be a photic zone, close to the illuminated surface and a dark zone, further away from this surface. The dark zone is due to light absorption by the cells and mutual shading. The hydrogen productivity of a photobioreactor is light limited and tends to decrease at higher light intensities (photosynthesis diverts the hydrogen production pathway) hence the light regime is determined by the light gradient (must be diluted and distributed as much as possible; absolute dark condition responsible for highest production). Liquid circulation time or aeration (hydrogen producing enzymes are oxygen susceptible; anaerobic condition or inert gas environment is preferred) rate has something to do with hydrogen productivity. It has followed that cyanobacteria absorb preferentially red light around 680 nm. To fulfill this demand, red light panels are constructed in specialized bioreactors to provide red light to the culture systems. As a result of mixing, cells will circulate between the light and the dark zone of the reactor at a certain frequency and regular intervals, which is dependent on reactor design and gas input. The position of the light source as well as gas liquid hydrodynamics also affects cyanobacterial growth as well as hydrogen production.

Several types of bioreactors have been used for hydrogen production. These can be mainly divided into three types of photobioreactors (PBRs): vertical column reactor, tubular type and flat panel photobioreactor. A reactor for photobiological hydrogen production must meet several conditions:

1) Photobioreactor should be an enclosed system so that the produced hydrogen may be collected without any loss.
2) The reactor design must allow sterilization with convenience and ease.
3) To maximize the area of incident light (thus allowing high growth and hydrogen production) photobioreactor design should provide high surface to volume ratio.

*3.1.1.4.1. Vertical column reactor (air-lift loop reactor and bubble column)*

Such PBRs consists of a transparent column usually made up of high quality glass and surrounded by a water jacket that while allowing maintenance of the temperature with circulating water allows adequate entry of light. Reactor top has provision for medium inlet and outlets for the gases such as argon and for the hydrogen. Fresh medium is added from a reservoir from above the PBR. Microorganisms are inoculated through a septum that helps maintenance of sterility and prevents contamination. Bottom part of the PBR column retains outlets for the culture and an inlet/outlet for argon gas. In bubble columns using sunlight as light source, the presence of gas bubbles enhances internal irradiance at sunset and sunrise. As the position of sun changes from low horizon to overhead at noon, the bubbles diminish the internal column irradiance relative to the ungassed state. The biomass productivity varies substantially, during the year, the peak productivity during summer may be several times greater than in the winter. An example of this type of PBR is the one used for hydrogen production using *Rhodobactor sp*. This reactor column was made up of a glass cylinder with an inner volume of 400 ml surrounded by a water jacket. The optimal dimensions of vertical column were about 0.2 m in diameter and 4 m in column height. The optimal column height depends on factors such as wind speed and strength of optically transparent materials like glass or thermoplastics.

*3.1.1.4.2. Flat panel Photobioreactor*

A typical flat-panel PBR consists of a stainless-steel frame and three polycarbonate panels. The reactor comprises of two compartments placed side by side. The front compartment contains the bacterial culture. Water is circulated via a temperature controlled water bath through the hind compartment in order to maintain the desired temperature of the culture. This design of PBR usually utilizes artificial light, the tungsten-halogen lamps (usually 500 W) are placed on one side of the reactor as light source. The average light intensity provided at the reactor surface is 175 W/m$^2$. Alternately red light emitting diode (LED peaking at 665 nm) is used as the light source on one side. A membrane gas pump circulates the gas through the spargers (hypodermic needles) at the bottom of the reactor. The produced gas is collected in a gasbag. In this reactor system, pressure vessels prevent pressure fluctuations in the gas recirculation system and a pressure valve maintains a constant input pressure to the mass flow controller. A condenser prevents water vapour from entering the gas recirculation system. The reactor is autoclaved prior to cyanobacterial cultivation and hydrogen production. The culture medium is autoclaved separately and fed to the reactor. Sampling is done through the sample port, attached to the outflow tube. Bacterial growth is monitored on-line. On the right-hand side of the reactor a small tube is attached to the reactor through which bacterial suspension flows due to an airlift effect.

*3.1.1.4.3. Tubular photobioreactors*

Tubular PBRs consists of long transparent tubes with diameters ranging from 3 to 6 cm, and lengths ranging from 10 to 100 m. The culture liquid is pumped through these tubes by means of mechanical or airlift pumps. The tubes can be positioned in many different ways: in a horizontal plane as straight tubes with a small or large number of U-bends; vertical, coiled as a cylinder or a cone; in a vertical plane, positioned in a fence-like structure using U-bends or connected by manifolds; horizontal or inclined, parallel tubes connected by manifolds. The predominant effect of the specific designs on the light regime is a difference in the photon flux density incident on the reactor surface. 'Tredici' PBR is special type of tubular PBR with internal gas exchange and consists of thin flexible tubular plastic sleeves filled with water and ganged together with top and bottom distribution pipes. The tubes are positioned in a corrugated plastic roofing sheet, which keeps them straight and even. The tubes can be of considerable length, with an optimal length in between 20 m and 50 m, depending on factor being optimized. The system is usually inclined at a slope allowing free rise of gas bubbles and a footer with compressed air line allows supply of air at the bottom of the reactor into selected tubes. The header serves as degasser to allow for containment of the fluid displaced during aeration. Every third or fourth tube is not gased, which serves as a fluid return tube and provides an efficient airlift type of recirculation. In this PBR, cooling is achieved by water spray. The mass transfer characteristics of the tubular photobioreactor vary with the shape of the reactor and type of mixing used. A summary of bioreactor types and their properties have been provided in Table 4.

**Table 4** Summary of photobioreactor properties

| PBR type | Cyanobacterial species used | Advantage | Disadvantage |
|---|---|---|---|
| Vertical Column | *Spirulina platensis* | 1. Simple and cost effective design<br>2. Greater rate of mass transfer in bubble columns | 1. Lack of control on irradiant light.<br>2. Wide fluctuations in productivity. |
| Flat Pannel | *Spirulina platensis* | 1. Greater control of incident light.<br>2. Effective control of gas pressure. | 1. Cost for production is high.<br>2. Complicated design and more maintenance |
| Tubular | *Arthrospira platensis, Anabaena variabilis PK84, Anabaena variabilis ATCC 29413, Anabaena variabilis PK84* | 1. Flexibility in volume to surface area ratio.<br>2. Flexibility in shifting the place receiving light.<br>3. Give higher biomass with internal static mixture. | 1. While provides flexibility to irradiant light but creates pockets of poor mass transfer<br>2. Mixing time is longer in internal static mixture TPBR |

*3.1.1.5. Future prospects for cyanobacterial hydrogen production*

Cyanobacterial hydrogen production is poised to be a very useful commodity, provided various effective utilization of the produced hydrogen is devised. There are various applications where the process of biological hydrogen production by cyanobacteria can be well utilized. The examples can be included from food and chemical industries, which employ the process of hydrogenation to produce derivatives that are used as food additives, commodities, and fine chemicals:

(a) *Hydrogenation of cheap carbohydrates into high valued derivatives:* High value-added $C_5$ and $C_4$ polyols can be obtained from cheap $C_6$ carbohydrates by oxidative decarboxylation followed by hydrogenation. These polyols blends are useful for the manufacture of polyesters or alkyl resins employed in the manufacture of paints. Sorbitol is another important polyol produced from hydrogenation of glucose, which is used industrially in a variety of physical and chemical processes, (for example, as humectants and softener in various food products, drugs and cosmetics). Derivatives are also used in protecting coatings, plasticizers, emulsifiers and detergents

(b) *Hydrogenation of fatty acids:* Hydrogenation of fatty acids is used to manufacture margarines and shortening oils. The hydrogenation of oils converts liquid oils into hard fats by adding hydrogen to the fat molecule. Oils can be hydrogenated to varying degrees, depending on the hardness. These fats are desirable for its melting point, allowing for high temperature cooking and frying. Hydrogenation involves the artificial saturation of unsaturated bond(s) present in fatty acids. In the process fatty acids are put under pressure, using hydrogen gas at temperatures of 120–210°C in the presence of a metal catalyst (nickel, platinum, or copper) for six to eight hours.

All these processes requires the rigorous hydrogenation. The hydrogen produced by cyanobacteria is very economical in comparison to the traditional large-scale hydrogen production. Cyanobacterial hydrogen produced in a photobioreactor can easily be directed to separate compartments containing the substrate for hydrogenation and specific catalysts. Provisions of external heating as well as purifying the products from different contaminants (catalyst and reactants) may be easily provided as well. Hydrogen generated in near vicinity in PBRs will reduce requirement for transport and hazards associated with transport. Apart from hydrogenation, hydrogen is combustible so it can be well applied as a substitute for conventional fuel or may be used in fuel cells to generate electricity.

#### *3.1.2.Photolysis by green algae*

Alike higher plants, the green algae also have two photosystems. There are two possible mechanisms for the generation of electron donors for hydrogen photoproduction. In the first type, hydrogen generation is linked with the photochemical dependent metabolism of endogenous stored compounds and the second one is hydrogen generation through photolysis of water. EMP, Kreb's cycle and pyruvate dehdrogenase reactions have been suggested as sources of reductants.

The ability of unicellular green algae to produce hydrogen gas upon illumination has been mostly, a biological curiosity. Historically, hydrogen evolution activity in green algae was induced upon a prior anaerobic incubation of the cells in the dark. A hydrogenase enzyme was expressed under such incubation and catalyzed, with high specific activity, a light-mediated hydrogen evolution. The monomeric form of the enzyme, reported to belong to the class of Fe hydrogenases is encoded in the nucleus of the unicellular green algae. However, the mature protein is localized and functions in the chloroplast stroma. Light absorption by the photosynthetic apparatus is essential for the generation of hydrogen gas because light energy facilitates the oxidation of water molecules, the release of electrons and protons, and the endergonic transport of these electrons to ferredoxin. The photosynthetic ferredoxin serves as the physiological electron donor to the Fe-hydrogenase and, thus, links the Fe hydrogenase to the electron transport chain in the chloroplast of the green algae.

Under these conditions, the activity of the hydrogenase is only transient (it lasts from several seconds to a few minutes) because, in addition to electrons and protons, the light-dependent oxidation of water entails the release of molecular $O_2$. Oxygen is a powerful inhibitor of the Fe hydrogenase. Current technological developments in this field have notyet succeeded in overcoming this mutually exclusive nature of the $O_2$ and $H_2$ photoproduction reactions. Thus, the physiological significance and role of the Fe hydrogenase in green algae, which normally grow under aerobic photosynthetic conditions, has long been a mystery. Given the $O_2$ sensitivity of the Fe hydrogenase and the prevailing oxidative environmental conditions on earth, questions have been asked as to whether the hydrogenase is anything more than a relic of the evolutionary past of the chloroplast in green algae, and whether this enzyme and the process of photosynthesis can ever be utilized to generate $H_2$ gas for commercial purposes. Nevertheless, the ability of green algae to photosynthetically generate $H_2$ gas has captivated the fascination and interest of the scientific community because of the fundamental and practical importance of the process. Below is an itemized list of the properties and promise of photosynthesis in green algal $H_2$ production, and the problems that are encountered with current technology:

(a) Photosynthesis in green algae can operate with a photon conversion efficiency of 80%.

(b) Micro-algae can produce $H_2$ photosynthetically, with a photon conversion efficiency of 80%.

(c) Molecular $O_2$ acts as a powerful and effective switch by which the $H_2$ production activity is turned off.

Aside from the above described photosystem II (PSII)-dependent $H_2$ photo-evolution, which involves water as a source of electrons and produces 2:1 stoichiometric amounts of $H_2$:$O_2$, an alternative mechanism has been described in the literature. Upon a dark anaerobic incubation of the algae and the ensuing induction of the hydrogenase, electrons for the photosynthetic apparatus are derived upon catabolism of endogenous substrate and the attendant oxidative carbon metabolism in the green algae. Electrons from such endogenous substrate catabolism feed into the photosynthetic electron transport chain between the two photosystems, and probably at the level of the plastoquinone pool. Light absorption by PSI and the ensuing electron transport elevates the redox potential of these electrons to the redox equivalent of ferredoxin and the hydrogenase, thus permitting the generation of molecular $H_2$ In the presence of the PSII inhibitor 3-(3, 4-dichlorophenyl)-1,1-dimethylurea (DCMU), this process generates 2:1 stoichiometric amounts of $H_2$:$CO_2$. Thus, following a sufficiently long dark anaerobic incubation of the culture, initially high rates of $H_2$ production can be detected upon illumination of the algae in the presence of DCMU, a PSII inhibitor.

#### *3.1.2.1. Two-stage photosynthesis and $H_2$ production in green algae*

Recent work has shown that lack of sulfur in the growth medium of *Chlamydomonas reinhardtii* causes a specific but reversible decline in the rate of oxygenic photosynthesis but does not affect the rate of mitochondrial respiration. In sealed cultures, imbalance in the photosynthesis-respiration relationship by S deprivation resulted in net consumption of oxygen by the cells causing anaerobiosis in the growth medium, a condition that automatically elicited $H_2$ production by the algae. In the course of this recent work, it was shown that expression of the Fe hydrogenase can be induced in the light, so long as anaerobiosis is maintained within the culture. Under such conditions, it was possible to photoproduce and to accumulate significant volumes of $H_2$ gas, using the green alga *C. reinhardtii,* in a sustainable process that could be employed continuously for several days. Thus, progress was achieved by circumventing the sensitivity of the Fe hydrogenase to $O_2$ through a temporal separation of the reactions of $O_2$ and $H_2$ photoproduction, i.e. by the so-called "two-stage photosynthesis and $H_2$ production" process. The novel application of this two-stage protocol revealed the occurrence of hitherto unknown metabolic, regulatory, and electron transport pathways in the green alga *C. reinhardtii* leading to the significant and sustainable light-dependent release of $H_2$ gas by the cells. The temporal sequence of events in this two-stage photosynthesis and $H_2$-production process is given below:

(a) Green algae are grown photosynthetically in the light (normal photosynthesis) until they reach a density of 3 to 6 million cells/ml in the culture.

(b) Sulfur deprivation is imposed upon the cells in the growth medium, either by carefully limiting sulfur supply in the medium so that it is consumed entirely, or by permitting cells to concentrate in the growth chamber prior to medium replacement with one that lacks sulfur nutrients. Cells respond to this S deprivation by fundamentally altering photosynthesis and cellular metabolism to survive.

(c) S deprivation exerts a distinctly different effect on the cellular activities of photosynthesis and respiration. The activity of oxygenic photosynthesis declines quasi-exponentially with a half-time of 15 to 20 h to a value less than 10% of its original rate. However, the capacity for cellular respiration remains fairly constant over the S deprivation period. As a consequence, the absolute activity of photosynthesis crosses below the level of respiration after about 24 h of S deprivation. Following this cross point between photosynthesis and respiration, sealed cultures of S-deprived *C. reinhardtii* quickly consume all dissolved oxygen and become anaerobic even though they are maintained under continuous illumination.

(d) Under S deprivation conditions, sealed (anaerobic) cultures of *C. reinhardtii* produce $H_2$ gas in the light but not in the dark. The volume and rate of photosynthetic $H_2$ production was monitored from the accumulating $H_2$ gas in an inverted burette, measured from the volume of water displacement. A rate of 2.0 to 2.5 ml, $H_2$ production / culture / h was sustained in the 24 to 70h period. The rate gradually declined thereafter.

(e) In the course of such $H_2$ gas production, cells consumed significant amounts of internal starch and protein. Such catabolic reactions apparently sustain, directly or indirectly, the $H_2$ production process.

(f) Profile analysis of selected photosynthetic proteins showed a precipitous decline in the amount of Rubisco (ribulose bisphosphate carboxylase / oxygenase) as a function of time in S deprivation, a more gradual decline in the level of PSII and PSI proteins, and change in the composition of the light-harvesting complex.

(g) Microscopic observations showed distinct morphological changes in *C. reinhardtii* during S deprivation and $H_2$ production. Ellipsoid-shaped cells (normal photosynthesis) gave way to larger and spherical cell shapes in the initial (0-24 h) stages of S deprivation and $H_2$ production, followed by cell mass reductions at longer (24-120 h) S deprivation and $H_2$ production times.

Upon exposure to hydrogen the anaerobically adapted green algae cells take up the molecular $H_2$ and reduce $CO_2$ in the dark. The reverse reaction, e.g. hydrogen production in the light. High rates of $H_2$ evolution could be measured in the light for short periods of time (from several seconds to a few minutes). Electrons were generated either upon the photochemical oxidation of water by PSII, which results in the simultaneous production of $O_2$ and $H_2$, or upon the oxidation of endogenous substrate (Figure 8), feeding electrons into the thylakoid membrane with the simultaneous release of $CO_2$ to the medium. It is known that *C. reinhardtii* can photoproduce hydrogen when PSII is blocked by DCMU, but no $H_2$ evolution occurs after an addition of 2,5-dibromo-3-methyl-6-isopropyl-p-benzoquinone, which blocks the function of the cytochrome b-f complex. Under anaerobic conditions in the presence of DCMU, accumulated reducing equivalents from the fermentative catabolism of the algae cannot be oxidized via respiration because the terminal electron acceptor $O_2$ is absent. An NAD(P)H reductase protein complex that feeds electrons into the plastoquinone pool recently has been identified in many vascular plant chloroplasts but so far only from the green alga *Nephroselmis olivacea*. Nevertheless, inhibitor experiments have yielded evidence in support of a thylakoid membrane-localized NAD(P)H reductase in *C. reinhardtii*, suggesting that electrons derived upon the oxidation of endogenous substrate may feed into the plastoquinone pool (Figure 8). Thereafter, electrons are driven upon light absorption by PSI to ferredoxin. The latter is an efficient electron donor to the Fe hydrogenase, which efficiently combines these electrons with protons to generate molecular $H_2$.

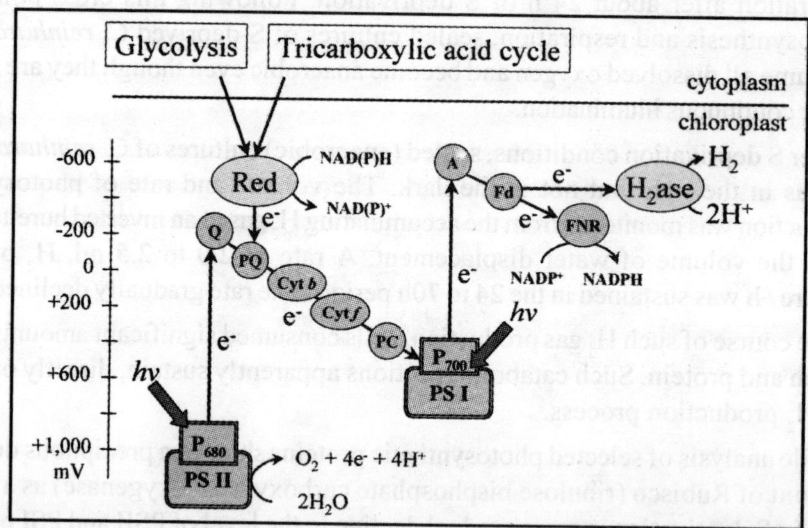


**Figure 8** Hydrogenase-related electron transport pathways in green algae. Electrons may originate either at PSII upon photooxidation of water, or at the plastoquinone pool upon oxidation of cellular endogenous substrate (e.g. via glycolysis and the tricarboxylic acid cycle). Electrons in the electron transport chain are transported via PSI to ferredoxin, which serves as the physiological electron donor to the Fe hydrogenase. $P_{680}$, Reaction center of PSII; $P_{700}$, reaction center of PSI; Q, primary electron acceptor of PS II; A, primary electron acceptor of PSI; PQ, plastoquinone; Cyt, cytochrome; PC, plastocyanin; Fd, ferredoxin; Red, NAD(P)H oxido-reductase; $H_2$ase, hydrogenase; FNR, ferredoxin-$NADP^+$ reductase

The physiology of $H_2$ production upon sulphur deprivation has many similarities and some distinct differences from the process described above. Sulfur-deprived and sealed cultures of *C. reinhardtii* become anaerobic in the light due to a significant and specific slowdown in the activity of the $O_2$-evolving PSII, which is followed by automatic induction of the Fe hydrogenase and by photosynthetic $H_2$ production. Biochemical analyses revealed that, concomitant with the $H_2$ production process, starch and protein content of the cells gradually declined . Such catabolic pathway(s) could be generating reductant that feeds electrons into the thylakoid membrane, perhaps via a chloroplast NAD(P)H-dependent process. More important, starch catabolism must also generate substrate for the cell's mitochondrial respiration. Mitochondrial respiration scavenges the small amounts of $O_2$ that evolve due to the residual activity of photosynthesis and thus ensures the maintenance of anaerobiosis in the culture (Figure9). Thus, the physiology of $H_2$ production by S deprivation involves a coordinated interaction between:

a) *Oxygenic photosynthesis:* The electrons are transported through the photosynthetic electron transport chain and eventually feed into the Fe hydrogenase, thereby contributing to $H_2$ production.

b) *Mitochondrial respiration:* This scavenges all oxygen generated by the residual photosynthesis and, thus, maintains anaerobiosis in the culture.

c) *Endogenous substrate catabolism:* This yields substrate suitable for the operation of oxidative phosphorylation in mitochondria, and possibly for an NAD(P)H-dependent electron transport in the chloroplast, both of which contribute to the generation of much-needed ATP.

d) *Electron transport via the hydrogenase pathway:* The release of $H_2$ gas by the algae sustains a baseline level of photosynthesis and, therefore, of respiratory electron transport for the generation of ATP and thus ensures the survival of the organism under protracted stress conditions.

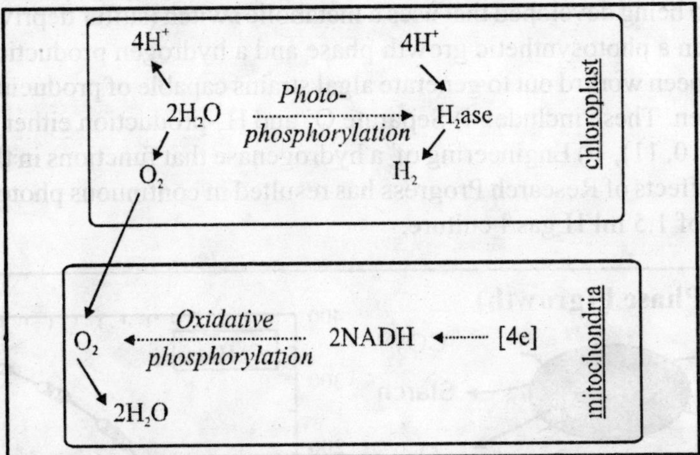


**Figure 9** Coordinated photosynthetic and respiratory electron transport and coupled phosphorylation during $H_2$ production. Photosynthetic electron transport delivers electrons upon photo-oxidation of water to the hydrogenase, leading to photophosphorylation and $H_2$ production. The oxygen generated by this process serves to drive the coordinate oxidative phosphorylation during mitochondrial respiration. Electrons for the latter ([4e]) are derived upon endogenous substrate catabolism, which yields reductant and $CO_2$. Release of molecular $H_2$ by the chloroplast enables the sustained operation of this coordinated photosynthesis-respiration function in green algae and permits the continuous generation of ATP by the two bioenergetic organelles in the cell

It is clear that more research is needed to dissect the four-way interplay and the intricate relationships between the processes of oxygenic photosynthesis, mitochondrial respiration, catabolism of endogenous substrate, and electron transport via the hydrogenase pathway leading to $H_2$ production. Nevertheless, the discovery of sustainable $H_2$ production that bypasses the sensitivity of the reversible hydrogenase to $O_2$ is a significant development in the field. It may lead to exploitation of green algae for the production of $H_2$ gas as a clean and renewable fuel. However, the actual rate of $H_2$ gas accumulation was at best 15% to 20% of the photosynthetic capacity of the cells, when the latter is based on the capacity for $O_2$ evolution under physiological conditions. The relatively slow rate of $H_2$ production suggests that there is room for significant improvement in the yield of the process, by as much as one order of magnitude. Similarly, other improvements must be made to optimize the process under conditions of mass culture of the algae. For example, optical problems associated with the size of the chlorophyll antenna and the light saturation curve of photosynthesis must be addressed before green algae can achieve high photosynthetic solar conversion efficiencies in mass culture. Moreover, the continuity of the process needs to be addressed because $H_2$ production by S deprivation of the algae cannot last forever. The yield begins to level off after about 70 h of S deprivation. After about 100 h of S deprivation, the algae need to go back to normal photosynthesis to be rejuvenated by replenishing endogenous substrate.

Photobiological technology holds great promise, but because oxygen is produced along with the hydrogen, the technology must overcome the limitation of oxygen sensitivity of the hydrogen-evolving enzyme systems. Researchers are addressing this issue by screening for naturally occurring organisms that are more tolerant to oxygen, and by creating new genetic forms of the organisms that can sustain hydrogen production in the presence of oxygen. A new system is also being developed that uses a metabolic switch (sulfur deprivation) to cycle algal cells between a photosynthetic growth phase and a hydrogen production phase. Two approaches have been workrd out to generate algal strains capable of producing hydrogen in presence of oxygen. These include: i) Separate $O_2$ and $H_2$-production either temporally or physically (Figs. 10, 11), ii) Engineering of a hydrogenase that functions in the presence of $O_2$ (Figure 12). Effects of Research Progress has resulted in continuous photoproduction of pure $H_2$ at a rate of 1.5 ml $H_2$gas/l culture.

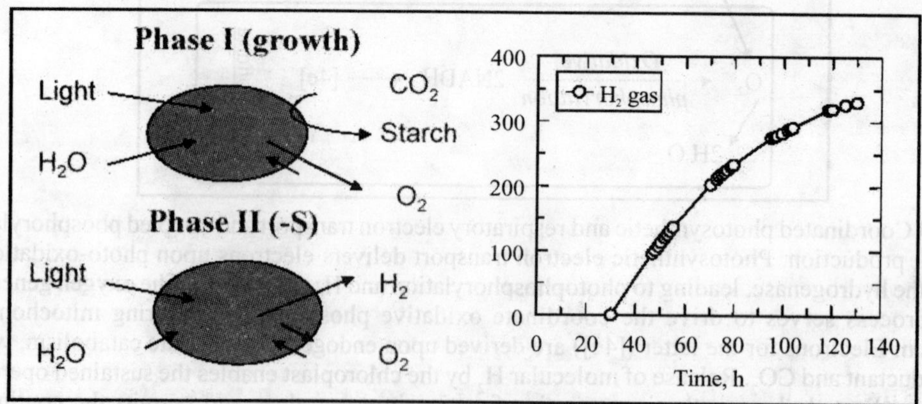


**Figure 10** Temporal separation of $O_2$ and $H_2$ production (batch system)

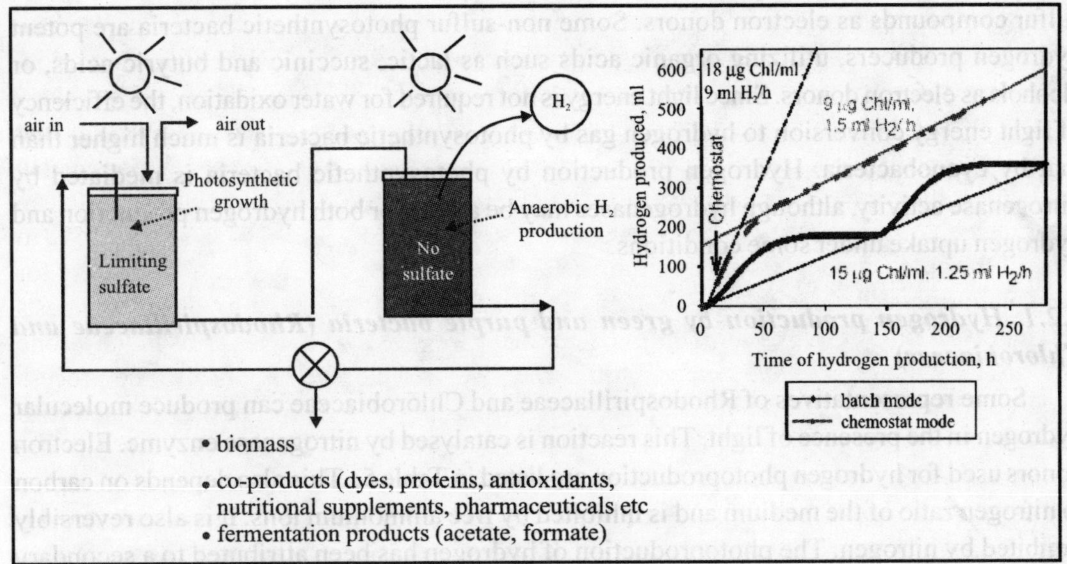


**Figure 11** Physical separation of $O_2$ and $H_2$ production (continuous system)

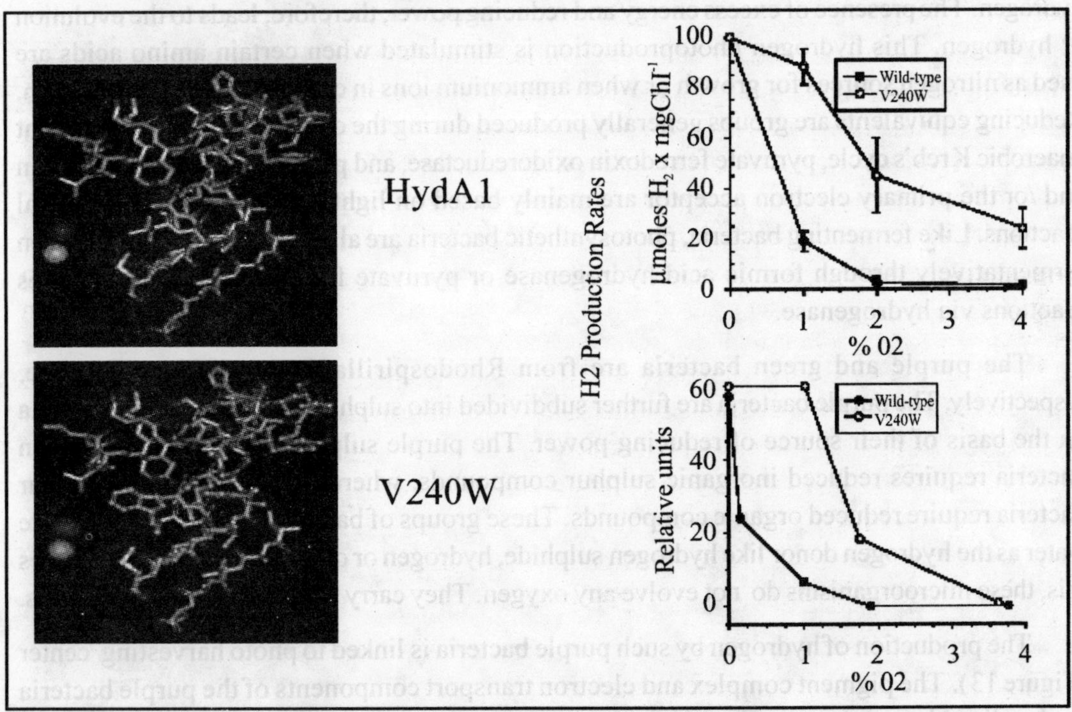


**Figure 12** Engineering the Hydrogenase for $O_2$-Tolerance: Cloning of two algal hydrogenases for generation of a $H_2$-channel mutant *(For a coloured version of this figure, see plate section, page 558)*

## 3.2. Photoreduction of organic compounds

This involves light dependent decomposition of organic compounds by photosynthetic bacteria where they undergo anaerobic photosynthesis with organic compounds or reduced sulfur compounds as electron donors. Some non-sulfur photosynthetic bacteria are potent hydrogen producers, utilizing organic acids such as lactic, succinic and butyric acids, or alcohols as electron donors. Since light energy is not required for water oxidation, the efficiency of light energy conversion to hydrogen gas by photosynthetic bacteria is much higher than that by cyanobacteria. Hydrogen production by photosynthetic bacteria is mediated by nitrogenase activity, although hydrogenases may be active for both hydrogen production and hydrogen uptake under some conditions.

### *3.2.1. Hydrogen production by green and purple bacteria (Rhodospirillaceae and Chlorobiaceae)*

Some representatives of Rhodospirillaceae and Chlorobiaceae can produce molecular hydrogen in the presence of light. This reaction is catalysed by nitrogenase enzyme. Electron donors used for hydrogen photoproduction are listed in Table 5. This also depends on carbon to nitrogen ratio of the medium and is inhibited by free ammonium ions. It is also reversibly inhibited by nitrogen. The photoproduction of hydrogen has been attributed to a secondary function of nitrogenase i.e., the ability to reduce proton as well as nitrogen and thus liberate hydrogen. The presence of excess energy and reducing power, therefore, leads to the evolution of hydrogen. This hydrogen photoproduction is stimulated when certain amino acids are used as nitrogen sources for growth or when ammonium ions in culture media limits growth. Reducing equivalents are groups generally produced during the operation of light dependent anaerobic Kreb's cycle, pyruvate ferredoxin oxidoreductase, and photoreduction of ferredoxin and /or the primary electron acceptor are mainly based on light dependent photochemical reactions. Like fermenting bacteria, photosynthetic bacteria are also able to produce hydrogen fermentatively through formic acid hydrogenase or pyruvate ferredoxin oxidoreductases reactions via hydrogenase.

The purple and green bacteria are from Rhodospirillaceae and Chlorobiaceae, respectively. The purple bacteria are further subdivided into sulphur and non- sulphur bacteria on the basis of their source of reducing power. The purple sulphur bacteria and all green bacteria requires reduced inorganic sulphur compounds, whereas the purple non sulphur bacteria require reduced organic compounds. These groups of bacteria are not able to utilize water as the hydrogen donor like hydrogen sulphide, hydrogen or organic compounds. Besides this, these microorganisms do not evolve any oxygen. They carry anoxygenic photosynthesis.

The production of hydrogen by such purple bacteria is linked to photo harvesting center (Figure 13). The pigment complex and electron transport components of the purple bacteria are localized in membranes. The pigment complex of the photochemical reaction center can be separated from antenna pigments. The energy absorbed by the antennae pigments is channeled to a reaction center. Isolated reaction centers consist of a protein complex containing bacterial chlorophyll a, bacteriophaeophytin, caroteniod, ubiquinone, and iron sulphur proteins. The reaction center pigment is designated as p870 according to the wavelength of the maximal absorption which decreases on illumination.

**Table 5** Various type of electron donors for biological monitoring of hydrogen production

| **Microorganism** | **Electron donor** |
|---|---|
| *Rhodospirillum rubrum* | Acetate, Lactate, Pyruvate, Succinate, Fumarate, Malate, Oxalacetate |
| Purple non sulphur bacteria | |
| *Rhodopseudomonas* | Lactate |
| *acidophila* | Propionate, Lactate, Pyruvate, |
| *R.capsulata* | Succinate, Fumarate, Malate, Butyrate, Glucose, Fructose, Sucrose |
| *R.palustris* | Formate, Pyruvate, α-Ketoglutarate, Succinate, Malate, Oxalacetate, Glucose, Thiosulfate |
| *R.sphaeroides* | Glucose |
| *Chrornatium D* | Thiosulfate, Pyruvate |
| *Purple sulphur bacteria* | |
| *Chromatium* sp. | Malate + CO, Sulphide, Thiosulfate, Succinate, Fumarate, Malate, Acetate, Succinate + Thiosulfate + Trace Metal |
| | Thiosulfate, Acetate, Pyruvate, |
| *Thiocapsa roseopersicina* | Oxalacetate, Lactate, Pyruvate, Citrate, α-Ketoglutarate, Xylose, Mannitol, Glucose |
| | Formate |
| *Chloropseudomonas sp.* | |

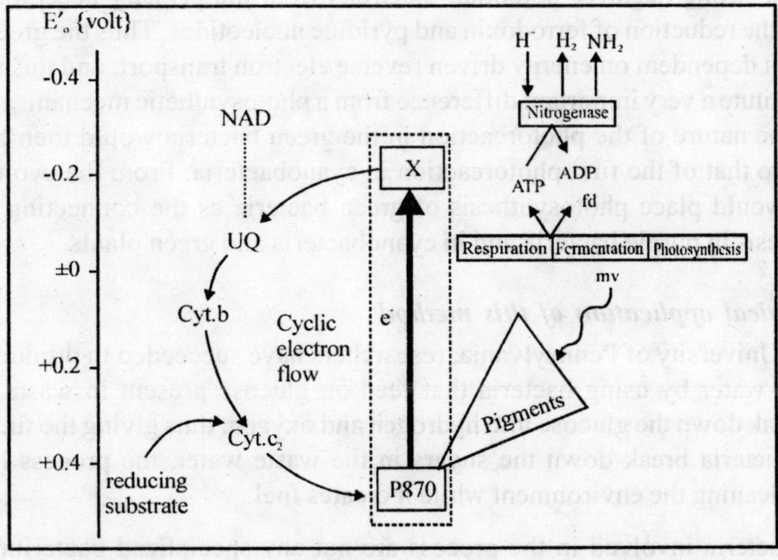


**Figure 13** Photoproduction of hydrogen in Rhodospirillaceae

The redox potential of this electron donor is in the region +4.50 to +4.90 mV. A complex of ubiquinone and Fe-S protein probably function as the primary electron acceptor. This complex seems to have a potential of only –100 mV, and it is therefore improbable that the electrons emitted by the light reaction can reduce NAD. This is an important difference from the otherwise analogous first photoreaction in oxygenic photosynthesis. To fill the electron gap in the cycle, the purple bacteria must rely on external electron donors: hydrogen sulphide, sulphur or thiosulphate , in the case of the purple sulpur bacteria and organic compounds (malate, succinate) or hydrogen in both groups of purple bacteria. The photoreaction of the green bacteria has not yet been fully understood (Figure 14).

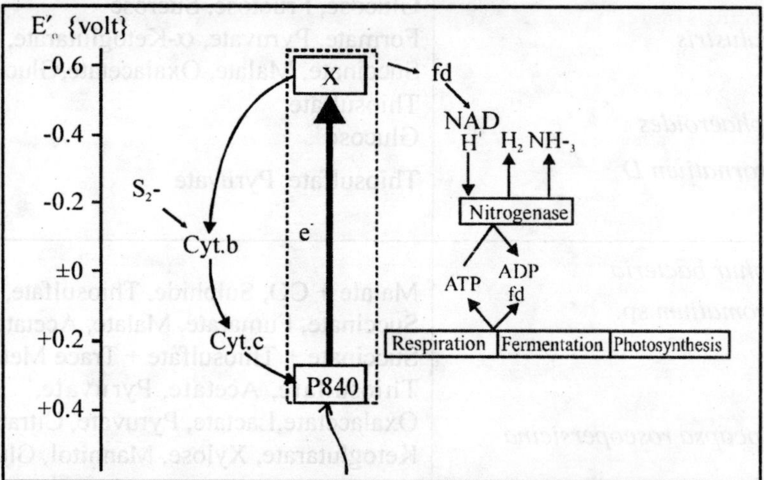


**Figure 14** Photoproduction of hydrogen in Chlorobiaceae

The primary electron acceptor of the light reaction has a potential of about –500 mV. With such a strong negative potential, electrons from the primary acceptor can be used directly for the reduction of ferrodoxin and pyridine nucleotides. Thus the green bacteria are probably not dependent on energy driven reverse electron transport, and this independence would constitute a very important difference from a photosynthetic mechanism of the purple bacteria. The nature of the photoreaction in the green bacteria would then be completely equivalent to that of the first photoreaction in cyanobacteria. From the evolution point of view, this would place photosynthesis of green bacteria as the connecting link between photosynthesis in purple bacteria and in cyanobacteria and green plants.

### *3.2.2. Practical application of this method*

In the University of Pennsylvania, researchers have succeeded in producing hydrogen from waste water by using bacteria that feed on glucose present in waste water. These bacteria break down the glucose into hydrogen and oxygen, thus giving the fuel. In addition, since the bacteria break down the sugars in the waste water, the process has the added benefit of cleaning the environment while it creates fuel.

The bacteria involved in the process are not any specialized bacterium, but can be found in any garden soil. For their experiment, the researchers took an ordinary soil sample from a local tomato plot. They segregated the hydrogen generating bacteria from those that consume hydrogen by heating the soil for 2 hours at a temperature just above water's boiling point. The hydrogen-consuming microbes died off, but bacteria that generate hydrogen survived as they could form heat-resistant spores. The researchers then mixed the tomato-plot dirt in an enclosed reactor with sugar water to represent wastewater from a food-production plant. The concoction generated gas that was about 60 percent hydrogen.

Bruce Logan, the Pennsylvania State University environmental engineering professor who manages the university's hydrogen research has found that similar fermentation experiments done by other research groups probably had hindered hydrogen generation. Those researchers had collected hydrogen from their reactors only intermittently rather than continuously as Logan's group had done. Letting the gas build up seems to suppress hydrogen production, says Logan. Taking it out continuously from a reactor yields 43 percent more hydrogen.

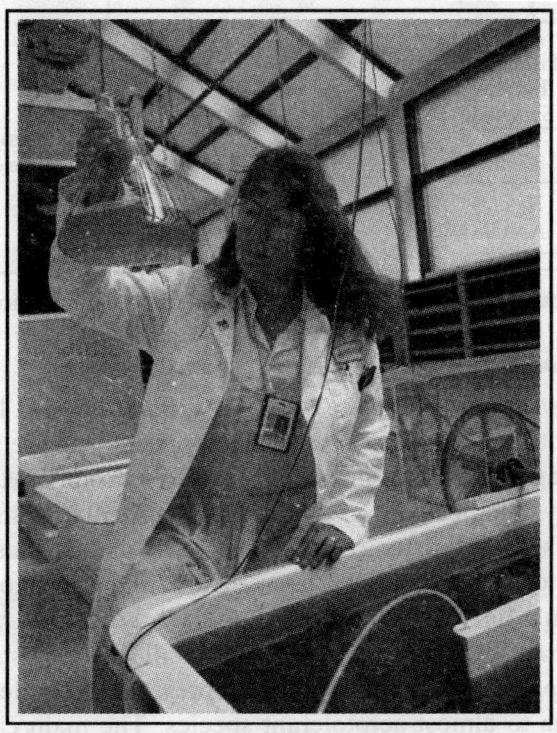

He says, this kind of biological method - which relies on bacteria and sugar-or starch-rich crops - has an advantage over, say, algae-based production, because it doesn't require large ponds for collecting the sunlight that drives the hydrogen-generating chemistry.

## 3.3. Fermentation of organic compounds

Fermentation is an ATP generating metabolic process in which degradation products of organic substrates serve as hydrogen donors as well as hydrogen acceptors. ATP is generated by photophosphorylation of ADP . The oxidized carbon compounds are finally released from the cell as carbon dioxide. The oxidizing steps are dehydrogenations in which the hydrogen is transferred to a cofactor like NAD. Intermediatary products of substrate dehydrogenation then serve as acceptors for the hydrogen from $NADH_2$.

During degradation of biomass (Fig 15), organic substrates or some other biowastes under anaerobic condition, besides the liberation of ATP like energetic compounds, many other products are also produced either singly or in various combinations like propionate, formate, ethanol, butyrate, succinate like value added chemicals and gases like carbon di oxide, hydrogen sulphide , methane or hydrogen. We will be restricted to fermentation where hydrogen gas is also a secondary product.

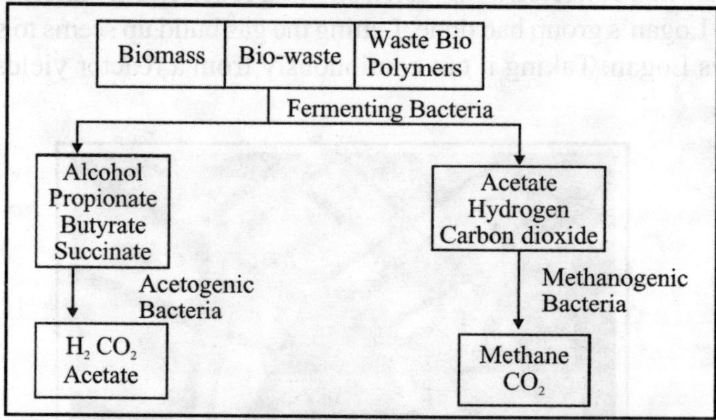


**Figure 15** Hydrogen production from biomass and biowaste

### *3.3.1. Hydrogen production by clostridial fermentation*

The genus *Clostridium* belongs to the family *Bacilliaceae.* They are gram positive bacteria and very motile by virtue of possessing peritrichous flagella. The vegetative cells are rods, but their shape is variable and influenced by environmental factors. The clostridia are characterized physiologically by their intensely fermentative metabolism and by their relation to oxygen. The optimum temperature for growth of most clostridia is in the range of 30-40°C. In addition to these mesophiles, there are also many thermophilic species that have temperature optima between 60–75°C, eg. *Clostridium thermoaceticum, Cl. thermohydrosulfuricum.* In common with other Bacillaceae, clostridia can grow only at neutral and alkaline pH. This group of bacteria represents a complete spectrum from strict anaerobes of bacteria to almost aerotolerant species. The main carbon sources of these bacteria are glucose, starch, lactate, glycerol, pyruvate and acetate and fermentation products are ethanol, acetate, butyrate, carbon dioxide, hydrogen etc. The bacterium *Clostridium kluyeri* metabolises ethanol and acetate to butyrate, acetate and hydrogen (Figure 16). Acetate serves as an additional hydrogen acceptor, though it is also formed during fermentation. During this process ATP is generated by the acetate kinase reaction. The fermentation of glucose by *Clostridium kutricum* and *Cl. acetobutylicum* can also be regarded as prototype clostridial fermentations. Their end products are butyrate, acetate, butanol, ethanol, acetone, 2-propanol, carbon dioxide and hydrogen. Their yield is variable and depends on conditions.

### 3.3.2. Hydrogen production during glucose fermentations

The rumen of cow and other grazing animals contains various types of bacteria, commonly known as rumen bacteria. These bacteria are mainly responsible for breaking down of the cellulose, complex carbohydrates and protein into simple fatty acids (acetate, propionate and butyrate) and gases. Representative bacteria types are shown in Table 6. *Ruminococcus albus* can convert one molecule of glucose to two molecules of hydrogen and carbon dioxide, but only if the concentrations of hydrogen can be kept low. This is possible in a mixed culture with hydrogen utilizing bacteria such as *Vibrio succinogenes* (Figs 17, 18).

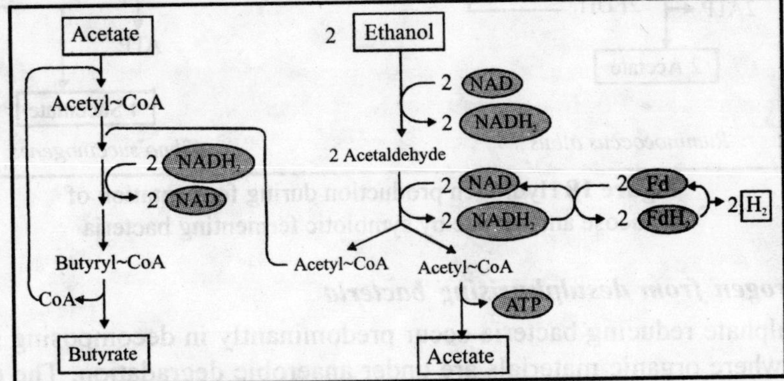


**Figure 16** Hydrogen generation during fermentation of ethanol and acetate by *Clostridium kluyveri*

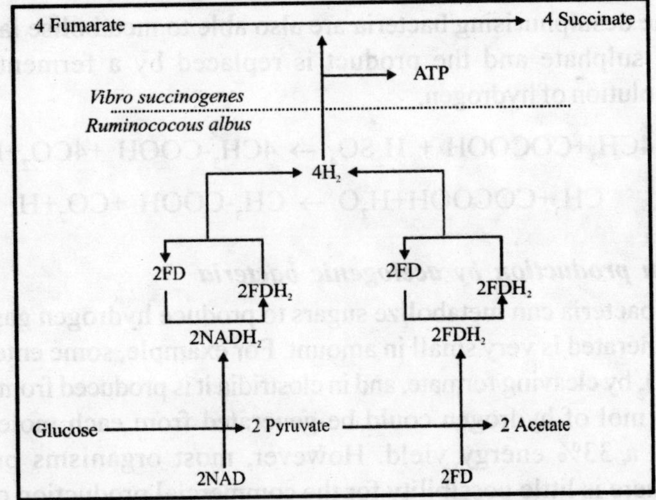


**Figure 17** Hydrogen production during fermentation of glucose

**Table 6** Different types of rumen bacteria

| Cellulose digesters | Starch digesters | Methane producers |
|---|---|---|
| *Butyrivibrio fibrisolvens* | *Selenomonas ruminantium* | ***Methanobacterium ruminantium*** |
| *Ruminococcus albus* | *Succinomonas amydolytica* | |
| *Bacteroides succinogenes* | *Bacteroides ruminicola* | |
| *Ruminococcus flavofaciens* | *Bacteroides amylophilus* | |

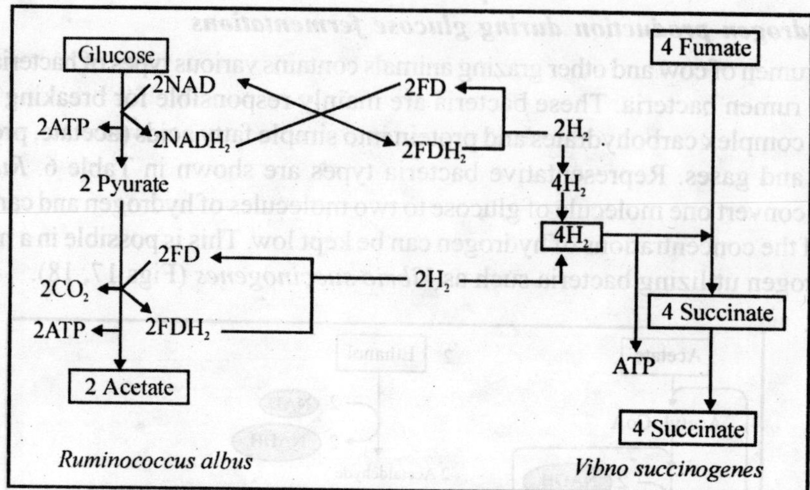


**Figure 18** Hydrogen production during fermentation of glucose and fumate by symbiotic fermenting bacteria

### 3.3.3. *Hydrogen from desulphurising bacteria*

The sulphate reducing bacteria occur predominantly in decomposing sediments and black mud where organic materials are under anaerobic degradation. The desulphurisers seem to be specially adapted to the products of incomplete carbohydrate metabolism i.e. fatty acids, oxyacids, alcohols and hydrogen from fermentations.

Some of the desulphurising bacteria are also able to metabolise lactate or pyruvate in the absence of sulphate and the product is replaced by a fermentative reaction and simultaneous evolution of hydrogen.

$$4CH_3{+}COCOOH + H_2SO_4 \rightarrow 4CH_3\text{-}COOH + 4CO_2 + H_2S$$

$$CH_3{+}COCOOH + H_2O \rightarrow CH_3\text{-}COOH + CO_2 + H_2$$

### 3.3.4. *Hydrogen production by acetogenic bacteria*

Anaerobic bacteria can metabolize sugars to produce hydrogen gas and organic acids, but hydrogen generated is very small in amount. For example, some enterobacteria produce hydrogen and $CO_2$ by cleaving formate, and in clostridia it is produced from reduced ferredoxin. Theoretically, 4 mol of hydrogen could be generated from each mole of glucose, which represents only a 33% energy yield. However, most organisms produce much less. Consequently, there is little possibility for the commercial production of hydrogen via this route in the near future.

$$C_6H_{12}O_6 + 2H_2O \rightarrow 2CH_3COO^- + 2H^+ + 2CO_2 + 4H_2$$

The efficiency of this process is low since anaerobic bacteria are incapable of further breaking down the organic acids formed. So photosynthetic and anaerobic bacteria are used together for the conversion of organic acids to hydrogen. Theoretically, one mole of glucose can be converted to 12 moles of hydrogen (Figure 19) through the use of photosynthetic bacteria capable of capturing light energy in such a combined system. Organic wastes frequently contain sugar or sugar polymers. It is not easy to obtain organic wastes containing organic acids as the main components. The combined use of photosynthetic and anaerobic bacteria potentially increases the likelihood of their application in photobiological hydrogen production.

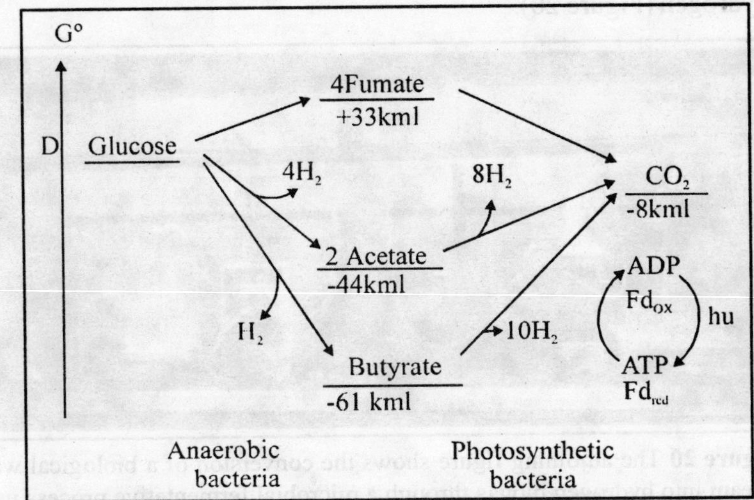


**Figure 19** Free energy changes in hydrogen-producing reactions by anaerobic an photosynthetic bacteria

### *3.3.5. Practical application of this method*

Researchers are developing a biological method of producing hydrogen from waste streams created by dining halls, kitchens, latrines. Anaerobic breakdown of glucose gives four molecules of hydrogen, two molecules of acetate and the process stops at the organic acid, which retains significant bound hydrogen (Figure 18). In order to fully extract the hydrogen from organic molecules, experts have developed a two-stage process a fermentative reactor with a photosynthetic bacterial hydrogen production process to yield the maximum hydrogen from organic carbon sources (Figure 19). The photosynthetic reaction uses the energy from sunlight to completely convert the organic acids to carbon dioxide and hydrogen molecules, releasing essentially all of the available hydrogen.

An up-flow, fixed-bed anaerobic reactor has been developed to begin testing biological hydrogen production. They filled the reactor with approximately 2 kg of diatomaceous earth pellets (naturally occurring substance comprised of the fossilized remains of microscopic hard shell marine creatures). Diatomaceous earth pellets have a high surface-area-to-mass ratio that provides abundant sites for the bacteria to attach and form biofilms. A large, fixed biofilm optimizes hydrogen production and prevents loss of bacteria from the reactor during high flow rates. They introduced a sucrose solution and soil sample, which contained the bacteria, to the earth pellets, then operated the reactor at low flow rates (1 ml per minute) for 2 months to allow the bacteria to colonize the earth pellets. As bacteria degraded the sucrose, the reaction produced carbon dioxide and hydrogen. The fixed-bed reactor consistently produced 50 to 60% hydrogen in the resulting biogas waste stream. In early experiments, the fixed-bed reactor did not properly channel the flow of the sucrose solution through the diatomaceous bed, thus decreasing the contact time the biofilm had with the substrate. To correct this, engineers adopted a fluidized bed reactor which eliminates the channeling problem and allows formation of a uniform biofilm. The fluidized bed reactor consistently produces 60 to 80% hydrogen (Figure 20).

**Figure 20** The adjoining figure shows the conversion of a biological waste stream into hydrogen biogas through a microbial fermentative process using a fluidized bed reactor

## 4. PROBLEMS ASSOCIATED WITH THE USE OF HYDROGEN AS A FUEL

*Storage:* Storage is the main technological problem of a viable hydrogen economy. Hydrocarbons are stored extensively at the point of use, however, hydrogen is quite expensive to store or transport. Hydrogen gas has good energy density per weight, but poor energy density per volume versus hydrocarbons, hence it requires a larger tank to store. A large hydrogen tank will be heavier than the small hydrocarbon tank used to store the same amount of energy, all other factors remaining equal. Increasing gas pressure would improve the energy density per volume, making for smaller, but not lighter container tanks. Compressing a gas will require energy to power the compressor. Higher compression will mean more energy lost to the compression step. Alternatively, higher volumetric energy density liquid hydrogen may be used as in the Space Shuttle. However liquid hydrogen is cryogenic and boils around –253 °C. Hence its liquefaction imposes a large energy loss, used to cool it down to that temperature. The tanks must also be well insulated to prevent boil off. Ice may form around the tank and help corrode it further if the insulation fails. Insulation for liquid hydrogen tanks is usually expensive and delicate.

*Leakage:* There have been some concerns over possible problems related to hydrogen gas leakage. One issue, that may present itself with widespread hydrogen usage, is permanent hydrogen loss. Molecular hydrogen is light enough to escape into space. With a continuous cycle of hydrogen being liberated and then combined with oxygen, some will leak from containment. Another issue is that hydrogen gas ($H_2$) may form water vapours as it reacts with oxygen and cool, or form free radicals (H) due to ultraviolet radiation, in the stratosphere. These free radicals can then act as a catalyst for ozone depletion. A large enough increase in stratospheric hydrogen from leaked $H_2$ could exacerbate the depletion process. However, the issues associated with hydrogen leakage may not really be as much of a problem for various reasons.

There are proposals to use metal hydrides as the carrier for hydrogen instead of pure hydrogen. Hydrides can be coherced, in varying degrees of ease, into releasing and absorbing hydrogen. Some are easy to fuel liquids at ambient temperature and pressure, others are solids which could be turned into pellets. Proposed hydrides for use in a hydrogen economy include boron and lithium hydrides. These have good energy density per volume, although their energy density per weight is often worse than the leading hydrocarbon fuels.

Hydride storage is a leading contender for automotive storage. A hydride tank is about three times larger and four times heavier than a gasoline tank holding the same energy. Often hydrides react by combusting rather violently upon exposure to moist air, and are quite toxic to humans in contact with the skin or eyes, hence cumbersome to handle. This is why such fuels, despite being proposed and vigorously researched by the space launch industry, have never been used in any actual launch vehicle. Certain metallic hydrides such as lanthanum hydride, however, may be somewhat more promising given their unique properties.

An alternative to hydrides is to use regular hydrocarbon fuels as the hydrogen carrier. Then a small hydrogen reformer would extract the hydrogen as needed by the fuel cell. The problem is reformers are slow and given the energy losses involved plus the extra cost of the fuel cell you were probably better off burning it in a cheap internal combustion engine to begin with.

More exotic hydrogen carriers based on nanotechnology have been proposed, such as carbon buckyballs and nanotubes, but these are still in the early research stage.

*Transmission:* By far the cheapest way to move energy around the planet is in the form of oil in a pipeline or supertanker, or coal on a barge or rail car. Natural gas pipelines and LPG tankers are much more expensive, in comparison. Hydrogen pipelines are unfortunately more expensive. Hydrogen is about three times bulkier than natural gas for the same energy delivered, and hydrogen accelerates the cracking of steel (hydrogen embrittlement), which increases maintenance costs, leakage rates, and material costs.

*End use:* The underlying promise of a hydrogen economy is that fuel cells will replace internal combustion engines and turbines as the primary way to convert chemical power into motive and electrical power. The reason to expect this changeover is that fuel cells, being electrochemical, can be more efficient than heat engines. Currently, fuel cells are very expensive, but there is active research to bring down fuel cell prices.

Fuel cells work with hydrocarbon fuels as well as pure hydrogen. If and when fuel cells become cost-competitive with internal combustion engines and turbines, one of the first adopters will be large gas-fired powerplants. These are currently being built in large numbers by a highly competitive industry, their owners can work with operational constraints (tight temperature ranges, low shock, slow power ramps, etc), power to weight is not an issue, and even small efficiency gains are worth quite a lot. If reforming natural gas into hydrogen and then using that hydrogen in a fuel cell is somehow more efficient than burning the natural gas, gas-fired powerplants will do that instead. But there is no serious discussion of fuel-cell powerplants.

Much of the popular interest in hydrogen seems to attach to the idea of using fuel cells in automobiles. The cells can have a good power-to-weight ratio, are more efficient than internal combustion engines, and produce no damaging emissions. If cheap fuel cells can be had, they may make sense in an advanced hybrid automobiles.

So long as methane is the primary source of hydrogen, it will make more sense to fill specialized car tanks with compressed methane and run the fuel cells directly off that. The resulting system uses the methane energy more efficiently, produces less total $CO_2$, and requires less new infrastructure. A further advantage is that methane is much easier to transport and handle than hydrogen. Methane used for fuel cells cannot have traces of methanethiol or ethanethiol, which are smelly chemicals injected into natural gas distributions to help users find leaks. The sulfur component of the odorant will destroy the membranes of the fuel cell. Since the technology for running internal combustion engines directly from methane is well developed, low polluting, and leads to long engine life, it is more likely that compressed natural gas (CNG) will be used for transportation in this way rather than in fuel cells for the near future.

## 5. FUTURE OUTLOOK

The oil crisis in 1973 prompted research on biological hydrogen production, including photosynthetic production, as part of the search for alternative energy technologies. Green algae were known as light-dependent, water-splitting catalysts, but the characteristics of their hydrogen production were not practical for exploitation. Hydrogenase is too oxygen-labile for sustainable hydrogen production: light-dependent hydrogen production ceases within a few to several tens of minutes since photosynthetically produced oxygen inhibits or inactivates hydrogenases. A continuous gas flow system designed to maintain low oxygen concentrations within the reaction vessel, was employed in basic studies, but has not been found practically applicable. The future of biological hydrogen production depends not only on research advances, i.e. improvement in efficiency through genetically engineering microorganisms and/or the development of bioreactors, but also on economic considerations (the cost of fossil fuels), social acceptance, and the development of hydrogen energy systems. There is a long way for this technology before we actually start driving microbe driven cars.

Today's world might run on fossil fuel, but many people predict that hydrogen will fuel the future—in cars, houses, and countless handheld electronic devices. Hydrogen-powered fuel cells can generate electricity much more efficiently than fossil fuel can and without spewing polluting byproducts such as nitrous oxides, which contribute to smog, and carbon dioxide, the most prevalent gas behind global warming. The car we drive in the future may well be powered by the breath of tiny living organisms. That's the promise, at least, of a new field of science known as biological energy production. Researchers in the field seek to find a cure for the world's dependence on fossil fuels by taking advantage of nature's ability to transform sunlight and simple sugars into usable fuels. Hydrogen is one such fuel that can be produced microbiologically and holds a great promise as the power fuel of the future. This project assignment looks into the various methods by which hydrogen can be produced microbially and the practical applications of these methods. Biological hydrogen production is the most challenging area of biotechnology with respect to environmental problems. Although hydrogen promises a much neater and efficient fuel for the future, but risks are also equally high.

## 6. FURTHER READING


Angenent, L.T., Karim, K., Al-Dahhan, M.H., Wrenn, B.A. and Domíguez-Espinosa, R. (2004). Production of bioenergy and biochemicals from industrial and agricultural wastewater. Trends Biotechnol. 22: 477-485.

Dutta, D., De, D., Chaudhuri, S., and Bhattacharya, S.K. (2005). Hydrogen production by Cyanobacteria. Microbial Cell Factories. 4:36-46.

Jacobson, M.Z., Colella, W.G., Golden, D.M. (2005). Cleaning the air and improving health with hydrogen fuel-cell vehicles. Science. 307: 1901-1905.

Logan , B., Oh , S., Kim , I., and Ginke , S.V. (2002). Biological hydrogen production measured in batch anaerobic respirometers. Environ. Sci. Technol. 36: 2530-2535.

McLaren, J.S. (2005). Crop biotechnology provides an opportunity to develop a sustainable future. Trends Biotechnol. 23: 339-342

Nishio, N. and Nakashimada, Y. (2004). High rate production of hydrogen/methane from various substrates and wastes. Adv. Biochem. Engin/Biotechnol. 90: 63-87.

Oh, S., Ginkel, S. V. and L ogan, B. (2003). The relative effectiveness of pH control and heat treatment for enhancing biohydrogen gas production. Environ. Sci. Technol. 37: 5186-5190

Rhoads, A., Beyenal, H. and Lewandowski, Z. (2005). Microbial Fuel Cell using Anaerobic Respiration as an Anodic Reaction and Biomineralized Manganese as a Cathodic Reactant. Environ. Sci. Technol. 39: 3401 -3408.

Varma, A. and Behera, B. (2003). Green biomass : processing and technology. In: Green Energy. Capital Publishing Company, New Delhi, India. pp. 63-83.

Venkataraman, C., Habib, G., Eiguren-Fernandez, A., Miguel, A.H. and Friedlander, S.K. (2005). Residential Biofuels in South Asia: Carbonaceous Aerosol Emissions and Climate Impacts. Science. 307: 1454-1456.


□□□

# 11

# Microbes and Electricity Generation

**S.K. SONI**
*Department of Microbiology, Panjab University, Chandigarh-160 014*

## 1. INTRODUCTION

Electricity has become indispensable in the modern era and the increasing dependency of the mankind on electrical energy has led to many discoveries for the development of portable devices for electricity generation by chemical reactions. The history of converting chemical energy into electrical energy dates back to 18$^{th}$ century when 'Volta' invented first battery which are used even these days to power a large number of portable electronic devices. The draw backs of batteries in the form of limited lifespan has made way for the invention of another portable device, powered by some fuel in a cell, for conversion into electricity by electro-chemical means. An electrochemical cell in which the energy of a reaction between a fuel, such as liquid hydrogen, and an oxidant, such as liquid oxygen, is converted directly and continuously into electrical energy is known as a fuel cell. Typical reactants used in a fuel cell are hydrogen on the anode side and oxygen on the cathode side (a hydrogen cell). In contrast, conventional batteries consume solid reactants and, once these reactants are depleted, must be discarded, recharged with electricity by running the chemical reaction backwards, or, at least in theory, by having their electrodes replaced. Typically in fuel cells, reactants flow in and reaction products flow out, and continuous long-term operation is feasible virtually as long as these flows are maintained. A typical fuel cell comprises two electrodes (anode and cathode) separated by a polymer exchange membrane. Each electrode is coated on one side with a platinum catalyst, which causes the hydrogen fuel to separate into free electrons and protons (positive hydrogen ions) at the anode. The free electrons are conducted in the form of usable electrical current through an external circuit. The protons migrate through the membrane electrolyte to the cathode, where the catalyst causes the protons to combine with oxygen from the air and electrons from the external circuit to form water and heat (Figure 1). There are several types of fuel cells: i) Proton-exchange fuel cell; ii) Reversible fuel cell; iii) Direct-methanol fuel cell; iv) Direct borohydride fuel cells; v) Solid-oxide fuel cells; vi) Molten-carbonate fuel cells; vii) Phosphoric-acid fuel cells; viii) Alkaline fuel cells ix) Biofuel cells

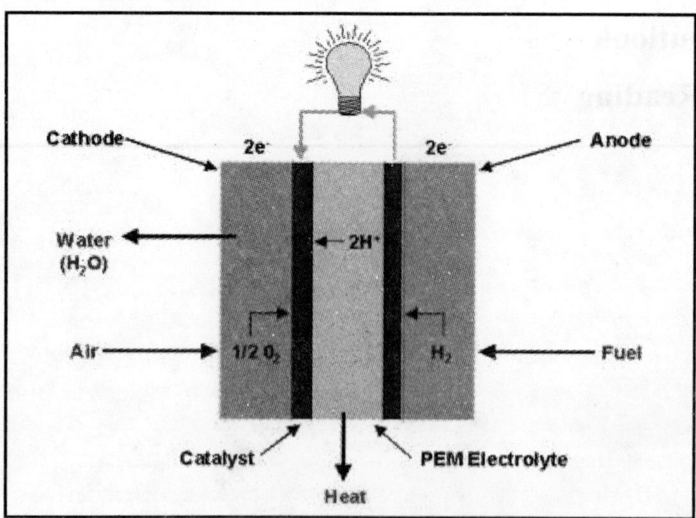


**Figure 1** A generalized fuel cell

Fuel cells offer many advantages over other portable devices of electricity generation. These include:

i) The electrical power output can be sustained almost indefinitely as long as the fuel can be continuously supplied.

ii) Fuel cells are usually more efficient than other types of energy converters and exhibit energy conversion efficiencies of 60-80%.

iii) They have a high energy efficiency suggesting that a smaller storage container as well as low fuel is needed for a specific energy requirement.

iv) Unlike a battery which contains only a limited amount of fuel material and oxidant, fuel cells are continuously supplied with fuel and air.

v) They donot emit pollutants and provide a truly sustainable power as the fuels used in these are produced from renewable energy sources.

Fuel cell industry is predicted to the largest, new industry of the 21st century, although there are many obstacles to overcome. It depends on which sources for hydrogen ultimately make sense. By itself, hydrogen is difficult to distribute and stockpile, and installing hydrogen pumps in every gas station would be a gigantic undertaking. Currently, Ballard Power Systems, Inc., Burnaby, British Columbia is the largest company making fuel cells. The principle of the fuel cell was discovered by Swiss scientist Christian Friedrich Schönbein in 1838. Based on this work, the first fuel cell was developed by Welsh scientist Sir William Grove. A sketch was published in 1843, but it wasn't until 1932 that British engineer Francis Thomas Bacon developed successful fuel cell devices. Twenty-seven years later in 1959, Bacon and his colleagues demonstrated a practical five-kilowatt unit capable of powering a welding machine. In the 1960s Bacon's patents were licensed by Pratt and Whitney from the U.S. where the concepts were used in the U.S. space program to supply electricity and drinking water (hydrogen and oxygen being readily available from the spacecraft tanks). Extremely expensive materials were used and the fuel cells required very pure hydrogen and oxygen. Early fuel cells tended to require inconveniently high operating temperatures that were a problem in many applications. However, fuel cells were seen to be desirable due to the large amounts of fuel available (hydrogen & oxygen). Further technological advances in the 1980s and 1990s, like the use of Nafion as the electrolyte, and reductions in the quantity of expensive platinum catalyst required, have made the prospect of fuel cells in consumer applications such as automobiles more realistic.

With the technological advancement in electronics, many energy-demanding applications, these days, require small, light power sources that are able to sustain operation over long periods of time, particularly in remote locations such as space and exploration. Furthermore, advances in the medical sciences are leading to an increasing number of implantable electrically-operated devices (e.g. pacemakers). These items need power supplies that will operate for extremely long durations as maintenance would necessitate surgery. Ideally, implanted devices would take advantage of the natural fuel substances found in the body, thus would continue to draw power as long as the person lives. Biofuel cells potentially offer solutions to all these problems, by taking nature's solutions to energy generation and tailoring them to our own needs. They take readily available substrates from renewable sources and convert them into benign by-products with the generation of electricity. Since they use concentrated sources of chemical energy, they can be small and light, and the fuel can even be taken from a living organism (e.g. glucose from the blood stream). Biofuel cells thus appear to an ideal source to use a renewable energy to effectively produce electricity or to convert other forms of energy to electricity. A biological fuel cell is a device that directly converts biochemical energy into electricity. The driving force of a biological fuel cell is the redox reaction of a carbohydrate substrate such as glucose and methanol using a microorganism or an enzyme as catalyst. Working principle is similar to that of chemical fuel cells. The main differences are that catalyst in the biological fuel cell is microorganism or enzyme, therefore noble metal is not needed, and its working conditions are mild: neutral solution and room temperature. For example, complete oxidation of one gram of methanol by an enzyme gives theoretically 5000 mAh of electric energy. The general characteristics of chemical and biofuel cells are given in Table 1.

**Table 1** General characteristics of chemical and biological fuel cell

| | Chemical fuel cell | Biological fuel cell |
|---|---|---|
| Catalyst | Noble Metals | Microorganism / enzyme |
| pH | Acidic Solution (pH<1) | Neutral Solution pH 7.0-9.0 |
| Temperature | over 200°C | Room Temperature 22-25°C |
| Electrolyte | Phosphoric-acid | Phosphate Solution |
| Capacity | High | Low |
| Efficiency | 40 - 60 % | over 40 % |
| Fuel Type | Natural gas, $H_2$, etc. | Any Carbohydrates and hydrocarbons |

## 2. BIOFUEL CELL

Biofuel cells use biocatalysts for the conversion of chemical energy to electrical energy. As most organic substrates undergo combustion with the evolution of energy, the biocatalyzed oxidation of organic substances by oxygen or other oxidizers at two-electrode interfaces provides a means for the conversion of chemical to electrical energy. Abundant organic raw materials such as methanol, organic acids or glucose can be used as substrates for the oxidation process, and molecular oxygen or $H_2O_2$ can act as the substrate being reduced. Typically the organic substrate is oxidized and the oxygen is reduced by the microbial cells or the enzymes:

$$\text{Anode: } C_6H_{12}O_6 + 6H_2O \xrightarrow{\text{Microbial cell}} 6CO_2 + 24H^+ + 24e^-$$

$$\text{Cathode: } 6O_2 + 24H^+ + 24e^- \xrightarrow{\text{Electron transport chain}} 12H_2O$$

$$\text{Anode: Glucose} \xrightarrow{\text{Glucose oxidase}} \text{Gluconate} + 2H^+ + 2e^-$$

$$\text{Cathode: } O_2 + 4H^+ + 4e^- \xrightarrow{\text{Laccase}} 2H_2O$$

The work on biofuel cells led to the development of the microbial fuel cell (MFC) in 1990 when it was proposed that the microorganisms including bacteria and yeasts can be exploited to produce electricity in a fuel cell. Later on it was discovered in 2002 that apart from pure strains, anaerobic sludge or marine sediments can be used to produce high currents. In the MFC, the microorganisms produce electrons during energy generating processes and donate these to a chemical electron mediator, which in turn transfers them to an electrode, producing electricity. Biofuel cells can use biocatalysts, enzymes or even whole cell organisms in one of two ways. Either i) the biocatalysts can generate the fuel substrates for the cell by biocatalytic transformations or metabolic processes, or ii) the biocatalysts may participate in the electron transfer chain between the fuel substrates and the electrode surfaces. Unfortunately, most redox enzymes do not take part in direct electron transfer with conductive supports, and therefore a variety of electron mediators (electron relays) are used for the electrical contacting of the biocatalyst and the electrode. Recently, novel approaches have been developed for the functionalization of electrode surfaces with monolayers and multilayers consisting of redox enzymes, electrocatalysts and bioelectrocatalysts that stimulate electrochemical transformations at the electrode interfaces. The assembly of electrically contacted bioactive monolayer electrodes could be advantageous for biofuel cell applications as the biocatalyst and electrode support are integrated. This chapter summarizes recent advances in the tailoring of conventional microbial-based biofuel cells and describes novel biofuel cell configurations based on biocatalytic interfaces structures integrated with the cathodes and anodes of biofuel cells. The extractable power of a fuel cell (Pcell) is the product of the cell voltage (Vcell) and the cell current (Icell)

$$\text{Pcell} = \text{Vcell} \times \text{Icell}$$

Although the ideal cell voltage is affected by the difference in the formal potentials of the oxidizer and fuel compounds.

$$\text{Vcell} = \text{Eox} - \text{Efuel}$$

Cell current controlled by electrode size, transport rates across membrane. But most redox enzymes do not transfer electrons directly. Therefore, one uses electron mediators (relays).

## 2.1. Microbial-based biofuel cells

All living organisms respire in order to obtain energy in the form of ATP. There are two types of respiration: aerobic and anaerobic. Some organisms respire using oxygen to produce carbon dioxide and water from glucose, while some microorganisms respire anaerobically. The electron transport chain provides a means of obtaining ATP at different levels on the chain. Electrons removed from glucose are donated to $NAD^+$, reducing it to NADH which then donates the electrons to a molecule of lower redox potential. When electrons are donated to a lower level, the released energy is captured in ATP. At the end of the chain, the electrons and protons react with oxygen to form water and carbon dioxide. Some microorganisms donate the electrons to pyruvate, acetaldehyde or even to some elements like Fe (III), sulphur and manganese etc. The transfer of electrons from a respiratory substrate to an electrode is the basis of the microbial fuel cell. Microbes can be exploited to donate electrons to an electron mediator or electrode and produce electricity. *Rhodoferax ferrireducens* is a marine sediment microbe capable of donating electrons directly to graphite electrode. Other microbes such as *Desulfuromonas acetoxidans* and *Proteus vulgaris*, require electron mediators in the form of some chemicals (eg. Neutral red and methylene blue) that accept the electrons and deliver them to electrode. There are two chambers in a microbial fuel cell. In the anode chamber, the microbes conduct anaerobic or aerobic respiration and the electrons extracted from the substrate are donated to an electron mediator while the protons ($H^+$) pass through a cation-selective membrane to the oxygenated cathode chamber. Once the electrons travel through the anode and reach the cathode, they react with oxygen and protons to form water. Besides oxygen, potassium ferricyanide can also be used as an electron acceptor in the cathode chamber. One of the important analysis conducted on MFCs is chemical oxygen demand (COD) which is the measure of how oxidisable a substance is. The main features of the cell are shown schematically in Figure 2. Scientists at NASA performed research on the microbial fuel cell in 1960s while the conversion of bacterial energy to electrical energy was first shown in 1976. In 1984 and 1985, the scientists described the ability of microbes to transfer electrons to metals present in soil and it was demonstrated in 1990 that microbes can be exploited to produce electricity in MFC. It was in 2003 when *Rhodoferax ferrireducens*, a bacterium was found to have the highest electron transfer efficiency.

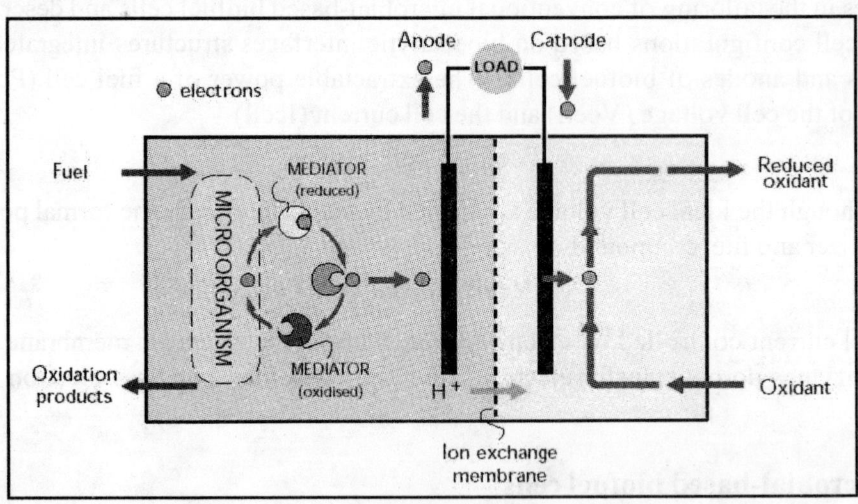


**Figure 2** Schematic representation of a microbial fuel cell

Three different types of microbial fuel cells have been designed. In 'Type-A' fuel cells, artificial redox mediators capable of penetrating bacterial cells are added to the culture medium within the anode fuel cell compartment, enabling electrons produced during fermentation or other metabolic processes to be shuttled to the anode. Type-B fuel cells incorporate metal-reducing bacteria including the members of the Geobacteriaceae and Schwanellaceae that partly exhibit special membrane-bound cytochromes capable of transferring electrons directly to the electrodes rather than dissolved or solid iron (III) or manganese (IV/III). Type-C fuel cells oxidize fermentation products, such as hydrogen and methanol, on electrocatalytic electrodes capable of effectively oxidizing such metabolites.

All these designs have some advantages and disadvantages. Type-A fuel cells show good current densities but have a disadvantage of using synthetic and toxic redox mediators that have to be added to the cell culture and are nonrecoverable. Type-B fuel cells can work in natural environments including sewage and sediments, however, the growth rate of bacteria and current densities are low. Type-C fuel cells, though produce the highest current densities and work with simple and easily available microorganisms including *E. coli*, electrocatalytic electrodes require rather expensive surface modifiers. *Rhodoferax ferrireducens*, a bacterium superior in its glucose catabolizing efficiency as compared to *Geobacter* and *Shewanella* species and capable of diverting 85% electron yield for electricity generation appears to be a good candidate for microbial fuel cells. In addition, electrons generated by *R. ferrireducens* are easily transferred to anode without the assistance of electron-shuttling mediators and the cells grow at a steady rate which guarantees a steady supply of electrons and therefore a consistent current density (Figure 3).

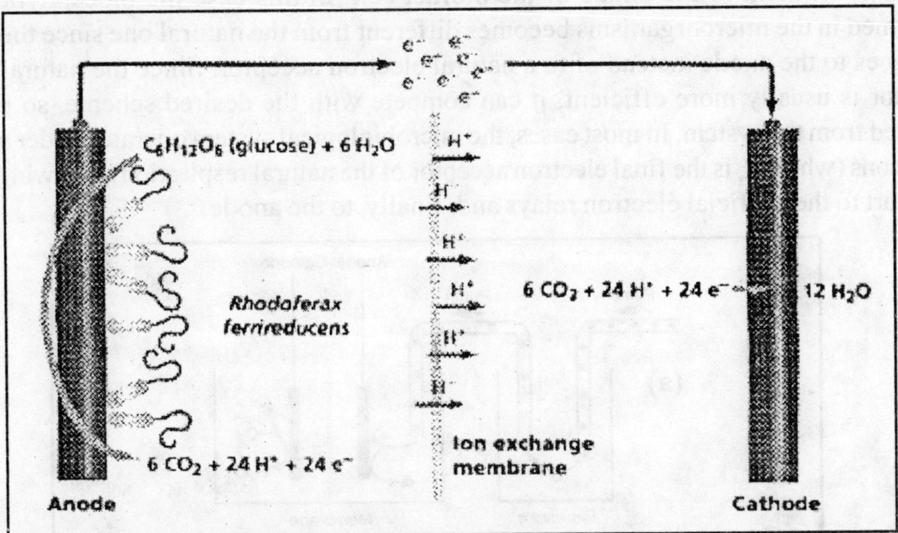


**Figure 3** *Rhodoferax ferrireducens* microbial fuel cell. *R. ferrireducens* burns carbohydrates to $CO_2$ in the anodic compartment, a process that produces free electronw which are directly captured by the anodes from where, these are channeled to the cathode, where they reduce oxygen to water

There are three approaches involving the use of microorganisms in fuel cells. In one of the approaches, microorganisms produce electrochemically active substances that may be metabolic intermediaries or final products of anaerobic respiration which can be produced in one place and transported to a biofuel cell to be used as a fuel. In this case the biocatalytic microbial reactor is producing the biofuel and the biological part of the device is not directly integrated with the electrochemical part (Figure 4a). This scheme allows the electrochemical part to operate under conditions that are not compatible with the biological part of the device. The two parts can even be separated in time, operating completely individually. The most widely used fuel in this scheme is hydrogen gas, allowing well-developed and highly efficient $H_2/O_2$ based fuel cells to be conjugated with a bioreactor.

In the second approach the microbiological fermentation process proceeds directly in the anodic compartment of a fuel cell, supplying the anode with the in situ produced fermentation products (Figure 4b). In this case the operational conditions in the anodic compartment are dictated by the biological system, so they are significantly different from those in the conventional fuel cells. At this point we have a real biofuel cell and not a simple combination of a bioreactor with a conventional fuel cell. This configuration also is often based on the biological production of hydrogen gas, but the electrochemical oxidation of $H_2$ is performed in the presence of the biological components under mild conditions. Other metabolic products (e.g. formate, $H_2S$) have also been used as fuels in this kind of system.

A third approach involves the application of artificial electron transfer relays that can shuttle electrons between the microbial biocatalytic system and the electrode. The mediator molecules take electrons from the biological electron transport chain of the microorganisms and transport them to the anode of the biofuel cell. In this case the biocatalytic process performed in the microorganisms becomes different from the natural one since the electron flow goes to the anode instead of to a natural electron acceptor. Since the natural electron acceptor is usually more efficient, it can compete with the desired scheme, so is usually removed from the system. In most cases, the microbiological system operates under anaerobic conditions (when $O_2$ is the final electron acceptor of the natural respiration), allowing electron transport to the artificial electron relays and, finally, to the anode.

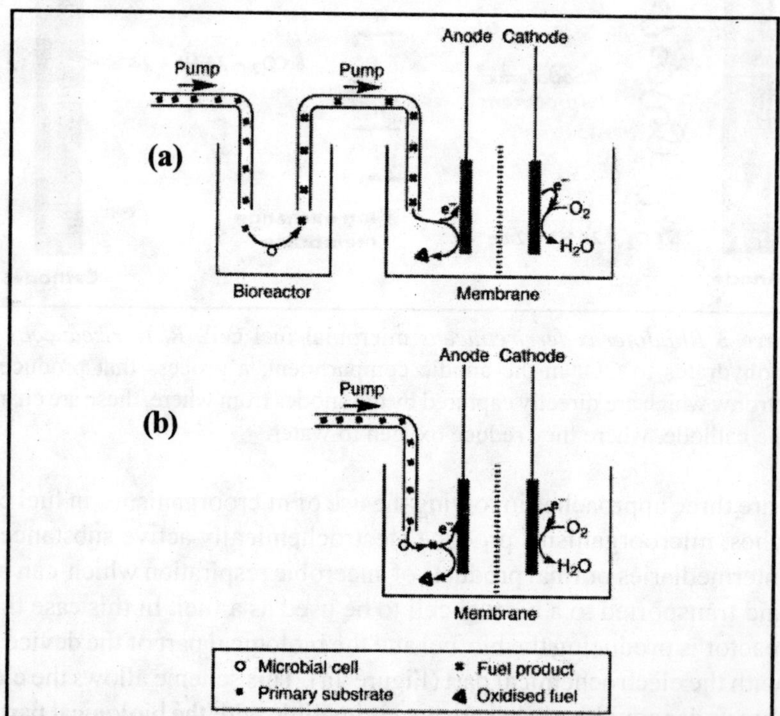


**Figure 4** Schematic configuration of a microbial biofuel cell: **(a)** With a microbial bioreactor providing fuel separated from the anodic compartment of the electrochemical cell. **(b)** With a microbial bioreactor providing fuel directly in the anodic compartment of the electrochemical cell

### 2.1.1. *Microbial bioreactors producing $H_2$ for conventional fuelcells*

Various bacteria and algae, for example *Escherichia coli*, *Enterobacter aerogenes*, *Clostridium butyricum*, *Clostridium acetobutylicum*, and *Clostridium perfringens* have been found to be active in hydrogen production under anaerobic conditions. The most effective $H_2$ production is observed upon fermentation of glucose in the presence of *Clostridium butyricum*. The Figure 3 shows a microbial fermentation process provided by *Escherichia coli* resulting in $H_2$ evolution as a typical example of the microbial fuel production.

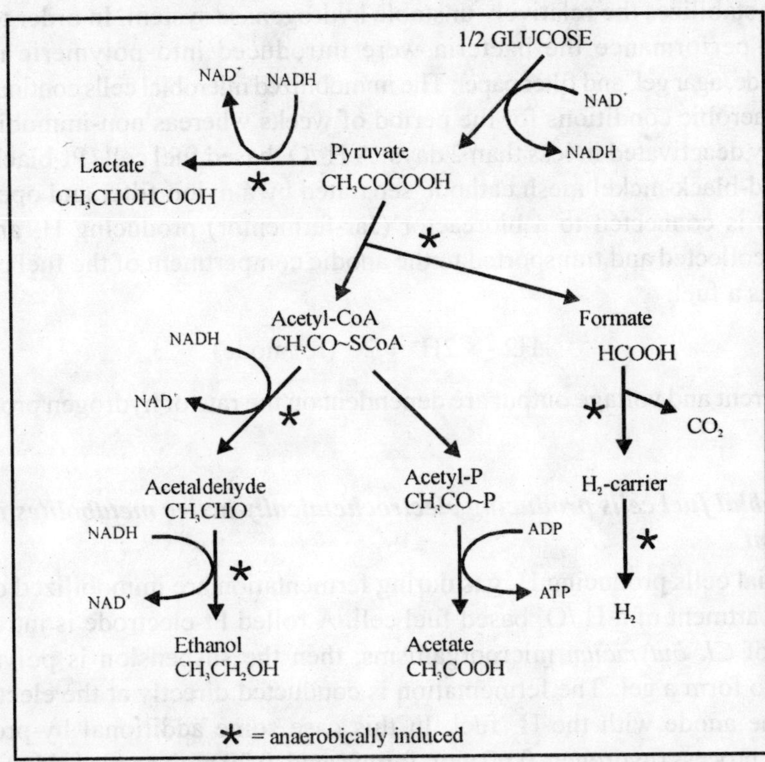


**Figure 5** Fermentative production of $H_2$ in *E. coli* for use in a biofuel cell

The conversion of carbohydrate to hydrogen is achieved by a multienzyme system. In bacteria the route is believed to involve glucose conversion to 2 mol of pyruvate and 2 mol of NADH formed by the Embden-Meyerhof pathway. The pyruvate is then oxidized through a pyruvate - ferredoxin oxidoreductase producing acetyl-CoA, $CO_2$, and reduced ferredoxin. NADH-ferredoxin oxidoreductase oxidizes NADH and reduces ferredoxin. The reduced ferredoxin is reoxidized to form hydrogen by the hydrogenase. As a result, 4 mol of hydrogen are produced from 1 mol of glucose under ideal conditions:

$$\text{Glucose} + 2\text{NAD} + \xrightarrow{\text{Multienzyme Embden-Meyerhof Pathway}} 2\text{Pyruvate} + 2\,\text{NADH}$$

$$\text{Pyruvate} + \text{Ferredoxin}_{ox} \xrightarrow{\text{Pyruvate-ferredoxin oxidoreductase}} \text{Acetyl-CoA} + CO_2$$

$$\text{NADH} + \text{Ferredoxin}_{ox} \xrightarrow{\text{NADH - ferredoxin oxidoreductase}} \text{NAD}^+ + \text{Ferredoxin}_{red}$$

$$\text{Ferredoxin}_{red} + 2H^+ \xrightarrow{\text{Hydrogenase}} \text{Ferredoxin}_{ox} + H_2$$

However, only 1 mol of $H_2$ per 1 mol of glucose is obtained under optimal conditions in a real system. Since the $H_2$ yield is only 25% of the theoretical yield, the improvement of hydrogen production by genetic engineering techniques and screening of new hydrogen-producing bacteria is possible for enhanced energy conversion. Glucose is an expensive substrate, and industrial wastewater containing nutritional substrates for $H_2$-producing bacteria have been successfully applied to produce hydrogen later used in a fuel cell.

The immobilization of hydrogen-producing bacteria, *Cl. butyricum*, has great value because this stabilizes the relatively unstable hydrogenase system. In order to stabilize the biocatalytic performance the bacteria were introduced into polymeric matrices, e.g. polyacrylamide, agar gel, and filter paper. The immobilized microbial cells continuously produce $H_2$ under anaerobic conditions for the period of weeks whereas non-immobilized bacteria cells are fully deactivated in less than 2 days. A $H_2/O_2$ based fuel cell (Pt-black-nickel mesh anode and Pd-black-nickel mesh cathode separated by a nylon filter and operated at room temperature) is connected to a bioreactor (Jar-fermentor) producing $H_2$ and the $H_2$ gas produced is collected and transported to the anodic compartment of the fuel cell, where the gas is used as a fuel:

$$H2 \rightarrow 2H^+ + 2e^- \text{ (to anode)}$$

The current and voltage output are dependent on the rate of hydrogen production in the fermentor.

### *2.1.2. Microbial fuel cells producing electrochemically active metabolites in the anodic compartment*

Microbial cells producing $H_2$ gas during fermentation are immobilized directly in the anodic compartment of a $H_2/O_2$ based fuel cell. A rolled Pt-electrode is introduced into a suspension of *Cl. butyricum* microorganisms, then the suspension is polymerized with acrylamide to form a gel. The fermentation is conducted directly at the electrode surface, supplying the anode with the $H_2$ fuel. In this case some additional by-products of the fermentation process (hydrogen, 0.60 mol; formic acid, 0.20 mol; acetic acid, 0.60 mol; lactic acid, 0.15 mol) can also be utilized as additional fuel components. For example, pyruvate produced can be alternatively oxidized to formate through a pyruvate-formate lyase, a central enzyme in bacterial anaerobic metabolism catalysing the reversible reaction of pyruvate and coenzyme A into acetyl-CoA and formate. The metabolically-produced formate is directly oxidized at the anode when the fermentation solution is passing the anode compartment:

$$\text{Pyruvate} \xrightarrow{\text{Pyruvate-formate lyase}} \text{Formate}$$

$$HCOO^- \rightarrow CO_2 + H^+ + 2e^- \text{ (to anode)}$$

The biofuel cell that includes. 0.4 g of wet microbial cells (0.1 g of dry material) yields upon optimal operating conditions the outputs $V_{cell} = 0.4$ V and $I_{cell} = 0.6$ mA. It should be noted that in the case that a Pt-black electrode is used as an anode, oxidation of the original substrate utilized by the microorganisms in the fermentation process (e.g. glucose) can contribute to the anodic current. Thus, the $H_2$ provided by the microorganisms is the main, but not the only source of the anodic current. Other fuels have also been produced microorganisms in the anodic compartments of biofuel cells. There are many microorganisms producing metabolically reduced sulfur-containing compounds (e.g. sulfides, $S^{2-}$, $HS^-$, sulfites, $SO_3^{2-}$). Sulfate-reducing bacteria (e.g. *Desulfovibrio desulfuricans*) form a specialized group of anaerobic microbes that use sulfate ($SO_4^{2-}$) as a terminal electron acceptor for respiration. These microorganisms yield $S^{2-}$ while using a substrate (e.g. lactate) as a source of electrons. This microbiological oxidation of lactate with the formation of sulfide has been used to drive an anodic process in biofuel cells. The metabolically produced sulfide is oxidized directly at an electrode, providing an anodic reaction that produces sulfate or thiosulfate:

$$\text{Lactate} + SO_4^{2-} + 8H^+ \xrightarrow{\text{Bacteria}} S^{2-} + 4\,H^2O + \text{Pyruvate}$$

$$S^{2-} + 4H_2O \rightarrow SO_4^{2-} + 8H^+ + 8^{e-} \text{ (to anode)}$$

$$2S^{2-} + 3H_2O \rightarrow S_2O_3^{2-} + 6H^+ + 8^{e-} \text{ (to anode)}$$

The fermentation solution is composed of a microbial suspension ($10^8$ non-immobilized cells per ml), with the nutritional substrates (mainly lactate) under anaerobic conditions. Accumulation of sulfides in the medium results in the inhibition of the metabolic process because of their interaction with iron containing proteins (e.g. cytochromes), causing the electron transport systems to be blocked. To prevent the toxic effect of $H_2S$, the anode should effectively oxidize it. However, many metallic electrodes are poisoned by sulfide because of its strong and irreversible adsorption. Thus, porous graphite electrodes were used (100 $cm^2$, impregnated with 10% w/w cobalt hydroxide, which in the presence of $S^{2-}$ undergoes a transition into a catalytically highly active cobalt oxide/cobalt sulfide mixture). The biocatalytic anode is combined with an oxygen cathode (porous graphite electrode, 100 $cm^2$ geometrical area, activated with iron(II)-phthalocyanine and vanadium(V) compounds) separated with a cation-exchange membrane in order to maintain anaerobic conditions in the anodic compartment.

Microbiological cultivation under aerobic conditions utilizes $O_2$ as a terminal electron acceptor. It has been shown that aerobic growth of *Saccharomyces cerevisiae* or *Micrococcus cerificans* bacteria in the presence of glucose as the nutritional substrate in an anodic compartment of a biofuel cell results in an anodic current. A biofuel cell in such a system works as an $O_2$-concentration cell utilizing the potential difference produced at the cathode and anode due to the oxygen consumption in the anodic compartment. Table 2 summarizes the electrical output obtained in biofuel cells operating without electron transfer mediators and using the natural products of microbial fermentation (e.g. $H_2$, $H_2S$) as the current providing species.

### *2.1.3. Microbial fuel cells operating in the presence of artificial electron relays*

The contact of the microbial cells with an electrode usually results in a very minute electron transfer across the membrane of the microbes. In some specific cases, however, the direct electron transfer from the microbial cells to an anode surface is still possible. The metal-reducing bacterium *Shewanella putrefaciens* has been reported to have cytochromes in its outer membrane. These electron carriers (i.e. cytochromes) are able to generate anodic current in the absence of terminal electron acceptors (under anaerobic conditions). However, this is a rather exceptional example.

**Table 2** Examples of microbial-based biofuel cells utilizing fermentation products for their oxidation at anodes

| Micro-organism | Nutritional substrate | Fermentation Product | Biofuel cell voltage | Biofuel cell current or current density | Anode |
|---|---|---|---|---|---|
| *Clostridium butyricum* | Waste water | $H_2$ | 0.62 V (at 1 Ω) | 0.8 A (at 2.2 V) | Pt-blackened Ni, 165 $cm^2$ (5 anodes in series) |
| *Clostridium butyricum* | Molasses | $H_2$ | 0.66 V (at 1 Ω) | 40 mA /$cm^2$ (at 1Ω) | Pt-blackened Ni, 85 $cm^2$ |
| *Clostridium butyricum* | Lactate | $H_2$ | 0.6 V (oc) | 120 μA /$cm^2$ (sc) | Pt-black, 50 $cm^2$ |
| *Enterobacter aerogenes* | Glucose | $H_2$ | 1.04 V (oc) | 60 μA /$cm^2$ (sc) short circuit | Pt-blackened stainless steel, 25 $cm^2$ |
| *Desulfovibrio desulfuricans* | Dextrose | $H_2S$ | 2.8 V (oc) open circuit | 1 A | Graphite, $Co(OH)_2$ impregnated (3 anodes in series) |

Low-molecular weight redox-species may assist the shuttling of electrons between the intracellular bacterial space and an electrode. However, there are many important requirements that such a mediator should satisfy in order to provide an efficient electron transport from the bacterial metabolites to the anode: (a) The oxidized state of the mediator should easily penetrate through the bacterial membrane to reach the reductive species inside the bacteria. (b) The redox-potential of the mediator should fit the potential of the reductive metabolite (the mediator potential should be positive enough to provide fast electron transfer from the metabolite, but it should not be too positive as to prevent significant loss of potential). (c) Neither oxidation state of the mediator should interfere with other metabolic processes (should not inhibit them or be decomposed by them). (d) The reduced state of the mediator should easily escape from the cell through the bacterial membrane. (e) Both oxidation states of the mediator should be chemically stable in the electrolyte solution, they should be well soluble, and they should not adsorb on the bacterial cells or electrode surface. (f) The electrochemical kinetics of the oxidation process of the mediator-reduced state at the electrode should be fast (electrochemically reversible).

Many different organic and organometallic compounds have been tested in combination with bacteria to test the efficiency of mediated electron transport from the internal bacterial metabolites to the anode of a biofuel cell. Thionine has been used extensively as a mediator of the electron transport from *Proteus vulgaris* and from *E. coli*. Other organic dyes that have been tested include benzylviologen, 2,6-dichlorophenolindophenol, 2-hydroxy-1,4-naphthoquinone, phenazines (phenazine ethosulfate, safranine), phenothiazines (alizarine brilliant blue, *N,N*-dimethyl-disulfonated thionine, methylene blue, phenothiazine, toluidine blue), and phenoxazines (brilliant cresyl blue, gallocyanine, resorufin). These organic dyes were tested with *Alcaligenes eutrophus*, *Anacystis nidulans*, *Azotobacter chroococcum*, *Bacillus subtilis*, *Clostridium butyricum*, *Escherichia coli*, *Proteus vulgaris*, *Pseudomonas aeruginosa*, *Pseudomonas putida*, and *Staphylococcus aureus* bacteria, usually using glucose and succinate as substrates. Among the dyes tested, phenoxazine, phenothiazine, phenazine, indophenol, bipyridilium derivatives, thionine and 2-hydroxy-1,4-naphthoquinone were found to be very efficient in maintaining relatively high cell voltage output when current was drawn from the biofuel cell. Some other dyes did not function as effective mediators because they are not rapidly reduced by microorganisms, or they lacked sufficient negative potentials. Ferric chelate complexes (e.g. Fe(III)EDTA) were successfully used with *Lactobacillus plantarum*, *Streptococcus lactis*, and *Erwinia dissolvens*, oxidizing glucose. Table 3 summarizes the structures and redox-potentials of electron transfer mediators as well as the rate constants of their reduction by microorganisms. It should be noted that the overall efficiency of the electron transfer mediators depends also on many other parameters, and in particular on the electrochemical rate constant of mediator re-oxidation, which depend on the electrode material.

**Table 3** Redox-potentials of electron relays used in microbial-based biofuel cells and the kinetics of their reduction by microbial cells

| Redox relay | Structural formula | Redox potential V (vs. NHE) | Rate of reduction μmol / g dry wt /s |
|---|---|---|---|
| 2,6-Dichlorophenol-indophenol | 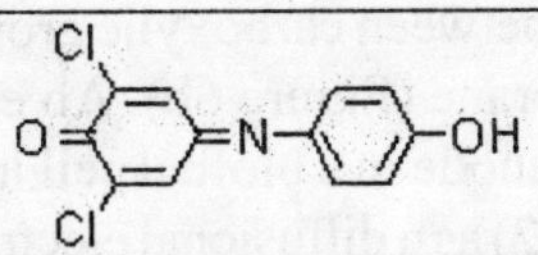 | +0.217 | 0.41 |
| Phenazine ethosulphate | 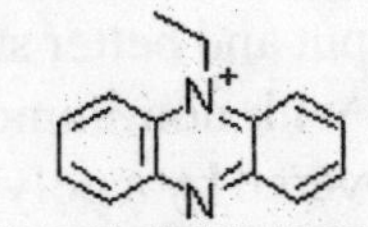 | +0.065 | 8.57 |
| Safranine-O | 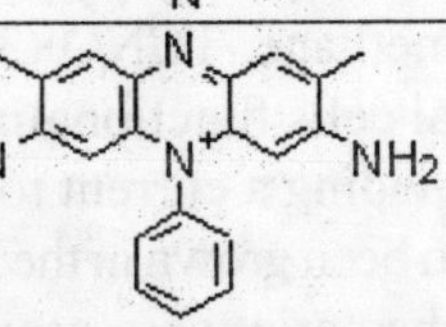 | -0.289 | 0.07 |
| *N,N*-Dimethyl-disulphonated thionine | 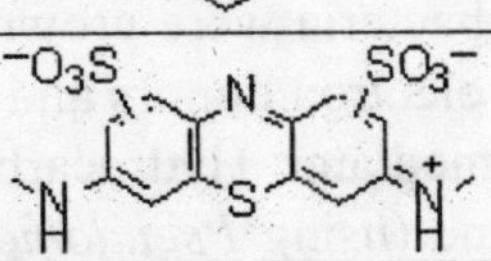 | +0.220 | 0.33 |

*Contd...*

*Table 3 Contd...*

| | | | |
|---|---|---|---|
| New Methylene Blue | 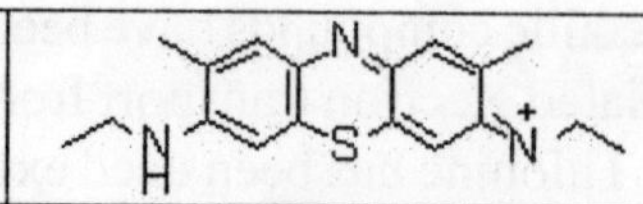 | -0.021 | 0.20 |
| Phenothiazinone | 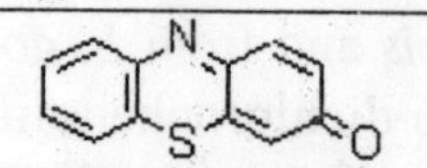 | +0.130 | 1.43 |
| Thionine | 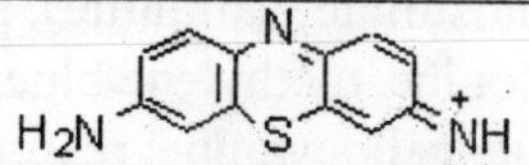 | +0.064 | 7.10 |
| Toluidine Blue-O | 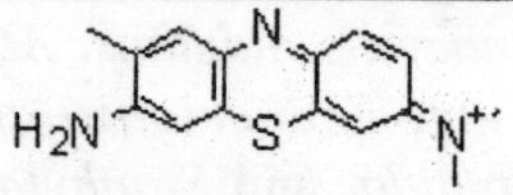 | +0.034 | 1.47 |
| Gallocyanine | 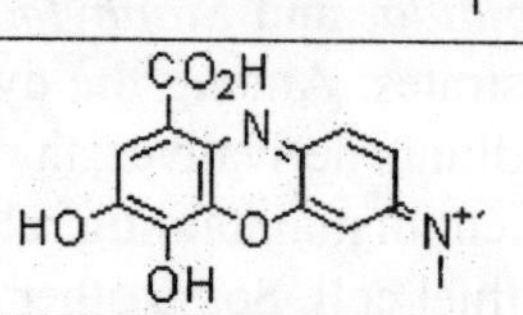 | +0.021 | 0.53 |
| Resorufin | 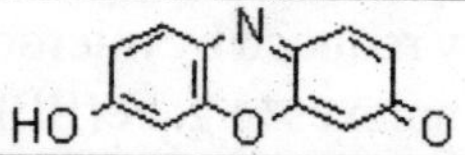 | -0.051 | 0.61 |

In order to organize an integrated biocatalytic assembly in the anode compartment of a biofuel cell, the microbial cells and the electron transfer mediator are co-immobilized at the anode surface using several methods. Neutral red (1), an organic dye known to be an active diffusional mediator for electron transport from *E. coli* has been covalently linked to a graphite electrode by making an amide bond between a carboxylic group on the electrode surface and an amino group of the dye (Figure 6a). The mediator-modified electrode when used as an anode in the presence of *E. coli*, the surface-bound dye provides electron transfer from the microbial cells to the conductive support under anaerobic conditions. In this case only those bacteria that reach the modified electrode surface are electrically contacted. *Proteus vulgaris* microbial cells have been covalently bound to an oxidized carbon electrode surface by making amide bonds between carboxylic groups on the electrode surface and amino groups of the microbial membrane (Figure 6b). An electrode modified with the attached microorganisms is applied as an anode in a biofuel cell in the presence of glucose as a primary reductive substrate and thionine (2) as a diffusional electron mediator. The microbial-modified anode shows an enhanced current output and better stability in comparison with a system composed of the same components but with non-immobilized cells. *Desulfovibrio desulfuricans* microbial cells have also been covered by a polymeric derivative of viologen or modified by 7,7',8,8'-tetracyanoquinodimethane, TCNQ (3) adsorbed on the surface of the microbial cell (Figure 6c). The microbial cells functionalized with the electron transfer mediators were applied in a biofuel cell providing a current to an anode in the absence of a diffusional mediator. Microbial cells have also been grown in the presence of various nutritional substrates. For example, *Proteus vulgaris* bacteria were grown using glucose, galactose, maltose, trehalose, and sucrose as primary electron donors and used in a biofuel cell with thionine as a diffusional electron transfer mediator. Hydrocarbons such as *n*-hexadecane (using *Micrococcus cerificans*) and methane (using *Pseudomonas methanica*) have also been used as fuels to maintain anodic current in the anodic compartment of a biofuel cell. It has been shown that biofuel cell performance depends heavily on the primary substrate used in the fermentation process. The metabolic process in the bacteria is very complex, involves many enzymes, and may proceed by many different routes. It has been shown that a mixture of nutritional substrates can result even in higher extractable current than any single component alone. It is possible to achieve maximum fuel cell efficiency just by changing the carbon source and thus inducing the various metabolic states inside the microorganism. Table 4 summarizes the electrical output of microbial biofuel cells operating with different electron transfer mediators as species providing anodic current, and using different nutrients.

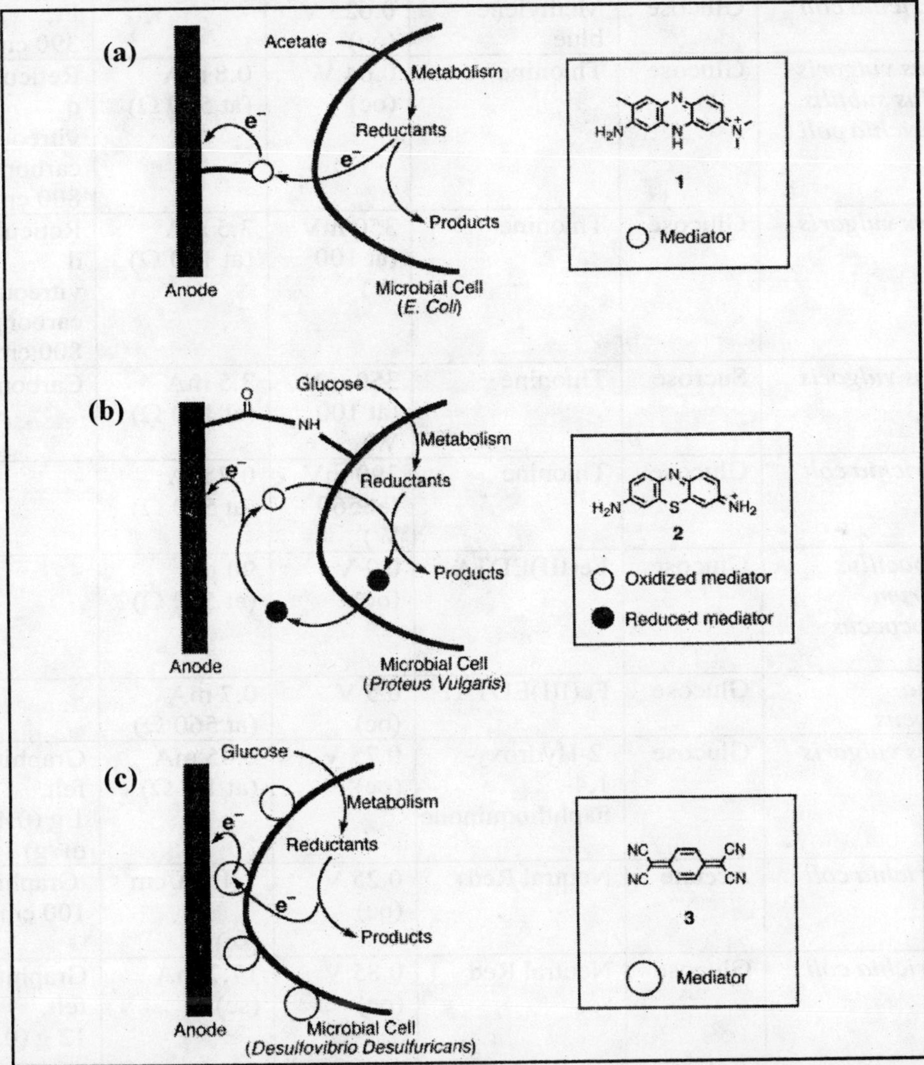


**Figure 6** Electrical wiring of microbial cells to the anode of the electrochemical cell using electron transfer meadiators: **(a)** A diffusional mediator shuttling between the microbial suspension and the anode surface. **(b)** A diffusional mediator shuttling between the anode and microbial cells covalently linked to the electrode. **(c)** A mediator adsorbed on the microbial cells providing the electron transport from the cells to the anode

**Table 4** Examples of microbial-based biofuel cells utilizing electron relays for coupling of the intracellular electron transfer processes with electrochemical reactions at anodes

| Microorganism | Nutritional substrate | Mediator | Biofuel cell voltage | Current or current density | Anode |
|---|---|---|---|---|---|
| *Pseudomonas methanica* | $CH_4$ | 1-Naphthol-2-sulfonate indo-2,6-dichloro-phenol | 05 - 06 V (oc) | 2.8 mA/cm$^2$ (at 0.35 V) | Pt-black, 12.6 cm$^2$ |
| *Escherichia coli* | Glucose | Methylene blue | 0.625 V (oc) | - | Pt, 390 cm$^2$ |
| *Proteus vulgaris* *Bacillus subtilis* *Escherichia coli* | Glucose | Thionine | 0.64 V (oc) | 0.8 mA (at 560 Ω) | Reticulated vitreous carbon, 800 cm$^2$ |
| *Proteus vulgaris* | Glucose | Thionine | 350 mV (at 100 W) | 3.5 mA (at 100 Ω) | Reticulated vitreous carbon, 800 cm$^2$ |
| *Proteus vulgaris* | Sucrose | Thionine | 350 mV (at 100 W) | 3.5 mA (at 100 Ω) | Carbon |
| *Escherichia coli* | Glucose | Thionine | 390 mV (at 560 W) | 0.7 mA (at 560 Ω) | - |
| *Lactobacillus plantarum* *Streptococcus lactis* | Glucose | Fe(III)EDTA | 0.2 V (oc) | 90 μA (at 560 Ω) | - |
| *Erwinia dissolvens* | Glucose | Fe(III)EDTA | 0.5 V (oc) | 0.7 mA (at 560 Ω) | - |
| *Proteus vulgaris* | Glucose | 2-Hydroxy-1,4-naphthoquinone | 0.75 V (oc) | 0.45 mA (at 1 k Ω) | Graphite felt, 1 g (0.47 m$^2$/g) |
| *Escherichia coli* | Acetate | Neutral Red | 0.25 V (oc) | 1.4 μA/cm$^2$ (sc) | Graphite, 100 cm$^2$ |
| *Escherichia coli* | Glucose | Neutral Red | 0.85 V (oc) | 17.7 mA (sc) | Graphite felt, 12 g (0.47 m$^2$ /g) |
| *Escherichia coli* | Glucose | 2-Hydroxy-1,4-naphthoquinone | 0.53 V (at 10 k W) | 0.18 mA cm$^{-2}$ (sc) | Glassy carbon, 12.5 cm$^2$ |

OC = open circuit, SC = short circuit

### *2.1.4. Microbial fuel cells producing electricity during wastewater treatment*

A new approach to fuel cells turns waste water into clean water and electricity. Microbial fuel cells (MFCs) have been used to produce electricity from different compounds, including acetate, lactate, and glucose. It is also possible to produce electricity in a MFC from domestic wastewater, while at the same time accomplishing biological wastewater treatment leading to removal of chemical oxygen demand. Organic water in waste water has energy value as these have high concentration of organic matter. Two types of MFCs have so far been used: small batch-fed laboratory systems primarily using defined substrates and MFCs developed to harvest energy from marine sediments. Laboratory systems typically use separate compartments for the anode and cathode connected by a proton exchange membrane (PEM). This two-compartment design is difficult to apply to larger systems for continuously treating wastewater. Marine sediment systems are based on the placement of a carbon pad into an anaerobic sediment (anode) and a second pad placed in overlaying oxygenated waters (cathode). The latter systems cannot be adapted for continuous treatment of a wastewater due to the need for large volumes of oxygenated water and the limited solubility of dissolved oxygen in water. Thus, changes in reactor design are needed to produce MFCs suitable for treating continuous streams of dissolved organic matter in waste waters.

To continuously generate electricity from organic matter in water, a MFC reactor must have both a large surface area for biofilm formation and a high void volume. Wastewater treatment reactors are designed to be highly porous to avoid clogging the reactor with high concentrations of particles in the wastewater. This requirement for high porosity is at odds with the need for a large surface area for biofilm formation. Thus, fixed-film wastewater reactors have been designed to produce thin fluid films and convective flow rates sufficiently high to shear off excess biofilm growth from surfaces. These reactors cannot be used without modification for energy production because MFCs have an additional requirement that the biomass be kept separated from any dissolved oxygen. These design challenges are similar to requirements for membrane bioreactors used to treat water and wastewater. In these reactors, a series of fibrous nanopore-sized membranes are inserted into a single, sheared compartment providing a high porosity tank containing the membrane. The high shear generated in the reactor by bubbles is used to help shear biofilms from the fibers.

The methods used to achieve a high porosity in membrane reactors and to improve the design of a MFC by providing a chamber that contains both the anode and the cathode in a single compartment MFC (SCMFC). A prototype reactor for the electricity generation accompanied by wastewater treatment demonstrated by the removal of organic matter in the form of chemical oxygen demand (COD) or biochemical oxygen demand (BOD) is revealed in Figure 7. This single chamber microbial fuel cell (SCMFC) contains eight graphite electrodes (anodes) and a single air cathode. The system is operated under continuous flow conditions with primary clarifier effluent obtained from the wastewater treatment plant. This is based on the sulphate reducing species, *Desulfovibrio desulfuricans* mixed with four other species, namely, Proteus vulgaris, *Escherichia coli, Pseudomonas aeruginosa* and *P. fluorescens*. The role of these organisms is to utilize a wide range of sugars and other organic substances and convert these into end products including lactate. *D. desulfuricans* is capable of utilizing lactate as its carbon energy source and used sulphate ($SO_4$) found in waste water and its end terminal electron acceptor which it reduces to sulphide ($S^{2-}$). Sulphide is electrochemically active at the anode and is oxidized at the electrode surface, giving electrons and sulphate. This fuel cell gives much higher power density than previous types and requires no synthetic exogenous mediators. The prototype SCMFC reactor generated electrical power (maximum of 26 mw/m$^2$) while removing up to 80% of the COD of the wastewater. Power output is proportional to the hydraulic retention time over a range of 3-33 h and to the influent wastewater strength over a range of 50-220 mg/l of COD. Current generation was controlled primarily by the efficiency of the cathode. Optimal cathode performance was obtained by allowing passive air flow rather than forced air flow (4.5-5.5 L/min). Bioreactors based on power generation in MFCs may represent a completely new approach to wastewater treatment. If power generation in these systems can be increased, MFC technology may provide a new method to offset wastewater treatment plant operating costs, making advanced wastewater treatment more affordable for both developing and industrialized nations.

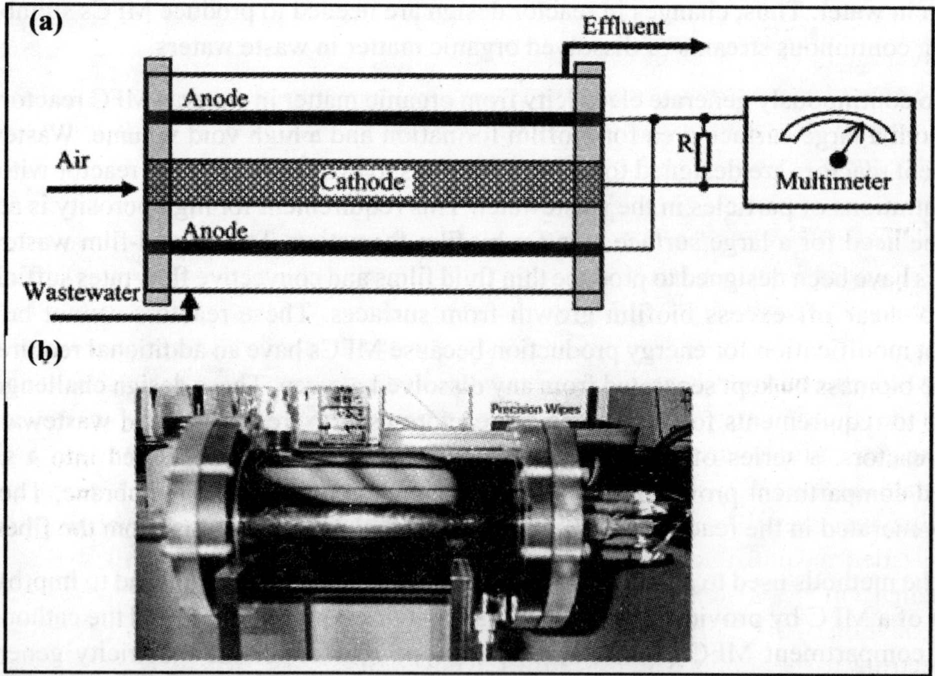


**Figure 7 (a)** Schematic and **(b)** laboratory-scale prototype of the SCMFC used to generate electricity from wastewater

### *2.1.5. Microbial fuel cell producing electricity from acetate or butyrate*

Hydrogen production from wastewater by biological fermentation has drawn much attention as a method of producing a valuable product during treatment of wastewaters containing high concentrations of carbohydrates. One mole of glucose can theoretically be converted into 12 mol of hydrogen, but the maximum yield via known fermentation routes is only 4 mol of hydrogen when acetate is the sole byproduct. While the maximum efficiency of hydrogen production is therefore 33%, typically only 15% of the energy is recovered as hydrogen with the remainder of the organic matter present as fatty acids and alcohols.

To improve the economics of hydrogen production from wastewater, additional processes are needed to recover the remaining energy. One approach is to link hydrogen production with methane production by using a two-stage process. Although two-stage anaerobic treatments have been used to make methane, it has not yet been proven outside of the laboratory that hydrogen can be recovered at high concentrations from the first stage using actual wastewaters. A second approach is to use phototrophic bacteria to recover additional hydrogen from the byproducts of hydrogen fermentation. Although solar energy is free, the availability of sufficient land area and the instability of sufficient solar energy at the plant would make such a process difficult for wastewater treatment applications. A third approach is to recover the remaining energy directly as electricity in a microbial fuel cell (MFC). A MFC can be linked to biohydrogen production where the power can be generated from degradation of these compounds. Power generation from acetate and butyrate, produced as byproducts of hydrogen production can be obtained in single as well as two-chambered MFCs. Such a cell uses *Geobacter sulfurreducens* which is capable of oxidizing acetate and hydrogen. This organism can produce electricity by forming a monolayer directly on the anode electrode surface and use this as their end terminal electron acceptor in anaerobic respiration. This is a unique ability that can exist in species termed as anodophiles such as *G. sulfurreducens* and *Rhodoferax ferrireducens*.

## 2.2. Enzymatic biofuel cells

Microbial biofuel cells require the continuous fermentation of whole living cells performing numerous physiological processes, and thus dictate stringent working conditions. In order to overcome this constraint, the redox-enzymes responsible for desired processes may be separated and purified from living organisms and applied as biocatalysts in biofuel cells rather the whole microbial cells. That is, rather than utilizing the entire microbial cell-apparatus for the generation of electrical energy, the specific enzyme(s) that oxidizes the target fuel-substrate may be electrically-contacted with the electrode of the biofuel cell element (Figure 8) . Enzymes are still sensitive and expensive chemicals, and thus special ways for their stabilization and utilization must be established. Upon utilizing enzymes as catalytically active ingredients in biofuel, one may apply oxidative biocatalysts in the anodic compartments for the oxidation of the fuel-substate which may be carbohydrates or alcohols etc and transfer of electrons to the anode, whereas reductive biocatalysts may participate in the reduction of the oxidizer in the cathodic compartment of the biofuel cell. Redox-enzymes lack, however, direct electrical communication with electrodes due to the insulation of the redox-center from the conductive support by the protein matrices. Several methods have been applied to electrically contact redox-enzymes and electrode supports. To understand the working principle of an enzyme fuel cell (EFC) based on methanol as a substrate, let's discuss the reactions occurring in anode chamber of the cell which contains the enzyme, mediator and methanol as the substrate which is oxidized by the enzyme and the release of electrons to anode are facilitated by the mediator which is an artificial electron acceptor (phenazine ethosulphate or phenazine methosulphate).

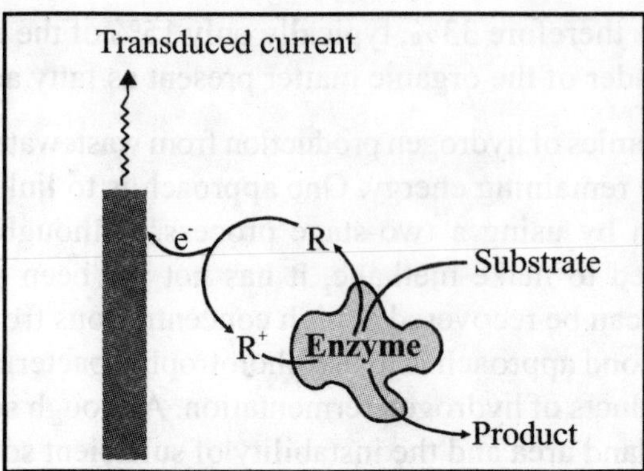


**Figure 8** A schematic representation of fuel-substrate oxidation by enzyme electrically-contacted with the electrode of the biofuel cell element

### *2.2.1. Anodes for biofuel cells based on enzyme-catalyzed oxidative reactions*

The electrochemical oxidation of fuels can be biocatalyzed by enzymes electrically communicating with electrodes. Different classes of oxidative enzymes (e.g. oxidases, dehydrogenases) require the application of different molecular tools to establish this electrical communication. Electron transfer mediators shuttling electrons between the enzyme active centers and electrodes are usually needed for the efficient electrical communication of FAD-containing oxidases (e.g., glucose oxidase, GOx). $NAD(P)^+$-dependent dehydrogenases (e.g. lactate dehydrogenase, LDH) require $NAD(P)^+$-cofactor and an electrode catalytically active for the oxidation of NAD(P)H and regeneration of $NAD(P)^+$ to establish an electrical contact with the electrode.

### *2.2.2. Anodes based on the bioelectrocatalyzed oxidation of NAD(P)H*

The nicotinamide redox cofactors ($NAD^+$ and $NADP^+$) play important roles in biological electron transport, acting as carriers of electrons and activating the biocatalytic functions of dehydrogenases, the vast majority of redox enzymes. The application of $NAD(P)^+$-dependent enzymes (e.g. lactate dehydrogenase; alcohol dehydrogenase, glucose dehydrogenase) in biofuel cells allows the use of many organic materials such as lactate, glucose and alcohols as fuels. The biocatalytic oxidation of these substrates requires the efficient electrochemical regeneration of $NAD(P)^+$-cofactors in the anodic compartment of the cells. The biocatalytically produced NAD(P)H cofactors participating in the anodic process transports electrons from the enzymes to the anode, and the subsequent electrochemical oxidation of the reduced cofactors regenerates the biocatalytic functions of the system. Electrochemistry of NAD(P)H has been studied extensively, and it has been demonstrated that the electrochemical oxidation process is highly irreversible and proceeds with large overpotentials at carbon, Pt and Au electrodes. Strong adsorption of NAD(P)H and $NAD(P)^+$ on Pt, Au, glassy carbon and pyrolytic graphite generally poisons the electrode surface and inhibits the oxidation process. Furthermore, $NAD(P)^+$ acts as an inhibitor for the direct oxidation of NAD(P)H, and adsorbed NAD(P)H can be oxidized to undesired products that lead to the degradation of the cofactor (e.g. to $NAD^+$-dimers). Thus, the non-catalyzed electrochemical oxidation of NAD(P)H is not appropriate for use in workable biofuel cells. For the efficient electrooxidation of NAD(P)H, mediated electrocatalysis is necessary. Several immobilization techniques have been applied for the preparation of mediator-modified electrodes: The mediator molecules can be adsorbed directly onto electrodes, incorporated into polymer layers, or covalently linked to functional groups on electrode surfaces.

A biofuel cell based on the electrocatalytic regeneration of $NAD^+$ at a modified anode has been developed in which glucose dehydrogenase is immobilized in a porous glass located in the anode compartment of the biofuel cell. The enzyme oxidizes the substrate (glucose) and produces the reduced state of the cofactor (NADH). The reduced cofactor reaches the anode surface diffusionally, where it gets oxidized to $NAD^+$. The covalent coupling of redox mediators to self-assembled monolayers on Au-electrode surfaces has an important advantage for the preparation of multi-component organized systems. Pyrroloquinoline quinone (PQQ) can be covalently attached to amino groups of a cystamine monolayer assembled on a Au-surface (Figure 9). The resulting electrode demonstrates good electrocatalytic activity for NAD(P)H oxidation, particularly in the presence of $Ca^{2+}$-cations as promoters.

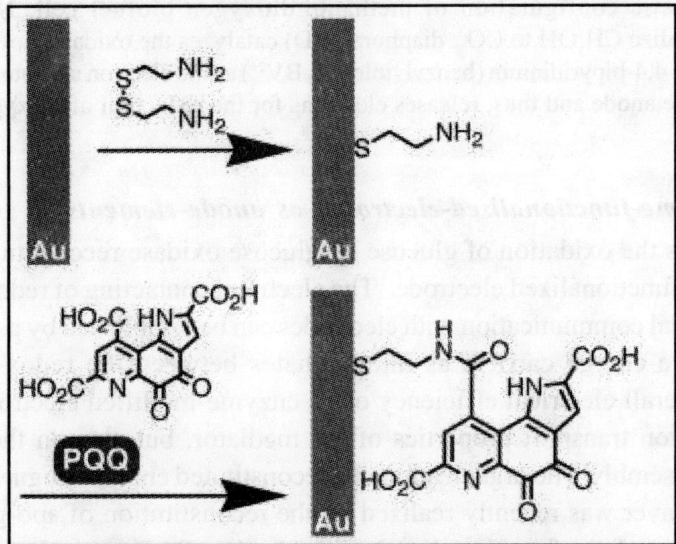


**Figure 9** Assembly of the PQQ-modified Au-electrode

NAD(P)$^+$-dependent enzymes electrically contacted with electrode surfaces can provide efficient bioelectrocatalysis for the NAD(P)H oxidation. For example, diaphorase can be applied to oxidize NADH using a variety of quinone compounds, several kinds of flavins, and viologens as mediators. A biofuel cell has been developed based on enzymes (producing NADH upon the biocatalytic oxidation of primary substrate) and diaphorase (electrically contacted via an electron relay and providing bioelectrocatalytic oxidation of the NADH to $NAD^+$. A number of $NAD^+$-dependent enzymes (alcohol dehydrogenase, aldehyde dehydrogenase, formate dehydrogenase) provide a sequence of biocatalytic reactions resulting in the oxidation of methanol to formaldehyde and finally $CO_2$ (Figure 10). The reduced cofactor (NADH) produced in all the steps is biocatalytically oxidized by diaphorase.

This diaphorase is electrochemically contacted by a diffusional electron relay (benzylviologen, $BV^{2+}$) that provides enzyme regeneration and anodic current. The biocatalytic anode is conjugated with an $O_2$-cathode to complete the biofuel cell.

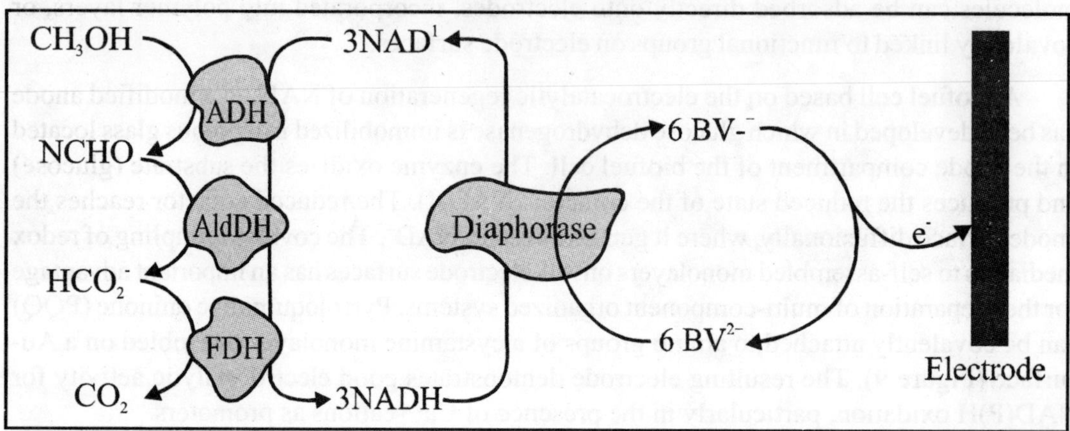


**Figure 10** Schematic configuration of methanol/dioxygen biofuel cell. $NAD^+$-dependent dehydrogenases oxidize $CH_3OH$ to $CO_2$; diaphorase (D) catalyzes the oxidation of NADH to $NAD^+$ using *N,N'*-dibenzyl-4,4-bipyridinium (benzylviologen, $BV^{2+}$) as the electron acceptor. $BV^{\cdot+}$ is oxidized to $BV^{2+}$ at a graphite anode and thus, releases electrons for the reduction of dioxygen at a platinum cathode

### *2.2.3. Flavoenzyme-functionalized-electrodes as anode-elements*

This involves the oxidation of glucose by glucose oxidase reconstituted on an FAD/ PQQ-monolayer-functionalized electrode. The electrical contacting of redox-enzymes that defy direct electrical communication with electrodes can be established by using synthetic or biologically-active charge carriers as intermediates between the redox-center and the electrode. The overall electrical efficiency of an enzyme-modified electrode depends not only on the electron transport properties of the mediator, but also on the transfer steps occurring in the assembly. The organization of a reconstituted enzyme aligned on an electron relay-FAD monolayer was recently realized by the reconstitution of apo-glucose oxidase (apo-GOx) on a surface functionalized with a relay-FAD monolayer (Figure 11). Pyrroloquinoline quinone (4) was covalently linked to a base cystamine monolayer at a Au-electrode, and $N^6$-(2-aminoethyl)-FAD (8) was then attached to the PQQ relay units. The resulting enzyme-reconstituted PQQ-FAD-functionalized electrode revealed bioelectrocatalytic properties. The electrode constantly oxidizes the PQQ site located at the protein periphery, and the PQQ-mediated oxidation of the FAD- center activates the bioelectrocatalytic oxidation of glucose:

$$FAD + glucose + 2H^+ \rightarrow FADH_2 + gluconic\ acid$$

$$FADH_2 + PQQ \rightarrow FAD + PQQH_2$$

$$PQQH_2 \rightarrow PQQ + 2H^+ + 2e^-\ (to\ anode)$$

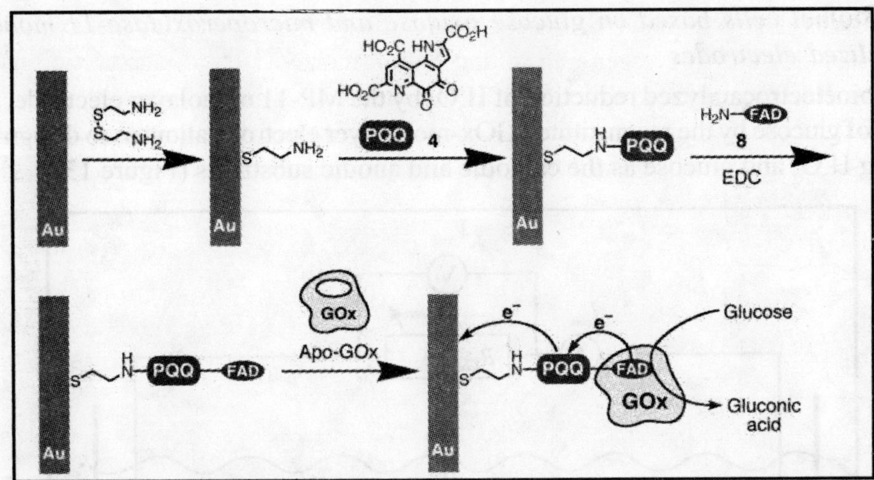


**Figure 11** The surface-reconstitution of apo-glucose oxidase on a PQQ-FAD monolayer assembled on a Au-electrode

### 2.2.4. *Biofuel cells based on layered enzyme-electrodes*

For the design of complete biofuel cells, it is essential to couple the cathode and anode units into integrated devices. Furthermore, for synchronous operation of the biofuel cell, charge compensation between the two electrodes must be attained, and the flow of electrons in the external circuit must be compensated by cation-transport in the electrolyte solution. To overcome these limitations, the catholyte and anolyte solutions may be compartmentalized.

#### *2.2.4.1. Biofuel cells based on pyrroloquinoline quinone and microperoxidase-11 monolayer-functionalized electrodes*

The bioelectrocatalyzed reduction of $H_2O_2$ by MP-11 and oxidation of NADH by PQQ has been used to design a biofuel cell using $H_2O_2$ and NADH as the cathodic and anodic substrates (Figure 12).

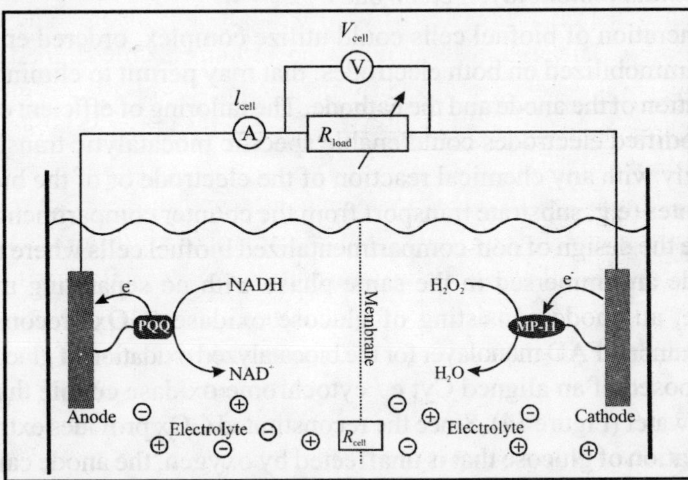


**Figure 12** Schematic configuration of a biofuel cell employing NADH and $H_2O_2$ as fuel and oxidizer substrates and PQQ- and MP-11-functionalized-electrodes as catalytic anode and cathode, respectively

*2.2.4.2. Biofuel cells based on glucose oxidase and microperoxidase-11 monolayer-functionalized electrodes*

The bioelectrocatalyzed reduction of $H_2O_2$ by the MP-11 monolayer electrode, and the oxidation of glucose by the reconstituted GOx-monolayer electrode allow us to design biofuel cells using $H_2O_2$ and glucose as the cathodic and anodic substrates (Figure 13).

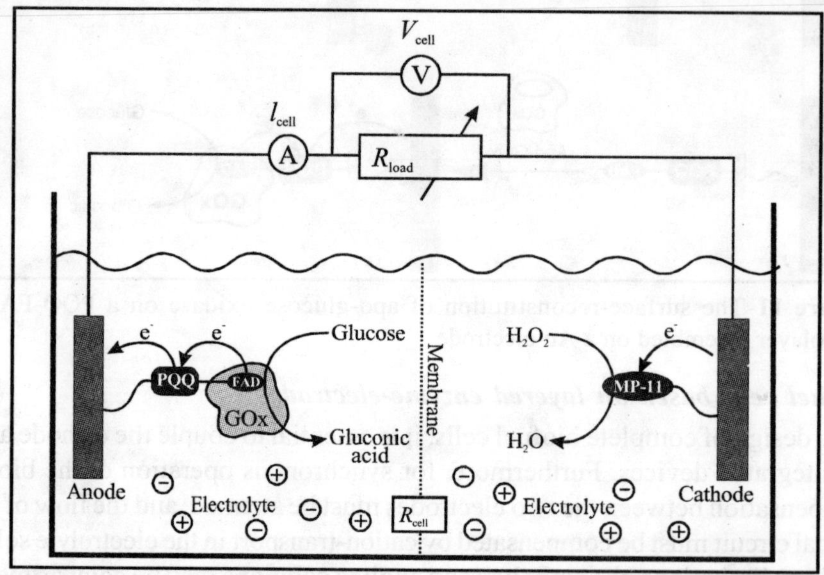


**Figure 13** Schematic configuration of a biofuel cell employing glucose and $H_2O_2$ as a fuel and an oxidizer, respectively. Glucose oxidase reconstituted onto a PQQ-FAD-monolayer and MP-11 - functionalized Au-electrodes act as the biocatalytic anode and cathode, respectively

*2.2.4.3. Non-compartmentalized biofuel cells based on glucose oxidase and cytochrome c / cytochrome oxidase monolayer-electrodes*

The next generation of biofuel cells could utilize complex, ordered enzyme or multi-enzyme systems immobilized on both electrodes, that may permit to eliminate the need for compartmentalization of the anode and the cathode. The tailoring of efficient electron transfer at the enzyme-modified electrodes could enable specific biocatalytic transformations that compete kinetically with any chemical reaction of the electrode or of the biocatalysts with interfering substrates (e.g. substrate transport from the counter compartment, oxygen, etc.). This would enable the design of non-compartmentalized biofuel cells where the biocatalytic anode and cathode are immersed in the same phase with no separating membrane. In a working example, an anode consisting of glucose oxidase (GOx) reconstituted onto a pyrroloquinoline quinone-FAD monolayer for the biocatalyzed oxidation of glucose was coupled to a cathode composed of an aligned Cyt c / cytochrome oxidase couple that catalyzes the reduction of $O_2$ to water (Figure 14). Since the reconstituted GOx provides extremely efficient biocatalyzed oxidation of glucose that is unaffected by oxygen, the anode can operate in the presence of oxygen. Thus, the biofuel cell uses $O_2$ as an oxidizer and glucose as a fuel without the need for compartmentalization of pacemakers or insulin pumps.

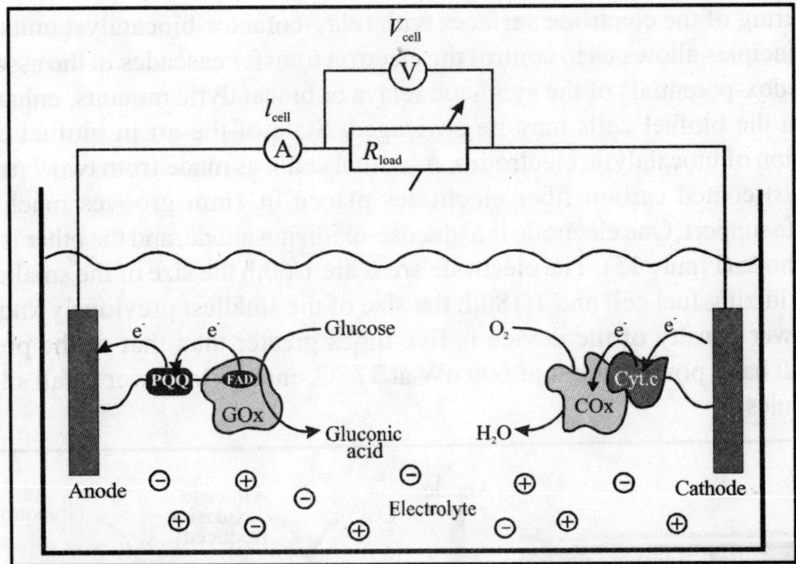


**Figure 14** Schematic configuration of a non-compartmentalized biofuel cell employing glucose and $O_2$ as fuel and oxidize, and using PQQ-FAD/GOx and Cyt.c / COx-functionalized Au-electrodes as biocatalytic anode and cathode, respectively

## 3. CONCLUSIONS

Biofuel cells for the generation of electrical energy from abundant organic substrates can be organized by different approaches. One approach includes the use of microorganisms as biological reactors for the fermentation of raw materials to fuel products, e.g. hydrogen, that are delivered into a conventional fuel cell. The second approach to utilize microorganisms in the assembly of biofuel cells includes the in situ electrical coupling of metabolites generated in the microbial cells with the electrode support using diffusional electron mediators. A further methodology to develop biofuel cells involves the application of redox-enzymes for the targeted oxidation and reduction of specific fuel and oxidizer substrates at the electrode supports and the generation of the electrical power output.

Towards this goal, it is essential to tailor integrated enzyme-electrodes that exhibit electrical contact and communication with the conductive supports. The detailed characterization of the interfacial electron transfer rates, biocatalytic rate-constants and cell resistances is essential upon the construction of the biofuel cells. The identification of the rate-limiting steps allows then the development of strategies to improve and enhance the cell output. The chemical modification of redox-enzymes with synthetic units that improve the electrical contact with the electrodes provides a general means to enhance the electrical output of biofuel cells. The site-specific modification of redox enzymes and the surface-reconstitution of enzymes represent novel and attractive means to align and orient biocatalysts on electrode surfaces. The effective electrical contacting of aligned proteins with electrodes suggests that future efforts might be directed towards the development of structural mutants of redox-proteins to enhance their electrical communication with electrodes. The stepwise nanoengineering of the electrode surfaces with relay-cofactor-biocatalyst units by organic synthesis principles allows us to control the electron transfer cascades in the assemblies. By tuning the redox-potentials of the synthetic relays or biocatalytic mutants, enhanced power outputs from the biofuel cells may be envisaged. State-of-the-art in biofuel cells allows miniaturization of biocatalytic electrodes. A biofuel cell was made from two 7 μm diameter, electrocatalyst-coated carbon fiber electrodes placed in 1mm grooves machined into a polycarbonate support. One electrode is a glucose-oxidizing anode, and the other is an oxygen-reducing cathode (Figure 15). The electrode areas are 1/60th the size of the smallest reported methanol-oxidizing fuel cell and 1/180th the size of the smallest previously known biofuel cell. The power density of the device is five times greater than that of the previous best biofuel cell. It has a power output of 600 nW at 37 °C, enough to power small silicon-based microelectronics.

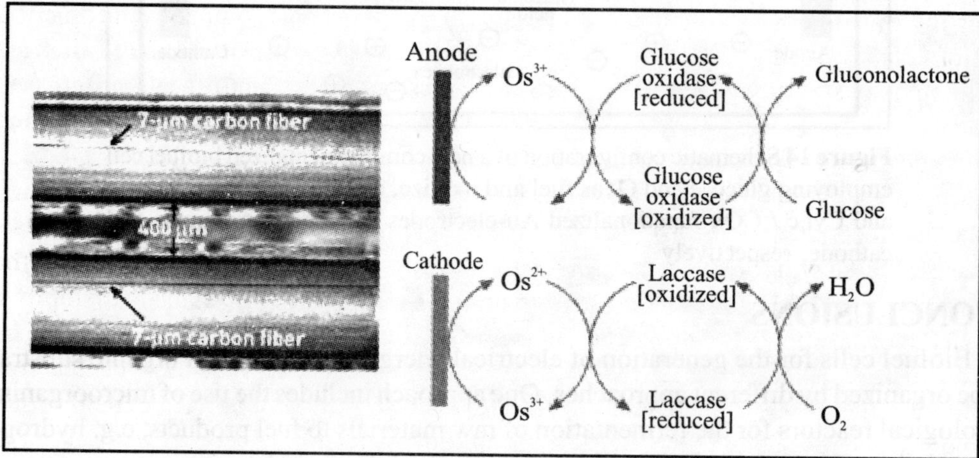


**Figure 15** Redox electron-transferring steps of glucose oxidation and $O_2$ reduction drive tiny biofuel cell

One more interesting example of a novel device is a robot powered by a biofuel. The robot, named "Gastronome", uses a microbial fuel cell (MFC) system to directly convert carbohydrate fuel to an electrical power source without combustion. This is achieved through tapping the metabolic pathways of microorganisms. Gastronome is thought to be the first robot of its kind to use such a novel biomass energy conversion method, and as such represents a new class of bioelectromechanical machines, dubbed Gastrobots (literally: robots with stomachs). Gastrobots can potentially be sustained from just an input of natural food, water and air, and are therefore ideal for a host of applications that demand "living off the land" during fully autonomous "start and forget" missions. Gastronome is built around a platform of three small wheeled wagons linked together like train cars. The wagons contain a "stomach", "lung", gastric pump, "heart" pump, and a six cell MFC stack. Exotic materials such as titanium terminal plates, reticulated vitreous carbon electrodes, carbon felt contact pads, and proton exchange membrane were required, together with a microbial biocatalyst and various chemical agents. Ni-Cd battery packs, kept charged by the MFC output, are drawn upon to run electronic controls and motors. The production of electrical energy from biomass substrates using biofuels could complement energy sources from chemical fuel cells. An important potential use of biofuel cells is their in situ assembly in human body fluids, e.g., blood. The extractable electrical power could then be used to activate implanted devices such as pacemakers, pumps (e.g., insulin pumps), sensors and prosthetic units.

## 4. FUTURE OUTLOOK

The electrical activity of micro-organisms is a fascinating and instructive area of science. Although much work remains to be done at the research level to elucidate the chemistry and biochemistry of such phenomena, recent studies of microbial fuel cells have greatly advanced our understanding of microbial electricity generation. Microbial fuel cells hold great promise as a sustainable biotechnological solution to future energy needs. Enzyme-based biofuel cells are attracting attention rapidly partially due to the promising advances reported recently. However, there are issues to be addressed before biofuel cells become competitive in practical applications. Two critical issues are short lifetime and poor power density, both of which are related to enzyme stability, electron transfer rate, and enzyme loading. Recent progress in nanobiocatalysis opens the possibility to improve in these aspects. Many nano-structured materials, such as mesoporous media, nanoparticles, nanofibers, and nanotubes, have been demonstrated as efficient hosts of enzyme immobilization. It is evident that, when nanostructure of conductive materials are used, the large surface area of these nanomaterials can increase the enzyme loading and facilitate reaction kinetics, and thus improving the power density of biofuel cells. In addition, research efforts have also been made to improve the activity and stability of immobilized enzymes by using nanostructures. It appears to be reasonable to us to expect that progress in nanostuctured biocatalysts will play a critical role in overcoming the major obstacles in the development of powerful biofuel cells. Further improvements can be made to such fuel cell configurations to improve energy recovery or to increase voltages by linking the fuel cells in series, resulting in new technologies that will make electricity generation using biofuel cells. Some day, it would replace rechargeable batteries. Instead of plugging into a fixed power outlet and waiting for a recharge, these new batteries would last up to a full month after they are charged instantly with a few milliliters of alcohol or glucose solution to power cell phones, laptops, robots, pacemakers etc.

## 5. FURTHER READING

Bennetto, H.P. (1990). Electricity generation by microorganisms. Biotechnol. Edu. 1: 163-168.

Bond, D.R. and Lovley, D.R. (2003). Electricity production by Geobacter sulfurreducens attached to electrodes. Applied and Environmental Microbiology. 69: 1548-1555.

Chaudhuri, S.K. and Lovley, D.R. (2003). Electricity generation by direct oxidation of glucose in mediatorless microbial fuel cells. Nature Biotechnology. 21: 1229-1232.

Ehrenman, G. (2005). From foul to fuel. Mechanical Engg. 21:1-16.

Ieropoulos, I.A., Greenman, J., Melhuish, C. and Hart, J. (2005). Comparative study of three types of microbial fuel cell. Enzyme and Microbial Technology. 37: 238-245.

Kim, J., Jia, H. and Wang, P. (2006). Challenges in biocatalysis for enzyme-based biofuel cells. Biotechnol Adv. 24: (in press).

Liu, H., Ramnarayanan, R. and Logan, B.E. (2004). Production of Electricity during Wastewater Treatment Using a Single Chamber Microbial Fuel Cell. Environmental Science Technology. 38 : 2281 -2285.

Logan, B.E. and Liu, H. (2004). Production of electricity during waste water treatment using a single chamber MFC. Environmental Science Technology. 38: 2281-2285.

Lovley, D.R. and Fertig, S.J. (2002). Harnessing microbially generated power of the sea floor. Nature Biotechnology. 20: 821-825.

Scholz, F. and Schroder, U. (2003). Bacterial batteries. Nature Biotechnology. 21: 1151-1152.

Service, R. (2002). Biofuel cells. Science. 296: 1223.

□□□

# 12

# Biotechnology for Developing Novel Microbes

**R. SONI[1], S.K. SONI[2] AND N. GOYAL[2]**

[1]*Department of Biotechnology, D.A.V. College, Chandigarh-160 011*

[2]*Department of Microbiology, Panjab University, Chandigarh-160 014*

# 1. INTRODUCTION

The term 'biotechnology' has come to mean different things to different people. The simplest definition of biotechnology is 'applied biology'. Another definition is 'the use of living organisms to make a product or run a process'. By this definition, the classic techniques used for plant and animal breeding to produce stock with enhanced qualities, fermentations to make alcohol, yogurt, cheese, vinegar, enzymes, antibiotics etc. would be considered biotechnology. Some people use the term biotechnology only to refer to the newer tools of genetic engineering developed since 1973. In this context, biotechnology may be defined as "the use of biotechnical methods to modify the genetic material of living cells so they will produce new substances or perform new functions". Examples include recombinant DNA (rDNA) technology, in which a copy of a piece of DNA containing one or a few genes is transferred between organisms or recombined within an organism, and rDNA or gene splicing, which may be likened to cutting a circle of tape, inserting a different piece, and rejoining both ends to the new piece

One of the important living organisms of interest in Biotechnology are the microorganisms which are exploited for the formation of a desired product. These are also recognized as the unacknowledged workers of the biotechnology industry. Microorganisms used are either the organisms already present in nature or the formation of new microbes that are able to show desired properties. Generation of microorganisms involves either screening or use of recombinant DNA technology. Screening involves obtaining organisms already present in nature. Recombinant DNA technology is the manipulation of existing microorganisms by insertion of foreign DNA into their genomes so as to obtain desirable characteristics of the donor organism in the host. It is however not very easy and is a slow process.

# 2. STRAIN DEVELOPMENT TECHNOLOGIES

The desired microbes can be modified genetically in many different ways. Some techniques alter limited regions of the genome, whereas other techniques are used to recombine or rearrange the entire genome. Techniques having the greatest potential in genetic programming include selection of variants, mutation and selection, hybridization, rare-mating, protoplast fusion, gene cloning and transformation. The combined use of replica plating, mutagenesis, hybridization and recombinant DNA methods have dramatically increased the genetic diversity that can be introduced into microbial cells.

## 2.1. Screening of natural variants of microorganisms

Screening involves the use of highly selective procedures to allow the detection and isolation of only those microbes of interest among a large microbial production. The various sources to obtain microorganisms are soil, sea water, marine mud, compost, rumen contents, domestic sewage, manure, spoiled foodstuff or feedstuff. For example, to obtain protease producing microorganisms, the various organisms are grown on opaque protein agar. Zones of clearance are found around protease producers. Further if we want to isolate alkalophilic protease producers, the organisms are grown in a alkaline pH medium . Screening is of two types: primary, to detect and isolate microbes that possess potentially interesting application and secondary screening, which allows further sorting out of those microbes that have real value for industrial process and the discarding of those lacking this potential. The following is an example of screening of microbes for novel enzymes.

The first stage of the screening procedure for commercial enzymes is to screen ideas, i.e. to determine the potential commercial need for a new enzyme, to estimate the size of the market and to decide, approximately, how much potential users of the enzyme will be able to afford to pay for it. The next stage involves the location of a source of the required enzyme screening of new microbial strains capable of performing the transformation required. This is not be a 'blind' screen but is usually a source of microbes that could have been exposed for countless generations to the conditions that the new enzyme should withstand. Hence, thermophiles are sought in hot springs, osmophiles in sugar factories, organisms capable of metabolizing wood preservatives in timber yards and so on. The identification of a microbial source of an enzyme is by no means the end of the story. The properties of the enzyme must be determined; i.e. temperature for optimum productivity, temperature stability profile, pH optimum and stability, kinetic constants ($K_m$ and $V_{max}$), whether there is substrate or product inhibition, and the ability to withstand components of the expected feedstock other than substrate. A team of scientists, engineers and accountants must then consider the next steps. If any of these parameters is unsatisfactory, the screen must continue until improved enzymes are located. Once an enzyme with suitable properties has been located, various decisions be made concerning the acceptability of the organism to the regulatory authorities, the productivity of the organism, and the way in which the enzyme is to be isolated, utilized (free or immobilized) and, if necessary, purified. If the organism is unacceptable from a regulatory viewpoint, two options exist; to eliminate that organism altogether and continue the screening operation, or to clone the enzyme into an acceptable organism. The latter approach is becoming increasingly attractive especially as cloning could also be used to increase the productivity of the fermentation process. The use of immobilised enzymes is now familiar to industry and their advantages are well recognised so the practicality of using the new enzymes in an immobilised form will be determined early in the screening procedure. If the enzyme is produced intracellularly, the feasibility of using it without isolation and purification will be considered very seriously and strains selected for their amenability to use in this way.

It should be emphasised that there should be a constant dialogue between laboratory scientists and biochemical process engineers from the earliest stages of the screening process. Once the biochemical engineers are satisfied that their initial criteria of productivity, activity and stability can be met, the selected strain(s) of microbe will be grown in pilot plant conditions. It is only by applying the type of equipment used in full scale plants that accurate costing of processes can be achieved. Pilot studies will probably reveal imperfections, or at least areas of ignorance, that must be corrected at the laboratory scale. If this proves possible, the pilot plant will produce samples of the enzyme preparation to be used by customers who may well also be at the pilot plant stage in the development of the enzyme-utilizing process.

Screening for new enzymes is expensive so that the intellectual property generated must be protected against copying by competitors. This is usually done by patenting the enzyme or its production method or, most usefully, the process in which it is to be used. Patenting will be initiated as soon as there is evidence that an innovative discovery has been made.

## 2.2. Genetic manipulation of microorganisms

The purpose of the genetic manipulation is to create a new strain with precisely defined genetic properties. The techniques used for the manipulations may be random or specific and are discussed below.

### *2.2.1. Random mutagenesis and selection*

Mutation of microbial strains can lead to improvement of certain traits with the simultaneous debilitation of other characteristics. Through mutations, the genetic constitution of microorganisms can be altered. Mutation for beneficial products was started since the time of Alexander Fleming during the end of 1920s when by using ultraviolet rays penicillin production by the mold *Penicillium chrysogenum* got increased by about 1000 fold. Mutations are permanent, sometimes transmissible changes to the genetic material of a cell which are caused by copying errors in the genetic material during cell division and by exposure to radiation, chemicals, or viruses. Although mutations are probably induced with the same frequency in haploid, diploid or polyploids, they are not easily detected in diploid and polyploid cells because of the presence of non-mutated alleles. Only if the mutation is dominant, is a phenotypic effect, it is detected without the need for additional alterations. Therefore, haploid strains of microorganisms are preferred, though not essential, when inducing mutations. Successful mutation and breeding is usually associated with mutations in meiotic segregants, where the two mating parents of a genetically stable hybrid provide a good basis for the introduction of recessive mutations. Mutagenesis has the potential to disrupt or eliminate undesirable characteristics and to enhance favourable properties microbes. Though the use of mutagens for directed strain development is limited, the method could be applied to isolate new variants of the microorganisms prior to further genetic manipulation. In multicellular organisms, mutations can be subdivided into germ line mutations, which can be passed on to progeny and somatic mutations, which often lead to the malfunction or death of a cell and can cause cancer. Mutations are considered the driving force of evolution, where less favorable (or deleterious) mutations are removed from the gene pool by natural selection, while more favorable (or beneficial) ones tend to accumulate. Neutral mutations do not affect the organism's chances of survival in its natural environment and can accumulate over time, which might result in what is known as punctuated equilibrium, the modern interpretation of classic evolutionary theory. Structurally, mutations can be classified as small-scale mutations and large-scale mutations in chromosomal structure.

#### *2.2.1.1. Small-scale mutations*

These affect one or a few nucleotides and may include point and frame shift mutations. Point mutations, often caused by chemicals or malfunction of DNA replication, exchange a single nucleotide for another. Most common is the transition that exchanges a purine for a purine or a pyrimidine for a pyrimidine ($A \leftrightarrow G$, $C \leftrightarrow T$). A transition can be caused by nitrous acid, base mispairing, or mutagenic base analogs such as 5-bromo-2-deoxyuridine (BrdU). Less common is a transversion, which exchanges a purine for a pyrimidine or a pyrimidine for a purine (C/T ↔ A/G). A point mutation can be reversed by another point mutation, in which the nucleotide is changed back to its original state (true reversion) or by second-site reversion (a complementary mutation elsewhere that results in regained gene functionality). These changes are classified as transitions or transversions. An example of a transversion is adenine being converted into a cytosine. There are also many other examples that can be found. There are three kinds of point mutations, depending upon what the erroneous codon codes for: i) silent mutations: codes for the same amino acid, so has no effect, ii) missense mutations: codes for a different amino acid, iii) nonsense mutations: codes for a stop, which can truncate the protein. Frameshift mutations involve the insertion or deletion of one or few nucleotides in the DNA molecule which shifts the reading frame of nucelotide sequences. These are usually caused by transposable elements, or errors during replication of repeating elements (e.g. AT repeats). Most insertions or deletions in a gene can either alter splicing of the mRNA, or cause a shift in the reading frame, both of which can significantly alter the gene product.

*2.2.1.2. Large-scale mutations*

These bring about a change in chromosomal structure and include i) amplifications or gene duplications, leading to multiple copies of chromosomal regions, increasing the dosage of the genes located within them ii) deletions of large chromosomal regions, leading to loss of the genes within those regions iii) chromosomal translocations attaching DNA from separate chromosomes, iv) interstitial deletions removing regions of DNA from a single chromosome, thereby apposing previously distant genes, v) chromosomal inversions switching the orientation of a segment of a chromosome, thereby apposing its ends to previously distant genes, vi) Loss of heterozygosity with loss of one allele, either by a deletion or recombination event, in organisms which previously had two.

Mutations can take place spontaneously or can be induced intentionally in various microorganisms. In nature, mutation appears as random or chance events with no regard to environmental conditions. Spontaneous mutations occur in the absence of any agent known to alter the genetic material. Induced mutations are produced by chemical and physical agents, termed as mutagens including base analogs, alkylating agents, deaminating agents, acridine derivatives, X-rays, gamma rays and UV rays.

The average spontaneous mutation frequency in microbes at any particular locus is approximately $10^{-6}$ per generation. The use of mutagen greatly increases the frequency of mutations in a wine yeast population. Mutation and selection (often through replica-plating on selective agar media) appear to be a rational approach to strain development when a large number of performance parameters are to be kept constant while only one is to be changed.

The phenotypic results of mutation can vary dramatically. If the changes occur in a non-functional area of DNA or, because of the redundant nature of the genetic code, do not change a gene at the level of the protein sequence, the mutation will be silent. Such a mutation will not have any phenotype although such mutation can accumulate and change the genetic structure of subsequent generations. Even if a mutation causes a change in the amino acid sequence of a protein, the consequences to the organism may be negligible. For example, conservative amino acid changes may not affect the activity of a mutated enzyme or the enzyme may not be required in the environmental niche in which the organism is growing. In the former case, such a change is an evolutionary step affecting the protein and may have consequences on protein activity at a latter stage of evolution change. In the latter case, if the environment changes, the lack of the enzyme may become critical. In general, changes at the amino acid level in a protein will have a consequential effect on the protein's activity, usually detrimental. Thus, an important metabolic pathway may be blocked by the lack of a functional enzyme resulting in identifiable phenotypes. These phenotypes can be classified into four main groups: auxotrophic mutations, substrate utilization mutations, resistance mutations and essential mutations.

Auxotrophic mutations are mutations that knock out genes that are involved in anabolic pathways inside the bacterial cells. The metabolic pathways within cells synthesize a wide range of amino acids, nucleic acids, lipids, cofactors etc that are central to the normal growth of the cell. Obviously, all bacterial species do not synthesize all such cellular requirements and for a particular species, some additions to a basic medium which provides a carbon, nitrogen and energy source are needed. Such a medium is called a defined minimal medium and it is the simplest medium on which bacteria can grow. An auxotrophic mutation, because it interrupts a central anabolic function, would now require an addition chemical added to its minimal medium. Thus, a mutation that inactivates an enzyme in the pathway that synthesizes histidine will result in a requirement for histidine to be added to the minimal medium for bacterial growth. Similarly, an interruption of the purine biosynthetic pathway would need the addition of purines. Such mutations are easily identified by the failure of such a strain to group on the minimal medium for that species but the strain will grow in the presence of the added factor.

Substrate utilization mutations are mutations that knock out genes that are involved in catabolic pathways inside and outside the bacterial cell. Different species of bacteria can utilize different complex substrates as sources of energy, carbon, nitrogen etc. Such utilization needs the production of enzymes capable of breaking down the complex substrates to simple compounds that can be taken up by the cell. A mutation in either an enzyme involved in the breakdown or involved in the uptake will result in the loss of the ability to utilize such a resource. For example, a mutation in β-galactosidase in *E. coli* stops utilization of lactose as a carbon and energy resource. Similarly, a mutation that affects a cellulase gene will reduce to some extent degradation of cellulase by that organism. Interestingly, in this case, because most cellulase utilizing bacteria contain a number of cellulase genes, the utilization will probably still continue, but at a lower rate.

Resistance mutations are a class of mutations which can be considered artificial for the most part although they can occur in nature. The presence of antibiotics, heavy metals and other toxic substances in the environment means that they do occur in specific environments. In general, they can be divided into three types: firstly, mutations that inhibit uptake of the toxic substance, usually by inactivating a gene involved in uptake; secondly, mutations that change the biological target of the toxic substance so that it is less affected by the substance; and finally, by changing an enzyme so that it can now degrade the toxic substance, either by switching the gene on, if it is already present or by natural selection of a new gene activity.

*2.2.1.3. DNA damage in mutations*

Although DNA is naturally quite resistant to change because of its structure, aspects of its structure can lead limited change. Figure 1 shows the tautometric changes that can occur with certain bases which allow miss-pairing during DNA replication. Such miss-pairing results in a change in the base on one strand of the daughter DNA molecule. A second round of replication will result in the fixing of such a mutation. Usually, however, the miss-pairing is picked up by the DNA repair systems of the organisms. At that point, there is a 50% chance that the repair mechanisms will fix the mutation.

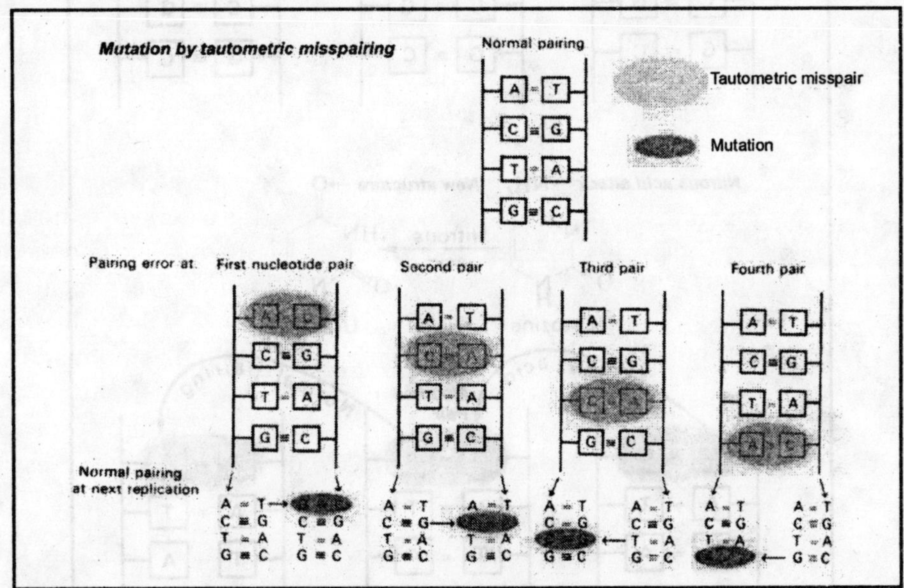


**Figure 1** Mutation by tautometric changes

Another form of DNA damage at the base level is the formation of thymidine dimers by ultra-violet light. As shown in Figure 2, two adjacent thymidines can be cross linked and such a complex stops DNA replication. Again, DNA repair comes in to remove the problem.

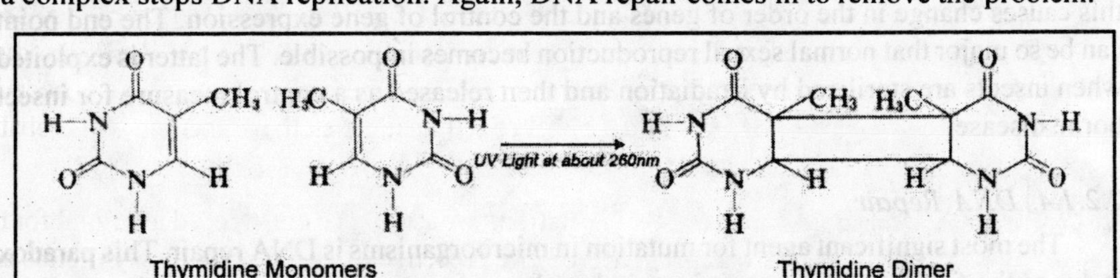

UV Light at about 260nm

Thymidine Monomers    Thymidine Dimer

**Figure 2** Formation of a thymidine dimer

Finally, at the base level, a variety of chemicals can damage DNA by modification of the base. This is shown in Figure 3. The resulting change may alter DNA pairing and result in a base sequence change.

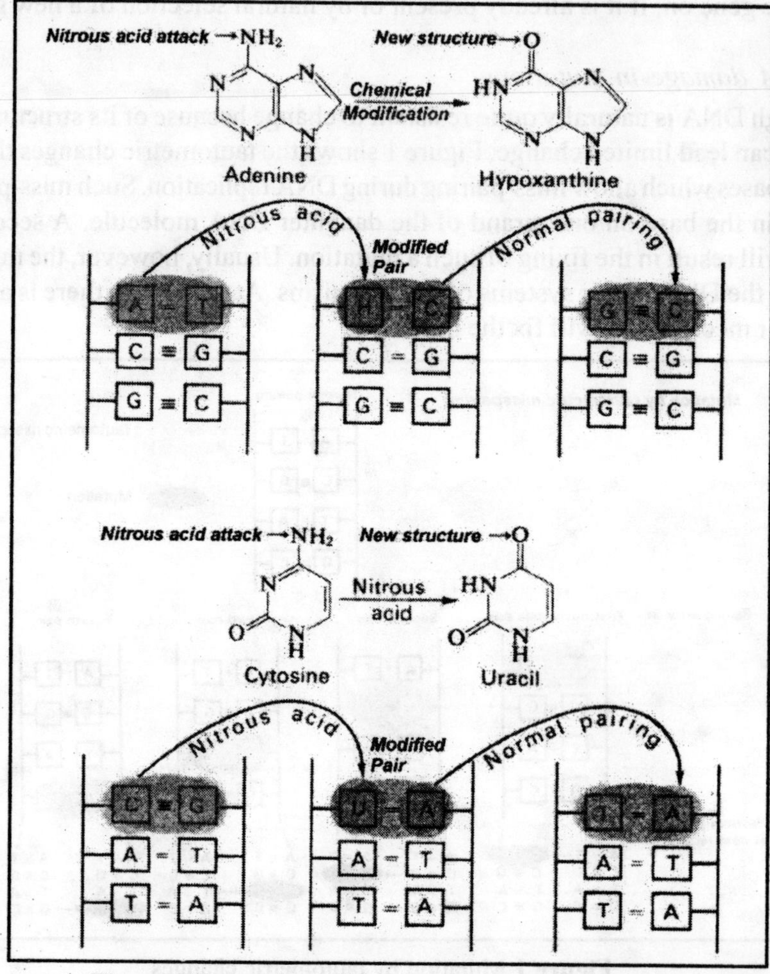


**Figure 3** Base pair modification caused by nitrous acid

Large scale changes can be induced by environmental factors such as ultra-violet light, nuclear radiation and chemicals. These can result in DNA strand breakage and rejoining; this causes change in the order of genes and the control of gene expression. The end point can be so major that normal sexual reproduction becomes impossible. The latter is exploited when insects are sterilized by irradiation and then released as a control measure for insect borne disease.

### *2.2.1.4. DNA Repair*

The most significant agent for mutation in microorganisms is DNA repair. This paradox is due to the fact that DNA repair is an induced emergency mechanism which has evolved to remove mutation that may cause the death of the cell. Because of this, some DNA repair mechanisms are error prone and as such cause further but less lethal DNA changes. An example of a DNA repair mechanism that is error free is the removal of thymidine dimers by photoreaction. Here, the enzyme photolyase uses visible light to split the linked bases and restore the ability of the DNA to replicate (Figure 4). However, error-prone repair in prokaryotes involves the enzyme DNA polymerase 1 which has a higher error rate than the DNA polymerase the cell uses for normal DNA replication. Thus, when DNA polymerase fills in the area where the mutation has been removed, as shown in Figure 5, the rate at which the wrong base is inserted is 100 to 1000 fold higher than in normal DNA replication. Because prokaryotes are haploid, such a mutation immediately become part of the genetic makeup of the cell unlike eukaryotic diploids where a normal second copy of the gene usually protects the cell from the potentially lethal effect of such a mutation.

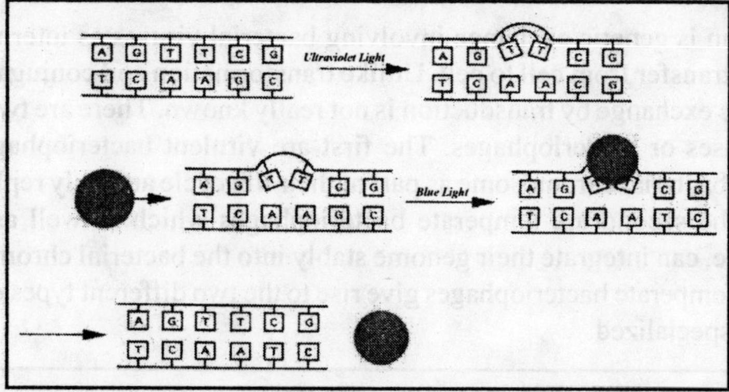


**Figure 4** Removal of thymidine dimers by photoreaction.

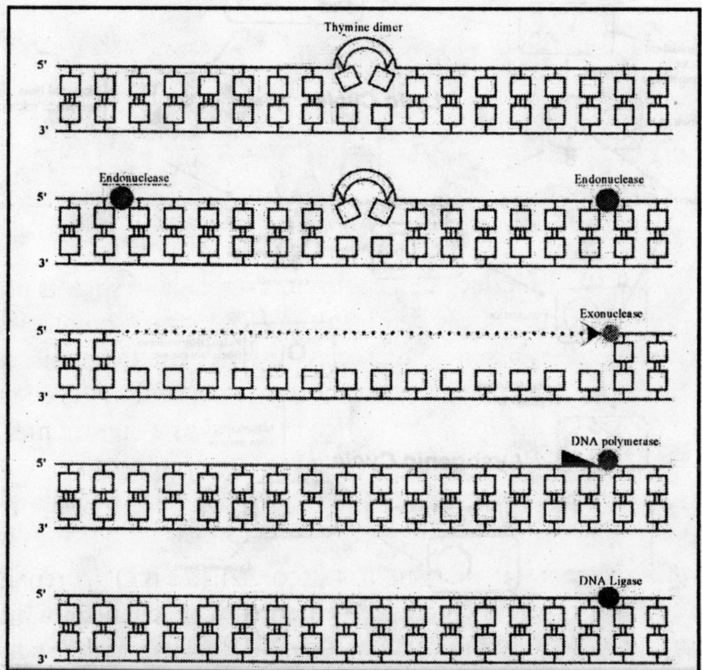


**Figure 5** Error-prone DNA repair in prokaryotes

### *2.2.2. Conjugation*

Conjugation is a method of genetic exchange that requires the physical connection of two bacterial cells for it to take place. Although not absolutely essential, conjugation usually involves a plasmid as the means of establishing the connection and causing the genetic exchange. In many cases, conjugation involves only plasmid exchange but chromosomal exchange also occurs at a significant rate. Conjugation, whereby chromosomal DNA is exchanged between bacteria, seems to occur as a byproduct of the presence of plasmids in bacteria although plasmid independent chromosome transfer is not unknown. Whether it is a circular or a linear plasmid, if the plasmid is capable of interacting with the main chromosome, it seems to be able to mobilize that chromosome.

### *2.2.3. Transduction*

Transduction is genetic exchange involving bacterial viruses as intermediate vectors that allow DNA transfer from cell to cell. Unlike transformation and conjugation, the extent of natural genetic exchange by transduction is not really known. There are two broad classes of bacterial viruses or bacteriophages. The first are virulent bacteriophages that do not interact with the bacterial chromosome as part of their life cycle and only replicate inside the bacterial cell. The second are temperate bacteriophages which as well as undergoing a virulent life cycle, can integrate their genome stably into the bacterial chromosome (Figure 6). Virulent and temperate bacteriophages give rise to the two different types of transduction, generalized and specialized.

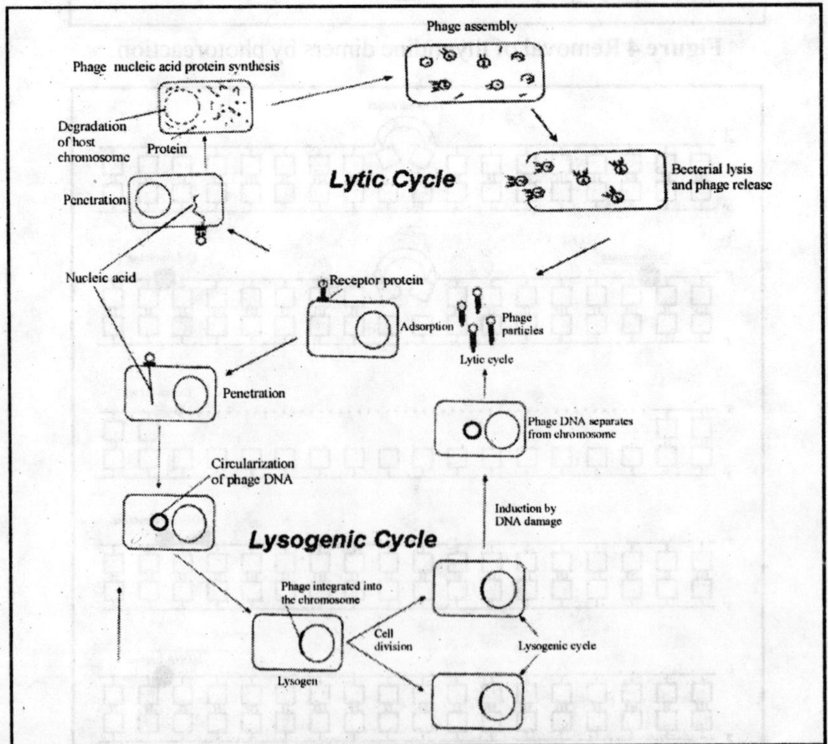


**Figure 6** Transduction by virulent and temperate bacteriophages

### 2.2.3.1. *Generalized transduction*

In this type of transduction, random pieces of chromosomal DNA are packaged by mistake into a bacteriophage head and can thus be transferred on infection into the new host DNA (Figure 7).

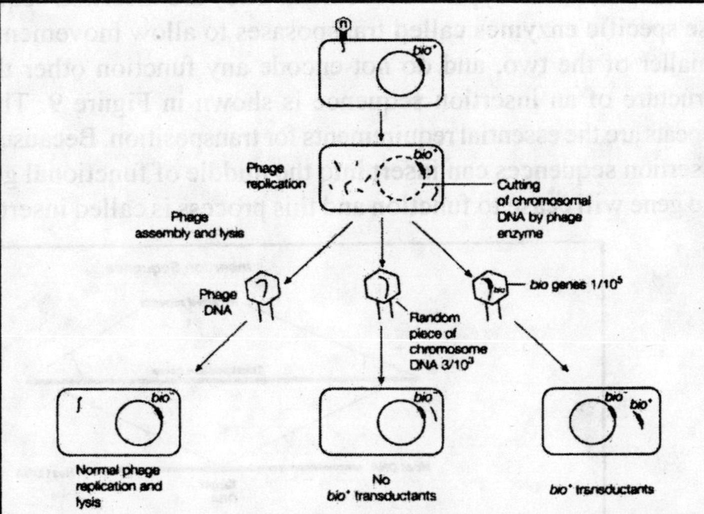


**Figure 7** Generalized transduction

### 2.2.3.2. *Specialized transduction*

Specialized transduction (Figure 8) involves the incorrect excision of DNA when lysogeny ends, with the result that the bacteriophage picks up a collinear piece of DNA from beside its insertion site. When lysogeny is re-established, the additional DNA is also inserted into the new host.

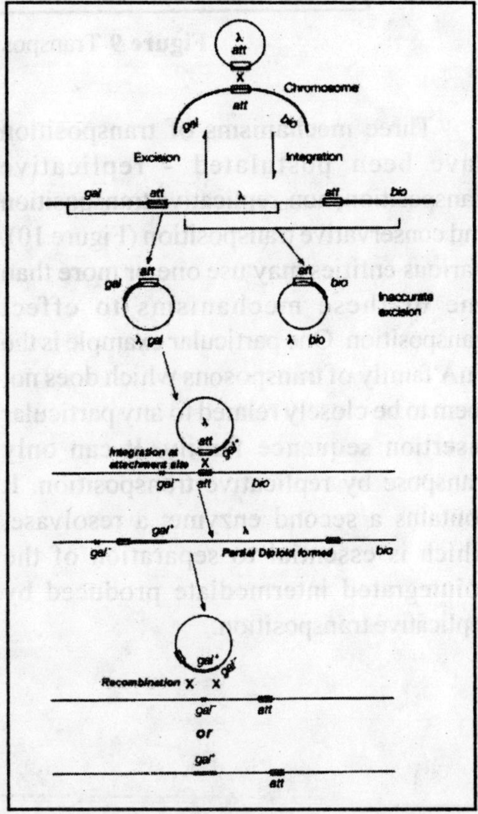


**Figure 8** Specialized transduction

### *2.2.4. Transposition*

Transposition is a mechanism by which pieces of DNA can move from one place to another on the DNA molecules within a cell. There are effectively two types of genetic unit that can cause this type of movement. They are insertion sequences and transposons. Both use specific enzymes called transposases to allow movement. Insertion sequences are the smaller of the two, and do not encode any function other than the ability to move. The structure of an insertion sequence is shown in Figure 9. The transposase and the invert repeats are the essential requirements for transposition. Because insertion is relatively random, insertion sequences can insert into the middle of functional genes. The result of this is that the gene will cease to function and this process is called insertional inactivation.

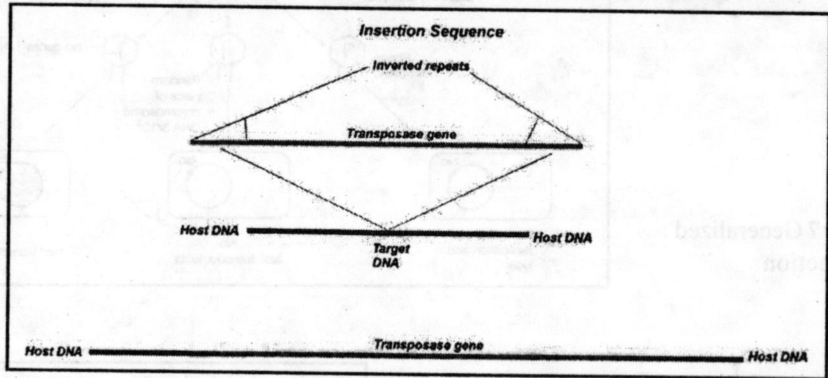


**Figure 9** Transposition by insertion sequence

Three mechanisms of transposition have been postulated - replicative transposition, non-replicative transposition and conservative transposition (Figure 10). Various entities may use one or more than one of these mechanisms to effect transposition. One particular example is the TnA family of transposons which does not seem to be closely related to any particular insertion sequence family. It can only transpose by replicative transposition. It contains a second enzyme, a resolvase, which is essential to separation of the cointegrated intermediate produced by replicative transposition.

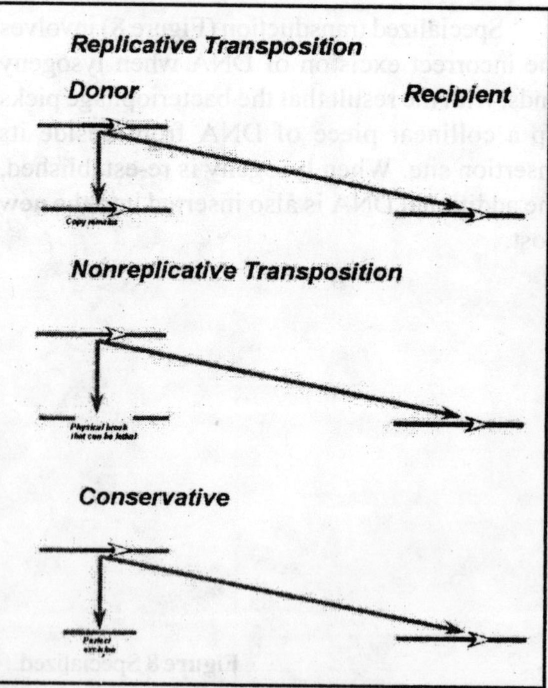


**Figure 10** Three mechanisms of transposition.

### 2.2.5. *Hybridization*

This is the first method, which is put into consideration when the desired property, discovered in another strain, is under multigenic control. Its main disadvantage is its imprecision and consequent disregard for the genome of the original strain. There are of course procedures, such as backcrossing, that can potentially regenerate the original genome. The regeneration is often only partial due to meiotic and mitotic recombination phenomena. Hybridization is usually only feasible between two haploid cells of opposite mating types belonging to the same species. The first attempt to produce new yeast types by hybridization was by recombination of existing variation between commercial strains. Others involved the production of new hybrids by conventional cross breeding but encountered problems related to the life cycle of commercial strains. Because of this, repeated backcrossing of a derived strain to its original parent to remove undesirable characteristics is both tedious and inefficient. The hybrids obtained by crossing the mating derivatives of commercial yeasts displayed intermediate properties when compared to the parental strains, a result which would be expected for characteristics under polygenic control e.g. fermentation rate in yeast.

Intra-species hybridization involves the mating of haploids of opposite mating-types to yield a heterozygous diploid (Figure 11). Recombinant progeny are recovered by sporulating the diploid, recovering individual haploid ascospores and repeating the mating/sporulation

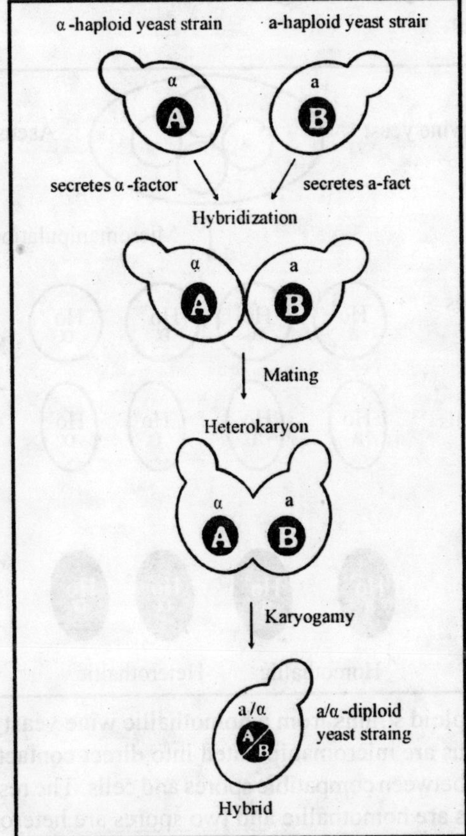


**Figure 11** Hybridization between haploids of two opposite mating-types of wine yeast

cycle as required. Haploid strains from different parental diploids, possessing different genotypes, can be mated to form a diploid strain with properties different from that of either parental strain. Thus, in theory, crossbreeding can permit the selection of desirable characteristics and the elimination of undesirable ones.

Generally, hybridization provides a solution for precise objectives, based on a correctly, defined genetic context. Non-$H_2S$ producing flocculant hybrids by conjugating spores of two homothallic commercial strains were constructed which on analysis had shown to be homozygous for the desired character; this combination is rarely found in indigenous populations.

Unfortunately, many wine yeasts are homothallic and the use of hybridization techniques for development of wine yeast strains has proved difficult. However, this problem can be circumvented by direct spore-cell mating using a micromanipulator, in which four homothallic ascospores from the same ascus are placed into direct contact with heterothallic haploid cells (Figure 12). Mating takes place between compatible ascospores and cells. The resulting diploid is sporulated. Since two spores in each ascus are homothallic and two are heterothallic, stable haploids can be isolated from the sporulated diploids. Elimination or inclusion of a specific property can thus, be achieved relatively quickly by hybridization, provided that it has a simple genetic basis, for example one or two genes. But, many desirable wine yeast characteristics are specified by several genes or are the result of several gene systems interacting with one another.

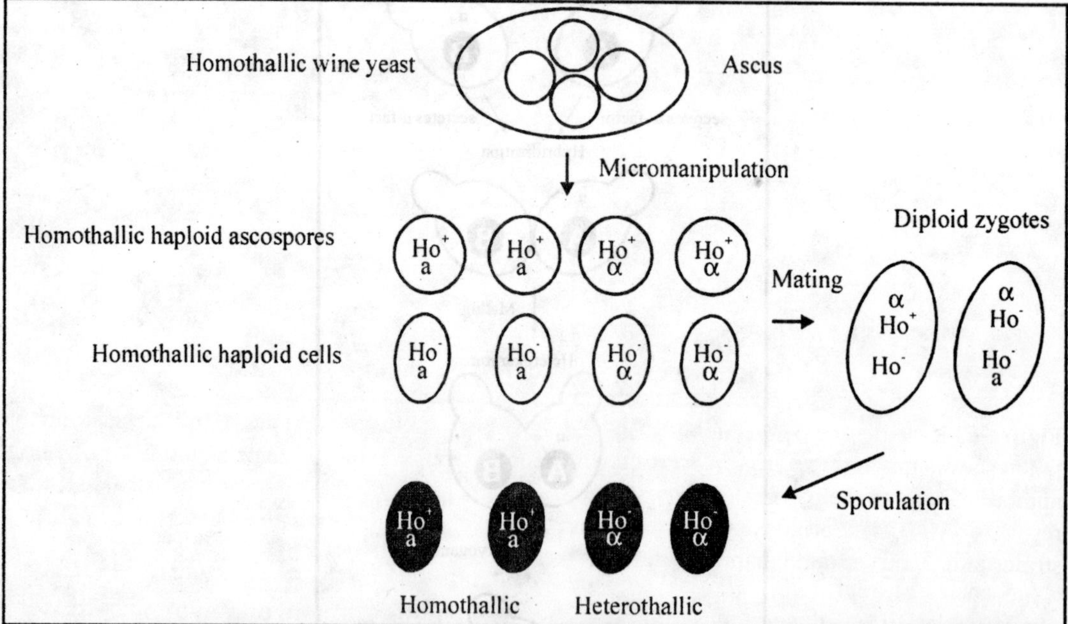


**Figure 12** The isolation of haploid strains from a homothallic wine yeast by spore-cell mating. Four ascospores from the same ascus are micromanipulated into direct contact with heterothallic haploid yeast cells. Mating takes place between compatible spores and cells. The resulting diploid is sporulated. Since two spores in each ascus are homothallic and two spores are heterothallic, stable haploids can be isolated from the sporulated diploids

### 2.2.6. *Rare mating*

Yeast strains that fail to express a mating-type can be force-mated (rare-mating) with haploid *MATa* and *MAT*α strains (Figure 13). Typically, a large number of cells of the parental strains are mixed and a strong positive selection procedure is applied to obtain the rare hybrids formed. For instance, industrial strains that have a defective form of, or lack, mtDNA (respiratory deficient mutants) can be force-mated with auxotrophic haploid strains having normal respiratory characteristics. Mixing of these non-mating strains at high cell density will generate only a few respiratory-sufficient prototrophs. These true hybrids with fused nuclei can then be induced to sporulate for further genetic analysis and crossbreeding.

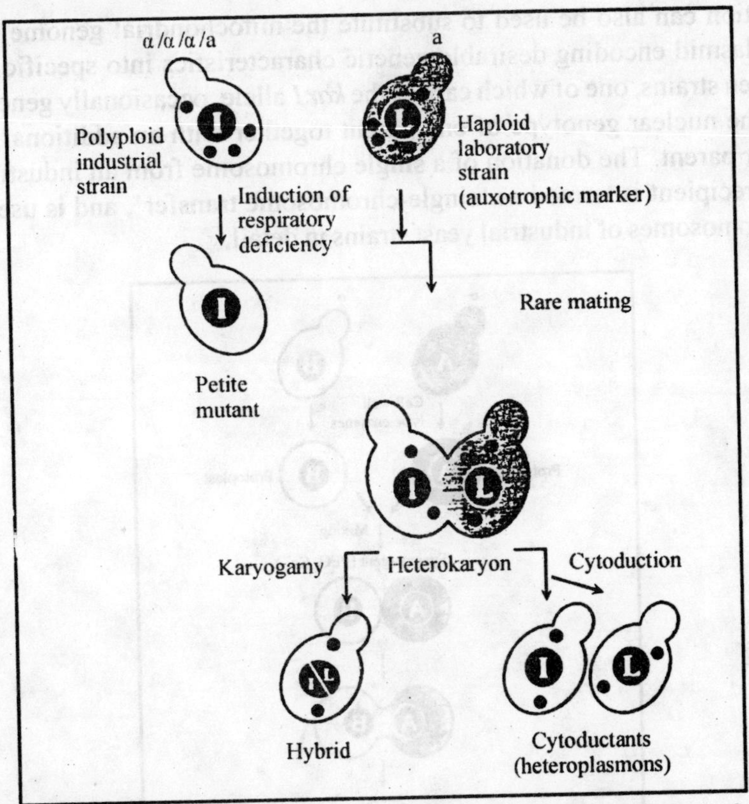


**Figure 13** Rare-mating between industrial and laboratory strains of wine yeast. Industrial strain that fail to show a mating type are force-mated with haploid strains, exhibiting a or α mating type. A large number of cells of the parental strains are mixed and the rare hybrids or cytoductants are selected as respiratory-sufficient prototrophs from crosses between a respiratory-deficient mutant of the industrial strains and an auxotrophic haploid laboratory strain

Rare mating is based on the fact that in a population of diploid or polyploid cells, some individuals become homozygous for the mating type and can, therefore, be crossed with a haploid cell of the opposite type. This cross is a rare event that can only be detected with a highly effective screen. The procedure has great advantage of allowing hybridization between strain of different ploidy and avoids the need for sporulation, which often does not work with commercial strains and runs the risk of genome alteration.

### *2.2.7. Cytoduction*

Rare-mating is also used to introduce cytoplasmic genetic elements into yeasts without the transfer of nuclear genes from the yeast parent (Figure 14). This method of strain development is termed as Cytoduction. Cytoductants, (heteroplasmons) receive cytoplasmic contributions from both parents but retain the nuclear integrity of only one. Cytoduction requires that a haploid mating strain carry the *kar1* mutation; i.e. a mutation that impedes karyogamy (nuclear fusion) after mating. This more specific form of strain construction can, for example, be used to introduce the dsRNA determinants for the $K_2$zymocin and associated immunity into particular wine yeast.

Cytoduction can also be used to substitute the mitochondrial genome of yeast or to introduce a plasmid encoding desirable genetic characteristics into specific yeast strains. Mating between strains, one of which carries the *kar1* allele, occasionally generates progeny that contain the nuclear genotype of one parent together with an additional chromosome from the other parent. The donation of a single chromosome from an industrial strain to a haploid *kar1* recipient is termed as 'single-chromosome transfer', and is used to examine individual chromosomes of industrial yeast strains in detail.

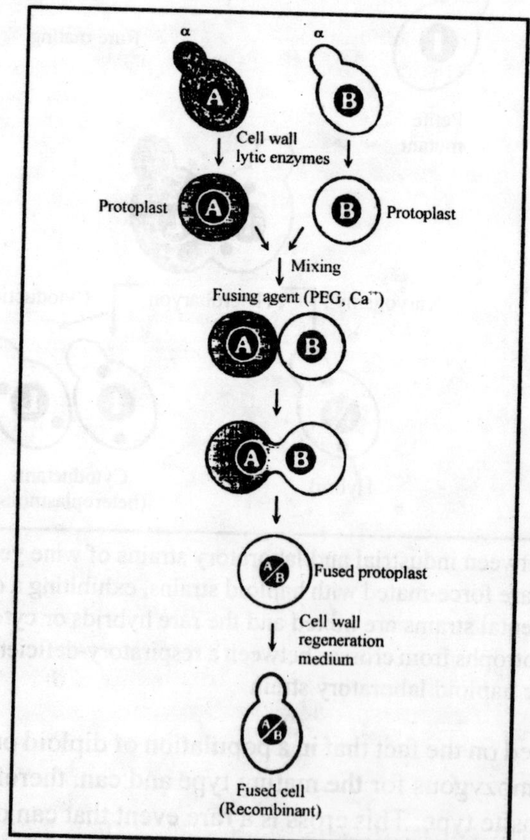


**Figure 14** Protoplast fusion between two different yeast cells is a direct asexual technique to produce either hybrids or cytoductants

### 2.2.8. *Protoplast fusion*

Protoplast fusion is a direct, asexual technique employed in crossbreeding as a supplement to mating. Like rare-mating, protoplast fusion can be used to produce either hybrids or cytoductants (Figure 14). Both these procedures overcome the requirement for opposite mating types to be crossed, thereby extending the number of crosses that can be done. Cell walls of yeasts can be removed by lytic enzymes in the presence of an osmotic stabilizer to prevent osmolysis of the resulting protoplasts. Spheroplasts from the different parental strains are mixed together in the presence of a fusing agent, polyethylene glycol and calcium ions, and then allowed to regenerate their cell walls in an osmotically-stabilized selective-agar medium.

Protoplast fusion of non-sporulating industrial yeast strains removes the natural barriers to hybridization. The desirable (and undesirable) characteristics of both parental strains will recombine in the offspring. Cells of different levels of polyploidy can be fused. For instance, a diploid wine yeast strain can be fused to a haploid strain to generate triploid strains. Alternatively, two diploid wine yeasts with complementing desirable characteristics can be fused to generate a tetraploid strain with all the genetic backgrounds of the two parental wine yeasts.

Protoplast fusion can be used to obtain interspecific and even intergeneric hybrids via a mechanism that has not been completely elucidated but is similar to somatic cell fusion. The final result is often a large predominance of the genome of one of the strains plus several genes of the other strain. However, considerable difficulties have been encountered in developing a coherent screen and in stabilizing the fusion products. The incomplete enzymatic removal of yeast cell walls and the polyethylene glycol-induced fusion of the resulting protoplast have been used to produce hybrids between what were previously non-mating commercial yeast strain. This technique allows the mating between the strains of the same species or across the species that normally do not mate. In general, protoplast fusion serves only to remove any natural barriers to hybridization, the products are still likely to combine both desirable and undesirable characteristics of the parents. Therefore, the parents must have some unique easily traceable traits.

### 2.2.9. *Transformation*

Transformation involves the genetic exchange by the uptake of naked DNA. In other words, the microbial cell either naturally or under artificial conditions, moves a foreign DNA molecule across its membrane(s) and then introduces it into its chromosome by recombination. Inherently, the naked DNA is highly vulnerable while floating free in the natural or artificial environment. The export of nucleases into the environment and the presence of restriction endonucleases in the bacterial cytoplasm both restrict transformation. Not withstanding this, transformation is probably the most important method of horizontal genetic exchange between highly divergent microorganisms.

Selection of variants, mutagenesis, conjugation, transduction, hybridization, rare-mating and protoplast fusion have value in strain development programmes, but these methods lack the specificity required to modify the organisms in a well-controlled fashion. It may not be possible to define precisely the change required using these genetic techniques, and a new strain may bring an improvement in some aspects, while compromising other desired characteristics. Transformation offers the possibility of altering the characteristics of the microorganisms with surgical precision: the modification of an existing property, the introduction of a new characteristic without adversely affecting other desirable properties, or the elimination of an unwanted trait. By using such procedures, it is possible to transform microbial cells with DNA fragments or purified genes from the same or other organisms. The transformation requires i) cell permeability to donor DNA, ii) an appropriate DNA vector, iii) an easily selectable gene marker, and iv) a suitable recipient strain.

The exogenous DNA normally does not enter the microbial cells due to the rigid cell wall which is generally partially removed by lytic enzymes which are commercially available as mixtures of α, β-glucanases (helicase, zymolyase, glusulase) for yeast cells and lysozyme for bacterial cells. The resulting spheroplasts are suspended in osmotically stabilized hypertonic solution alongwith donor DNA in presence of fusing agents, polyethylene glycol (PEG) and calcium chloride. The spheroplasts are then embedded in 3% agar to allow the regeneration of cell wall and placed on selective medium that supports the growth of transformed cells (Figure 15).

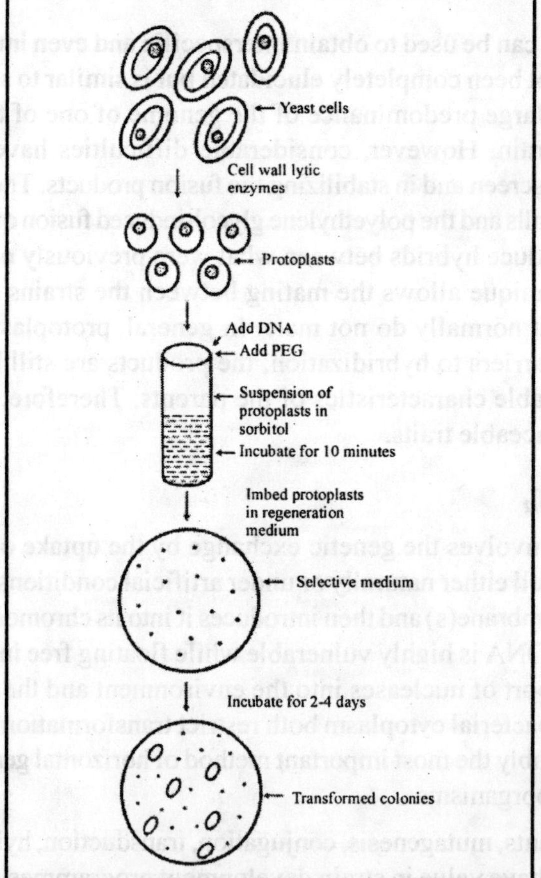


**Figure 15** The method of yeast transformation

### *2.2.10. Gene cloning and transformation (Recombinant DNA technology)*

Recombinant DNA consists of a DNA sequence that has been artificially produced by joining pieces of DNA from different organisms. This technique can also be used to develop "useful" micro-organisms that have the ability to degrade toxic wastes. New strains of agriculturally essential plants such as wheat, corn, and rice represent byproducts of rDNA technology. Recombinant DNA technology can be used in a wide range of industrial enterprises to develop: i) Microorganisms that produce new products, ii) more efficient methods of producing currently available products, iii) large quantities of products which are scarce in the natural environment.

Recombinant DNA technology comprises a mixture of techniques, some new and some borrowed from other fields, such as microbial genetics. These techniques consist of 1) specific cleavage of DNA by restriction nuclease, 2) DNA cloning, whereby a specific DNA fragment is integrated into rapidly replicating vectors, so that it can be amplified in bacteria or yeast cells (Figure 16), 3) nucleic acid hybridization, which makes it possible to identify specific sequences of DNA or RNA with great accuracy and sensitivity by their ability to bind a complementary nucleic acid sequence and 4) DNA sequencing of the nucleotides in a cloned DNA fragment.

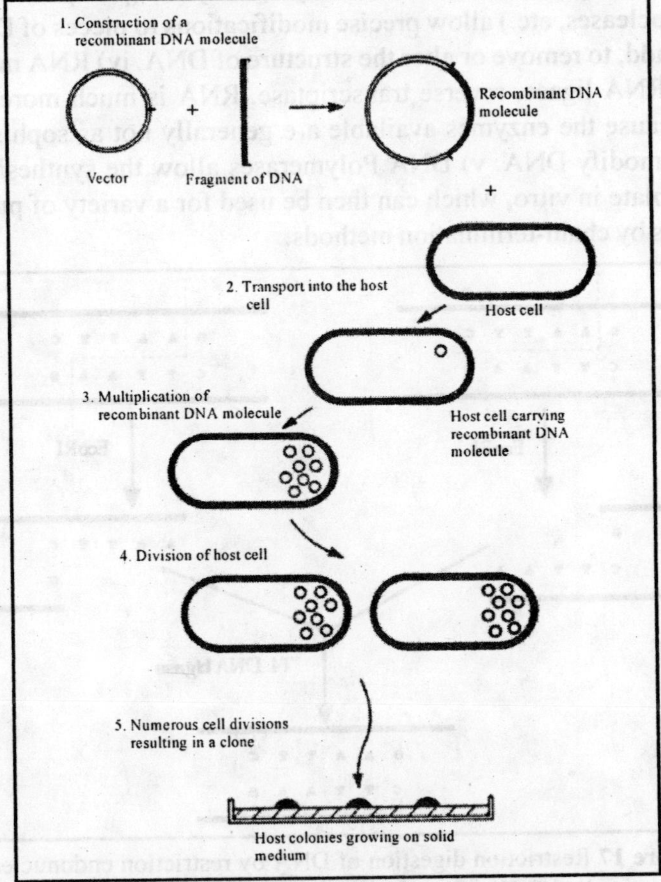


**Figure 16** The basic steps in gene cloning

*2.2.10.1. Specific enzymes used in recombinant DNA technology*

The specific cleavage of DNA is carried out by the use of naturally-occurring enzymes, most of which are derived from bacteria or viruses (Figure 17). These include restriction endonucleases produced by bacteria in natural environments where they are continually exposed to a dilute 'soup' of foreign DNA released from other organisms which have died or lysed as well as the bacteriophage DNA. They protect themselves from becoming infected or killed by producing many different restriction-modification systems, called restriction enzymes which are named after the organisms from which they were originally isolated, e.g.: EcoRI from *Escherichia coli,* BamHI from *Bacillus amyloliquefaciens*. These systems operate by enzymes which recognize specific short regions of DNA sequence, which are usually palindromic, e.g: 5' GGATCC 3', 3' CCTAGG 5' . Molecular biologists use highly purified preparations of R.E.s to cut DNA into specific fragments. ii) DNA ligase which provides the ability to rejoin cut fragments of DNA and form artificial recombinant molecules. The enzyme utilizes ATP to create new phosphodiester bonds, covalently linking together the ends of DNA strands. The ligase encoded by bacteriophage T4 is particularly valuable since it has very good properties in vitro, such as the ability to join together 'blunt-ended' DNA fragments which allows the joining of otherwise non-compatible fragments, e.g. different restriction enzymes (Figure 18). iii) Other modifying enzymes (phosphatases, kinases, single-strand specific nucleases, etc.) allow precise modifications to pieces of DNA to be made in vitro in order to add, to remove or alter the structure of DNA. iv) RNA modifying enzymes - exonucleases, RNA ligase, reverse transcriptase. RNA is much more difficult to work with in vitro because the enzymes available are generally not as sophisticated as the set available which modify DNA. v) DNA Polymerases allow the synthesis of DNA from a pre-existing template in vitro, which can then be used for a variety of purposes, including sequence analysis by chain-termination methods.

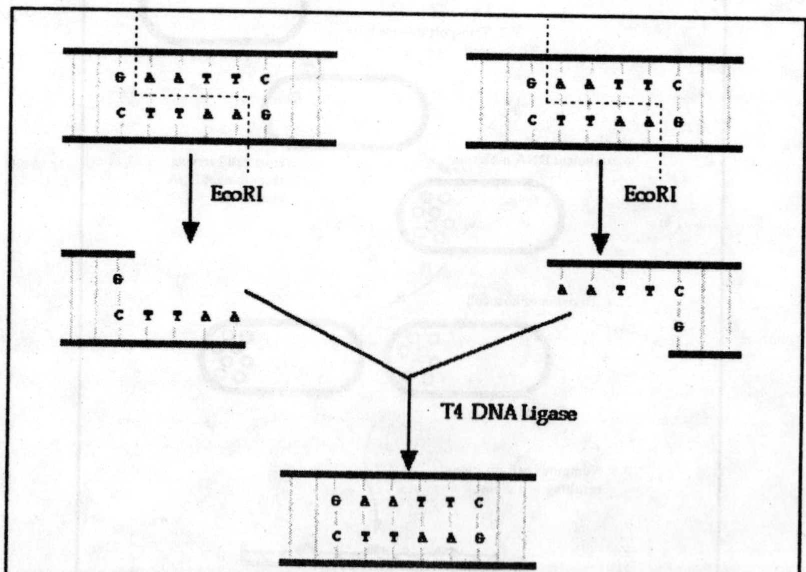


**Figure 17** Restriction digestion of DNA by restriction endonucleases and rejoining of cut fragments of DNA by DNA ligase

Recently, thermostable polymerases have become important, e.g. Taq DNA polymerase from *Thermus aquaticus*. This bacterium has evolved to grow in hot springs at temperatures which kill most other species. These enzymes allow the amplification of as little as one molecule of DNA into a large amount by means of repeated cycles of melting, primer annealing and extension by the enzyme which is not destroyed by the high temperatures used in this process. This is known as the polymerase chain reaction (Figure 18).

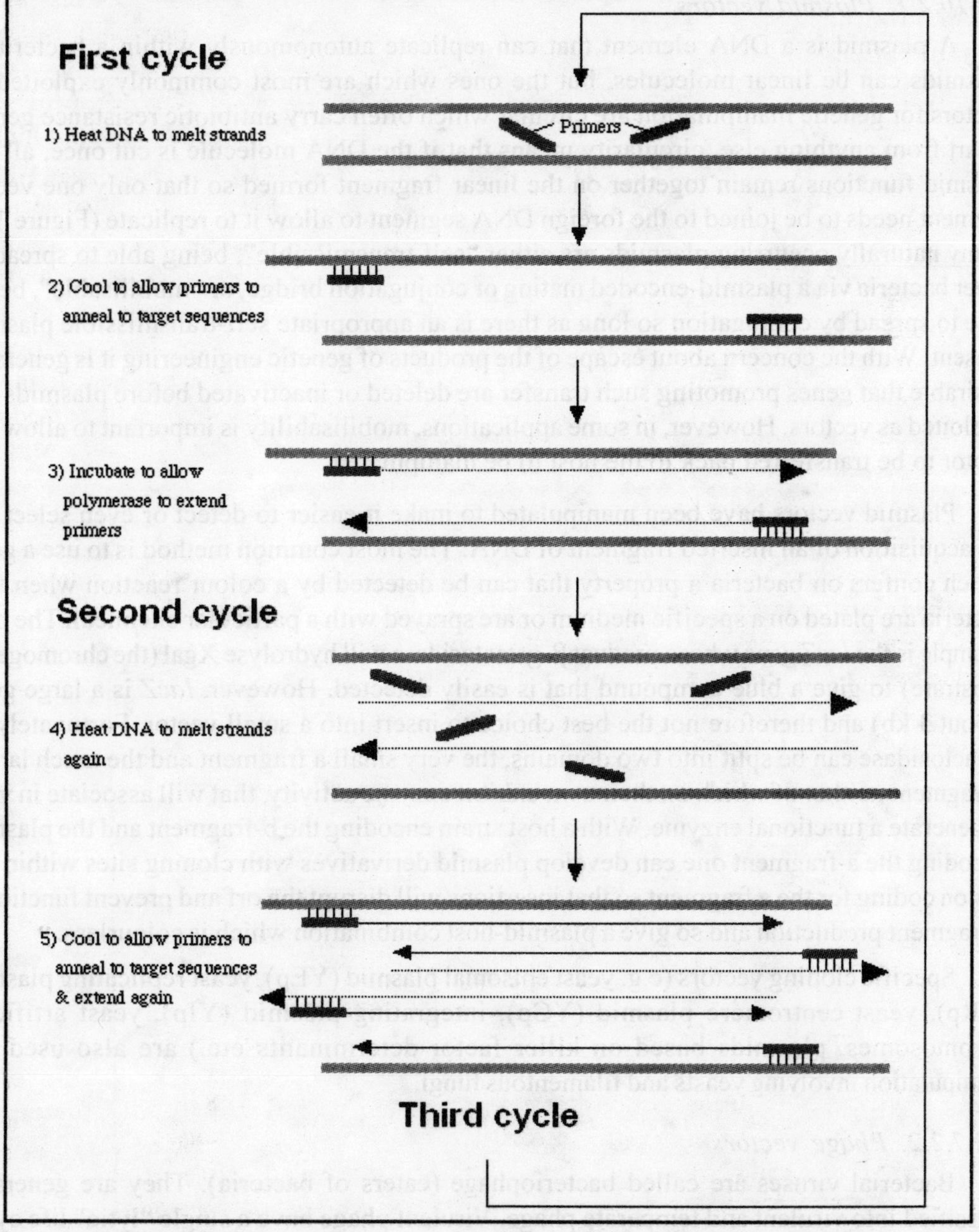


**Figure 18** A schematic representation of polymerase chain reaction

*2.2.10.2. Vectors used in DNA cloning*

Because DNA fragments obtained after the action of restriction endonucleases donot generally replicate on their own when introduced into a bacterium they need to be joined to a DNA molecule that can replicate-a vector. Vectors need to be small enough to be manipulated conveniently and also to allow the joined DNA to be easily analysed. Vectors are derived either from plasmids or bacterial viruses-bacteriophage.

*2.2.10.2.1. Plasmid vectors*

A plasmid is a DNA element that can replicate autonomously within a bacterium. Plasmids can be linear molecules, but the ones which are most commonly exploited as vectors for genetic manipulation are circular which often carry antibiotic resistance genes. Apart from anything else, circularity means that if the DNA molecule is cut once, all the plasmid functions remain together on the linear fragment formed so that only one vector segment needs to be joined to the foreign DNA segment to allow it to replicate (Figure 19). Many naturally occurring plasmids are either "self-transmissible", being able to spread to other bacteria via a plasmid-encoded mating or conjugation bridge, or "mobilisable", being able to spread by conjugation so long as there is an appropriate self-transmissible plasmid present. With the concern about escape of the products of genetic engineering it is generally desirable that genes promoting such transfer are deleted or inactivated before plasmids are exploited as vectors. However, in some applications, mobilisability is important to allow the vector to be transferred back to the host to be manipulated.

Plasmid vectors have been manipulated to make it easier to detect or even select for the acquisition of an inserted fragment of DNA. The most common method is to use a gene which confers on bacteria a property that can be detected by a colour reaction when that bacteria are plated on a specific medium or are sprayed with a particular chemical. The best example is the *lacZ* gene whose product β-gatactosidase will hydrolyse Xgal (the chromogenic substrate) to give a blue compound that is easily detected. However, *lacZ* is a large gene (about 3 kb) and therefore not the best choice to insert into a small vector. Fortunately β-galactosidase can be split into two domains, the very small a fragment and the much larger b fragment neither of which on their own exhibit enzyme activity, that will associate in vivo to generate a functional enzyme. With a host strain encoding the b-fragment and the plasmid encoding the a-fragment one can develop plasmid derivatives with cloning sites within the region coding for the a fragment so that insertions will disrupt the orf and prevent functional a-fragment production and so give a plasmid-host combination which is colourless.

Specific cloning vectors (e.g. yeast episomal plasmid (YEp), yeast replicating plasmid (YRp), yeast centromere plasmid (YCp), integrating plasmid (YIp), yeast artificial chromosomes, plasmids based on killer factor determinants etc.) are also used for manipulation involving yeasts and filamentous fungi.

*2.2.7.2.2. Phage vectors*

Bacterial viruses are called bacteriophage (eaters of bacteria). They are generally classified into virulent and temperate phage. Virulent phage have a single "lytic" life cycle which involves infection, multiplication and host death (lysis) with the release of new viruses.

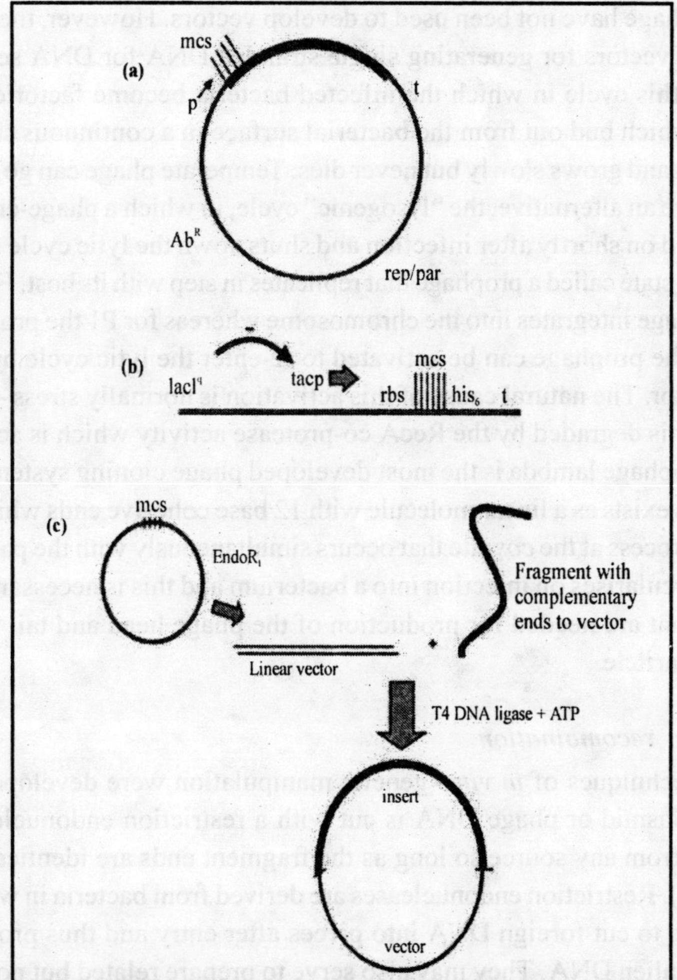


**Figure 19. (a)** Simplified scheme of the basic plasmid vectors. The essential components are a selectable marker (Ab$^R$ - conferring resistance to antibiotics such as ampicillin, tetracycline, kanamycin or chloramphenicol); a replication and stability system providing a high or a low copy number; one or more sites that can be cut by restriction endonucleases to allow insertion of foreign DNA fragments – a cluster of such sites is called a multiple cloning site - mcs; and a gene running across the mcs whose disruption can be detected easily. Inclusion of a strong transcriptional terminator (t) downstream of the mcs can prevent strong transcription from interfering with plasmid replication **(b)** Below the vector is shown an alternative vector segment which includes a strong regulated promoter - the tac promoter - that can be used to direct controlled expression of the coding region for any gene inserted into the mcs. If the gene lacks its own translation signals then the presence of a ribosome binding site (rbs) can direct translation on the mRNA produced so long as the distance between the rbs and the start codon for the coding region is approximately 7-8 nucleotides in length. The *lacI*$^q$ gene encodes a repressor protein that binds to the tac promoter and prevents transcription except when an "inducer" such as isopropyl-β-D-thiogalactoside (IPTG), which binds to the repressor and alters its conformation so that it no longer binds to the DNA, is present **(c)** Simple scheme showing the ligation of a foreign fragment into the vector by cutting with a restriction enzyme and using T4 DNA ligase to join the fragments together before introducing them into bacteria by transformation or electroporation

In general such phage have not been used to develop vectors. However, the M13 phage that form the basis of vectors for generating single stranded DNA for DNA sequencing uses a slight variant of this cycle in which the infected bacteria become factories for producing phage particles which bud out from the bacterial surface in a continuous stream so that the host is debilitated and grows slowly but never dies. Temperate phage can go through the lytic cycle but also have an alternative, the "lysogenic" cycle, in which a phage-encoded repressor system is switched on shortly after infection and shuts down the lytic cycle so that the phage enters a quiescent state called a prophage that replicates in step with its host. For bacteriophage lambda the prophage integrates into the chromosome whereas for P1 the prophage multiplies like a plasmid. The prophage can be activated to re-enter the lytic cycle by inactivation of the phage repressor. The natural cause of this activation is normally stress - for example the lambda repressor is degraded by the RecA co-protease activity which is activated by DNA damage. Bacteriophage lambda is the most developed phage cloning system. Lambda DNA in phage particles exists as a linear molecule with 12 base cohesive ends which are generated by the cleavage process at the *cos* site that occurs simultaneously with the packaging process. This molecule circularises on injection into a bacterium and this is necessary to generate the series of genes that are needed for production of the phage head and tail that assemble to form the phage particle.

### *2.2.10.3. In vitro recombination*

The basic techniques of *in vitro* genetic manipulation were developed in bacteria in which isolated plasmid or phage DNA is cut with a restriction endonuclease is joined to DNA fragments from any source so long as the fragment ends are identical to those of the vector (Figure 19). Restriction endonucleases are derived from bacteria in which they appear to act as a system to cut foreign DNA into pieces after entry and thus protect the bacteria from completely alien DNA. They may also serve to prepare related but not identical DNA for homologous recombination. Self DNA is protected from cutting by methylation of the DNA at the sites which would otherwise be recognised by the endonuclease. Restriction systems must be considered during genetic manipulation because they can be a reason why the transfer of DNA from one bacterial strain to another is not successful: if unmodified vector or "insert" DNA is used with a strain carrying a restriction system, then the efficiency can be significantly reduced. One strategy that has been used to reduce the efficiency of the restriction barrier to DNA entry is to grow the bacteria at high temperature for a number of generations. The original source of DNA ligase is the phage T4 where it plays a key role in phage DNA replication. Normally the enzyme is derived from bacteria that have been manipulated to over-produce this enzyme.

Restriction enzymes can be used to generate fragments either by complete or partial digestion of the DNA. Complete digestion is the simpler strategy in which DNA is digested with excess enzyme (1 unit = enough to cut all sites in 1 mg DNA at 37°C) and individual fragments which can be as small as 100 bp for (enzymes with a 4 bp recognition sequence) to >10 kb (enzymes with a 6 bp recognition sequence) are joined to vector. Partial digestion involves limited amount of enzyme for a limited amount of time and gives random segments running from one restriction site to another but generating fragments consisting of two or more "fragments" still joined together. A given gene may thus end up on more than one type of fragment. The advantages of this are that if the genetic region of interest has a target site for the restriction endonuclease to be used within it then the only way that the region can be obtained intact is by partial digestion. The other advantage is that a library of random fragments produced by partial digestion will contain fragments that overlap with each other. Therefore a long DNA segment can be built up as a series of different cloned segments and identified as such by hybridisation of the fragments with each other, even though they are not identical fragments. Subsequent mapping of restriction sites along the DNA of each related clone can identify those segments that are identical segments in different clones.

*2.2.10.4. Introduction of DNA into bacteria*

Transformation, mixing naked DNA with bacteria that are "competent" to take up DNA, is the commonest and most straight forward way of introducing DNA into bacteria. Some bacteria become competent naturally towards the end of exponential phase or upon starvation (for example *Bacillus subtilis*, *Streptococcus pneumoniae* and *Haemophilus influenzae*), but the active uptake process involved often converts the DNA during uptake into single stranded fragments that do not regenerate the desired combination of vector and cloned fragment. For uptaking the products of *in vitro* recombination, bacteria are normally treated with a mixture of divalent cations, the most basic requirement being calcium ($Ca^{2+}$), that allow DNA to bind to the negatively charged outer surface of the bacteria in such a way that during a short period at 42°C after holding on ice the DNA is taken up into the cytoplasm. When naked phage DNA is taken up this is called transfection. Not all bacteria can be easily made competent for transformation, presumably due to the structure of the outer surfaces of the bacterial cell. Electroporation has provided an alternative, more general method in which bacteria and naked DNA are placed in a small chamber between two electrodes and then subject to a short but intense electric surge which temporarily creates pores through which DNA can enter the cell. A variety of application notes and publications are available providing details for the optimisation of this process for most bacterial genera currently studied. A third way of introducing DNA is by injection after *in vitro* packaging into phage particles, but this is more specialised and only applies to vectors that carry the DNA sequence signals that allow the DNA to be recognised by the proteins that initiate the process. In the case of bacteriophage lambda this is the cos site, recognised by gene product A, an endonuclease that introduces nicks 12 bp apart to create an end that start the wrapping of the DNA into the protein head of the phage. Extracts from different bacterial strains carrying mutant phage that have accumulated different components of the packaging process, can be mixed *in vitro* so as to provide all that is necessary, and ligated DNA added to this mix will be packaged. Production of stable phage particles depends on the DNA being approximately the same size as the normal phage genome, so it provides a selection for clones that have substantial inserts.

## 3. CURRENT STATUS AND FUTURE PROSPECTS OF GENETICALLY MODIFIED MICROBES

In theory, almost any gene can be transferred from one organism to another and carry out its function, provided that the proper signals are present to enable the host recognise the gene. The strain improvement involves the re-routing of metabolic pathways, and engineering of the flux of a certain route in industrial micro-organisms. Some of examples of genetic manipulation in microorganisms for developing novel strains are discussed below:

### 3.1. Manipulation of bacterial metabolism

In trying to exploit bacteria for particular applications it is likely that the condition under which the bacteria are grown may be different from those under which they have evolved. It may be therefore realistic to try to alter the bacterial metabolism so as to make it more fit for the new growth regime. This may have advantage if the changes increase the energetic efficiency of bacterial growth so that there is a greater biomass or product yield for a given amount of input medium. A classical example in which this was achieved was for the species *Methylotrophus methylophilus*, a methane/methanol utilising organism that was exploited in the late 70s and early 80s by the UK chemical company ICI as a source of single cell protein (SCP). The organism was used to scavenging its environment for nitrogen-rich compounds and used an energetically expensive but high affinity system for assimilating ammonium ions. The strategy adopted was to knock out the endogenous pathway by inactivating the gene encoding oxo-glutarate amino transferase and replacing it with glutamine synthase from *E.coli* which is less energetically expensive, but has lower affinity for ammonium. Thus in non-nitrogen-limited conditions the manipulated strain would yield more biomass per unit carbon source input.

### 3.2. Manipulation of biodegradative capacity

Bacteria are capable of degrading a vast range of organic compounds and are a major force in the recovery of natural environments from pollution such as oil spills or release of industrial effluent or pesticides and herbicides. However, this natural ability tends to be limited to compounds that can already be found in nature, so there are many structures that have been created by modern chemical industry (for example polysubstituted aromatic compounds) which are recalcitrant to degradation by natural isolates. In some cases it may be that only simple evolutionary steps are needed to develop strains that are capable of the degradation and application of selective pressure on a pure strain or a mixture of strains is all that is needed to engineer the desired change. A simple example is the transfer of the TOL plasmid pWW0 into a naturally occurring strain that uses TNT (trinitrotoluene) as its nitrogen source. Removal of the nitro groups produces toluene which is normally discarded but the presence of pWW0 allows this to be degraded. This recombinant strain is therefore able to degrade TNT to harmless compounds which are used for bacterial growth. In other cases intervention with modern methods of genetic manipulation has short circuited this process. For example, *Pseudomonas putida* carrying the TOL plasmid pWW0 can degrade toluene, m- and p-xylenes and m-ethyltoluene, but not p-ethyltoluene - for two reasons. First, the latter does not activate the positive control protein XylS needed for transcription of the genes. Second, the catechol 2,3-oxygenase enzyme is inactivated when it interacts with this abnormal substrate. Both of these problems were overcome by point mutants isolated by appropriate selection and then both mutations were introduced into the same strain, resulting in retention of the ability to utilise the normal range of compounds but now also able to degrade p-ethyltoluene. Selecting for both changes to happen simultaneously would result in a very low frequency and neither of the mutations on their own would create a mutant with useful properties.

## 3.3. Bacteria as biosensors

An extension of the fact that many bacteria carry resistance to and /or the ability to degrade toxic compounds which are harmful to the environment and the living organisms in it, is the ability to manipulate and exploit the gene expression signals and regulatory proteins that control expression of the resistance or degradation genes. Once these signals and regulators have been defined they can be joined to a reporter gene such that coding for the green fluorescent protein (GFP) from the jelly fish *Aequoria victoria* or the lux genes from *Vibrio fischeri*, so that when the gene is switched on the bacteria will glow green when illuminated with ultraviolet light, or simply emit light. With such a construction the emission of a light signal will indicate that the chosen gene system is switched on and this can be interpreted in light of the properties of the regulator. For example the *mer* gene system can be used to detect the presence of $Hg^{2+}$ ions above the critical levels. Similarly, the *xylR/S* genes can be used to detect the presence of a variety of toluene-related aromatic compounds. Manipulation of these regulators as referred to above can results in a broadening of the spectrum of compounds that can be detected.

## 3.4. Manipulation of antibiotic biosynthesis

Two excellent examples of genetic manipulation concern the manipulation of biosynthetic pathways responsible for production of polyketide antibiotic production. The essence of the biosynthetic pathway is the cyclic addition of acyl units like acetate to a growing chain. The resultant polyketide may be cyclised to produce multiple aromatic rings, or may be modified during the rounds of acetate addition followed by modification of the keto group through reduction, dehydration and a second reduction. In this second approach, variations in the product can be achieved by performing only the first step or only the first two steps. Two general types of biosynthetic pathway have been discovered by molecular genetic studies. In type II pathways there is a basic set of separate enzymes, acyl carrier protein, acyl transferase, keto-synthase, keto-reductase, which catalyse the essentially uniform chain growth followed by cyclisation to produce a series of aromatic rings which are then further modified in a variety of ways determined by other enzymes encoded within the cluster. The chain length and the way that it cyclises are determined by chain length factor and cyclases particular to each pathway. However, the basic pathway is similar for each type II system. Therefore creation of hybrid gene mixtures from two or more different pathways can generate novel compounds. This has been achieved by mixing the genes from two different pathways so that the backbone modifications reflect some of the steps from each pathway and therefore create a structure different from either of the original products.

An alternative strategy is presented by the paradigm of the erythromycin biosynthetic pathway which is the best studied type I pathway. Such pathways encode related polyketide biosynthesis proteins but are rolled up into large, multifunctional proteins. Each protein may code for one or more cycles of addition and modification, so that there is one enzymic domain for each step in the biosynthetic pathway. Whereas in the type II systems the basis for each cycle is the same, in type I systems there is the potential for each cycle to be different. The growing chain is carried by an acyl carrier protein and eventually released from this by a thioesterase which may also help the chain to fold correctly. By placing the thioesterase at different points in the system and deleting the downstream modules one can generate compounds of different length or ring size. By adding or deleting a modification step at particular steps in the pathway one can also manipulate to produce new compounds.

### 3.5. Microbial production of 1,3-propanediol

1,3-Propanediol (1,3-PD) production by fermentation of glycerol was described in 1881 but little attention was paid to this microbial route for over a century. Glycerol conversion to 1,3- PD can be carried out by clostridia as well as enterobacteriaceae. The main intermediate of the oxidative pathway is pyruvate, the further utilization of which produces $CO_2$, $H_2$, acetate, butyrate, ethanol, butanol and 2,3-butanediol. In addition, lactate and succinate are generated. The yield of 1,3-PD per glycerol is determined by the availability of $NADH_2$, which is mainly affected by the product distribution (of the oxidative pathway) and depends first of all on the microorganism used but also on the process conditions (type of fermentation, substrate excess, various inhibitions). In the past decade, research to produce 1,3-PD microbially was considerably expanded as the diol can be used for various polycondensates.

In particular, polyesters with useful properties can be manufactured. A prerequisite for making a "green" polyester is a more cost-effective production of 1,3-PD, which, in practical terms, can only be achieved by using an alternative substrate, such as glucose instead of glycerol. Therefore, great efforts are now being made to combine the pathway from glucose to glycerol successfully with the bacterial route from glycerol to 1,3-PD. Thus, 1,3-PD may become the first bulk chemical produced by a genetically engineered microorganism.

### 3.6. Biodegradable plastics using recombinant bacteria

From water bottles to artificial heart valves, plastics are essential to our everyday lives. Plastics are synthetic resins made of large, organic polymers refined from valuable natural fossil fuels (oil, coal, petroleum, or natural gas). Normally, plastics take years to degrade and are a major component of landfill and waste areas. However, Dr. Chris Somerville, director of the Carnegie Institute of Washington's Department of Plant Biology, and other institute scientists have discovered a method for engineering plants to make plastic that would degrade more easily-and not use up natural resources in the production process.

*Alcaligenes eutrophus,* as well as other bacterial species, produce the organic polymer polyhydroxyalkanoate as a carbon reserve. Moreover, these same microorganisms produce the enzyme necessary to break down the polymer into monomers, metabolized as a carbon source. This natural source of plastic polymer is therefore biodegradable. With the aid of modern molecular tools, three genes within the DNA recombinant bacteria and plasmid isolation sequence of *A. eutrophus* have been identified as enabling this organism to manufacture the plastic polymer. An additional gene was eventually revealed to code for the polymers to sequester specifically in plastids. Armed with the genes to code for the making and storing of the plastic polymer, the scientists used gene-splicing (genetic engineering) techniques to insert these codes into the DNA of the plant *Arabidopsis thaliana.* Until the organelle-specific gene was revealed, plastic polymers were detected in the nucleus, vacuole, and cytoplasm of *A. thaliana.* These polymer accumulations were associated with decrease in both plant growth and proper production. However, polymer accumulation in the plastids had no obvious negative effects on either the growth or fertility of the plants.

Bacteria in general lend themselves to this type of genetic transfer because they are a kingdom of prokaryotes (Monera). As such, their DNA is not housed within a nucleus but is found in a single strand and in easier-to-work-with circular fragments called plasmids. The plant-produced plastics are harvested in a series of chloroform extractions in which the plant material is separated from the plastic polymer. Transferring the genes into plants, as opposed to harvesting the plastic from *A. eutrophus,* is more cost-effective.

### 3.7. Re-routing of carbon metabolism in *Lactococcus lactis*

Lactic acid bacteria display a relatively simple metabolism wherein the sugar is converted mainly to lactic acid. The extensive knowledge of metabolic pathways and the increasing information of the genes involved allows for the re-routing of natural metabolic pathways by genetic and physiological engineering. We discuss several examples of metabolic engineering of *Lactococcus lactis* for the production of important compounds, including diacetyl, alanine and exopolysaccharides. Lactic acid bacteria (LAB) are used world-wide in the industrial manufacture of fermented food products. Their most important application in this respect is undoubtedly in the dairy industry, where these micro-organisms are used to convert milk or milk-derived products to an enormous variety of fermented dairy products. However, next to milk also other unprocessed food materials, like meat and vegetables, are subjected to fermentation by LAB on an industrial scale. In milk these bacteria encounter lactose as the major carbon source, but they have the capacity to use a number of other mono- and disaccharide substrates. The major product of these fermentations is lactic acid, which plays a crucial role in protection of the final fermented product against spoilage. Besides this acidification that acts as a natural preservative effect, LAB metabolism is essential for development of desired product properties like flavor, shelf-life and texture. The metabolic conversions that generate these end-products vary widely, depending on the lactic acid bacterium. Metabolic engineering strategies that involve inactivation of undesired genes and/or overexpression of existing or novel ones have been used to create rerouting of the metabolic fluxes by changing the energy metabolism or the concentrations of metabolic intermediates or of existing or completely new end-products.

### 3.8. Genetic Engineering for improving brewing, wine-making and baking yeasts

Yeasts have been used to produce food and beverages since the Neolithic age. Their use in fermentation was recognized in 1836-1838; and Louis Pasteur demonstrated their unequivocal role in the conversion of sugar to ethanol and carbon dioxide in 1861. The genetic improvement of industrial strains traditionally relied on classical genetic techniques (mutagenesis, hybridization, protoplast fusion, cytoduction), followed by selection for broad traits such as fermentation capacity, ethanol tolerance, absence of off-flavors (e.g. $H_2S$ for wine strains), fast dough fermentation, osmotolerance, rehydration tolerance, organic acid resistance (baker's strains), flocculation and carbohydrate utilization (brewer's strains). Despite considerable work, a major limitation of these classical genetic techniques was the difficulties of adding or removing features from a strain without altering its performance. One major advantage of gene technology over classical genetic techniques is that just one characteristic can be precisely modified, without affecting other desirable properties. In addition, molecular biology approaches have introduced a new dimension. The expression of heterologous genes has substantially increased the possibilities. In the past 20 years, impressive progress has been made in the development of molecular techniques for *Saccharomyces cerevisiae*. These advances have been successfully applied to industrial strains in the past decade, allowing the development of a new generation of specialized industrial yeast strains. The principal targets for strain development fall into two broad categories: (1) improvement of fermentation performance and simplification of the process and (2) improvement of product quality, e.g. organoleptic and hygienic characteristics. Selected examples of the most advanced applications of yeast genetic engineering in the fields of wine-making, brewing and baking have been reviewed in this chapter. Many targets for yeast improvement are relevant to several of these fields and will be presented in a sole section. Conversely, each of these fields has specific demands, which will be reviewed in separate sections.

### *3.8.1. Relevant targets for brewer's and wine-making yeasts*

#### *3.8.1.1. Flocculation*

Yeast flocculation, the asexual aggregation of cells into flocs and their subsequent removal from the fermentation medium by sedimentation, is of major interest for some fermentation processes. In brewing, especially bottom fermentation, good flocculation towards the end of primary fermentation is essential for a bright beer with sufficient aroma. The onset of cell sedimentation is critical. If it occurs too late, the beer will be cloudy and hard to filter; and if it occurs too early, maturation will be inefficient. Flocculation is also of interest for the elaboration of sparkling wines. In the classic Champagne method, a secondary fermentation (the so-called "prise de mousse") is conducted in the bottle to develop specific organoleptic characteristics. After 1 year, the yeast cells are removed from the bottle by a time-consuming and expensive procedure ("remuage"). This process could be advantageously simplified by the sedimentation of flocculent yeast cells. Yeast flocculation is an asexual, calcium-dependent, reversible aggregation of cells into flocs, involving a protein-sugar interaction between a specific cell-surface lectin and cell wall mannan. Several attempts have been made to construct flocculent industrial strains by classical genetic approaches. Electrofusion was used to convert a non-flocculent brewer's yeast into a flocculent one with adequate brewing performance. Flocculation is also introduced into wine yeast strains by hybridization or cytoduction. Extensive work in the past decade has led to a better understanding of the molecular and biochemical basis of flocculation. Several dominant flocculation genes, FLO1,

FLO5 and FLO8, are involved in flocculation. The dominant FLO1 gene has been isolated, characterized and shown to encode a cell-wall protein containing a lectin domain. Transfer of the flocculation characteristics into a wine yeast strain was achieved by transforming the strains with a multicopy vector containing FLO1, isolated from a flocculent *S. cerevisiae* strain. A strong flocculation phenotype is achieved in a non-flocculent brewer's yeast by integration of the ADH1-regulated FLO1 gene into the ADH1 locus.

*3.8.1.2. Glycerol overproduction*

Glycerol is the most abundant by-product of alcoholic fermentation, after ethanol and carbon dioxide. This polyol is thought to contribute to the sensory quality to beverages. In particular, high levels of glycerol may contribute to the perceived sweetness of wine. Therefore, many attempts have been made to increase the glycerol yield during fermentation. Another interest for re-routing the carbon flux towards glycerol, is the expected decrease in ethanol yield. This may be an alternative approach to the current methods for removing ethanol from beverages (e.g. dialysis), which are expensive and detrimental for aroma compounds. In *S. cerevisiae*, glycerol is produced by reduction of dihydroxyacetone phosphate, catalyzed by the glycerol-3-phosphate dehydrogenase (GPDH), followed by a subsequent dephosphorylation realized by a glycerol-3-phosphatase. Glycerol export and glycerol-3-phosphate dehydrogenase, but not glycerol phosphatase, are rate-limiting for glycerol production during fermentation in *S. cerevisiae*. The most direct approach for overproducing glycerol is to overproduce the GPDH. Overexpression of GPD1 in laboratory strains of wine and brewer's yeasts resulted in a 2- to 3-fold increase in glycerol production and a lower ethanol yield. Wine yeast strains transformed with multicopy plasmids carrying GPD1 under the control of ADH1 promoter produced 12-18 g of glycerol/l and about 1% (v/v) less ethanol. As a result of carbon re-routing and altered NADH metabolism, these strains exhibit increased production of by-products, mainly acetate, 2,3-butanediol and succinate. The high production of acetate is a major disadvantage; and this problem is solved by deleting ALD6, which encodes the $NADP^+$- dependent, $Mg2^+$-activated cytosolic acetaldehyde dehydrogenase isoform in the strains over expressing GPD1. High amounts of glycerol are obtained without increasing acetate formation and are accompanied by the reorientation of carbon flux towards the formation of succinate and 2,3-butanediol. Attention must be paid to acetoin, the production of which was shown to increase in strains producing very high amount of glycerol. The same approach was applied to brewer's yeast. Over expression of GPD1 in brewer's yeast resulted in a lower ethanol yield (80% compared to the reference strain) and a 4-fold increase in glycerol yield. Although the concentration of higher alcohols and esters was not strongly affected, there was a marked increase in the production of acetaldehyde, diacetyl and acetoin.

*3.8.1.3. Increased production of acetate esters*

Isoamyl acetate (banana-like aroma) is an important determinant for beer, sake and young wines. This ester is produced by yeast from isoamyl alcohol, which is itself a by-product of leucine synthesis. Overexpression of one of the genes responsible for leucine synthesis (LEU4 encoding-isopropylmalate synthase) in a sake yeast strain resulted in a very slight increase in isoamyl alcohol concentrations and the corresponding ester. Another strategy developed in a brewing strain was to increase the formation of acetate esters by overexpression of the ATF1 gene, which encodes the alcohol acetyltransferase catalyzing the formation of esters from acetyl CoA and the relevant alcohols. This led to a 27-fold increase in the production of isoamyl-acetate during fermentation on a laboratory scale. The strategy has been also successfully applied to wine yeast strains. However, a major disadvantage of this approach is that it generates a simultaneous increase in the formation of esters, e.g. ethyl acetate, which can be undesirable in wine above a critical amount. Another approach developed on sake yeast strains was to disrupt the EST2 gene, which codes for the major esterase hydrolyzing isoamyl acetate. However, this only resulted in a limited increase (approximately 2-fold) in isoamyl acetate production.

#### *3.8.1.4. Elimination of ethylcarbamate*

Ethylcarbamate is a suspected carcinogen found in many fermented foods and beverages, e.g. wine, sherry, brandy and sake. It is mainly formed by the spontaneous chemical reaction of ethanol and urea at elevated temperatures in acidic media. Urea is produced mainly from the cleavage of arginine by arginase. To reduce the formation of urea in sake, the two copies of the CAR1 gene, coding for arginase, were disrupted in an industrial sake yeast. This resulted in the elimination of urea and ethylcarbamate formation during sake brewing. It may be also of interest to reduce the formation of ethylcarbamate in wines. Arginine is one of the most abundant amino acids in grape must and is quickly assimilated by yeast. Therefore, wines made from arginine-rich grape musts may sometimes contain amounts of ethylcarbamate that exceed the authorized concentrations. Although the approach described above might be conceivable for wine yeast, it has the major drawback of reducing the amount of nitrogen available during fermentation. Another possibility is to express an acidic urease that degrades urea into ammonia and carbon dioxide in yeast. The acidic urease from *Lactobacillus fermentum* was recently expressed in *S. cerevisiae*, with the aim of degrading urea during wine fermentation. However, this approach was ineffective, because *S. cerevisiae* lacks several auxiliary proteins and appropriate cofactors required for the correct folding of urease.

#### *3.8.1.5. Antimicrobial properties*

In brewing, sake-brewing and wine-making, there is always a risk that wild yeast becomes predominant. Many indigenous yeasts secrete zymocin, a toxin that kills sensitive strains. Zymocin production and immunity to this toxin are determined by a cytosolic, doublestranded RNA. There are three major types of killer toxins (K1-K3). As most of the yeast strains found in grape musts produce the K2 zymocin, the K1 dsRNA has been integrated into the genome of a K2 wine yeast. A brewing strain with increased resistance to contamination was also constructed by expressing the genes coding for zymocin production and immunity. The modified yeast was resistant to the zymocin which can be produced by contaminating yeast and can kill yeast flora that is sensitive to the toxin. Although the idea of achieving a better control of yeast implantation is relevant, the utilization of starter strains with a selective advantage is questionable. Industrial yeasts, for example wine yeasts, are classically used without containment and are released into the environment. Therefore, great attention must be paid to the potential impact of yeast strains possessing a competitive advantage over the natural microflora.

The fermentation media are also susceptible to bacterial contamination. A brewer's strain carrying both the killer factor and antibacterial properties has been constructed. This strain was obtained by mating a respiratory-deficient strain with antimicrobial properties and a killer strain, followed by fusion of the resulting hybrid with a brewer's strain. Attempts to develop bactericidal wine yeast strains have been recently described. Two bacteriocin genes, encoding a pediocin and a leucocin gene from *Pediococcus acidilactici* and *Leuconostoc carnosum* respectively, have been expressed in *S. cerevisiae*. As these bacteriocins have a rather narrow spectrum, this approach should be extended to other toxins to warrant a satisfying stability of wines. The use of bacteriocidal yeasts would be useful for the production of wine with reduced levels of sulfur dioxide and other chemical preservatives. However, as sulfur dioxide has other interesting properties as an antioxidant, its addition to wine will probably remain necessary.

### 3.8.2. *Specific targets for brewer's yeast*

#### 3.8.2.1. *Fermentation of dextrins*

Brewer's yeast strains cannot utilize dextrins, which represent about 25% of malt wort sugars and contribute significantly to the calorific content of beer. Crude commercial preparations containing glucoamylases (usually from *Aspergillus*) are used to produce light beers. An early target for yeast genetic improvement was the development of amylolytic yeast strains that hydrolyze residual starch or dextrin in wort, producing low carbohydrate beers. Although classical genetic approaches have succeeded in transferring the amyloglucosidase activity of *S. diastaticus* (which hydrolyzes the α-1,4 linkages of dextrins) to brewer's strains, a major problem was the co-transfer of the POF1 gene, causing phenolic off-flavor. This side-effect was overcome by the expression of the *S. diastaticus* STA2 gene encoding amyloglucosidase. To overcome plasmid instability problems, an all-yeast, multicopy plasmid carrying STA2 under the control of the PGK promoter, was used to transform a commercial lager-brewing yeast strain. The recombinant brewer's yeast secreting amyloglucosidase exhibited improved carbohydrate degradation and produced approximately 1% more ethanol, without altering the fermentation performance or the quality of the beer. The degree of dextrin degradation was further increased by expressing amyloglucosidases able to hydrolyze α-1,6 links. The corresponding genes from *A. niger* or *A. awamori* have been integrated into the genome of brewer's yeast strains, to give recombinant strains with high levels of secreted enzymes and dextrin degradation. Good quality, superattenuated beers have been produced on a pilot scale by brewer's yeast secreting the *A. niger* enzyme. Increased dextrin degradation has also been obtained with a brewer's yeast expressing the gene coding for the amyloglucosidase of *Schwanniomyces occidentalis*. This enzyme has both α-1,4 and α-1,6 activities and has the additional advantage of being thermolabile, thus being inactivated by pasteurization. This ensures that the beer does not become sweet during storage. To increase starch degradation further, various *Saccharomyces cerevisae* yeasts coexpressing the STA2 gene from *S. distaticus*, coding for a glucoamylase, and the AMY gene from *Bacillus amyloliquefaciens* were constructed. These strains were very efficient at starch degradation in laboratory conditions.

*3.8.2.2. Beer viscosity and filtration*

The β-1,4 and β-1,6 linkages of β-glucans, a polysaccharide found in barley cell walls, are cleaved by a specific endo-β-glucanase during malting. This thermolabile enzyme is only present in small amounts in the wort, due to the elevated temperature during malt drying. Consequently, β-glucan degradation is often insufficient, which leads to increased viscosity, reduced filterability and the formation of gels and hazes in the beer. The filtration performance of beer can be significantly improved by the addition of a commercial preparation of bacterial β-glucanase to the fermenter. A cheaper solution is the use of yeast strains secreting glucanase. The secretion of *B. subtillis* β-glucanase in brewer's yeast strains was obtained by expressing the appropriate gene fused to the promoter and signal sequence of factor, in a multicopy plasmid. More efficient β-glucan degradation was subsequently achieved by expressing a β-glucanase of *Trichoderma reesei* with a lower optimum pH, between 4 and 5, instead of 6.7 for the *B. subtillis* enzyme. Brewer's strains carrying one copy of the *T. reesei* EG1 gene integrated into the chromosome (ADH1 or PGK1 locus) and devoid of bacterial sequences have been constructed and tested on a pilot scale. Almost all β-glucans were digested by some of the integrants during fermentation, resulting in a significant reduction of beer viscosity. The growth and brewing properties of the strains were unaltered.

*3.8.2.3. Diacetyl elimination*

Brewer's yeasts positively or negatively influence the flavor of the beer by producing aroma compounds during primary fermentation. Vicinal diketones, in particular pentanedione and diacetyl, are considered unpleasant. The main reason for lagering (secondary fermentation) is to decrease the concentration of diacetyl, which can be tasted at concentrations of below 0.02-0.10 mg/l. Diacetyl is formed outside the cell by the chemical oxidation of acetolactate, an intermediate in the biosynthesis of valine, which diffuses into the fermenting wort. Once diacetyl has been formed, it is enzymatically reduced inside the yeast cell to acetoin and finally to 2,3-butanediol, which has no impact on the flavor of the beer. The complete removal of diacetyl sometimes requires lengthy maturation (1-3 weeks). Various approaches based on the engineering of enzymes of the valine biosynthetic pathway have been tested to reduce yeast diacetyl formation. Diacetyl formation was reduced in ILV2 mutants and in strains in which the expression of ILV5, which codes for acetohydroxy acid reductoisomerase, was increased. Another strategy, which has the advantage of not interfering with the biosynthesis of valine or isoleucine, consisted of expressing the acetolactate decarboxylase gene from *Enterobacter aerogenes* in brewer's yeast. Brewer's strains with the ADHI-controlled ALDC gene integrated into the chromosome produced considerably less diacetyl than the parent yeast in fermentation tests.

*3.8.2.4. Reduction of hydrogen sulfide production*

Low levels of hydrogen sulfide are produced by *S. cerevisiae*, by the reduction of sulfate during methionine synthesis. Due to its very low sensory threshold, trace amounts of hydrogen sulfide can alter the organoleptic characteristics of beer. This compound must be eliminated during beer maturation. An alternative strategyis to develop brewing yeasts with reduced hydrogen sulfide formation. Increased expression of NHS5, encoding the cystathionine-synthase in brewing yeast, has been shown to suppress the formation of hydrogen sulfide in beer produced on the laboratory scale, without affecting other fermentation characteristics. Another approach, based on the overexpression of MET25, which encodes o-acetylhomoserine and o-acetylserine sulfhydrylase, resulted in a 10-fold decrease in hydrogen sulfide production in beer, in pilot-scale experiments. This can be explained by increased consumption of the substrate by the overproduced enzyme and by a decrease in sulfate uptake. An alternative approach was to partially or fully eliminate MET10, which encodes a putative sulfite reductase subunit. This led to a substantial reduction in hydrogen sulfide and the accumulation of sulfite. The beer produced showed increased flavor stability.

Another sulfur compound that may cause organoleptic problems, especially for some lager beers, is dimethyl sulfide (DMS) which is produced during fermentation by reduction of dimethyl sulfoxide (DMSO). The MXR1 gene was recently shown to encode a methionine sufoxide reductase. A MXR1 disruption mutant was reported to be unable to reduce DMSO in laboratory conditions. This work opens up the construction of brewing strains that do not produce DMS.

*3.8.2.5. Increased production of sulfur dioxide*

Sulfite is an antioxidant and a key compound for the flavor stability of beer, because it combines with aldehydes. There has been extensive work to control the accumulation of sulfite by brewer's yeast. One approach, based on the overexpression of MET3 and MET14 in brewer's yeast, led to increased sulfite production. Alternatively, brewer's yeast with several of the four copies of MET2 inactivated was shown to produce higher amounts of sulfite. However, hydrogen sulfide production was also increased. As mentioned above, the inactivation of MET10 was shown to be an efficient strategy for increasing sulfite formation and decreasing hydrogen sulfide production.

### *3.8.3. Specific targets for wine yeast improvement*

*3.8.3.1. Malolactic fermentation*

Alcoholic fermentation by pure-cultured wine yeast strains has become an increasingly well controlled process. In contrast, malolactic fermentation remains unreliable in numerous situations, due to the poor development of lactic acid bacteria in wine. This secondary fermentation (decarboxylation of malate to lactate by the malolactic enzyme) has an essential role for the deacidification and stabilization of wine. Delayed or stuck malolactic fermentation leads to scheduling problems in cellars and increases the risks of wine alteration. To circumvent the problems of the unreliability of malolactic fermentation, *S. cerevisiae* strains able to degrade malic acid completely into lactic acid and carbon dioxide have been constructed. The malolactic gene from *Lactococcus lactis* has been cloned and efficiently expressed in *S. cerevisiae*. Due to the inability of *S. cerevisiae* to transport malate efficiently, the malate permease from *Schizosaccharomyces pombe* was coexpressed with the malolactic enzyme in *S. cerevisiae*. The recombinant strains fully degraded up to 7 g of malate/l in 4 days, simultaneously with alcoholic fermentation and without affecting the growth properties and fermentation rate. A careful comparison of wines, obtained on a pilot scale with engineered industrial strains and traditional malolactic fermentation, will be necessary. Attention will have to be paid to the impact on flavors, because bacterial malolactic fermentation has been described in some instances to improve the organoleptic complexity of wine. Similarly, it will be interesting to compare the advantages of both yeast and bacterial malolactic fermentation in terms of wine stabilization, because the composition of residual micronutrients might differ in the wines obtained by the two systems. Although the use of engineered yeast strains for malolactic fermentation to produce safer wines of high quality might compete with bacterial starter cultures in the future, the use of malolactic-engineered yeast would provide an additional advantage. Indeed, the elimination of secondary fermentation limits the risk of developing bacterial flora responsible for the alteration of wine and/or the production of undesirable metabolites.

*3.8.3.2. Production of lactic acid*

Achieving the correct balance between sugar and acidity is a major requirement for wine quality. In some hot regions, grape musts are often insufficiently acidic and the balance must be corrected. Wine acidification is usually accomplished by adding tartaric acid, which is authorized in certain specific situations and the efficiency of which is limited by the instability of the K-tartrate. Biological acidification using a lactic acid-producing *S. cerevisiae* strain is a promising alternative, due to the high stability and organoleptic properties of this organic acid. Moreover, lactic acid is naturally found in wines after malolactic fermentation. Therefore, *S. cerevisiae* strains performing a dual fermentation (ethanol and lactate) have been constructed by expressing a lactate dehydrogenase ( LDH) gene from *Lactobacillus casei.* These strains can efficiently adjust the acidity of grape must: production of lactate at around 5 g/l increased total acidity by 0.2-0.3 pH units. Consistent with the diversion of sugars towards lactate, the ethanol content of the wines obtained is slightly lower (approximately 0.25% (v/v) for 5 g lactate/l). It is technically possible to construct a set of wine yeast strains producing different amounts of lactic acid across a range suitable to correct grape musts lacking in acidity. Lactate-producing yeasts are one of the most promising examples of wine-yeast improvement and may also be of interest in other fermentation fields (production of acidic doughs for baking, industrial lactate production).

*3.8.3.3. Decrease of volatile acidity*

The level of acetic acid, the main component of volatile acidity, is critical for the quality of wines. The concentration of acetic acid in wines is usually approximately 0.5 g/l and must remain below 0.8 g/l. Yeasts sometimes produce excessive acetic acid, due either to the genetic background of the yeast or to the wine-making processes (e.g. excessive clarification). Despite the importance of this organic acid, the genetic basis of acetate production during alcohol fermentation is still largely unknown. It has recently been shown that the level of acetate produced by *S. cerevisiae* can be effectively controlled by increasing or decreasing the level of expression of the cytosolic acetaldehyde dehydrogenase encoded by ALD6. Inactivation of the two alleles of this gene in a wine yeast leads to a 2-fold reduction in the amount of acetate produced during wine fermentation. As aconsequence of the resulting redox imbalance, glycerol, succinate and 2,3-butanedediol production is slightly increased.

### *3.8.3.4. Improved polysaccharide degradation*

The use of pectinases and glucanases in winemaking has several advantages. These enzymes can be used to facilitate wine clarification and to improve liquefaction of grapes, thereby increasing the juice yield. Polysaccharide-degrading enzymes may also enhance the liberation of various compounds trapped in grape skins, thereby improving the bouquet and color of the wine. As these commercial enzyme preparations are expensive, much effort has been made to develop yeast strains that secrete heterologous pectinases, glucanases, xylanases or a combination of these enzymes. For example, polypectate degradation was increased by a wine yeast co-expressing the *Erwinia chrysanthemi* pectate lysate gene and the *E. carotovora* polygalacturonase gene. Another approach was based on the expression of the *Fusarium solani* pectate lyase gene. Glucanolytic wine yeast strains have been constructed by the expression of fungal and bacterial endo-β-1,4-glucanases, or a combination of endo- and exoglucanases.

### *3.8.3.5. Improved liberation of varietal aroma*

The bouquet of a wine is determined by a combination of many compounds, the production of which is influenced by grape variety, terroir, viticulture practices, enological practices and yeast strain fermentation conditions. Due to this complexity, only a small number of suitable targets can be defined. A well identified target for aroma produced by yeast metabolism is isoamyl acetate (described above). In addition to producing aromatic compounds, yeast can increase the liberation of varietal aroma by producing enzymes that hydrolyze non-aromatic precursors in the grape must. For example, various grape monoterpenols of (e.g. linalol, geraniol) are linked to diglycosides, which can be converted to monoglucosides by cleaving of the 1,6-glucosidic linkage. The flavor compound is then liberated by the action of a β-glucosidase. However, grape and yeast β-glucosidases are inhibited by glucose, are unstable at low pH and therefore poorly hydrolyze monoterpenyl-glucosides during wine fermentation. The addition of commercial preparations of fungal β-glucosidases after fermentation has been shown to improve the hydrolysis of the glyco-conjugated aroma compounds. Progress in this area relies on the isolation of new glucose-tolerant b-glucosidases that are stable at low pH. A highly glucose-tolerant β-glucosidase has been purified from *Candida peltata*. A β- glucosidase from *A. oryzae* that is highly resistant to inhibition by glucose and is stable at low pH was recently described and the corresponding gene cloned. This enzyme has a broad specificity, because it can hydrolyze 1,3-, 1,4-, and 1,6-β-diglycosidase and can release flavor compounds such as geraniol, nerol and linalol from the corresponding monoglucosides in a rich-glucose medium at pH 2.9. An alternative approach for increasing the flavor of wine involves the modification of existing *S. cerevisiae* metabolic pathways associated with the production of aromatic compounds. Interestingly, mutants of the ergosterol biosynthetic pathways have been shown to produce monoterpenes (geraniol, citronelol, linalool) similar to those of the floral grape cultivars.

### *3.8.3.6. Reduced sulfide*

Hydrogen sulfide is a highly undesirable compound in wine and is a product of the yeast's sulfur metabolism. Hydrogen sulfide is produced during wine fermentation, mainly in response to the depletion of nitrogen and possibly certain vitamins; and its production is influenced by many environmental factors and by the yeast strain. In conditions of nitrogen starvation, hydrogen sulfide may accumulate and diffuse out of the cells. Wine yeast strains with low hydrogen sulfide production have been obtained by hybridization. Alternatively, reduced sulfide production might be achieved by manipulating the sulfate metabolic pathway, as described above for brewer's yeast. However, strategies such as the elimination of MET10 cannot be used for wine yeast, because they result in an increased sulfite production. The final amount of sulfur dioxide in wine is regulated; and sulfur dioxide is currently added to grape must and wine as a preservative. However, due to the current unfavorable public opinion, the tendency is to reduce its content in the wine.

*3.8.3.7. Melibiose-utilizing strains*

Molasses, a commonly used raw material for the production of baker's yeast, contains up to 8% raffinose in addition to sucrose. This trisaccharide (fructose/glucose/galactose) is hydrolyzed by yeast invertase to fructose and the disaccharide melibiose. Bakers' yeast cannot utilize melibiose, because it does not have α-galactosidase (melibiase), the enzyme responsible for the hydrolysis of melibiose into the fermentable sugars, galactose and glucose. However, the α-galactosidase enzyme is found in bottom-fermenting brewers' yeast strains. The MEL1 gene encoding this enzyme has been cloned and transferred into baking strains. Strains expressing MEL1 on multicopy plasmids or integrated into the LEU2 locus have been reported to secrete significant amounts of α-galactosidase. All available melibiose was utilized in a beet molasses medium, resulting in higher yeast yields. Recombinant MEL+ bakers' yeast strains devoid of plasmid sequences have been recently developed. These strains exhibit an 8% increase in biomass yield without alteration of the growth rate.

*3.8.3.8. Maltose utilization*

A high fermentation rate is a prerequisite for bakers' yeast strains. The free sugars present in the flour (sucrose, glucose, fructose, maltose) are sequentially consumed. Fermentation continues due to the action of amylases present in the dough, which release maltose from starch. Maltose utilization by yeast requires a maltose permease and a maltase; and both are induced in the presence of maltose. A major factor limiting the dough fermentation rate is the repression of the synthesis of maltose-utilizing enzymes and the inactivation of the maltase enzyme by glucose. Low concentrations (1-2%) of free sugars (mainly glucose and fructose) in the dough repress maltose utilization, causing a lag phase in carbon dioxide production. To avoid this lag phase, maltose-utilizing enzymes have been derepressed by replacing the native promoters of the maltase and maltose permease with constitutive promoters. Baker's yeast strains display intrinsic differences in their rates of maltose utilization. Non-lagging bakers' strains are characterized by rapid maltose fermentation, unlike lagging strains, which exhibit a lag phase. It has been reported that the non-lagging phenotype depends on a high level of maltase expression in the presence of glucose. More recently, it has been suggested that non-lagging strains display higher constitutive basal levels of expression of MAL genes than lagging strains, under non-inducing and non-repressing conditions. It has been proposed that these differences are due to divergences in the genetic structure of the

MALX3 gene, which codes for an activator of the other MAL genes. Therefore, expression of a MALX3 gene isolated from a non-lagging strain in a lagging strain might improve maltose metabolism in unsweetened dough.

### *3.8.4. Other approaches to increase the fermentative capacity*

Several attempts have been made to increase the glycolytic flux. Various control steps have been suggested, including sugar uptake, the hexokinase, the phosphofructokinase and the pyruvate kinase. However, overexpression of a combination of these enzymes and other glycolytic enzymes failed to increase the glycolytic flux. An alternative strategy was based on the hypothesis that a decrease in the intracellular ATP concentration would result in an increased carbondioxide production rate. This was tested by derepression of the ATP consuming gluconeogenic enzymes, fructose 1,6-diphosphatase and phosphoenolpyruvate carboxykinase. Although no evidence could be obtained of functional futile cycling, this resulted in a 25% increase in the rate of glucose consumption. This strain also displayed a significant reduction in biomass yield. Recently, recombinant strains, simultaneously overexpressing a set of seven enzymes involved in the lower part of glycolysis, were constructed. The recombinant strain exhibited increased glycolytic flux, but only under conditions of increased ATP demand.

## 3.9. Pathogenic potential

Microorganisms used for the production of fermented food (e.g. acetic, propionic, and lactic acid bacteria, yeasts and certain filamentous fungi) have a long history of safe use. Although some enteric lactic acid bacteria have, in rare instances, been identified as the cause of bacteremia or endocarditis in patients with severe underlying disease, they can by no means be regarded as food-borne pathogens. Foodborne pathogens are either invasive and/or toxinogenic in the food or in the human intestine. Opportunistic pathogens in food may not be hazardous for the healthy consumer, but may pose a threat to some health-compromised persons. The genomes of many of the important food-borne pathogens and of some opportunistic pathogens have been fully sequenced, and the genes responsible for pathogenicity traits have been identified .This opens the way to identify similar genetic information in the genomes of microorganisms used in food fermentation. Of the several genomes of microorganisms used in food fermentation that have been completely sequenced, two examples (*Saccharomyces cerevisiae* and *Lactococcus lactis*) have been published and reported to be free of known pathogenicity traits. If strains of species known to carry potential toxin genes are subjected to genetic modification, they must not carry such genetic information. The long history of safe use and the available genetic evidence suggests that the genetic background of the majority of microorganisms used for food fermentation is free from pathogenicity islands and other pathogenicity determinants.

In addition to this, the following need to be considered: i) The genetic modification could produce a metabolic imbalance that may enhance the level of common metabolites that are normally not toxicologically significant in food to levels that are unacceptable (e.g. formic acid, acetaldehyde, biogenic amines in lactic acid bacteria or yeast; cyclopiazonic acid or roquefortin in *Penicillium camemberti/roqueforti*). ii) The genetic modification could switch on genes coding for normally unexpressed toxins in the microorganism. iii) The genetic modification (i.e. the expression of new protein(s)) could change the "cross-talk" between the microbe and the intestinal immune system of the consumer leading, for example, to an undesirable immune reaction or undesirable reactions with other cells (e.g. enterocytes) of the GI tract. The first and second of these points may be addressed with *in vitro* studies. The third point may need to be addressed by *in vivo* studies in suitable animal models or human volunteers. If required, such studies will needed to be carried out in accordance with good practice guidelines and ethical standards.

## 4. WHOLE-GENOME SHUFFLING - A MILESTONE IN STRAIN IMPROVEMENT

Recent studies describe a major advance in the construction of mutants with an apparently significantly higher probability of improved phenotype. The method involves whole-genome shuffling (with protoplast fusion) of a small number of parental strains exhibiting subtle improvements in the phenotype of interest obtained by classical strain improvement methods. The researchers demonstrated this method in two cases of industrial importance: the improvement of tylosin production (a complex polyketide antibiotic) by *Streptomyces fradiae* and the isolation of acid-tolerant strains of *Lactobacillus* suitable for the production of lactic acid in low pH fermentations. In both cases, significant phenotypic improvements were obtained after only a few rounds of genome shuffling, at least comparable to those resulting from multi-year strain improvement efforts by mutation and screening/selection. Although random mutation and selection methods have succeeded in generating many industrial strains, today they must compete with recombinant technologies and direct genetic manipulation of strains through the introduction and control of specific genes. However, the difficulty with such "rational" methods is that the profile of an ideal cell depends on a multitude of genes that are rather poorly understood, mostly unknown, and broadly distributed throughout the genome. This raises serious obstacles in the direct application of genetic engineering for strain improvement and invites combinatorial approaches for the determination of the optimal genetic configuration in microbes for industrial applications. In combinatorial approaches, cells that are the most useful with respect to the phenotype of interest are screened or selected from a genotypically diverse collection (or library) of candidates. Such libraries have been generated in the past by random mutation of cells, successive rounds of mutation and screening/selection of individual genes (directed evolution), or shuffling of gene homologs from different strains. The success of combinatorial approaches depends on the initial selection of variants, the efficiency of the genetic recombination process, and the power of the selection method. The two reports described here demonstrate that protoplast fusion allows recombination events to occur throughout the genome and at much longer range relative to other methods. Genome shuffling is in principle applicable to the improvement of most production pathways and phenotypes. It is rather straightforward to implement and well suited for industrial applications, where time is of the essence. As such, it constitutes a major milestone in strain-improvement technology and metabolic engineering because of the opportunities, along with new challenges, that emerge with this technology.

## 5. MICROBIAL GENOMICS RESEARCH

By some estimates, microbes make up about 60% of the earth's biomass, yet less than 1% of microbial species have been identified. Microbes play a critical role in natural biogeochemical cycles. Because most do not cause disease in humans, animals, or plants and are difficult to culture, they have received little attention. Microbes have been found surviving and thriving in an amazing diversity of habitats, in extremes of heat, cold, radiation, pressure, salinity, and acidity, often where no other life forms could exist. Identifying and harnessing their unique capabilities, which have evolved over 3.8 billion years, will offer us new solutions to longstanding challenges in environmental and waste cleanup, energy production and use, medicine, industrial processes, agriculture, and other areas. Scientists also are starting to appreciate the role played by microbes in global climate processes, and we can expect insights about both the biological underpinnings of climate change and the contributions of microbes to earth's biosphere. Their capabilities soon will be added to the list of traditional commercial uses for microbes in the brewing, baking, dairy, and other industries. Microbes may, for example, contain enzymes that are effective in driving chemical reactions in extreme environments. Some may provide enzymes useful in research; one such "extremozyme" derived from a bacterium living in hot springs in Yellowstone National Park has become critical to current protocols for sequencing any genome, including that of humans. Other microbes have metabolic processes with potential for breaking down toxic waste or even producing methane, an energy source. Knowledge about the enormous range of microbial capacities has broad and far-reaching implications for environmental, energy, health, and industrial applications.

Genomic analyses are badly needed of microbial consortia and species for biotechnological exploitations. The establishment in 1994 of the US Department of Environment's Microbial Genome Program put the microbial genomic data into the public databases and over the last 10 years, sequencing of a range of microorganisms that live in a wide diversity of environments has provided a considerable information base. Researchers estimate that more than 200 million additional microbial DNA bases will be sequenced in the next 2 to 3 years, with some 200,000 predicted genes which will be useful in generating various new genetically improved microorganisms. The genetically engineered microorganisms or the biotechnological processes involving them are patentable.

## 6. STATUTORY REGULATION FOR GENETICALLY MODIFIED ORGANISMS AND CONSUMER DEMANDS

*S. cerevisiae* and lactic acid bacteria have been playing a major role in the food and beverage industries since long and are mankind's oldest domesticated organisms, generally regarded as safe. Their metabolism has made possible various biotechnological processes like beer brewing, leavening bread doughs, wine making, cheese making, yoghurt production and many indigenous fermented foods. *S. cerevisiae* was also the first genetically modified organism (GMO) to be cleared for food use, as a baking-and brewing strain. Although considerable progress has been made in genetic manipulation of wine yeasts for specific wine making processes but these strains have not yet been cleared for commercial use probably due to the public fear and perception of risk with regard to genetically modified food. It has been generally agreed that these fears are baseless as no scientific test has shown any of the GM foods currently being used to be toxic.

Fears about food and environmental safety has evoked a plethora of strict legislation and regulatory guidelines. Such Regulations, although differing in detail, are broadly similar in most countries. Guidelines for approval of GM products and the release of GMOs usually require a number of obvious guarantees. These include a complete definition of the DNA sequence introduced, and the elimination of any sequence that is not indispensable for expression of the desired property, the absence of any selective advantage conferred on the transgenic organismthat could allow it to become dominant in natural habitats, no danger to human health and/or the environment from the transformed DNA, and a clear advantage to both the producer and the consumer.

The regulatory authorities appear more willing to approve the use of GMOs than the public as the latter suspects that GM foods will prove unhealthy in the long term and that the escape of GMOs with transplanted genes will damage the environment and result in the loss of biodiversity. As the fears of the general public are baseless, consumer education is essential to remove their fear of the unknown. Scientists must consistently inform the public and remain open about experiments, research and products. The consumer should be reassured of transparent regulatory systems and the meticulous. implementation of biosafety legislation with clear technical standards and definitions with respect to GM products.

Successful application of recombinant DNA technology in the wine industry will depend on assuring commercial users of genetically modified wine yeasts that existing desirable characteristics have not been damaged, that the requirements of beverage legislation are met, and that the engineered strain will be stable in practice.

## 7. CONCLUSIONS AND FUTURE OUTLOOK

Considerable progress has been made during the past two decades in the development of industrial strains possessing optimized and new characteristics. However, the wine industry, two decades after the first successful yeast transformation, has entered the third millennium without a transgenic wine yeast used on a commercial scale to produce wine. For the most advanced examples, genetically modified industrial strains have been constructed according to the general requirements for genetically modified organisms (GMOs), particularly the absence of resistance markers and stability (usually obtained by chromosomal integration); and the new properties have been confirmed on a pilot-scale). However, despite the remarkable progress during the past 20 years, only two have so far received official approval (both from the British government) for commercial use, but they are not currently used commercially. The first is a baker's yeast derepressed for maltase and maltose permease. The second is a brewer's yeast gene expressing the STA2 gene and producing exocellular glucoamylase.

Public acceptance considerations remain the major obstacle to the commercialization of genetically modified industrial yeast strains. One of the major difficulties is that the benefits of genetically modified yeast strains are in most cases not perceptible to consumers, except for some nutritional or hygiene advantages. Efforts to inform and to discuss with the general public have been limited and need to be developed to increase public awareness of the potential benefits (safe production, high quality/low cost) of recombinant DNA technology. The acceptability of GMOs in food will also depend on the presence (e.g. bread, wine) or absence (e.g. beer, filtered wines) of the GMO in the product. Detection methods will have to be developed to differentiate these two types of product. The debate should also include aspects concerning the practical consequences of the introduction of this new technology for the industry. For example, the risks associated with GMO release. Each industrial field may have to be considered separately and specific approaches may have to be defined and implemented for each. The end of the twentieth century has been marked by the explosion of life technologies, in particular those related to genomics. *S. cerevisiae* recently became the first eukaryotic microorganism whose entire genome has been sequenced. The challenge for the next years will be to expand the knowledge and data obtained for laboratory strains of *S. cerevisiae* in laboratory conditions to industrial strains and conditions. The genome of industrial yeast is more complex (e.g. alloploidy, polyploidy, aneuploidy, chromosomal rearrangements) than that of laboratory strains. Furthermore, these strains are exposed to many environmental stresses, e.g. alcohol concentration, desiccation, high osmolarity, freezing and nitrogen depletion, either simultaneously or sequentially. Their adaptation and response to these various stresses is largely unknown. Several commercial traits (fermentation capacity, metabolite production, stress tolerance, etc.) are under multigenic control and their molecular bases are largely unknown. New possibilities are now available for exploring the metabolic and genetic control of gene expression on a genomic scale. The development of functional genomic approaches in relevant industrial conditions will obviously help us to identify genes that are regulated coordinately, transcription factors that control metabolism and relevant technological properties. Besides a huge increase in the knowledge of the adaptation mechanisms developed by industrial strains, new targets for genetic engineering can be expected.

## 8. FURTHER READING

Lal, R., Khanna, R., Kaur, H., Khanna, M., Dhingra, N., Lal, S., Gartemann, K.H., Eichenlaub, R. and Ghosh, P.K. (1996). Engineering antibiotic producers to overcome the limitations of classical strain improvement programs. Crit. Rev. Microbiol.22: 201-255.

Maraz, A. (2002). From yeast genetics to biotechnology. Acta Microbiol. Immunol. Hung. 49: 483-491.

Baltz, R.H. (2001). Genetic methods and strategies for secondary metabolite yield improvement in actinomycetes. Antonie Van Leeuwenhoek. 79:251-259.

McDaniel, R., Licari, P. and Khosla, C. (2001). Process development and metabolic engineering for the overproduction of natural and unnatural polyketides. Adv. Biochem. Eng. Biotechnol. 73:31-52

Schuller, D. and Casal, M. (2005). The use of genetically modified *Saccharomyces cerevisiae* strains in the wine industry. Appl. Microbiol. Biotechnol. 68: 292-293.

Chiang, S.J. (2004). Strain improvement for fermentation and biocatalysis processes by genetic engineering technology. J. Indust. Microbiol. Biotechnol. 31:99-108.

Nielsen, J. (1998). Metabolic engineering: techniques for analysis of targets for genetic manipulations. Biotechnol. Bioeng. 58:125-132.

Parekh, S., Vinci, V.A. and Strobel, R.J. (2000). Improvement of microbial strains and fermentation processes. Appl. Microbiol. Biotechnol. 54:287-301.

Poggeler, S. (2001). Mating-type genes for classical strain improvements of ascomycetes. Appl. Microbiol. Biotechnol. 56:589-601.

Popov, B.K. and Konstantinova, R.D. (1986). Production and regeneration of yeast protoplasts: a review. Zentralbl Mikrobiol. 141:217-224.

Pretorius, I.S. (2000). Tailoring wine yeasts for the new millennium: novel approaches to the ancient art of winemaking. Yeast. 16: 675.

Pretorius, I.S. and Bauer, F.F. (2002). Meeting the consumer challenge through genetically customized wine yeast strains. Trends. Biotechnol. 20:426.

Pretorius, I.S. and Van der Westhuizen, T.J. (1991). The impact of yeast genetics and recombinant DNA technology on the wine industry – a review. S. Afr. J. Enol. Vitic. 12: 3.

Stafford, D.E, and Stephanopoulos, G. (2001). Metabolic engineering as an integrating platform for strain development. Curr. Opin. Microbiol. 4:336-340.

Yanase, H., Kato, N. and Tonomura, K. (1994). Strain improvement of *Zymomonas mobilis* for ethanol production. Bioprocess. Technol. 19:723-739.

□□□

# 13

# Patenting of Biotechnological Inventions

S.K. SONI[1], R. SONI[2] AND N. GOYAL[1]
[1]*Department of Microbiology, Panjab University, Chandigarh-160 014*
[2]*Department of Biotechnology, D.A.V. College, Chandigarh-160 011*

## 1. INTRODUCTION TO PATENT LAWS

A patent is an intellectual property right (IPR) relating to inventions and is the grant of exclusive right, for limited period, provided by the Government to the patentee, in exchange of full disclosure of his invention, for excluding others, from making, using, selling, importing the patented product or process producing that product for those purposes. The purpose of this system is to encourage inventions by promoting their protection and utilization so as to contribute to the development of industries, which in turn, contributes to the promotion of technological innovation and to the transfer and dissemination of technology. Under the system, patents ensure property rights (legal title) for the invention for which patent have been granted, which may be extremely valuable to an individual or a company. One should make the fullest possible use of the patent system and the benefits it provides. Patent right is territorial in nature and a patent obtained in one country is not enforceable in other country. The inventors/their assignees are required to file separate patent applications in different countries for obtaining the patent in those countries.

A patent provides protection for the invention to the owner of the patent. The protection is granted for a limited period, generally 20 years. A patent also provides protection to the invention that the invention cannot be commercially made, used, distributed or sold without the patent owners consent. These patent rights are usually enforced in a court, which, in most systems, holds the authority to stop patent infringement. Conversely, a court can also declare a patent invalid upon a successful challenge by a third party. A patent owner has the right to decide who may - or may not - use the patented invention for the period in which the invention is protected. The patent owner may give permission to, or license, other parties to use the invention on mutually agreed terms. The owner may also sell the right to the invention to someone else, who will then become the new owner of the patent. Once a patent expires, the protection ends, and an invention enters the public domain, that is, the owner no longer holds exclusive rights to the invention, which becomes available to commercial exploitation by others. Patents provide incentives to individuals by offering them recognition for their creativity and material reward for their marketable inventions. These incentives encourage innovation, which assures that the quality of human life is continuously enhanced. Patented inventions have, in fact, pervaded every aspect of human life, from electric lighting (patents held by Edison and Swan) and plastic (patents held by Baekeland), to ballpoint pens (patents held by Biro) and microprocessors (patents held by Intel, for example).

All patent owners are obliged, in return for patent protection, to publicly disclose information on their invention in order to enrich the total body of technical knowledge in the world. Such an ever-increasing body of public knowledge promotes further creativity and innovation in others. In this way, patents provide not only protection for the owner but valuable information and inspiration for future generations of researchers and inventors.

The first step in securing a patent is the filing of a patent application. The patent application generally contains the title of the invention, as well as, an indication of its technical field; it must include the background and a description of the invention, in clear language and enough detail that an individual with an average understanding of the field could use or reproduce the invention. Such descriptions are usually accompanied by visual materials such as drawings, plans, or diagrams to better describe the invention. The application also contains various "claims", that is, information which determines the extent of protection granted by the patent.

An invention must, in general, fulfill the following conditions to be protected by a patent. It must be new - the invention must not have been disclosed in any way, by word of mouth, journal article or abstract. An invention may be completely new or an improvement on an existing invention. It must involve an inventive step - the invention should not be obvious to another scientist with experience of the field, though in practice this is often difficult to judge, particularly if the invention involves an incremental improvement. It must have industrial application - the invention should be capable of use within some kind of industry, where industry has the broad definition of anything distinct from purely intellectual or aesthetic activity. In practical terms this means that the invention takes the form of a device or apparatus, a product such as a new material or substance, or a new industrial process or method of application. Finally, its subject matter must be accepted as "patentable" under law. In many countries, scientific theories, mathematical methods, plant or animal varieties, discoveries of natural substances, commercial methods, or methods for medical treatment (as opposed to medical products) are generally not patentable.

A patent is granted by a national patent office or by a regional office that does the work for a number of countries, such as the European Patent Office and the African Regional Industrial Property Organization. Under such regional systems, an applicant requests protection for the invention in one or more countries, and each country decides as to whether to offer patent protection within its borders. The WIPO-administered Patent Cooperation Treaty (PCT) provides for the filing of a single international patent application which has the same effect as national applications filed in the designated countries. An applicant seeking protection may file one application and request protection in as many signatory states as needed.

Patenting is done to protect the creative work of people. The generation of new microorganisms by recombinant DNA technology (rDNA technology) are patentable.

## 2. AN OVERVIEW OF THE PATENTING PROCESS

A patent is an exclusive right of its owner to exclude others from making, using, or selling the invention as defined in the claims of the patent for a period of time, which in the United States is 20 years from the date of filing the patent application. The typical process of obtaining a patent on an invention is shown in Figure 1.

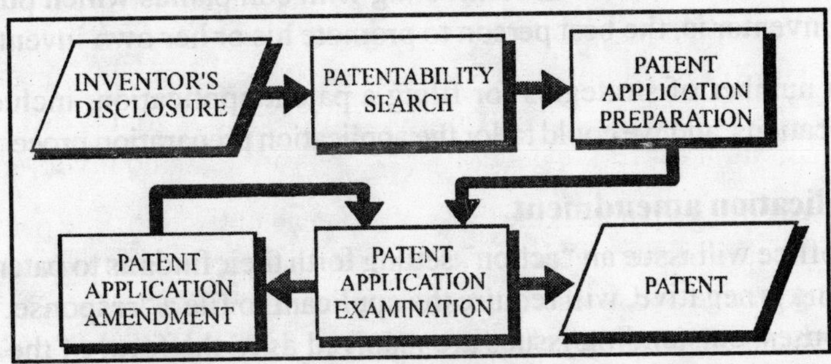


**Figure1** Typical process of obtaining a patent

## 2.1. Inventor's disclosure

If you believe that your invention may be protectable, then you should prepare a disclosure of the invention in written form. Your disclosure should indicate what your invention is, how it works, how it can be used, and how it is an improvement over known articles or methods. Preferably, your disclosure information should include one or more drawings of your invention, and your written description should also make reference to your drawings as appropriate. Your disclosure should be signed and dated by you and witnessed by at least two individuals who are not co inventors. It is recommended that you maintain a diary or laboratory notebook to keep track of the dates when you initially conceived your invention and the steps taken to reduce it to practice.

## 2.2. Patentability search

We have found that the best method of evaluating whether to proceed with a patent application is to conduct a patentability search in the patent office of the country. You disclose your invention to the office which, will conduct the search in which all properly classified and filed patents which are pertinent to the invention are reviewed and copies are ordered for your records. The office then drafts a patentability opinion letter indicating the scope of the patent protection. The results of the search may help you to better define the invention or to identify alternative embodiments of the invention, and may help to draft the text of the patent application and to draft claims in a way which would give the broadest possible protection to the invention.

## 2.3. Patent application preparation

The office requires authorization to proceed with preparing of the patent application. Preparation of the application depends upon the technical complexity of the subject matter, the quality of the written description provided by the inventor, and the number of revisions of the application necessitated by the redefining of the invention by the inventor during the application drafting process. The application may be filed with "informal drawings." Once the application is filed, it is "patent pending."

You may be interested to note that licensing can take place any time after the application is filed. Some manufacturers prefer to license an invention while it is still in the patent pending stage. Furthermore, it is advisable to delay any marketing efforts until after you have filed your application. The office normally does not get involved in such marketing efforts. Additionally, care should be taken in contracting with companies which purport to market inventions. The inventor is, the best person to promote his or her own invention.

There are a number of strategies for filing a patent application, including the use of provisional applications, and we could tailor the application preparation process to your needs.

## 2.4. Patent application amendment

The patent office will issue an "action" setting forth their finds as to patentability and, in the case the finding is negative, will require the applicant to file a "response." Usually, after one or two responses, outstanding issues are resolved as to the form of the claims and the scope of protection.

Patent application prosecution is dependent upon the complexity of the rejection, the reasonableness of the examiner, the closeness of the prior art as compared to the scope of patent protection desired by applicant, and the quality of the applicant's comments for responding to the examiner. The period of prosecution, and the interval over which these added costs are incurred, averages one to two years. After allowance, and upon payment of issuance fees, the application issues into a patent.

## 2.5. Patent application examination

When a patent application is filed with the patent office, a patent examiner, in determining patentability, must find that the invention has the following features:

i) *Usefulness:* Having some practical utility, fit for some desirable, practical or commercial purpose.

ii) *Novelty:* An invention will be considered novel if it does not form a part of the global state of the art. Information appearing in magazines, technical journals, books, newspapers etc. constitutes the state of the art. Oral description of the invention in a seminar/conference can also spoil novelty. Novelty is assessed in a global context. An invention will cease to be novel if it has been disclosed in the public through any type of publications anywhere in the world before filing a patent application in respect of the invention. Prior use of the invention in the country of interest before the filing date can also destroy the novelty. Novelty is determined through extensive literature and patent searches. It should be realized that patent search is essential and critical for ascertaining novelty as most of the information reported in patent documents does not get published any where else.

iii) *Non-obviousness:* A patent application involves an inventive step if the proposed invention is not obvious to a person skilled in the art i.e., skilled in the subject matter of the patent application. The prior art should not point towards the invention implying that the practitioner of the subject matter could not have thought about the invention prior to filing of the patent application. Inventiveness cannot be decided on the material contained in unpublished patents. The complexity or the simplicity of an inventive step does not have any bearing on the grant of a patent. In other words a very simple invention can qualify for a patent. If there is an inventive step between the proposed patent and the prior art at that point of time, then an invention has taken place. A mere 'scintilla' of invention is sufficient to found a valid patent.

## 2.6. Main parts of a patent

Front page information:

| Patent number | Title | Date of publication |
|---|---|---|
| Inventor: | Assignee: | Application number: |
| Date of application: | Cited references: | Classification numbers: |
| International classification code: | Abstract: | |
| Descriptions: | Claims: | Drawings: |

To be awarded a patent, the inventor should not have earlier abandoned, suppressed or concealed the invention. This requires that the inventor be diligent in both completing the invention and in filing for patent protection. Further, an inventor should always keep in mind that an earlier filing date is helpful where the patent office has to decide upon which of two pending applications for the same invention should be allowed to issue.

**2.7. Granting of patent**

If the results of the patent examination are favorable, a patent will be granted. A patent is an exclusive right of its owner to exclude others from making, using, or selling your invention as defined in the claims of your patent for a period of time, which in most of the countries is 20 years from the date of filing your patent application. After your patent has issued, maintenance fees must be paid at 3½ , 7½ , and 11½ years from the date of issue to keep your patent in force. Publication of an invention in any form by the inventor before filing of a patent application would disqualify the invention to be patentable. Hence, inventors should not disclose their inventions before filing the patent application. The invention should be considered for publication after a patent application has been filed. Thus, it can be seen that there is no contradiction between publishing an inventive work and filing of patent application in respect of the invention.

**2.8. Date and term of patent**

The date of patent is the date of filing the complete specification. This is an important date because it is from this date that the legal protection of an invention covered in the patent takes effect. The term of the patent is counted from this date:

a) Five years from the date of sealing of the patent or seven years from the date of the patent (i.e. the date of filing the complete specification), whichever period is shorter, for an invention claiming the method or process of manufacture of a substance, where the substance is intended or capable of being used as a drug, medicine or food.

b) Fourteen years from the date of patent in respect of any other patentable invention.

There are two types of patent documents usually known as patent specification, namely (i) Provisional specification and (ii) Complete specification.

**2.9. Provisional specification**

A provisional specification is usually filed to establish priority of the invention in case the disclosed invention is only at a conceptual stage and a delay is expected in submitting full and specific description of the invention. Although, a patent application accompanied with provisional specification does not confer any legal patent rights to the applicants, it is, however, a very important document to establish the earliest ownership of an invention. The provisional specification is a permanent and independent scientific cum legal document and no amendment is allowed in this. No patent is granted on the basis of a provisional specification. It has to be followed by a complete specification for obtaining a patent for the said invention. Complete specification must be submitted within 12 months of filing the provisional specification. This period can be extended by 3 months. It is not necessary to file an application with provisional specification before the complete specification. An application with complete specification can be filed right at the first instance.

## 2.10. Complete specification

Submission of complete specification is necessary to obtain a patent. The contents of a complete specification would include the following

1. Title of the invention.
2. Field to which the invention belongs.
3. Background of the invention including prior art giving drawbacks of the known inventions and practices.
4. Complete description of the invention along with experimental results.
5. Drawings etc. essential for understanding the invention.
6. Claims, which are statements, related to the invention on which legal proprietorship is being sought. Therefore the claims have to be drafted very carefully.

## 2.11. Cost of patenting

The Government fee for filing a patent application (complete/provisional) in India is Rs. 1,500 for individuals and Rs. 5,000 for legal entities. A sealing fee of Rs. 1,500 for individuals and Rs. 5,000 for legal entities has to be paid at the time of grant of patent. A patent granted in one country is not enforceable in other countries. There is nothing like a global patent or a world patent. Patent rights are essentially territorial in nature and are protected only in a country (or countries) which, has (have) granted these rights. In other words, for obtaining patent rights in different countries one has to submit patent applications in all the countries of interest for grant of patents. This would entail payment of official fees and associated expenses, like the attorney fees, essential for obtaining patent rights in each country. However, there are some regional systems where by filing one application one could simultaneously obtain patents in the member countries of a regional system; European Patent Office is an example of a similar system. Each country is free to grant or refuse a patent on the bases of scrutiny by its patent office. This means that granting a patent in one country of the union does not force other countries to grant the patent for the same invention. Also, the refusal of the patent in one country does not mean that it will be terminated in all the countries.

To be patentable the invention must fall within one of the following statutory classes:

a) *Utility patents:* May be granted to anyone who invents or discovers any new and useful process, machine, article of manufacture, or compositions of matters, or any new useful improvement thereof. Pertains to chemical, mechanical, electrical or physical patents. Statutory period lasts typically 20 years from filing date (prior to 1994 only 17 years were granted). Be aware of the fact that companies, especially pharmaceutical companies, file for extensions of the patent life. Upto 3 years can be granted.

b) *Plant patents:* Covers asexually or sexually reproducible plants. Statuary period is 17 years from issuance.

c) *Design patents:* Covers the unique, ornamental, nonfunctional or visible shape or design of an object. Statutory period is 14 years from issuance.

## 3. PATENTING OF BIOTECHNOLOGICAL INVENTIONS

Patenting of biotechnology inventions has become front-page news with each big breakthrough in this burgeoning field, elevating the patenting of inventions to a higher profile. Biotechnological research and developments involve an increasing amount of time, money and energy. Patent protection encourages research, allowing inventors to profit from their inventions. This economic and intellectual investment has made the relationship between intellectual property and biotechnological inventions an issue of immediate interest.

The rules and regulations governing biotechnology patents are country-specific. Each country has set its own standards for granting biotechnological patents. Some countries place stricter standards than others regarding the patentability of biotechnological inventions. Given the great potential value of such inventions, these country-specific nuances need to be understood if inventors are to profit from their inventions.

Exciting inventions in the field of biotechnology have been made in recent years. Following the examples of the owners of the biotech patents, companies with high investments in the field of biotechnology now recognize the advantages of protecting and enforcing their intellectual property rights.

### 3.1. Specific patent needs in the case of biotechnology inventions

The current controversy in biotechnology is what is patentable and what is not. Generally, an invention is patentable while a discovery is not. While this rule may, in other areas, appear well defined, in biotechnology, it is often the cause of differences in regulations between countries. Discovery is merely making available what already exists in nature. A substance freely occurring in nature, if merely found or discovered, is not patentable. However, if the substance found in nature has first to be isolated from its surroundings, and a process for obtaining it is developed, that process is patentable.

The granting of patent is subject to strict criteria of novelty, inventiveness and industrial application. In the case of biotechnology, often involving living entities like microorganisms, there is an additional requirement of sufficient disclosure, for which the invention is required to be deposited at an authorized "depository authority". Each country has set its own standards for granting patents to biotechnological inventions. This chapter focuses on biotechnological inventions — what is patentable, where and how, with focus on patentability of biotechnology.

### 3.2. Biotechnological patents in Australia, USA and Europe

Even given these broad guidelines, across the globe, each country has taken a different approach to biotech patent regulation. As mentioned, countries differ even in the subject matter they view as "inventable" and "discoverable".

Australia's approach to biotech patents is one of the most liberal. In 1976, the Australian Patent Office (APO), in Rank Hovis McDougall Ltd.'s Application, held that living organisms are patentable, implying that they are inventable. The Australian Patent Act contains no express prohibition against the patenting of life forms (aside from human beings). The APO considers all living organisms excluding human beings as potentially patentable subject matter.

The US Patent Law has followed Australia's liberal approach towards biotech patents. The US Supreme Court decision in Diamond vs Chakraborty (1980) opened the way for inventions relating to genetic engineering and living organisms, declaring them as inventable. The United States Patent Office (USPTO) has since issued patents for over 6,000 genes, and about 1,000 of these relate to human genes. Currently, there are more than 20,000 patent applications related to genes pending in the US. However, in response to criticism that their gene patents are too liberal in defining what is invented, the US recently issued "utility guidelines" requiring stricter applicability standards.

In contrast, Europe adopts a more cautious approach towards granting biotech patents, which is addressed by the European Union directive on the Legal Protection of Biotechnological Inventions. The directives include a non-exclusive list of unpatentable processes, for example, cloning, germ-line modifications, embryo processes, transgenic processes, etc. The discovery/invention distinction also referred to as "attack of obviousness," has been one which has featured strongly in biotech patents in Europe. In Genetech Inc.'s patent, a claim to an isolated and characteristic protein produced by r-DNA technology was held by the English Court of Appeal to lack this required "attack of obviousness".

## 3.3 Indian perspective

India is a storehouse of biological resources and one of the world's richest biodiversity countries. In the past two years, there has been a rise in the investment in the biotech-oriented industries. It is estimated that over the next five years, biotechnology can offer opportunities for fresh investment of Rs 7-8 billion in India. Biotechnology is poised to take India to a different playing field where it can dominate the world market. However, to achieve its full potential, the biotech sector requires a facilitative environment. The government needs to make a conceded effort to amend its Patent Laws, strengthen its IPR regime, so as to protect the economic interests of those who innovate. Additionally, India needs to sign the "Budapest Treaty on the International Recognition of the Deposit of Microorganisms for the Purposes of Patent Procedure" to assist in the standardization process of biotechnology patenting.

## 3.4 What is patentable in biotechnology?

This is the question that India must decide in regards to new biotechnology including genomics, proteomics, bioinformatics, genetic engineering. As per the Indian Patent Act, the following are patentable:

- *Microorganisms:* under the Indian Patent Act, microbiological processes can be patented. Also patentable are processes for producing new-microorganisms through genetic engineering and the products that result out of this process, such as microorganisms including plasmids and viruses if they are non-living.

- Cell lines: A cell line is patentable if artificially produced.
- r-DNA, RNA, Amino acid: If the end result is non-living, it is patentable.
- Hybridoma technology: Patents are also allowed on hybridoma technology, but not on protoplast fusion.
- Expressed sequence tag's, or ESTs, are small fragments of genetic material obtained by reverse transcriptions of messenger RNA (mRNA) from expressed genes. The gene sequence, or expressed sequence tags (ESTs), can be patented if it has a use, such as if it works as a probe.

## 3.5. What is not patentable in biotechnology?

Additionally, the Indian Patent Act defines what is not patentable biotechnology: inventability does not apply to plant or animals. Accordingly, a method of producing a new form of a known plant or tissue culture method for production of plant variety is not patentable, nor is a method of treatment of a human body by surgery or operation for diagnosis. Nor is a method of improving or changing the appearance of the human body or parts of it patentable. Still, these categories are not as clear-cut as they appear. Skillful wording may decide whether a finding is an "invention" or a "discovery". Thus, it is best to consult a patent attorney prior to filing a patent. An invention may satisfy the conditions of novelty, inventiveness and usefulness but it may not qualify for a patent under the following situations:

- An invention which is frivolous or which claims anything obviously contrary to well established natural laws e.g. different types of perpetual motion machines.
- An invention, the primary or intended use of which would be contrary to law or morality or injurious to public health e.g. a process for the preparation of a beverage which involves use of a carcinogenic substance, although the beverage may have higher nourishment value.
- The mere discovery of a scientific principal or formulation of an abstract theory e.g., Raman effect.
- The mere discovery of any new property or new use of known substance or the mere use of a known process, machine or apparatus unless such a known process results in a new product or employs at least one new reactant.
- A substance obtained by a mere admixture resulting only in the aggregation of the properties of the components thereof or a process for producing such substance.
- The mere arrangement or rearrangement or duplication of features of known devices each functioning independently of one another in a known way.
- A method or process of testing applicable during the process of manufacture for rendering the machine, apparatus or other equipment more efficient.
- A method of agriculture or horticulture.
- Any process for medical, surgical, curative, prophylactic or other treatment of human beings, or any process for a similar treatment of animals or plants.
- Inventions relating to atomic energy.

To compete worldwide, India must decide whether to remain with the more conservative European approach, or if the Australia/US or some other approach better suits the needs of its emerging economy. In either case, Indian companies, inventors, or investors venturing into the biotech sector must be well-informed and well-aware of Indian laws, as well as the laws of other countries as they seek to join the biotechnology headlines. More importantly and immediately, these companies and investors venturing into the biotech sector need to fully realize the significant role intellectual property plays in the commercialization process for biotechnological innovation.

## 3.6. TRIPs and patent law

The need for a uniform law to harmonise the different laws related to the protection of intellectual property prevailing in the various countries resulted in the Trade related aspects of Intellectual Property Rights (TRIPs) agreement that came into force in 1995. It provides the minimum standards of patent protection that require mandatory compliance by all the member countries. In India the patent law was amended so as to bring it in compliance with the TRIPs agreement. This has led to the grant of product patents in the field of chemicals, pharmaceuticals, food and biotechnology which was not possible previously.

Article 27 of the TRIPs agreement forms the basis of the changes in the patent law. Article 27.1 states that "patents shall be available for all inventions, whether products or processes, in all fields of technology, provided that they are new, involved an inventive step and are capable of industrial application. Patents shall be available and patent rights enjoyable without discrimination as to the place of invention, the field of technology and whether products are imported or locally produced".

Article 27.2 states that "members may exclude from patentability inventions, the prevention within their territory of the commercial exploitation of which is necessary to protect public order or morality, including to protect human, animal or plant life or health or to avoid serious prejudice to the environment, provided that such exclusion is not made merely because the exploitation is prohibited by their law". This provision has also been incorporated into the Indian Patents Act. This subsection excludes inventions, the use of which would be contrary to public order and morality or would be seriously jeopardizing animal and plant life.

Article 27.3(b) states that "members may also exclude from patentability: plants and animals other than microorganisms, and essentially biological processes for the production of plants and animals other than non-biological and microbiological processes. However, members shall provide for the protection of plant varieties either by patents or by an effective sui generic system or by any combination thereof". This forms the basis for the provisions on the exclusions from patentability under the Indian law.

The patent act now also allows for the opposition against the grant of patent. As per the act, any person within three months from the date of publication can apply and oppose the grant of a patent. The Patent act also states that the use of the patented invention, without the consent of the inventor, would be considered as an infringement if the use of the invention is for the purposes of research and development only.

The present age of technology arguably belongs to biotechnological innovations involving the direct use of living organisms or parts or products of living organisms in their natural or modified form. Biotechnology also involves recombinant DNA technology which can be used to modify the genetic material of living cells to produce new substances or perform new functions.

The patent protection is obtainable for most of the biological innovations. The protection thus provided serves as an incentive for further development and technical innovation. Biological innovations in the patent system are similar to chemical inventions. These include such inventions as biological, microbiological, biotechnological, genetic engineering and medical inventions. Biotechnological inventions which may be patentable can be broadly categorized as:

i) products in the form of chemicals, organisms, plant extracts, fermented material;
ii) processes or methods for producing useful products; and
iii) compositions or formulations of products such as vaccines, proteins, hormones.

The frontier areas in biotechnological inventions include DNA sequencing methods, genetically modified organisms, stem cell research, gene therapy, diagnostic kits, vaccines, bioinformatics and cloning.

## 3.7. Patenting of microorganisms

Microorganism patents are now routinely granted by the US, European and Japanese Patent Offices. Although a US patent had been granted in 1873 to Pasteur for "yeast free from germs of disease as an article of manufacture", the US courts later held that the "discovery of some of the handiwork of nature" was unpatentable.

Article 27.3(b), of the TRIPs agreement forms the basis for the provisions on the patentability of microorganisms. Nevertheless, microorganisms also have to satisfy the novelty, utility and non-obviousness criteria to be patentable. However, TRIPs agreement does not provide a precise definition of microorganisms. The term microorganisms is generally understood to include viruses, bacteria, yeast and other forms of fungi, protozoa and unicellular algae. Even though microorganisms can be patented as per TRIPs agreement: one is often faced with the dilemma whether at all microorganism constitute a patentable subject matter since they are real "life forms". The law which opened the gates for the inventions in the field of biotechnology, particularly microorganisms, was by a landmark judgement of the US supreme court in 1980 in Diamond vs Chakraborty. The US Supreme Court decided that a microorganism was not precluded from patentability solely because it was alive. Thus a *Pseudomonas* bacterium manipulated to contain more than one plasmid controlling the breakdown of hydrocarbons (therefore more useful in dispersing oil slicks than the natural organism containing only one such plasmid) was "a new bacterium with markedly different characteristics from any found in nature" and hence not nature's handiwork but that of the inventor. The "product of nature" objection therefore failed and the modified organism was held patentable. The landmark judgement paved the way for the grant of a number of such biotechnology related patents.

The UK Patents Act further mentions that innovations involving discoveries, scientific theories, mathematical methods, literary works and presentation of information are not inventions. Moreover the Act specifies inventions for which patent shall not be granted. This includes inventions which are contrary to the public morality. The UK Act allows the patentability of microbial processes or the microorganism per se. Going by US and European precedents, it would appear that only such microorganisms that are a result of human intervention would be patentable. Naturally occurring microorganisms are likely to be excluded from patentability, unless the microorganism loses its natural characteristics as a result of human intervention. However, subsequent to the human intervention, the involved microorganism should satisfy the criteria of possessing an industrial applicability, that is the organism should not be devoid of a utility.

The Act also requires that the sample of the microorganism which forms the subject matter of the claimed invention be deposited not later than the date of filing the application. The deposit of the sample of the biological material is required to ensure that the specification is sufficiently enabled so as to render it workable. In India the position of patentability of microorganisms is parallel to that of the UK and Europe.

Microbiological processes are patentable subject matter after the implementation of the second amendment in May 2003. Earlier, the deposit of the culture samples were required to be made at an institution authorized by the Government. After the third amendment the deposit is now to be made as per the Budapest Treaty to/at an International depository authority. The deposit of the culture sample in India can be made at the Microbial Type Culture Collection (MTCC) and Gene Bank at the Institute of Microbial Technology, Chandigarh, India. The act also provides other conditions to be fulfilled at the time of the deposit of the sample as follows:

- All characteristics for the identification of the microbial sample.
- Access to material allowed after publication of the application.
- Disclosing the geographical source of the biological material.

Even though in the absence of a working definition various countries are allowing the patenting of microorganisms. However, there are debates regarding the patentability of microorganisms in many countries. Some countries consider that there are disadvantages, hence it is imperative to clearly differentiate the microorganisms claimed from the naturally occurring microbes and show that biological material is an outcome of invention and not a mere discovery. Clearly, the recombinant, mutated and adapted microorganisms are patentable as these are not naturally occurring organisms. Therefore, the position that India adapts regarding the patentable matter will become clear since the act has been amended recently, to include product patents for microbes.

The Indian Patent Act has no specific provision for patenting of microorganisms and microbiological processes. However, as a matter of practice microorganisms per se are not patentable in India, however, a recent decision of the Kolkata High Court has held that microbiological processes are patentable in India. In order to meet the obligation under

TRIPS agreement, India is required to introduce a patenting of microorganisms. Draft laws in this regards have been formulated. It may, however, be noted that many countries allow both process and product patents in regard to microbiological inventions and microorganism per se. All such countries allow patenting of genetically modified microorganisms but a few also allow patenting of naturally occurring microorganisms if isolated from nature for the first time and if other conditions of patentability are satisfied.

## 3.8. Patenting of genes, expressed sequence tags and single nucleotide polymorphisms

The patentability of various forms of genetic information is dependent on knowledge of the functional significance of that information. Assuming that the genetic information is new and not obvious, patentability may depend on the usefulness of the information. Patents on genes that code for medically useful proteins such as erythropoeitin and thrombopoietin have been granted. The situation is less clear for open reading frames of unknown function or for expressed sequence tags (ESTs) or single nucleotide polymorphisms (SNPs). If it can be shown that genes, ESTs or SNPs are associated with a particular disease in a way that enables diagnosis of that disease or susceptibility to disease, then this may enable patenting. Similarly, if a gene encodes a protein that possesses beneficial function or is a drug target for the treatment of a disease, this may also allow patenting.

The European Patent Office (EPO) and the US Patent Office take different approaches to the patenting of nucleic acids. In the US, patent applications for genes without functional relevance are rejected for lack of utility. In the EPO, the simple discovery of the sequence or partial sequence of a gene is not patentable because it is a mere discovery. A gene isolated from the human body by a technical process may be patentable but the industrial application must be disclosed in the patent application. However, in practical terms the outcomes of the determinations of patentability are roughly the same.

## 3.9. Patenting of computer-related (bioinformatic) inventions

Computer programs are patentable so long as they generate the ability for the computer to perform a novel technical function. For example, a program that simply sorted something into alphabetical order, which can be done manually, is unlikely to be patentable. Applications of computation are also patentable. For instance, patents have been granted on the following:

- 'Computer-aided visualization & analysis system for sequence evaluation (US5795716)'
- 'Computerized method and system for analysis of an electrophoresis gel test (US5754524)'
- 'Computer-aided probability base calling for arrays of nucleic acid probes on chips (US5733729)'
- 'Database and system for storing, comparing and displaying genomic information (WO/98/26407)'.

## 3.10. Who owns the patent?

If an employee makes an invention, the rights usually belong to the employer (unless specific contractual arrangements are in place to the contrary). This means that intellectual property developed by a university researcher in the UK is generally owned by the university, with the researcher named as an inventor. In the USA the patent applicant must be the inventor, although it is usual for the patent to be assigned to the employing institution.

## 4. CONCLUSIONS AND FUTURE OUTLOOK

A patent is a legal device that grants an inventor market exclusivity over a new invention or medication. Market exclusivity can mean tremendous economic rewards for the patent holder because it provides the inventor with a monopoly over the invention for the 20-year patent term. The patenting of biotechnological inventions is controversial. It raises broad social and moral concerns and puts a strain on traditional patentability criteria. To understand the debate that is currently taking place, it is useful to be reminded of the nature of patent protection and the benefits that can be derived from the grant of a patent.

In general terms, patents enable the patentee to prevent others using his invention; they grant a monopoly in exchange for disclosure of the invention. Patents can be granted in respect of a wide range of inventions, from the simplest to the most sophisticated. The criteria for patentability do not vary; they are the same for all. To be patentable, an invention must be novel (it must not have existed before), it must be inventive (not obvious to the skilled worker in the field), and it must be industrially applicable (able to be put to practical use). The invention must also be described in sufficient detail to enable it to be carried out by a skilled worker in the field.

Patents are national rights. A patent covers only the country (or in some cases the group of countries) for which it is granted. Thus, a US patent applies only in the US and a German patent applies only in Germany. Issues of validity and infringement are decided by the local courts. In Europe, a patent may be granted for several European countries together, but it can only be enforced in each country separately. There are plans to introduce a Community patent which would change this situation, but it will take some time.

Obtaining a patent and retaining market exclusivity can be a treacherous process, especially in the arena of biotechnology patents. It is probably not the case that patenting in the field of biotechnology has any greater hurdles to overcome than patenting in any other field. There are, however, three aspects that make biotechnology patenting seem very different. First, biotechnology is a science in which ultimate goals are achieved step by step, and each step might contribute something inventive. If each of those steps is patented, a single product may use several separate patented inventions. This is unusual. Secondly, this science is moving at such a rate that it is difficult for patent examiners to keep up. The speed of development also means that by the time the patent is granted (often several years after the work was done), the patented invention may be commonplace, even though it was inventive at the time the application was made which is when the invention has to be judged. This can give the impression that a patent has not been properly scrutinised. Thirdly, biotechnology patents bring with them their own special questions of morality, as discussed at the start of this article. Resolution of these issues is not, however, so much a problem for the patent system as for society generally.

The patent system is being challenged by this relatively new and fast-moving science. Scientific, legal, and practical considerations must be carefully weighed to best protect an inventor's rights. To obtain market exclusivity, biotech companies need to be aware of how science and patent law interact. Scientific issues affecting patentability, competent legal counsel, and inconsistencies in the way courts apply and interpret biotechnology patent law can all affect a company's ability to obtain, and retain, market exclusivity. Therefore, judicial developments will continue to define the scope of patent protection and guide the future of biotechnological inventions.

## 5. FURTHER READING

Barker, R. (2005). Changes to India's Patent Law: integrated approach is needed. BMJ. 330:692.

Crespi, R.S. (2004). Patenting for the research scientist: an update. Trends Biotechnol. 22:638-42.

Farnley, S. Morey-Nase, P. and Sternfeld, D. (2004). Biotechnology - a challenge to the patent system. Current Opinion in Biotechnology. 15: 254-257

Gersten, D.M. (2005), The quest for market exclusivity in biotechnology: navigating the patent minefield. NeuroRx. 2:572-578.

Jayaraman, K.S. (1999). Indian Patent Law will protect crops 'unless injurious to health'. Nature. 402:566.

Kintisch, E. (2005). U.S. Patent Law. Case probes what's fair game in the search for new drugs. Science. 308:174.

Mudur, G. (2005). Changes to India's Patent Law may deny cheap drugs to millions. BMJ. 330:1025.

Normile, D. (2005), Patent Law. Inventor knocks Japan's system after settlement. Science.307:337.

Plomer, A. (2005). European Patent Law and ethics. Drug Discov. Today.10: 947-948.

Witek, R. (2005). Ethics and patentability in biotechnology. Sci. Eng. Ethics. 11:105-111.

□□□

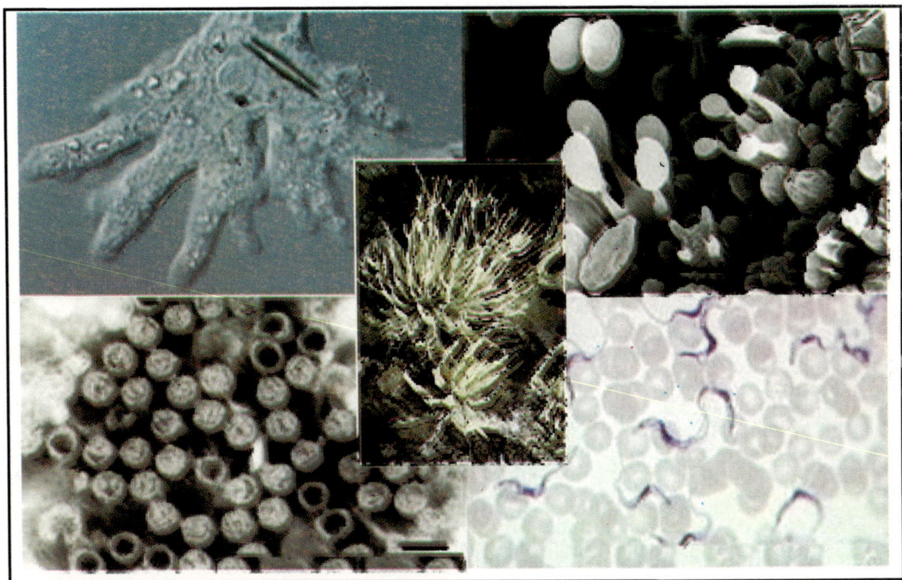

**Figure 3** Images of some representative microbial types Clockwise from upper left: *Amoeba proteus*, a member of the protista; a close-up view of a fog-desert *Niebla,* lichen [mergers]; basidiospores of the shiitake mushroom, *Lentinula edodes* [fungi]; *Trypanosoma brucei rhodesiense* [bacteria]; rotavirus [viruses]. *(See page 14)*

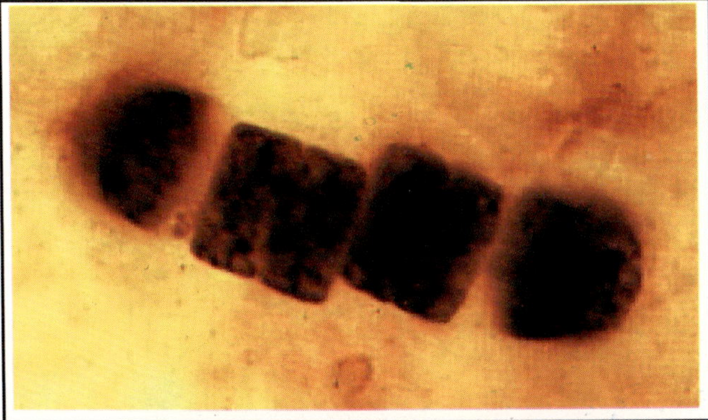

**Figure 4** Cyanobacteria fossil *(See page 17)*

**Figure 8. (a)** The only above-ground signs of the humongous fungus are patches of dead trees and **(b)** the mushrooms that form at the base of infected trees *(See page 24)*

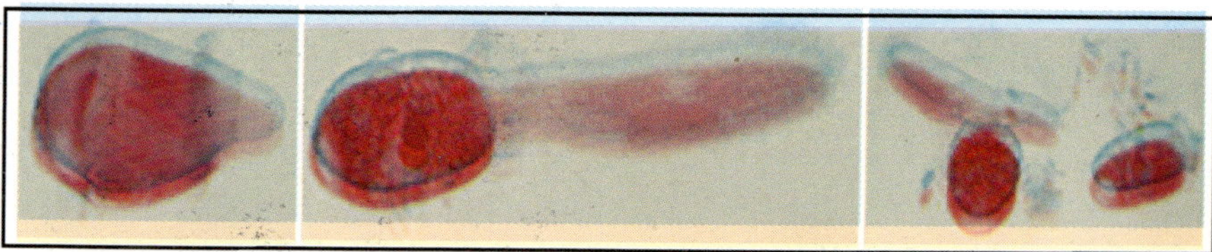

**Figure 16** *Didinium*, a ciliate that lives in fresh water, is a voracious hunter of live food such as Paramecium. A *Didinium* makes contact with a *Paramecium* and begins to ingest it *(See page 33)*

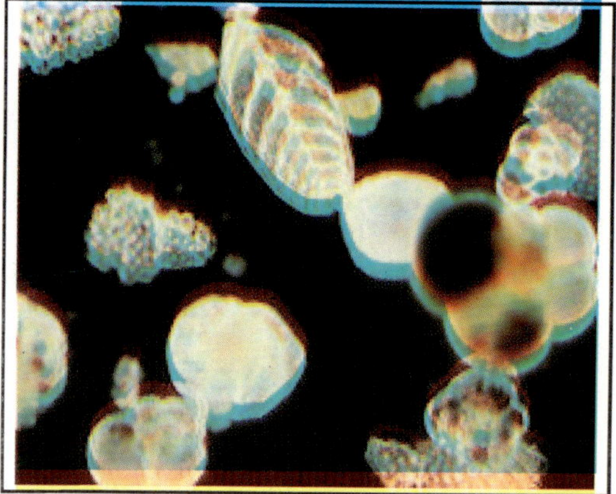

**Figure 18** Foraminifera bodies made up of chalk extracted from sea water *(See page 34)*

**Figure 19** Slime molds spend most of their lives independently, but during food shortages, they swarm and aggregate into an enormous single cell *(See page 35)*

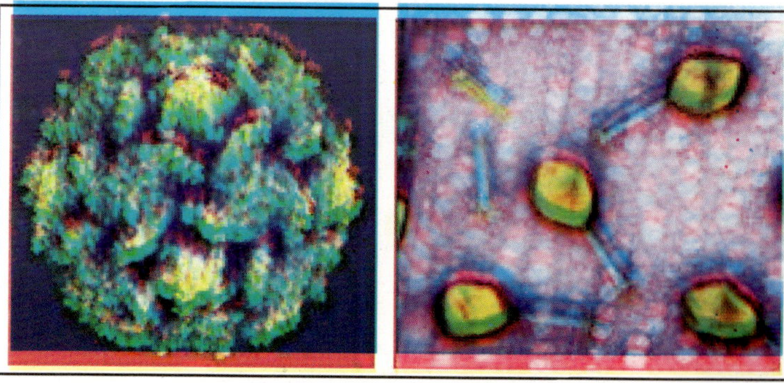

**Figure 20** The *polio virus* (left) once crippled millions. The *T4 bacteriophage* (right), is a virus that invades bacterial cells *(See page 37)*

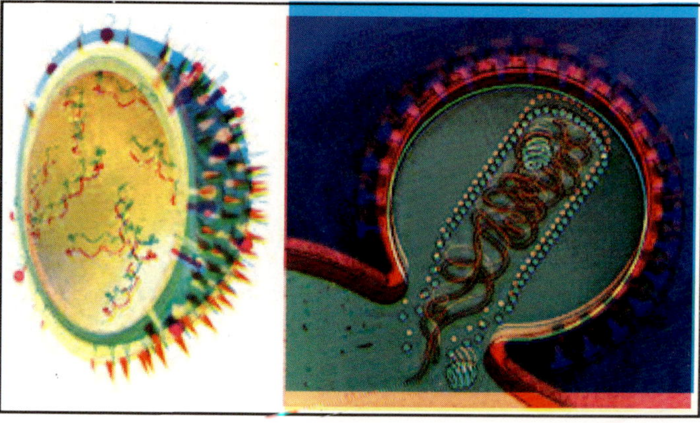

**Figure 21** An illustration of influenza virus (left) and human Immuno-deficiency Virus (HIV) (right) *(See page 38)*

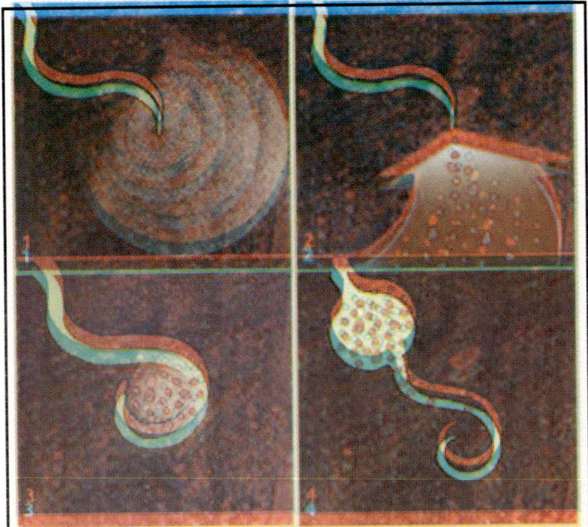

**Figure 22** The pea root sends a chemical message that attracts *Rhizobia*, then surrounds the bacteria, which set up housekeeping inside the root's cells *(See page 41)*

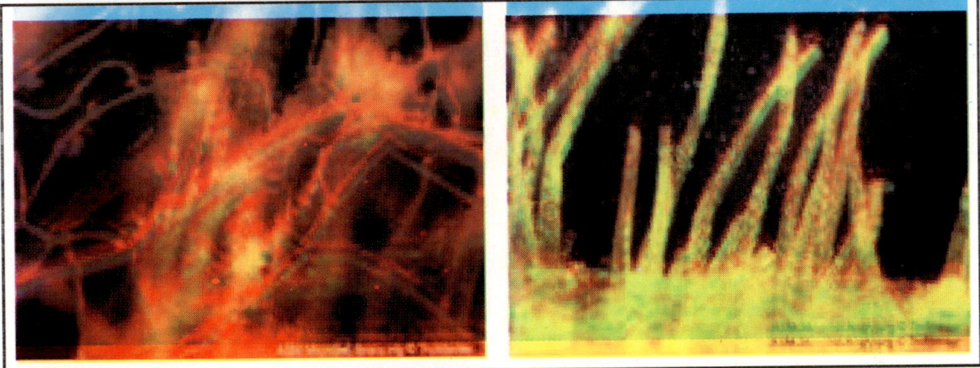

**Figure 23** Rhizobia living as a merger agreement with legumes (including beans and peas as well as clover and alfalfa) *(See page 42)*

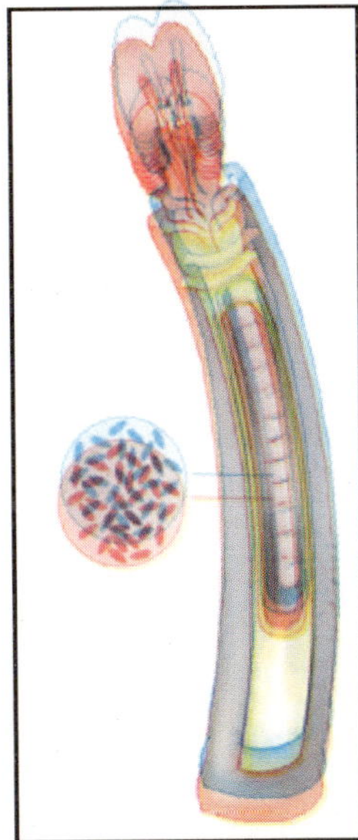

**Figure 24** Bacteria and their tubeworm host *(See page 43)*

**Figure 25** Leafcutter ant *(See page 44)*

**Figure 26** A close-up view of a fog-desert *Niebla* lichen (left), one of the common kinds of lichens found on rock and a lichen following cracks in the rock (right) *(See page 45)*

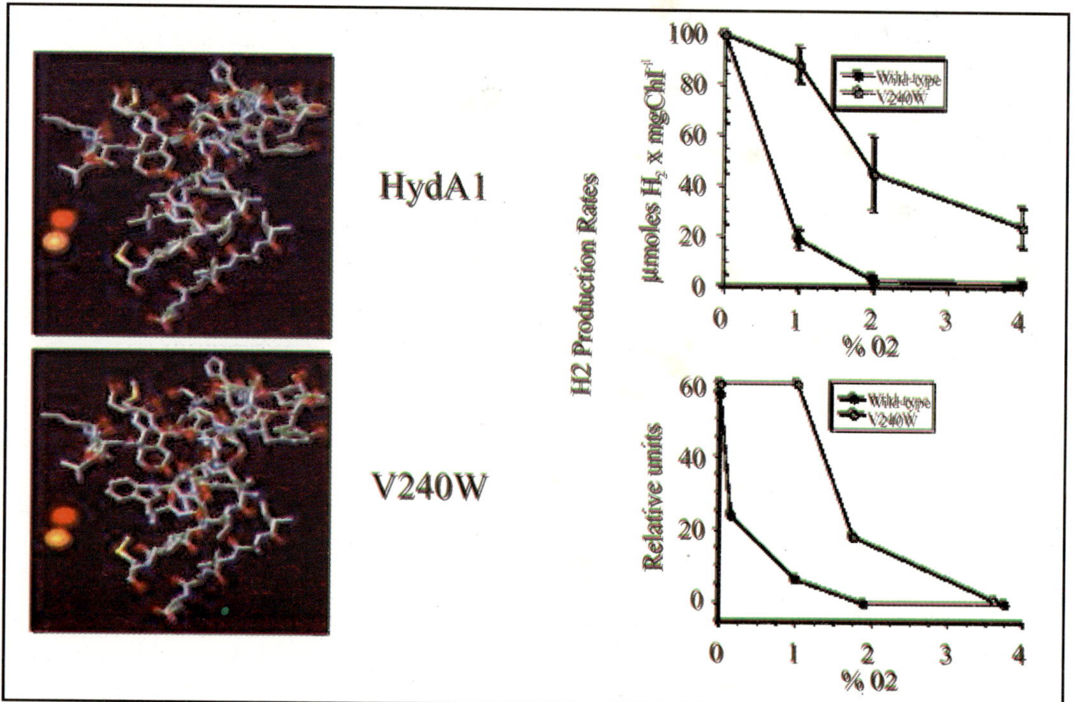


**Figure 12** Engineering the Hydrogenase for $O_2$-Tolerance: Cloning of two algal hydrogenases for generation of a $H_2$-channel mutant *(See page 453)*

# Index

## B

## D

## E

## L



## M

## N

## O



## P

## Q



## R

## T



## U



## V

□□□